ANNUAL REVIEW OF ASTRONOMY AND ASTROPHYSICS

EDITORIAL COMMITTEE (1990)

GEOFFREY BURBIDGE
PETER S. CONTI
HUGH S. HUDSON
DAVID LAYZER
MORTON S. ROBERTS
ALLAN SANDAGE
FRANK H. SHU
BARUCH T. SOIFER

Responsible for the organization of Volume 28
(Editorial Committee, 1988)

JOHN N. BAHCALL
GEOFFREY BURBIDGE
PETER S. CONTI
CARL HEILES
HUGH S. HUDSON
DAVID LAYZER
JOHN G. PHILLIPS
ALLAN SANDAGE
JOHN FAULKNER (Guest)
DONALD E. OSTERBROCK (Guest)

Production Editor   KEITH DODSON
Subject Indexer   KYRA GORDENEV

ANNUAL REVIEW OF ASTRONOMY AND ASTROPHYSICS

VOLUME 28, 1990

GEOFFREY BURBIDGE, *Editor*
University of California, San Diego

DAVID LAZYER, *Associate Editor*
Harvard College Observatory

ALLAN SANDAGE, *Associate Editor*
Observatories of the Carnegie Institution of Washington

ANNUAL REVIEWS INC   4139 EL CAMINO WAY   P.O. BOX 10139   PALO ALTO, CALIFORNIA 94303-0897

ANNUAL REVIEWS INC.
Palo Alto, California, USA

COPYRIGHT © 1990 BY ANNUAL REVIEWS INC., PALO ALTO, CALIFORNIA, USA. ALL RIGHTS RESERVED. The appearance of the code at the bottom of the first page of an article in this serial indicates the copyright owner's consent that copies of the article may be made for personal or internal use, or for the personal or internal use of specific clients. This consent is given on the conditions, however, that the copier pay the stated per-copy fee of $2.00 per article through the Copyright Clearance Center, Inc. (21 Congress Street, Salem, MA 01970) for copying beyond that permitted by Section 107 or 108 of the US Copyright Law. The per-copy fee of $2.00 per article also applies to the copying, under the stated conditions, of articles published in any *Annual Review* serial before January 1, 1978. Individual readers, and nonprofit libraries acting for them, are permitted to make a single copy of an article without charge for use in research or teaching. This consent does not extend to other kinds of copying, such as copying for general distribution, for advertising or promotional purposes, for creating new collective works, or for resale. For such uses, written permission is required. Write to Permissions Dept., Annual Reviews Inc., 4139 El Camino Way, P.O. Box 10139, Palo Alto, CA 94303-0897 USA.

International Standard Serial Number: 0066-4146
International Standard Book Number: 0-8243-0928-6
Library of Congress Catalog Card Number: 63-8846

Annual Review and publication titles are registered trademarks of Annual Reviews Inc.

∞ The paper used in this publication meets the minimum requirements of American National Standard for Information Sciences—Permanence of Paper for Printed Library Materials, ANSI Z39.48-1984.

Annual Reviews Inc. and the Editors of its publications assume no responsibility for the statements expressed by the contributors to this *Review*.

TYPESET BY AUP TYPESETTERS (GLASGOW) LTD., SCOTLAND
PRINTED AND BOUND IN THE UNITED STATES OF AMERICA

PREFACE

This volume was planned at a meeting held in San Francisco on May 7, 1988. Those who attended the meeting included Geoffrey Burbidge (Editor), John Phillips (Associate Editor), Keith Dodson (Production Editor), and Committee members John Bahcall, Peter Conti, Carl Heiles, Hugh Hudson, and Allan Sandage. Guests were John Faulkner and Donald Osterbrock.

In the preface to Volume 27, I pointed out that 32 articles were planned for this volume. In fact we have 17, or only slightly over one half of the articles originally planned for this volume. For Volume 29, at the time of writing (February 1990), 31 articles are on the list. The first-year default rate remains high.

Once again, I would like to thank all the authors, the Associate Editors, and especially Keith Dodson. It is a pleasure to work with such a group.

THE EDITOR

SOME RELATED ARTICLES IN OTHER *ANNUAL REVIEWS*

From the *Annual Review of Earth and Planetary Sciences*, Volume 18 (1990):

What is a Space Scientist? An Autobiographical Example, James A. Van Allen

Formation of the Earth, George W. Wetherill

The Nature of the Earth's Core, Raymond Jeanloz

From the *Annual Review of Fluid Mechanics*, Volume 22 (1990):

Mixing, Chaotic Advection, and Turbulence, J. M. Ottino

Plasma Turbulence, P. L. Similon and R. N. Sudan

From the *Annual Review of Nuclear and Particle Science*, Volume 40 (1990):

Neutrinos From Supernova Explosions, A. Burrows

Radiative Capture Reactions in Nuclear Astrophysics, C. Rolfs and C. A. Barnes

Accelerator Mass Spectrometry in Nuclear Physics and Astrophysics, Walter Kutschera and Michael Paul

Annual Review of Astronomy and Astrophysics
Volume 28, 1990

CONTENTS

NOTES OF AN AMATEUR ASTROPHYSICIST, *Vitaly L. Ginzburg*	1
INTERSTELLAR DUST AND EXTINCTION, *John S. Mathis*	37
THEORIES OF THE HOT INTERSTELLAR GAS, *Lyman Spitzer, Jr.*	71
MASSES AND EVOLUTIONARY STATUS OF WHITE DWARFS AND THEIR PROGENITORS, *Volker Weidemann*	103
COOLING OF WHITE DWARFS, *Francesca D'Antona and Italo Mazzitelli*	139
THE ORIGIN OF NEUTRON STARS IN BINARY SYSTEMS, *Ramon Canal, Jordi Isern, and Javier Labay*	183
H I IN THE GALAXY, *John M. Dickey and Felix J. Lockman*	215
SOLAR CONVECTION, *H. C. Spruit, Å. Nordlund, and A. M. Title*	263
QUANTITATIVE SPECTROSCOPY OF HOT STARS, *R. P. Kudritzki and D. G. Hummer*	303
RADIO IMAGES OF THE PLANETS, *Imke de Pater*	347
GAMMA-RAY BURSTS, *James C. Higdon and Richard E. Lingenfelter*	401
THE SPACE DISTRIBUTION OF QUASARS, *F. D. A. Hartwick and David Schade*	437
EQUILIBRIUM AND DYNAMICS OF CORONAL MAGNETIC FIELDS, *B. C. Low*	491
EXTRAGALACTIC H II REGIONS, *G. A. Shields*	525
RADIO PROPAGATION THROUGH THE TURBULENT INTERSTELLAR PLASMA, *B. J. Rickett*	561
RAPIDLY OSCILLATING Ap STARS, *D. W. Kurtz*	607
THE SOFT X-RAY BACKGROUND AND ITS ORIGINS, *Dan McCammon and Wilton T. Sanders*	657
INDEXES	
Subject Index	689
Cumulative Index of Contributing Authors, Volumes 18–28	699
Cumulative Index of Chapter Titles, Volumes 18–28	701

ANNUAL REVIEWS INC. is a nonprofit scientific publisher established to promote the advancement of the sciences. Beginning in 1932 with the *Annual Review of Biochemistry*, the Company has pursued as its principal function the publication of high quality, reasonably priced *Annual Review* volumes. The volumes are organized by Editors and Editorial Committees who invite qualified authors to contribute critical articles reviewing significant developments within each major discipline. The Editor-in-Chief invites those interested in serving as future Editorial Committee members to communicate directly with him. Annual Reviews Inc. is administered by a Board of Directors, whose members serve without compensation.

1990 Board of Directors, Annual Reviews Inc.

J. Murray Luck, Founder and Director Emeritus of Annual Reviews Inc.
  Professor Emeritus of Chemistry, Stanford University
Joshua Lederberg, President of Annual Reviews Inc.
  President, The Rockefeller University
James E. Howell, Vice President of Annual Reviews Inc.
  Professor of Economics, Stanford University
Winslow R. Briggs, *Director, Carnegie Institution of Washington, Stanford*
Sidney D. Drell, *Deputy Director, Stanford Linear Accelerator Center*
Sandra M. Faber, *Professor of Astronomy, University of California, Santa Cruz*
Eugene Garfield, *President, Institute for Scientific Information*
William Kaufmann, *President, William Kaufmann, Inc.*
D. E. Koshland, Jr., *Professor of Biochemistry, University of California, Berkeley*
Donald A. B. Lindberg, *Director, National Library of Medicine*
Gardner Lindzey, *Director, Center for Advanced Study in the Behavioral Sciences, Stanford*
William F. Miller, *President, SRI International*
Charles Yanofsky, *Professor of Biological Sciences, Stanford University*
Richard N. Zare, *Professor of Physical Chemistry, Stanford University*
Harriet A. Zuckerman, *Professor of Sociology, Columbia University*

Management of Annual Reviews Inc.

John S. McNeil, Publisher and Secretary-Treasurer
William Kaufmann, Editor-in-Chief
Mickey G. Hamilton, Promotion Manager
Donald S. Svedeman, Business Manager

ANNUAL REVIEWS OF
Anthropology
Astronomy and Astrophysics
Biochemistry
Biophysics and Biophysical Chemistry
Cell Biology
Computer Science
Earth and Planetary Sciences
Ecology and Systematics
Energy
Entomology
Fluid Mechanics
Genetics
Immunology
Materials Science
Medicine
Microbiology
Neuroscience
Nuclear and Particle Science
Nutrition
Pharmacology and Toxicology
Physical Chemistry
Physiology
Phytopathology
Plant Physiology and
  Plant Molecular Biology
Psychology
Public Health
Sociology

SPECIAL PUBLICATIONS

Excitement and Fascination
  of Science, Vols 1, 2
  and 3

Intelligence and Affectivity,
  by Jean Piaget

A detachable order form/envelope is bound into the back of this volume.

NOTES OF AN AMATEUR ASTROPHYSICIST

Vitaly L. Ginzburg

P. N. Lebedev Physical Institute of the Academy of Sciences USSR, Moscow, USSR

KEY WORDS: radio astronomy, synchrotron radio emission, cosmic-ray astrophysics, gamma-ray astronomy.

INTRODUCTION

I was surprised to receive the kind invitation to write the Prefatory Chapter for this volume of the *Annual Review of Astronomy and Astrophysics*, in that I do not consider myself a "real" astronomer. Rather, I am an amateur astrophysicist or, possibly better to say, amateur astronomer. In any case, I'm somewhere on the periphery of the astronomical community. For this and other reasons I did not think that one day I would be asked to write such a paper. On receiving the invitation, however, I immediately decided to make an attempt because I have always liked the idea of publishing in review volumes autobiographical papers or other nonstandard reviews including autobiographical elements.

The content of scientific knowledge does not, of course, depend on who has established some facts or how they have carried out observations and measurements. However, it is human beings, with their tastes, passions, and fates, that produce this knowledge, and it is much more difficult to understand a person than to decipher the spectrum of a star. At the same time, isn't it a natural temptation to desire insight into the inward life of a "comrade-in-battle" whom you meet at conferences and/or on journal pages? This problem is solved to some extent by obituaries and by other posthumous publications, but they can in no way substitute for autobiographical narration.

Below I shall try to outline the contributions due to my astrophysical attempts [which has been partially done already in (1, 2)]. I shall also dwell on my biography, which I hope will be of interest to my colleagues in the West and help them better understand the living conditions in my country, especially during those now already remote years when my generation of physicists and astronomers was brought up.

AUTOBIOGRAPHY I

I was born on October 4, 1916, in Moscow, where I have lived all my life (with the exception of two war years that I spent in Kazan'). My father was an engineer engaged in water purification and held several patents. He married for the first time in 1914 at the age of 51. My mother, a physician, was then 28. I was the only child in the family. My mother died of typhoid fever in 1920, and I have almost no memories of her. After she died, her younger sister came to live with us and was as good a mother to me as she could be. It was a very hard time—first World War I and then the Civil War. Moscow became the capital and was in general in a privileged position, but all the same there was little food and diseases raged. My memory is generally poor or, at any rate, has a high threshold. Above this threshold I recall one scene that I witnessed not far from our home in the center of Moscow in about 1920: a cart carrying coffins, a carter dragging himself along, and arms and legs sticking out of the coffins. Another reminiscence, not so ghastly, but still typical, also comes to mind: We managed to buy fresh meat somewhere, but it appeared to be dog meat; and under normal conditions, dog meat was never used as food in Russia. Nevertheless, my family suffered much less from what are called practical difficulties, I think, than did most people in the country at that time—we had a roof above our heads, Moscow was not occupied by troops, and there was no real hunger. What I had in excess was loneliness. It was aggravated by the fact that I did not go to elementary school and went only to the fourth form at age 11. I do not remember why this was so. The schools, like almost everything in the country at that time, were being subjected to all kinds of reorganization, and my parents probably thought it better for me to get my education at home. I believe this was legal at that time. (I don't know of such cases now, except for sick children.) It was undoubtedly a mistake, since when I finally did go to school, I found that it was not so bad. My school was a former gymnasium where many of the old teachers worked. But bad luck is bad luck. When in 1931 I finished at this seven-year school, it was decided, somewhere by somebody, that 7 years were quite enough, and thus the remainder of the secondary schooling was liquidated. (Secondary school has been different at different times; some time later, they changed their minds and returned to the ten-year school, as it is today.) My total school education, however, amounted to only four years.

After seven years of school one was supposed to enter a factory-plant school, where one was simultaneously to advance in education and to train for a worker's qualifications. But I did not take this route and instead became a laboratory assistant at an X-ray diffraction laboratory at one of the high schools. There I mainly communicated with two other laboratory workers who were three years my senior and took a great interest in physics and invention. (By the way, both my fellow lab workers later became good physicists.) I did not learn much but was imbued with something more important than knowledge—enthusiasm and an interest in work. In 1933, for the first time in many years, one could become a student at Moscow State University through open competitions at the entrance exams, and I decided to attempt to enter the physical faculty. It took me three months to "cover" the three-year course corresponding to the eighth, ninth, and tenth years of school. I passed the entrance exams but was not admitted—preference was given to those with better biographical particulars (i.e. more proletarian social origins and occupations of parents). But there was no special discrimination (for instance, related to the fact I am a Jew)—the results of my exams were not brilliant. I decided not to wait until the next year, left my work, entered a correspondence course at the university, again studied almost without assistance, and finally, in 1934, at the age of 18, started attending lectures and studying normally "like other students."

AMATEUR ASTROPHYSICIST

Why all these details? My aim is only to warn people against this route because I am deeply convinced that one shouldn't follow my example if given the choice. Of course, the school curriculum can be mastered in much less than ten years, but for this one generally must pay a high price. In my case, this price was a deficient background in Russian grammar, which to some extent has hampered my ability to write in this fairly rich language. Mastering a language requires exercise, exercise, and more exercise. I hated cramming and did not do it without pressure. The same also applies to mathematics. To understand something, one need only solve a dozen or so problems. But at school, I would have solved ten, if not a hundred, times more such problems, which would have provided the necessary automatism. All this is obviously so clear that further explanations are not needed here, and, in any case, I have already written about it in more detail elsewhere (3). However, I shall give a less trivial example. First-year students of the direct department (i.e. not including students taking correspondence courses) were taught astronomy, and my schoolmates recalled it with pleasure. However, I somehow contrived to become a second-year student without passing exams in astronomy or in chemistry. For this reason and because of my lack of corresponding school knowledge, even on becoming a professor of physics I remained illiterate in astronomy and chemistry. In both cases, I'm sure, I have paid a high price. As far as chemistry is concerned, I am still paying the price even now, because since 1964 I have been engaged in research on high-temperature superconductivity, which is today the center of gravity of my work (see e.g. 4), and this research appears to be closely connected with some chemical problems of which I am ignorant. As for astronomy, I'll begin with a curious illustration. Many times, especially when I took a great interest in the "new astronomy" with its quasars, pulsars, etc., I have spoken in lectures or with friends about various astronomical discoveries, about the radio, X-ray, and gamma-ray "sky." But the ordinary stellar sky is unfamiliar to me, and when asked what star or constellation is it, I must honestly say that I don't know. And if I have called this situation funny rather than shameful, it is because I do not consider myself to be a professional astronomer. I'm somewhat ashamed nonetheless, but such is my life.

When in 1946, at the age of 30, I wrote my first paper in astronomy, I had already authored many papers in physics, and an even greater number were waiting their turn. I had neither spare time nor extra strength, and life was hard. How could I learn the map of the stellar sky, remember it, and get used to it?

This may seem strange, but the facts show that a lack of elementary knowledge in one or another field—I mean now astronomy and physics—is not yet an obstacle for obtaining interesting and important results in these fields. Examples are especially numerous among mathematicians engaged in solving (and very successfully) various physical problems in spite of their ignorance of physics as a whole, to say nothing of numerous details. Analogously, quite a number of physicists, myself included, have written papers in astronomy that are of interest and were published without any allowances in the astronomical literature in spite of a very poor general astronomical background of the authors.

This is my personal opinion, and although I haven't had a chance to discuss it with anyone, it would be of interest to know what other people think. This is the first reason why I have entered upon this discourse here. The second is my desire to explain why I consider myself to be an amateur astrophysicist, as can be seen from the title of this article. Finally, and of most interest to me, I am curious as to how my work as astronomer would have been influenced if I had a "normal" astronomical education, i.e. the same as any astronomer who chose this profession as a university student, if not earlier. Unfortunately, it is extremely difficult, indeed almost impossible, to answer such a question. Alas, one cannot start one's life anew. The best of all imaginable ways of obtaining the answer would be to trace the life of identical twins developing under different conditions. But given the absence of not only a twin brother but even analogous examples of other people, one can only make assumptions. I do not, of course, mean a fairly trivial general assertion concerning the benefit of being well educated and well informed. I refer to quite concrete hypotheses, results, and even discoveries for which I've asked myself whether I could have been the author of some of them. Sometimes the answer was negative, whereas in certain cases I'm sure I would have immediately given a correct answer if asked, or if at least acquainted with the corresponding astronomical material, I would have asked myself this or that question. But can a man who has never heard of neutron stars ask why they can rotate rapidly, possess a huge magnetic field, and be superfluid in some part of them? In short, and this is of course common knowledge, once a question is formulated, the work is sometimes half done. And in order to formulate a question concerning astronomical objects or effects, one must be acquainted with them.

Overall, I'm sure that it was a mistake that when I was first entering the field of astronomy, I restricted myself only to the material that was directly connected with what I was working on [at first it was the Sun; see (5)]. I should have eliminated my astronomical illiteracy in spite of all the obstacles that I have already mentioned. I was only 30 years old then, and of course I could, if I had realized this, have postponed some $n+1$ paper in physics and instead studied a textbook in astronomy. Today, as in the past, some physicists and engineers who are far from astronomy make attempts, say, to measure cosmic gamma-ray emission or to detect gravitational waves. It seems to them that they may remain unacquainted with astronomy as a whole. They are wrong. I know, of course, that such advice usually is not taken, but all the same I try to convince people that extending their astronomical horizons and advancing in their general level of knowledge is justified even within an exclusively pragmatic approach.

AUTOBIOGRAPHY II

I shall, however, return to my biography. From 1934 to 1938 I studied quite conscientiously at the physical faculty of Moscow State University. This was a flourishing period for the physical faculty (one that ended with the beginning of the Second World War and the evacuation from Moscow). My sympathy was from the very beginning with L. I. Mandelstam and his school. (Representatives of this school were I. E. Tamm, G. S. Landsberg, and others, including A. A. Andronov, although by that time he had already moved to Gorky and I only made his acquaintance later). The name of L. I. Mandelstam (1879–1944) is little known in the West, although he made great scientific achievements in optics and radio physics. (Suffice it to say that Mandelstam and Landsberg simultaneously and, of course, absolutely independently of C. V. Raman discovered the effect named after Raman; in the USSR, however, we prefer the term "combinational light scattering.") Mandelstam delivered lecture courses on various topics that had a wide audience.[1] These lectures, as well as subsequent discussions and occasional seminars, were a perfect school of true physics.

A third- or fourth-year student (the whole course consisted of five years, after which a student obtained the degree of "diploma") had to make a precise choice of his speciality. The Department of Theoretical Physics was headed by I. E. Tamm (1895–1971) and seemed most attractive to me. But my mathematical aptitude is rather modest, whereas in theoretical physics mathematics is considered to be not simply important (this is undoubted) but very essential. Therefore I chose experimental optics (the department was headed by Landsberg) and started working under the guidance of S. M. Levi. We tried to study the spectrum of canal rays, but in hindsight it is now clear that our experimental potentialities were not adequate to this difficult task. Nonetheless I received my degree. More interesting were my contacts with Levi. He was a Jew from Lithuania who had worked for a long time in Berlin, if I am not mistaken, in R. Landenburg's laboratory. The advent of fascism to power impelled him to leave for Moscow. Jumping ahead, I shall mention that later (in 1937 or 1938) Levi was dismissed from the physical faculty but, fortunately, not imprisoned. He could move abroad and found himself in the USA. In the 1960s I visited the USA three times (1965, 1967, 1969) and tried to find Levi. Miss Helen Dukas, the former secretary of Einstein, and some other people tried to help me, but all in vain. Later, when back in Moscow, I learned Levi's address when I complained to an acquaintance of mine of the bad luck in my attempts to find Levi in the USA. It turned out that this man had for a long time exchanged New Year's greetings with Levi. Such tangles are, of course, not surprising. It is more interesting that as far back as the 1930s Levi (and obviously many others) understood clearly the sense and the possible role of induced emission. Levi told me straightforwardly, "Create an overpopulation at higher atomic levels and you will obtain an amplifier; the whole trouble is that it is difficult to create a

---

[1] These lectures have been published, although I do not know whether they have been translated into English. In such cases, I shall henceforth not refer to the Russian editions, which are practically inaccessible to the readers of the present volume.

substantial overpopulation of levels." As is now known, it is not so difficult to create an overpopulation, and, more importantly, by using mirrors the optical pathlength can be extended so that we obtain a laser. Why lasers were not created as far back as the 1920s, I do not understand. Much becomes obvious in hindsight. Perhaps there were some obstacles or simply nobody thought of using mirrors. This idea did not occur to me either, and here I definitely cannot explain this by my lack of knowledge. But the grains dropped by Levi have produced shoots. (I do not mention the general influence upon me of this pleasant and educated man.) As mentioned below, at the beginning of the Second World War, I was engaged in work on radio wave propagation. In particular, I paid attention to the decisive role of induced emission for propagation of radio waves in some ranges of the Earth's atmosphere (6). O_2 molecules possess magnetic moment, and therefore in the Earth's magnetic field the O_2 molecule levels (the lower electron levels) split. But the difference between the magnetic sublevels is $\hbar\omega \sim e\hbar H/mc \sim 10^{-19}$ erg. Therefore, at a temperature $T \sim 100\text{--}300$ K, the energy is $kT \gg \hbar\omega$ and the sublevels are filled almost equally. As a result, waves with frequencies corresponding to the energy difference between sublevels propagate by means of a continuous alternation of absorption of a wave (radio photon) from a lower sublevel and an induced emission of the same photon from the upper sublevel. If the induced emission is ignored in this situation, the resulting absorption will be $kT/\hbar\omega \sim 10^5$ times stronger than the actual absorption. It is now, of course, a well-known fact that it is necessary in astrophysics that allowance be made for induced emission. To the best of my knowledge, my paper (6) was the pioneering application of this to the Earth's atmosphere. However, radio wave propagation in the ionosphere and atmosphere has long been beyond the scope of my interests, and I shall not dare estimate the role of the paper (6). This role, I think, was negligible because the resulting effect is small and the paper (6) itself most likely remained unnoticed. In 1942, people had other troubles.

On graduating from the university in 1938, I became a postgraduate student,[2] but not everything went smoothly. The war was coming, and deferments for postgraduates were rescinded. I was called up and remember receiving a document in which I was called an "espirant" [because, I think, the word "aspirant" (a postgraduate student) was associated with the then-popular language Esperanto]. The physical faculty still managed to grant deferments to its postgraduate students, but in 1938 this happened for the last time. I do not doubt that had I been then inducted into the

---

[2] This is a three-year period intended for preparation of a candidate dissertation, which generally corresponds to a PhD dissertation.

Army, I wouldn't be writing this paper now. Only a few of my university fellows who entered the army survived the war. The situation on the whole, however, was more complicated. Three times I quite accidentally failed to get in the army; two of the three times I tried to volunteer. All of this, as well as much else that I mention only casually, is not without interest, but I'm afraid that I have already paid too much attention to my biography and further details would be out of place. (Some of these details, however, will become clear from what follows.)

While awaiting the summons (or more precisely the call-up "to present myself with things") I did not rush to the black-walled windowless room where I conducted measurements. Instead, I tried to somehow explain the strange angular dependence of the canal-ray radiation that we were attempting to find. (The indications of strange angular dependence found in the old literature were erroneous, as I am now sure.) It occurred to me that the electromagnetic field of a moving charge may play the same role as a photon flux and, in particular, induce radiation. This idea is erroneous, since a charge field is not equivalent to a free field. But in those times the situation was less clear. At any rate, when in the fall of 1938 (if I am not mistaken, this historical event in my life happened on September 13) I asked Tamm a corresponding question, he got interested, advised me to look through the literature, and wished me well. I soon found out that the difference between the methods used in classical and quantum electrodynamics gave rise to misunderstandings and obscurities that I partially managed to clarify, and that what was important was not the application of a complicated mathematical technique but in understanding the formulation of the problem. This is how I became a theoretician and left experimentation for life. Obtaining on my own these (maybe modest) results, I realized that I could fruitfully work in theoretical physics. This had a great effect on me, and my new life began. In this respect I am much obliged to Tamm and his attitude toward people. I do not write about Tamm in more detail here, since the interested reader may refer to my reminiscences about him (7). Here I only emphasize how important a friendly atmosphere and mere kindness are for many (although not all) beginners [for more details, see (7)]. Neither do I dwell on the content of my first three papers devoted to quantum electrodynamics (published in 1939). The corresponding references can be found in my book (8) and paper (9), where the subject itself[3] is also considered, of course.

In 1939 S. I. Vavilov and P. A. Cherenkov discovered the phenomenon that is now known as the Vavilov-Cherenkov effect or radiation. [True, in

---

[3] Here and below I try to refer to more recent and accessible literature, especially in what does not concern astronomy.

the West it is most often referred to as Cherenkov radiation, but many Soviet authors (I among them) who know well the history of this discovery use only the term Vavilov-Cherenkov (VC) radiation.] The classical theory of the effect was formulated in 1937 by I. M. Frank and Tamm, who showed that a charge uniformly moving in a medium at a "super-light" velocity $v > c/n$ (where n is the refractive index of the medium) must radiate. This theory was already within my memory in 1939, and it was thus very natural that while studying quantum electrodynamics at this time, I constructed the quantum theory of VC radiation as well as considered this effect in an anisotropic medium. Since then, the study of radiation of uniformly moving sources (including, besides the VC effect, the transition radiation and the Doppler effect) has become one of my favorite fields. All of these effects seem beautiful to me. Of course, the definition of "beautiful" is always subjective. For example, L. D. Landau (1908–1968) was almost absolutely indifferent to the above-mentioned effects. By the way, I regard L. D. Landau, as well as Tamm, to be one of my teachers. Like Tamm, Landau was an outstanding personality [see my contribution in the book of reminiscences about Landau (10)]. Therefore, it may not be out of place to narrate, as an illustration of the above-said, how we first met (and, in fact, got acquainted) on scientific grounds. At one of the joint seminars [for more details, see (10, 11)] Tamm mentioned the quantum theory of VC radiation formulated by me, to which Landau immediately made a remark to the effect that this is absolutely unnecessary, since the effect is classical. As was usual in such cases, Landau's criticism had serious grounds—the quantum corrections in the theory of VC radiation are of the order of $\hbar\omega/mc^2$ (ω is the radiation frequency, and m is the mass of a radiating particle) and are small because at high frequencies ω, the refractive index $n(\omega)$ tends to unity and VC radiation disappears. However, the quantum theory of VC radiation is in fact not of narrow methodical interest only, since it also substantially clarifies the situation when applied to the Doppler effect in a medium [for a discussion of this and practically all my papers devoted to the theory of uniformly moving sources, see (8, 11)].

The papers mentioned above were written within about a year and a half and were the basis for my candidate dissertation, defended at Moscow State University in May 1940. A candidate dissertation and the corresponding candidate degree in the USSR correspond to the PhD in the West. Of course, the level of candidate dissertations is sometimes very low, but in physics and at good institutes and universities, the level of candidate dissertations seems to me to be rather high. We also have another degree— the Doctor of Sciences (Dr. Sc.). In this case, and again in the proper places, a Doctor's thesis is usually defended by an author of dozens of papers, quite an independent research worker not younger than 30 to 40. The Doctor's degree gives one the right to hold a professor's post, although there are not a few professors in our country without the Doctor's degree. At that time (1940) the USSR Academy of Sciences had special vacancies for those who were to prepare a Doctor's dissertation within three years. There was one such vacancy in the Department of Theoretical Physics at the P. N. Lebedev Physical Institute of the USSR Academy of Sciences (FIAN), which I filled on September 1, 1940. Ever since, I have been working in this department, which was founded by Tamm in 1934. (Until mid-1941 he had also been head of the theoretical department at the physical faculty of Moscow State University.) Before entering FIAN, I had already been engaged in the theory of particles with higher spin states. This work continued for many years (partially in collaboration with Tamm and young colleagues). I shall restrict myself to mentioning only my most recent papers (12, 13) that touch upon this problem.

As is well known, Germany invaded the USSR on June 22, 1941. Several cities had already been under bombing attacks for many hours when, at about midday, Molotov announced over the radio that war had broken out. I clearly remember listening to his speech with my two-year-old daughter on my lap. I also remember Stalin's speech on July 3, when this dictator, who had poured the country with blood and tears, called his audience "brothers and sisters" for the first and last time. I was a "private untrained" and therefore was not mobilized. But for people like me and, it seems to me, for all those who were not mobilized, a people's volunteer corps was organized. Soon a great many of these people were killed or captured near Moscow. Of course, I immediately joined the volunteer corps and spent the whole first day in a school building, where we were "formed," and in the evening left with an order to come "with things" at the first call. But soon there came a decision to evacuate the Academy of Sciences, and, if I am not mistaken, exactly one month after the war broke out, my elderly father (he was 78 years old), aunt, wife, and I left for Kazan'. (My only daughter had been evacuated with her grandmother some time before.) Before our departure, Moscow underwent only one air attack by the Germans. When we heard the sound of alert, my wife and I happened to be near a metro station, where we had to spend that night. In the morning I saw only traces of the attack. I shall note that in Kazan' I had to join the "labor front"—we dug trenches not far from the town, which fortunately were not used. I had no deferment ("reserve") because in those times it was intended only for people of higher standing. But having had no military training I was of no value to the army, and in any event I was not called up. While waiting for the summons, I wished to finish my Doctor's dissertation as soon as possible, and in May 1942 I defended my thesis on the theory of higher spin particles. I have no reason for considering my thesis to be weak, and indeed it was highly estimated. Under peaceful conditions, however, I would not have hurried to get my Doctor's degree. But in the situation described, and taking into account the fact that I changed my subject of research (see below), my haste was justified. By the way, the defense of a thesis did not have any influence, as far as I know, upon one's being called to military service, at least in those hard times—by that time, the Germans had reached the Volga. Furthermore, I was invited to volunteer to become a paratrooper, and I tried to do so but was rejected because of the state of my health by virtue of an anecdotic set of circumstances. In Kazan' our institute occupied a floor in the wing of the Kazan' University building, and there were no more than 100–200 people in the institute. We very often were sent to fulfill different kinds of jobs, and I remember unloading a barge of logs on the Volga. Everybody worked, including Tamm, Landsberg, and others who then seemed to me to be very old although they were no more than fifty. As for me, I carried big logs "on crate"—this is a setup in which straps are put on like a rucksack and a "ground" is then fixed on one's back, on which a log or something else is put. Such a contrivance was widely used in Russia for carrying bricks and other heavy objects. I was not at all mighty (180 cm tall and weighing about 60 kg at that time). Using this contrivance, however, I carried rather heavy short logs that two men could hardly put on the "ground." But the load was apparently too heavy, and one time I found blood in my mouth—a small vessel must have broken. Tuberculosis was suspected and I was sent to the dispensary, where they found some "petrificated foci," after which I was registered. Therefore I was regarded as unfit for the landing force. I am not going to play the hypocrite and say that I regret this. But at that time, I felt that it was better to die in a battle than to find myself under German occupation and then in a death camp.

Life in Kazan' was not a honey, of course. My family lived in a small room in the university hostel where the temperature in winter fell below $0°C$. My father could not endure such a life and died in mid-1942. The Academy was supplied with some food and thus we did not actually starve, but we did feel hungry all the time. I remember a canteen where for a corresponding recompense waitresses served pea porridge. People came in with a tin stuck into a briefcase, filled the tin with porridge or soup, and then carried away the briefcase as if it were filled with papers.

At the end of 1943 the members of FIAN began returning to Moscow as victory approached. Below I narrate another important chapter of my biography, and thus now I shall only note that on returning to Moscow the theoretical department started growing. For several years I was Tamm's deputy, and after his death in 1971 I had to become head of the department. I say "had to become," for I did not want this position, but in the Academy it was necessary "for the welfare of the department staff" that its head be a man of position. In 1971 there were only two full members of the Academy (academicians) in the department: A. D. Sakharov and myself. Sakharov had already become a dissident and was not suitable as a chief, so I had to be head of the department for 17 years. A new rule has recently been introduced obliging all members of the Academy beyond the age of 70 to resign their administrative posts with the right to become a councillor (adviser) without a decrease in salary. Of course, to be head of the department that now numbers nearly 60 workers, including three full and four corresponding members of the USSR Academy of Sciences (the same number as in the rest of our great institute), required time and strength. But I do not complain, for our department is known for its friendly terms and very kind-hearted atmosphere, and during the lifetime of the department (54 years!) we have not had a single serious conflict.

Since 1943 I have been engaged in the theory of superconductivity and have written a number of papers in this field; the most well-known among them (14) (in collaboration with Landau) was published in 1950. I have also been concerned with the theory of ferroelectrics, superfluidity, and many other things, including astrophysics, which will be a special subject here. But all this work came some time later. In spite of all the warnings, the coming war was not openly regarded as inevitable in the USSR. Therefore, to the best of my knowledge, no direct preparation for the war was conducted by the Academy. In any event, when the war broke out on June 22, 1941, physicists were in general out of it. Suffice it to recall that even I. V. Kurchatov set himself to the problem of demagnetization of ships against magnetic mines, and it was only at the end of 1942 that he was appointed to head the nuclear program. As has already been mentioned, I was occupied in an absolutely abstract analysis of equations for higher spin particles. As the war began, we all started searching for applications of our abilities closer to practice, if not to defense. Accidentally I was given advice to concern myself with radio-wave propagation in the ionosphere, which was supposedly an important problem for defense. I followed this advice and spent much time on this subject.

SIDE ASSOCIATIONS

There is a special type of person who usually has "side associations." I belong to this type, in that I am constantly developing various associations that require comment.

I have mentioned above the title "academician," the use of which in the

West has always surprised me and aroused an unpleasant feeling. I first experienced this feeling many years ago, when in the list of participants of some congress I came across the names of Prof. P. Dirac, Prof. N. Bohr, and Academician X. Why a special title for a Soviet scientist? Does it not suffice to be a professor, especially in the same row with Bohr and Dirac?

In the USSR Academy of Sciences there exist two "degrees" (titles): full member (or academician) and corresponding member. Unfortunately, both these titles often are used in the Soviet literature. I think that this is the influence of the prerevolutionary habit not to forget to mention the title of count or duke, if not grand duke. Strong, also, was the German influence, whereby a colleague is thanked as, say, Herr Geheimrat Professor Doctor X. The progressive tendency in our country is now to omit titles, say, in acknowledgments (acknowledge X, Y, or Z but not professor, doctor, or academician X, Y, or Z). In the West, as far as I know, such a style has already become conventional. In any case, I am a resolute opponent to the use of the academician title in the English literature. However, one cannot totally abandon titles in the USSR, because otherwise some authority will not answer the telephone, among other things.

Since I have already touched upon the subject, it seems reasonable to say some more about the USSR Academy of Sciences. (We have numerous other academies as well.) This academy numbers about 800–900 members, of which about one third are academicians and the rest are corresponding members. Such a structure was created before the revolution of 1917. Merely for the title, an academician now receives 500 roubles a month (without any duties), and a corresponding member receives half that. These sums are rather large if one takes into account that my salary as head of a department or councillor is also 500 roubles, while a young candidate of sciences gets 150–200 roubles a month. Members of the Academy have some other privileges as well; for instance, an academician may call a car with a driver for several hours a day. This may greatly surprise some readers, but one should remember that the salary of 500 roubles a month is often much less than the salary of a man of equal qualifications in the West. As for the use of a car, some time ago the members of the Academy were fewer in number than now, the majority of them were elderly people (the average age of academicians in 1985 was still 69.9), and there were very few private cars.

I am convinced that in principle the only correct thing to do is to abolish all material privileges for members of the Academy. But in our society this is evidently a long way off,

Academic privileges have, naturally, a very pernicious effect upon the elections: Not only scientists, but also designers (engineers), high-ranking officials, etc., try to join the Academy. Despite this, the USSR Academy of Sciences, especially in the exact sciences, is on the whole a focus of the most qualified people of this country. Besides, the Academy does not act exclusively on the basis of arbitrariness but is still guided by a charter that provides for a competitive ballot and not "one place–one candidate" type elections. However, until recently the charter has, in essence (although not formally), been repeatedly violated. (For example, special vacancies were allotted for candidates welcome to the authorities.) But under the conditions of the deep changes now taking place in my country, I hope that the charter will be made more precise and, most importantly, observed. The Academy and all of Soviet science have, however, problems that are no less important. One of them is the necessity to fight against fantastic bureaucratism. It would be out of place to dwell on this subject here, the more so as I can refer the interested reader to my paper (15) that elucidates this problem [for those who wish to obtain insight into the state of Soviet society in mid-1988, I advise them to get acquainted with the whole of the volume containing (15)].

Another "association" arises in connection with the above-mentioned fact that I am the author of a large number of works and a still larger number of papers. By "work," I mean some scientific results in the form of a paper or papers that I have registered in a special list. Long ago, when I was defending my theses, by some formal requirement I had to start such a list, and I have since continued it voluntarily because it is convenient to do so for references. The list now numbers 325 items, and one item often includes several papers devoted to one problem. A number of notes and reviews, popular scientific papers, etc., are not included, nor will the present paper figure in the list, although it was more difficult for me to write this one than a whole number of scientific papers. The total number of my publications probably approaches one thousand, and the number of my "works" is also large. What does this prodigality signify? Is it good or bad? I have had to face such or similar questions, and opinions in this connection are very diverse. For example, I have been reproached for publishing papers very easily. There also exists an opinion that authors who publish many papers do so for the sake of glory, for increasing the number of references to their papers, etc.

It is quite obvious that a large number of publications or even "works" cannot in itself be put to one's credit. One essential work may, of course, be more valuable than a thousand weak papers. So in estimating the contribution of one or another author, the decisive role is undoubtedly played by the content of his publications. All other conditions being equal, the number of publications characterizes basically the style of the activities and tastes of the author. Personally, I write rather easily if, of course, I have an idea of what I am going to write about. Moreover, I do not feel satisfied until the paper is written; of course, dissatisfaction often remains afterward, but nonetheless the very process of writing is, at least for me (and I think for many people), an important element of the work itself. With some exceptions, a paper is written not in the way of a duty but as a result of some interest. And if the author himself is interested, why not share this with others and send a paper or a note to press? As for those who write little, there exist different cases. Some merely have nothing to write about; others find it difficult to write or believe that they have already reached such a level that publication of a not very important result, a popular paper, etc., will add nothing to their scientific reputation and may even arouse ironical smiles. I am well aware of the fact that many of my publications only give rise to criticism. But I have no fear for the "clamours of Boeotians," for I do not want to be led by snobs. Sometimes there are mistakes and I regret having sent a paper to press, but idlers alone never make mistakes. Finally, tastes and opinions are diverse [a very typical example is given in (3)]. The author must heed his inner voice and not try to please everybody. If a paper is not interesting to someone, he can choose not to read it. This, by the way, consoles me with respect to this paper, in spite of the fact that I realize that some may not like it.

The value of the works and their quality is an important question in scientific activities. For young people who are only beginning their scientific life it is even of vital importance, but irrespective of age and position, one wants to hear a just opinion. An unambiguous and reliable way of judging the merits of works probably does not exist. In the West, one common way of estimating papers is based on the citation index. This method is seldom used in the USSR and is in general unpopular, but I do not think that there is enough reason for such a negative attitude. The citation index may, of course, absolutely distort the picture. For instance, an erroneous but sensational paper may give rise to a lot of references. Another source of errors is that only the first author's name[4] is taken into account. Finally, papers published in Russian are not as frequently cited as those appearing in English in well-known journals. But information from the citation index is nonetheless interesting and significant. True, I do not look in the index. (The library of our institute does not have it, and

---

[4] By the way, I, along with the majority of my colleagues in the USSR, use only an alphabetical ordering of names for my publications. (I have written only two or three papers in which this order was more or less accidentally violated.) One should bear in mind, however, that the Russian alphabet differs from the Latin one. In my case it is particularly important that the letter Г (G) is the fourth in the Russian alphabet, and to V or W there corresponds the third letter of the Russian alphabet (В). Also, the Russian Ф (F) is toward the end of the alphabet. Therefore, when a paper is translated from Russian into English, the alphabetical ordering often appears to be violated.

if I am not mistaken it is available only in two places in the whole of Moscow.) But once in a Russian publication I came across a listing (16) of the 249 most frequently cited scientists in the world, based on the overall number of references to their papers in 1961–75, i.e. within 15 years. In this list, L. D. Landau was in second place (18,888 references), and five other Soviet authors were mentioned (four physicists and one chemist), of which I had the most citations (6834 references, or an average of 456 references a year); overall, I was in 66th place. In subsequent years the number of references to my papers has become somewhat smaller, which is natural. [In recent years the papers that I have published have been less in number and maybe also in quality; I do note, however, that some of my old papers, especially (14), are now most frequently cited in the text without any mention in the list of references.] Unfortunately, I do not know which place is taken by references to my astronomical papers (including cosmic-ray astrophysics).

In spite of all the limitations of the conclusions that can be drawn from the citation index, it would nonetheless be of interest to have such information with respect to astronomical themes. (True, it is not always easy to decide to what branch of science this or that paper should be ascribed.)

AUTOBIOGRAPHY III

In 1985 the Soviet Union entered a new epoch, of which one of the most important and distinctive features is *glasnost* (openness). We now (especially since 1987) read books and articles in magazines and newspapers whose publication was unthinkable some time ago. Possessing, to say nothing of spreading, such literature used to lead to arrest and many years of imprisonment. *Glasnost* also allows me to dwell here on the part of my biography of which I could not write or speak publicly of before. The ignorance of some foreign colleagues with respect to life in the Soviet Union has often amazed us, especially since quite a lot of materials fairly sufficient to remove such "illiteracy" for anyone who can read have already been published. But not everybody is fond of reading, and for astronomers a concrete example concerning an "amateur" astronomer, but still one of their colleagues, may be of some additional interest.

When the war ended in 1945, physicists in the USSR were already held in great esteem. The existing institutes grew rapidly, and new scientific institutions and high schools appeared. As far back as the 1930s, there was at Gorky University a strong group of physicists and mathematicians, the most outstanding of which was A. A. Andronov. This group decided to organize a special radio physical faculty. Since they were short of their own experts in this field, they invited three professors from Moscow (from FIAN) for whom it was to become their "part-time" job, i.e. they were expected to remain working in Moscow but at the same time come to Gorky from time to time to deliver lectures. I was one of these three and was put in charge of organizing and heading the department of radio-wave propagation because, as I have already mentioned, I had been occupied with precisely this problem since the beginning of the war and had already published a number of papers. I remember, at the end of 1945, arriving in Gorky by train (it is a night's journey) in far from comfortable conditions and hiring a fellow to carry my suitcase by sledge (we were walking) from the station to the central part of town, where the university and hotel were located. At the beginning, if I am not mistaken, there was only one colleague (M. M. Kobrin) in "my department" and one strange student, who later left the university. But soon after there appeared capable students, followed by postgraduates. Many of those who graduated from the department have long since obtained their Doctor's degree, and the department continues to exist. That I headed this department under hard postwar conditions now seems to me to have been a venturesome enterprise. But I was 29 then, and as a doctor of sciences had the right to be at the head of a department. I was willing to teach youngsters, but in Moscow it was difficult for me to realize this ambition. I think I would have left Gorky in a couple of years, as did my colleagues from Moscow, had fate not willed things differently. I met there my present (second) wife Nina Ermakova, and in 1946 we got married. I would not, of course, even mention here my family life if it were not for some special circumstances—namely, that Nina was, in fact, in exile in Gorky and had no right to move to Moscow. In short, the story (which has even been mentioned in some publications) is this. Her father was arrested in 1938 and died in Saratov's prison in 1942. (He was in the same prison as the well-known geneticist N. I. Vavilov, who died there during the same period.) A student of the mathematical faculty of Moscow State University, Nina was arrested in July 1944 and along with her fellows (some of them, as she herself, were the children of repressed parents) was accused of plotting an attempt on Stalin's life—Stalin was supposedly to be shot from the window of her flat in the Arbat. But the "scriptwriters" from the KGB had never taken the trouble before the arrest to check everything, and only later was it established that the windows of the room where Nina lived with her mother did not face the Arbat. For all the nonsense of the accusations brought by the KGB, the investigators somehow tried not to make easily refutable assertions. In any event, the accusation of terror was remitted, and there remained "only" the accusation of counterrevolutionary group anti-Soviet activities (Articles 58.10 and 58.11 of the then Criminal Code). She spent

9 months in prison and in March 1945, without any trial and by the decision of the so-called special assembly, was sentenced to three years' confinement in a camp. This was a very "short" term for a person convicted under Article 58. Maybe it was for this reason that when an amnesty was announced on the occasion of the end of the war, even prisoners charged under Article 58 were given amnesty if their terms were three years or less. (The majority of those charged under Article 58 had longer terms and were not included in the amnesty.) Thus, in September 1945 Nina was released, but without the right of residing in a number of large cities.[5] She had an aunt in Gorky, and therefore she chose Gorky, but nonetheless she could get registered only on the other bank of the river Volga in the village Bor. Despite this, she managed (and this was nontrivial—kind people helped her) to enter the polytechnical institute in Gorky, from which she graduated in 1947. Until 1949 she lived illegally in my room, but at the end of 1949 she was registered in Gorky by Andronov's application. (This happened after a large accident on the river Volga on October 29, 1949, in which a ship carrying people from Gorky to Bor was wrecked; Nina was among some dozen or so out of about 250 passengers who survived.) I, naturally, handed in an application each year (it could not be done more frequently) with a request to allow my wife to move to Moscow, but all in vain. It was only in 1953, after Stalin's death, that a new amnesty was announced and Nina could leave for Moscow at last. In 1956 she, as well as all her fellows from the "anti-Soviet group," was completely rehabilitated—that is, the accusation was recognized as fully groundless. To characterize the rehabilitation process, I will say that an inspector came to Nina's mother's place to draw up a statement, in the presence of witnesses, that the windows of the room did not face the Arbat.

In addition to the reasons that I have already mentioned, I narrated this story also to explain why I taught students and conducted research work in Gorky for so long. I had students, postgraduates, and colleagues there, and much research was conducted. Therefore, after 1953 I continued, although more and more seldom, to visit Gorky. I headed the department, I believe until 1961, and the last two times (in 1980 and at the end of 1983) I went to Gorky it was mainly to see Sakharov, who, by irony of fate (at least I feel so) was also exiled to Gorky.

In 1942 I became a candidate member of the CPSU (Communist Party of the Soviet Union) and in 1944 a CPSU member. The war was in full swing then, and today, when we know what was Stalin's true face, it is still hard to believe that many millions of people, including myself, were absolutely blind about him. Even when I learned the "criminal" story of

---

[5] In order to live in a given place, Soviet citizens must be registered there by the police and get visas in their passport. Amnestied people were given passports with limitations for visas.

my wife and her fellows, I did not think yet that everything came from the "great leader." My eyes were opened wide only after N. S. Khrushchev's report on February 25, 1956, at the 20th Congress of the CPSU.

If I continued writing my biography in further detail, it would, of course, overstep the admissible page limits of the present paper. Therefore, I shall restrict myself only to a few more remarks. By unwritten rule, CPSU members were not, of course, supposed to marry "counterrevolutionaries." In addition, on October 4, 1947, there appeared in the Literary Gazette an article in which a certain physicist D. Ivanenko (he wrote denunciations of many others as well, including Tamm) accused me of "servility" and "cosmopolitism," i.e. admiration for bourgeois science. This gave me much trouble, and still more was to be faced. I think it would have cost me my head if it were not for I. V. Kurchatov, who invited Tamm in 1947 to take part in nuclear research. Tamm, in turn, enlisted me, Sakharov, and a number of other workers of our department. Soon I made an important proposal; another was made by Sakharov. Unfortunately, even now, at a time when the USSR and the USA exchange military observers and the Secretary of Defense of the USA has recently examined the new Soviet armaments, these 40-year-old works still remain classified. Therefore, so as not to get into new troubles (I have had enough of them in my life), I shall only say the following. Tamm, Sakharov, and some others left for far-off lands sometime in 1948. As one that aroused suspicion (the wife, you know, was in exile), I remained in Moscow, but a sentry was posted at the door of my office. I knew no real secrets and was not at all interested in them, but in 1950, on the initiative of Sakharov and Tamm, research work in the field of controlled thermonuclear synthesis was begun. At the first stage this was, of course, a physics problem, and I was also engaged in it and obtained some results. By 1952 (or late 1951) this work was considered so important and secret that I was altogether removed from it. Maybe it was for this reason that after this problem was declassified (this was due to Kurchatov, who made his well-known report on the thermonuclear problem in England in 1956), I decided in 1962 to publish my old reports (17).

As is well known, in 1952 and early 1953 the situation in my country became still more heated. Stalin went absolutely mad, and vivid evidence of this was the notorious "case of the physicians." New monstrous repressions were coming. Fortunately, on March 5, 1953, "the greatest man of genius of all times" was finally gathered to his fathers. There followed many rapid changes in the country, the "case of the physicians" was disavowed and they were released, the next amnesty was announced (I have mentioned it above), and Beria—one of the closest bloody assistants of Stalin—was shot. It was noted that the Academy of Sciences of the USSR had not called elections since 1946, and late in 1953 the elections were finally held. I was then elected a corresponding member. At the end of the year I was awarded the order of Lenin (the highest civil order in this country) and the Stalin Prize of 1st degree (later this prize was renamed the State Prize, and three degrees were abolished and reduced to only one) for the above-mentioned "closed" works. In general, by 1954, from a half-disgraced man I had become a person. Since then, my life has been generally normal. From 1962 to 1970 I even could rather often go abroad, sometimes with my wife (this was a rare thing for Soviet citizens, but an exception was made for full members of the Academy; I was elected in 1966). Since 1971, true, I have gotten into new troubles, but these were not a threat to my life and, therefore, I shall say only a few words about them. First, I was almost never allowed to go abroad. For instance, I could not deliver my Darwin Lecture myself. [It was delivered at a meeting of the Royal Astronomical Society from a text that I had submitted beforehand (18).] Only after several indignant letters sent by me "to the very top" did I manage to go to several international conferences, but each time with such nervous strain that I would not wish it on my enemy. Only since 1987 have I been going abroad without extreme difficulties, although, as is quite a usual thing with us, with a great many bureaucratic impediments (see 15). Second, complications arose owing to the fact that in 1969, A. D. Sakharov again started working in our department. I refused to sign any letters condemning his activities, and generally the attitude toward him was quite loyal in the department, even when in 1980 he was exiled to Gorky. On our initiative he remained a worker in our department, and, with permission from the administrators, his colleagues from the department went to see him there.[6] When, at the very end of 1986, Sakharov could finally return to Moscow, he appeared in the department on the day of his arrival and was heartily greeted. He is working in our department now. Being head of the department, I had, of course, to deal with "the Sakharov case." I think that I have done my best and I have no reason to reproach myself, but on the whole it is not up to me to judge.

In general, as can be seen from what has been said, it was only in 1987 that I got out (is it forever?) of "external" difficulties. In return, by virtue of what is called the "law of conservation," so to speak, my "internal" difficulties increased. The point is that beginning at the age of about 65, it has become more and more difficult for me to work fruitfully, although I try to overcome this obstacle (see 3).

In concluding this section, I would like to answer one, obviously natural question: What was the influence of my severe trials upon my scientific

---

[6] About this period, see, in particular, my article (written 17 December 1989, after A. D. Sakharov's death, on behalf of the editor) in the literary magazine *Znamja* (1990, no. 2, p. 3) (noted added in proof).

activities? A simple answer suggests itself—under better conditions I would have managed to do more. But, frankly speaking, I am not sure of that. During all of my "scientific life" (since 1938) I could almost without interruption be engaged in anything I wanted and not worry about earning my bread, even though it was not always with butter. This was first. Second, and this is also typical of many of my colleagues, science and scientific activities occupied a predominant place in our lives; they were simultaneously work, hobby, and rest (and even a narcotic). I think that if I had lived under better conditions, I would have probably been happier and have rested and seen more. But the integral of my scientific activity, if I may say so, most probably would not be larger than it is.

RADIO ASTRONOMY[7]

As has already been mentioned, in the middle of 1941 I turned to the study of radio-wave propagation in the ionosphere and generally to plasma physics. This activity is reflected in the monograph (19) as well as in (8). It was, in fact, the ionosphere that was the starting point for my astronomical and, more precisely, radio astronomical studies.

The well-known Soviet physicists and radio specialists L. I. Mandelstam and N. D. Papaleksi contemplated the problem of Moon radiolocation (radar) long before the Second World War. In 1944, stimulated by the progress in radiolocation, Papaleksi returned to this idea by considering, naturally, radiolocation of the planets and the Sun. In this connection, at the end of 1945 or the beginning of 1946, he asked me to clarify the conditions of radio-wave reflection from the Sun. As it turned out, this was a typical ionosphere problem, and I had all the formulas at hand. The results of the calculations did not seem to be very optimistic, since for a large set of parameters, such as electron concentration and temperature in the corona and chromosphere (which then remained unknown in many respects), radio waves should be strongly absorbed in the corona or chromosphere and not even reach the level of reflection. [The reflection due to inhomogeneities was not considered; the "point" $n^2(\omega) = 1 - \omega_p^2/\omega^2 = 0$ was, roughly speaking, taken as the level of reflection.) But this was immediately followed by a more interesting conclusion: The source of solar radio emission must not be the photosphere, but rather the chromosphere and, for longer waves, the corona. At the same time, the corona was already assumed to be heated up to hundreds of thousands or even a million

---

[7] In this and the following sections I have used material from, and sometimes even the exact texts of, my papers (1, 2). Since these papers are easily accessible, I refer those (probably very few) interested in the details to them. A number of references omitted here can also be found in these papers.

degrees. Thus, even under equilibrium conditions, i.e. in the absence of perturbations and sporadic processes, the temperature of emission from the corona at waves longer than about a meter must reach approximately 10^6 K at a photospheric temperature $T_{ph} \approx 6000$ K. All this was presented in my first astronomical paper (5). In the same year, analogous conclusions were published by Martyn (20) and Shklovsky (21). [I can only say that my paper (5) was submitted for publication on March 27, 1946, whereas the dates of submission of the other two papers are not indicated; they appeared, respectively, in *Nature* on November 2, 1946, and in the November–December 1946 volume of *Astronomicheskii Zhurnal*.] In the calculations, I used absolutely clear and reliable formulas known from the theory of ionospheric wave propagation (19). Martyn did not present the formulas, but he evidently acted in the same way. As for Shklovsky, he believed (21) that the absorption due to "free-free" transitions and the absorption due to electron-proton collisions should be taken into account separately and then summed up, whereas, in reality, these are one and the same mechanism (19). However, this circumstance was of no importance, since the parameters of the corona were not exactly known at that time.

The existence of thermal radio emission of the corona at $T \sim 10^6$ K was confirmed in the paper by Pawsey published immediately after Martyn's paper (20), in which such emission was shown to play the role of a lower limit that is reached as soon as the sporadic component of solar radio emission becomes sufficiently weak.

A weak point of radio astronomy of that period was a low angular resolution that prevented investigations of the Sun even for regions of sizes of arcminutes—it is difficult to believe this today when the angular resolution of radio interferometers far surpasses that of the best optical telescopes. In this connection, Papaleksi suggested measuring solar radio emission during the total solar eclipse of May 20, 1947, with the help of a 1.5-m wavelength antenna that was installed on board a ship and had a wide directivity pattern of several degrees. These measurements were successful [see the literature cited in the book containing (1)] and appeared to be the first of their kind. Whereas the intensity of optical emission from the Sun during a total eclipse decreases by several orders of magnitude, at a wavelength of 1.5 m the intensity during the eclipse decreased no more than by 60%. Thus, meter-wave radio emission was proven to come from the corona, which remains uncovered by the Moon even during a total optical eclipse.

In 1947 I took part in the Brazil expedition of the USSR Academy of Sciences, on board the ship *Griboyedov*, in which radio observations of the Sun were conducted. It seems that I was included on the staff of the expedition in recognition of my early work in the field of radio astronomy, which was only then beginning in the USSR. I did not take part, however, in the measurements themselves—they were carried out on board the ship while the main part of the expedition made its way to Brazil to take optical measurements, which were, unfortunately, unsuccessful because of bad weather. This main part of the expedition also included a small ionospheric group headed by Ya. L. Al'pert. (I was in this group; weather could not, certainly, prevent ionospheric measurements.)

The above-mentioned activities drew me further into radio astronomy, and I became for a while almost a professional radio astronomer. (I tried to acquaint myself with all the available material, methods of measurement, etc.). As a result, I wrote two reviews of radio astronomy (22, 23), among the first in the world literature. (I cannot, however, vouch for this.)[8] Now that 40 years have passed, it is difficult for me to judge the value of these papers, and I do not feel like analyzing them in detail. I shall only mention the suggestion made by me in (23) that the radio-wave diffraction on the Moon edge should be used to increase angular resolution of details on the Sun during eclipses. This question I considered in further detail in my 1950 paper with G. G. Getmantsev [see the reference in (1)], where we were already thinking more of discrete sources of cosmic radio emission than of the Sun. The diffraction of radio emission on the Moon edge has since been widely used, and therefore I need only add here that I also discussed in (23) the possibility of enhancing the angular resolution still more by observing a source located on the line that joins the Moon center with the point of observation. (Here we are evidently dealing with the Arago-Poisson light spot.)[9] The nonspherical shape of the Moon and the necessity of having a source on or very near to the indicated line strongly hamper such observations, of course, and I do not know of any attempts to exploit this method. But maybe one should nonetheless analyze such a possibility in more detail, not only for the Moon but also for the planets

---

[8] How far I still remained from astronomy as a whole, in spite of what has been said, can be seen from my note (24) in the volume dedicated to the 80th birthday of J. Oort. There, I recount that on the way back from Brazil, the participants of our expedition were lucky enough to visit Leiden, although quite accidentally. And while there, instead of taking the opportunity of meeting Oort and generally taking part in the discussion of astronomical problems, I rushed to the Kamerling Onnes Cryogenic Laboratory because I was then most interested in low-temperature physics.

[9] As is known, as an objection to the wave theory of light, Poisson pointed out a consequence of this theory that seemed to him quite absurd: On the axis of the geometrical shadow of a round nontransparent screen a light spot must be observed. (The source is considered to be pointlike and is positioned behind the screen on the axis perpendicular to its plane.) The experiment conducted by Arago immediately after this confirmed the existence of the light spot. For an opaque sphere as screen, the conditions for observing a central peak are facilitated, as compared with for a flat screen.

and their satellites and artificial screens (both flat and spherical). Perhaps this has already been done and I do not know about it, for I have not looked through the corresponding literature.

If I tried to dwell on my subsequent papers in so much detail, it would take too much space. My work in astrophysics has been rather sporadic and chaotic, and the portion closest to radio astronomy may somewhat conditionally be divided into three main trends:

1. Twinkling of radio sources in and beyond the ionosphere, oscillations of the intensity of solar emission, the use of polarization measurements, and satellite measurements.
2. The theory of sporadic emission from the Sun. V. V. Zheleznyakov and I began studying this range of questions in 1958 (25). A number of other papers followed, but it would be out of place to refer to them here, since the corresponding results (with references) are fully given in the book by Zheleznyakov (26).
3. The theory of cosmic synchrotron radio emission, and its connection with the problems of the cosmic-ray origin and with high-energy astrophysics as a whole. This was the subject of my main astrophysical interest, and I still work in this field, although not so much as before. From the point of view of historical information that I can offer, these problems also play the most important role. Therefore, they are discussed in the next section [for more details, see (1, 2)].

I shall now dare to mention some more of my papers close to radio astronomy. The origin of radio galaxies and, specifically, the question of the energy source of their radio emission were not immediately clear. For instance, one can mention the hypothesis of colliding galaxies, which proved, as a rule, incapable of explaining the mechanism of energy output in radio galaxies. Another suggested idea was that of a sharp increase in the number of supernova flares in radio galaxies, an idea that has always been rather groundless. Therefore, in my opinion, my paper (27) was not without value, since it stated that the required energy output and cosmic-ray acceleration in radio galaxies are, in principle, easily provided by gravitational energy—in particular, in the process of star formation. I pointed out that "it seems more attractive to associate the galactic flares not with supernova flares, but with another large-scale mechanism, for example, gravitational instability of a galaxy or of its central part." This trend of thought continued to a certain extent in later papers devoted to quasars [see (28) and the literature cited therein, as well as (29)]. I would also like to mention paper (71) about heating of intergalactic gas.

The discovery of pulsars gave rise to the temptation to clarify the mechanism of their radio emission. Zheleznyakov and I (and partially V.

V. Zaytsev) published several articles on this subject, the last being the review (30) published in these volumes. But the problem proved to be much more complicated than it at first seemed. (It happens sometimes that a problem turns out to be particularly sophisticated when this is not suspected in advance.) For this reason I decided long ago to leave this field, and I am sure I was right to do so. Only representatives of the younger generation (or even generations) appeared to be able to gain a real understanding (though not to the bottom) of the very interesting but complicated and many-sided range of questions concerning pulsars, including the mechanism of their radiation and the whole theory of pulsar magnetospheres (see 31).

SYNCHROTRON RADIO EMISSION, COSMIC-RAY ASTROPHYSICS, AND GAMMA-RAY ASTRONOMY

Now I turn to cosmic-ray astrophysics and gamma-ray astronomy, which are very important areas of astronomy in which I have taken part. My activity in these fields started with the theory of synchrotron radio emission.

It is reasonable to comment first on the terminology, which has not yet become conventional. Cosmic rays are now usually understood to be charged particles (protons, other nuclei, electrons, etc.) of cosmic origin that possess high energy (say, kinetic energy $E_k \gtrsim 100$ MeV, but this limit is rather conditional). The whole range of questions connected with the origin and role of cosmic rays in space may be called cosmic-ray astrophysics, although it is frequently referred to as the problem of cosmic-ray origin. However, the origin (in the literal sense of the word) of cosmic rays is only a part of cosmic-ray astrophysics. The term high-energy astrophysics is also used and includes, besides cosmic-ray astrophysics, gamma-ray astronomy, X-ray astronomy, and high-energy neutrino astrophysics. Many authors, I among them, have written much on these subjects. I refer to the reviews (32–34) and particularly to the Proceedings of the International Cosmic-Ray Conferences (ICRC); the 20th ICRC was held in 1987 (35). [Paper (36) is my introductory talk at this conference.] Therefore I shall only briefly comment on the early stages in the development of the theory of cosmic synchrotron radiation and the appearance of cosmic-ray astrophysics [for more details, see (1, 2)].

Approximately in 1947–49 it became quite clear that comparatively long-wave, nonsolar cosmic radio emission (including, in particular, the very first radio astronomical measurements made by K. Jansky at a wavelength of about 15 m) possesses an effective temperature T_{eff} exceeding 10^4 K. Therefore, it was impossible to interpret such radio emission as thermal radiation from interstellar gas, since this gas, generally speaking, has a temperature $T \lesssim 10^4$ K, and in any case radiation with $T \gg 10^4$ K cannot be explained by thermal radiation of gas. Thus, one had to assume the existence of some nonthermal source analogous, for example, to the sporadic sources of nonthermal solar radio emission. This was how the "radio star hypothesis" arose in a quite natural way. According to this idea, some stars are anomalously powerful radio sources responsible for the nonthermal, cosmic radio emission with its continuous spectrum and diffuse directional distribution (37–39). The radio star hypothesis, however, faced many difficulties, mainly involving assumptions (sometimes arbitrary and unrealistic) concerning the hypothetical radio stars. An alternative soon appeared that with time became stronger and stronger—the synchrotron hypothesis of the origin of nonthermal radio emission, which in the end proved to be valid.

The competition or struggle between these two hypotheses took several years. From the physical point of view, synchrotron radiation had been known and understood for many years (40), and in the 1940s it was especially widely discussed in the physical literature in connection with the analysis of synchrotrons. But it was only in 1950 that the first papers (41, 42) appeared in which the synchrotron mechanism was considered as related to cosmic radio emission: Alfvén & Herlofson (41) discussed radiation from radio stars and Keipenheuer (42) that from interstellar space. It seems rather strange to me that these papers appeared in a journal of physics, and only then in the form of short letters. In any case, as far as I know, these papers did not attract the attention of astronomers. However, I at once believed that the synchrotron mechanism was responsible for nonthermal cosmic radio emission. I do not attribute this to any keen insight on my part, but rather to the above-mentioned fact that I was closer to physics and far from classical astronomy. In this situation the synchrotron mechanism seemed clear and realistic, whereas hypothetical, strange radio stars remained purely speculative. I immediately verified the calculations in these two papers, but, if I am not mistaken, I did not add anything essential. [I have not now compared all the expressions and estimates from (41), (42), and my own paper (43), submitted for publication on October 31, 1950, because it does not seem essential here, especially in that I have never claimed priority.] But the fact remains that my paper (43) was the first, and for some time the only one, that responded to the proposals of Kiepenheuer and of Alfvén & Herlofson to use the synchrotron mechanism in astronomy. The reaction of astronomers was probably quite the opposite, i.e. that the synchrotron mechanism seemed mysterious and speculative, whereas radio stars, although posing riddles, were more acceptable, for what kinds of stars cannot exist? In this respect, I. S.

Shklovsky was not an exception. He not only developed the radio star hypothesis (39), but also positively denied the synchrotron hypothesis ["which for a number of reasons does not seem to us to be acceptable" (39)]. This quotation from Shklovsky is presented in more detail in (2). I dare make this remark here only because it is just Shklovsky's paper (39) that has been repeatedly cited in the world literature as the first and principal discussion of the application of the synchrotron hypothesis. A similar error is often spread concerning the question of exploiting polarization of synchrotron radiation as a criterion for establishing the validity of a synchrotron origin of cosmic radiation [for more details, see (1, 2)]. After the appearance of a very important paper by Pikelner (44), who emphasized that the interstellar magnetic field exists in the entire galactic volume, Shklovsky (45) realized that his earlier objection (39) to the efficiency of the synchrotron mechanism had been groundless. For reasons that will become clear from the section to follow, I do not here dwell in more detail on Shklovsky's paper (45) and his subsequent papers [see (1, 2)].

Judging from the proceedings of the 1955 Manchester IAU symposium on radio astronomy (46), which included a paper on the radio star hypothesis while my paper sent to the symposium was not even published, astronomical "public opinion" in 1955 was still on the side of the radio star model. However, by the next IAU radio astronomy symposium, in Paris in 1958, the synchrotron mechanism was unconditionally accepted as the dominant production mechanism of nonthermal cosmic radio emission. (I do not mean, of course, the radiation coming from the solar atmosphere and generally from relatively dense regions.) This time my report "Radio Astronomy and the Origin of Cosmic Rays" was included in the proceedings of the symposium (47), although I myself was not able to participate.

The establishment of a connection between radio astronomy and cosmic rays has led, as a matter of fact, to the appearance of a new field of research in astronomy—namely, cosmic-ray astrophysics (now usually included in the term high-energy astrophysics). Although cosmic rays were recognized earlier (prior to 1950–53) to be cosmic objects, they were investigated only on the Earth (in the atmosphere and on its boundaries) exclusively for the purposes of high-energy physics. As is well known, the cosmic-ray results have led to exceedingly important discoveries in physics [$\mu^{\pm}$-leptons, $\pi^{\pm}$-mesons, and some other particles were discovered; see the sources cited in (2)]. But since cosmic rays are highly isotropic, studying them near the Earth is similar to making a spectral analysis of the light of all stars taken together. It is clear that under such conditions, without additional information about celestial bodies, astronomy could not have developed. The reception of synchrotron radio emission from cosmic rays (more precisely, from their electron component) has sharply changed this situation. It has become clear that cosmic rays are a universal phenomenon: They are present in interstellar space, in supernova remnants, and in galaxies. (Radio galaxies and then quasars have been discovered.) Radio astronomical data, together with the information on primary cosmic rays near the Earth and with the available astronomical concepts, have promoted further advances. I note that as far back as 1934, Baade & Zwicky (48) associated cosmic-ray generation and neutron star formation with supernova flares. In 1949, in considering cosmic rays as a gas of charged particles, Fermi (49) discussed a possible mechanism of their acceleration. These ideas, together with radio astronomical data, have been the basis for constructing the galactic model of the origin of the main part of cosmic rays observed near the Earth. Since 1953, I have supported only the galactic model with halo. The halo problem aroused many discussions (see e.g. 50), which seem to me to be the results of misunderstanding [see (50, 51)]. It is impossible to write here about this and many other problems of cosmic-ray astrophysics in further detail, and I once again refer the reader to the literature [see (32–36, 51)]. True, it is necessary to lay special emphasis on the principal achievement of recent years or, more precisely, of the last two decades. I refer to gamma-ray astronomy and, more specifically, to the reception of gamma-ray emission due to π^0-meson decay, with the energy of photons $E_\gamma > 30$–50 MeV. π^0-mesons are produced, in turn, during collisions of protons and nuclei [which make up nearly 99% (by number of particles) of cosmic rays] with gas nuclei in the interstellar medium. Therefore, identification of gamma-ray photons generated by π^0-meson decay provides information on the proton-nuclear component of cosmic rays far from the Earth. In this respect, gamma-ray astronomy plays the same role that radio astronomy plays in the study of the electron (or, more precisely, electron-positron) component of cosmic rays. Thus, only through the combination of radio and gamma-ray astronomical methods can we, in principle, obtain rather complete information on cosmic rays far from the Earth. Unfortunately, gamma-ray astronomy is not yet sufficiently developed. The reason is that it is necessary to use special satellites (gamma-ray observatories) to detect photons in the range 30 MeV $< E_\gamma < $ (1–5) GeV, but not a single gamma-ray observatory has been in operation, at the time of this writing (1989), since the very successful satellite *COS-B* finished its work in 1982.

The problems to be solved by gamma-ray astronomy cover the whole energy range from $E_\gamma \gtrsim 10^5$ eV to $E_\gamma \gtrsim 10^{16}$ eV [gamma-ray bursts, gamma-ray lines, diffuse gamma-ray emission, radiation from discrete sources—in particular, gamma-ray emission from SN 1987A (52)]. Here, too, I restrict myself only to references (34–36, 53).

The proton-nuclear component of cosmic rays generates neutrinos as well, and the potentialities of high-energy neutrino astrophysics are very large. [Since I have participated in a single work (54) in this field, I shall refer only to it and the review contained in (34).] The prospects for the development of cosmic-ray astrophysics and the whole of high-energy astrophysics have been repeatedly discussed, by me among others [most recently in (36)].

Summarizing, I can say that within the memory of my generation, astronomy has undergone a profound transformation—from optical to "all-wave"; in the course of this process, cosmic-ray astrophysics has been added, and in the near future it will also include high-energy neutrino astrophysics. Along with physics, I was lucky to plunge rather early into astronomical problems as well.

A FEW WORDS ABOUT PRIORITY

Questions of priority have long played an important role in the scientific community. Suffice it to recall how much time and strength the great Newton spent on priority questions (see e.g. 55) even in the period when he had already obtained general recognition. His was, of course, another era, but in not so remote times and even now the priority passions have flared and do flare up, although people have learned to hide them. At the same time, a great increase in the number of scientific publications and of various conferences has created additional difficulties.

On August 3, 1987, I gave an introductory talk at the 20th International Cosmic Ray Conference in Moscow; in the talk itself I did not mention the names of the authors for the majority of the papers that I touched upon. That is why I gave some explanations that were absent in the published version (36). For this reason and for purposes of further discussion, I briefly dwell on these explanations here.

First, the gleam of names in a talk (and, say, on overhead transparencies) hinders discussion of the material itself in that it diverts attention from the content. Second, papers and authors now are so numerous that it is impossible to mention everybody, while selection often brings displeasure and offense. In this connection I showed a transparency with the following two phrases:

> "Priority questions are a dirty business. Priority mania or supersensitivity is a disease."

I then advised not to attach much importance to the absence of references. In most cases, this omission is explained not by evil intent but rather by the abundance of literature, by ignorance, by the desire to refer only to the most recent review, etc. It is only a negligible minority of authors who avoid citations deliberately and act from ill-intentioned motives. Such people and such actions, with some exceptions, do not deserve attention. As I understood, my transparency and comments were taken exactly as I expected, i.e. as advice given half-jokingly, inspired by my experiences of many years in the fields of physics and astrophysics. By the way, two weeks after (on August 21, 1987) I made a remark in the same spirit at the 18th International Conference on Low-Temperature Physics but in quite different conditions and atmosphere, due to which some people misunderstood me [see (4)]. But, of course, my way of thinking is exactly as I have tried to explain it above. My point is not that I am somehow indifferent to the question of priority. Yes, I do notice when my papers are cited and when not, but I never lay claims to authors who do not refer to my works. Besides feeling that such claims (which are, unfortunately, not rare) do not seem to be very decent, I think that in most cases, as I have already mentioned, the reasons for which a citation is sometimes omitted are not ill intentioned.

I decided to touch upon this question of priority for two reasons. First, it seems to me that it deserves attention because some people are deeply agitated about it and get upset. Therefore, one should not pretend that such a problem does not exist. Second, from time to time I happen to be, to this or that extent, involved in priority disputes or some collisions. I may state that I have never tried to prove or defend my priority. In all cases my indignation was provoked by the tactless behavior of my opponents. It is, of course, up to the reader to decide whether or not I am sincere—a man may be sophisticated and often play the hypocrite or distort, maybe unconsciously (subconsciously?), the truth. But I also have the right to express my opinion, which is what I have done above.

I would not do this, however, if I did not think it necessary to somehow explain here my relations with I. S. Shklovsky (1916–1985). It has been well known in my country and possibly abroad that since the end of 1967 I was on bad terms with Shklovsky, that we did not speak to nor greet each other. I do not know about the West, but in our country such a fantastic form of relations is sometimes encountered. By the way, it does not necessarily show some special enmity. In this case it simply happened, and neither of us wanted afterward to break the ice of estrangement. In the opinion of some colleagues, the break-off of my relations with Shklovsky was due to my priority claims. In my view, this is absolutely wrong. Fortunately, I need not prove this assertion here, since it is quite clear from papers (1, 2) written while Shklovsky was alive.

Of course, he made use of my papers and took into account my results, but I did the same with respect to his papers. True, we had different manners of citation, but this question is of secondary importance and therefore is not the point here.

We were of the same age and studied almost simultaneously at Moscow State University, but we made a closer acquaintance only during the Soviet expedition to Brazil, aimed at observing the total solar eclipse of May 20, 1947, in which we both participated. From then our relations, if not very close, were quite friendly up to 1966. By the way, all those "claims" that I could, in principle, lay to Shklovsky refer to the pre-1966 period [see references in (1, 2)]. But I did not pay attention to his tactlessness and attached no importance to it. But from 1966 on, beginning with the IAU conference on radio astronomy in Nordwijk, Shklovsky was ill-disposed toward me for some reason, of which I can only guess. This had no consequences until the fall of 1967, when Shklovsky published his paper (56) in which he associated powerful X-ray radiation from a number of sources with binary systems, in which one of the components is a neutron star onto which plasma accretes. Meanwhile, all of this had, as a matter of fact, been said at a round-table discussion held during the Nordwijk conference the year before (50). The results of the discussion were reported at the conference by G. Burbidge, who was the chairman of this discussion [see (50, p. 463)]. I was present at the discussion, and so was Shklovsky, who, as far as I remember, did not utter a word. In his paper (56) he did not refer to this discussion, which was published some time later (50). The history of the interpretation of powerful X-ray radiation has been described in more detail by Burbidge (57), who also narrated what I have said above. I spoke at the round-table discussion and after Burbidge's report [see (50)] as well, but I have not published anything on this subject and have no special priority claims. But as far back as 1966, Ya. B. Zel'dovich, together with I. D. Novikov, had published a paper (58) containing the idea of accretion onto a neutron star. According to Zel'dovich, Shklovsky knew about it but made no reference to it in (56). Zel'dovich was generally very sensitive to priority questions, and with respect to Shklovsky's paper (56) he was indignant and told me about this. I, in turn, told him about the Nordwijk round-table discussion. Eventually, we wrote (of course, with both our signatures) a corresponding letter to Shklovsky, and I added there the examples presented in (1, 2). I have made no attempts now to find the copy of this letter among my papers. The letter was not distributed—it was sent to Shklovsky only. However, his close colleagues read this letter. Zel'dovich and I were led by the naive idea that Shklovsky's eyes would be opened, but nothing of the sort happened. He wrote an answer, but I was not shown it because I left for England for three months and when I returned, Shklovsky either was ill or fell ill soon thereafter, and thus his friends and coworkers decided to "damp down" this matter.

After that I did not communicate with Shklovsky, but he remained on diplomatic terms with Zel'dovich. I think that in sending this letter, Zel'dovich and I made a mistake, the more so as it did not change in any way Shklovsky's position. In particular, in his paper written in 1982 on the occasion of the 20th anniversary of X-ray astronomy, and even in spite of the paper by Burbidge (57), Shklovsky insisted on the priority of his paper (56).

Shklovsky was a very capable man and had a number of actual achievements in the field of astrophysics. As regards his unfortunate statements concerning radio astronomy and X-ray astronomy, they were harmful, first of all, to himself; I will not take the liberty of interpreting his motives. In his last years, Shklovsky wrote autobiographical short stories. At the end of 1987 I was given a volume of these short stories, entitled *Echelon*. I found this book very interesting and hope it will someday be published. Influenced by Shklovsky's short stories, I wrote a rather long letter (addressed to L. S. Marochnik) in which I narrated the story of my relations with Shklovsky and the cause of our quarrel. This letter was read by Shklovsky's former coworkers from the Institute for Space Research, and I asked one of them (B. V. Komberg) to keep this letter and, if possible, other documents. Historians of astronomy (or psychology?) may one day be interested in these documents.

There remains only to answer the following question: Was this worth writing? Shklovsky is unfortunately not with us now, and so my position is rather delicate—he cannot object to me, and how can I avoid suspicions of ignoble posthumous settling a score with him? I understand this, and at first hesitated, but I still decided not to delete this section. First, all my remarks are completely based on already published materials, which are cited in (1, 2) and above. Thus, those interested may check everything. Second, some unpublished material (including the letter—Shklovsky's reply—that I have not read) is with his former coworkers, and I will only be too glad if they now publish everything that they may care to, including their own comments. Finally, as I write this paper (in fall 1988) I am already 72 years old, and I am not at all sure I shall live to see it published. Under such conditions it seems to me simply preposterous to write not with the goal of clarifying the truth but from some other considerations. I hope nobody will doubt that.

CONCLUDING REMARKS

In this paper I have not intended to enumerate all my papers that refer to astronomy. The main references, however, were made, directly or indirectly, when I referred the reader to papers (1, 2) and others. In view of the fact that this paper is to be published in the *Annual Review of Astronomy and Astrophysics*, it seems not out of place to refer separately to my papers (59, 60), which were published in this series and devoted to the theory of synchrotron radiation. I shall also mention papers (61, 62) on X-ray and gamma-ray astronomy. In (63) I discussed the problem of superfluidity and superconductivity in astrophysics.

Besides these, I shall dwell only on the papers connected with the general theory of relativity (GTR). GTR is, of course, a physical theory, but it is so closely connected with astronomy that it can be referred to as astrophysics as well. This, in particular, is what I did in my popular book (64).

In general, I did not work hard on GTR and do not belong to the so-called relativists. I shall mention the study of the collapse of a magnetic star (29, 65) which led to the conclusion that the magnetic moment of the star tends to zero as the Schwarzschild radius is reached. This was the first indication of an effect that later was described by the phrase "a black hole has no hair." How large the magnetic field of collapsing stars can be was pointed out in (29), several years prior to the discovery of pulsars and, therefore, neutron stars.

The application of the equivalence principle, the fundamental principle of GTR, in taking into account the quantum effects and, specifically, zero-point fluctuations of different fields is discussed in (66). Finally, several of my papers were devoted to experimental verification of GTR [my last review on this subject is (67); see also (64)] and to the limits of its applicability. [I refer only to (68, 69).] The latter problem is associated with the question of limitations on certain physical concepts when applying them to astronomy, a question that has attracted the attention of some astronomers. My Darwin Lecture (18) was devoted to this subject. Today I am of the same opinion as that expressed in this lecture—namely, that one can give no definite guarantees, but it is most probable and realistic that if some deviations from the known physical laws and, specifically, from Einstein's original, classical GTR can be expected, they may happen only at very early stages of cosmological evolution. [I mean the observable, expanding Universe, which, from the point of view of some inflationary cosmological models, may appear to be part of a larger or even infinite system—see the review (70).] True, I am not going to touch upon black holes that are deep below the horizon (i.e. for a Schwarzschild black hole for $r \ll r_g = 2GM/c^2$), nor upon mini–black holes, cosmic strings, etc.

One of the possible ways in which GTR might be violated is connected with the existence (purely hypothetical) of a fundamental length $l_f \gg l_g = (G\hbar/c^3)^{1/2} \sim 10^{-33}$ cm [for more details, see (69)]. At later stages of cosmological evolution (for example, for the redshift parameter $z < 10^3$), there is absolutely no reason to expect that the well-known laws of macroscopic physics do not apply. Of course, this is on the whole a large and complicated problem, and it is impossible to consider it here more thoroughly.

In concluding, I would like to ask this final question: Have I contributed anything essential to astronomy? This is undoubtedly up to other people to decide, but the author himself also has the right to have his own opinion, even if it is erroneous. For instance, I believe that I have made a noticeable contribution to the theory of superconductivity, but with respect to astrophysics I do not have any definite opinion. I have done some things, of course, but how important are they, and what role have they played? It would be interesting to have the answer, but unfortunately I do not know a single work that considers in detail and reliably assesses the history of "new astronomy"—i.e. the development of astronomy over the last 40–50 years. On the contrary, I have come across rather many papers and even booklets or books containing errors concerning this history. Absurd assertions, adapted by repetition, roam from one such paper or book to another. It is surprising how credulous some authors who elucidate history are not to consult all the original literature and check their assertions. I hope that the situation will be clarified with time and believe that the prefatory chapters published in the Annual Reviews will make their contribution to this process. I hope the present paper is not an exception, although I have not tried [with some exceptions; see (1, 2)] to take on the role of historian—that is, to compare, verify, and assess the results and assertions of various authors. Besides, against my own expectations and original aim, this paper proved to be devoted not so much to astrophysics as to autobiographical and other remarks. But maybe this is not so bad.

Acknowledgment

I would like to warmly thank Kip Thorne and Keith Dodson for reading and commenting on the manuscript.

Literature Cited

1. Ginzburg, V. L. 1984. In *The Early Years of Radio Astronomy*, ed. W. Sullivan, pp. 289–302. Cambridge: Univ. Press
2. Ginzburg, V. L. 1985. In *The Early History of Cosmic Ray Studies*, ed. Y. Sekido, H. Elliot, pp. 411–26. Dordrecht: Reidel
3. Ginzburg, V. L. 1986. *Priroda* 1986(10): 80–94 (In Russian)
4. Ginzburg, V. L. 1988. *Prog. Low-Temp. Phys.* 12: 1–44
5. Ginzburg, V. L. 1946. *C. R. (Dokl.) Acad. Sci. URSS* 52: 487–90
6. Ginzburg, V. L. 1942. *C. R. (Dokl.) Acad. Sci. URSS* 36: 8–13
7. Ginzburg, V. L. 1987. In *Reminiscences About I. E. Tamm*, p. 190. Moscow: Nauka (In Russian)
8. Ginzburg, V. L. 1989. *Applications of Electrodynamics in Theoretical Physics and Astrophysics*. New York: Gordon & Breach

9. Ginzburg, V. L. 1984. *Sov. Phys. Usp.* 26: 713
10. Ginzburg, V. L. 1989. In *Reminiscences About L. D. Landau.* Oxford: Pergamon. In press
11. Ginzburg, V. L. 1986. In *The Lesson of Quantum Theory*, p. 113. Elsevier [Unfortunately, my paper was cut down by the editors (evidently for want of space). The complete version is published in Russian in 1986. *Tr. Fiz. Inst. P. N. Lebedev Akad. Nauk SSSR* 176: 3. This volume is being translated into English by Nova Science Publ., New York)
12. Ginzburg, V. L., Man'ko, V. I. 1976. *Sov. J. Part. Nucl.* 7: 1
13. Ginzburg, V. L. 1987. In *Quantum Field Theory and Quantum Statistics*, 2: 15–33. Bristol, Engl: A. Hilger
14. Ginzburg, V. L., Landau, L. D. 1950. *Zh. Eksp. Teor. Fiz.* 20: 1064–82
15. Ginzburg, V. L. 1988. In *Ynogo ne dano (One Way)*, pp. 135–53. Moscow: Progress Publ. (In Russian)
16. Garfield, E. 1977. *Current Contents*, 1977(49): 5–15
17. Ginzburg, V. L. 1962. *Tr. FIAN (Proc. P. N. Lebedev Phys. Inst.)* 18: 55–104
18. Ginzburg, V. L. 1975. *Q. J. R. Astron. Soc.* 16: 265–85
19. Ginzburg, V. L. 1970. *Propagation of Electromagnetic Waves in Plasmas.* Oxford: Pergamon
20. Martyn, D. F. 1946. *Nature* 158: 632
21. Shklovsky, I. S. 1946. *Astron. Zh.* 23: 333
22. Ginzburg, V. L. 1947. *Usp. Fiz. Nauk* 32: 26
23. Ginzburg, V. L. 1948. *Usp. Fiz. Nauk* 34: 13
24. Ginzburg, V. L. 1980. In *Oort and the Universe*, p. 129. Dordrecht: Reidel
25. Ginzburg, V. L., Zhelezhnyakov, V. V. 1958. *Astron. Zh.* 35: 694 (*Sov. Astron. AJ* 2: 653)
26. Zhelezhnyakov, V. V. 1970. *Radio Emission of the Sun and Planets.* Oxford: Pergamon
27. Ginzburg, V. L. 1961. *Astron. Zh.* 38: 380
28. Ginzburg, V. L., Ozernoy, L. M. 1977. *Astrophys. Space Sci.* 50: 23 (see also 1970. *Astrophys. Space Sci.* 9: 116)
29. Ginzburg, V. L. 1964. *Dokl. Akad. Nauk SSSR* 156: 43–46 (*Sov. Phys. Dokl.* 9: 329)
30. Ginzburg, V. L., Zheleznyakov, V. V. 1975. *Annu. Rev. Astron. Astrophys.* 13: 511–35
31. Beskin, V. S., Gurevich, A. V., Istomin, Ya. N. 1986. *Sov. Phys. Usp.* 29: 946 (see also 1988. *Astrophys. Space Sci.* 146: 205)
32. Ginzburg, V. L., Syrovatskii, S. I. 1964. *The Origin of Cosmic Rays.* Oxford: Pergamon
33. Ginzburg, V. L., Ptuskin, V. S. 1976. *Rev. Mod. Phys.* 48: 161, 675
34. Ginzburg, V. L., ed. 1990. *Astrofizika kosmicheskych luchey (Cosmic Ray Astrophysics).* Moscow: Nauka. 2nd ed. (expected to be translated into English by North-Holland Publ., Amsterdam)
35. 1987. *Proceedings of International Cosmic Ray Conference (ICRC), 20th, Moscow*
36. Ginzburg, V. L. See Ref. 35, 7: 7 (see also 1988. *Usp. Fiz. Nauk* 155: 185; 1988. *Sov. Phys. Usp.* 31: 491–510)
37. Ryle, M. 1949. *Proc. Phys. Soc. London Sect. A* 62: 491
38. Unsöld, A. 1949. *Z. Astrophys.* 26: 176
39. Shklovsky, I. S. 1952. *Astron. Zh.* 29: 418
40. Schott, G. A. 1912. *Electromagnetic Radiation.* Cambridge: Univ. Press
41. Alfvén, H. A., Herlofson, N. 1950. *Phys. Rev.* 78: 616
42. Kipenheuer, K. O. 1950. *Phys. Rev.* 79: 738
43. Ginzburg, V. L. 1951. *Dokl. Akad. Nauk SSSR* 76: 377
44. Pikelner, S. B. 1953. *Dokl. Akad. Nauk SSSR* 88: 229
45. Shklovsky, I. S. 1953. *Astron. Zh.* 30: 15
46. 1957. *Radio Astronomy, IAU Symp. No. 4.* Cambridge: Univ. Press
47. 1959. *Paris Symposium on Radioastronomy, IAU Symp. No. 9.* Stanford, Calif: Stanford Univ. Press
48. Baade, W., Zwicky, F. 1934. *Phys. Rev.* 46: 76 (also 1934. *Proc. Natl. Acad. Sci. USA* 20: 259)
49. Fermi, E. 1949. *Phys. Rev.* 75: 1169
50. 1967. *Radio Astronomy and the Galactic System, IAU Symp. No. 31.* Academic
51. Ginzburg, V. L. 1989. In *Essays on Particles and Fields*, p. 103. Bangalore: Indian Acad. Sci.
52. Beresinsky, V. S., Ginzburg, V. L. 1987. *Nature* 329: 807
53. Dogiel, V. A., Ginzburg, V. L. 1989. *Space Sci. Rev.* 49: 311–83
54. Beresinsky, V. S., Ginzburg, V. L. 1981. *MNRAS* 194: 3
55. Westfall, R. 1982. *Never at Rest: A Biography of Isaac Newton.* Cambridge: Univ. Press
56. Shklovsky, I. S. 1967. *Ap. J. Lett.* 148: L1
57. Burbidge, G. 1972. *Comments Astrophys. Space Sci.* 4(4): 105
58. Novikov, I. D., Zel'dovich, Ya. B. 1966. *Nuovo Cimento Suppl.* 4: 810

59. Ginzburg, V. L., Syrovatskii, S. I. 1965. *Annu. Rev. Astron. Astrophys.* 3: 297–350
60. Ginzburg, V. L., Syrovatskii, S. I. 1969. *Annu. Rev. Astron. Astrophys.* 7: 375–420
61. Ginzburg, V. L., Syrovatskii, S. I. 1965. *Space Sci. Rev.* 4: 267–312 (see also 1964. *Dokl. Akad. Sci. SSSR* 154: 557)
62. Ginzburg, V. L. 1967. *Sov. Phys. Usp.* 9: 543–50
63. Ginzburg, V. L. 1969. *Sov. Phys. Usp.* 12: 241–51 (see also 1971. *Physica* 55: 207–12)
64. Ginzburg, V. L. 1985. *Physics and Astrophysics (A Selection of Key Problems)*. Oxford: Pergamon
65. Ginzburg, V. L., Ozernoy, L. M. 1965. *Sov. Phys. JETP* 20: 689–96
66. Ginzburg, V. L., Frolov, V. P. 1987. *Usp. Fiz. Nauk* 153: 633–74 (Engl. transl., *Sov. Phys. Usp.*)
67. Ginzburg, V. L. 1980. *Sov. Phys. Usp.* 29: 514–27
68. Ginzburg, V. L., Kirzhnits, D. A., Lyubushin, A. A. 1971. *Sov. Phys. JETP* 33: 242–46
69. Ginzburg, V. L., Mukhanov, V. F., Frolov, V. P. 1988. *Zh. Eksp. Teor. Fiz.* 94: 1 (Engl. transl., *Sov. Phys. JETP.* 1988. 67: 649)
70. Barrow, J. D. 1988. *Q. J. R. Astron. Soc.* 29: 101
71. Ginzburg, V. L., Ozernoy, L. M. 1966. *Astron. J.* 9: 726

INTERSTELLAR DUST AND EXTINCTION

John S. Mathis

Washburn Observatory, University of Wisconsin-Madison, 475 North Charter Street, Madison, Wisconsin 53706

KEY WORDS: interstellar medium, molecular clouds, interstellar grains

1. INTRODUCTION

Interstellar dust is an important constituent of the Galaxy. It obscures all but the relatively nearby regions at visual and ultraviolet (UV) wavelengths and reradiates the absorbed energy in the far-infrared part of the spectrum, thereby providing a major part ($\sim 30\%$) of the total luminosity of the Galaxy. The far-infrared radiation from dust removes the gravitational energy of collapsing clouds, allowing star formation to occur. Dust is crucial for interstellar chemistry, in that it reduces the UV radiation, which causes molecular dissociations, and provides the site for the formation of the most abundant interstellar molecule, H_2. Probably grain surfaces are responsible for other chemistry as well. Dust controls the temperature of the interstellar medium (ISM) by accounting for most of the elements that provide cooling, as well as by providing heating through electrons ejected photoelectrically from grains.

The past decade has seen an increase in interest in interstellar dust because of the discovery of spectroscopic features in both emission and absorption, along with laboratory studies of candidate materials. There have been good observations of the extinction law of dust in many directions. Probably the most important feature to emerge from these studies is that "interstellar dust" refers to a variety of materials of widely varying properties.

Many studies of interstellar dust have involved lines of sight through the diffuse, low-density ISM, including some clouds having densities of up to several hundred H atoms per cubic centimeter. This material is referred to herein as *diffuse dust*. In the literature, most references to "interstellar dust" apply to diffuse dust. Dust in the outer parts of molecular clouds that can be studied by optical and UV observations is called *outer-cloud dust*. Finally, there have been many studies of sources embedded so deeply within molecular clouds that only the near-infrared or perhaps optical part of the spectrum can be studied. This type of dust is referred to as *inner-cloud dust*. There is, of course, a continuous gradation of properties from diffuse dust to inner-cloud dust, but these three designations allow us to emphasize the rather different properties of interstellar dust in the various regions. This review is confined to diffuse dust and outer-cloud dust; for excellent reviews of inner-cloud dust, see (156, 157, 165).

Recent general references regarding interstellar dust are the proceedings of (*a*) a 1985 workshop held at Wye, Maryland (123); (*b*) the 1987 conference on "Dust in the Universe," held in Manchester, England (7); and (*c*) IAU Symposium No. 135 on "Interstellar Dust," held in Santa Clara, California, in July, 1988 (3). In general, reviews in these volumes on specialized aspects are not referenced here specifically. In addition, recent references are generally given in preference to older but important papers.

Extinction (absorption plus scattering) is by far the best-studied property of diffuse dust and outer-cloud dust because it can be determined accurately over a wide range of wavelengths and for lines of sight sampling different physical conditions in the ISM. Both continuous extinction and certain spectral features (narrow-wavelength regions over which the extinction varies appreciably) are discussed in Section 2. Another very important diagnostic is the emission from dust (Section 3), both in spectral features (which provide clues as to specific materials) and in the far-infrared (representing the emission from grains warmed by incident radiation or particles). Scattering (Section 4), polarization (Section 5), and other diagnostics (Section 6) also provide information. The evolution of dust is outlined in Section 7, and theories are discussed in Section 8. A summary is given in Section 9.

In this article I refer to the wavelength region 0.9 μm $< \lambda <$ 10 μm as the near-infrared (NIR), 10 μm $\leq \lambda \leq$ 30 μm as the mid-infrared (MIR), and $\lambda >$ 30 μm as the far-infrared (FIR).

2. INTERSTELLAR EXTINCTION

2.1 *Continuous Extinction*

Each line of sight has its own "extinction law," or variation of extinction with wavelength, usually expressed by $A(\lambda)/A(V)$ in this article. This means of expressing the extinction law is not unique; it has been common practice to use instead the ratios of two colors, $E(\lambda-V)/E(B-V)$, where

$E(\lambda - V) = A(\lambda) - A(V)$. The use of $A(V)$ as the reference extinction is arbitrary, and it might be preferable to use instead a longer wavelength, such as the J bandpass (≈ 1.25 μm), because the extinction law would then be virtually independent of the line of sight (see Section 2.1.3). The information content of the extinction law is independent of how the law is expressed, but certain relations (see Section 2.1.1) become more obvious when $A(\lambda)/A(V)$ is used in place of $E(\lambda - V)/E(B - V)$.

2.1.1 OPTICAL/ULTRAVIOLET EXTINCTION OF DIFFUSE DUST AND OUTER-CLOUD DUST   There have been several studies of the spatial distribution of extinction in the optical part of the spectrum (e.g. 51, 101, 122) that might be useful in estimating the amount of extinction in a particular direction, but this review concentrates instead upon the form of the extinction law and the physical nature of dust. Ardeberg & Virdefors (5) discuss the optical extinction law, with references.

Many authors have utilized the *International Ultraviolet Explorer* (*IUE*) to make detailed studies of the UV extinction law of diffuse dust and outer-cloud dust. There are considerable differences among the various lines of sight. Cardelli, Clayton & Mathis [19, 20 (hereinafter CCM89)] have used the UV observations of Fitzpatrick & Massa [54 (hereinafter FM86), 55, 109], with optical/NMR observations of the same stars, to explore the relationships between various extinction laws over the entire available interval of wavelengths. These observations were spread over the sky and included both diffuse dust and lines of sight to the Ophiuchus, Orion, and other molecular clouds, as well as to H II regions. CCM89 used the optical total-to-selective extinction ratio $R_V = A(V)/E(B-V)$ as a parameter. (R_V is determined by extrapolating NIR extinction to infinite wavelength.) Figure 1 shows the observed extinction laws of many lines of sight, plotted against R_V^{-1}, for several values of λ ranging from the red to almost the limit of the *IUE* spacecraft (1200 Å). There are fairly tight linear relationships between $A(\lambda)/A(V)$ and R_V^{-1} in each case, including the UV wavelengths.

The value of R_V depends upon the environment along the line of sight. A direction through low-density ISM usually has a rather low value of R_V (about 3.1). Lines of sight penetrating into a dense cloud, such as the Ophiuchus or Taurus molecular clouds, usually show $4 < R_V < 6$. However, it is not possible to estimate R_V quantitatively from the environment of a line of sight; for example, the star VI Cyg 12 lies behind a dense cloud of dust but has an R_V of 3.1 (78a), a value appropriate for the diffuse ISM. As a further example, $R_V \approx 3.0$–3.5 in parts of the Taurus cloud (159).

CCM89 fitted the slopes of the various $A(\lambda)/A(V)$–R_V^{-1} relations, examples of which are shown in Figure 1, by an analytic formula that

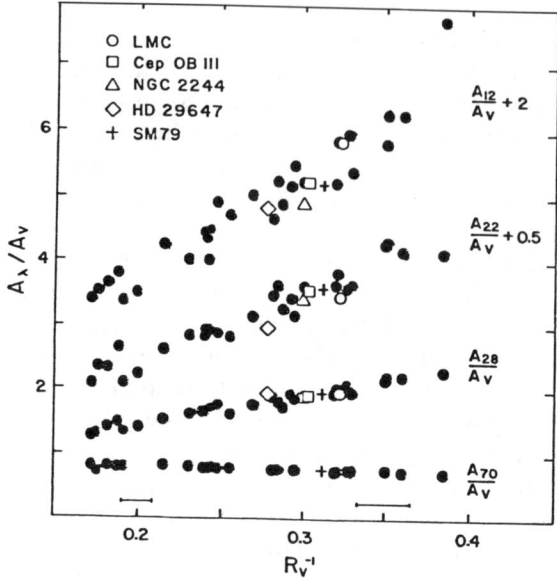

Figure 1 The observations of $A(\lambda)/A(V)$ plotted against R_V^{-1}, where $R_V = A(V)/E(B-V)$ (from CCM89). A_{12} refers to 1200 Å, A_{22} to 2175 Å, A_{28} to 2800 Å, and A_{70} to 7000 Å (the standard R filter). The black-dot observational values are from Fitzpatrick (53). The regularity of the observations and the scatter about the mean relationship are also shown.

represents the mean extinction law as a function of R_V. The expression is not reproduced here. Figure 2 (from CCM89) shows the mean extinction law for three values of R_V as calculated from the formula. Also shown are actual observations with the same values of R_V. The central curve is about as discrepant as actual observations are from the mean relationship. The dispersion of individual extinction laws around that mean law is shown in Figure 1 (from the spread in the individual observed points) and in the panel in Figure 2 (as error bars giving the standard deviation). The lowest set of curves plotted in Figure 2 are for Herschel 36, an exciting star of the H II region M8 and considered to have very "peculiar" extinction. Rather, Herschel 36 has a peculiar value of R_V (≈ 5.3) but a "normal" extinction law for that value of R_V. Note, however, that there are real deviations from the mean extinction law for any particular value of R_V. These deviations are especially large at 1200 Å, where the standard deviation of $A(\lambda)/A(V)$ from the mean extinction law is about 0.3. Extreme deviations might be found in the future and will provide valuable information regarding the processes that modify the grains.

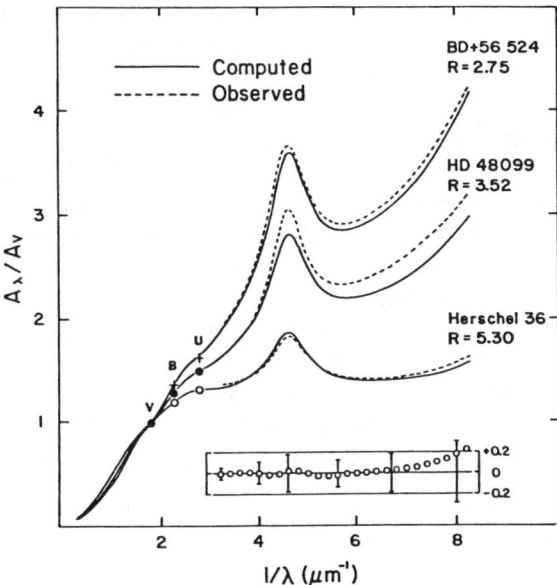

Figure 2 Three cases of a mean extinction law. Solid lines are obtained by fitting the slopes of the $A(\lambda)/A(V)$–R_V^{-1} relationship (Figure 1) by an analytic formula (CCM89), and dashed lines are actual extinction laws of stars (FM86) with the appropriate values of R_V. The error bars in the lower panel show the standard deviations of the observations of the entire sample (54) from the extinction law obtained from the formula. The open circles in the panel show the deviations from the "standard" mean extinction law (149) for the value $R_V = 3.2$, appropriate for the diffuse ISM.

The "mean" extinction laws of Savage & Mathis (144) and Seaton (149) are commonly used to correct observations for the presence of dust. Both laws are reproduced closely if $R_V = 3.2$ is substituted into the R_V-dependent mean extinction law given in CCM89. The panel in Figure 2 shows the deviations of the CCM89 mean extinction law, with $R_V = 3.2$ [from Seaton (149)]. It is not surprising that mean extinction laws correspond to an R_V slightly higher than 3.1; some lines of sight used in the averaging penetrate fairly dense regions.

Figure 1 shows that there is a continuous change between properties of diffuse dust, with $R_V \approx 3.1$, and outer-cloud dust, with large values of R_V. The two designations merely contrast one end of the observed range of R_V with the other.

The differences in the extinctions between diffuse dust and outer-cloud dust strongly affect any predictions concerning physical conditions inside clouds. Figure 2 shows that outer-cloud extinction laws, i.e. those with $R \gtrsim 4$, rise much less steeply at shorter wavelengths than diffuse dust. Consequently, interstellar radiation incident upon a cloud can penetrate the cloud much more easily than would be predicted from the Seaton or Savage-Mathis extinction laws. Figure 3 shows the mean radiation intensity at the center of a cloud with a radial extinction of $A(V) = 5$ magnitudes (a typical value) computed using $R_V = 3.1$, and also with a typical outer-cloud dust value of $R_V = 5$. The difference in the predicted mean intensities has large implications for the physical processes in the cloud.

The fact that the extinction law depends so regularly on R_V suggests that the processes that modify the sizes and/or the compositions of grains must operate on all sizes simultaneously and quite efficiently. One could imagine that the small and large grains would be modified independently along various lines of sight, but such is not the case. The consequences of the regularity of the various extinction laws are considered in Section 7.

2.1.2 THE 2175-Å "BUMP" The strongest spectroscopic feature in the entire observed spectrum, in terms of equivalent width in frequency units, is the "bump" situated at 2175 Å, or 4.6 μm^{-1} (see Figure 2). The bump is present for all values of R_V. Its origin is not well understood (see ref. 38

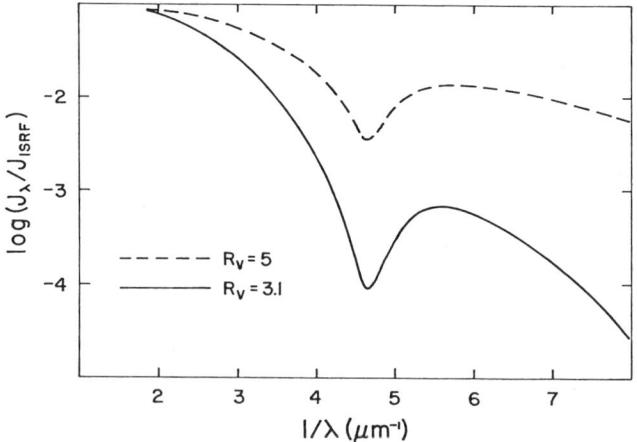

Figure 3 The mean intensity of radiation at the center of a cloud with a radial extinction of $A(V) = 5$ mag and no internal sources, expressed in terms of the mean intensity of the incident interstellar radiation field. Two values of $R_V [= A(V)/E(B-V)]$ are shown: one corresponding to the mean value for diffuse dust ($R_V = 3.1$), and the other for a typical observed value for lines of sight penetrating clouds ($R_V = 5$). The use of a mean extinction law corresponding to diffuse dust for predicting the radiation field within clouds leads to a gross error in the predicted radiation field within the cloud.

for an excellent review), although there is general (but not unanimous) agreement that it is probably caused by graphite or a slightly less well ordered form of carbon.

Let $A_{bump}(\lambda^{-1})$ be the extinction at a wave number λ^{-1} between 3.3 and 6 μm^{-1} in excess of a linear extinction interpolated between the end points. The important properties of $A_{bump}(\lambda^{-1})$ are as follows (176, FM86, CCM89):

1. The bump is extremely strong and must be produced by a very abundant material (which is why most theories attribute it to carbon). The equivalent width of the bump per $A(V)$, known from CCM89 or Seaton (149), can be expressed in terms of the oscillator strength, f_{bump}, times the number of absorbing atoms, N_{bump}. Bohlin et al (9) determined the mean $N(H)/E(B-V)$, from which $N(H)/A(V)$ follows. Dividing the expressions yields $N_{bump} f_{bump} = 9.3 \times 10^{-6} N(H)$. Only the elements C, N, O, Ne, Mg, Si, and Fe (excluding noble gases) can provide enough absorption strength, even if the transition is exceedingly strong ($f_{bump} \approx 1$). Each of the elements Fe, Si, and Mg require $f_{bump} = 0.3$ even if the entire cosmic abundance is responsible for producing the bump, while 8% of the carbon would be required for the same f_{bump}.

2. The central wave number λ_0^{-1} is surprisingly constant. For the stars in the FM86 sample, it is given by $\lambda_0^{-1} = 4.599 \pm 0.019$ μm^{-1}, corresponding to $\lambda_0 = 2174 \pm 9$ Å. This amounts to a mean deviation of 0.4% in λ_0^{-1}, whereas other properties of the extinctions vary considerably. However, there are real variations in λ_0. The spreads of λ_0^{-1} for stars within a given cluster (101) are significantly less than among the field stars and serve to establish an upper limit to the observational errors. The stars HD 29647 and HD 62542 have especially peculiar extinctions (18), with $\lambda_0 = 2128$ Å and 2110 Å, respectively, that are smaller than the mean λ_0 by many standard deviations. These stars have the broadest bumps known (see below) but a rather small central absorption $A_{bump}(\lambda_0)$, so that the integrated strengths of their bumps are about average. Their environments are very different, one being in a quiescent region in Taurus and the other in the Gum Nebula.

It is remarkable that graphite, a completely ordered and stable form of carbon, has a resonance very close to 2175 Å of about the right width and strength to produce the bump. Small graphite particles (radii < 0.005 μm) of various sizes and similar shapes would have the resonance at a common wavelength independent of size, but larger ones would have λ_0 shifted to longer wavelengths. Almost all theories suggest that the bump is produced by graphitic carbon in this way, along the lines suggested by Hecht (66).

3. The width of the bump, expressed as the full width at half-maximum

(FWHM), varies widely, currently with extremes of 0.768 μm^{-1} (HD 93028) and 1.62 μm^{-1} (HD 29647). The only significant correlation of the FWHM is with the mean gas density along the line of sight (FM86). The lack of correlation between the FWHM and λ_0^{-1} suggests that the variations in width are not caused by coatings of varying thicknesses upon a common carrier particle. Such coatings should produce a shift in λ_0^{-1} that is related to the change in width. This variation in width, unrelated to the position of the resonance, is difficult to explain with graphite of any size. The total area of the bump relative to $A(V)$, $\int A_{\text{bump}}(\lambda^{-1})/A(V)\,d\lambda^{-1}$, varies by over a factor of two among the FM86 stars.

4. The albedo of the bump probably has not been determined reliably. Lillie & Witt (98), using the observations of the diffuse galactic light from *Orbiting Astronomical Observatory 2*, suggested that the albedo drops across the bump. This is expected if the particles producing the bump are small (as suggested by the constancy of λ_0^{-1}, which can most easily be explained by absorption from small particles). However, plane-parallel axisymmetric geometry was used for interpreting the data, while the actual sky brightness in the UV is now considered to be uncertain and almost surely quite asymmetrical (see Section 6).

5. Two reflection nebulae show evidence of scattering in the bump, with a different profile in each (177), suggesting that some carriers of the bump can be large. (Small grains do not scatter.) This scattering is not observed in other nebulae. The expected shift in λ_0 to longer wavelengths (because of the relatively large particles causing the scattering) is not observed in the extinction of the exciting stars of these nebulae.

6. Observations of carbon stars (see Section 6.2) suggest that amorphous carbon, not graphite, is injected into the ISM. Perhaps *small* graphite particles can be produced later by annealing, but it is difficult to see how large graphite flakes can be made.

2.1.3 NIR CONTINUOUS EXTINCTION Figure 2 shows great differences in extinction laws among various lines of sight throughout the optical and UV portions of the spectrum. We might expect a corresponding variation at somewhat longer wavelengths, but apparently there is little if any.

Most NIR photometry is at the Johnson filters J (1.25 μm), H (1.65 μm), and K (2.2 μm). Whittet (166) tabulates $E(J-H)/E(H-K)$ as determined by several studies, in diffuse dust and outer-cloud dust alike, and finds them to be consistent with the value $E(J-H)/E(H-K) = 1.61 \pm 0.04$. Koornneef (87) considered a large body of data and suggests an extinction law that has a value of 1.70 for this ratio. Jones & Hyland (79) also concluded that NIR extinction is the same for both diffuse dust and outer-cloud dust, although they found $E(J-H)/E(H-K) =$

2.09 ± 0.10. The constancy of their ratio between lines of sight through diffuse dust and outer-cloud dust is more significant than the difference in the numerical value itself, which depends upon reduction to a standard photometric system.

The NIR extinction law is well fitted by the form $A(\lambda)/A(J) = (\lambda/1.25 \, \mu m)^{-\alpha}$. Recent values of α are 1.70 ± 0.08 (166), 1.61 (136), 1.75 (37), and ~ 1.8 (108). The value 1.70 seems a reasonable compromise for both diffuse dust and outer-cloud dust and implies that $E(J-H)/E(H-K) \sim 1.6$.

The constancy of the NIR extinction law implies that the size distributions of the largest particles are almost the same in all directions. This conclusion was also reached (111) on the basis of the interstellar polarization law, which involves only the largest particles.

2.1.4 FAR-UV EXTINCTION Martin & Rouleau (107) have extended the Draine & Lee (42; hereinafter DL84) opacities through the ionizing-UV range to X-ray energies (3.5 keV), assuming that grains are composed of silicates and graphite. The opacity rises to a maximum of 2.8×10^{-21} cm^2 (H atom)$^{-1}$ at 730 Å ($= 17$ eV) and declines to 7.4×10^{-22} cm^2 (H atom)$^{-1}$ at 124 Å ($= 100$ eV). At energies > 300 eV, the absorption law of the dust is approximately the same as if all of its atoms were neutral in the gas phase. At lower energies, especially just above the thresholds of abundant elements such as carbon, the large grains are opaque and the effective cross section per H atom is reduced. At 24 eV, the reduction amounts to a factor of four. The major effects of the dust as regards high-energy radiation are (a) to keep its constituents absorbing as neutral atoms, rather than possibly being ionized; and (b) to scatter the radiation, with a cross section about equal to the absorption cross section. This scattering can be observed as an X-ray halo around point sources (117). Dust does not affect the ionization equilibrium of H II regions very significantly (112) because its absorption, peaking at 17 eV, resembles hydrogen absorption too closely.

2.1.5 EXTRAGALACTIC EXTINCTION Reasonably reliable measurements for the extinction laws and dust/gas ratios exist only for the Magellanic Clouds. In the Large Magellanic Cloud (LMC), it is found that $R_V \approx 3.2 \pm 0.2$, virtually the galactic value (24, 86, 119). For the UV, the stars near the giant H II region 30 Doradus have weak bumps and extinctions rising steeply at the shortest IUE wavelengths, a behavior unfortunately known as "the LMC extinction law." However, the stars well away from 30 Dor (>500 pc projected distance), spread throughout the Galaxy, have approximately galactic extinction laws (24, 53). The $N(H)/E(B-V)$ is 2×10^{22} atoms mag^{-1} (53, 86), about four times the galactic value (9) and roughly proportional to the gaseous carbon abundance in the LMC.

In the Small Magellanic Cloud (SMC), there are almost no suitably reddened stars. In general there seems to be a low value of R_V, almost no bump, and a very steep far-UV rise (11, 121), as might be expected from a small R_V. One star, though, shows an extinction law similar to that of the Galaxy (96). The $N(H)/E(B-V)$ value is 4.6×10^{22} atoms mag^{-1}, about 10 times the galactic value and consistent with the gaseous C abundance in the SMC (104).

2.2 The 9.7- and 18-μm Silicate Features

Spectral absorption features can provide a great deal of information regarding the compositions and nature of the dust grains. Such features have not been detected in the UV (55, 148, 184), except, of course, for the bump. However, many have been found in the NIR, especially from the icy mantles within molecular clouds. In this review, we concentrate on diffuse-dust and outer-cloud dust and do not discuss the bands of molecular ices in general.

There is a broad, smooth absorption feature peaking at 9.7 μm, attributed to the Si–O stretch in silicates, which is always seen in interstellar dust if $A(V)$ is suitably large. The 9.7-μm band is found in emission from warm circumstellar dust surrounding oxygen-rich stars. In these objects the heavy elements (Fe, Mg, etc) are in silicates in the expanding envelope, while the carbon combines almost completely with oxygen to form CO. The 9.7-μm feature is not seen in circumstellar dust surrounding carbon-rich objects, except for some dusty planetary nebulae in which the grains were expelled by an earlier, and possibly oxygen-rich, phase of stellar evolution. A somewhat weaker and even broader feature peaking at 18 μm, the Si–O–Si bending mode, has been detected in circumstellar dust, in stars near the galactic center, and in molecular clouds (2, 118, 128, 160). The 18-μm band is much less well studied but is found with a strength relative to that of the 9.7-μm feature expected for silicates (about 0.4). The bands are polarized at the same angle (2, 84) with the amplitudes expected for silicates.

The strength and profile of the 9.7-μm band, relative to $A(V)$, have been determined (137) from several Wolf-Rayet type WC stars. These objects have the advantages of being luminous and of having no intrinsic spectral features in the 10-μm region (because they are carbon rich and therefore contain no circumstellar silicate dust). If $\tau_{9.7}$ is the optical depth of the silicate feature above the underlying continuum, the mean value of $A(V)/\tau_{9.7}$ in diffuse dust is 18.5 ± 1.0. However, there are substantial variations of $A(V)/\tau_{9.7}$ within the Taurus molecular cloud (169). The line of sight toward the galactic center has $A(V)/\tau_{9.7} \approx 9$, about half the local value (138), although the strong emission band at 7.7 μm confuses the determination of the continuum underlying the band (28). The dust seems to be diffuse rather than in dense clouds; there are only weak radio lines from molecules commonly seen in molecular clouds. However, conditions in the inner Galaxy (chemical composition, for instance) certainly differ from those in our neighborhood.

The derived shape of the 9.7-μm band is uncertain because the band is only about half as strong at maximum as the continuous extinction at the K band (2.2 μm). In diffuse dust toward WC stars, the shape of the band is similar to the emission seen in dusty oxygen-rich stars such as μ Cep (99, 137). The emission profile near the Trapezium in the Orion Nebula is $\approx 40\%$ broader (60) than in μ Cep, and the profile in dense clouds appears to be 10% narrower (128). It is not surprising that the shape of the band is different in various types of objects; very probably the silicates are in different states of order, with different degrees of impurities. I recommend the Roche & Aitken (137) profile as being typical of diffuse dust.

The 9.7-μm profile shows that interstellar silicate is not crystalline (16). Crystalline silicate absorption peaks at about 10.5 μm, rather than at 9.7 μm (see spectra in ref. 143). Laboratory measurements of amorphous or radiation-damaged silicates (33, 88) show a satisfactory, but not perfect, fit to the observed profile. Stars embedded in molecular clouds warm and partially anneal nearby grains (2).

The silicate band is weak for observational purposes, but it is difficult to account for its strength even if all of the silicon is in silicates and if a rather large opacity (3000 cm^2 g^{-1}) is assumed for the maximum absorptivity (DL84, 157). This value is larger than most laboratory measurements of amorphous or lunar silicates. The total equivalent width of the band is rather constant from one type of silicate to another, so the broad profile of the interstellar band limits the maximum strength. The fundamental Kramers-Krönig relations (9a) limit the strength of the band (DL84) if astronomical silicates are similar to the minerals studied in the laboratory, but heavy contamination with other materials and perhaps a porous nature greatly complicate the issue.

There are suggestions that interstellar silicates are hydrated (68, 85), based on a 6.00-μm H–O–H bending band. However, in general the 6.00-μm feature is not correlated with the 9.7-μm band (172).

2.3 Mean Extinction Laws

Table 1 is an estimate of the extinction law for the observable range of wavelengths, normalized to J (≈ 1.25 μm) because the extinction law is assumed to be independent of environment for $\lambda > 0.9$ μm. The reader can convert to $A(\lambda)/A(V)$ by means of the tabulated $A(V)/A(J)$ values. There are two columns for $\lambda < 0.9$ μm, representing the mean for diffuse

Table 1 Interstellar extinction and $A(\lambda)/A(J)$, where $J \approx 1.25$ μm[a]

λ (μm)	$A(\lambda)/A(J)$	λ (μm)	$A(\lambda)/A(J)$ $R_V = 3.1$	$A(\lambda)/A(J)$ $R_V = 5.0$	λ (μm)	$A(\lambda)/A(J)$ $R_V = 3.1$	$A(\lambda)/A(J)$ $R_V = 5.0$
250[b]	0.0015	5	0.095	0.095	0.24	9.03	5.13
100	0.0041	3.4	0.182	0.182	0.218	11.29	6.03
60	0.0071	2.2	0.382	0.382	0.20	10.08	5.32
35	0.013	1.65	0.624	0.624	0.18	8.93	4.66
25	0.048	1.25	1.00	1.00	0.15	9.44	4.57
20	0.075	0.9	1.70	1.70	0.13	11.09	4.89
18	0.083	0.7	2.66	2.43	0.12	12.71	5.32
15	0.053	0.55	3.55	3.06	0.091[c]	17.2	—
12	0.098	0.44	4.70	3.67	0.073	19.1	—
10	0.192	0.365	5.53	4.07	0.041	9.15	—
9.7	0.208	0.33	5.87	4.12	0.023	7.31	—
9.0	0.157	0.28	6.90	4.34	0.004	3.39	—
7	0.070	0.26	7.63	4.59	0.002	1.35	—

[a] $A(\lambda)/A(J)$ is the same for $\lambda > 0.9$ μm for all lines of sight, to within present errors. To estimate $A(\lambda)/N(H)$, multiply tabulated entry for $R_V = 3.1$ by 1.51×10^{-22} cm^2 (H atom)$^{-1}$. Except as noted below, entries are calculated from CCM89. Other values of R_V can be determined from that paper.
[b] For $\lambda > 250$ μm, multiply entry for 250 μm by $(250 \, \mu m/\lambda)^2$.
[c] For $\lambda < 0.1$ μm, entries are from (107), increased by 1.15 for continuity with the CCM89 extinction value at 0.12 μm.

dust ($R_V = 3.1$) and for outer-cloud dust ($R_V = 5$), both calculated from CCM89. The difference between the two laws is striking. The 9.7- and 18-μm feature profiles, as given by "astronomical silicate" (DL84), have been added to a power-law interpolation of an underlying continuum fitted between 250 and 7 μm. The profile of the silicate band was truncated at 25 μm, as is appropriate for circumstellar dust (128) but perhaps not for interstellar dust. The FIR opacity should be extrapolated to longer wavelengths with a λ^{-2} dependence (see Section 3.2.3); the value in the table is determined by the estimate of Hildebrand (70). The ionizing-UV cross sections are from Martin & Rouleau (107), adjusted by a factor of 1.15 to make them continuous with the extinction value of CCM89 at 0.12 μm. Both the "astronomical silicate" and ionizing-UV opacities are based upon the bare-silicate/graphite grain model (see Section 8) and depend (through the Kramers-Krönig relations) upon the assumption that interstellar grains have densities of ~ 3 g cm^{-3}.

The entries in Table 1 for $\lambda > 15$ μm are uncertain by at least a factor of two. There are as yet very few observational constraints upon the extinction law between the longer wavelengths of the silicate feature (which probably varies from one line of sight to another) and wavelengths $\lambda > 100$ μm, for which the energy is produced by steady-state emission from large grains. The opacity at $\lambda > 100$ μm is somewhat constrained by the emission from isolated clouds warmed by the interstellar radiation field, which can be estimated.

3. EMISSION FROM DUST

3.1 *The Unidentified Infrared Bands*

The realization (150, 151) that diffuse dust produces strong unidentified infrared emission bands (UIBs) in the 3.3–11.3 μm range, as well as an associated continuum, has stimulated much research within the last five years. The carriers of the UIBs are surely important components of the ISM. The UIBs have been discussed extensively (3, 94, 133, 157), and some of their main features are the following:

1. The strongest bands are at 3.3, 6.2, 7.7, 8.6, and 11.3 μm. These wavelengths all closely correspond to the C–H or C–C bond vibrations in aromatic (benzene-ring) structures. The simplest substances that can produce these bands are simple, planar molecules called polycyclic aromatic hydrocarbons (PAHs), but other less well ordered configurations of carbon and hydrogen can also produce them (10, 142). A suggestive fit to the bands is provided by absorption from vitrinite (125), partially ordered graphite from coal. A mixture of PAHs can reproduce all of the UIBs, both weak and strong (57, 180).
2. Diffuse UIB emission, found throughout the Galaxy (59), is responsible for 10–20% of the total radiation from dust. UIBs and the associated continuum dominate the *Infrared Astronomical Satellite* (*IRAS*) filter responses at 12 and 25 μm (141), and are presumably responsible for the galactic "cirrus" emission seen with these filters (12).
3. The bands are also found in planetary nebulae, "reflection" nebulae, H II regions, extragalactic objects (27, 29, 171, and references therein), and carbon-rich or interstellar dust environments, *but not in dust produced by oxygen-rich objects*. There is a direct relationship between the C/O ratio in planetary nebulae and the strength of the UIBs (29).
4. The wavelength of the 11.3-μm UIB shows that the hydrocarbons are not saturated with H. This band is due to the out-of-plane C–H bending and occurs at 11.6–12.5 μm if there are two C–H bonds on the same aromatic ring, and at 12.4–13.3 μm for three bonds (94). The indicated amount of H coverage on the outer rings is 20–30%. Observational selection of relatively intense emission regions has meant that rather high radiation fields and subsequent dehydrogenation are favored; perhaps the 11–13 μm emission from low-radiation environments will indicate more than one C–H bond on the same ring.

5. The bands are excited by the absorption of a single UV photon by the carrier. This is easy to understand (133) if the carriers (planar PAHs or three-dimensional carbon structures no larger than about 5 Å) float freely in space, so that a single photon can provide the energy required to emit the UIBs. The degree of excitation suggests that roughly 50 carbon atoms are required, with an upward size range. If the carriers are attached to larger grains, the absorbed energy must be localized within a 5-Å region for the time required for the emission (of the order of a second). This process requires an *exceedingly* small thermal coupling.
6. The carriers of the UIBs are modified significantly by environment and history. The *IRAS* 12-μm response shows that the UIBs are not present in regions of very high radiation fields (13, 141), demonstrating that the carrier can be modified or destroyed by intense radiation. The wavelength of at least the 7.7-μm UIB is significantly different in planetary nebulae (where the carriers are newly produced by the carbon-rich material from the star) than in H II regions and reflection nebulae (where the carrier was presumably in the ISM before any interactions with the star presently causing the excitation).
7. PAHs would be mostly ionized in the diffuse ISM, since their first ionization potential is <13.6 eV. Up to now, many laboratory studies of PAHs have, necessarily, involved only neutral molecules.
8. An individual PAH has strong discrete absorption bands in the visual through the UV, *and there are no such features observed in interstellar extinction*. However, a mixture of PAHs of varying sizes and structural arrangements produces continuous absorption.

3.2 *Continuum Emission*

Continuum radiation from dust arises from two mechanisms: (*a*) fluorescence, giving rise to a red continuum; and (*b*) thermal radiation. The latter is observed either in the 1–60 μm range following the transient heating of a small grain by a single UV photon; or for $\lambda > 100$ μm, where the energy is reprocessed by steady-state emission of larger grains; or in the intermediate-wavelength range, where the effects compete.

3.2.1 RED CONTINUUM Grains in the "reflection" nebula NGC 2023 were found to produce a red continuum that peaks at about 6800 Å (178). Subsequent studies of many such nebulae confirm that there is an extended red emission in many of them (175, and references therein), typically contributing 30–50% of the flux in the *I* band (0.88 μm). In NGC 2023, the red emission is found in filamentary structures that spatially coincide with patches of H_2 infrared fluorescence and with the 3.3-μm UIB emission

(56) but not necessarily with the intensity of reflected light, which is determined by total dust density. IC 435, in the same molecular cloud, shows no red luminescence (174), nor does the Merope reflection nebula. A small patch of red fluorescent emission near the star γ Cas also shows a strong H_2 fluorescence in the UV (179). The "Red Rectangle" (HD 44179), a bipolar outflow excited by a central star, shows both strong UIBs and the red emission (140, 146).

The simplest interpretation of the red fluorescent emission is that it is excited by a strong UV flux incident upon hydrogenated carbon particles, either amorphous (46) or PAHs (32), thereby producing both the fluorescent red emission and the H_2 fluorescence. A strong enough radiation field will alter the carbon particles and dissociate the H_2, as might be the case in the Merope nebula.

3.2.2 3-30 μm CONTINUUM In addition to the UIBs, there is a continuous emission in the NIR/MIR range that accounts for a large part of the radiation from reflection nebulae (151) and, presumably, from the Galaxy as a whole. A survey at 11 and 20 μm (132) found unexpectedly high diffuse galactic emission. *IRAS* mapped almost the entire sky at 12, 25, 60, and 100 μm and found that the 12- and 25-μm intensities vary considerably relative to the 100-μm intensity (90, 91), which is roughly proportional to N(H I). The NIR/MIR emission arises from warm grains; simple considerations of the peak wavelength of the Planck function (and hence the emissivity of a grain) as a function of temperature show that this radiation must come from grains with temperatures of hundreds of Kelvins. By contrast, the mean local interstellar radiation field and optical constants for likely grain materials (carbon, silicates, organic refractory mantles) all suggest equilibrium temperatures of $\sim$ 20 K for the grains having the sizes (>0.01 μm) needed to account for the optical extinction (133). Grains of this size have large enough heat capacities so that their temperatures do not fluctuate appreciably after absorbing a single UV photon.

The NIR/MIR emission must be produced by grains in the size range 5–50 Å; such grains are large enough to have an almost continuous density of energy states. In this case they radiate a continuum rather than emission bands. A single UV photon heats the grain to a high peak temperature that depends upon the size of the grain and the energy of the photon (150). The grain emits the NIR/MIR radiation and cools to very low temperatures between photon absorptions. Calculations of thermal fluctuations (34, 41) have shown that very small grains can account for the spectrum of the emission. For clear discussions on the properties of small grains, see (4, 133).

3.2.3 FIR EMISSION Grains with sizes of 0.01 μm or more are cold

(~20 K) and reradiate most of the energy that they absorb into the FIR part of the spectrum. Hildebrand (70) has clearly explained the process of determining grain properties and cloud masses from FIR observations. Cox & Mezger (30), in a recent review of the galactic FIR/submillimeter radiation from dust, estimate that about 10^{10} $L_\odot$, or 30% of the total luminosity of the Galaxy, is radiated in the FIR, mostly from dust heated in H I regions by the interstellar radiation field of early-type stars (12).

The FIR provides important information regarding the spatial variations of the interstellar radiation field throughout the Galaxy and on how molecular clouds form stars and evolve into H II regions, but it is of limited use as a diagnostic of dust because each line of sight samples grains with temperatures that depend upon their local radiation fields. However, for wavelengths much longer than about 150 μm (the peak of the emitted radiation), the FIR can be used to determine the relative opacities of the emitting grains. The Galaxy is optically thin at submillimeter wavelengths, in which case the intensity of emission I_λ is proportional to the opacity κ_λ times the Planck function $B_\lambda(T)$. For long wavelengths, $B_\lambda(T)$ varies linearly with the temperature, and thus the wavelength dependence of I_λ provides relative values of κ_λ. The results for diffuse radiation from the Galaxy (115) and from other galaxies (22) show that for $\lambda > 100$ μm, κ_λ is proportional to λ^{-2}, which is predicted by theory (e.g. DL84). The constant in the proportionality is difficult to determine because it depends on an estimation of the column density of grains, or of hydrogen, along the line of sight of the observation. The theoretical opacities of DL84, based on a graphite-silicate model for grains, are approximately half of Hildebrand's estimate (70) based on a calibration in local dense globules (cf 124). The uncertainties are probably at least a factor of two, if not more. Table 1 uses the Hildebrand value.

The mass of interstellar dust, and thereby an estimate for the mass of the ISM, is often determined from the FIR intensity, together with the opacity of grains per unit mass and an estimated grain temperature. It is important to realize (40) that this procedure is reasonably safe *only* if the observations are all on the long-wavelength side of the Planck function of the coldest grains—normally, in the submillimeter range. For instance, the "temperature" obtained from the ratio of the 60- and 100-μm intensities from *IRAS* is biased by a few warm grains that can provide almost all of the 60-μm emission. One can easily be off by a substantial factor (>3) in the resulting mass estimate!

In dust surrounding very young objects, bipolar outflows, or the cores of giant molecular clouds, the flux suggests that the opacity even in the submillimeter range (400–1300 μm) might vary as λ^{-1} or $\lambda^{-1.5}$ (145, 162, 181) instead of as λ^{-2}. The nebulae might be so optically thick that radiative transfer effects are important at submillimeter wavelengths. However, the grains in these extremely dense objects might not be extremely small in comparison to 1 mm. Cometary grains also extend up to this size range. The growth of fractal grains (182; see Section 8) would explain the observations.

4. SCATTERING FROM DUST

Scattering from grains provides another diagnostic for their nature and composition, since cross sections for scattering at any angle (the "phase function") can be computed for a given material in much the same way as for extinction. However, the relative locations of the illuminating sources, the scattering grains, and the observer are very important in determining the actual intensity of scattered radiation. The presence of fluorescent emission (Section 3.2.1) complicates the interpretation of red and NIR scattering, but there seems to be little fluorescence in the range $\lambda < 0.5$ μm (139).

In practice, the geometry is so uncertain that one attempts to determine only two quantities characterizing the scattering process: the albedo, or the fraction of the extinction that is scattering; and g, the mean value of the cosine of the angle of scattering. For isotropic scattering, $g = 0$; for completely forward-throwing scattering, $g = 1$.

Scattering can be observed in three general situations: (*a*) the "diffuse galactic light" (DGL)—i.e. scattering by the diffuse dust of the general incident interstellar radiation field—which is strongly concentrated in the galactic plane, since the dust has a relatively small scale height; (*b*) reflection nebulae, with a known source of illumination (usually a B or A star because of their favorable luminosities); and (*c*) scattering of the general interstellar radiation field by a dark cloud, seen at high enough latitudes so that it contrasts with a relatively dark sky background.

The DGL in the optical part of the spectrum has been analyzed (110, 173). It is quite faint and asymmetric in its angular distribution and requires careful correction for the contribution of faint stars, airglow, and (especially) zodiacal light. Its advantage is that the geometry of the sources and scatterers is well known, in contrast to reflection nebulae. Witt (174) quotes a study (158) from *Pioneer 10* at 3 AU (so that both the airglow and zodiacal light are negligible) in which Toller (158) finds the albedo at 0.44 μm to be 0.61 ± 0.07 and g equal to 0.60 ± 0.22. Grains exhibit strong forward throwing scattering in the optical.

In the UV, there are surprising spatial fluctuations in the diffuse brightness of the sky, with background intensities ranging over an order of magnitude at the same wavelength (cf 77, 103, 120, 127, 155). The minimum

$N(H\ I)$ presently observed in the sky corresponds to $A(V) \approx 0.05$ mag (100), or a scattering optical depth of ~ 0.06 at 0.16 μm. This translates into a sky brightness that is comparable to that observed. There may be an extragalactic component (which, in fact, is a common interpretation of the observed brightness). Clearly, the properties of grains or the intensity of any extragalactic component cannot be analyzed until the scattered intensity is known.

Reflection nebulae, in which a bright star illuminates a nearby scattering cloud, have been analyzed in some detail by various authors. Their advantages over the DGL and scattering by clouds are that they are much brighter, and that they are reflecting light from well-studied early-type stars, whereas the DGL and the clouds (discussed below) are illuminated by the much less well known interstellar radiation field (especially at UV wavelengths). Reflection nebulae suffer from three disadvantages: (a) The geometry of the star and scattering particles is not well known, and the placement of a given grain relative to the star and observer is crucial for determining how much light it scatters into the observer's line of sight. Simple geometries used in modeling the scattering, such as plane-parallel or spherical, are not so appropriate as one would like. (b) The patchiness of the dust is more serious for the reflection nebulae than for the DGL or the clouds. For clumpy objects, mean values of the albedo and g determined for the clump may not represent the values for a single scattering from an individual grain. (c) Some of the scattering is at large angles, in which case the assumed form of the phase function becomes important in the analysis. Astronomical observations of scattering have been interpreted by theoretical analyses employing the simple analytical Henyey-Greenstein (HG) phase function (69), which is computationally convenient but has no physical basis. The HG function is suitable for analyzing scattering at small to modest angles (<1 rad), in which case the important quantity is the fraction of the light thrown almost in the forward direction, but the HG function is not realistic for large-angle scattering.

Whether the advantages of reflection nebulae for determining the scattering properties of dust outweigh the disadvantages is a matter of opinion. I feel the geometrical uncertainties are such that results should be taken with considerable caution. Witt (174, and references therein) disagrees, arguing that the brightest reflection nebulae must have a common geometry in which the intensity of scattered light is maximized. He also points out that in the UV there might be three wavelengths at which the extinction optical depth is the same (two on each side of the bump and one in the steeply rising part of the extinction law at very short wavelengths; see Figure 2). In this case differences in the scattered light relative to the illuminating star's luminosity directly reflect changes in the grains' albedo.

His conclusion that g is smaller (more isotropic scattering) at $\lambda < 2000$ Å than for longer wavelengths seems correct.

Possibly small, relatively isolated interstellar clouds ("globules") contain dust that is more like outer-cloud dust than diffuse dust, but the determination of their optical properties is still of considerable interest. The geometry of scattering from globules at high enough latitude to be seen against a dark background (free from the DGL in the plane) is better known than for other reflection nebulae. At optical wavelengths, a cloud is seen strongly limb brightened from scattered radiation from behind (52), which is a direct indication of highly forward-throwing scattering by the grains in the optical. If the scattering were isotropic, the center of the cloud (with the greatest optical depth) would be brightest. Mattila (116) determined an albedo of ~ 0.6 and $g \approx 0.75$ by comparing the brightnesses of two globules at different latitudes and assuming that their dust is intrinsically similar.

5. POLARIZATION

Interstellar polarization arises from the propagation of radiation through the ISM containing aligned, elongated interstellar grains. The galactic magnetic field is responsible for aligning the grains (71), which spin with their long axes perpendicular to the field. Under these conditions, radiation is subjected to extinction, to linear dichroism (differential linear extinction for the two waves polarized along and perpendicular to the direction of alignment), and to linear birefringence (differential phase shift between the two waves). The birefringence produces circular polarization from the linearly polarized light, but the quantitative interpretation of the circular polarization to date has not been very fruitful because it depends upon the unknown geometry of the change of direction of grain alignment along the line of sight.

In principle, polarization is a diagnostic that provides another integral of a grain property over the size distribution, similar to extinction and scattering. However, it involves an additional function that is poorly understood: the alignment of grains of various sizes (reviewed in ref. 71). Even so, polarization is important because it provides information regarding the optical properties of grains and the conditions under which grains can be aligned.

5.1 *Continuum Polarization*

5.1.1 OPTICAL AND NIR The empirical wavelength dependence of optical/NIR polarization (151a) follows "Serkowski's law," i.e. $p(\lambda)/p(\lambda_{max}) = \exp[-K\ln^2(\lambda/\lambda_{max})]$, where λ_{max} is the wavelength of the maximum polarization $p(\lambda_{max})$. The quantity K was originally taken to be 1.15, but an improved fit (170) is $K = -0.10 + 1.86\lambda_{max}$. This law is entirely empirical, and it would be important to determine deviations at large $|\ln(\lambda/\lambda_{max})|$. Salient features of the optical polarization law are the following:

1. The averaged value of λ_{max} is 0.55 μm (151a), with extremes from about 0.34 to about 1 μm. The values of λ_{max}, determined by a least-squares fitting to Serkowski's law rather than a direct search for the maximum, depend mainly upon the observations at extreme wavelengths.

2. The polarization typically rises with wavelength through the ground-based UV to a maximum in the optical and then falling slowly through the NIR. Such behavior bears little resemblance to the extinction law, which keeps rising monotonically, except for the bump, toward shorter wavelengths throughout the observable UV. The grains responsible for the extinction in the ground-based UV do not participate in polarization because they are not elongated and/or not aligned.

3. The value of λ_{max} is almost proportional to R_V (25, 167, 168), although there is more scatter in the relationship than for the extinction laws. To a large extent, optical polarization measurements can substitute for NIR extinction in obtaining R_V.

4. The polarization law $p(\lambda)$ varies as $p(\lambda) \propto \lambda^{-1.8}$ for both diffuse dust and outer-cloud dust in the range 0.9 μm $< \lambda <$ 5 μm (108). The polarization law exponent is less well determined than for extinction, varying between -1.5 and -2.0 for various samples, but it is certainly similar to the value for extinction (-1.7---1.8; see Section 2.1.3). Note that this relation involves the absolute polarization, not relative to $p(\lambda_{max})$. The optical $p(\lambda)$ does vary strongly with R_V (see point 2 above), and the silicate feature has strong polarization that dominates for $\lambda > 5$ μm. The independence of $p(\lambda)$ from R_V again suggests that the size distribution of large grains is similar for clouds and the diffuse ISM.

5. The maximum value of $p(\lambda_{max})/A(\lambda_{max})$ is about 0.03 mag^{-1}, far less than from perfectly aligned spinning cylinders [0.22 mag^{-1} (111)]. This is interesting because the polarization direction closely follows the contours of the edges of several molecular clouds, presumably in regions where hydrogen changes its state from molecular to atomic relatively abruptly in space and perhaps in time. If the alignment mechanism keeps grains aligned under these adverse conditions, one would expect almost complete alignment when conditions are favorable, and a larger value of $p(\lambda_{max})/A(\lambda_{max})$ than is observed, in directions where the line of sight is perpendicular to the field. Perhaps there are two or more separate types of grains, only some of which are aligned. Alternatively, all grains might be

well aligned but have shapes that are less efficient than a spinning cylinder for producing polarization. A third possibility is that there is always a randomly oriented component to the field.

6. Polarization in the UV is unknown except for two stars (58). These limited data suggest that the bump is unpolarized. Upcoming observations from space (the WUPPE experiment on ASTRO missions) should provide many data. The polarization of the $\lambda 2175$ bump has been predicted if the bump is produced by aligned graphite (36).

An explanation for the form of the polarization law (111) assumes that grains can be aligned only if they contain one or more "superparamagnetic" particles (magnetite or other magnetic materials), which dissipate rotational energy as heat. Large grains are preferentially aligned because they are relatively likely to contain inclusions. Polarization is not specific to any particular grain model; if the large grains are aligned and a model predicts the extinction correctly, it will do well for the polarization also.

5.1.2 FIR POLARIZATION Polarization is observed in the emission from grains deep within the Orion Molecular Cloud and Sgr A, near the galactic center (35, 72, 163). The direction of the FIR polarization is perpendicular to the optical, exactly as expected: Light transmitted in the optical is polarized in the direction of smaller absorption. Emission, on the other hand, is largest in the direction of largest absorption.

Grain alignment is even more difficult for dense clouds than for the diffuse ISM (71). Alignment depends upon the grain being far from equilibrium with its surroundings, in which case there is no preferred axis because of the equipartition of energy. Deep inside a cloud, a grain should come to thermal equilibrium with the dense surrounding gas. The fact that polarization is observed shows that the rotation of aligned grains within clouds is not thermalized; they are presumably kept spinning in a particular direction because of the ejection of particles (H_2 after formation, or electrons) from particular sites (134), so the momentum of the ejected particles is not random. However, deep inside a cloud one would expect the gas impinging upon the grain to already be overwhelmingly H_2. Probably there also needs to be an enhanced dissipation of energy by superparamagnetic inclusions in the grains.

5.2 *Polarization in Spectral Features*

Inner-cloud dust provides information about diffuse dust through the polarization in the 3.08-μm ice band and the 9.7- and 18-μm silicate bands. If the polarization arises from aligned grains, the profiles of polarization, as compared with the corresponding extinction, are potentially important diagnostics for both the shapes and the types of grains (DL84, 105). If the sizes and optical properties of interstellar grains are such that the light wave acts as a uniform electromagnetic field across the particle (the "Rayleigh approximation"), the extinction and polarization cross sections are both simple functions of the optical constants of the material. The Kramers-Krönig relations force a relationship between the optical constants and, therefore, the extinction and polarization. When there is a strong frequency variation of the optical constants, as across a spectral band, the polarization peaks at longer wavelengths than does the extinction. Indeed, the silicate polarization peaks at about 10.5 μm, whereas the extinction peak is at 9.7 μm (1, 2). The amount of shift depends upon the shapes of the grains as well as upon the optical constants. If the band is strong enough, there is a polarization reversal at wavelengths on the short side of the maximum polarization. The presence of coatings also affects the shape of the polarization relative to the extinction, even if the coatings have a weak wavelength dependence of optical constants across the band.

Polarization can be produced by scattering in the NIR as well as by extinction due to aligned grains. Such scattering, commonly observed in reflection nebulae around sources in dense, star-forming regions (95, 129, 130), shows that grains in very dense regions have grown to sizes of at least the order of 1 μm, partly by acquiring the ice coatings producing the 3.08-μm band.

In the Becklin-Neugebauer (BN) object in Orion (see ref. 93 for references), the linear polarization is strong (16%) in the 3.08-μm ice band, as opposed to $\sim$10% in the neighboring continuum. The polarization is $\sim$15% in the 9.7-μm silicate feature but only $\sim$1% in the continuum. The position angle is constant across the bands and is the same as in the nebula in general, as expected from grain alignment but not from scattering. However, there is a reflection nebula around BN (see references in ref. 129) that complicates the interpretation of the polarization. By assuming that the polarization in BN is entirely from aligned grains, Lee & Draine (93) found grains are oblate (disk shaped) rather than prolate, with modest (2:1) axial ratios. One consequence of oblate grains is that the degree of alignment required is dropped by a factor of two, which is the ratio of the mean polarizing power of oblate and prolate grains.

A potentially powerful diagnostic for grains is the comparison of the polarizations in the 9.7- and 18-μm silicate bands, for which many of the geometrical uncertainties, such as angles of the magnetic field relative to the line of sight to the source, cancel because the same particles produce both bands.

6. OTHER DIAGNOSTICS OF INTERSTELLAR DUST

6.1 *Depletions in Interstellar Gas*

The strengths of the interstellar absorption lines of various ions in the spectra of background stars indicate the ionic column densities in the gas phase and, by inference, the amounts of the elements in grains. Jenkins (78) reviews the method of analysis, results, and several pitfalls. The principal results of depletions as regards grains are as follows:

1. The elements O, N, and Zn are only slightly depleted with respect to solar abundances, and their depletion does not vary measurably between dense and diffuse gas. The errors are such that these elements could be either undepleted or depleted by a factor of two.

2. Depletions of other elements increase significantly with average gas density along the line of sight. The elements P, Mg, and Cl are almost undepleted in diffuse gas ($\langle n(H) \rangle \approx 0.1$ cm^{-3}) and are depleted by about an order of magnitude when $\langle n(H) \rangle \approx 10$ cm^{-3}.

3. The elements Fe, Cr, and Si are depleted by about one order of magnitude in the diffuse ISM, and the depletions go roughly as the square root of the mean gas density, so that their depletion is about a factor of 10 more than P, Mg, and Cl, over a wide range in mean gas density (80). Thus, these depletions are two orders of magnitude along lines of sight through dense gas. This difference in depletion between outer-cloud dust and diffuse dust implies that grains evolve as they go from one environment to another. The elements Ca, Ti, and Al have similar depletions in low-density gas but a steeper dependence of depletion on mean gas density.

4. The depletion of C is, unfortunately, not reliably determined except for one line of sight: toward δ Sco with the *Copernicus* satellite (73). For observations with *IUE*, all lines of C$^+$, the dominant ionization stage in H I regions, lie on the "flat" (insensitive) portion of the curve of growth. For δ Sco, about 25% of the C is in the gas phase, which is about the amount found in CO in molecular clouds. The gas-phase carbon in the diffuse ISM is apparently simply converted to CO in molecular clouds, whereas the solid fraction remains approximately fixed.

There is some confusion in the literature because the C depletion can be *crudely* estimated from the well-determined gas-phase abundance of neutral C by estimating C^0/C$^+$. Unfortunately, the ionization corrections are factors of hundreds to thousands!

5. At a given mean gas density, there is a surprisingly small dispersion of the depletions (about ± 0.3 dex), whereas some depletions vary by one

or two orders of magnitude. Much of this dispersion must arise from averaging various local conditions along the line of sight. Again we see that the state of grains must be quite well described by only one parameter (perhaps local gas density).

6. Depletions are a function of z, the height above the plane of the Galaxy (50). Probably the extinction law also depends on z (83).

6.2 Sites of Dust Formation

Probably most dust is injected into the ISM from stars on the asymptotic giant branch, either C rich or O rich, although supernovae, despite a low *rate* of mass injection, might be important because of their large heavy-element composition. Carbon stars show an 11.15-μm feature of SiC with a quite uniform profile (26, 99). There is also a featureless optical/NIR continuum that can be modeled by amorphous carbon but not graphite (92, 106), unless the graphite is in small particles with a loose fractal structure (183). Oxygen-rich stars show the silicate feature with a profile that varies appreciably from star to star (99, 126). Differing physical conditions in the atmosphere affect the nature of the grains, and the silicate band in one star (2) shows clear signs of annealing. A few circumstellar shells show the 3.08-μm ice band as well.

The UV bump is found in C stars at 0.24 μm rather than at 0.2175 μm (67), and in one H-poor, C-rich planetary nebula (Abell 30) at 0.25 μm (63). These shifted wavelengths are consistent with amorphous C or fractal graphite grains. There is *no bump* in the circumstellar dust of oxygen-rich α Sco (153). The bump is seen in circumstellar dust surrounding a few hot stars (152) at the normal λ_{max}, but the dust might be interstellar, remaining from the epoch of the star's formation.

Novae and Wolf-Rayet stars (late-type WCs) are minor sources of dust because of their low mass input into the ISM (58a). Both presumably inject carbon-rich dust without silicates. Planetary nebulae represent a considerable source of the return of gas to the ISM (102) but are not a large source of dust because they have a low dust/gas ratio (131).

Isotopic anomalies in meteorites (see Section 6.3) prove that some dust forms in expanding supernova shells. Dust is also found within hot supernova remnants (48, 49), but it might have been produced by a presupernova red supergiant. How much dust is actually formed in supernovae is not known.

6.3 Solar System Dust

Primitive meteorites, interplanetary dust particles (IDPs), and cometary dust provide some information regarding interstellar grains, although all solar system dust has been significantly processed, both chemically and physically, since having been in the pre–solar system molecular cloud. For reviews on the various types of objects, see (14, 15, 82).

Meteorites are almost entirely asteroidal in origin, since *cometary* meteoroids cannot survive entry into the Earth's atmosphere. Most meteorites have undergone obvious metamorphism, with carbonaceous chondrites being the most primitive. One of this class, the Murray meteorite, has tiny SiC inclusions showing isotopic anomalies (8, 186), proving an interstellar origin. The small amount of carbon in primitive meteorites is mostly poorly ordered, not graphitic, but tiny (ca. 25 Å) diamonds are present (97). It is difficult to imagine a solar system origin for the high temperatures and/or pressures needed to produce diamonds, so diamond bonding in interstellar carbon is strongly indicated. Meteoritic silicates are much more crystalline than the interstellar varieties, showing that meteorites must have been much warmer than interstellar temperatures. Graphite, if originally present, could have been lost by chemical reactions with water and hydrogen.

IDPs are both cometary and asteroidal in origin, with the cometary variety being more primitive and, therefore, relevant to interstellar dust. Some cometary IDPs, collected from high-flying aircraft, are a few microns in size, with a very fragile, open structure consisting of submicron mineral grains stuck together into an open matrix (see, e.g., 14, 15). Their silicates have been annealed to crystalline forms (143), with the types of minerals similar to those observed in cometary dust. Some of the carbon, poorly ordered but aromatic in nature, occurs as submicron grains. There is also carbon coated onto the silicate materials. Some IDPs have relatively large D/H ratios (185), suggestive of molecular clouds.

The crystalline structure of cometary silicates shows that even cometary dust has been fairly heavily modified from the original interstellar dust. There are whole particles consisting of volatile material containing C, H, O, and N ("CHON"), with little refractory material within; there are also low-density silicate and carbonaceous grains. Meteoroids (mostly cometary) entering the upper atmosphere have densities of $0.01-1$ g cm^{-3}, so the fluffy refractory particles are common in comets and are probably present in interstellar dust, at least deep within molecular clouds.

In summary, solar system material shows that (*a*) small particles can survive all of the rigors of the ISM after formation in a supernova; (*b*) grains form large structures deep inside molecular clouds, with voids possibly packed with ices; and (*c*) heating and associated chemical processing took place before the formation of comets, possibly in the molecular cloud material.

7. EVOLUTION OF DUST

Relevant time scales show that the present form of interstellar dust must be more a reflection of the processing it has received within the ISM than a sample of the conditions at its origin. A typical parcel of gas and dust is cycled back and forth through molecular clouds several times during its lifetime, changing its grain properties significantly each time.

The lifetime of a grain against incorporation into stars can be estimated by dividing the surface density of the ISM by the rate of star formation. The local H I surface density is $\sim 10^7 \, M_\odot \, \text{kpc}^{-2}$ (89), and for H_2 is $3 \times 10^6 \, M_\odot \, \text{kpc}^{-2}$ (147). A mean rate of star formation of $3.4 \times 10^{-3} \, M_\odot \, \text{kpc}^{-2} \, \text{yr}^{-1}$ would account for the present surface density of low-mass ($M \leq M_\odot$) stars over 10^{10} yr (6); presumably the present rate is lower. High-mass stars contribute about $1.1 \times 10^{-3} \, M_\odot \, \text{kpc}^{-2} \, \text{yr}^{-1}$ (135). Therefore, the corresponding mean lifetime for a parcel of gas/dust in the ISM is more than 3 Gyr. On the other hand, about 30% of the local ISM is in molecular clouds, each with a lifetime of perhaps 10^8 yr (the time for the gas to proceed from one spiral arm to the next) or less. These numbers imply that a given parcel of gas has been into and out of a molecular cloud at least every 3×10^8 yr, or more than 10 times during its mean lifetime. Each time, the differences in extinction laws between diffuse dust and inner-cloud dust require that the grains be heavily modified.

Let us follow the state of a typical parcel of gas/dust near the Sun (see also 39). Since most of the mass of the local ISM is in H I, the parcel must spend the bulk of its time outside of molecular clouds. During this phase, about 10% of its mass is returned to the ISM from stars. Perhaps 10–20% of the returned gas is from hot stellar winds with no dust or from planetary nebulae with a dust/gas ratio lower than the ISM, providing substantial amounts of *gaseous* Fe, Al, and the other strongly depleted elements (81). Refractory elements must encounter grains frequently and stick to them very efficiently in order to achieve the observed strong depletions in the denser parts of the diffuse ISM.

Each time the gas/dust mixture is incorporated into the outer regions of a molecular cloud, many things happen to the grains: (*a*) The extinction law changes from diffuse dust to outer-cloud dust in such a way that the relative numbers of small, medium, and large grains depend primarily upon only one parameter (local gas density?), regardless of the local environment or past history. (*b*) The refractory elements are more strongly depleted onto the grains than in the diffuse phase. (*c*) The atomic H becomes molecular. Gaseous carbon recombines from C^+ to C^0 and finally to CO. (*d*) The grains almost certainly coagulate in the outer parts of molecular clouds before there is much coating of icy mantles. Much deeper in the cloud, the fluffy micron-sized cometary and interplanetary dust particles can be produced by further coagulation of the silicate and carbonaceous parts of the grains, with ices probably filling the voids and producing the observed molecular absorption features. It is difficult to see how fluffy cometary mineral grains can form within the cloud if icy or organic refractory mantles envelop the minerals before the coagulation.

Coagulation seems to dominate the change in the size distribution in going from the diffuse ISM to outer-cloud dust. At least two well-observed stars in outer-cloud dust have $A(V)/N(H)$ smaller than in diffuse dust (CCM89). Since $\int [A(\lambda^{-1})/A(V)] d\lambda^{-1}$ is much lower in the outer-cloud dust than in diffuse dust (Figure 2), in these stars the integrated extinction per H atom is substantially smaller than in diffuse dust. This reduction in grain cross section per H atom, despite the accretion of small amounts of the refractory elements, can only be achieved by coagulation, which prevents the inner parts of the larger grains from participating efficiently in the extinction. Adding substantial coatings would increase, rather than reduce, the extinction cross section per H atom for the dust in the cloud.

There are real deviations of the various extinction laws about the mean. These differences must reflect somewhat different histories and present local environments (radiation fields, shocks, magnetic fields, etc) of the grains along each line of sight. Studying these deviations should lead to a better understanding of the factors influencing extinction laws.

8. SOME THEORIES, AND PROBLEMS WITH EACH

Space does not allow even a superficial discussion of all the many individual theories interpreting the observational evidence summarized above. Thus, we summarize a few.

1. *Bare silicate/graphite grains*: DL84, in a very careful discussion of the optical constants of both graphite and silicates, greatly extended in wavelength an older theory of Mathis, Rumpl & Nordsieck (114; often referred to as MRN). The features of the theory are that (*a*) individual grains are bare and homogeneous, composed of either silicate or graphite; (*b*) the size distribution is a power law, where $n(a)$, the number of grains of radius a, is proportional to $a^{-3.5}$; and (*c*) the size distribution is truncated at the upper end at 0.25 μm, with the lower end of sizes extending downward to a few angstroms to fit the *IRAS* data (41, 133), and very likely all the way to PAHs in order to produce the UIBs.

DL84 subjected their theory to a much more exacting comparison with observations than did most other authors. The fit to the extinction law for diffuse dust, over the entire observed wavelength range from 0.1 μm to

1000 μm, is impressively good. On the other hand, DL84 required that large graphite particles be produced from the amorphous carbon that late-type stars inject into the ISM. It is very difficult to understand how the necessary annealing can take place under interstellar conditions. Furthermore, the materials in the two types of grains (silicates and graphite) must be kept separate, despite the many cycles of coagulation and rearrangement of the size distributions that take place as the grains cycle into and out of clouds.

2. *Core/mantle grains*: Greenberg and his collaborators (see references in ref. 62) believe that the bulk of interstellar grains have refractory silicate cores covered with an organic refractory mantle. This mantle is produced from the processing by both UV photons and cosmic rays deep inside molecular clouds, after the icy mantles observed in such clouds are deposited upon the grain surfaces. Laboratory studies show that molecules in such mantles can be partially converted into free radicals that are chemically active enough to react violently when warmed, producing complex molecules (31). Such runaway reactions could be triggered on inner-cloud grains by cosmic rays. After such an event, an organic polymeric substance known as "organic refractory" material, stable at room temperatures remains. In this theory, organic refractory mantles on silicate cores produce the optical/NIR extinction, while the bump is produced by small graphite particles, PAHs produce the UIBs, and the shortest wavelength extinction is from small particles (probably silicates). Chlewicki & Laureijs (23) suggest that an additional component of iron will produce most of the 60-μm emission observed by *IRAS*.

These ideas have a great deal of appeal as regards events deep within clouds. The Greenberg scenario explains why interstellar molecules are found in the gas phase within dense clouds, where they should freeze onto grains in a short time—the runaway reactions drive off the molecules, and gas-phase chemistry takes place before refreezing onto grain surfaces. An alternative explanation for gas-phase molecules inside dense clouds is that a very rapid circulation of grains takes place between the surface and the center of the cloud (e.g. 21).

There are four problems as regards extending these ideas into diffuse dust: (*a*) Grains are larger in outer-cloud dust because of coagulation, not accretion of mantles, as shown by the reduced extinction per H atom in some cases. (*b*) Organic refractory mantles, which are less refractory than silicates and solid carbon, would be more readily destroyed by shocks in the diffuse ISM. The destruction rate of materials depends sensitively upon the binding energy (43, 44). (*c*) Solar system dust particles suggest that the silicate and carbon materials coagulate into large structures before icy mantles envelop them. (*d*) The 3.4-μm C–H absorption band, seen in organic refractory material in the laboratory along with the 3.08-μm ice band, is locally weak or absent. The object with the strongest 3.4-μm band, IRS 7 near the galactic center (17), is not typical of local dust. The 3.08- and 3.4-μm bands are not yet seen toward the local star VI Cyg 12 (61) [for which $A(V) \approx 10$ mag], limiting these bands to less than 0.3 the strength, per $A(V)$, of those for IRS 7. The 3.4-μm band is seen in Lynga 8/IRS 3 [154; $A(V) = 17$], about 0.4 times as strong as for IRS 7. The 3.4-μm band is always accompanied by a stronger 3.08-μm ice feature, and there are some lines of sight (64) with no ice band for $A(V) < 20$ mag.

It is possible (156) that the organic refractory mantles are so heavily processed that they lose almost all of their N and O, becoming essentially amorphous carbon. This material would be difficult to distinguish from amorphous C injected directly into the ISM from carbon stars. In this case, the Greenberg theory is very similar to composite-grain theories (see below).

3. *Silicate cores with amorphous carbon mantles*: Duley et al (47) suggest that grains are silicates with mantles of hydrogenated amorphous carbon (HAC). One population is very small and produces the bump by (OH)$^-$ ion absorption in the presence of Si atoms. (All other theories produce the bump from well-ordered carbon.) The UIBs are caused by absorption of UV photons by "islands" of HAC on the silicate core surfaces, so thermally isolated that they can radiate like free particles [for about 1 second! (133)]. The rapid increase of extinction with wave number for $\lambda^{-1} > 6$ μm^{-1} is produced by diamondlike bonding in the "amorphous" C.

This theory makes several predictions that can be tested. It explains the differences between diffuse dust and outer-cloud dust rather naturally as arising from different depletions of carbon onto the silicate cores. However, it requires a very large fraction of the Si atoms to have OH$^-$ ions nearby, near the surfaces of small grains, even if the bump transition in OH$^-$ has an oscillator strength of unity. The thermal isolation of the "islands" of HAC is difficult to achieve.

4. *Composite grains*: Mathis & Whiffen (113) and Tielens (156) suggest that interstellar grains consist of an assembly of small particles of carbon and silicates, jumbled together loosely. These grains are the natural result of coagulation and disruption of grains as they cycle into clouds. The particles inside the porous structure are protected from shocks and might well be covered with highly processed organic material. The bump is provided by small graphitic particles; PAHs can produce the UIBs. The rise in extinction for $\lambda < 0.16$ μm is provided by the diamond bonding in "amorphous" C. The composite grains are mostly open, in analogy with interplanetary dust particles. However, too much porosity provides too large an opacity in the FIR, making the grains too cold because they

radiate efficiently. It is difficult to calculate the extinction of composite grains, so the calculated fit should be taken as provisional.

5. *Fractal grains*: Wright (182) suggested that interstellar grains are the product of coagulation into very large fractal structures resembling twisted branches (65). If one defines the fractal dimension α by $M \propto R^{\alpha}$, then α depends upon the sequence of coagulation (and the probability of the fractal grains' breaking up, which is neglected in the calculations). In general $\alpha < 3$, and in some cases $\alpha < 2$. One of the major features of fractal grains is a FIR absorption per unit mass larger by an order of magnitude or more than that for solid grains.

Fractal grains can explain very large radar backscattering in comets (183) without large masses of dust. They also explain the very shallow (λ^{-1}) wavelength dependence of the submillimeter opacity observed in some very dense nebulae (Section 3.2.3). However, the FIR opacity of fractal grains is so large that the grains would be too cold to explain the observed FIR spectrum of galactic dust ($T \approx 20$ K).

6. *Biological grains*: Hoyle, N. C. Wickramasinghe, and others (76, 161, and references therein) have suggested that the grains producing visual extinction have a biological origin, with the bump provided by graphite. The extinction and polarization laws are fitted reasonably well. However, there are two problems with the model: (*a*) There is not enough cosmic phosphorus to accommodate the amount found in organisms (45, 164; but see 74). The cosmic abundance of P is low, and most of it is in the gas phase for low-density lines of sight, so this criticism seems valid. (*b*) Organisms, even when dried, show strong O–H and C–H stretch absorptions (75), which are not seen except deep within molecular clouds.

9. SUMMARY

"Interstellar dust" refers to materials with rather different properties, and the "mean extinction law" of Seaton (149) or Savage & Mathis (144) should be replaced by the expression given in CCM89, using the appropriate value of total-to-selective extinction R_V. The older laws were appropriate for the diffuse ISM, but dust in clouds differs dramatically in its extinction law (Figure 2). However, there are certainly real deviations from the mean extinction law (see error bars in the inset in Figure 2). The extinction law for $\lambda > 0.9$ μm seems to be independent of environment, to within the present observational errors. Other diagnostics of dust, especially the depletions from the gas phase, confirm that properties of the grains vary along different lines of sight, but only one parameter, probably related to the local gas density, determines the grain properties surprisingly well.

Dust is heavily processed while in the ISM by being included within clouds and cycled back into the diffuse ISM many times during its lifetime. Consequently, grains probably reflect only a trace of their origin, although meteoritic inclusions with isotopic anomalies prove that some tiny particles survive intact from a supernova origin to the present. Grains apparently grow by coagulation while in clouds. Cometary and interplanetary dust suggests that very large sized grains are produced before extensive icy mantles are formed. Within the dark clouds, there is likely processing of the icy mantles by cosmic rays or by the UV radiation produced by cosmic rays, and heavy molecules are released by runaway reactions. If there is an organic refractory mantle remaining after this processing, it is probably converted to almost pure amorphous carbon by the continued processing to which the grains are subjected.

There are several theories that explain the extinction law for diffuse dust, but a much more challenging problem is to understand the relation between dust of all types. The evolution of dust is probably the next theoretical challenge.

ACKNOWLEDGMENTS

This review has been partially supported by contract 957996 with the Jet Propulsion Laboratory and grant NAGW-1768 with NASA. Comments and assistance from L. J. Allamandola, J. A. Cardelli, G. C. Clayton, B. T. Draine, P. G. Martin, and B. D. Savage are appreciated.

Literature Cited

1. Aitken, D. K., Bailey, J. E., Roche, P. F., Hough, J. H. 1985. *MNRAS* 215: 815
2. Aitken, D. K., Roche, P. F., Smith, C. H., James, S. D., Hough, J. H. 1988. *MNRAS* 230: 629
3. Allamandola, L. J., Tielens, A. G. G. M., eds. 1989. *Interstellar Dust, IAU Symp. No. 135*. Dordrecht: Kluwer. 530 pp.
4. Allamandola, L. J., Tielens, A. G. G. M., Barker, J. R. 1989. *Ap. J. Suppl.* 71: 733
5. Ardeberg, A., Virdefors, B. 1982. *Astron. Astrophys.* 115: 347
6. Bahcall, J. N., Soneira, R. M. 1980. *Ap. J. Suppl.* 44: 73
7. Bailey, M. E., Williams, D. A., eds. 1988. *Dust in the Universe*. Cambridge: Univ. Press. 573 pp.
8. Bernatowicz, T., Fraundorf, G., Tang, M., Anders, E., Wopenka, B., et al. 1987. *Nature* 330: 728
9. Bohlin, R. C., Savage, B. D., Drake, J. F. 1978. *Ap. J.* 224: 132
9a. Bohren, C. E., Huffman, D. R. 1983. *Absorption and Scattering of Light by Small Particles*. New York: Wiley. 530 pp.
10. Borghesi, A., Bussoletti, E., Colangeli, L. 1987. *Ap. J.* 314: 422
11. Bouchet, P., Lequeux, J., Maurice, E., Prévot, L., Prévot-Burnichon, M. L. 1985. *Astron. Astrophys.* 149: 330
12. Boulanger, F., Pérault, M. 1988. *Ap. J.* 330: 964
13. Boulanger, F., Beichman, C., Désert, F. X., Helou, G., Pérault, M., Ryter, C. 1988. *Ap. J.* 332: 328
14. Brownlee, D. E. 1985. *Annu. Rev. Earth Planet. Sci.* 13: 147
15. Brownlee, D. E. 1989. In *Highlights of Astronomy*, ed. D. McNally, 8: 281. Dordrecht: Kluwer
16. Butchart, I., Whittet, D. C. B. 1983. *MNRAS* 202: 971
17. Butchart, I., McFadzean, A. D., Whittet, D. C. B., Geballe, T. R., Greenberg, J. M. 1986. *Astron. Astrophys.* 154: L5

18. Cardelli, J. A., Savage, B. D. 1988. *Ap. J.* 325: 864
19. Cardelli, J. A., Clayton, G. C., Mathis, J. S. 1988. *Ap. J. Lett.* 329: L33
20. Cardelli, J. A., Clayton, G. C., Mathis, J. S. 1989. *Ap. J.* 345: 245 (CCM89)
21. Chièze, J. P., Pineau des Forêts, G. 1989. *Astron. Astrophys.* 221: 89
22. Chini, R., Krügel, E., Kreysa, E. 1989. *Astron. Astrophys.* 216: L5
23. Chlewicki, G., Laureijs, R. J. 1988. *Astron. Astrophys.* 207: L11
24. Clayton, G. C., Martin, P. G. 1985. *Ap. J.* 288: 558
25. Clayton, G. C., Mathis, J. S. 1988. *Ap. J.* 327: 911
26. Cohen, M. 1984. *MNRAS* 206: 137
27. Cohen, M., Allamandola, L. J., Tielens, A. G. G. M., Bregman, J., Simpson, J., et al 1986. *Ap. J.* 302: 737
28. Cohen, M., Tielens, A. G. G. M., Bregman, J. D. 1989. *Ap. J. Lett.* 344: L13
29. Cohen, M., Tielens, A. G. G. M., Bregman, J., Witteborn, F. C., Rank, D. M., et al. 1989. *Ap. J.* 341: 246
30. Cox, P., Mezger, P. G. 1989. *Astron. Astrophys. Rev.* 1: 49
31. d'Hendecourt, L. B., Allamandola, L. J., Grim, R. J. A., Greenberg, J. M. 1986. *Astron. Astrophys.* 158: 119
32. d'Hendecourt, L. B., Léger, A., Olofsson, G., Schmidt, W. 1986. *Astron. Astrophys.* 170: 91
33. Day, K. L. 1979. *Ap. J.* 234: 158
34. Désert, F. X., Boulanger, F., Shore, S. N. 1986. *Astron. Astrophys.* 160: 295
35. Dragovan, M. 1986. *Ap. J.* 308: 270
36. Draine, B. T. 1988. *Ap. J.* 333: 848
37. Draine, B. T. 1989. In *Infrared Spectroscopy in Astronomy. Proc. ESLAB Symp., 22nd.* In press
38. Draine, B. T. 1989. See Ref. 3, p. 313
39. Draine, B. T. 1990. In *The Evolution of the Interstellar Medium*, ed. L. Blitz. San Francisco: Astron. Soc. Pac. Press. In press
40. Draine, B. T. 1990. In *The Interstellar Medium in Galaxies*, ed. H. A. Thronson, J. M. Shull. Dordrecht: Kluwer. In press
41. Draine, B. T., Anderson, N. 1985. *Ap. J.* 292: 494
42. Draine, B. T., Lee, H. M. 1984. *Ap. J.* 285: 89 (DL84)
43. Draine, B. T., Salpeter, E. E. 1979. *Ap. J.* 231: 77
44. Draine, B. T., Salpeter, E. E. 1979. *Ap. J.* 231: 438
45. Duley, W. W. 1984. *Q. J. R. Astron. Soc.* 25: 109
46. Duley, W. W. 1985. *MNRAS* 215: 259
47. Duley, W. W., Jones, A. P., Williams, D. A. 1989. *MNRAS* 236: 709
48. Dwek, E., Dinerstein, H. L., Gillett, F. C., Hauser, M. G., Rice, W. L. 1987. *Ap. J.* 315: 571
49. Dwek, E., Petre, R., Szymkowiak, A., Rice, W. L. 1987. *Ap. J. Lett.* 320: L27
50. Edgar, R. J., Savage, B. D. 1989. *Ap. J.* 340: 762
51. FitzGerald, M. P. 1968. *Astron. J.* 73: 983
52. FitzGerald, M. P., Stephens, T. C., Witt, A. N. 1976. *Ap. J.* 208: 709
53. Fitzpatrick, E. L. 1985. *Ap. J.* 299: 219
54. Fitzpatrick, E. L., Massa, D. 1986. *Ap. J.* 307: 286 (FM86)
55. Fitzpatrick, E. L., Massa, D. 1988. *Ap. J.* 328: 734
56. Gatley, I., Hasegawa, T., Suzuki, H., Garden, R., Brand, P. W. J. L., et al. 1987. *Ap. J. Lett.* 318: L73
57. Geballe, T. R., Tielens, A. G. G. M., Allamandola, L. J., Moorhouse, A., Brand, P. W. J. L. 1989. *Ap. J.* 341: 278
58. Gehrels, T. 1974. *Astron. J.* 79: 590
58a. Gehrz, R. 1989. See Ref. 3, p. 445
59. Giard, M., Pajot, F., Lamarre, J. M., Serra, G., Caux, E. 1989. *Astron. Astrophys.* 215: 92
60. Gillett, F. C., Forrest, W. J., Merrill, K. M., Capps, R. W., Soifer, B. T. 1975. *Ap. J.* 200: 609
61. Gillett, F. C., Jones, T. W., Merrill, K. M., Stein, W. A. 1975. *Astron. Astrophys.* 45: 77
62. Greenberg, J. M. 1989. In *Highlights of Astronomy*, ed. D. McNally, 8: 241. Dordrecht: Kluwer
63. Greenstein, J. L. 1981. *Ap. J.* 245: 124
64. Harris, D. H., Woolf, N. J., Rieke, G. H. 1978. *Ap. J.* 226: 829
65. Hawkins, I., Wright, E. L. 1988. *Ap. J.* 324: 46
66. Hecht, J. H. 1987. *Ap. J.* 314: 429
67. Hecht, J. H., Holm, A. V., Donn, B., Wu, C.-C. 1984. *Ap. J.* 280: 228
68. Hecht, J. H., Russell, R. W., Stephens, J. R., Grieve, P. R. 1986. *Ap. J.* 309: 90
69. Henyey, L. G., Greenstein, J. L. 1941. *Ap. J.* 93: 70
70. Hildebrand, R. H. 1983. *Q. J. R. Astron. Soc.* 24: 267
71. Hildebrand, R. H. 1988. *Q. J. R. Astron. Soc.* 29: 327
72. Hildebrand, R. H., Dragovan, M., Novak, G. 1984. *Ap. J. Lett.* 284: L51
73. Hobbs, L. M., York, D. G., Oegerle, W. 1982. *Ap. J. Lett.* 252: L21
73a. Hollenbach, D. J., Thronson, H. A. Jr. 1987. *Interstellar Processes. Astrophys. Space Sci. Libr.*, Vol. 134. Dordrecht: Reidel
74. Hoyle, F., Wickramasinghe, N. C. 1984. *Astrophys. Space Sci.* 103: 189
75. Hoyle, F., Wickramasinghe, N. C., Al-

Mufti, S., Olavesen, A. H., Wickramasinghe, D. T. 1982. *Astrophys. Space Sci.* 83: 405
76. Jabbir, N. L., Jabbar, S. R., Salih, A. H., Majeed, Q. S. 1986. *Astrophys. Space Sci.* 123: 351
77. Jakobsen, P., Bowyer, S., Kimble, R., Jelinsky, J., Grewing, M., et al. 1984. *Astron. Astrophys.* 139: 481
78. Jenkins, E. B. 1987. See Ref. 73a, p. 533
79. Jones, T. J., Hyland, H. 1980. *MNRAS* 192: 354
80. Joseph, C. L. 1988. *Ap. J.* 335: 157
81. Jura, M. 1987. See Ref. 73a, p. 3
82. Kerridge, J. F., Mathews, M. S., eds. 1988. *Meteorites and the Early Solar System*. Tucson: Univ. Ariz. Press
83. Kiszkurno-Koziej, E., Lequeux, J. 1987. *Astron. Astrophys.* 185: 291
84. Knacke, R. F., Capps, R. W. 1979. *Astron. J.* 84: 1705
85. Knacke, R. F., Krätchmer, W. 1980. *Astron. Astrophys.* 80: 281
86. Koornneef, J. 1982. *Astron. Astrophys.* 107: 247
87. Koornneef, J. 1983. *Astron. Astrophys.* 128: 84
88. Krätchmer, W., Huffman, D. R. 1979. *Astrophys. Space Sci.* 61: 195
89. Kulkarni, S. R., Heiles, C. 1987. See Ref. 73a, p. 87
90. Laureijs, R. J., Mattila, K., Schnur, G. 1987. *Astron. Astrophys.* 184: 269
91. Laureijs, R. J., Chlewicki, G., Clark, F. O. 1988. *Astron. Astrophys.* 192: L13
92. Le Bertre, T. 1987. *Astron. Astrophys.* 176: 107
93. Lee, H. M., Draine, B. T. 1985. *Ap. J.* 290: 211
94. Léger, A., d'Hendecourt, L., Défourneau, D. 1989. *Astron. Astrophys.* 216: 148
95. Lenzen, R., Hodapp, K.-W., Solf, J. 1984. *Astron. Astrophys.* 137: 202
96. Lequeux, J., Maurice, E., Prévot-Burnichon, M. L., Prévot, L., Rocca-Volmerange, B. 1982. *Astron. Astrophys.* 113: L15
97. Lewis, R. S., Tang, M., Wacker, J. F., Anders, E., Steel, E. 1987. *Nature* 326: 160
98. Lillie, C. F., Witt, A. N. 1976. *Ap. J.* 208: 64
99. Little-Marenin, I. R. 1986. *Ap. J. Lett.* 307: L15
100. Lockman, F. J., Jahoda, K., McCammon, D. 1986. *Ap. J.* 302: 432
101. Lucke, P. B. 1978. *Astron. Astrophys.* 64: 367
102. Maciel, W. J. 1981. *Astron. Astrophys.* 98: 406
103. Martin, C., Bowyer, S. 1989. *Ap. J.* 338: 677
104. Martin, N., Maurice, E., Lequeux, J. 1989. *Astron. Astrophys.* 215: 219
105. Martin, P. G. 1975. *Ap. J.* 202: 393
106. Martin, P. G., Rogers, C. 1987. *Ap. J.* 322: 374
107. Martin, P. G., Rouleau, F. 1989. *Berkeley EUV Colloq.*
108. Martin, P. G., Whittet, D. C. B. 1990. *Ap. J.* Submitted for publication
109. Massa, D., Fitzpatrick, E. L. 1986. *Ap. J. Suppl.* 60: 305
110. Mathis, J. S. 1973. *Ap. J.* 173: 815
111. Mathis, J. S. 1986. *Ap. J.* 308: 281
112. Mathis, J. S. 1986. *Publ. Astron. Soc. Pac.* 98: 995
113. Mathis, J. S., Whiffen, G. 1989. *Ap. J.* 341: 808
114. Mathis, J. S., Rumpl, W., Nordsieck, K. H. 1977. *Ap. J.* 217: 425
115. Matsumoto, T., Hayakawa, S., Matsuo, H., Murakami, H., Sato, S., et al. 1988. *Ap. J.* 329: 567
116. Mattila, K. 1979. *Astron. Astrophys.* 78: 253
117. Mauche, C. W., Gorenstein, P. 1986. *Ap. J.* 302: 371
118. McCarthy, J. F., Forrest, W. J., Briotta, D. A., Houck, J. R. 1980. *Ap. J.* 242: 965
119. Morgan, D. H., Nandy, K. 1982. *MNRAS* 199: 979
120. Murthy, J., Henry, R. C., Feldman, P. D., Tennyson, P. D. 1988. *Ap. J.* 336: 954
121. Nandy, K., Morgan, D. H., Willis, A. J., Wilson, R., Gondhalekar, P. M. 1981. *MNRAS* 196: 955
122. Neckel, Th., Klare, G. 1980. *Astron. Astrophys. Suppl.* 42: 251
123. Nuth, J. A. III, Stencel, R. E., eds. 1986. *Interrelationships Among Circumstellar, Interstellar, and Interplanetary Dust*. NASA Conf. Publ. 2403. Washington, DC: US Gov. Print. Off.
124. Pajot, F., Gispert, R., Lamarre, J. M., Pomerantz, M. A., Puget, J. L., Serra, G. 1989. *Astron. Astrophys.* 224: 107
125. Papoular, R., Conard, J., Giuliano, M., Kister, J., Mille, G. 1989. *Astron. Astrophys.* 217: 204
126. Papoular, R., Pegourie, B. 1983. *Astron. Astrophys.* 128: 335
127. Paresce, F., McKee, C., Bowyer, S. 1980. *Ap. J.* 240: 387
128. Pegourie, B., Papoular, R. 1985. *Astron. Astrophys.* 142: 451
129. Pendleton, Y., Tielens, A. G. G. M., Werner, M. 1990. *Ap. J.* In press
130. Pendleton, Y., Werner, M., Capps, R., Lester, D. 1986. *Ap. J.* 311: 360
131. Pottasch, S. R., Baud, B., Beitema, D., Emerson, J., Habing, H. J., et al. 1984. *Astron. Astrophys.* 138: 10

132. Price, S. D. 1981. *Astron. J.* 86: 193
133. Puget, J. L., Léger, A. 1989. *Annu. Rev. Astron. Astrophys.* 27: 161
134. Purcell, E. M. 1979. *Ap. J.* 231: 404
135. Ratnatunga, K. U., van den Bergh, S. 1989. *Ap. J.* 343: 713
136. Rieke, G. H., Lebofsky, M. J. 1985. *Ap. J.* 288: 618
137. Roche, P. F., Aitken, D. K. 1984. *MNRAS* 208: 481
138. Roche, P. F., Aitken, D. K. 1985. *MNRAS* 215: 425
139. Rush, W. F., Witt, A. N. 1975. *Astron. J.* 80: 31
140. Russell, R. W., Soifer, B. T., Willner, S. P. 1978. *Ap. J.* 220: 568
141. Ryter, C., Puget, J. L., Pérault, M. 1987. *Astron. Astrophys.* 186: 312
142. Sakata, A., Wada, S., Tanabe, T., Onaka, T. 1984. *Ap. J. Lett.* 287: L51
143. Sandford, S. A., Walker, R. M. 1985. *Ap. J.* 291: 838
144. Savage, B. D., Mathis, J. S. 1979. *Annu. Rev. Astron. Astrophys.* 17: 73
145. Schloerb, F. P., Snell, R. L., Schwartz, P. R. 1987. *Ap. J.* 319: 426
146. Schmidt, G. D., Cohen, M., Margon, B. 1980. *Ap. J. Lett.* 239: L133
147. Scoville, N. Z., Sanders, D. B. 1987. See Ref. 73a, p. 21
148. Seab, C. G., Snow, T. P. 1985. *Ap. J.* 295: 485
149. Seaton, M. J. 1979. *MNRAS* 187: 73P
150. Sellgren, K. 1984. *Ap. J.* 277: 623
151. Sellgren, K., Allamandola, L. J., Bregman, J. D., Werner, M. W., Wooden, D. H. 1985. *Ap. J.* 299: 416
151a. Serkowski, K., Mathewson, D. S., Ford, V. L. 1975. *Ap. J.* 196: 261
152. Sitko, M. L., Savage, B. D., Meade, M. R. 1981. *Ap. J.* 246: 161
153. Snow, T. P., Buss, R. H., Gilra, D. P., Swing, J.-P. 1987. *Ap. J.* 321: 921
154. Tapia, M., Persi, P., Roth, M., Ferrari-Toniolo, M. 1989. *Astron. Astrophys.* 225: 488
155. Tennyson, P. D., Henry, R. C., Feldman, P. D., Hartig, G. F. 1988. *Ap. J.* 330: 435
156. Tielens, A. G. G. M. 1989. See Ref. 3, p. 239
157. Tielens, A. G. G. M., Allamandola, L. J. 1987. See Ref. 73a, p. 397
158. Toller, G. N. 1981. PhD thesis. State Univ. N.Y., Stony Brook
159. Vrba, F. J., Rydgren, A. E. 1985. *Astron. J.* 90: 1490
160. Volk, K., Kwok, S. 1988. *Ap. J.* 331: 435
161. Wallis, M. K., Wickramasinghe, N. C., Hoyle, F., Rabilizirov, R. 1989. *MNRAS* 238: 1165
162. Weintraub, D., Sandell, G., Duncan, W. D. 1989. *Ap. J. Lett.* 340: L69
163. Werner, M. W., Davidson, J. A., Hildebrand, R. H., Morris, M. R., Novak, G., Platt, S. R. 1988. *Ap. J.* 333: 729
164. Whittet, D. C. B. 1984. *MNRAS* 210: 479
165. Whittet, D. C. B. 1987. *Q. J. R. Astron. Soc.* 28: 303
166. Whittet, D. C. B. 1988. See Ref. 7, p. 25
167. Whittet, D. C. B., van Breda, I. G. 1978. *Astron. Astrophys.* 66: 57
168. Whittet, D. C. B., van Breda, I. G. 1980. *MNRAS* 192: 467
169. Whittet, D. C. B., Bode, M. F., Longmore, A. J., Adamson, A. J., McFadzean, A. D., et al. 1988. *MNRAS* 233: 321
170. Wilking, B. A., Lebofsky, M. J., Rieke, G. H. 1982. *Astron. J.* 87: 695
171. Willner, S. P. 1984. In *Galactic and Extragalactic Infrared Spectroscopy*, ed. M. F. Kessler, J. P. Phillips, p. 37. Dordrecht: Reidel
172. Willner, S. P., Gillett, F. C., Herter, T. L., Jones, B., Krassner, J., et al. 1982. *Ap. J.* 253: 174
173. Witt, A. N. 1968. *Ap. J.* 152: 59
174. Witt, A. N. 1988. See Ref. 3, p. 1
175. Witt, A. N., Schild, R. E. 1988. *Ap. J.* 325: 837
176. Witt, A. N., Bohlin, R. C., Stecher, T. P. 1984. *Ap. J.* 279: 698
177. Witt, A. N., Bohlin, R. C., Stecher, T. P. 1986. *Ap. J. Lett.* 305: L23
178. Witt, A. N., Schild, R. E., Kraiman, J. B. 1984. *Ap. J.* 262: 708
179. Witt, A. N., Stecher, T. P., Boroson, T. A., Bohlin, R. C. 1989. *Ap. J. Lett.* 336: L21
180. Witteborn, F. C., Sandford, S. A., Bregman, J. D., Allamandola, L. J., Cohen, M., et al. 1989. *Ap. J.* 341: 270
181. Woody, D. P., Scott, S. L., Scoville, N. Z., Mundy, L. G., Sargent, A. I., et al. 1989. *Ap. J. Lett.* 337: L41
182. Wright, E. L. 1987. *Ap. J.* 320: 818
183. Wright, E. L. 1989. *Ap. J. Lett.* 346: L89
184. York, D., Drake, J., Jenkins, E., Morton, D., Rogerson, J., Spitzer, L. 1973. *Ap. J. Lett.* 182: L1
185. Zinner, E., McKeegan, K. D., Walker, R. M. 1983. *Nature* 305: 119
186. Zinner, E., Tang, M., Anders, E. 1987. *Nature* 330: 730

THEORIES OF THE HOT INTERSTELLAR GAS

Lyman Spitzer, Jr.

Princeton University Observatory, Princeton, New Jersey 08544

KEY WORDS: supernova remnants, superbubbles, galactic fountains, coronal gas in Galaxy

1. INTRODUCTION

Research on the hot component of the interstellar gas, with a temperature of about 10^6 K or more, has developed rapidly during the last two decades as part of space astronomy. A few theoretical papers on this topic had appeared somewhat earlier. In 1956 there was a somewhat speculative suggestion (89) that a hot gas, extending a kiloparsec or so from the galactic plane, would confine the diffuse clouds observed far above the plane and would prevent their expansion and dissipation. A major milestone was the 1962 paper by Shklovsky (87), who applied to a nonradiating supernova envelope, expanding into the interstellar medium, the self-similar solution developed several years earlier by Sedov (85). Since supernovae provide a major galactic source of hot gas, this detailed quantitative result on the temperature within model supernova (SN) remnants provided an important advance. The basic information needed to compute the temperature immediately behind a supernova shock had been known as early as the 1940s, but the high temperature resulting was not much discussed, perhaps because there seemed to be no prospect of observational data on this topic.

Observational research on the hot galactic gas began in 1968 with sounding rocket measurements of the diffuse X rays softer than 1 keV (12). Data were obtained rapidly by several groups, but the source of the radiation was still unclear in 1973 (36, 88); the possibilities discussed included numerous point sources and hot gas, either galactic or intergalactic. The following year conclusive observations of apparently interstellar O VI absorption lines (52, 105) made the presence of widespread hot gas in the Galaxy very likely. This result gave strong support (102) to a corresponding interstellar origin of the soft X-ray background, a point of view now generally accepted. Another important milestone in 1974 was the detection (104) of the $\lambda 5303$ Å emission line of [Fe XIV] from the X-ray-emitting region of the Cygnus Loop, verifying the thermal origin of the observed X rays in a gas at about 2×10^6 K; by that time, soft X-ray spectra (0.1 to about 1 keV) from some half-dozen supernovae were known (38) to be consistent with thermal radiation (mostly line emission from ionized atoms) at gas temperatures between 2 and 15×10^6 K, though some nonthermal emission could also be present. Thus, in one year three different lines of evidence converged to show that a hot gas formed an important constituent of the interstellar medium.

The extensive observational material that by 1987 had accumulated on this hot gas has been clearly summarized in the following broad reviews: of thermal X rays from SN remnants by Aschenbach (2), of diffuse soft X rays by Cox & Reynolds (29), and of absorption lines from highly ionized atoms by Jenkins (51) and by Savage (82). From the latter two of these three topics the general picture emerging is that the observed soft X-ray background is largely produced in a "Local Bubble" of hot gas within some 100 pc of the Sun. The temperature of this gas is about 10^6 K, somewhat less than in old SN remnants. Absorption of O VI lines in the disk and of C IV and N V in the halo, especially at $z > 1$ kpc, is produced by a more extended gas, probably cooling to lower temperatures than in the Local Bubble. The properties of this halo gas are consistent with those required to produce the recently observed C IV and O III] emission lines at high galactic latitude (65), though some emission may also be produced by cooling gas nearer to the Sun.

We make no attempt here to deal with the many observational details covered in the numerous papers summarized in the above reviews. Instead, attention is focused chiefly on the origin and development of the hot interstellar gas in our Galaxy, starting with the origin for much of it in SN remnants. In particular, the present paper describes theoretical work on this topic, especially that carried out during the last few years. In a few areas, such as the production of highly ionized atoms and the equilibrium of the interstellar gas perpendicular to the galactic plane, observations that are of particular relevance to recent theories are briefly discussed. Hot gas in other galaxies, in galaxy clusters, and in intergalactic space is not considered here.

Understanding the processes that occur as the hot interstellar gas evolves in our Galaxy is an ambitious goal that we are far from achieving. The dynamics of a compressible gas, subject to the photons and cosmic rays in interstellar space, is a complex topic. Some progress has been made through the development of idealized models, which are so simplified that one can hope to understand them and to compute their properties. In terms of such models one can distinguish three scenes in the unfolding drama of the hot interstellar gas. First, the explosion of a supernova ejects a rapidly expanding envelope, whose interaction with the surrounding medium creates the hot gas in which we are interested. Next, as this heated gas expands, it encounters regions whose internal density is well above the average. These regions, which we call clouds, are compressed by the hot gas, are heated by conduction, and sometimes evaporate or are disrupted. In the final scene, the remnant of heated gas surrounding one or more supernovae can rise to appreciable distances from the galactic plane and may produce a hot galactic corona before it falls back down or escapes the Galaxy entirely.

In actuality, these three scenes overlap so much that their mutual interactions are important. In most theoretical models these scenes have been kept somewhat separate to simplify the theory and to clarify what happens in at least a few highly idealized situations.

During the last few years theorists have constructed a number of such simplified models. To describe the details of all these models would require a substantial monograph. The present paper comments briefly on a few models, indicating the assumptions made and the general character of the results, together with some of the chief problems remaining. After discussions of the three evolutionary scenes listed above, a final section treats the vertical structure of the interstellar medium, again through discussion of simple models. This paper provides only a cursory introduction to the complex dynamical processes in which the hot gas participates, often as the primary driving force, and which must play a major role in any overall account of the interstellar medium, particularly of its structure and evolution.

2. EXPANSION OF SUPERNOVA REMNANTS

The phenomena that follow a supernova explosion can in principle be followed in rather full theoretical detail if spherical symmetry is assumed. Such a spherical model is applicable if the stellar explosion itself produces this symmetry, if also the initial properties of the surrounding interstellar gas are functions only of r (the distance from the supernova), and if finally the magnetic field **B** is ignored. In addition, one must assume that nonspherical instabilities will not arise. Under these conditions, all quantities are functions of radius r and time t, and the relevant differential equations can be integrated. An initial particle density independent of r is usually assumed. While spherical symmetry is not likely to be realized in detail, it may provide an adequate first approximation, especially in those regions where the interstellar gas is reasonably homogeneous. The approximately circular shape observed for some SN remnants, especially the younger ones (2), is consistent with this assumption, though cylindrical symmetry may provide a better overall fit (15, 55) for the radio emission observed from most remnants.

Most theoretical models also make the restrictive "hydrodynamic" assumption that the mean free path of all particles is much less than r. This assumption has the great advantage that it yields the familiar equations of fluid dynamics, for which the techniques of numerical solution have been much studied. Physically, the hydrodynamic assumption leads to a thin shock wave and to negligible conductive heat flow. In fact the mean free path of protons for $90°$ deflections in encounters with other protons is many parsecs for a newly born supernova remnant, decreasing to about 1 pc when $T = 10^7$ K, if the proton density is 0.1 cm^{-3}. A magnetic field restricts the travel of charged particles transverse to **B** but does not yield the hydrodynamic approximation for motions parallel to **B**. Other processes have been suggested that can reduce the effective mean free path. Energetic particles moving through an ionized gas are sometimes slowed down by the plasma instabilities that they excite. This complex effect can perhaps provide full justification for the hydrodynamic assumption, which is also consistent (72) with the relatively sharp boundaries observed for the X-ray emission from some young supernova remnants, notably around the entire circumference of Cas A and Tycho. However, definite confirmation is lacking.

We ignore initially here both the magnetic field, which is almost certainly present, and the uncertainties associated with the hydrodynamic assumption. The effects of thermal conductivity and of a **B** field are discussed briefly at the end of this section. The processes that result when the initial distribution of the ambient gas is cloudy or has a vertical density gradient are treated in subsequent sections. Effects produced by relativistic particles or by any type of energy source within the SN remnants have not been much considered in theoretical models of such remnants and are ignored here. Hence, radiation from a rapidly rotating neutron star, produced by the collapse of a stellar core, is not taken into account. In the observed "filled-center" (or "plerion") remnants (99), such as the Crab Nebula, such effects may be highly important.

2.1 *Results for Spherical Models*

The spherical models based on these assumptions have yielded a substantial body of knowledge on supernova remnants. Three familiar evolutionary stages are distinguished. First, there is free expansion of the ejected material, a stage that lasts as long as the ejected mass is large compared with the mass of the swept-up interstellar gas. Subsequently, the interstellar mass swept up by the outward-moving shock, of radius r_s, much exceeds the ejected mass. As long as the kinetic temperature is high enough throughout the remnant that radiative cooling is slight, this is the well-known "Sedov-Taylor" stage [see the monograph by Ostriker & McKee (77)]; the density decreases rather steeply inward behind the shock front, with half the mass in the outer 6% of the radius. In this stage the shock velocity V_s varies as $r_s^{-3/2}$, giving $r_s \propto t^{2/5}$.

With increasing r_s, the postshock temperature T_s decreases as V_s^2. When T_s falls below roughly 10^6 K, the increased rate of radiation then cools the postshock gas to a temperature below 10^4 K, and the density increases by a large factor behind the shock, forming a cold shell much thinner than before; this is the third, or "snowplow," stage. As long as the internal pressure deep within the remnant remains high, the outward momentum of the cold shell gradually increases. In this third stage, V_s varies about as $r_s^{-5/2}$ (25, 73), giving $r_s \propto t^{2/7}$.

For an actual remnant these stages are approximations, since the transitions between successive stages produce additional effects. In particular, interaction between the freely expanding ejected SN gas and the surrounding medium produces a "reverse shock" that progresses back through the ejecta, an effect pointed out (17) independently in three papers presented in 1973. The ejected gas, moving out at a velocity v_e through the ambient medium of density ρ_a, produces an outward moving shock, with a postshock pressure of about $\rho_a v_e^2$. This pressure tends to propagate back into the ejecta; as these continue to move outward they cool from adiabatic expansion, and when their internal pressure falls significantly below the postshock pressure ahead, the pressure disturbance that is propagating back through the ejecta steepens and forms a reverse shock (68). The observed emission from the SN ejecta in young remnants is largely produced by this shock.

The reverse shock is only one example of the spherical pressure disturbances that result in large part from transitions between successive stages in a SN remnant. An exact numerical solution of the fluid-dynamical equations (21), including radiative emission in the appropriate energy equation, shows a variety of such disturbances moving radially inward and outward. As a result of these disturbances and the limited duration of each stage, there is only rough agreement with the predictions for the various stages in isolation. However, the combination of numerical calculations with approximate analytic results gives a reasonably complete understanding of the idealized spherical model, based on the hydrodynamic assumption and an initially uniform interstellar gas density.

In subsequent discussions we take as typical parameters for SN remnants the results obtained in this numerical model, with an ambient density of 0.1 H atom cm^{-3} and an initial kinetic energy of 0.93×10^{51} ergs in an ejected envelope of mass 3 $M_\odot$. In this model the swept-up and ejected masses are equal at roughly 10^3 yr. The cold shell forms during the interval from 1.2×10^5 to 1.7×10^5 yr as the shock velocity drops below about 150 km s^{-1}; the shock radius r_s is about 55 pc during this interval.

Other spherical models have emphasized the very early stages of expansion, when SN ejecta, with a density varying as r^{-n}, expand into an ambient interstellar medium of constant density or into a circumstellar medium with a density varying as r^{-2}. For $n > 5$, self-similar solutions exist (18) that include both the outward-moving shock and the reverse shock traveling inward through the ejecta. A recent numerical fluid-dynamics calculation (7) shows that the flow within the SN remnant is initially close to the self-similar solution and gradually evolves to the Sedov-Taylor blast wave. Rayleigh-Taylor instabilities in the flow between the outer shock and the reverse shock are likely to produce turbulent motion.

2.2 *Effects of Thermal Conduction and Magnetic Fields*

We turn now to the effects that thermal conduction and a magnetic field can produce in supernova remnants. A model that takes into account thermal conduction by electrons as well as energy exchange between ions and electrons shows (22) that after the initial free-expansion stage the electron temperature T_e becomes nearly constant with radius inside the remnant, a marked change from the Sedov-Taylor solution. When the electrons are nearly isothermal, the positive ion temperature T_i, assumed to equal T_e immediately behind the shock, is found to increase inward. For the particular case treated, the rate of expansion is not much altered by such effects, with r_s at each time increased by not more than 8%. These detailed results depend on the assumption that no heat conduction occurs through the moving shock front. More detailed computations (28, 35) show the effects of thermal conduction and of electron-ion energy exchange on various properties of an evolving supernova remnant.

If a magnetic field **B** is present, as seems very likely, thermal conduction transverse to **B** will be almost completely suppressed. Thus, T_e will tend to be constant along **B** but to show a variation of Sedov-Taylor type in planes transverse to **B**. The positive-ion temperature T_i will exceed T_e deep within the remnant, since approach to equipartition through electron-ion encounters is relatively slow (see references in preceding paragraph); conduction of heat by positive ions contributes significantly to holding down dT_i/dr along **B**.

The effects that a magnetic field produces on a supernova remnant are more conspicuous if the energy density $B^2/8\pi$ is at least comparable with the material pressure nkT, where n is the total number of particles per cubic centimeter. An equivalent condition is that the Alfvén speed $V_A = B/(4\pi\rho)^{1/2}$ be at least comparable with the isothermal sound speed $C_s = (kT/m)^{1/2}$, where $m = \rho/n$ is the mean mass per particle. This situation can arise even if the magnetic field in the gas surrounding the supernova is initially very weak.

One such case for which detailed calculations have been made (58) is the hydromagnetic flow around a conducting spherical shell, or "piston," assumed to be expanding at a constant rate into a conducting gas permeated by an initially weak and uniform magnetic field. The lines of force that have been pushed outward by the piston tend to accumulate in a thin boundary layer, where the magnetic field grows steadily until its energy density becomes comparable with that of the streaming gas. In an actual remnant the ionized ejected gases from the supernova may take the place of the expanding piston. The high magnetic pressure in the surrounding boundary layer would then tend to decelerate the inner ejected gases, and the Rayleigh-Taylor instability should occur. While the analysis is evidently idealized and other effects will certainly be present, such a process may play a part in producing the filaments of high magnetic field observed in the inner region of the Crab supernova.

A uniform interstellar **B** field can produce important effects during the snowplow stage of supernova expansion. With representative parameters for the warm interstellar medium ($n_H = 0.15$ cm^{-3}, $T = 6000$ K, $B = 3$ μG) the Alfvén speed V_A is 14 km s^{-1}, substantially greater than the isothermal sound speed C_s of 6.1 km s^{-1}. As pointed out above, when the cold radiative shell starts to form, the shock velocity V_s is about 150 km s^{-1}; we adopt 40 km s^{-1} as a representative value of V_s during the snowplow phase. For these parameters the relative increase of density across an isothermal shock (90A) is a factor $(V_s/C_s)^2 = 43$ if **B** vanishes or is parallel to V_s but is only 3.4 (approximately $2^{1/2} V_s/V_A$) for propagation transverse to the assumed magnetic field. This latter compression is not only weak but nearly reversible, since the energy stored in compressing the magnetic field can drive a reexpansion when the pressure falls. If the high-compression "parallel shocks" were assumed to be relatively infrequent and the low-compression transverse ones were regarded as dominant, one would conclude (26, 27) that the late expansion stages of a supernova remnant produced only a very modest compression of the interstellar medium.

To evaluate the compression expected in an actual interstellar situation one must consider oblique shocks, with a shock-front normal at some arbitrary angle to the magnetic field. The physical principles governing such shocks are well known (54, 60). It turns out that for the conditions previously specified, the parallel shock discussed above cannot occur. Instead, a single shock, moving parallel to the magnetic field (i.e. V_s parallel to **B**), produces (92) a relatively low compression (a factor of 8.2) and leaves the postshock magnetic field inclined at a substantial angle (75°) with respect to the shock front normal. This configuration is called a "switch-on shock." More generally, if the shock velocity V_s exceeds V_A and V_A exceeds C_s, the switch-on shock replaces the parallel shock that one might expect to occur when V_s is parallel to the preshock **B**. A corresponding shock that bends the magnetic field back and leaves the postshock **B** perpendicular to the shock front is called a "switch-off shock." Remarkably enough, a combination of two successive shocks, first a switch-on followed by a switch-off front, produces (54) the same overall effect as does the parallel shock described above, and thus in one sense nature contrives to elude its ban on parallel shocks.

One may expect that similar results, with relatively high compressions, are possible for two successive shocks within some range of directions relative to **B**. Until such possibilities have been explored, the average compression in the late stage of a supernova shock is essentially unknown, except that it presumably lies between $(V_s/C_s)^2$ and about $2^{1/2} V_s/V_A$.

The amount of this compression has important effects on the structure and dynamics of the interstellar gas. A familiar picture of the interstellar medium, often used as a standard of reference, is based on the sweeping synthesis by McKee & Ostriker (73), which brings into one theoretical framework many different aspects of this medium, including particularly the interpenetration of hot and cooler gas proposed by Cox & Smith (31). If the compression in an isothermal shock during the snowplow phase were much reduced by magnetic forces, with the compression ratio decreasing from 4 to 1 as V_s decreases from 40 down to about 14 km s^{-1}, some aspects of this picture would require modification (26). In particular, the warm neutral medium (WNM) would not be swept up into dense shells but would occupy an appreciable part of the remnant's volume, reducing the fraction of this volume occupied by the hot gas. The overall filling factor f_h of the hot gas (the fraction of the volume of the galactic disk occupied by this gas) would be correspondingly reduced.

Such effects are shown by a SN remnant model (30) that exaggerates the dynamical influence of the magnetic field; in this model $p_B = B^2/8\pi$ is assumed isotropic, with B increasing through the shock in direct proportion to the density. In the ambient medium $B = 5$ μG, n_H is taken to be 0.1 cm^{-3}, and $T = 10^4$ K. When radiative cooling becomes important, the postshock density increase in this model is at most a factor of four and decreases to two after some 6×10^5 yr, when the maximum velocity of this gas is only about 30 km s^{-1}. The radius of the inner hot bubble ($T > 10^5$ K) is then 60 pc, and it does not increase above 65 pc. After about 3×10^6 yr the bubble disappears, compressed by the external p_B and cooled by radiation. This model differs markedly from a nonmagnetic one but does not necessarily yield a small f_h, which can lie between 0.1 and nearly 1 depending (30) on the rate and clustering (see Section 4) of SN explosions.

In a more realistic model, compression would be likely to occur along some lines of magnetic force. Coupling between motions parallel and transverse to **B** might conceivably convert the enhanced magnetic energy of the compressed WNM into radiation from dense clumps of cooling gas. In any case, magnetic tensions and pressures in the complex interstellar medium are likely to produce unexpected consequences. Computers are reaching the power necessary to follow such hydromagnetic processes approximately. Until more knowledge is available either from analysis or from simulations, theory cannot indicate the value of f_h, the hot-gas filling factor. The topology of this gas—isolated hot bubbles at one extreme and isolated cooler clouds (either warm or cold) at the other—is equally unclear.

There have been several attempts to determine f_h observationally. The number of observed SN remnants with a radio surface brightness above a limiting value, together with an assumed SN rate of once every 30 yr in the Galaxy, is consistent (46) with a filling factor f_h of 0.9 for the rarefied and presumably hot gas into which the SN remnant expands rapidly; however, for a Type II SN this rapid expansion probably occurs (15) into the cavity produced by the stellar wind or by UV radiation from the progenitor star, and a low density in this cavity need not be representative of the interstellar medium in general. In another approach, the lack of anticorrelation between the column densities $N(O^{+5})$ and E_{B-V} has been used (50) to derive an upper limit of 0.2 for f_h; this limit does not apply if the cold clouds responsible for most of the interstellar extinction are assumed unaffected by the hot gas, as seems likely sufficiently far from a SN. An upper limit of 0.2 has also been found (41) for the volume fraction of the galactic disk occupied by observed cavities in the H I distribution, considering those with a minimum extension of roughly 50 pc. A similar low value of f_h is suggested by a comparison of the observed $N(O^{+5})$ values with a simple model of conductive evaporating envelopes around cold clouds—see Section 3.2. However, another recent survey of the observations concluded (69) that f_h is between 0.4 and 0.7. Evidently these observational results are about as inconclusive as the theoretical ones.

3. INTERACTION OF REMNANTS WITH CLOUDS

When a star explodes, the expanding remnant may sweep through a gas quite different from the uniform interstellar medium discussed above. It is well known (57, 91) that the gas between stars has inhomogeneities with a wide variety of sizes, ranging from filaments less than a parsec across to giant molecular clouds and cloud complexes 100 pc in size. Similarly, the X-ray emission from SN remnants generally shows much complex structure (2). If the exploding star was particularly luminous or some of its companions had exploded a short time before, the local gas may have been greatly modified and perhaps somewhat homogenized by ultraviolet photons and by expanding hot gases. We do not treat here the details of these complex environments but rather consider some of the physical processes that can occur when a supernova remnant engulfs a cloud. Since a full review of such processes has recently been given by McKee (70), the present discussion is brief and highly selective.

Except in a brief discussion near the end of Section 3.2, we regard clouds as spherical. This assumption, which vastly simplifies the analysis, has been a basic part of most theoretical models of the interstellar gas. It is well established (45) that the 21-cm emission observed at intermediate galactic latitudes ($10° < b < 50°$) emanates mostly from conspicuous filaments (each generally clumpy in structure), oriented along the magnetic field direction indicated by the optical polarization. This result presumably applies to warm H I gas at appreciable height above the galactic plane. For cold H I gas in the galactic disk the evidence is less clear, and the assumption of spherical cold clouds may provide a reasonable first approximation.

Strong compression of clouds is believed to be a major effect produced by SN remnants. A combination of analytic theory and numerical simulation gives an approximate indication of how this "cloud crushing" might proceed. A passing supernova shock generates a slower shock in the denser cloud material. Behind this cloud shock the velocity field leads both to compression and shear and can produce instabilities that may disrupt the cloud, at least in part. An analysis of cloud compression when a magnetic field is present indicates (75), as expected, that if the magnetic pressure is dominant, motions transverse to **B** are suppressed. The effect of instabilities may also be less when the compression is one dimensional, though further study would be needed to establish this conclusion.

The development of a SN remnant must be strongly affected by the various interactions with clouds (70). A detailed evolutionary calculation for such a remnant, including the effects of cloud compression and evaporation, was carried out numerically some years ago (24), based on the clouds present in the galactic disk generally. Recently, a different hydrodynamic technique has been applied (103) to this same situation with general overall agreement. In particular, vigorous evaporation from clouds, which is most rapid nearest to the supernova, keeps the density nearly independent of r. As a result, the cold shell that is formed by radiative cooling, at a remnant age of typically some 10^5 yr, first appears well inside the remnant rather than immediately behind the shock layer.

3.1 Conducting Envelopes of Clouds

We consider now in more detail the evaporation process that results when a cold cloud is surrounded by hot gas, whose heat energy flows into the cloud by thermal conduction. The shock itself is not directly involved in this process, which is usually modeled with the hot gas in pressure equilibrium with the cloud and with no systematic velocity of the cloud with respect to the gas. The nature of this process depends on a global saturation parameter σ_0 (23), which is essentially the ratio of the electron mean free path to the cloud radius a, and is equal to $0.4(T_f \times 10^{-7})^2/n_{ef}a_{pc}$; here T_f is the asymptotic temperature of the hot gas, far from the cloud, n_{ef} is the asymptotic electron density, and a_{pc} is the cloud radius in parsecs. When σ_0 is small compared with unity, the thermal conductivity is given by its classical value, the temperature distribution (in this three-dimensional situation) has a quasi-steady state, and the velocity of the gas in the expanding envelope is subsonic everywhere. However, if σ_0 is too small, less than about 0.03 (corresponding to a_{pc} exceeding about 10 pc in a typical situation), the heat flow is inadequate to offset radiative cooling, and condensation of the hot gas replaces evaporation of the cloud (71). On the other hand, when $\sigma_0 > 1$ the physical situation is more complicated; the heat flow in this "saturated" condition can be estimated from solar wind data (23) and the resultant mass loss computed.

A recent investigation (33) indicates that for large σ_0 the effect of viscosity, produced by atomic ions, must also be considered. Calculations based on the simplifying assumption that $T_i = T_e$ show that for $\sigma_0 \geq 100$ the pressure in the cold cloud much exceeds that in the hot gas, and that the viscosity increases the mass loss rate by at least an order of magnitude above the inviscid case; the viscosity requires a pressure increase within the cloud to drive the flow, and at the resultant higher density the saturated heat flow and the resultant mass loss rate are increased. If the positive ions are heated only by two-body encounters with electrons, T_i will be much less than T_e and these viscous effects will be much reduced.

A further modification in the theory is needed if a magnetic field is present; as pointed out above, there is virtually no conductive flow of heat across an interstellar magnetic field. Limitation of the conductive flux to the direction of **B** has been taken into account (4) for the evaporative flow from an infinite plane surface, initially separating a cold gas on one side from a hot gas extending infinitely far on the other. In this model all variables depend only on t and on z, the distance from the plane; both $B^2/8\pi$ and ρv^2 are assumed small compared with p. No steady state is possible in this one-dimensional situation (unless a surface at some fixed temperature is located a finite distance away). Instead, a self-similar solution is obtained, with a similarity variable proportional to $z/t^{1/2}$; thus, as t increases, the conductive front thickens and the mass loss rate decreases.

An interesting result of this analysis is that the initial heat flux varies as $\cos^2 \theta_\infty$, where θ_∞ is the angle between **B** and the z axis at large z. This may be understood physically, since the temperature gradient parallel to **B** equals $\cos \theta \, dT/dz$, and the component of the heat flow parallel to z varies as $\cos^2 \theta \, dT/dz$. However, as the front evolves, the ablation rate is proportional more nearly to $\cos \theta_\infty$, since dT/dz within the front is greater for larger θ_∞ (see ref. 10).

Since the magnetic fields within diffuse clouds (determined from the Zeeman effect of 21-cm lines) are apparently about the same as those in the warm ionized medium (43), a model of straight, parallel lines of force extending into the hot gas may, perhaps, provide a reasonable approximation. On this basis the thickening of the conduction front with time, resulting from the one-dimensional character of the heat flow, should be an important effect, probably more so than the dependence of the flow on θ_∞.

3.2 *Highly Ionized Atoms*

An important attribute of these various conductive evaporating envelopes, with or without magnetic fields, is that they contain highly ionized atoms such as O^{+5}, whose O VI absorption features have been observed along numerous lines of sight through the interstellar gas. At a kinetic electron temperature above 10^5 K, oxygen atoms will be highly ionized by electron collisions. We consider here the theories of this process that have been developed and the numerical calculations that have been carried out for the resultant column densities.

In these envelopes the temperature of each fluid element in the moving gas increases with time, and hence collisional ionization equilibrium is not fully reached. Since the ionization lags behind the temperature, the fraction of oxygen atoms in O^{+5} ions will reach its peak value at temperatures exceeding the value of 3×10^5 K for the O^{+5} peak in collisional equilibrium. The relative numbers of atoms in different stages of ionization must be computed from the relevant differential equations, including the rates of ionization and recombination. Such computations (5) for a spherical conductive envelope, expanding radially outward according to the model described above (23), give the total number of O^{+5} ions in the envelope. We denote by F the ratio of this number to its value in collisional equilibrium; this factor F depends on $n_{ef}a_{pc}$, where again n_{ef} is the asymptotic electron density, and a_{pc} is the cloud radius in parsecs. The ratio F generally exceeds unity. The detailed calculations show that for an asymptotic temperature T_f equal to 10^6 K, F increases from 2.2 at $n_{ef}a_{pc} = 0.10$ pc cm^{-3} to 40 at $n_{ef}a_{pc} = 0.01$ pc cm^{-3}. Later computations (6) have considered the non-Maxwellian distribution of electron velocities resulting from the rapid increase of mean free path with velocity. While the ionization rates are somewhat increased, the relative number of highly ionized atoms in the evaporative flow is reduced by at most 20%.

These results, combined with an assumed cosmic composition, give a theoretical value of $\langle n(O^{+5}) \rangle$ (the mean particle density of O^{+5} ions in the galactic disk) equal to about 2×10^{-7} cm^{-3} for $a = 5$ pc, $n_{ef} = 0.01$ cm^{-3}, and $T = 10^6$ K; the filling factors f_c and f_h for the cold clouds and the hot gas in the galactic disk are set equal to 0.02 and 1, respectively. If n_{ef} is decreased to 0.001 cm^{-3}, the computed O^{+5} density rises to 5×10^{-7} cm^{-3}. These densities exceed by an order of magnitude or somewhat more the observed mean value of about 2×10^{-8} cm^{-3} (51).

Later computations for the ionization distribution in expanding spherical conductive envelopes (9), taking radiative losses also into account, give similar results. These were used to compute $N(O^{+5})$, the column density of O^{+5} ions in a radial line of sight through a single envelope, from $r = a$ to $r = 10a$. The resultant values were found to be clustered around 10^{13} cm^{-2}, with a relatively slow dependence on $n_{ef}a$; as this parameter increases, the increased mass loss rate is offset by a decrease in F. This theoretical column density of 10^{13} cm^{-2} agrees with the observed mean value (50) for two thirds of the individual components constituting the O^{+5} gas. [The remaining components have larger column densities and were not included in the value of $\langle n(O^{+5}) \rangle$ cited above.] However, the observed values should exceed these theoretical ones by a geometrical factor. For a line of sight through the cloud center the column density is twice the theoretical one. For lines passing tangentially through the shell of highest $n(O^{+5})$, the increase will be somewhat greater; an average increase by a factor of two should provide a rough approximation.

While one would not expect such idealized models to correspond closely with reality, it is of interest to note that plausible changes in the assumed parameters can bring the theoretical value of $\langle n(O^{+5}) \rangle$ into rough agreement with the observed value. Absorption-line data indicate (106) that oxygen is depleted in interstellar clouds, with a depletion factor δ_O between

0.4 and 0.7. If we set $\delta_O = 1/2$, this factor compensates for the geometrical factor discussed above, leaving the observed and computed component column densities in agreement. If also f_h, the filling factor of the hot gas, is set equal to 0.2, the theoretical value for $\langle n(O^{+5}) \rangle$ is then reduced by an order of magnitude and agrees with the observations to within the many uncertainties involved. About this same value of f_h is needed to give the observed average number (50) of about six O^{+5} conductive envelopes (each with a radius typically of about $2a$) in the line of sight per kiloparsec.

As another test of the conductive envelope theory, the O^{+5} velocity distribution computed for the model may be compared with the observations. For the thermal velocity dispersion of these ions within a single envelope, the model yields a value between 14 and 18 km s^{-1} "in most cases" (9), corresponding to kinetic temperatures between 4×10^5 and 6×10^5 K. Because of overlapping components, the observed line profiles can give only the minimum values of v_m, the rms velocity dispersion. The envelope expansion velocity can increase v_m appreciably for some lines of sight but will have little effect on the minimum values, corresponding to lines of sight passing tangentially through the outer layers. Thus, one would expect from the theory that for envelopes in general the minimum v_m should be about 14 km s^{-1}.

While a number of the observed O VI profiles (49) have values of v_m less than 14 km s^{-1}, ranging down to 10 km s^{-1}, the effect of observational errors must be considered, and because of these one can infer only that values as low as 14 km s^{-1} are "indeed rather common" (50). One may conclude that the observed line widths are not inconsistent with the conductive envelope theory, though more precise data would be required for definitive results.

We have seen in Section 3.1 that thermal conduction in an ionized interstellar gas is necessarily parallel to **B**, and that in consequence a one-dimensional (or "slab") expanding conduction front is likely to be a more realistic model than a spherical, radially expanding front. The temperature and ionization levels in such a slab front have been computed in a series of models (10), taking radiative heat losses and nonequilibrium ionization into account. The initial hot-gas temperature T_h and total particle density n_h were set equal to 7.5×10^5 K and 5×10^{-3} cm^{-3}, respectively; for the evaporating H II cloud considered, the corresponding initial values were 10^4 K and 0.375 cm^{-3}. Several values were considered for θ_∞, the asymptotic angle between **B** and the outward normal to the cloud surface.

At $t = 0$ the hot gas has an interface with the slab cloud. The models show that at later t the column density of each highly ionized species first increases about as t^2 and then flattens off at a reasonably constant value. The time required to reach this value varies from about 4×10^3 yr for Si^{+3} to about 4×10^5 yr for O^{+5}. For $\theta_\infty < 60°$ the quasi-constant column densities for the different species are equal, within a factor of two, to those found for spherical conduction fronts under comparable conditions. The kinetic temperatures of the absorbing O^{+5} ions are also about the same as in the spherical models. Hence, the one-dimensional slab fronts seem consistent with the observed O VI lines as regards both the component column densities and the minimum line widths.

It is not clear under what conditions this theory will give the observed number of about six O^{+5} components per kiloparsec and thus yield the observed $\langle n(O^{+5})\rangle$. The evaporation of both cold gas and warm gas must be considered in this connection, though possibly the warm low-density gas evaporates so rapidly that it makes only a minor contribution to $\langle (O^{+5})\rangle$. In addition, if the cloud is a thin filament parallel to **B**, conductive evaporation will occur only at the two ends. If the diameter-to-length ratio of the filament is small, the value of f_h required to fit the O VI data may be relatively high. Until such topics have been explored, it would be premature to take expanding conductive envelopes as the primary locations of observed interstellar O^{+5} ions.

Absorption by O^{+5} ions should also be produced in the inner regions of an idealized spherical supernova remnant when some of the remnant gas cools through temperatures of about 3×10^5 K. Since the initial temperature of the remnant decreases outward, the outer layers cool first and a radiative cooling front gradually eats its way into the inner hot bubble, surrounded by the cold shell of the snowplow stage. Studies of this effect (73) in the absence of any magnetic field showed that the mean density of the resulting O^{+5} ions could exceed the observed values, but that the line widths should be 100–200 km s^{-1}, much exceeding the observed widths.

In the magnetic remnant model (30) discussed in Section 2.2 the magnetic field decreases the expansion velocity of the hot inner bubble with time; thus, the younger bubbles will produce wider lines, and the older bubbles narrower ones. When radiative cooling first sets in and the column densities of the highly ionized species are at their maxima, the expansion velocities of these hot shells during some 10^5 yr are about 100 km s^{-1}, significantly higher than the observed O VI rms line widths, which range up to about 30 km s^{-1} with only one value as high as 40 km s^{-1} (49). Wide lines are more difficult to detect than narrow ones, but it seems unlikely that observational selection effects can explain the complete absence of rms line widths exceeding 40 km s^{-1}. Some 10^6 yr later, when the O^{+5} column densities are less by a factor of five, these expansion velocities have virtually vanished.

As with the conductive envelope models, the magnetic bubble model can yield agreement with observed O^{+5} column densities for individual components if f_h has a small value, estimated as about 0.1 (30). It is noteworthy that this magnetic model gives relative intensities of the various ion species that agree with the observations to within a factor of two, significantly better than achieved with other models. If effects of initial inhomogeneities were also included in such a magnetic model, these inner regions of hot cooling gas might have a somewhat similar appearance, though a different course of development, as the conductive envelopes between cold clouds and hot gas discussed above.

4. EFFECTS FAR FROM THE GALACTIC PLANE

The possibility that a hot gas might form a galactic corona at kiloparsec distances from the galactic plane (89) has given particular interest to the role of supernova remnants in this connection, since the hot gases in such remnants provide an obvious heat source. While the remnant from a single stellar explosion can provide heating far from the galactic plane if the supernova is located in the halo or if the ambient particle density is 10^{-2} cm^{-3} or less, more energetic events provide a better source. It has long been clear (19) that adjacent explosions of several supernovae, as might be expected in young stellar groups, would more easily break out of the gaseous galactic disk and pervade the halo. Recent studies have emphasized such sequential explosions of supernovae and the "superbubbles" that they produce in this and other galaxies.

A primary reason for this emphasis has been the accumulating observational evidence (40, 42, 66, 67, 95) for the existence of superbubbles not only in our own Galaxy but in other Local Group spiral and irregular systems. Extensive arcs and filaments seen in 21-cm emission reveal "supershells" of neutral H. X rays are detected from the hot gas within some of these supershells, and surveys of Balmer emission lines confirm such extended structures. The observed radii range from 100 to 1000 pc. The kinetic energies E_k found for 17 supershells range (40) from $10^{51.3}$ to $10^{53.5}$ ergs.[1] If the three values with the lowest confidence level are omitted, the mean log E_k is 52.3, corresponding to about 20 supernovae, with only two values exceeding 53 (53.1 and 53.4).

A second reason for the recent focus on superbubbles is the strong theoretical expectation that such concentrations of supernovae in time and space should in fact exist. Some supernovae (the Type Ia class) originate in old stars, which in the galactic disk show virtually no concentration in groups. However, the Type II supernovae and also those of Type Ib (99, 101) result from core collapse in young, massive stars, which are apparently

---

[1] I am indebted to Dr. Carl Heiles for informing me that the values of log E_k given (40) for GS 123+07−127 and GS 139−03−69 should each be corrected by subtracting 0.6.

formed to a large extent in groups, including clusters and associations. Some of these stars escape as runaways before they explode. According to a recent survey (37), about 70% of the O stars are now in groups; most of the supernovae produced when these massive stars in each group die will be found within a radius of a few times 10 pc and a time interval of several times 10^7 yr (66). If several stellar groups are formed within the same cloud complex, the various superbubbles produced may overlap, producing an even more spectacular explosion. The number of supernovae participating in one explosion is uncertain but probably has a wide distribution, with values between 10 and 100 most likely but possibly extending up to 10^3 or even more (44, 66).

4.1 Dynamical Models of Superbubbles

Theoretical models of these superbubbles may be constructed, based on most of the assumptions made for a one-supernova remnant. For a superbubble, however, the disturbance can spread so far from the galactic plane that all physical quantities must be functions of two spatial dimensions: cylindrical radius r and vertical height z. Offsetting somewhat this complication is the initial homogeneity that may characterize the ambient interstellar gas, thanks to the processing of this medium by energetic stellar winds and intense ultraviolet stellar radiation emitted from the massive bright stars before these die and explode.

The nature of blast waves and outgoing winds under these conditions has received much study (84). An important aspect of such disturbances in an exponential atmosphere, for example, is that they can lead to "blowout," in which the segment of the outgoing shock at the greatest height starts to accelerate and attains a much increased velocity in the high layers of very low density.

Numerical computations, recently reviewed in some detail by Tenorio-Tagle & Bodenheimer (93), have shown the development of galactic superbubbles produced by consecutive supernovae (63, 96). While some details are still controversial, the general outline seems clear. If the ambient density is taken to be a function of z only and set about equal to the observed smoothed value (61), blowout does not occur unless the center of the superbubble is about 100 pc or more from the galactic plane. In any case, an energetic superbubble expands rapidly in z and rises far above the galactic plane. In one example (63), 80 supernovae occur uniformly during 10^7 yr at $z = 0$; the ambient particle density at $z = 0$ equals 1.0 cm^{-3} (0.19 cm^{-3} of warm gas and 0.81 cm^{-3} of cold)[2] and for large z varies as

---

[2] I am indebted to R. H. McCray for providing me with the values actually used in the computations.

$0.19 \times \exp(-z/H)$ cm^{-3}, where $H = 500$ pc. After 10^7 yr the hot expanding gas has reached $z \approx 600$ pc, as compared with $r \approx 300$ pc reached at $z = 0$. The temperature of the rising gas is of order 10^6 K.

In a second example (96), with 50 supernovae in 10^7 yr, again centered at $z = 0$, the total mass of the ambient gas in a column 1 cm^2 in cross section is less by an order of magnitude; the interstellar particle density for large z equals $0.035 \exp(-z/H)$ cm^{-3}, where $H = 250$ pc. In this case the superbubble expands more rapidly, reaching $z \approx 1000$ pc in 10^7 yr, when $r \approx 400$ pc at $z = 0$.

The actual ambient gas density at high z is probably between the values assumed in these two examples. Hence, one may conclude that some hot gas from superbubbles reaches z values of at least 500 pc and probably substantially more. This conclusion is strengthened by calculations of superbubbles centered initially at $z = 100$ pc. Such computations were made for each of the two examples cited above, and even for the case with the higher ambient density show hot gas rising to some 1500 pc, where blowout begins.

Since different calculations disagree (64), these results on superbubble development are somewhat uncertain. Definite information is particularly lacking on the later stages of superbubble development, in view of the complexities involved and the idealizations introduced in the dynamical calculations; in particular, the likely presence both of density inhomogeneities and of magnetic fields is ignored. As a result, it is not clear under what conditions hot gas will escape as a galactic wind and when it will fall back to the galactic plane (see Section 4.2).

There is also uncertainty as to whether multiple supernovae can account for the most energetic supershells observed. The model calculations suggest that superbubble kinetic energies as great as 10^{53} ergs are not easily obtained (93), even if the number of supernovae participating has the relatively high value of 10^3 cited above. While multiple supernovae should certainly produce some conspicuous extended structures, the most energetic supershells, with E_k perhaps comparable to 10^{53} ergs (see above), may require some different energy source. Since there are many uncertainties affecting not only the dynamical calculations (64) but also the overall energy budget of superbubbles (42), definitive conclusions are not yet to be expected.

4.2 *Galactic Fountains*

If the superbubble gas rising to kiloparsec heights is mostly confined to the Galaxy, the material will recirculate, falling toward the galactic plane as cooler gas, presumably concentrated in clouds formed through thermal instabilities. Such a model has been called a galactic fountain (86); in a later model, the superbubbles, spewing out hot gas at great altitudes, have been likened to smoking chimneys (48).

The distribution of the falling clouds in velocity and in space when they reach the base of the corona has been computed with an idealized model (13). First the formation of clouds in the uprushing hot gas was analyzed, using two-dimensional hydrodynamic calculations with symmetry assumed about the z axis through the galactic center; condensation into cold clouds was assumed whenever the gas temperature fell below 10^4 K as a result of adiabatic expansion and radiative cooling. As boundary conditions, in a thin layer at $z = 0$ and at all times after $t = 0$, the gas temperature was taken to be about 10^6 K, with a particle density of about 10^{-3} cm^{-3}, varying slowly with distance from the galactic center. Since the initial density at high z was negligibly small, the boundary layer generates outward winds, which expand into a vacuum; such models can be regarded as exploding gaseous disks, which approach a quasi-steady state as a result of cloud condensation.

After the clouds form, they can be assumed to move ballistically (86), independently of the ambient pressure and density, falling back to $z = 0$. For the values of initial temperature and density cited above, the clouds form at a height of some 5 kpc, with each fluid element at a galactocentric radius r_G appreciably exceeding its initial value. This outward motion results from an assumed radial pressure gradient in the hot gas. As the clouds fall back toward the galactic plane, their r_G values will decrease, since their azimuthal velocities provide too small a centrifugal force for equilibrium. Some 10^8 yr after its formation a cloud returns to the galactic plane at a galactocentric radius about equal to its initial value, when the gas now in the cloud was expelled at high temperature from the disk.

As a result of these effects, the returning clouds, at a height of a few kiloparsecs, will have velocities in the downward z direction and also in the r_G direction, toward the galactic center. In addition, conservation of angular momentum will produce an azimuthal velocity in the θ direction, relative to the local standard of rest (LSR); the cloud will lag behind the circular velocity of rotation. These predictions of the dynamical fountain model (13) can be compared with 21-cm observations of high-velocity clouds, which may be defined (97) as those with $|v| > 80$ km s^{-1} in the LSR. A simple kinematic model shows (53) that infall velocities of up to 100 km s^{-1} both in the z and r_G direction, with a lag typically of 100 km s^{-1} in rotational velocity with respect to the LSR, provide a reasonable fit for most of the high-velocity clouds, if those in the Magellanic Stream ($v < -250$ km s^{-1}, $70° < l < 180°$) are excluded.

These parameters are entirely consistent with those found from the fountain model (13). However, this model would seem too idealized to permit definitive conclusions. As noted above, density inhomogeneities and magnetic fields complicate the picture. In addition, it is not clear to what extent the lateral accelerations of the assumed uprushing hot gas, whose initial properties vary slowly with galactocentric distance, are applicable to the rising columns of initially separate superbubbles. While the general agreement of the fountain model with available 21-cm data on cloud velocities seems impressive, other explanations must also be considered (76, 97, 98); probably at least some of the high-velocity clouds may be explained as extragalactic gas, or as parts of supernova or superbubble shells, or as high-z extensions of a highly warped galactic disk.

It is possible that a galactic fountain can account also for the observed presence (1, 56) of some warm gas with z velocities in the intermediate range, between 20 and 80 km s^{-1} (see Section 5.1). To investigate this possibility, detailed fluid-dynamical calculations of uprushing gas have been made (47), showing the dependence of the flow on T_0 and n_0, the temperature and particle density at $z = 0$. For a model that best fits the warm gas, these take on values $T_0 = 3 \times 10^5$ K and $n_0 = 10^{-3}$ cm^{-3}; the gas cools steadily as it rises, with clouds condensing out at about 1.5 kpc and then falling down ballistically. The upward velocities of these clouds when first formed are about 30 km s^{-1}; the free-fall cloud velocity attained at $z = 0$ is about -100 km s^{-1}. The atomic column density $N(v)$ per unit velocity interval for all halo clouds in this model is essentially constant for v between 30 and -100 km s^{-1}, vanishing for other v. This distribution is very different from the $1/v^2$ relation suggested by the data (1, 56) for intermediate-velocity clouds, but various modifications of the simple theory might give improved agreement. In any case, a "low-temperature" galactic fountain can produce clouds up to 1 kpc with intermediate velocities; again, the relation of such models to superbubbles and transient phenomena generally is unclear.

The circulation of gas in superbubbles and galactic fountains must have a major influence on the structure and evolution of the galactic interstellar medium. A preliminary theory (74) of such global relationships points out that the hot gas filling factor f_h should depend on the superbubble fluxes of mass and energy, which in turn depend on the rate and clumpiness of supernovae. In the solar neighborhood a value of about 0.1 is suggested for f_h, with the hot gas largely confined to the uprushing fluid.

4.3 Ionization of Coronal Gas

Though detailed theoretical arguments are lacking, it seems not unlikely that the hot gas rising in superbubbles may provide a substantial source of thermal energy to the halo and thus maintain a more extensive coronal gas at a high temperature. While the existence of a widespread hot gas in the galactic disk seems conclusively established by measures of O VI absorption and soft X-ray emission, the presence of a similar hot gas at kiloparsec distances from the galactic plane is somewhat less certain. We summarize briefly here the evidence that makes the existence of such a hot coronal gas highly probable.

The presence of Si^{+3}, C^{+3}, and N^{+4} ions at high z seems definite (83). The ionization leading to the production of the first two of these species could probably be produced by photons (14, 51, 82) either from hot O-type stars and from the nuclei of planetary nebulae or from extragalactic sources. However, such photons can apparently not produce sufficient N^{+4} to account for the observed N V absorption lines. If no unknown source of far-UV radiation is present, collisional processes seem required. Unless nonthermal electrons are more abundant at high z than is generally supposed, a gas with a high kinetic temperature seems the plausible ionizing mechanism.

Recent detection of diffuse line emission from several highly ionized atomic species, particularly C^{+3} ions (65), strongly suggests the existence of a coronal gas. The observed flux in the C IV lines agrees quantitatively with that expected from a halo gas cooling radiatively through a temperature of some 10^5 K, with n_e in the range from 0.004 to 0.02 cm^{-3} and with a C IV column density taken from the absorption-line data. A warm gas, at about 10^4 K, could not readily fit the data, since electrons at this temperature would have a mean energy of only 1.2 eV, far short of the 8.0 eV required to excite C^{+3} ions. While inhomogeneities in the interstellar gas complicate the interpretation, it seems probable that much the same coronal gas produces both the emission and the absorption of C IV as well as of other highly ionized species.

The coronal gas provides a thermal pressure that can confine the high-z clouds, preventing the free expansion of the warm gas along the magnetic lines of force. Successive shock waves may also play an important part in the overall equilibrium of hot gas and warm clouds far from the galactic plane.

If the existence of a coronal gas is accepted, one must take into account that a gas in radiative equilibrium at some 10^5 K is thermally unstable and tends to heat up or cool down. A cooling flow back toward the galactic plane was assumed in fitting the C IV emission data discussed above. A similar flow has been assumed to fit the observed N^{+4} column densities (34); this calculation assumed a mass flow rate (uniformly into the galactic disk, with a radius r_G set equal to 15 kpc) of 4 $M_\odot$ yr^{-1} on each side of the plane. The predicted column densities of C^{+3} and Si^{+3} were too low by factors of about 2 and 5, respectively. [A flow rate of 8 $M_\odot$ yr^{-1} is consistent with the range of values found (65) in fitting the C IV emission.] This discrepancy between N^{+4} on the one hand and C^{+3} and Si^{+3} on the other is in the direction to be expected if photons also contribute to ion production, especially of Si^{+3}, but other effects may be involved. In any case, the scatter in the observed column densities is comparable with these discrepancies. As usual, the physical situation is doubtless more complicated than envisaged in our idealized models.

5. STRUCTURE OF THE HALO GAS

The injection of a superbubble into the halo will certainly have dramatic effects on the properties of the local gas, with consequences that are difficult to predict. After such an eruption is over, the local halo will presumably relax to some physical state that may even remain somewhat constant, at least in a statistical sense, until the next great explosion nearby. Thus, one may reasonably ask what sort of steady state may be physically possible, subject to what little we know about conditions in the halo.

In view of the many complications, we consider here one of the simplest questions we can ask about the halo gas—how it is held up in the galactic gravitational field (i.e. how on the average it satisfies the equation of hydrostatic equilibrium). To make this question theoretically tractable we assume that such properties as gas pressure, density, and rms velocity are functions only of z, the distance from the galactic plane.

Thus we ignore here most of the interesting physical problems associated with the maintenance of the halo gas. A survey of several such problems shows (20) that heating of the hot coronal gas in the halo can perhaps be attributed to supernovae of Type Ia; these old stars are present in the halo, and their remnants can provide shock heating in extended regions of the low-density gas. Further study is required to indicate the relative importance that superbubbles and individual Type Ia SN remnants have for the halo gas. Another result of this survey (20) confirms that heat input from supernovae can under some circumstances lead to a galactic wind rather than the downward flow envisaged in a galactic fountain. We ignore this possibility, since the intermediate-velocity clouds systematically show an infall velocity (56). Evidently, further analyses of supernova and superbubble interaction with the halo gas are strongly needed.

The balance of this section is devoted to two recent one-dimensional models of the halo gas, which investigate how hydrostatic equilibrium is maintained. A basic element in each model is the assumed topography of **B**, the magnetic field. The magnetic pressure p_B, equal to $B^2/8\pi$, makes a major contribution to the pressure in the galactic disk if $B_z = 0$. On the other hand, cosmic rays can effectively stream only parallel to the magnetic lines of force, and hence the escape of these energetic particles from the Galaxy is most simply explained if **B** has an appreciable z component. The first model considered here is a configuration supported in part by the magnetic pressure, with **B** assumed everywhere parallel to the galactic plane; this approach is consistent with the observed direction of **B** in the galactic disk, as determined both from pulsar rotation measures and from the optical polarization of starlight, resulting from alignment of interstellar grains. The second model is a cosmic-ray-supported configuration, which is based on the escape of cosmic rays, with B_z the only nonvanishing component of **B**; this approach is consistent with the escape of relativistic particles, which seems required by the observed composition of cosmic rays.

5.1 Models With **B** Perpendicular to z

In hydrostatic equilibrium, all physical properties are independent of time. In the one-dimensional case, where all quantities are functions of z only, the conservation of momentum (or "momentum balance") requires that the total pressure $p_{TOT}(z)$ at height z equal the weight of the column of gas above z plus any external pressure p_{ext}; here we ignore p_{ext}. The total pressure equals the sum of all the pressures acting on the gas—the gas pressure $p_G(z)$, the magnetic pressure $p_B(z)$, and the cosmic-ray pressure $p_R(z)$. We shall usually denote these pressures simply by p_{TOT}, p_G, etc, except that values at $z = 0$ will be denoted by $p_{TOT}(0)$, $p_G(0)$, etc.

We give here a detailed discussion of one recent such model (8) and then point out the various problems associated with some of the principal assumptions. This model includes various components of the interstellar gas—cold, warm, and hot. A significant new feature is inclusion of a condition for hydromagnetic stability, particularly against perturbations of Parker type (78). In these perturbations each line of magnetic force is raised in some regions and lowered in others, with the gas flowing along **B** into the lowered regions. This stability condition (59) requires that p_G (including the turbulent pressure that is produced by cloud motions) exceed $(dp_{TOT}/dz)/(\gamma d\ln \rho/dz)$. As usual, γ is the ratio $\delta \ln p_G/\delta \ln \rho$ during perturbations of a gaseous element; the cosmic-ray pressure is assumed to remain uniform along **B**. This criterion, which is derived only for the idealized hydrostatic equilibrium model, is a necessary condition for stability but may not be sufficient to guarantee complete stability.

If we ignore for the moment the uncertain contribution from the hot gas, the condition of momentum balance at $z = 0$ is satisfied in this model with the quantities given (8) in Table 1. Listed here are the adopted pressure at $z = 0$ for the relevant components of the gas; for the cold gas (including molecular clouds), this pressure results almost entirely from turbulent motions rather than from thermal velocities. Also shown are the weights

Table 1 Properties of the interstellar medium at $z = 0$

Component	Pressure (10^{-12} dyne cm^{-2})	Weight (10^{-12} dyne cm^{-2})
Cold gas	0.6	1.4
Warm gas	0.3	1.4
Cosmic rays	0.5	—
Magnetic field	1.4	—
Total	2.8	2.8

of the gas components in a column 1 cm^2 in cross section. Each weight is defined as the integral of $g(z)\rho_j(z)\,dz$ from $z = 0$ to ∞, where $g(z)$ is the gravitational acceleration in the z direction and $\rho_j(z)$ is the gas density in component j. Another important new feature of the model is the assumption of two components for the warm gas: the familiar one confined to the galactic disk, the other much more extended with a scale height taken to be 400 pc, consistent with recent observations (61). Since $g(z)$ increases markedly with z out to about 500 pc (3), the total weight of this extended component is relatively large (27). As a result, the total weight of the gas is 2.8×10^{-12} dyne cm^{-2}, substantially exceeding the value 1.7×10^{-12} dyne cm^{-2} typically assumed in earlier such models (90B). The gas pressures at $z = 0$, on the other hand, are about the same as those assumed in earlier work.

The other two quantities needed for momentum balance are p_R and p_B, also listed in Table 1. The cosmic-ray pressure $p_R(0)$ is reasonably well known. The value of $p_B(0)$ has been adjusted to give momentum balance, with $p_{TOT}(0)$ equal to the weight of all the gas. The corresponding rms B field is about 6 μG, about twice the value thought necessary (90B) when the extended warm gas was not included. While so large a field agrees with a variety of estimates (8, 43), its consistency with the precise measures of B from pulsar data is a primary requirement. These data give a mean field B_m in the solar neighborhood that ranges from about 1.6 to 3.5 μG, depending on the choice of region averaged and on other features of the data analysis (43, 62, 80). We set B_m equal to 2.5 μG.

While the rms dispersion of measured fields is less than B_m, the rms deviation of **B** from **B**$_m$, which we denote by B_r, may exceed B_m if the scale size is substantially less than the pulsar distances. Such a small-scale field will have dynamical consequences, producing oscillations of the lines of force, together with the clouds attached to these lines. If we assume equipartition of kinetic and magnetic energy in these oscillations, then B_r^2

equals $4\pi\rho\langle v^2\rangle$. Since we have assumed that $B_z = 0$ in this model, we ignore the z velocity of the clouds and take the two-dimensional velocity dispersion, assumed to equal $2^{0.5}$ times the one-dimensional value. Summing over the components in Table 1, with the values of $\rho(0)$ and $\langle v^2\rangle$ given in (8), we find that $B_r = 4.8$ μG. This value is consistent with the observed dispersion of pulsar measures if the field variations parallel to the galactic plane are assumed to be random with a scale size in the range from about 100 to 200 pc (80, 94), depending on the relative importance of fluctuations in n_e and **B**. This model for B_r is doubtless much too idealized (80) but may provide a useful first approximation. The quadratic sum of B_m and B_r then gives about 5 μG, not far below the value obtained from $p_{TOT}(0)$.

The detailed discussion (8) of this halo model suggests that an increased $p_{TOT}(0)$, together with the hydromagnetic stability criterion, promises to fit together various aspects of the interstellar medium at high z, including particularly the origin of the synchrotron radiation observed at radio frequencies. To provide sufficient p_G to satisfy the stability criterion, a gas of high pressure but low density apparently suffices. For an assumed halo exponential scale height H_h of 6 kpc the halo temperature needed at high z is about 10^6 K, with lower values possible at about a kiloparsec. The value of $n_h(0)$, defined as the atomic particle density of the hot gas extrapolated to $z = 0$, is determined from the added constraint that the observed synchrotron radio emission places on p_R and p_B; it turns out that $n_h(0)$ varies about as H_h^{-2}. If H_h is assumed to equal 6 kpc, a value of 5×10^{-3} cm^{-3} provides a "best estimate" for $n_h(0)$.

Such a hot gas will affect the momentum balance at $z = 0$, discussed in connection with Table 1. With these values of H_h and $n_h(0)$, the model calculations (8) show that the hot gas contributes about 1.1×10^{-12} dyne cm^{-2} to the total weight of the gas above the galactic midplane. Since other values of H_h are possible and the model becomes very approximate at high z, this result is particularly uncertain. This halo gas may be fully self-supporting in that its contribution to the average interstellar gas pressure at $z = 0$ may about equal 1.5×10^{-12} dyne cm^{-2} if $f_h \approx 1$. If the hot gas does not extend into the disk the necessary support at $z = 0$ is provided by $p_B(0)$, and the rms B approaches about 8 μG, an appreciable increase.

We turn now to the problems and questions raised by this hydrostatic model. As pointed out at the beginning of this section, a model in pressure equilibrium, stratified in plane-parallel layers, is of uncertain relevance to a halo dominated by superbubbles and galactic fountains. This uncertainty is reinforced by the thermal instability of a hot halo (see Section 4.3), which would seem to require transient conditions and cooling flows. Under such time-dependent conditions a momentum balance equation still applies to averages taken over time, but the input of momentum from super-bubbles and the output to a galactic wind must also be included.

The model described above, although it includes a variety of gas components, has recently turned out to be incomplete in an important sense. A determination of the electron density at high z, making use of the dispersion measures found for pulsars in several globular clusters, suggests (81) that $n_e \approx 0.025 \exp(-z/1500 \text{ pc}) \text{ cm}^{-3}$ for large z. This ionized gas is presumably the source of the diffuse Hα emission at high latitudes. Such an extended warm ionized gas has a weight at $z = 0$ of about 1.2×10^{-12} dyne cm^{-2}, about the same as for the extended component of the warm neutral gas. If we assumed that the weight of this warm gas at $z = 0$, and also that of the hot gas discussed above, were balanced by $p_B(0)$, then the total field required at $z = 0$ would increase to some 10 μG, an embarrassingly high value.

An important effect in the opposite direction may result from an excess of high random velocities in warm high-z gas. Profiles of 21-cm emission observed at high latitude show extended wings, corresponding to velocities up to about ± 100 km s^{-1}. Moreover, studies of various absorption lines show (1, 32) that these large velocities are found preferentially at roughly a kiloparsec or more from the galactic plane. If some large local variations [especially (56, 100) an extended complex of infalling gas near the north galactic pole] are ignored, the 21-cm data can be fitted (56) with a model in which at most a fifth of the neutral H has a velocity dispersion of 35 km s^{-1} and a corresponding scale height of about 500 pc, the value observed (61) for the high-z, presumably warm gas of neutral H. Removal of stray radiation (61) does not significantly alter these high-v wings. Thus, a considerable fraction of the high-z neutral gas, possibly all of it, can be entirely supported by its own turbulent pressure, with no gradient of p_B or p_R required for this purpose. Preliminary model calculations, based on observed 21-cm line profiles and on an assumed gravitational acceleration $g(z)$, confirm this possibility for the solar neighborhood.[3] The velocity dispersion of the warm ionized gas at high z may be similarly large, though this is uncertain.

At low z, the velocity dispersion of the warm neutral gas may be reduced by collisions with the low-velocity gases in the disk, in which case $p_G(0)$ (summed over all components of the gas) has about the same relatively low value as that given in Table 1, and a high $B(0)$ is again required to balance the total weight of the gas.

---

[3] I am indebted to F. J. Lockman for sending me the results obtained with his models and also some of his average 21-cm profiles (see ref. 61).

The assumption that $B_z = 0$ in this plane-stratified model is probably a poor approximation for the actual galactic halo. If random motions of cold clouds in the disk are responsible for small-scale magnetic fields parallel to the galactic disk, it seems inevitable that a similar small-scale field will be produced in the z direction. If $\langle B_x B_z \rangle$ and $\langle B_x B_y \rangle$, averaged over x or y on scales larger than the scale size of B_r, are functions of z only, independent of x and y, then the average magnetic force in the z direction [denoted by $F_{Bz}(z)$] is (11, 79) the z derivative of $(\langle B_x^2 \rangle + \langle B_y^2 \rangle - \langle B_z^2 \rangle)/8\pi$. The physical process involved here is that the B_z field component gives a tension in the z direction, tending to offset the pressure resulting from the gradient of the B_x and B_y components. Given the lack of any information on the relative magnitude of $\langle B_z^2 \rangle$ relative to the mean square fields in the x and y directions, it is not clear how much, if at all, F_{Bz} is increased by the random component of the magnetic field.

With so many uncertainties involved, a hydrostatic equilibrium model can readily be made consistent with the data. As further observations reduce these uncertainties, such a model may indicate whether all the processes affecting the momentum balance of the gas have in fact been taken into account.

A basic assumption of all these models is that, apart from small-scale random fields, **B** is parallel to the galactic plane, and the smoothed distribution of the gas is uniform in horizontal planes. With such a model one must consider how cosmic rays escape from the Galaxy. Streaming along spiral arms to large distances and then spraying out is conceptually possible but might require diffusion over too great a distance and would complicate explanations of the observed cosmic-ray composition. Diffusion of cosmic-ray particles transverse to **B** can be produced by small-scale magnetic turbulence (16), but the magnitude of this effect in the halo is not clear.

5.2 Models With **B** Parallel to **z**

We turn to the second theoretical model, in which the escape of cosmic rays is a basic element of the picture, with **B** taken parallel to **z**. This model (39) takes into account the detailed physical processes involved when energetic ions stream outward along lines of magnetic force, exciting Alfvén waves. These waves scatter the ions, which drift along **B** by a combination of diffusion and convection. The model calculations assume hydrostatic equilibrium, with dp_R/dz balancing $-g\rho$, and two energy equations—one for p_R (which includes convection, diffusion, and conversion of cosmic-ray energy into wave energy) and one for the wave energy density. In this latter equation the wave damping normally provided by collisions of ions with neutral atoms is ineffective because of low density and high ionization. Instead, damping is attributed to interactions between upward- and downward-traveling Alfvén waves. It is not clear what physical process will produce downward waves of the necessary strength. The stability of such an equilibrium model is also unclear.

Apart from these problems, the model presents a full and self-consistent picture for the particular magnetic topography assumed. Since $p_R(0)$ is only a fraction of $p_{TOT}(0)$, this model applies only to the halo gas, at values of z where the particle density does not exceed about 10^{-3} cm^{-3}. The energy input from the diffusing cosmic rays into Alfvén waves and into thermal energy via wave-wave damping provides a heat source for the gas. Depending on the parameters assumed, such models would be consistent either with equilibrium kinetic temperatures somewhat below 10^5 K and photon ionization as a source of C^{+3} and Si^{+3} or with higher temperatures and collisional ionization. Since thermal instability is probable, a transient solution, with the temperature of a fluid element rising or falling, may again be needed to explain the observed C IV and Si IV line strengths.

A time-dependent solution would also be helpful in reconciling the magnetic field perpendicular to the galactic plane, which is suggested by the cosmic-ray data, with the parallel field, which is observed in the solar neighborhood and which may be needed to support the cold and warm components of the interstellar gas. Such a situation is provided, of course, by the occasional eruption of superbubbles. If some fraction of the rising hot gas escapes from the Galaxy, the magnetic lines of force will be stretched far out, with some reconnection perhaps recurring. As pointed out by several researchers (16), such a time-dependent sequence might permit the intermittent escape of cosmic-ray particles.

Evidently, progress in understanding the hot gas takes place through a succession of idealized models, a familiar state of affairs in much of modern astrophysics. Many of these models have been proposed, some have been developed in considerable detail, and a few have been described here. For the most part they are not very realistic, but they may provide a tentative picture of the hot gas, of how it arises, and of what range of effects it produces. In any case, these models are helpful in suggesting new theoretical questions and new observational programs.

ACKNOWLEDGMENTS

This review is an expansion of a paper (92) prepared for a meeting in Copenhagen in May 1988, organized by the Royal Danish Academy of Sciences and Letters in honor of Bengt Strömgren. In this work I have been greatly aided by comments and suggestions from a number of astronomers, including S. A. Balbus, J. B. G. M. Bloemen, J. N. Bregman, B. T. Draine,

L. L. Cowie, G. B. Field, T. W. Hartquist, F. J. Lockman, R. McCray, B. D. Savage, and M. Wardle. Several very detailed discussions with D. P. Cox, E. B. Jenkins, C. F. McKee, and J. P. Ostriker have been particularly helpful.

Literature Cited

1. Albert, C. E. 1983. *Ap. J.* 272: 509
2. Aschenbach, B. 1988. In *Supernova Remnants and the Interstellar Medium, IAU Colloq. No. 101*, ed. R. S. Roger, T. L. Landecker, p. 99. Cambridge: Univ. Press
3. Bahcall, J. N. 1984. *Ap. J.* 287: 926
4. Balbus, S. A. 1986. *Ap. J.* 304: 787
5. Ballet, J., Arnaud, M., Rothenflug, R. 1986. *Astron. Astrophys.* 161: 12
6. Ballet, J., Luciani, J. F., Mora, P. 1989. *Astron. Astrophys.* 218: 292
7. Band, D. L., Liang, E. P. 1988. *Ap. J.* 334: 266
8. Bloemen, J. B. G. M. 1987. *Ap. J.* 322: 694
9. Böhringer, H., Hartquist, T. W. 1987. *MNRAS* 228: 915
10. Borkowski, K. J., Balbus, S. A., Fristrom, C. C. 1990. *Ap. J.* 355: 501
11. Boulares, A., Cox, D. P. 1990. Submitted for publication
12. Bowyer, C. S., Field, G. B., Mack, J. E. 1968. *Nature* 217: 32
13. Bregman, J. N. 1980. *Ap. J.* 236: 577
14. Bregman, J. N., Harrington, J. P. 1986. *Ap. J.* 309: 833
15. Caswell, J. L. 1988. In *Supernova Remnants and the Interstellar Medium, IAU Colloq. No. 101*, ed. R. S. Roger, T. L. Landecker, p. 269. Cambridge: Univ. Press
16. Cesarsky, C. J. 1980. *Annu. Rev. Astron. Astrophys.* 18: 289
17. Chevalier, R. A. 1977. *Annu. Rev. Astron. Astrophys.* 15: 175
18. Chevalier, R. A. 1982. *Ap. J.* 258: 790
19. Chevalier, R. A., Gardner, J. 1974. *Ap. J.* 192: 457
20. Chevalier, R. A., Oegerle, W. R. 1979. *Ap. J.* 227: 398
21. Cioffi, D. F., McKee, C. F., Bertschinger, E. 1988. *Ap. J.* 334: 252
22. Cowie, L. L. 1977. *Ap. J.* 215: 226
23. Cowie, L. L., McKee, C. F. 1977. *Ap. J.* 211: 135
24. Cowie, L. L., McKee, C. F., Ostriker, J. P. 1981. *Ap. J.* 247: 908
25. Cox, D. P. 1972. *Ap. J.* 178: 159
26. Cox, D. P. 1986. In *Workshop on Model Nebulae*, ed. D. Pequinot, p. 11. Paris: Obs. Meudon
27. Cox, D. P. 1988. In *Supernova Remnants and the Interstellar Medium, IAU Colloq. No. 101*, ed. R. S. Roger, T. L. Landecker, p. 73. Cambridge: Univ. Press
28. Cox, D. P., Edgar, R. J. 1983. *Ap. J.* 265: 443
29. Cox, D. P., Reynolds, R. J. 1987. *Annu. Rev. Astron. Astrophys.* 25: 303
30. Cox, D. P., Slavin, J. D. 1990. In *Extreme Ultraviolet Astronomy*, ed. R. F. Malina, S. Bowyer. New York: Pergamon. In press
31. Cox, D. P., Smith, B. W. 1974. *Ap. J. Lett.* 189: L105
32. Danly, L. 1989. *Ap. J.* 342: 785
33. Draine, B. T., Giuliani, J. L. Jr. 1984. *Ap. J.* 281: 690
34. Edgar, R. J., Chevalier, R. A. 1986. *Ap. J. Lett.* 310: L27
35. Edgar, R. J., Cox, D. P. 1984. *Ap. J.* 283: 833
36. Friedman, H., Fritz, G., Shulman, S. D., Henry, R. C. 1973. In *X- and γ-Ray Astronomy, IAU Symp. No. 55*, ed. H. Bradt, R. Giacconi, p. 215. Dordrecht: Reidel
37. Gies, D. R. 1987. *Ap. J. Suppl.* 64: 545
38. Gorenstein, P., Harnden, F. R. Jr., Tucker, W. H. 1974. *Ap. J.* 192: 661
39. Hartquist, T. W., Morfill, G. E. 1986. *Ap. J.* 311: 518
40. Heiles, C. 1979. *Ap. J.* 229: 533
41. Heiles, C. 1980. *Ap. J.* 235: 833
42. Heiles, C. 1987. *Ap. J.* 315: 555
43. Heiles, C. 1987. In *Interstellar Processes*, ed. D. J. Hollenbach, H. A. Thronson, Jr., p. 171. Dordrecht: Reidel
44. Heiles, C. 1990. *Ap. J.* 354: 483
45. Heiles, C., Jenkins, E. B. 1976. *Astron. Astrophys.* 46: 333
46. Higdon, J. C., Lingenfelter, R. E. 1980. *Ap. J.* 239: 867
47. Houck, J. C., Bregman, J. N. 1990. *Ap. J.* 352: 506
48. Ikeuchi, S. 1987. In *Starbursts and Galaxy Evolution. Moriond Astrophys. Meet.*, ed. T. X. Thuan, T. Montmerle, J. T. Thanh Van, p. 27. Paris: Editions Frontières
49. Jenkins, E. B. 1978. *Ap. J.* 219: 845

50. Jenkins, E. B. 1978. *Ap. J.* 220: 107
51. Jenkins, E. B. 1987. In *Exploring the Universe With the IUE Satellite*, ed. Y. Kondo, p. 531. Dordrecht: Reidel
52. Jenkins, E. B., Meloy, D. A. 1974. *Ap. J. Lett.* 193: L121
53. Kaelble, A., de Boer, K. S., Grewing, M. 1985. *Astron. Astrophys.* 143: 408
54. Kantrowitz, A., Petschek, H. E. 1966. In *Plasma Physics in Theory and Application*, ed. W. B. Kunkel, p. 148. New York: McGraw-Hill
55. Kesteven, M. J., Caswell, J. L. 1987. *Astron. Astrophys.* 183: 118
56. Kulkarni, S. R., Fich, M. 1985. *Ap. J.* 289: 792
57. Kulkarni, S. R., Heiles, C. 1987. In *Interstellar Processes*, ed. D. J. Hollenbach, H. A. Thronson, Jr., p. 87. Dordrecht: Reidel
58. Kulsrud, R. M., Bernstein, I. B., Kruskal, M., Fanucci, J., Ness, N. 1965. *Ap. J.* 142: 491
59. Lachièze-Rey, M., Asséo, E., Cesarsky, C. J., Pellat, R. 1980. *Ap. J.* 238: 175
60. Landau, L. D., Lifschitz, E. M., Pitaevskii, L. P. 1984. *Electrodynamics of Continuous Media*, Sects. 72, 73. New York: Pergamon. 2nd ed.
61. Lockman, F. J., Hobbs, L. M., Shull, J. M. 1986. *Ap. J.* 301: 380
62. Lyne, A. G., Smith, F. G. 1989. *MNRAS* 237: 533
63. Mac Low, M.-M., McCray, R. 1988. *Ap. J.* 324: 776
64. Mac Low, M.-M., McCray, R., Norman, M. L. 1989. *Ap. J.* 337: 141
65. Martin, C., Bowyer, S. 1990. *Ap. J.* 350: 242
66. McCray, R., Kafatos, M. 1987. *Ap. J.* 317: 190
67. McCray, R., Snow, T. P. Jr. 1979. *Annu. Rev. Astron. Astrophys.* 17: 213
68. McKee, C. F. 1974. *Ap. J.* 188: 335
69. McKee, C. F. 1987. In *Interstellar Processes*, ed. D. J. Hollenbach, H. A. Thronson, Jr., p. 237. Dordrecht: Reidel
70. McKee, C. F. 1988. In *Supernova Remnants and the Interstellar Medium, IAU Colloq. No. 101*, ed. R. S. Roger, T. L. Landecker, p. 205. Cambridge: Univ. Press
71. McKee, C. F., Cowie, L. L. 1977. *Ap. J.* 215: 213
72. McKee, C. F., Hollenbach, D. J. 1980. *Annu. Rev. Astron. Astrophys.* 18: 219
73. McKee, C. F., Ostriker, J. P. 1977. *Ap. J.* 218: 148
74. Norman, C. A., Ikeuchi, S. 1989. *Ap. J.* 345: 372
75. Oettl, R., Hillebrandt, W., Miller, E. 1985. *Astron. Astrophys.* 151: 33
76. Oort, J. H. 1966. *Bull. Astron. Inst. Neth.* 18: 481
77. Ostriker, J. P., McKee, C. F. 1988. *Rev. Mod. Phys.* 60: 1
78. Parker, E. N. 1966. *Ap. J.* 145: 811
79. Parker, E. N. 1969. *Space Sci. Rev.* 9: 651
80. Rand, R. J., Kulkarni, S. R. 1989. *Ap. J.* 343: 760
81. Reynolds, R. J. 1989. *Ap. J. Lett.* 339: L29
82. Savage, B. D. 1987. In *Interstellar Processes*, ed. D. J. Hollenbach, H. A. Thronson, Jr., p. 123. Dordrecht: Reidel
83. Savage, B. D., Massa, D. 1987. *Ap. J.* 314: 380
84. Schiano, A. V. R. 1985. *Ap. J.* 299: 24
85. Sedov, L. I. 1957. *Similarity and Dimensional Methods in Mechanics*. Moscow: Gostekhizdat. 375 pp. 5th ed. (In Russian). Transl., 1959. New York: Academic; see ref. 77
86. Shapiro, P. R., Field, G. B. 1976. *Ap. J.* 205: 762
87. Shklovsky, I. S. 1962. *Astron. Zh.* 39: 209. Transl., 1962, in *Sov. Astron. AJ* 6: 162
88. Silk, J. 1973. *Annu. Rev. Astron. Astrophys.* 11: 269
89. Spitzer, L. 1956. *Ap. J.* 124: 20
90. Spitzer, L. 1978. *Physical Processes in the Interstellar Medium*. New York: Wiley [see Chapters 10 (90A) and 11 (90B)]
91. Spitzer, L. 1985. *Phys. Scr.* T11: 5
92. Spitzer, L. 1990. In *Astrophysics: Recent Progress and Future Possibilities*. K. Dan. Vidensk. Selsk. Mat.-Fys. Medd., ed. B. Gustafsson, P. E. Nissen, 42(4): 153. Copenhagen: R. Dan. Acad. Sci. Lett.
93. Tenorio-Tagle, G., Bodenheimer, P. 1988. *Annu. Rev. Astron. Astrophys.* 26: 145
94. Thomson, R. C., Nelson, A. H. 1980. *MNRAS* 191: 863
95. Tomisaka, K., Habe, A., Ikeuchi, S. 1981. *Astrophys. Space Sci.* 78: 273
96. Tomisaka, K., Ikeuchi, S. 1986. *Publ. Astron. Soc. Jpn.* 38: 697
97. van Woerden, H., Schwarz, U. J., Hulsbosch, A. N. M. 1985. In *The Milky Way Galaxy, IAU Symp. No. 106*, ed. H. van Woerden, R. J. Allen, W. B. Burton, p. 387. Dordrecht: Reidel
98. Verschuur, G. L. 1975. *Annu. Rev. Astron. Astrophys.* 13: 257
99. Weiler, K. W., Sramek, R. A. 1988. *Annu. Rev. Astron. Astrophys.* 26: 295
100. Wesselius, P. R., Fejes, I. 1973. *Astron. Astrophys.* 24: 15

101. Wheeler, J. C., Levreault, R. 1985. *Ap. J. Lett.* 294: L17
102. Williamson, F. O., Sanders, W. T., Kraushaar, W. L., McCammon, D., Borken, R., Bunner, A. N. 1974. *Ap. J. Lett.* 193: L133
103. Wolff, M. T., Durisen, R. H. 1987. *MNRAS* 224: 701
104. Woodgate, B. E., Stockman, H. S. Jr., Angel, J. R. P., Kirshner, R. P. 1974. *Ap. J. Lett.* 188: L79
105. York, D. G. 1974. *Ap. J. Lett.* 193: L127
106. York, D. G., Spitzer, L., Bohlin, R. C., Hill, J., Jenkins, E. B., Savage, B. D., Snow, T. P. 1983. *Ap. J. Lett.* 266: L55

MASSES AND EVOLUTIONARY STATUS OF WHITE DWARFS AND THEIR PROGENITORS[1]

Volker Weidemann

Institut für Theoretische Physik und Sternwarte, Universität Kiel, 2300 Kiel 1, Federal Republic of Germany

KEY WORDS: central stars in planetary nebulae, hot subdwarfs, late stages of stellar evolution

1. INTRODUCTION

During the last decade it has become firmly established that the white dwarf number–mass distribution is sharply peaked around an average mass of $\approx 0.60\ M_\odot$. At the same time it has become clear that white dwarfs are the final stage of stellar evolution for low- *and* intermediate-mass stars all the way up to the lower limiting mass for carbon ignition, i.e. for all stars that pass through the double-shell-burning phase and prefabricate the ensuing white dwarf by the growth of a degenerate core while they are on the asymptotic giant branch (AGB; see Section 4.1). Taken together, these results imply that heavy mass loss occurs during pre–white dwarf evolution. Since normal stellar winds can only explain this mass loss, if at all, for low-mass stars, additional mass loss on the AGB has to be invoked and has indeed been inferred from observations of circumstellar shells (Section 4.1.5). It is thus a natural assumption to identify the planetary nebula stage as post-AGB and the central stars of the planetary nebulae as immediate progenitors of white dwarfs.

This evolutionary connection is now also firmly established and is

---

[1] Submitted October 15, 1989, on J. L. Greenstein's 80th birthday, to whom this review is dedicated.

strongly supported by a similarly sharply peaked mass distribution for the central stars of planetary nebulae (Sections 3.1, 4.3.2).

However there are other evolutionary channels toward the white dwarf stage (Section 3.2) that have not yet been fully explored, but that are statistically less important (Section 4.2.3). Calculations of synthetic population models and comparisons with observed data for white dwarfs and their progenitors are an important tool for constraining our assumptions about stellar and galactic evolution.

Here we examine the most important pieces of evidence for this evolutionary picture. We do not discuss white dwarfs as such and refer the interested reader to past reviews (Weidemann 1968, 1975, 1979, 1988, Ostriker 1971, Liebert 1980, Sion 1986) and/or to the proceedings of recent conferences in Tucson, Arizona (Philip et al 1987), and Hanover, New Hampshire (Wegner 1989b).

2. WHITE DWARF MASSES

Aside from a few cases in which dynamical masses can be determined (Section 2.4), most information on the masses of white dwarfs has been obtained by spectroscopic and photometric methods. For stars with known distances, it is sufficient to determine the effective temperature in order to calculate the radius, from which the mass $M(R)$ can be derived under the assumption of a mass-radius (M-R) relation (Section 2.1.1). In general the M-R relation for zero-temperature degenerate carbon configurations (Hamada & Salpeter 1961) has been used, although it is now known that for finite temperatures and different thicknesses of the outer hydrogen and helium layers the relation must be modified (Koester & Schönberner 1986), in the sense that the radius for a given mass is larger at higher temperatures and only slowly approaches the zero-temperature value.

In many cases where distances are not known, spectrophotometric analysis with model atmospheres allows one to determine the surface gravity, from which a mass $M(g)$ can be determined using a M-R (M-g) relation. This method has been widely used in the case of DA stars, where g-sensitive Balmer lines and well-established theoretical continua are available (Section 2.1.2). For non-DA stars (Section 2.2) the g-determination is less reliable. In cases where both R and g are determined, one can of course derive masses directly $[M(R,g)]$ without recourse to a mass-radius relation: Conversely, it is possible to check if and how well theoretical M-R relations are reproduced (Section 2.1.4). A third method, also specific to white dwarfs, consists in the evaluation of spectroscopically determined gravitational redshifts v_R, which, according to general relativity, are a function of M/R. Thus, it is possible to obtain masses either by assuming a mass-radius relation [$M(v_R)$] or by combining v_R with a radius or gravity determination [$M(R, v_R)$ or $M(g, v_R)$]. Modern detectors have improved the observational determination of redshifts considerably, and these improved redshifts yield more reliable individual masses. However, the method is restricted to DA white dwarfs within binaries or systems with known radial velocities (Section 2.1.3).

2.1 DA Stars

2.1.1 MASSES DETERMINED FROM RADII

Radii and masses for stars with known parallaxes have been derived primarily by Shipman (1979), Koester et al (1979; hereinafter KSW), and McMahan (1989). The resulting H-R diagrams show comparatively small scatter around a constant-radius line that corresponds to about 0.6 $M_\odot$ for a Hamada-Salpeter zero-temperature C-configuration. Since the effective temperatures needed to calculate radii depend on model atmospheres and calibration of flux distributions, minor differences can be expected (apart from different observational sources). As demonstrated by KSW, the scatter—and thus the width of the mass distribution—becomes smaller with better parallaxes, reducing the interval width within which two thirds of the stars are found from about 0.4 to 0.2 $M_\odot$.

A narrow sequence has also been established by Greenstein (1984, 1985) using improved parallaxes and multichannel indices for color-luminosity diagrams. From 25 DAs with known parallaxes Leggett (1989), using infrared photometry for the determination of temperatures, also confirmed the restricted range of radii and masses (0.6 ± 0.2 $M_\odot$).

2.1.2 DA MASSES DETERMINED FROM GRAVITY

Strömgren photometry and multichannel photometry have been most helpful in the determination of masses for larger ensembles of DA stars. The *uvby* observations of Graham (1972) convincingly outlined a single cooling sequence of DA white dwarfs and thereby eliminated the earlier hypothesis of two distinct DA "crystallization" sequences. As demonstrated by Weidemann (1971), Strömgren photometry is superior to *UBV* photometry for the determination of atmospheric parameters of DA stars. The Palomar multichannel spectrophotometry of Greenstein (1976, 1984) and Oke (1974) provided even better material for evaluation. It is the basis for the least-squares analysis by KSW, who demonstrated that the resulting $M(g)$ distributions are as narrow as those for $M(R)$, especially when the listed monochromatic multichannel indices are replaced by broadband indices, which use the full scans. Extending the KSW study with newer observations but restricting the evaluation to the temperature range 8000–16,000 K (where the g-sensitivity of the continuous flux is largest owing to the pressure dependence of H^-), Weidemann & Koester (1984; hereinafter WK) obtained a $M(g)$ distribution for 70 DA stars that—even with unweighted data—is remarkably narrow [$\sigma(M) = 0.13\ M_\odot$] and shows a steep increase at 0.45 $M_\odot$ as well as a flatter tail up to 1 $M_\odot$, a shape that is to be expected from calculations of stellar and galactic evolution (Section 4.3.1). The average mass value, but not the shape of the distribution, depends somewhat on the calibration: for Hayes & Latham (1975) (also used in the KSW study) it is 0.58 $M_\odot$; for Oke & Gunn (1983) it is 0.62 $M_\odot$. Comparison with Strömgren data, however, demonstrated these results to be less reliable, owing to larger observational scatter introduced by inclusion of fainter stars from different sources (e.g. Wegner 1983), than those of Graham (1972). This also holds for the larger scatter found by Shipman & Sass (1980) using color-color diagrams with monochromatic Greenstein multichannel or Strömgren indices, from which they derived a DA mean mass of 0.60 or 0.45 $M_\odot$, respectively, and a dependence on effective temperature. For a more homogeneous sample of 63 DAs, however, Fontaine et al (1985) found no temperature dependence and $\log\langle g\rangle = 7.98 \pm 0.31$ (corresponding to $M = 0.58 \pm 0.17\ M_\odot$).

If one takes into account selection effects operating against larger mass (smaller radius) stars, the true average mass could be larger—by an amount depending on the intrinsic width of the distribution. Shipman (1979) corrected the mean mass of his sample from 0.55 to 0.75 $M_\odot$, and Guseinov et al (1983a), in a general compilation of data for DA stars, arrived at a similar high average mass of $0.75 \pm 0.20\ M_\odot$. However, Koester (1984) studied the selection effects in a magnitude-limited sample in detail and found them to be of minor importance for DA stars, with corrections amounting to only about 0.03 $M_\odot$. The recent study of 53 DA stars by McMahan (1989) arrives at a similar selection-effect correction of 0.04 $M_\odot$ and otherwise confirms the narrow mass distribution found by KSW [$\sigma = \pm 0.10\ M_\odot$], as well as a fairly low average $M(g) = 0.52\ M_\odot$. As shown by KSW, $M(g)$ in general is more sensitive to calibration than $M(R)$. The smaller $\langle M(g)\rangle = 0.52\ M_\odot$ of KSW compared with their value of 0.58 $M_\odot$ for $M(R)$, however, is due to systematic effects for hotter stars; WK found that the discrepancy disappears for equal calibration. McMahan uses the older calibration of Oke & Schild (1970) and does not try to explain the differences quantitatively. The mass distribution of McMahan differs from that of WK in showing a less steep increase at small masses, but this is probably not significant in view of other uncertainties.

2.1.3 DA MASSES FROM GRAVITATIONAL REDSHIFTS Following the pioneering work by Greenstein & Trimble (1967) and Trimble & Greenstein (1972), the determination of masses by using gravitational redshifts has recently been revived by the availability of high-S/N detectors. The surprisingly high average mass, derived from the K-term for 74 DA stars, found by Trimble & Greenstein (1972) [$v_R = 53$ km s^{-1}, $M(v_R) \approx 0.8\ M_\odot$] gave much impetus to further studies.

Wegner (1974) obtained $v_R = 43 \pm 14$ km s^{-1} for the statistical average of southern DA stars. Schulz (1977) tried to explain this large value by means of asymmetries (see also Grabowski et al 1987) and concluded that future measurements should be restricted to the line cores, which had been found to be very sharp, especially for Hα (Greenstein et al 1977). The latter study reduced the K-term to 45 km s^{-1} and the range for the mean mass to 0.60–0.75 $M_\odot$. Individual gravitational redshifts can only be determined if the true systematic radial velocity is known, i.e. for wide binaries (common proper-motion stars) and cluster members. This method was first applied by Wegner (1973). More recently, it was put on a more reliable basis by Koester (1987), Wegner & Reid (1987), and Wegner et al (1989) by use of CCD detectors with a resolution of 0.2 Å at Hα to obtain radial velocities with an accuracy of typically 3 km s^{-1}, corresponding to an uncertainty of about 0.03 $M_\odot$ for $M(v_R)$.

The first results of ongoing studies with this method yield an average mass $M(v_R)$ of 0.58 $M_\odot$ for 9 objects (Koester 1987), 0.57 $\pm$ 0.03 $M_\odot$ for 14 DA field stars (excluding Sirius B; Wegner 1989a), and 0.66 $\pm$ 0.05 $M_\odot$ for 7 Hyades (Wegner et al 1989). The slightly higher average in the Hyades case might be due to higher progenitor masses (Section 4.1.1).

Special cases are 40 Eri B (the brightest DA), which is in a well observable triple system, and Sirius B, whose closeness to Sirius A makes observations difficult. For 40 Eri B the following values have been found: $v_R = 21 \pm 4$ km s^{-1} (Popper 1954), $v_R = 28$ km s^{-1} (Greenstein et al 1977), $v_R = 20 \pm 3$ km s^{-1} (Wegner 1979), and $v_R = 24.2 \pm 0.7$ km s^{-1} (Wegner 1980). The latter value corresponds to a mass of $M(v_R) = 0.50\ M_\odot$, uncomfortably higher than the dynamical mass of $0.43 \pm 0.02\ M_\odot$ (Heintz 1974; see Section 2.4). For Sirius B Greenstein et al (1971) measured $v_R = 89 \pm 16$ km s^{-1}, with $M(v_R)$ in agreement with the dynamical mass. However, the problems in this case are still not solved, as is pointed out by Greenstein et al (1985; see Section 2.4). An unsettled case is the hot DA CD $-38°10980$. The redshift for this object determined by Holberg et al (1985) from sharp Si II and Si III UV lines is 28.4 ± 4.8 km s^{-1}, whereas Koester (1987) arrives at a redshift value of 37.9 km s^{-1}, compatible with Wegner's (1978) value (44 ± 7 km s^{-1}) and corresponding to $M(v_R) = 0.67\ M_\odot$; combining Koester's value with the radius gives $M(R, v_R) = 0.87\ M_\odot$ and $M(R) = 0.42\ M_\odot$.

2.1.4 MASS-RADIUS RELATION FOR DA STARS   It has been demonstrated

(Weidemann & Yuan 1989) that even with the best determined atmospheric parameters [log g and $R(T_{\text{eff}}, \pi'')$ for 22 DA stars from WK], the theoretical mass-radius relation is poorly fitted. Indeed, it is impossible to discriminate between Hamada-Salpeter predictions for different interior compositions (e.g. He vs C). In fact, the historical struggle between Eddington and Chandrasekhar on the validity of relativistic degeneracy would even today remain undecided if not for the presence of Sirius B (the only reliable high-mass DA) and of three newly found faint white dwarfs in the young cluster NGC 2516 with less reliable $M(g)$ and $M(R)$ around or above 1 $M_\odot$ (see Section 4.1.1), all of which confirm Chandrasekhar's theoretical prediction. A very slight improvement is achieved if one uses $M(v_R, R)$ instead of $M(g, R)$ (see Wegner 1989a, Figure 2), but still the scattering is uncomfortably large. The major uncertainty stems from distances entering the R-determination, rather than from errors in effective temperatures. Even in cases where dynamical masses are known, such as 40 Eri B and Sirius B, the theoretical relations are not fulfilled. For 40 Eri B the redshift mass (0.50 $M_\odot$) and the dynamical mass (0.43 $M_\odot$) are discrepant, and R is comparatively too small, to the point that the star is almost located on the Fe-relation in the M-R diagram. For Sirius B the opposite is true: The radius is significantly larger than expected for a C/O configuration with the dynamical mass if one assumes that the lower effective temperature derived from *IUE* and *Voyager* observations (Holberg et al 1984) is correct. This problem has recently been discussed by Thejll & Shipman (1986).

2.2 *Non-DA Stars*

For non-DA stars, spectroscopic mass determinations are more uncertain. The best values are obtained for stars with good parallaxes and well-determined effective temperatures [i.e. $M(R)$]; these are the cool ($T_{\text{eff}} < 10,000$ K) DC and DQ stars, for which a histogram with 16 objects has been published (Weidemann 1987c) based on analysis by Koester et al (1982). The average mass (0.55 $M_\odot$) and the shape of the mass distribution agree within the errors with those of the DA stars. The DB stars, which have hotter helium-rich atmospheres, are generally farther away; since parallaxes are uncertain or not available, only $M(g)$ can be determined from analysis of the He I line-dominated spectra. Wickramasinghe & Reid (1983) arrive at $M = 0.58$ $M_\odot$ from high-resolution line spectra for 7 stars, whereas multichannel observations evaluated by Oke et al (1984) for 25 stars yield DB masses in the range $M = 0.55 \pm 10\,M_\odot$. Equality of DA and non-DA masses and mass distributions has thus been established within the error bounds: This is an important result bearing on the still-unsettled question of the origin of the two white dwarf varieties, which have different atmospheric compositions. A slightly lower average mass for the DB stars, however, cannot be excluded if the results of Oke et al (1984), calibrated on AB 79, are compared with the DA stars for the same calibration (average mass $M = 0.62\ M_\odot$) (Weidemann & Koester 1984). Guseinov et al (1983b), in a comparative study of 81 He-rich white dwarfs, also conclude that the DB masses are on average lower, with 70% less than $0.55\ M_\odot$. Additional information would thus be very valuable. Strömgren colors, however, are much less sensitively dependent on surface gravity for DBs than for DAs, as demonstrated by Koester et al (1982) and Oke et al (1984). Gravitational redshifts are hard to obtain, since the intrinsic and varying pressure shifts (mostly to the blue) for DB stars are not known, and since sharp pressure-unaffected cores like $H\alpha$ for the DA stars do not exist for DB stars (Koester 1987).

As for the origin of DA vs non-DA atmospheres, considerable efforts have been made to explain the differences by diffusion and/or accretion *during* white dwarf evolution rather than by different progenitor channels. The reader is referred to the proceedings of recent conferences in Tucson, Arizona (Philip et al 1987), and Hanover, New Hampshire (Wegner 1989b). A remarkable attempt to trace the differences back to progenitor evolution, especially to the phase in the He shell flash cycle at which the progenitor leaves the AGB, has been made by Iben (1984). If this scenario is correct, the average masses of DA and non-DA stars should be exactly equal. On the other hand, hypotheses in which DB masses are significantly different and higher than DA masses (Wood & Faulkner 1986) can be ruled out.

2.3 *Population II White Dwarfs*

Very little is known about the masses of Population II white dwarfs. Repeated attempts have been made to detect white dwarfs in globular clusters (Ortolani & Rosino 1987). With the presently most reliable results obtained for M71 by Richer & Fahlman (1988), an indication of a faint (22–25 mag) blue sequence is interpreted to be DB stars, which are $0.1\ M_\odot$ more massive than disk white dwarfs. However, this result must still be considered preliminary. Hubble Space Telescope observations will certainly be necessary in order to provide additional information.

From evolutionary considerations, one expects to find a single mass cooling sequence; however, this assumes the existence of a single-valued initial- to final-mass relation, which must be doubted (Section 4.1.6). A theoretical scenario has been proposed by Renzini (1989) (see also Renzini & Fusi Pecci 1988), according to which low-mass Population II progenitors lose their entire hydrogen-rich envelope during a post–shell flash luminosity peak and thus produce DB stars of mass $\sim 0.54\ M_\odot$. Population II white dwarfs are very rare in the solar neighborhood: From tangential velocity distributions only a few cooler degenerates are known, for which it is not possible to determine masses (Liebert et al 1989).

2.4 Masses of White Dwarfs in Binaries

White dwarfs in wide binaries are extensively discussed by Greenstein (1986a). It appears that these do not differ from single white dwarf stars as far as average radius and therefore mass is concerned, but many pairs remain undetected. The number of closer pairs for which dynamical masses can be determined is restricted to nearby objects with periods in years; the most famous cases are Sirius B, for which the latest mass determination (Gatewood & Gatewood 1978) is $M = 1.053 \pm 0.028\ M_\odot$, and 40 Eri B with $M = 0.43 \pm 0.02\ M_\odot$ (Heintz 1974). Other more uncertain dynamical masses are those for Procyon B [$0.68\ M_\odot$ (Cester 1965)], Stein 2051B [$0.60\ M_\odot$ (Strand & Kallerkal 1989)], the cool double degenerate G107-70 [$0.46 \pm 0.10\ M_\odot$ (Harrington et al 1981)], Case 1 [$0.38\ M_\odot$ (Stauffer 1987)] and HZ 9 [$0.51\ M_\odot$ (Stauffer 1987)]. The latter two are short-period composite spectrum systems that must have evolved through phases with mass loss in common envelopes. A similar case is the eclipsing DA+dK2 V pair V471 Tau, with M between 0.6 and $0.8\ M_\odot$ (Nelson & Young 1970). These are considered to be pre-cataclysmic variable detached systems, for 10 of which Ritter (1986) finds $\langle M \rangle = 0.62\ M_\odot$. A spectroscopic double degenerate of similar dimensions is the formerly "overluminous" DA white dwarf L870-2 (Saffer et al 1988), with $P = 1.56$ days and most probable masses of 0.52 and $0.47\ M_\odot$ (Bergeron et al 1989). The masses of white dwarfs in cataclysmic variables, on the other hand, seemed for a long time to be larger on average [around $1\ M_\odot$ (Robinson 1976) or somewhat lower (Weidemann 1982)], but this has been found to be caused by selection effects acting against low-mass degenerate companions. If these effects are taken into account, the average mass is reduced to $0.62\ M_\odot$ (Ritter & Burkert 1986). Predicted mass distributions (Politano & Webbink 1989, Iben & Webbink 1989, Politano et al 1989) give similar values.

Finally, there are the ultrashort-interaction binary systems GP Com = G61-29 [$P = 46$ min (Nather et al 1981)] and PG 1346+082 [$P = 25$ min (M. A. Wood et al 1987)], which like AM CVn = HZ 29 ($P = 18$ min) are modeled to consist of a very low mass ($M \approx 0.02\ M_\odot$), large-radius helium degenerate and a more massive accreting white dwarf primary. In the case of AM CVn, however, the period is still debated, and other models seem possible (Solheim 1989). Recently, V803 Cen (AE-1) has been shown by O'Donoghue et al (1987) also to belong to this class of interacting binary white dwarfs.

The few noninteracting double degenerates are in general spectroscopic twins (Greenstein 1986b), with two remarkable exceptions: the Sanduleak-

Pesch pair, for which Greenstein et al (1983) find unequal masses (0.80 and 0.43 $M_\odot$); and L151-8A/B, a wide common proper motion pair with unequal spectral types (DA and DB) for which masses have not been determined (Oswalt et al 1988).

3. MASSES OF WHITE DWARF PROGENITORS

It has been known for a long time that the central stars of planetary nebulae (CSPN) are the immediate progenitors of white dwarfs (Section 3.1). However, there are evidently other evolutionary channels toward the white dwarf stage, e.g. the hot subdwarfs (sdO, sdB, and other blue subluminous stars) that may not have reached the AGB (Section 3.3) or that have already lost their nebulae. Probable examples of the latter class are the very hot PG 1159 pre–white dwarfs (Section 3.2.3). Since planetary nebulae become not visible before they are ionized when the central star reaches $T_{\rm eff} \approx 30{,}000$ K, stars leaving the AGB, at low $T_{\rm eff}$ (≈ 5000 K), must spend some time in transition (Schönberner 1988); this is the so-called proto–planetary nebula stage (Section 3.2.1). Recently, many candidates have been found in this temperature range, but it is not clear if all of them will become CSPN. For close binaries, theory predicts the possibilities for reaching the white dwarf stage, both with or without merging and with or without a foregoing AGB phase, depending on initial parameters and common envelope evolution. Since these stages are not easily identified and empirical mass determinations are restricted to pre-cataclysmic variables (Section 2.4), we confine the discussion in this section mainly to single stars.

3.1 Central Stars of Planetary Nebulae

Mass determinations for CSPN rely heavily on the existence of evolutionary tracks, which are available mainly as a result of calculations by Schönberner (1979, 1981, 1983) and Wood & Faulkner (1986) covering the mass range from 0.54 to 0.89 $M_\odot$. For higher masses, one still uses the tracks of Paczyński (1971). The exact shape depends on the phase in the double-shell-burning cycle when the star leaves the AGB. If the star leaves the AGB during quiet hydrogen burning, and is sufficiently separated from a He shell flash, which extinguishes the H shell and forces the star out of thermal equilibrium, it proceeds from the AGB toward higher $T_{\rm eff}$ through a plateau phase at constant luminosity. When the H shell dies out, the star fades—first quickly, then slower—and appears as a hot white dwarf after the nebula is dispersed.

If the star leaves the AGB close to a shell flash, a variety of possibilities can change the tracks and the time scale of evolution (Schönberner 1979,

Iben 1984, Wood & Faulkner 1986). For helium-burning post–shell flash stars, the luminosity decreases during the evolution toward higher temperatures, but the time spent in the high-luminosity region increases sufficiently so as to let the nebula disperse before the fading phase toward the white dwarf region begins.

For hydrogen-burning CSPN, the luminosity in the plateau phase is given by the core mass–luminosity relation that was valid during the preceding AGB evolution for quiet interflash hydrogen burning. The mass of these CSPN is thus easily determined if the star can be located in the H-R diagram. For known distances this can be done in two different ways: either by using effective temperatures derived by a variety of methods (Zanstra, Stoy, *IUE*, optical spectroscopy), whose uncertainties enter the luminosity primarily via the strong dependence on bolometric corrections; or, without recourse to effective temperatures, by equating a nebular expansion age with evolutionary ages on the tracks. The latter method was first applied by Schönberner (1981) and Schönberner & Weidemann (1981, 1983). It yielded the astonishing result that the derived CSPN mass distribution was sharply peaked at 0.58 $M_\odot$, with only a very few objects at higher masses. The first method has been applied by many other investigators [e.g. for large, optically thin nebulae by Kaler (1983), who finds a much wider mass distribution]. Similarly, Stecher et al (1982) claimed to have detected three high-mass CSPN in the Magellanic Clouds, a finding that prompted Maran (1983) to discuss the mass distribution discrepancies in an article in *Nature*. After criticism of this claim (Tylenda 1984, Barlow et al 1986, Heap & Augensen 1987), reinvestigations by Barlow et al (1986) and Aller et al (1987) now place the masses of these stars close to 0.6 $M_\odot$. More observations of planetary nebulae in the Magellanic Clouds (Barlow 1987, P. R. Wood et al 1986, 1987, Monk et al 1988, Henry et al 1989) confirm the near-absence of more massive CSPN in the observed sample. Masses for 16 CSPN are confined to the interval 0.56–0.64 $M_\odot$, with a mean mass of 0.595 ± 0.022 $M_\odot$ (Barlow 1989). Of course, these results depend slightly on the adopted distances to the Magellanic Clouds, which are still debated.

Investigations of planetary nebulae (PN) in other extragalactic systems (see Ford et al 1989) show that few if any high-luminosity CSPN exist. Since CSPN are not resolved, nebular modeling enters the results. Jacoby (1989) concludes that the observed bright-end cutoff in the [O III]–luminosity functions can only be reproduced if the mass distribution of CSPN is narrow, with σ near 0.02 $M_\odot$ to the high-mass side of 0.61 $M_\odot$. Another case in which a presumed high mass has been revised downward is the central star of NGC 7027. Whereas Walton et al (1988) derived $T_{\rm eff} = 310,000$ K, corresponding to $M = 0.89$ $M_\odot$ and in agreement with earlier estimates (Pottasch et al 1982), Jacoby (1988) now finds $T_{\rm eff} = 180{,}000$ K and $M = 0.65$ or 0.70 $M_\odot$ for hydrogen- or helium-burning star tracks, respectively. The latter value is confirmed by evaluation of CCD detections of several central stars in the presumably hottest CSPN (Heap & Hintzen 1990, Heap 1990).

A somewhat broader mass distribution for CSPN had been obtained by an extensive study of 68 stars for which *IUE* magnitudes were used instead of visual magnitudes with the Schönberner method (Heap & Augensen 1987). However, it has been demonstrated by Weidemann (1989) that the differences are entirely due to the use of smaller (Daub) distances. If the distances are increased according to the results of other investigations, the *IUE* mass distribution becomes identical with that derived from optical data. In particular, the strong peak around 0.6 $M_\odot$ is recovered. The fraction of CSPN with masses smaller than 0.64 $M_\odot$ increases from 44% to 77%.

Up to this point the derived CSPN masses have been dependent on often doubtful distances and have been obtained by using the evolutionary tracks for hydrogen-burning stars. It has become possible, however, to use distance-independent methods. The first such method relies on spectroscopic analysis of CSPN, which were observed at high resolution and evaluated with the help of non-LTE atmospheres (Méndez et al 1988a). The resulting atmospheric parameters ($\log g$ and $T_{\rm eff}$) were used to place the stars in a distance-independent $\log g$–$\log T_{\rm eff}$ diagram, where theoretical evolutionary tracks allowed mass determinations. The derived masses for 21 CSPN scatter about equally between 0.55 and 0.9 $M_\odot$. The higher masses, on average, may partly be due to remaining uncertainties in the gravity determination and partly due to selection effects. (Naturally the brightest CSPN were chosen.) For the same reason, not too much emphasis should be put on the failure to reconcile expansion ages with evolutionary ages, which decrease rapidly for higher masses.

The Schönberner method has also been criticized by Pottasch (1989) and Renzini (1989). Pottasch's criticism concerns mainly the use of Shklovsky distances and the derivation of the mass distribution for optically thin PN, whereas Pottasch, following Gathier (1984) and collecting independent material, considers most of his objects to be optically thick and finds a large scatter in the luminosity–expansion age diagrams. The discrepancies can at least partly be traced back to selection effects favoring young bright PN, which are nearly absent in a locally complete ensemble (Schönberner 1981). Renzini (1989) bases his criticism on the uncertainty of the transition time from the AGB to the PN phase, and on Schönberner's use of a fixed zero point for the nebular expansion. However, recent hydrodynamical models for nebular expansion with CSPN evolving on the theoretical tracks (Schmidt-Voigt & Köppen 1987) demonstrate that kinematical ages differ only slightly from evolutionary ages. The observations of expansion velocity, He II/Hβ line ratios, Zanstra temperatures, and electron densities agree best with CSPN having masses of 0.60–0.64 $M_\odot$. Using nebular properties that are distance independent, Szczerba (1987) also concludes that a histogram of 194 PN is best fitted by a correlation between PN and CSPN masses, in the sense that less massive CSPN have less massive nebulae, and that the mass distribution is very narrowly peaked at 0.6 $M_\odot$. A similar conclusion is reached by Tylenda & Stasińska (1989) in their comparison of models and observations.

We note at this point that the narrow mass distribution around 0.6 $M_\odot$ does not exclude the *existence* of more massive CSPN, for two reasons: First, the above-mentioned selection effects operate strongly against detection in the short-lived high-luminosity phase, as was originally pointed out by Renzini (1979) and later by Schönberner & Weidemann (1983) and also quantitatively shown by Shaw (1989). As demonstrated in an H-R distribution (see Figure 6 in Weidemann & Yuan 1989) most massive CSPN are expected to be found at very low luminosities, where observational efforts must be concentrated (e.g. Ishida & Weinberger 1987, Kwitter & Jacoby 1989).

Second, the rather flat initial- to final-mass relation (Section 4.1) predicts more massive ($>0.65\ M_\odot$) CSPN only for initial masses larger than about 3 $M_\odot$; judging from the steepness of the initial-mass function, such large initial masses are rare. This also explains why the vast majority of CSPN evolve similarly, as evidenced by the dust behavior in PN (Lenzuni et al 1987), by the excitation behavior (Schmidt-Voigt & Köppen 1987, Szczerba 1987), and by the appearance of gaps in the luminosity function and the excitation distribution (Schönberner 1981, 1986a). The latter evidence has also been taken as a strong argument for the fact that the majority of CSPN evolve on hydrogen-burning tracks: Schönberner (1981) has shown that these gaps would disappear if the evolution follows helium-burning CSPN, which left the AGB close to a helium shell flash.

Although some hydrogen-deficient CSPN exist, such as NGC 246 or the 15% Wolf-Rayet-type CSPN found in the Magellanic Clouds (Monk et al 1988), the masses seem not to be systematically different.

The fraction of presently undetected massive CSPN can be estimated from models of stellar and galactic evolution (Section 4.3), which reproduce the white dwarf mass distribution. As Figure 4 in Weidemann & Yuan (1989) demonstrates, it amounts to about 50% of the CSPN between 0.7 and 0.9 $M_\odot$, and less than 20% of the total.

Another problem is the occurrence of low-mass CSPN. Renzini's (1979) estimates and Schönberner's (1981) calculations predict that below a critical luminosity of log $L/L_\odot \approx 3.2$, or a mass of 0.55 $M_\odot$, the transition time from the AGB to $T_{\text{eff}} \approx 30{,}000$ K becomes too large, so that a nebula created on the AGB would disperse before ionization occurs and not be detected as a PN. Although this cutoff is very well confirmed if one uses the Schönberner method [see Tylenda & Stasińska (1989) for a compilation of data for 200 PN], several CSPN are placed below the 0.55-$M_\odot$ track if one uses the conventional method via a T_{eff} determination (e.g. Kaler 1983, Gathier & Pottasch 1989). Even if in many cases distances and temperatures are too uncertain, there may remain a few cases in which lower luminosity CSPN occur; two extreme examples are EGB 5 (Méndez et al 1988a) and PHL 932 (Méndez et al 1988b), which are most probably the results of close binary common envelope evolution on the first giant branch. Other less extreme examples are given by investigations of PN in the galactic bulge (Pottasch & Acker 1989, Zijlstra & Pottasch 1989). These cases could perhaps be understood if the transition time is shortened owing to continuing mass loss (Trams et al 1989). In any case, the fraction of CSPN with masses below 0.55 $M_\odot$ must be extremely small, and thus white dwarfs with smaller masses (like 40 Eri B) have probably not evolved through the PN channel (Section 3.2).

Few direct mass determinations of CSPN in close binaries have been attempted. The most convincing case has been provided by Drilling (1985), who uses the reflection properties of the low-mass main sequence companion to determine the mass of LSS 2018, the CSPN of DS1, to be 0.4–0.7 $M_\odot$. Similarly, Grauer & Bond (1983) conclude for Abell 41 that all observations of the ultrashort binary nucleus are consistent with a mass of 0.6 $M_\odot$ for the CSPN. Data for other CSPN in close binaries are collected by Bond (1989).

3.2 Other Post-AGB Stars

There are several classes of stars and individual objects that must be placed between the AGB and the white dwarf region. Most of these are probably post-AGB stars, although the white dwarf stage can also be reached by non-post-AGB stars, such as single post–horizontal branch stars that do not reach the AGB or binaries evolving from the first giant branch (Section 3.3).

Infrared observations, especially by *IRAS*, have recently led to the detection of many objects that, on the basis of their dust, IR, and optical spectra, must be considered to be in transition between AGB and CSPN stars, in the so-called proto–planetary nebula (PPN) stage.

3.2.1 PROTO–PLANETARY NEBULAE Examples of this class are nonvariable OH/IR sources for which the 9.7-μm silicate absorption goes to emission, indicative of shell thinning after the stars have left the AGB and mass loss has stopped (Volk & Kwók 1989, Kwok & Volk 1989). Often the stars remain invisible, completely hidden by dust. When the nebula becomes transparent, the central stars appear predominantly with F or G supergiant spectra.

These post-AGB objects have been at the focus of several recent conferences [Calgary 1986 (Kwok & Pottasch 1987), Vulcano 1986 (Preite-Martinez 1987), Mexico City 1987 (Torres-Peimbert 1989), Bloomington 1988 (Johnson & Zuckerman 1989), Montpellier 1989 (Mennessier & Omont 1990)], to which the reader is referred for details. A recommendable introduction is given by Kwok (1989a). Although in the cases of F and G supergiants (e.g. Waelkens et al 1987, Pottasch & Parthasarathy 1988, Hvrinak et al 1988, 1989) the similarity of IR flux distributions and implied dust properties to those of PN makes the identification as PPN almost certain, the distances, and thus the luminosities and (by implication) the masses, are not well known. However, Schönberner (1990) concludes from a comparison of estimated post-AGB ages to evolutionary tracks that the masses are consistent with $M \approx 0.60 \pm 0.05 \, M_\odot$.

3.2.2 sdO STARS Hot subluminous subdwarfs populate a wide range between the horizontal branch, the CSPN, and the white dwarf region. The pioneering study by Greenstein & Sargent (1974) showed most of them to be pre–white dwarfs. Many more have been found in the Palomar-Green survey (Green et al 1986), at higher temperatures (Drilling 1983, Schönberner & Drilling 1984), and at low galactic latitudes (Drilling 1983, Downes 1986). Although in surveys they appear to be as numerous as white dwarfs, their space density is small (Section 4.2.1) because their luminosities and average distances are much larger. With individual distances unknown, it is difficult to find positions in the H-R diagram. However, from spectroscopic analysis it is possible to use the g-T_{eff} diagram (Groth et al 1985), which indicates in general positions below the CSPN region. These sdO's are not necessarily post-AGB stars, since for acceptable masses (0.45–0.6 $M_\odot$) the luminosities may be too low. Others are located in the region of CSPN tracks (Heber & Hunger 1987) but show no trace of a nebula. These objects could be either (*a*) post-AGB stars that evolve very slowly, as is predicted for stars with $M < 0.55 \, M_\odot$ ("laziness"; Renzini 1981), or (*b*) "born-again" stars that originally evolved as normal CSPN but experienced a late helium shell flash (Schönberner 1979, Iben et al 1983) that brought them back to the AGB, from where they evolved a second time on a helium-burning track. Méndez et al (1988c) did not find any traces of old nebulae in a deep search around 12 of these sdO's, and this lack of detection casts doubt on the born-again interpretation. Since evolutionary post–horizontal branch tracks show loops, sometimes reach or miss the AGB, and are dependent on composition (Sweigart et al 1974, Gingold 1976), it is not possible to obtain a one-to-one mapping between H-R and g-$T_{\rm eff}$ diagrams. Therefore, unlike the CSPN case, masses cannot be determined by spectroscopic methods.

3.2.3 PG 1159 STARS These stars are very hot pre–white dwarfs characterized by spectra that are similar to those of sdO's but do not show any hydrogen. C IV and O VI absorption lines indicate carbon and oxygen enrichment. The prototype, PG 1159−035, is a pulsating variable (GW Vir), which (like several other members of this class) displays complex, low-amplitude nonradial modes with periods between 100 and 1000 s (McGraw et al 1979, Bond et al 1984, Bond & Grauer 1987, Demers et al 1989). From comparison with other hot white dwarfs with He II lines, Wesemael et al (1985) estimate surface gravities of log $g \approx 7$, which place the stars in the region of hydrogen-burning CSPN in the g-$T_{\rm eff}$ diagram (and H-R diagram, if one assumes a mass). However, no nebulae could be detected. This fact and the absence of hydrogen in the spectra indicate that the PG 1159 stars probably follow helium-burning post-AGB tracks. Such tracks predict longer lifetimes at higher luminosities (Section 3.1) and also raise the detection probability such as to allow an estimate that 20% of all pre–white dwarfs evolve through this path (Weidemann 1987c), which is about the fraction of DB stars. Recently, detailed spectroscopic analysis with sophisticated non-LTE atmospheres (Werner et al 1989a,b) for four PG 1159 stars has yielded $T_{\rm eff} = 110{,}000$ or $140{,}000$ K (two stars at each value), log $g \approx 7$, and photospheric abundances dominated by a carbon mass fraction of 55%, followed by helium (25%) and oxygen (15%). The g-$T_{\rm eff}$ position remains in the region of CSPN. Kawaler (1987, 1988) has evaluated the observed period spacing that is most sensitive to mass for PG 1159−035 and concludes that if it is an $l = 1$ or 3 pulsator, the mass is (to high accuracy) 0.60 $M_\odot$. For $l = 2$ or 4, the mass would be 0.74 or 0.48 $M_\odot$, respectively, values that are less probable if PG 1159 stars are DB progenitors with typical masses of 0.55 $M_\odot$ (Oke et al 1984). Combined with the spectroscopic atmospheric parameters, the luminosities can be estimated to be in the range $2.3 \leq \log L/L_\odot \leq 2.7$. For further discussion of the pulsational properties and several related problems, we refer the reader to the proceedings of a conference held in Tucson (Philip et al 1987) and to a review by Winget (1988). The situation is complicated by the facts that not all of the spectroscopically classified PG 1159 stars are pulsators, although theory predicts them to be if the pulsations are excited by nuclear burning, and that pulsations should be damped away if the excitation mechanism is partial ionization of carbon and oxygen and the helium content is larger than 20% (Kawaler & Hansen 1989). A nonpulsating, even hotter pre–white dwarf that is not classified as a PG 1159 star is H1504+65 (*HEAO-1*), for which Nousek et al (1986) estimates a temperature of 160,000 K. This star is exceptional in showing no trace of either hydrogen or helium but only faint O VI and C IV absorption lines in ultraviolet spectra. Since spectroscopic analysis is not yet possible and the distance is unknown, a mass determination for this star is not possible.

3.2.4 UV-BRIGHT STARS As many studies have shown, there are blue stars above the horizontal branch in globular clusters that are probably Population II post-AGB stars. Harris et al (1983) have collected data for 129 UV-bright stars that also comprise Population II (low-luminosity) Cepheids. Spectroscopic analysis, also using *IUE* data, by de Boer (1985) for 10 UV-bright stars in globular clusters allowed reliable temperature, gravity, and (with known distances) luminosity determinations that place the stars in the H-R diagram with log $L/L_\odot$ between 3.1 and 3.4, mostly just below the Schönberner 0.54-$M_\odot$ post-AGB track. Planetary nebulae have not been detected, with the exception of K648, for which the luminosity might be somewhat higher. Schönberner (1987) estimates the masses of the UV-bright stars to be between 0.53 and 0.55 $M_\odot$ under the assumption that they are on horizontal tracks that obey the core mass–luminosity relation on the AGB.

One can, however, question whether these stars are really on post-AGB tracks, since the predicted hotter objects—the temperatures of the detected UV-bright stars are generally between 8000 and 30,000 K—have not (yet?) been found. Some of them could also be post–horizontal branch (PHB) stars that evolve on tracks as calculated by Sweigart et al (1974) or Gingold (1976) and do not reach the AGB. The masses should then be larger than $\sim 0.50\ M_\odot$ and smaller than 0.52 $M_\odot$. (For larger masses PHB tracks reach the AGB without crossing the UV-bright star region.) Additional masses have been derived for one sdO star in M22 [0.44 $M_\odot$ (Glaspey et al 1985)]; for a central bright star of M30 without a PN [$T_{\rm eff} \approx 9000$ K, log $L/L_\odot \approx 3.62$, corresponding to $M = 0.57\ M_\odot$ (Caloi et al 1984)]; and for ROB 162 in the globular cluster NGC 6397 [for which NLTE analysis by Heber & Kudritzki (1986) yields $T_{\rm eff} = 51,000$ K, log $g = 4.5$, and a mass of 0.55–0.60 $M_\odot$ on post-AGB tracks].

3.3 Pre–White Dwarfs Evolving Through Other Evolutionary Channels

Stars on the extended horizontal branch evolve directly to the white dwarf region if their mass is smaller than about 0.50 $M_\odot$ (Caloi 1989). As spectroscopic analysis has convincingly shown, these are the sdB and sdOB stars (Heber 1986), which are burning helium in their centers like other horizontal branch (HB) stars but have so little hydrogen on their surfaces that their hydrogen shells are not ignited. The Palomar-Green survey has shown them to be three times as numerous as sdO stars, a fact that can be explained by their comparatively long, central helium-burning lifetimes. The sdOB stars are located at higher effective temperatures (>30,000 K) than the sdB stars and are closer to the white dwarf region. The masses thus should be determined by the core mass at helium ignition on top of the first red giant branch, i.e. between 0.45 and 0.48 $M_\odot$ (Sweigart & Gross 1978, Sweigart et al 1989). While observations have been mainly restricted to field sdB's and sdOB's, Heber et al (1986a) succeeded in obtaining spectra for sdB's in the globular cluster NGC 6752. Analysis showed them to be strikingly similar to the field stars and yielded a mean mass of $0.45^{+0.36}_{-0.22}$ $M_\odot$. For further discussion, the reader is referred to Heber (1987).

Horizontal branch blue stars beyond the second Newell gap, with $T_{\text{eff}} < 20,000$ K, evolve with a hydrogen-burning shell first toward the red for zero-age horizontal branch total masses around 0.52 $M_\odot$ (Caloi 1989), but they do not reach the AGB and instead return toward the white dwarf region.

Since few tracks have been calculated so far, and those that have are not only dependent on composition but also on the details of helium exhaustion (breathing pulses versus overshooting), the mass limits against post-AGB evolution are not yet established. It should be kept in mind that differential mass loss on the first giant branch is responsible for the extension of the HB, and that HB stars also exist for Population I, albeit in a smaller fraction (Section 4.1.6).

Another group of stars that must be considered to be pre–white dwarfs comprises the R CrB and extreme helium stars. Most probably they are the result of binary evolution (Iben & Tutukov 1985); however, their evolutionary status is still not clear (Schönberner 1986b), and thus masses cannot be obtained by using evolutionary tracks.

Saio (1986) determined the mass of R CrB and RY Sgr to be 0.8 ± 0.1 $M_\odot$ from observed pulsation periods. The rapid period decrease of RY Sgr is taken as evidence of shrinking and supports evolution toward the white dwarf region.

4. EVOLUTIONARY STATUS

In this section we trace the masses of white dwarfs and their immediate progenitors back to masses in earlier stages (e.g. on the AGB) and especially to initial masses on the main sequence. Initial- to final-mass relations can be obtained by different methods, and these are discussed in Section 4.1. By combining these relations with stellar statistical data on white dwarfs and their progenitors, one can estimate the fraction of stars that have evolved through each of the different evolutionary channels that lead to the white dwarf stage (Section 4.2). Finally, we demonstrate how galactic evolution models can be used to predict mass distributions of white dwarfs and their progenitors, and how comparison with observed data provides a powerful tool to constrain these models (Section 4.3).

4.1 Initial- to Final-Mass Relations

4.1.1 WHITE DWARFS IN CLUSTERS If white dwarfs are established as members of clusters with known ages and main sequence turnoff masses, it is possible to use the cluster main sequence luminosity function and the number of evolved stars to estimate the upper limiting mass for white dwarf versus supernova production, M_{WD}. Alternatively, if spectroscopic analysis allows the masses and cooling ages to be determined for individual white dwarfs, one can calculate their progenitor masses by equating cluster age minus cooling age with theoretical evolutionary ages to the AGB tip as a function of initial mass. Both methods have been extensively applied during recent years. More simply, one obtains lower limits to M_{WD} by postulating that the white dwarf progenitor mass should be larger than the turnoff mass. The presence of a white dwarf in the Pleiades has thus been used to estimate that $M_{WD} \approx 6\ M_\odot$ (Woolf 1974). Early attempts to use white dwarf cooling ages to derive initial masses were made by Sweeney (1976) for the Hyades; a more thorough discussion in this case was made by Weidemann (1977a). The first method was then applied to southern open clusters by Romanishin & Angel (1980), who found the faint blue white dwarf candidates by searching deep red and blue photographic plates and based their membership estimate on excess numbers in clusters compared with field areas. Their results showed that $M_{WD} \approx 7\ M_\odot$. Koester & Reimers (1981) confirmed the white dwarf cluster members by spectroscopic observations and extended this method to additional clusters. The most important result was obtained for NGC 2516 (Reimers & Koester 1982) in which three high-mass white dwarfs were found ($M > 0.9\ M_\odot$). Using the luminosity function and the number of evolved stars, Reimers & Koester estimated that $M_{WD} = 8^{+3}_{-2}\ M_\odot$. In a subsequent investigation, Weidemann & Koester (1983) determined individual progenitor masses by the second method and established initial- to final-mass relations $M_f(M_i)$, which appear flat in the range $M_i = 1-3\ M_\odot$ and then bend upward to $M_{WD} = \sim 8\ M_\odot$. This result indicates that heavy mass loss occurs for intermediate-mass stars, definitely stronger than that according to linear $M_f(M_i)$ relations (Iben & Renzini 1983) parameterized by a Reimers wind mass loss and planetary nebula efficiency parameter. Observations by Koester & Reimers (1985) of an even younger cluster with turnoff mass $\approx 5.3\ M_\odot$ yielded another white dwarf with $M_f = 0.2$–$0.6\ M_\odot$ and $M_i = 5.9\ M_\odot$, whereas two more recent investigations (Reimers & Koester 1988, 1989) give (for NGC 2168) white dwarf masses of 0.6–0.8 $M_\odot$ for $M_i > 5\ M_\odot$, and (for NGC 3532) $M_f = 0.5$, 0.9, and 0.9 $M_\odot$ for $M_i = 3.9$, 4.1, and 4.1 $M_\odot$, respectively. Since uncertainties in $M_f(g)$ are not small for these generally very faint stars (19–20 mag), it is important that $M_f(R)$, which is based mainly on effective temperature and distance, in general yield the same results. In particular, the case of NGC 2516 has provided the first convincing proof that stars with larger initial masses produce white dwarfs with larger masses.

However, as far as the numerical value of M_{WD} is concerned, two facts must be kept in mind. The figures for M_i given above were derived using pre–white dwarf evolutionary ages (from the main sequence up to the AGB), which were calculated for stellar models with a moderate amount of core overshooting [measured by a parameter $\alpha_c = 0.5$, using the terminology of Maeder & Mermilliod (1981)]. In the meantime, the Padua group (Bertelli et al 1985, 1986, Chiosi et al 1989) has promoted a case for stronger overshooting. In this case, main sequence lifetimes would be considerably increased, turnoff masses lowered, and M_i for individual white dwarfs reduced, since the unchanged cooling age becomes fractionally smaller. The influence of overshooting assumptions on M_f/M_i relations and on M_{WD} has been demonstrated and extensively discussed by Weidemann (1987a,b). If overshooting considerably prolongs the total age of a star of given M_i (see Maeder & Meynet 1989), the semiempirical value of M_{WD} will be correspondingly lowered, from $\approx 8\ M_\odot$ for $\alpha_c = 0.5$ to about 5.5 $M_\odot$ for $\alpha_c = 1$. Since the theoretical mass limit for nondegenerate carbon ignition, M_{up}, goes down similarly, the idea that $M_{WD} = M_{up}$ (Mengel 1976, Weidemann 1979) can be upheld. However, the case of no overshooting can be excluded, since the empirically derived M_{WD} would be much larger than M_{up}.

A second modification of M_{WD} occurs because the values quoted were obtained by extrapolation of the empirical $M_f(M_i)$ relations to $M_f = 1.4\ M_\odot$ (the Chandrasekhar mass), whereas it is more consistent to use an upper mass limit for single white dwarfs around 1.1 $M_\odot$ [the core mass for which nondegenerate carbon ignition occurs (see Iben 1985)].

4.1.2 STELLAR EVOLUTION WITH MASS LOSS Theoretical initial- to final-mass relations can easily be obtained for single-star evolution with mass loss. A simple example has been given by Paczyński (1970), whose envelope ejection scheme leads to $M_{WD} = 3.5\ M_\odot$. Stellar wind mass loss has been incorporated by Fusi Pecci & Renzini (1976), Mengel (1976), Scalo (1976), and Wood & Cahn (1977), leading to $M_{WD} \approx 5\ M_\odot$ for mass loss with a Reimers parameter $\eta = 1$. It soon became clear that, in addition to the Reimers wind, a superwind (Renzini 1981) must be invoked in order to understand the creation of planetary nebulae. This is in line with recent determinations of high mass loss rates on the AGB for OH/IR stars or other sources with circumstellar dust or gas shells. Since mass loss can still not be calculated from first physical principles, some parameterization is necessary. Cases for accelerated mass loss on the AGB and formulas have been put forward by Baud & Habing (1983), Bedijn (1988), and Van der Veen (1989). These formulas predict that initial- to final-mass relations will be in approximate agreement with the empirical relation for white dwarfs—although this is not accidental in the study by Van der Veen, who uses a free parameter to achieve this result. For more discussion on mass loss laws, the reader is referred to the proceedings of recent conferences (Kwok & Pottasch 1987, Mennessier & Omont 1990).

A theoretical relation between the initial and the minimum final mass, which is assumed to be given by the core mass at the beginning of the thermal pulsing phase, has been calculated by Mazzitelli & D'Antona (1986) for Population I stars with masses ranging from 1 to 5 $M_\odot$. The $M_f(M_i)$ relation is rather flat and thus predicts a sharply peaked white dwarf mass distribution, although at higher masses than are observed. More recently, corrections have been made that lower M_f for $M_i < 2.5$ $M_\odot$ (Mazzitelli 1989).

4.1.3 AGB TIP LUMINOSITIES   Additional information about initial- to final-mass relations can be obtained from maximum AGB luminosities reached in clusters of known age. Since galactic clusters are not sufficiently populated to outline the AGB, one must turn to the Magellanic Clouds, which contain clusters covering a wide range of ages and turnoff masses. A standard case is NGC 1866, for which the bright AGB stars predicted by stellar evolution with moderate (Reimers) mass loss (Iben & Truran 1978, Renzini & Voli 1981, Iben & Renzini 1983) were found to be missing. Inclusion of data for other clusters and determination of ages from turnoffs whose luminosities had been reached (Hodge 1983) enabled the construction of an M_i–M(tip AGB) diagram (Weidemann 1984) that reproduced the $M_f(M_i)$ relation derived from white dwarfs and galactic CSPN. A similar construction with newer data for some clusters was presented by Aaronson & Mould (1985), and this study confirmed that the AGB tip luminosities and the corresponding core masses—interpreted as final masses—were considerably below earlier estimates. Weidemann (1987a) has reconsidered the problem, especially the dependence of the $M_f(M_i)$ distribution on the distance to the Magellanic Clouds [which is now considered to be shorter (Reid & Strugnell 1986)] and on an overshooting parameter, which again affects the determination of the initial mass for a given turnoff magnitude.

It appears that for the older, low-turnoff-mass and low-metallicity clusters, the final masses are somewhat above the galactic $M_f(M_i)$ relation (see also Frantsman 1986a), in agreement with the theoretical expectation that mass loss will increase with metallicity and thus that M_{WD} will be smaller for Population II (Jura 1986). Since young clusters like NGC 1866 have nearly a Population I metallicity, the $M_f(M_i)$ relation for the Magellanic Clouds is expected to bend, as is seen in a tabulation by Weidemann (1987a). Increased mass loss for intermediate-mass stars, and not a lowering of M_{WD} by overshooting (which also shifts the turnoff masses!), is thus the explanation for the absence of bright AGB stars. A similar conclusion is reached by Frantsman (1988).

4.1.4 AGB LUMINOSITY FUNCTIONS It is possible to use synthetic population calculations (Section 4.3) to model the AGB luminosity function of field stars. The essential ingredients are the assumptions of an initial-mass function (IMF), a star formation rate (SFR), and an $M_f(M_i)$ relation. However, in order to take the different climbing rates on the AGB before and after the start of the thermal pulsing phase (E-AGB versus TP-AGB) into account, one must know the "start of TP-AGB" versus initial-mass relation. Recent investigations (Weidemann 1987d, Lattanzio 1987) have shown that this relation must be revised from that of Iben & Renzini (1983). The synthetic method was first applied by Reid & Mould (1984, 1985), who succeeded in reproducing their Large Magellanic Cloud field luminosity functions using the $M_f(M_i)$ relation from Aaronson & Mould (1985). That one can use the synthetic method the other way around —i.e. to derive an $M_f(M_i)$ relation from a given luminosity function— has since been demonstrated by Weidemann (1987a) for the same observational material. He showed how the derived $M_f(M_i)$ relations depend on Magellanic Cloud distance and start of TP-AGB relations. The results confirm the $M_f(M_i)$ shape obtained by the AGB tip luminosity method. Synthetic integrated AGB populations have also been calculated by Frantsman (1986b) for different mass loss assumptions.

4.1.5 OH/IR SOURCES, *IRAS* SOURCES In recent years it has become evident that AGB evolution is terminated by a phase of heavy mass loss that leads to dust formation and radiative acceleration of gas and dust. Mass loss rates increase to about 10^{-4} $M_\odot$ yr^{-1}, or rates so high that the envelope is lost in a time scale short compared with that of core mass growth by nuclear burning. Stars reaching this final phase of AGB evolution are long-period variables, a fact that is thought to cause the lifting of the surface material by shock waves produced by a pistonlike mechanism (Bowen 1988).

The dust shells increase in thickness, as shown by the appearance of strong silicate absorption at 9.7 μm (for oxygen-rich AGB stars). From their infrared spectra—obtained primarily by *IRAS*—the stars can be found even if there is no visible counterpart.

A special subgroup of the *IRAS* sources comprises the OH/IR stars, which are detected by line emission of their masering shell. For these (oxygen-rich) AGB stars it is possible to determine distances by the phase-lag method and therefore to obtain a luminosity function (Herman & Habing 1985), the reproduction of which was used to obtain a formula for smoothly accelerated mass loss on the AGB (Baud & Habing 1983). [A similar study was made by De Jong (1983).] Bedijn (1988) extended this approach by modeling the dust shell formation and derived a formula for mass loss as a function of mass, radius, luminosity, and effective temperature, with free parameters fixed by fitting to some observations. Using this formula, he obtained an $M_f(M_i)$ relation with which, after incorporating it into a model of galactic evolution, he was able to reproduce the period distribution of classical Miras, the OH luminosity function in the solar neighborhood, and the white dwarf mass distribution. The $M_f(M_i)$ relation in his study is linear in the range $0.86 < M_i < 8\ M_\odot$. The deviation from the somewhat bended shape of the relation found by Weidemann (1987a) is minor; however, Bedijn needed to assume a special star formation history in order to obtain a narrow white dwarf mass distribution (Section 4.3.1).

As for the long-period variables in general, Wood et al (1983) were able to separate AGB from supergiant pulsators by IR photometry in the Magellanic Clouds and determined final masses as a function of pulsational masses. Their conclusion that $M_{WD} \approx 5\ M_\odot$, however, depends on the assumed pulsational mode and on the occurrence of mass loss before onset of pulsation.

Habing (1988) used the *IRAS* point source catalogue to investigate the luminosity distribution of AGB stars with thick circumstellar shells, selected according to their positions in *IRAS* broadband two-color diagrams, which should reflect the luminosity distribution in the AGB termination phase. Combining these data with information about the galactic distribution, he found two populations that either (*a*) have the same spatial distribution but different luminosities or (*b*) have 80% in a thin-disk component and 20% in a thick-disk component with equal luminosity peaks at 4000 $L_\odot$, corresponding to $M \approx 0.55\ M_\odot$ on the core mass–luminosity relation. The luminosity distribution in the thin disk is about equal to that of the bulk distribution, and the implied mass distribution is very similar to that of the white dwarfs. The conclusion is that the selected *IRAS* sources are indeed the immediate predecessors of the white dwarfs. Of course, the next stage for these objects will be the proto–planetary (nebula) phase (Section 3.2.1). Kwok (1989b) has attempted to outline the evolution from the AGB to PN (Miras, OH/IR and CO emission variable stars, post-AGB stars and proto-PN, young PN), and Van der Veen (1989) has modeled the evolution by using counts in the different regions of the *IRAS* two-color diagram. However, it is still debated if this interpretation of the observed well-outlined two-color sequence from the blue to the red, or from normal Miras to PN, is followed by individual stars or is built up piecewise by stars with different initial masses. For details, we refer the reader to Mennessier & Omont (1990). In any event, there can be little doubt that most pre–white dwarfs go through a phase with thick circumstellar dust envelopes, whose presence is so convincingly revealed by infrared and radio observations.

4.1.6 DIFFERENTIAL MASS LOSS Up to now it has been assumed that the decisive parameter for pre–white dwarf evolution is the initial mass, reflected in the existence of a single-valued $M_f(M_i)$ relation. However, we know from the interpretation of the horizontal branch for Population II or from the existence of Population I RR Lyrae stars (Taam et al 1976) and of field horizontal branch stars that stars lose different amounts of mass already on the first giant branch. Prior evidence for this has been collected and discussed with a view on rotation as a second parameter by Weidemann (1977a). Similarly, Peterson (1985) concludes from an investigation of rotational velocities in horizontal branch stars that rotation, coupled with increased mass loss, might be responsible for the anomalous increase in the proportion of blue stars on the HB in some globular clusters. That also the He flash mass, and thus the core mass at the horizontal branch (or at the clump for Population I), might be influenced has been shown by evolutionary calculations (Mengel & Gross 1976). The recent interpretation of the sdB and sdOB field stars (Section 3.3) as objects evolving from an extended horizontal branch and their identification as old disk objects (Heber 1986) show that even if the fraction of Population I stars evolving through this pre–white dwarf evolutionary channel is small [of the order of 2% (Section 4.2.3)], differential mass loss on the first giant branch is important. Most probably, differential mass loss will also occur on the AGB. Therefore, a single-valued $M_f(M_i)$ relation is only a first approximation.

However, differential mass loss cannot be too large, since otherwise the narrow mass distribution of white dwarfs could not be reproduced by reasonable models of galactic evolution (Section 4.3.1). It is another question if M_{WD} is also affected by differential mass loss (see Weidemann 1981).

4.1.7 BINARIES   In view of the many possibilities of binary evolution, it is not yet possible to outline a comprehensive picture. For wide binaries there are no indications of differences compared with single-star evolution, with the exception of the as-yet unexplained observation that white dwarfs with helium-rich atmospheres are overabundant (Sion & Oswalt 1988). For closer binaries, if and at which evolutionary phase mass transfer or exchange occurs depend on masses and separation. In general, the growth of the core mass of the primary [along the red giant branch (RGB) or AGB] will be reduced compared with single-star evolution: We thus expect smaller final masses for a given initial mass. It is possible, however, that the mass of the helium white dwarf (mass loss on the RGB) or C/O white dwarf (on the AGB) will be increased by a second phase of mass transfer. Most common is Roche lobe overflow on the AGB, when the radius is largest. It is agreed that this leads to a common envelope phase in which most of the mass and angular momentum are lost from the system, thus bringing the stars closer together. This is considered to be the way to precataclysmic and cataclysmic variables, i.e. to close binary stages. However, a variety of evolutionary histories are possible. For a discussion of these, the reader is referred to papers by Webbink (1979) and, especially, by Iben & Tutukov (e.g. 1984a, 1985, 1986a,b, 1987). Concerning the evolutionary status of white dwarfs, the most interesting suggestion is that degenerates are created through mergers of close binaries involving either low-mass helium white dwarfs with a predicted peak at $0.2\,M_\odot$ or more massive C/O degenerates, leading to supernova Type I explosions [see Webbink & Iben (1987) for a survey]. Tutukov & Yungelson (1987) have tried to explain nearly all subluminous blue stars and pre–white dwarfs by means of binary evolution. However, attempts to detect close double degenerates—which are predicted in fairly large numbers—have widely failed (Robinson & Shafter 1987, 1989, Bragaglia et al 1989, Saffer & Liebert 1989). Furthermore, since there is no evidence in the white dwarf number–mass distribution for a secondary peak, these evolutionary channels for white dwarf production have probably been overestimated. Although a somewhat revised picture for $M_f(M_i)$ relations and mass distributions has recently been presented by Iben & Webbink (1989), numerical results can only be obtained by making several assumptions and thus must be considered preliminary.

4.2   *Stellar Statistics, Birth Rates*

Now that different white dwarf progenitors and progenitor stages have been identified, we must try to obtain a coherent picture, especially in order to estimate the fraction (and mass range) of white dwarfs that have evolved through the different evolutionary channels. This task cannot be achieved without recourse to stellar statistical data, such as space densities, scale heights, velocity distributions, and birth rates.

4.2.1 WHITE DWARFS White dwarfs are known only in the immediate neighborhood of the Sun, mainly within 100 pc. Counts demonstrate (Jahreiss 1987) that completeness is restricted to even smaller distances (about 10 pc), and that white dwarfs like Sirius B are hidden in binaries at larger distances. Although recent surveys (e.g. Palomar-Green) reach out to fainter magnitudes, estimates of the space density are still uncertain. Indeed, the derived space densities depend on the red limit—or, more generally, on color limits—for color-selected white dwarfs or on proper-motion limits for cooler degenerates.

The most comprehensive study—that by Fleming et al (1986)—is restricted to Palomar-Green objects with $U-B < -0.44$ and yielded a space density of 0.49 ± 0.05 DA white dwarfs with $M_v < 12.75$ within 1000 pc^3, from which birth rates can be derived if the cooling time to $M_v = 12.75$ is known. Depending on different evolutionary cooling models (see D'Antona & Mazzitelli 1990), the resulting birth rates are between 3.9×10^{-13} and 6.1×10^{-13} yr^{-1} pc^{-3}. In order to obtain a total birth rate, one must correct for the non-DA contribution. The non-DA/DA ratio is still arguable, since it is not yet known if both spectral varieties are due to different pre–white dwarf evolutionary channels or to changes of surface composition during white dwarf evolution itself [see Shipman (1989) and many other contributions at recent conferences (Philip et al 1987, Wegner 1989b)]. For the more restricted temperature range between 30,000 and 10,000 K, the DB/DA ratio is of order 10% (Oke et al 1984). Correcting for the non-DA fraction, Fleming et al (1986) obtain birth rates of 4.9×10^{-13} and 7.5×10^{-13} yr^{-1} pc^{-3} for the cooling models of Iben & Tutukov (1984b) and Koester & Schönberner (1986), respectively, values that are considerably lower than earlier estimates (Weidemann 1977b). The results of Fleming et al (1986) will probably need to be revised upward in view of problems of incompleteness. Weidemann (1990) obtains a birth rate range of 0.9–1.4×10^{-12} yr^{-1} pc^{-3} for single white dwarfs, which is corrected to 1.5–2.3×10^{-12} yr^{-1} pc^{-3} if white dwarfs in close binaries are included. The latter value is important for calculating the mass contribution of white dwarfs to the local galactic matter density (Section 4.3.4) and for a comparison with the birth rates of CSPN.

The white dwarf scale height is still not observationally established, but it is of the order of 300 pc (Fleming et al 1986). Similarly, the velocity distribution is not sufficiently well determined as to allow conclusions about different pre–white dwarf evolutionary channels (Sion et al 1988).

4.2.2 CENTRAL STARS OF PLANETARY NEBULAE The space densities and birth rates of PN have been the subject of many investigations. In a recent tabulation Phillips (1989) lists 24 birth rate determinations yielding values between 0.1×10^{-12} and 11×10^{-12} PN yr^{-1} pc^{-3}, mostly depending on the distance scale used. Relatively recent values derived by Phillips (1984) based on the Daub distance scale (4.4×10^{-12} PN yr^{-1} pc^{-3}) and by Ishida & Weinberger (1987) from locally detected old PN (8.0×10^{-12} PN yr^{-1} pc^{-3}) caused concern in that these figures are too high compared with the white dwarf birth rate. Phillips (1989) now arrives at a best value of $2.39 \pm 0.32 \times 10^{-12}$ PN yr^{-1} pc^{-3}, and Weidemann (1989), after demonstrating that increased planetary nebula distances compared with the Daub distance scale reconcile the Schönberner and Heap/Augensen mass distributions of CSPN (Section 3.1), has pointed out that these larger distances also reduce the birth rate and scale height discrepancies between white dwarfs and planetary nebulae. However, one must note that not all white dwarfs go through the planetary nebula evolutionary channel (see Section 4.2.3). Therefore, a slight discrepancy still remains, which also serves as an indication that the above-favored upward revision of the white dwarf birth rate is reasonable.

4.2.3 WHITE DWARFS FROM OTHER EVOLUTIONARY CHANNELS Few attempts have been made to determine the birth rates for white dwarfs evolving through other progenitor stages. For sdO stars the problem is not yet solved because not only are distances and therefore space densities uncertain, but also they are produced in different ways, both as post-AGB stars in binary evolution (10% according to Ferguson et al 1984) and as post–horizontal branch stars that do not reach the AGB. Evolutionary time scales, which are necessary in order to transform space densities into birth rates, are also unknown. The only estimate that can be made is that about 20% extreme helium-rich sdO's evolve into DB white dwarfs (Heber et al 1986b).

The situation is much better for the sdB stars, but only if their interpretation as helium-burning extended horizontal branch stars with a non-burning hydrogen skin is correct. Evolutionary times can then be obtained from tracks such as that of Caloi (1989). Heber (1986) arrives at a birth rate of 2×10^{-14} yr^{-1} pc^{-3} for the sdOB/sdB stars, a figure indicating that this evolutionary channel feeds at most 2% into the white dwarf region, especially into the low-mass tail (0.45–0.50 $M_\odot$).

A similar result has been obtained by Drilling & Schönberner (1985), who used a complete ensemble of very hot low-galactic-latitude sdO's for which effective temperatures had been determined from UV continua, and distances from color excesses (Schönberner & Drilling 1984). Dividing the ensemble into CSPN, post-AGB stars, and post–horizontal branch stars, they obtained birth rates of 1, 0.6, and 0.02×10^{-12} yr^{-1} pc^{-3}, respectively. Discussing the uncertainties, especially for the second group, they arrived at fractions of 30–90%, 5–70%, and 0.2–3%, respectively, entering the white dwarf regions through the different evolutionary channels. This should be compared with the results of Weidemann & Koester (1984), who found from the DA mass distribution (Section 2.1) that 28–44% (depending on calibration) have masses smaller than 0.55 $M_\odot$, the lower limit for visible CSPN according to Schönberner's tracks (see, however, Section 3.1). In summary, it appears that about 30% of the white dwarfs reach this stage via non-PN evolutionary channels. Only about 2% evolve directly from the extended horizontal branch and the remainder probably evolve about equally through the post-AGB and binary evolutionary channels.

4.3 Galactic Evolution Models

Having learned much about white dwarfs and their progenitors, we can try to integrate our knowledge into a general scheme of galactic evolution and—in the case of the white dwarfs—more specifically into a model of evolution of the stellar population in the solar neighborhood. Using numerical population synthesis, we can test our concepts of stellar pre–white dwarf evolution and, since the white dwarfs that we observe today evolved from progenitors that were born long ago, also gain some information about the history of the Galaxy. We shall see that the observed narrow white dwarf mass distribution puts severe constraints on these models (Section 4.3.1). Similarly, we can use the observed white dwarf luminosity function as a tool to constrain models of galactic evolution, e.g. to restrict the galactic age (Section 4.3.3). Vice versa, it is possible to predict past rates of white dwarf production, especially in order to estimate the fraction of cooled-down degenerates contributing to local dark matter (Section 4.3.4). Up to now, the galactic evolution models used for this purpose have been comparatively simple and restricted to the evolution of a single population characterized by an initial-mass function (IMF), a star formation rate (SFR), a zero point at the assumed age of the galactic disk, and an initial- to final-mass relation. Also entering the calculations are evolutionary ages from the main sequence up to the AGB, and down the white dwarf cooling sequence [see Weidemann (1979) for a flow chart and further details]. With so much emphasis now being put on galactic evolution models—primarily on chemical and dynamical evolution rather than white dwarf evolution (see Hensler 1987, Pagel 1990)—modeling will soon be extended to incorporate, for example, different populations and bursts of star formation. With increasing computational possibilities, the simple models applied in the white dwarf case will thus soon be outdated. However, the knowledge gained—especially about initial- to final-mass relations, limiting progenitor mass ranges, and birth rates—will remain an important ingredient for all future investigations.

4.3.1 WHITE DWARF MASS DISTRIBUTION The first models using steep $M_f(M_i)$ relations were calculated by Thuan et al (1975). After it became clear that mass loss plays a larger role in stellar evolution than initially thought, Weidemann (1977a) predicted white dwarf mass distributions [for different $M_f(M_i)$ relations, a Salpeter IMF, and an exponentially declining SFR] that showed a steep increase at $M_f(M_i = 1\ M_\odot)$ and a slower decrease toward higher masses. When the KSW study showed that the mass distribution is in reality much narrower, a comprehensive model calculation with varying input assumptions was carried out by Koester & Weidemann (1980). They demonstrated that the majority of the models could be ruled out, since the width and median value of the mass distribution and the birth rate ratio between white dwarfs and supernovae were not compatible with observed data. After the establishment of a more reliable $M_f(M_i)$ relation from cluster data, especially from massive white dwarfs found in NGC 2516 (Section 4.1.1), new calculations that also took the scale height evolution into account were carried out by Weidemann & Koester (1983). Compared with the use of linear parameterized $M_f(M_i)$ relations (Iben & Renzini 1983), these showed better agreement with the observed mass distribution. Recently the Koester & Weidemann (1980) study has been repeated by Yuan (1987), who incorporated newer IMFs, SFRs, and different $M_f(M_i)$ relations. Her results showed that a constant SFR (at least through the past 5×10^9 yr) and $M_f(M_i)$ relations from cluster data (Section 4.1.1) yielded the best agreement with the observed distribution, whereas many other combinations [e.g. $M_f(M_i)$ from Mazzitelli & D'Antona (1986) or Bedijn (1988)] could be ruled out for constant SFR. (Bedijn achieved better agreement by assuming a special shape for the SFR.) However, the influence of the overshooting parameter, which enters into the evolutionary age and strongly influences the limiting mass for white dwarf versus supernova production, M_{WD} (Section 4.1.1), turned out to be minor. This result is indicative of the fact that the vast majority of white dwarfs arise from low-mass progenitors. The supernova/white dwarf ratio, however, is not expected to be changed with M_{WD} as a function of overshooting, since the shape of the IMF, derived from an observed present-day mass function and main sequence evolutionary ages, varies in such a way as to compensate the more extended mass range for supernova progenitors (Weidemann 1987b).

4.3.2 MASS DISTRIBUTION OF CSPN The numerical galactic evolution models can both be used to predict white dwarf mass distributions (which must be obtained by integrating the white dwarf birth rates as a function of M_f over a cooling time down to a specified luminosity limit) and to predict the *present* white dwarf birth rates, which are identical with the CSPN birth rates. In connection with the evolutionary tracks and time scales, these birth rates can be used to predict the H-R distribution of CSPN expected for a complete (local) ensemble (Section 3.1) and the true (selection-free) mass distribution of CSPN. Yuan's (1987) calculations [which assume the same input parameters that reproduced the white dwarf mass distribution best, but begin with $M_f(M_i = 1\ M_\odot) = 0.55\ M_\odot$, the lower limit for CSPN] give a distribution that is even sharper than the white dwarf (integrated) mass distribution in agreement with Schönberner's (1981) and Weidemann's (1989) findings. This confirms the conclusion that the vast majority of white dwarfs with $M > 0.55\ M_\odot$ evolve through the planetary nebula evolutionary channel.

4.3.3 WHITE DWARF MASSES AND LUMINOSITY FUNCTION Following improved empirical determinations of the DA luminosity function (LF) (Fleming et al 1986, Liebert et al 1988, 1989), efforts have been made to reproduce the observed shape using theoretical cooling curves for white dwarfs. Besides depending on other physical parameters, these curves differ as a function of mass: For example, at lower luminosities, massive degenerates cool down faster below the Debye temperature (see reviews by D'Antona 1989, D'Antona & Mazzitelli 1990). Theoretical cooling curves for different masses have been combined to an integrated LF using the observed (present) DA mass distribution in an attempt by Winget et al (1987) to determine the age of the Galaxy (more precisely, of the galactic disk) from a fit especially at the observed low-luminosity cutoff. Since degenerates at the faint end of the LF have cooled for several billion years, the assumptions by Winget et al (constant mass distribution, constant SFR) needed revision (Weidemann & Yuan 1989). Iben & Laughlin (1989) improved the model calculations by using semianalytical expressions for the mass dependence of cooling curves. They conclude that the LF at the cool end is strongly dependent on the early mass distribution. Yuan (1989) used instead the fully numerical galactic evolution model—which allowed easy variation of assumptions regarding SFR, IMF, and $M_f(M_i)$—in order to check the compatibility range with the observed LF. For example, she shows how the assumption of bimodal star formation—according to which many massive white dwarfs are produced in the early stages of galactic evolution (Larson 1986)—would steepen the cool end of the LF beyond the observational error limits.

4.3.4 DARK MATTER CONTRIBUTION The models described in the preceding sections can be used to estimate the contribution of cooled-down white dwarfs to dark matter, an important figure in the context of the discussion of missing mass. Although this problem has been reduced owing to revised determinations of the local dynamical mass (Bienaymé et al 1987, Kuijken & Gilmore 1989), the white dwarf contribution to invisible matter is still debated and is of interest in itself. Yuan's galactic evolution models have thus been used to calculate space and column mass densities for stellar remnants under varying assumptions for the galactic disk age, SFR, IMF, and $M_f(M_i)$ relation. The results show that the invisible mass fraction in the solar neighborhood is of the order of 2 $M_\odot$ pc^{-2}, or 4% of the revised dynamical mass, a value that is somewhat higher than earlier estimates. For a tabulation and further discussion, see Weidemann (1990). For these calculations the most important ingredients are the initial- to final-mass relations and the progenitor mass ranges, which certainly vary with chemical composition and in time and space through the Galaxy. With more comprehensive models of galactic evolution under construction at many places, one should ensure that knowledge on white dwarfs and their progenitors—as outlined in this review—is adequately incorporated.

ACKNOWLEDGMENTS

I am grateful to my many colleagues and, over the years, friends who supported my work with preprints, reprints, letters, and visits in Kiel. In particular, I am deeply grateful to Jesse L. Greenstein, who started me on the white dwarf track in 1957 and continued to provide me with research possibilities and observational material during later stays in Pasadena, up to 1981/82. I furthermore thank those who attended the European Workshops on White Dwarfs from 1974 to 1984 and shared their knowledge in this growing research field. Finally, I thank my local supporters, especially A. Unsöld, D. Koester (now at Louisiana State University), D. Schönberner, and U. Heber, as well as I. Schmidt for typing and patiently correcting the manuscript. Finally, I am much indebted to the Production Editor, Keith Dodson, for many improvements in style and clarifying suggestions.

Literature Cited

Aaronson, M., Mould, J. 1985. *Ap. J.* 288: 551
Aller, L. H., Keyes, C. D., Maran, S. P., Gull, T. R., Michalitsianos, A. G., Stecher, T. P. 1987. *Ap. J.* 320: 159
Barlow, M. J. 1987. *MNRAS* 227: 161
Barlow, M. J. 1989. See Torres-Peimbert 1989, p. 319
Barlow, M. J., Morgan, B. L., Standley, C., Vine, H. 1986. *MNRAS* 223: 151
Baud, B., Habing, H. J. 1983. *Astron. Astrophys.* 127: 73
Bedijn, P. J. 1988. *Astron. Astrophys.* 205: 105
Bergeron, P., Wesemael, F., Liebert, J., Fontaine, G. 1989. *Ap. J. Lett.* 345: L91

Bertelli, G., Bressan, A. G., Chiosi, C. 1985. *Astron. Astrophys.* 150: 33
Bertelli, G., Bressan, A., Chiosi, C., Angerer, K. 1986. *Astron. Astrophys. Suppl.* 66: 191
Bienaymé, O., Robin, A. C., Creze, M. 1987. *Astron. Astrophys.* 180: 94
Bond, H. E. 1989. See Torres-Peimbert 1989, p. 215
Bond, H. E., Grauer, A. D. 1987. *Ap. J. Lett.* 321: L123
Bond, H. E., Grauer, A. D., Green, R. F., Liebert, J. 1984. *Ap. J.* 279: 751
Bowen, G. H. 1988. *Ap. J.* 329: 299
Bragaglia, A., Greggio, L., Renzini, A., D'Odorico, S. 1989. See Wegner 1989b, p. 138
Caloi, V. 1989. *Astron. Astrophys.* 221: 27
Caloi, V., Castellani, V., Galluccio, D., Wamsteker, W. 1984. *Astron. Astrophys.* 138: 485
Cester, B. 1965. *Mem. Soc. Astron. Ital.* 36 (3): 1
Chiosi, C., Bertelli, G., Meylan, G., Ortolani, S. 1989. *Astron. Astrophys.* 219: 167
D'Antona, F. 1989. See Wegner 1989b, p. 44
D'Antona, F., Mazzitelli, I. 1990. *Annu. Rev. Astron. Astrophys.* 28: 139
de Boer, K. S. 1985. *Astron. Astrophys.* 142: 321
De Jong, T. 1983. *Ap. J.* 274: 252
Demers, S., Wesemael, F., Irwin, M. J., Fontaine, G., Lamontagne, R., et al. 1989. *Publ. Astron. Soc. Pac.* In press
Downes, R. A. 1986. *Ap. J. Suppl.* 61: 569
Drilling, J. S. 1983. *Ap. J. Lett.* 270: L13
Drilling, J. S. 1985. *Ap. J. Lett.* 294: L107
Drilling, J. S., Schönberner, D. 1985. *Astron. Astrophys.* 146: L23
Ferguson, D. H., Green, R. F., Liebert, J. 1984. *Ap. J.* 287: 320
Fleming, T. A., Liebert, J., Green, R. F. 1986. *Ap. J.* 308: 176
Fontaine, G., Bergeron, P., Lacombe, P., Lamontagne, R., Talon, A. 1985. *Astron. J.* 90: 1094
Ford, H. C., Ciardullo, R., Jacoby, G. H., Hui, X. 1989. See Torres-Peimbert 1989, p. 335
Frantsman, Ju. L. 1986a. *Astrofizika* 25: 517
Frantsman, Ju. L. 1986b. *Sov. Astron. Lett.* 12: 94
Frantsman, Ju. L. 1988. *Astrophys. Space Sci.* 145: 251
Fusi Pecci, F., Renzini, A. 1976. *Astron. Astrophys.* 46: 447
Gatewood, G. D., Gatewood, C. V. 1978. *Ap. J.* 225: 191
Gathier, R. 1984. Thesis. Univ. Groningen, Neth.
Gathier, R., Pottasch, S. R. 1989. *Astron. Astrophys.* 209: 369
Gingold, A. 1976. *Ap. J.* 204: 116
Glaspey, J. W., Demers, S., Moffat, A. F. J., Shara, M. 1985. *Ap. J.* 289: 326
Grabowski, B., Madej, J., Halenka, J. 1987. *Ap. J.* 313: 750
Graham, J. 1972. *Astron. J.* 77: 144
Grauer, A. D., Bond, H. E. 1983. *Ap. J.* 271: 259
Green, R. F., Schmidt, M., Liebert, J. 1986. *Ap. J. Suppl.* 61: 305
Greenstein, J. L. 1976. *Astron. J.* 81: 323
Greenstein, J. L. 1984. *Ap. J.* 276: 602
Greenstein, J. L. 1985. *Publ. Astron. Soc. Pac.* 97: 827
Greenstein, J. L. 1986a. *Astron. J.* 92: 859
Greenstein, J. L. 1986b. *Astron. J.* 92: 867
Greenstein, J. L., Trimble, V. 1967. *Ap. J.* 149: 283
Greenstein, J. L., Oke, J. B., Shipman, H. L. 1971. *Ap. J.* 169: 563
Greenstein, J. L., Sargent, A. I. 1974. *Ap. J. Suppl.* 28: 157
Greenstein, J. L., Boksenberg, A., Carswell, R., Shortridge, K. 1977. *Ap. J.* 212: 186
Greenstein, J. L., Dolez, N., Vauclair, G. 1983. *Astron. Astrophys.* 127: 25
Greenstein, J. L., Oke, J. B., Shipman, H. L. 1985. *Q. J. R. Astron. Soc.* 26: 279
Groth, H. G., Kudritzki, R. P., Heber, U. 1985. *Astron. Astrophys.* 152: 107
Guseinov, O. H., Novruzova, H. I., Rustamov, Y. S. 1983a. *Astrophys. Space Sci.* 96: 1
Guseinov, O. H., Novruzova, H. I., Rustamov, Y. S. 1983b. *Astrophys. Space Sci.* 97: 305
Habing, H. J. 1988. *Astron. Astrophys.* 200: 40
Hamada, T., Salpeter, E. E. 1961. *Ap. J.* 134: 683
Harrington, R. S., Christy, J. W., Strand, K. Aa. 1981. *Astron. J.* 86: 909
Harris, H. C., Nemec, J. M., Hesser, J. E. 1983. *Publ. Astron. Soc. Pac.* 95: 256
Hayes, D. S., Latham, D. W. 1975. *Ap. J.* 197: 593
Heap, S. R. 1990. In Mennesier & Omont 1990. In press
Heap, S. R., Augensen, H. J. 1987. *Ap. J.* 313: 268
Heap, S. R., Hintzen, P. M. 1990. *Ap. J.* In press
Heber, U. 1986. *Astron. Astrophys.* 155: 33
Heber, U. 1987. In Philip et al 1987, p. 79
Heber, U., Kudritzki, R. P. 1986. *Astron. Astrophys.* 169: 244
Heber, U., Hunger, K. 1987. See Philip et al 1987, p. 599
Heber, U., Kudritzki, R. P., Caloi, V., Castellani, V., Danziger, J., Gilmozzi, R. 1986a. *Astron. Astrophys.* 162: 171
Heber, U., Drilling, J. S., Husfeld, D. 1986b. See Hunger et al 1986, p. 345

Heintz, W. D. 1974. *Astron. J.* 79: 819
Henry, R. B. C., Liebert, J., Boroson, T. A. 1989. *Ap. J.* 339: 872
Hensler, G. 1987. *Mitt. Astron. Ges.* 70: 141
Herman, J., Habing, H. J. 1985. *Astron. Astrophys. Suppl.* 59: 523
Hodge, P. W. 1983. *Ap. J.* 264: 470
Holberg, J. B., Wesemael, F., Hubeny, I. 1984. *Ap. J.* 280: 679
Holberg, J. B., Wesemael, F., Wegner, G., Bruhweiler, F. C. 1985. *Ap. J.* 293: 294
Hunger, K., Schönberner, D., Rao, N. K., eds. 1986. *Hydrogen-Deficient Stars and Related Objects. Proc. IAU Colloq. No. 87*. Dordrecht: Reidel
Hvrinak, B. J., Kwok, S., Volk, K. M. 1988. *Ap. J.* 331: 832
Hvrinak, B. J., Kwok, S., Volk, K. M. 1989. *Ap. J.* 346: 265
Iben, I. Jr. 1984. *Ap. J.* 277: 333
Iben, I. Jr. 1985. In *Mass Loss from Red Giants*, ed. B. Zuckerman, M. Morris, p. 1. Dordrecht: Reidel
Iben, I. Jr., Laughlin, G. 1989. *Ap. J.* 341: 312
Iben, I. Jr., Renzini, A. 1983. *Annu. Rev. Astron. Astrophys.* 21: 271
Iben, I. Jr., Truran, J. W. 1978. *Ap. J.* 220: 980
Iben, I. Jr., Tutukov, A. V. 1984a. *Ap. J. Suppl.* 54: 335
Iben, I. Jr., Tutukov, A. V. 1984b. *Ap. J.* 282: 615
Iben, I. Jr., Tutukov, A. V. 1985. *Ap. J. Suppl.* 58: 661
Iben, I. Jr., Tutukov, A. V. 1986a. *Ap. J.* 311: 742
Iben, I. Jr., Tutukov, A. V. 1986b. *Ap. J.* 311: 753
Iben, I. Jr., Tutukov, A. V. 1987. *Ap. J.* 313: 727
Iben, I. Jr., Webbink, R. F. 1989. See Wegner 1989b, p. 477
Iben, I. Jr., Kaler, J. B., Truran, J. W., Renzini, A. 1983. *Ap. J.* 264: 605
Ishida, K., Weinberger, R. 1987. *Astron. Astrophys.* 178: 227
Jacoby, G. H. 1988. *Ap. J.* 333: 193
Jacoby, G. H. 1989. *Ap. J.* 339: 39
Jahreiss, H. 1987. *Mem. Soc. Astron. Ital.* 58: 53
Johnson, H., Zuckerman, B., eds. 1989. *Evolution of Peculiar Red Giant Stars. Proc. IAU Colloq. No. 106*. Cambridge: Univ. Press
Jura, M. 1986. *Ap. J.* 301: 624
Kaler, J. B. 1983. *Ap. J.* 271: 188
Kawaler, S. D. 1987. See Philip et al 1987, p. 297
Kawaler, S. D. 1988. In *Advances in Helio- and Asteroseismology, IAU Symp. No. 123*, ed. J. Christensen-Dalsgaard, S. Frandsen, p. 329. Dordrecht: Reidel
Kawaler, S. D., Hansen, C. J. 1989. See Wegner 1989b, p. 97
Koester, D. 1984. *Astrophys. Space Sci.* 100: 471
Koester, D. 1987. *Ap. J.* 322: 852
Koester, D., Reimers, D. 1981. *Astron. Astrophys.* 99: L8
Koester, D., Reimers, D. 1985. *Astron. Astrophys.* 153: 260
Koester, D., Schönberner, D. 1986. *Astron. Astrophys.* 154: 125
Koester, D., Weidemann, V. 1980. *Astron. Astrophys.* 81: 145
Koester, D., Schulz, H., Weidemann, V. 1979. *Astron. Astrophys.* 76: 262 (KSW)
Koester, D., Weidemann, V., Zeidler-K. T., E. 1982. *Astron. Astrophys.* 116: 147
Kuijken, K., Gilmore, G. 1989. *MNRAS* 239: 605
Kwitter, K. B., Jacoby, G. H. 1989. See Torres-Peimbert 1989, p. 303
Kwok, S. 1989a. See Torres-Peimbert 1989, p. 401
Kwok, S. 1989b. *MNRAS*. In press
Kwok, S., Pottasch, S. R., eds. 1987. *Late Stages of Stellar Evolution. Astrophys. Space Sci. Libr. No. 132*. Dordrecht: Reidel
Kwok, S., Volk, K. 1989. See Torres-Peimbert 1989, p. 452
Larson, R. B. 1986. *MNRAS* 218: 409
Lattanzio, J. C. 1987. *Ap. J. Lett.* 313: L15. Correction in *Ap. J. Suppl.* In press
Leggett, S. K. 1989. *Astron. Astrophys.* 208: 41
Lenzuni, P., Natta, A., Panagia, N. 1987. See Preite-Martinez 1987, p. 249. See also 1989. *Ap. J.* 345: 306
Liebert, J. 1980. *Annu. Rev. Astron. Astrophys.* 18: 363
Liebert, J., Dahn, C. C., Monet, D. G. 1988. *Ap. J.* 332: 891
Liebert, J., Dahn, C. C., Monet, D. G. 1989. See Wegner 1989b, p. 15
Maeder, A., Mermilliod, J. C. 1981. *Astron. Astrophys.* 93: 136
Maeder, A., Meynet, G. 1989. *Astron. Astrophys.* 210: 155
Maran, S. P. 1983. *Nature* 305: 470
Mazzitelli, I. 1989. See Wegner 1989b, p. 29
Mazzitelli, I., D'Antona, F. 1986. *Ap. J.* 311: 762
McGraw, J. T., Starrfield, S. G., Liebert, J., Green, R. F. 1979. In *White Dwarfs and Variable Degenerate Stars, IAU Colloq. No. 53*, ed. H. M. Van Horn, V. Weidemann, p. 377. Rochester, NY: Univ. Rochester
McMahan, R. K. 1989. *Ap. J.* 336: 409
Méndez, R. H., Kudritzki, R. P., Herrero, D., Groth, H. G. 1988a. *Astron. Astrophys.* 190: 113

Méndez, R. H., Groth, H. G., Kudritzki, R. P., Herrero, A. 1988b. *Astron. Astrophys.* 197: L25
Méndez, R. H., Gathier, R., Simon, K. P., Kwitter, K. B. 1988c. *Astron. Astrophys.* 198: 287
Mengel, J. G. 1976. *Astron. Astrophys.* 48: 83
Mengel, J. G., Gross, P. G. 1976. *Astrophys. Space Sci.* 41: 407
Mennessier, M. O., Omont, A., eds. 1990. *From Miras to Planetary Nebulae*. Gif sur Yvette, Fr: Frontières. In press
Monk, D. J., Barlow, M. J., Clegg, R. E. S. 1988. *MNRAS* 234: 583
Nather, R. E., Robinson, E. L., Stover, R. J. 1981. *Ap. J.* 244: 269
Nelson, D., Young, A. 1970. *Publ. Astron. Soc. Pac.* 82: 699
Nousek, J. A., Shipman, H. L., Holberg, J. B., Liebert, J., Pravdo, S. H., et al. 1986. *Ap. J.* 309: 230
O'Donoghue, D., Menzies, J. W., Hill, P. W. 1987. *MNRAS* 227: 347
Oke, J. B. 1974. *Ap. J. Suppl.* 27: 21
Oke, J. B., Gunn, J. E. 1983. *Ap. J.* 266: 713
Oke, J. B., Schild, R. E. 1970. *Ap. J.* 161: 1015
Ortolani, S., Rosino, L. 1987. *Astron. Astrophys.* 185: 102
Ostriker, J. P. 1971. *Annu. Rev. Astron. Astrophys.* 9: 353
Oswalt, T. D., Hintzen, P. M., Liebert, J. W., Sion, E. M. 1988. *Ap. J. Lett.* 333: L87
Paczyński, B. 1970. *Acta Astron.* 20: 47
Paczyński, B. 1971. *Acta Astron.* 21: 417
Pagel, B. E. J. 1990. In *Baryonic Dark Matter*, ed. G. Gilmore, D. Lynden-Bell. Dordrecht: Kluwer. In press
Peterson, R. C. 1985. *Ap. J.* 289: 320
Philip, A. G. D., Hayes, D. S., Liebert, J. W., eds. 1987. *The Second Conference on Faint Blue Stars.* Proc. IAU Colloq. No. 95. Schenectady, NY: L. Davis
Phillips, J. P. 1984. *Astron. Astrophys.* 137: 92
Phillips, J. P. 1989. See Torres-Peimbert 1989, p. 425
Politano, M., Webbink, R. F. 1989. See Wegner 1989b, p. 440
Politano, M., Ritter, H., Webbink, R. F. 1989. See Wegner 1989b, p. 465
Popper, D. M. 1954. *Ap. J.* 120: 316
Pottasch, S. R. 1989. See Torres-Peimbert 1989, p. 481
Pottasch, S. R., Acker, A. 1989. *Astron. Astrophys.* 221: 123
Pottasch, S. R., Parthasarathy, M. 1988. *Astron. Astrophys.* 192: 182
Pottasch, S. R., Goss, W. M., Arnal, E. M., Gathier, R. 1982. *Astron. Astrophys.* 106: 229
Preite-Martinez, A., ed. 1987. *Planetary and Protoplanetary Nebulae: From IRAS to ISO.* Astrophys. Space Sci. Libr. No. 135. Dordrecht: Reidel
Reid, N., Mould, J. 1984. *Ap. J.* 284: 98
Reid, N., Mould, J. 1985. *Ap. J.* 299: 236
Reid, N., Strugnell, P. R. 1986. *MNRAS* 221: 887
Reimers, D., Koester, D. 1982. *Astron. Astrophys.* 116: 341
Reimers, D., Koester, D. 1988. *Astron. Astrophys.* 202: 77
Reimers, D., Koester, D. 1989. *Astron. Astrophys.* 218: 118
Renzini, A. 1979. In *Stars and Star Systems*, ed. B. E. Westerlund, p. 155. Dordrecht: Reidel
Renzini, A. 1981. In *Physical Processes in Red Giants*, ed. I. Iben, Jr., A. Renzini, p. 431. Dordrecht: Reidel
Renzini, A. 1989. See Torres-Peimbert 1989, p. 391
Renzini, A., Fusi Pecci, F. 1988. *Annu. Rev. Astron. Astrophys.* 26: 199
Renzini, A., Voli, M. 1981. *Astron. Astrophys.* 94: 175
Richer, H. B., Fahlman, G. G. 1988. *Ap. J.* 325: 218
Ritter, H. 1986. *Astron. Astrophys.* 169: 139
Ritter, H., Burkert, A. 1986. *Astron. Astrophys.* 158: 161
Robinson, E. L. 1976. *Annu. Rev. Astron. Astrophys.* 14: 119
Robinson, E. L., Shafter, A. W. 1987. *Ap. J.* 322: 296
Robinson, E. L., Shafter, A. W. 1989. See Wegner 1989b, p. 492
Romanishin, W., Angel, J. R. P. 1980. *Ap. J.* 235: 992
Saffer, R. A., Liebert, J. 1989. See Wegner 1989b, p. 408
Saffer, R. A., Liebert, J., Olszewski, E. 1988. *Ap. J.* 334: 947
Saio, H. 1986. See Hunger et al 1986, p. 425
Scalo, J. M. 1976. *Ap. J.* 206: 215
Schmidt-Voigt, M., Köppen, J. 1987. *Astron. Astrophys.* 174: 223
Schönberner, D. 1979. *Astron. Astrophys.* 79: 108
Schönberner, D. 1981. *Astron. Astrophys.* 103: 119
Schönberner, D. 1983. *Ap. J.* 272: 708
Schönberner, D. 1986a. *Astron. Astrophys.* 169: 189
Schönberner, D. 1986b. See Hunger et al. 1986, p. 471
Schönberner, D. 1987. In *Stellar Evolution and Dynamics in the Outer Halo of the Galaxy*, ed. M. Azzopardi, F. Matteucci, p. 519. München: ESO
Schönberner, D. 1988. In *Mass Outflows From Stars and Galactic Nuclei*, ed. L.

Bianchi, R. Gilmozzi, p. 137. Dordrecht: Kluwer
Schönberner, D. 1990. See Mennessier & Omont 1990. In press
Schönberner, D., Drilling, J. S. 1984. *Ap. J.* 278: 702
Schönberner, D., Weidemann, V. 1981. In *Physical Processes in Red Giants*, ed. I. Iben, Jr., A. Renzini, p. 463. Dordrecht: Reidel
Schönberner, D., Weidemann, V. 1983. In *Planetary Nebulae, IAU Symp. No. 103*, ed. D. R. Flower, p. 359. Dordrecht: Reidel
Schulz, H. 1977. *Astron. Astrophys.* 54: 315
Shaw, R. E. 1989. See Torres-Peimbert 1989, p. 473
Shipman, H. L. 1979. *Ap. J.* 228: 240
Shipman, H. L. 1989. See Wegner 1989b, p. 220
Shipman, H. L., Sass, C. A. 1980. *Ap. J.* 235: 177
Sion, E. M. 1986. *Publ. Astron. Soc. Pac.* 98: 821
Sion, E. M., Oswalt, T. D. 1988. *Ap. J.* 326: 249
Sion, E. M., Fritz, M. L., McMullin, J. P., Lallo, M. D. 1988. *Astron. J.* 96: 251
Solheim, J.-E. 1989. See Wegner 1989b, p. 446
Stauffer, J. R. 1987. *Astron. J.* 94: 996
Stecher, T. P., Maran, S. P., Gull, T. R., Aller, L. H., Savedoff, M. P. 1982. *Ap. J. Lett.* 262: L41
Strand, K. Aa., Kallerkal, V. V. 1989. See Wegner 1989b, p. 413
Sweeney, M. A. 1976. *Astron. Astrophys.* 49: 375
Sweigart, A. V., Gross, P. G. 1978. *Ap. J. Suppl.* 36: 405
Sweigart, A. V., Mengel, J. G., Demarque, P. 1974. *Astron. Astrophys.* 30: 13
Sweigart, A. V., Greggio, L., Renzini, A. 1989. *Ap. J. Suppl.* 69: 911
Szczerba, R. 1987. *Astron. Astrophys.* 181: 365
Taam, R. E., Kraft, R. P., Suntzeff, N. 1976. *Ap. J.* 207: 201
Thejll, P., Shipman, H. L. 1986. *Publ. Astron. Soc. Pac.* 98: 922
Thuan, T. X., Hart, M. H., Ostriker, J. P. 1975. *Ap. J.* 201: 756
Torres-Peimbert, S., ed. 1989. *Planetary Nebulae, Proc. IAU Symp. No. 131*. Dordrecht: Kluwer
Trams, N. R., Waters, L. B. F. M., Waelkens, C., Lamers, H. J. G. L. M., Van der Veen, W. E. C. J. 1989. *Astron. Astrophys.* 218: L1
Trimble, V., Greenstein, J. L. 1972. *Ap. J.* 177: 441
Tutukov, A. V., Yungelson, L. 1987. See Philip et al 1987, p. 435
Tylenda, R. 1984. *Astron. Astrophys.* 138: 317
Tylenda, R., Stasińska, G. 1989. *Astron. Astrophys.* 217: 209
Van der Veen, W. E. C. J. 1989. *Astron. Astrophys.* 210: 127
Volk, K., Kwok, S. 1989. *Ap. J.* 342: 345
Waelkens, C., Waters, L. B. F. M., Cassatella, A., LeBertre, T., Lamers, H. J. G. L. M. 1987. *Astron. Astrophys.* 181: L5
Walton, N. A., Pottasch, S. R., Reay, N. K., Taylor, A. R. 1988. *Astron. Astrophys.* 200: L21
Webbink, R. F. 1979. In *White Dwarfs and Variable Degenerate Stars*, ed. H. M. Van Horn, V. Weidemann, p. 426. Rochester, NY: Univ. Rochester
Webbink, R. F., Iben, I. Jr. 1987. See Philip et al 1987, p. 445
Wegner, G. 1973. *MNRAS* 165: 271
Wegner, G. 1974. *MNRAS* 166: 271
Wegner, G. 1978. *MNRAS* 187: 17
Wegner, G. 1979. *Astron. J.* 84: 650
Wegner, G. 1980. *Astron. J.* 85: 1255
Wegner, G. 1983. *Astron. J.* 94: 272
Wegner, G. 1989a. See Wegner 1989b, p. 401
Wegner, G., ed. 1989b. *White Dwarfs. Proc IAU Colloq. No. 114. Lect. Notes Phys.*, Vol. 328. Heidelberg: Springer-Verlag
Wegner, G., Reid, I. N. 1987. See Philip et al 1987, p. 649
Wegner, G., Reid, I. N., McMahan, R. K. 1989. See Wegner 1989b, p. 378
Weidemann, V. 1968. *Annu. Rev. Astron. Astrophys.* 6: 351
Weidemann, V. 1971. In *White Dwarfs, IAU Symp. No. 42*, ed. W. J. Luyten, p. 81. Dordrecht: Reidel
Weidemann, V. 1975. In *Problems in Stellar Atmospheres and Envelopes*, ed. B. Baschek, W. Kegel, G. Traving, p. 173. Heidelberg: Springer-Verlag
Weidemann, V. 1977a. *Astron. Astrophys.* 59: 411
Weidemann, V. 1977b. *Astron. Astrophys.* 61: L27
Weidemann, V. 1979. In *White Dwarfs and Variable Degenerate Stars*, ed. H. M. Van Horn, V. Weidemann, p. 206. Rochester, NY: Univ. Rochester
Weidemann, V. 1981. In *Effects of Mass Loss on Stellar Evolution*, ed. C. Chiosi, R. Stalio, p. 339. Dordrecht: Reidel
Weidemann, V. 1982. In *Binary and Multiple Stars as Tracers of Stellar Evolution*, ed. Z. Kopal, J. Rahe, p. 403. Dordrecht: Reidel
Weidemann, V. 1984. *Astron. Astrophys.* 134: L1
Weidemann, V. 1987a. *Astron. Astrophys.* 188: 74
Weidemann, V. 1987b. *Mem. Soc. Astron. Ital.* 58: 33

Weidemann, V. 1987c. See Philip et al 1987, p. 19
Weidemann, V. 1987d. See Kwok & Pottasch 1987, p. 347
Weidemann, V. 1988. In *A Decade of UV Astronomy With the IUE Satellite. ESA SP-281*, ed. E. Rolfe, 1: 17. Paris: ESA
Weidemann, V. 1989. *Astron. Astrophys.* 213: 155
Weidemann, V. 1990. In *Baryonic Dark Matter*, ed. G. Gilmore, D. Lynden-Bell. Dordrecht: Kluwer. In press
Weidemann, V., Koester, D. 1983. *Astron. Astrophys.* 121: 77
Weidemann, V., Koester, D. 1984. *Astron. Astrophys.* 132: 195
Weidemann, V., Yuan, J. W. 1989. See Wegner 1989b, p. 1
Werner, K., Heber, U., Hunger, K. 1989a. See Wegner 1989b, p. 194
Werner, K., Heber, U., Hunger, K. 1989b. In *Intrinsic Properties of Hot Luminous Stars. Proc. Boulder Conf.* San Francisco: Astron. Soc. Pac. In press
Wesemael, F., Green, R. F., Liebert, J. 1985. *Ap. J. Suppl.* 58: 379
Wickramasinghe, D. T., Reid, N. 1983. *MNRAS* 203: 887
Winget, D. E. 1988. In *Advances in Helio- and Asteroseismology, IAU Symp. No. 123*, ed. J. Christensen-Dalsgaard, S. Frandsen, p. 305. Dordrecht: Reidel
Winget, D. E., Hansen, C. J., Liebert, J., Van Horn, H. M., Fontaine, G., et al. 1987. *Ap. J. Lett.* 315: L77
Wood, M. A., Winget, D. E., Nather, R. E., Hessmann, F. V., Liebert, J., Kurtz, D. W. 1987. *Ap. J.* 313: 757
Wood, P. R., Cahn, J. 1977. *Ap. J.* 211: 499
Wood, P. R., Faulkner, D. J. 1986. *Ap. J.* 307: 659
Wood, P. R., Bessell, M. S., Fox, M. W. 1983. *Ap. J.* 272: 99
Wood, P. R., Bessell, M. S., Dopita, M. A. 1986. *Ap. J.* 311: 632
Wood, P. R., Meatheringham, S. J., Dopita, M. A., Morgan, D. H. 1987. *Ap. J.* 320: 178
Woolf, N. J. 1974. In *Late Stages of Stellar Evolution*, ed. R. J. Tayler, J. E. Hesser, p. 43. Dordrecht: Reidel
Yuan, J. W. 1987. Diplomarbeit. Univ. Kiel, Germ. Unpublished
Yuan, J. W. 1989. *Astron. Astrophys.* 224: 108
Zijlstra, A. A., Pottasch, S. R. 1989. *Astron. Astrophys.* 216: 245

COOLING OF WHITE DWARFS

Francesca D'Antona

Osservatorio Astronomico di Roma, I-00040 Monte Porzio, Italy

Italo Mazzitelli

Istituto di Astrofisica Spaziale CNR, CP 67, I-00044 Frascati, Italy

KEY WORDS: stellar structure, stellar evolution, degenerate matter, galactic age

1. INTRODUCTION

The problem of the surface luminosity and temperature evolution of white dwarfs (WDs), which cannot be completely decoupled from the problem of the surface chemical composition of WDs, consists of two main aspects:

1. the determination of the total energy content E of the WD;
2. the determination of the rate at which this energy is radiated away, which gives the total luminosity L.

The equation

$$L(t) = -\frac{dE(t)}{dt} \qquad 1.$$

defines the *cooling* of a WD. The terminology arose from the early recognition that by far the largest contribution to the energy radiated from the surface is due to the decrease in *thermal* energy E_{th} of the star (86, 133, 167). As most of the WD is isothermal, this can be stated, as a first approximation, by $L(t) = -(dE_{th}/dT_c)(dT_c/dt)$. The basic underlying picture is that the WD has a degenerate, nearly isothermal core at temperature T_c that gradually cools down as its residual heat escapes through the nondegenerate envelope. In this approximation, the small internal density adjustments during the cooling (and their influence on E_{th}) can be neglected, since it can be easily shown that when degeneracy is very large, the gravitational energy released by compression is completely absorbed to force the degenerate electrons into higher energy levels.

Thanks to the formal simplicity of this picture and to the long since well-understood thermodynamic properties of a fully degenerate ideal plasma, the basic theory of WD cooling was developed some 40 years ago in a work by Mestel (133), who found a simple power-law relation

$$t_{cool} \propto L^{-5/7} \qquad 2.$$

between the age and luminosity of a WD. It is worth noting that the most recent works on WD cooling, which take into account several more energy terms and give a careful treatment of the surface nondegenerate layers (75, 102, 126), lead to an amazing similarity of behavior (partly due to chance) with the simple Mestel theory given by Equation 2, *for most of the WD life*, as first realized by Iben & Tutukov (75). Nevertheless, if we wish to extract from WDs more than coarse information on their overall significance for the chemical and dynamical evolution of the Galaxy, even to the point of putting constraints on the age of the galactic disk, we must treat in more detail those phases in which the above cooling law does not well describe the WD evolution.

In the following, we continue to use the historical term *cooling* to indicate WD evolution, even if we account for *all* the main forms of energy release. Neutrino cooling (L_ν) was first included in the computations during the 1960s (e.g. 21, 157), as was the latent heat of crystallization [a further contribution to the release of thermal energy, L_{th} (104, 134, 183)]. The release of nuclear energy (L_{nuc}) has been considered more recently (75, 102, 126). Finally, gravitational energy release (L_g) from surface nondegenerate layers also is very important in both the early and the late stages (36), and thus we have the more complete equation

$$L(t) = L_{th} + L_g + L_{nuc} + L_\nu. \qquad 3.$$

In this review we consider *single* WDs, i.e. remnants of single-star stellar evolution; furthermore, most of the results given refer to WDs with mass $M \sim 0.6\ M_\odot$, the "typical" mass remnant of evolution. The problem of cooling in connection with accretion onto a WD from a companion in a binary system is not discussed. For general reference to the theory of cooling, the reader is referred to Cox & Giuli (29). An early theoretical review is that of Ostriker (145), while Liebert (112) has presented the status of observations. The problem of the relation between progenitor mass M_{in} and WD remnant mass is discussed by Weidemann (194) in this volume. In what follows, we first discuss (Section 2) the fundamentals of the formation of WDs, whose previous phases of evolution give rise to the initial chemical and physical stratifications. Here we provide an operational definition of pre-WD and WD stages. In Section 3 we describe the various terms contributing to the energy release from WDs and discuss at which stage each of these terms is dominant or important. Section 4 deals with heat transfer, mainly in the surface layers, and discusses its dependence on crucial parameters such as the opacity and the equation of state. In Section 5 we discuss cooling rates as a whole, and in Section 6 we review the problems of comparison of the WD cooling theory with the observed luminosity function. Finally, we conclude by summarizing the work that remains to be done (Section 7).

2. THE PRE–WHITE DWARF EVOLUTION

2.1 *Which Stars are White Dwarf Progenitors?*

WDs are the remnants of the evolution with mass loss of low- and intermediate-mass stars. The progenitors have a mass M such that

$$M_{\min} \leq M \leq M_{\mathrm{up}}, \qquad 4.$$

where $M_{\min}$ is constrained by

$$t_{\mathrm{ev}}(M_{\min}, Y, Z) \leq t_{\mathrm{sample}}. \qquad 5.$$

Here t_{ev} is the time spent prior to the WD phase by a star of given initial mass $M_{\min}$, helium content Y, and metal abundance Z; t_{sample} is the maximum age of the sample considered; and M_{up} is the maximum mass for a star that does not explode as a supernova (74). In the case of the nearby WDs, for instance, t_{sample} is the age of the disk, so that $0.8 \leq M_{\min}/M_\odot \leq 1.0$, depending on the chemical composition; for a young cluster, $M_{\min}$ can be much larger (up to several solar masses).

While $M_{\min}$ can be determined relatively well, the value of M_{up} not only is a function of the chemical composition, but also is still subject to debate because it is strongly affected by some not yet well-understood physical phenomena, such as convective overshooting in both core hydrogen and helium burning (9, 11), semiconvection, and helium spikes (19). The currently accepted values range between 7 and 9 $M_\odot$ when making conservative hypotheses about overshooting (7, 8), and they can be as low as 4–5 $M_\odot$ if large overshooting is included in the computations. Observations cannot help in deciding which is the case; for example, the existence of a WD in the Pleiades cluster is consistent both with $M_{\mathrm{up}} = 8\ M_\odot$, $M_{\mathrm{turnoff}} = 6\ M_\odot$ as derived without overshooting, and with $M_{\mathrm{up}} = 5\ M_\odot$, $M_{\mathrm{turnoff}} = 4\ M_\odot$ when overshooting is taken into account. Mass loss by wind and planetary nebula ejection (e.g. 152, 153) are further complications, since they influence M_{up} and also determine the relation between the initial mass of the star and the WD remnant mass. In addition, the physical mechanisms at the base of these processes are presently uncertain enough that it is better to rely on semiempirical formulations (125, 194).

It is useful to define here what we mean by WD and pre-WD. The latter term describes a star that has already departed from the asymptotic giant branch (AGB), since its H-rich envelope is very thin. It can no longer stay close to the Hayashi track, but there is still enough fuel to burn hydrogen via the CNO cycle and to ignite hydrogen-shell flashes, also at large T_{eff}. In contrast, a WD is a star in which CNO-cycle burning is definitely ruled out. A pre-WD of $\sim 0.6\,M_\odot$ has a hydrogen envelope of mass $\leq 10^{-3}\,M_\odot$ for solar chemical composition, and an even larger mass if the metal abundance or the core mass is lower.

2.2 *The Chemical Constitution of the Core*

If we consider WD remnants of single-star evolution and ignore brown dwarfs, which become degenerate before igniting hydrogen, three main cases seem to be possible for the inner chemical composition.

1. Some WD remnants of the evolution of low-mass stars ($<1\,M_\odot$) may have a helium core if mass loss has been so efficient as to stop the nuclear evolution before the helium flash (58). (This probably does not happen in the Galaxy.) The total mass of these WDs should be smaller than that required for degenerate helium ignition [$\sim 0.5\,M_\odot$ (124)].
2. In a few stars with mass close to M_{up}, carbon ignition may be able to remove the core degeneracy. Impulsive but not disruptive carbon burning follows, and the star forms an oxygen-neon-magnesium core of mass $\sim 1.2\,M_\odot$. Further evolution may lead to core collapse, but mass loss during the precollapse phase may leave an O-Ne-Mg WD. This is a possible outcome in single-star evolution (143), although it is more likely to occur in binaries (76, 111), and it is relevant to the formation of neutron star compact binaries (142). O-Ne-Mg WDs are known to be present in some nova systems (181, 199) based on observations of the ejecta.
3. In the vast majority of cases, WDs are remnants of the evolution following helium burning. Mass loss and/or a sudden ejection of the envelope can occur after a phase of helium thermal pulses in the AGB evolution (e.g. 74) or even at the second dredge-up stage (125). These stars are thus composed of carbon and oxygen.

We deal in the following with the evolution of carbon-oxygen WDs. Typical stratifications for these WDs can be found in (126-128), while Figure 1 (Section 2.3.1) shows the chemistry in the most external fraction of WDs having masses of 0.6 and $0.84\,M_\odot$. The inner part of the core of carbon-oxygen WDs emerges from the convective helium core burning.

The final $^{12}C/^{16}O$ ratio is primarily dependent on the ratio of the cross sections for the 3α and $^{12}C+\alpha$ processes. The latter reaction is favored at lower burning temperatures, so that more oxygen is produced during the core helium burning than during helium shell burning, which occurs at higher temperatures. For the same reason, the larger the WD progenitor mass, the larger is the carbon abundance in the WD core. According to the $^{12}C+\alpha$ cross section by Fowler et al (48), corrected by a factor of 2–3 according to Kettner et al (96), large oxygen abundances (70–80%) are expected in the vast majority of WDs, coming from stars in the mass range $1\ M_\odot \leq M \leq 3\ M_\odot$. The oxygen abundance can be even larger (80–90%) if convective overshooting, semiconvection, or helium spikes inject fresh ^{4}He into the core during the final helium-burning phases (126). For WDs produced from stars of larger mass, the central oxygen abundance is somewhat lower [$\sim$50% in WD remnants of 5-$M_\odot$ progenitors (127)]. Surrounding the remnant of the former convective core, there is a region left from steady thick-helium-shell burning, in which the $^{12}C/^{16}O$ ratio also is affected by the possible occurrence of large-scale mixing due to previous helium spikes (126). If these spikes indeed occur, the final oxygen abundance here is nearly the same as in the core. Inhibition of the helium spikes leads to a carbon mass fraction of about 50%. The carbon abundance left by the final phases of steady helium-shell burning is even larger, and the same is true for matter processed during the thermal pulse phase (see Figure 1), owing to the large temperature at which helium burns in a thin shell. Finally, in the He-rich region close to the surface left by the *last* helium thermal pulse, the abundance of carbon (mixed with helium) can be 10–30%, but little or no oxygen is produced (see Section 2.3.1).

2.3 *The Envelope Chemical Composition*

The envelope remnant from the preceding phases of evolution suffers, during the WD stage, a first important chemical rearrangement due to the powerful action of gravitational separation (42, 139, 149, 158, 187). This eventually leads to a simplified structure consisting of an almost pure hydrogen layer (if any) on top of an almost pure helium envelope on top of the carbon mantle. At the interfaces between these layers, the main constituents are not neatly separated, as gravitational sorting is counteracted by the large concentration gradients that produce ordinary diffusion. Thus, inside the He layer there is a tail of hydrogen from the top and a tail of carbon from the bottom.

While we refer the reader to the recent spectral classification by Sion et al (173), we recall that WDs are broadly divided into two main classes according to the presence (DAs) or absence (non-DAs) of an almost pure hydrogen atmosphere. The thickness of the H and/or He layers determines the ultimate evolution of WDs, since these layers act as an insulating blanket on top of the isothermal core, and the energy loss rate is a function of the opacity (i.e. of the chemical composition) of the blanket; consider also that only at $T_{\text{eff}} \geq 30,000$ K can radiative support allow metals to remain suspended in the atmosphere of WDs (e.g. 174, 184, 186, 187). At lower T_{eff}, gravitational settling of heavy elements frees the atmosphere from a metallic contribution to the opacity.

The onset of other mechanisms that can further perturb the monoelemental composition of the surface layers induced by gravitational settling depends on the thickness of the H and/or He layers. For instance, convective mixing occurs if the H and He layers are thin enough ($\leq 10^{-7} M_\odot$ and $\leq 10^{-4} M_\odot$, respectively) and leads to dilution of H in He or to the dredge-up of carbon from the core.

Finally, interstellar matter accretion (if any) can also help to alter the surface composition, but this process is not yet clearly understood.

2.3.1 EVOLUTIONARY CONSTRAINTS The theory of the latest phases of stellar evolution is, unfortunately, not refined enough (due to the unknown details of mass loss and planetary nebula ejection) to put very strong constraints on the envelope stratification (128). A first point to be considered is the following: Is the separation of WDs into DAs and non-DAs due to their pre-WD evolution? The theory provides many reasonable ways by which some pre-WDs can get rid of the hydrogen envelope:

1. Wind mass loss may be very important during the phases in which stationary helium burning, following a last helium thermal pulse in the pre-WD, completely supports the star luminosity, and the hydrogen shell is in the meantime turned off (68, 160, 206).
2. A late He thermal pulse, occurring at very large T_{eff}, may be responsible for mixing the flash-driven He convective shell with the H envelope (49, 67, 160), bringing hydrogen down to regions where it quickly burns (17, 70, 78).
3. A last hydrogen-shell flash occurring when the star has apparently already settled upon its cooling track (71) can in some cases be responsible for a temporary reexpansion to the AGB and a complete loss of the H-rich envelope due to wind (see Section 3.2).

From the preceding discussion it is easy to conclude that the hydrogen envelopes one should expect for DAs if their progenitors are hydrogen burners are only upper limits; in principle, the mass of the hydrogen envelope of H-rich WDs can be anywhere between $\sim 10^{-15} M_\odot$ [which is the lower limit required to provide the observed pure hydrogen spectrum (101)] and the *maximum* mass remnant from the previous evolution including hydrogen-burning [which is a function of the metallicity of the envelope (76) and of the total WD mass, ranging from $\sim 10^{-4}\,M_\odot$ for $0.6\,M_\odot$ to $\sim 4 \times 10^{-6}\,M_\odot$ for $1.2\,M_\odot$ (146)].

Considering now the thickness of the helium layers, let us define as the "buffer mass" (M_b) the mass contained in the pure helium region below the bottom of the hydrogen shell. Below this region there is a zone (of mass M_m) in which convection, at the peak of the last thermal pulse, has raised the carbon abundance up to ~ 10–30%, depending mainly on the number of pulses suffered by the progenitor. The mass of hydrogen processed between two consecutive thermal pulses (M_{tp}) ranges between 0.005 and $0.01\,M_\odot$; the latter value is the proper one for a total WD mass of $0.6\,M_\odot$. The minimum value of M_b is $\sim 10^{-7}\,M_\odot$ if the WD has formed at the end of a thermal pulse, since in this case convection has driven the C-enriched zone very close to the base of the hydrogen shell. The upper limit of M_b is M_{tp} if the formation of the WD closely precedes the onset of the next pulse. The value of M_m is again a function of both core mass and number of previous thermal pulses, and ranges between ~ 0.005 and $0.02\,M_\odot$. The latter value is most likely for a 0.6-$M_\odot$ WD that has suffered only a few thermal pulses on the AGB. In many thermally pulsing giants the "third dredge-up" may occur (e.g. 50, 65, 74): If surface convection sinks down into the carbon-rich region, M_b can be reduced to zero and M_m to half (or even less) of the values given above.

Thus, the helium envelope expected on top of a 0.6-$M_\odot$ WD at its birth is composed of a pure helium shell ranging between zero and $\sim 0.01\,M_\odot$ (depending on the interpulse phase at which the hydrogen-shell burning has ceased) and a deeper helium- and carbon-rich shell of $M_m \sim 0.01\,M_\odot$. Consideration of the ratio of the pulse to interpulse duration times should provide the statistical distribution of the He- and C-enriched layers expected in WDs.

Figure 1 presents the *initial* stratification of the envelope of two WDs: The upper part of the figure describes a typical Pop I $0.6\,M_\odot$ remnant of the evolution with mass loss of a 1-$M_\odot$ model star that has suffered only a few thermal pulses before the superwind phase; the lower part describes a 0.84-$M_\odot$ WD remnant of a 3-$M_\odot$ model star after a long series of thermal pulses (29, 123). The difference in the thicknesses of the intershell (including both the helium buffer and the He-C layer) between the two cases is evident.

Later in the evolution of the WDs, the chemical profiles will be modified by diffusion and gravitational settling. For WDs at $T_{\text{eff}} \sim 13,000$ K (where the carbon-rich spectral type DQ occurs), the structure of the WD should have a pure helium layer with mass $0.005\,M_\odot < M_{\text{He}} < 0.02\,M_\odot$ on top of the carbon-rich region. Notice, also, that despite the large cross sections

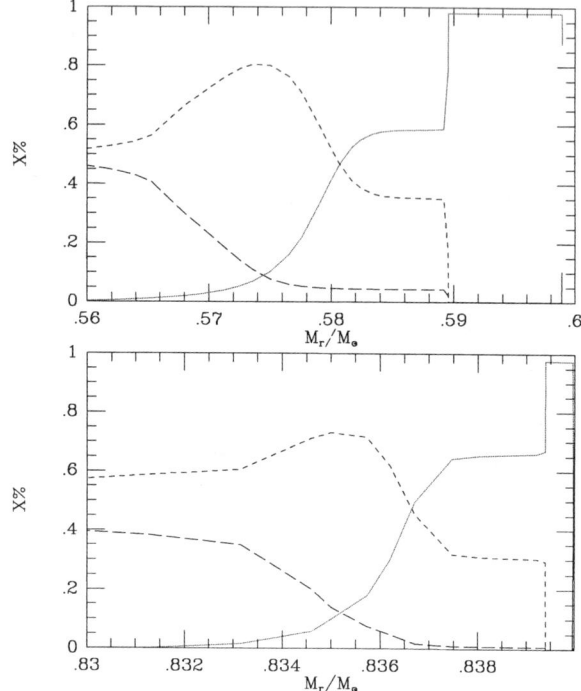

Figure 1 External initial chemical stratification for a $0.6\text{-}M_\odot$ remnant (*top*) of $1\text{-}M_\odot$ evolution (35) and for a $0.84\text{-}M_\odot$ remnant (*bottom*) of $3\text{-}M_\odot$ evolution (123). Dotted curve: ^{4}He; short-dashed curve: ^{12}C; long-dashed curve: ^{16}O.

adopted for the ^{12}C$+\alpha$ reaction, the abundance of oxygen below the helium is quite small (a few percent). We do not expect that oxygen will have much more chance of being "dredged up" in WDs than other metals. When the WD is very hot, the time scale for gravitational settling of carbon through the He-C-rich region is longer than the cooling time ($\sim 10^5$ yr): Outward chemical diffusion of carbon through M_b can thus lead to contact with the hydrogen envelope, leading to explosive burning (see Section 3.2).

A general warning must be made concerning the given values of the helium envelope mass. In fact, the third dredge-up is thought to be responsible for bringing carbon to the surface of red giants, but the observed location in the HR diagram of carbon stars (e.g. 1) agrees only marginally with the theory. Carbon stars appear at *lower* luminosity than is predicted by even the most recent models (10, 109, 110), unless semiconvection (72, 73) and convective overshooting are properly included (61), and they are not present at the large luminosities predicted by the earliest studies (66). As for the first problem (i.e. the lower luminosity of carbon stars compared with predictions), dredge-up for small core masses is probably more efficient than is provided for by the standard models, and thus the relation between helium mass and WD mass is probably somewhat different from what we believe. The second problem (i.e. the lack of carbon stars at large luminosities) probably results from a termination of the AGB phase (193, 195), for somewhat massive progenitor stars, that is early, even with respect to the predictions including reasonable rates of mass loss. In principle, this should affect mainly the relation between M_{in} and M_{WD}.

2.3.2 OBSERVATIONAL CONSTRAINTS Since the present status of the theory cannot give us unambiguous answers, comparison with the observations remains the most reliable means for further constraining the external stratification of WDs. Schönberner (161, 163) and Schönberner & Weidemann (166) identify with the sudden turnoff of the hydrogen burning shell an observed drop of the luminosity function of planetary nebula nuclei (PNN), and they conclude that most pre-WDs are indeed hydrogen burners. Furthermore, $\sim 35\%$ of PNN show extreme hydrogen deficiency (131), so that observations indicate a well-defined "primordial evolution" link between DAs (descending from hydrogen-burning PNN and thus having hydrogen envelopes of $M_H \sim 10^{-4} M_\odot$) and non-DAs (descending from hydrogen-poor PNN).

Unfortunately, other observational signatures of WDs do not agree with such a simple, clear-cut view. Pulsations in DA WDs (see Section 5.4) seem to imply hydrogen mass values of $M_H \leq 10^{-8} M_\odot$ (203; but see also 25). The strongest constraint from spectral evolution comes from the interpretation of the number-density ratios of DAs over non-DAs in different ranges of T_{eff}. Recent reviews of this problem are given in (29, 44, 170). The transition of several DAs to non-DAs at $T_{eff} \leq 8000$ K is generally accepted as statistically significant in the literature and attributed to the joining and full mixing of the H and He convective regions below this value of T_{eff}. The nearly complete disappearance of any hydrogen feature in the spectra requires large H dilution in He, and this is possible only if $M_H \leq 10^{-8} M_\odot$ in all the DAs that become non-DAs [$\sim 70\%$ (55)].

At larger luminosities and surface temperatures, another possible constraint lies in the transition DO → DA → DB if we ascribe it to the initial interplay of He sedimentation (DO → DA), followed by convective overshooting of the He below (DA → DB) (44, 118). Although this interpretation is more subject to doubt than the former one (101), it has some observational support (113). The resulting constraint is $M_H \sim 10^{-13}-10^{-14} M_\odot$ for the group of DAs that are turned into non-DAs at $T_{eff} = 30,000$ K.

It must be mentioned, however, that variations in the number ratios of different subgroups can be due not only to transitions of spectral types but also to differences in the cooling times of the subgroups themselves. This can be particularly important at large $T_{\rm eff}$ and at very low $T_{\rm eff}$.

A possible way to qualitatively reconcile the apparent discrepancies in the interpretation of observations is as follows: Hydrogen-shell burning is active during the first pre-WD phases, consistent with the observed drop in the PNN luminosity function. Then, at lower luminosities, diffusion-induced explosive hydrogen-burning (77; see Section 3.2) frees the star from most of its H-rich envelope, consistent with the requirements of the pulsation theory and with the observed spectral evolution. Possibly in contrast to this scheme, the presence of a burning shell is expected to produce pulsational instability among the PNN and hot DA WDs (93), but no hot DAs (60) are pulsating.

As for the helium envelopes masses, it has been shown (33, 185) that if $M_{\rm He} \leq 10^{-6}\ M_\odot$, then the helium layer of typical WDs becomes fully convective and merges with the convective carbon region beneath, producing C-dominated atmospheres; however, such atmospheres have never been detected. A more straightforward constraint for the He mass is given by the appearance of weak carbon features in a class of WDs (DQs) at $T_{\rm eff} \leq 13{,}000$ K (103, 188, 189). The observed relation between carbon abundance and $T_{\rm eff}$ can be achieved by the assumption that the carbon tail, in diffusive equilibrium, is dredged up into the He envelope when He convection in the external layers reaches its maximum depth. With present-day calculations of diffusion coefficients of carbon into helium (148), $M_{\rm He}$ is constrained at $\sim 2 \times 10^{-4}\ M_\odot$ in DQs (150) and, consequently, at $M_{\rm He} > 2 \times 10^{-4}\ M_\odot$ in spectral types without carbon.

We conclude that the *initial* chemical stratification of the WD envelope is *not* known in sufficient detail from stellar evolution theory or from the observations, not even in a statistical sense. This in turn leads to uncertainties in the evaluation of cooling times.

3. THE ENERGY CONTENT OF WHITE DWARFS

Figure 2 shows schematically the percentage of luminosity due to various energy contributions as a function of the total luminosity for a 0.564-$M_\odot$ WD having $M_{\rm H} \sim 4 \times 10^{-4}\ M_\odot$ (case B in ref. 36). We now discuss these terms in detail.

3.1 *The Gravitational Energy*

For historical reasons, it is commonly assumed that WDs begin their lives at an already constant radius. Actually, the radius at the beginning of the WD phase can be up to twice the zero-temperature fully degenerate radius,

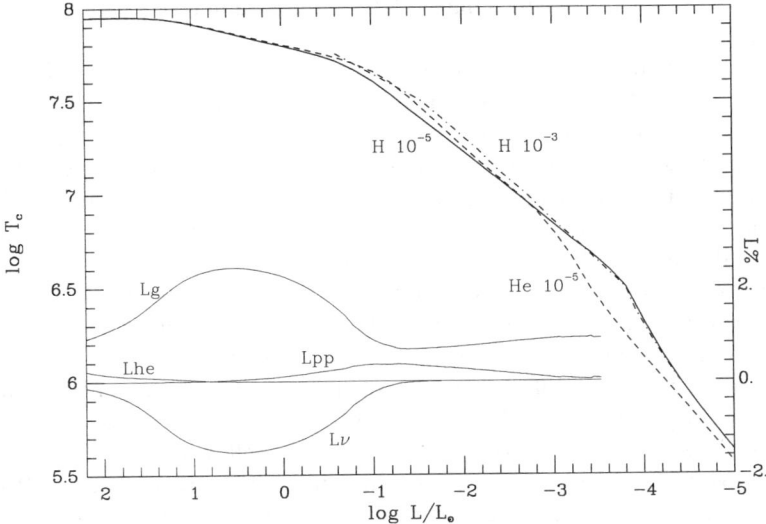

Figure 2 The percentages of total luminosity due to p-p burning (L_{pp}), He burning (L_{he}), neutrinos (L_ν), and gravitational plus thermal energy (L_g) are shown as a function of the total luminosity for the 0.564-$M_\odot$ remnant of a 1-$M_\odot$ model star having $M_H \sim 4 \times 10^{-4} \, M_\odot$. The run of central temperature (T_c) for this evolution is also given (solid curve). The dashed curve is T_c for the same structure, but with no hydrogen, and $Z = 10^{-5}$. The dash-dotted curve is T_c for a model of mass 0.6 $M_\odot$ by Tassoul et al (180) having $M_H = 10^{-10} \, M$.

and contraction is large just where degeneracy is low, mainly in the more external parts of the star; young WDs can even show contraction and a slight increase of temperature in the core. A proper treatment of WD evolution must then include release of gravitational energy at all stages. In particular, the gravitational contraction makes an important contribution to the energetics during the first evolutionary phases, whereas it becomes relatively less important during the phases of intermediate to low luminosity. Even if it is less obvious, its relative contribution grows again in the *final* phases. When Debye cooling has largely depleted the thermal energy content, the residual contraction of the thin subatmospheric layers, which are not yet fully degenerate, may provide up to 30% of the star luminosity output (36).

3.2 The Nuclear Energy

Nuclear energy generation by residual helium burning is important in the pre-WD phases, particularly if the star has suffered a last thermal pulse after leaving the AGB. By the time that $L \leq 100 \, L_\odot$, the helium shell quickly turns off. Its main influence is to extend the lifetime at large luminosity, where the winds may still be active and may lower the remnant hydrogen envelope.

Hydrogen burning is much more complicated: In particular, it must be recalled that proton-proton and CNO burning have very different temperature dependences. The hydrogen envelope is consumed in a pre-WD by CNO burning until the efficiency of the shell suddenly drops, whereupon the H-rich envelope rapidly contracts. As the pre-WD approaches its final, degenerate radius, the contraction of the external envelope slows down, and the surface temperature cannot increase any more. In these conditions, the H-rich envelope is already so thin ($\sim 10^{-4} M_\odot$) that the readjustment of the radiative gradients cannot maintain a sufficiently large temperature at the H-burning shell, and, owing to the high temperature dependence of the CNO cycle [$\varepsilon_{CNO} \sim T^{14-16}$ (75, 147)], hydrogen burning switches off. This occurs at $\log L/L_\odot \sim 3.2$ for $M = 0.6 M_\odot$. According to Schönberner (161, 165), the following rapid drop in luminosity is seen as a gap in the absolute magnitude distribution of PNN near $M_v = 5$. In the absence of mass loss, the hydrogen-envelope mass remains close to the upper limits discussed in the previous section, and at later stages, nuclear burning by proton-proton reactions may become the dominant source of energy support for the star [down to $\log L/L_\odot \sim -2.5-$ -3 (75)], providing up to 50% or more of the luminosity and depleting the hydrogen mass by a further factor of 2. The lower the metal abundance of the parent star, the larger is the proton-proton contribution, since M_H increases with decreasing Z (70).

Gravitational separation complicates the burning picture because the accumulation of CNO elements in the H shell can make CNO burning efficient for a somewhat longer time than when diffusion is neglected.

Stationary, diffusion-induced CNO hydrogen burning was first suggested (135) as a possible mechanism to reduce the H envelope of the evolutionary remnant from $M_H \sim 10^{-4} M_\odot$ to $M_H \lesssim 10^{-7} M_\odot$, as suggested by observations (Section 2.3.2). The situation is not so simple, however, because of the nonlinear response of the structure to hydrogen burning in a very thin shell. Iben & McDonald's (70) fully evolutionary computations, in fact, have shown that the ^{12}C- to ^{14}N-burning time scale becomes longer than the cooling time scale ($T_c/\dot{T}_c$) when not much hydrogen has been burnt (after about 10^8 yr of evolution). In the scenario suggested by Michaud & Fontaine (135), if the buffer helium mass between the carbon-oxygen core and the hydrogen layer is thin enough, the tail of the hydrogen layer diffusing inward meets the top of the carbon tail diffusing outward. In this case, Iben & McDonald (71) show that diffusion-induced hydrogen burning is *explosive*. A powerful hydrogen-shell flash, which may also occur several times, is ignited (self-induced nova). The flash may be responsible for depletion of the remnant H envelope by *orders of magnitude*. The depletion is not due to the burning itself, but rather to the fact that the surface of the star reaches AGB conditions, where presumably mass loss starts again.

From an observational point of view, a probable constraint to the theoretical framework discussed above is the following: As observations indicate the presence of mass loss during the PNN phase (20, 59) or even during the final pre-WD phase (115), proton-proton burning is probably never significant, since a depletion of M_H by a mere factor of 2 with respect to the upper theoretical limit is sufficient to severely inhibit it.

3.3 *Energy Loss by Neutrinos*

Neutrinos were first recognized to be a powerful sink of energy from the cores of giants and white dwarfs in the 1960s (21, 157). The energy losses due to neutrinos remain almost constant during the latest phases of AGB and PNN evolution and during the first phases of WD evolution. When CNO nuclear burning is no longer able to support the star, however, the surface luminosity of the star drops quickly, whereas the central temperature changes on a much longer time scale. This implies that the neutrino rates and the total neutrino luminosity also decrease more slowly than the surface luminosity. The ratio of neutrino losses to surface luminosity increases from a few percent up to a maximum of 200–300% when $\log L/L_\odot \sim 1.5$, and thus neutrino losses become the driving mechanism for cooling of the core. The neutrino luminosity then steadily decreases until, at $\log L/L_\odot \sim -1.5$ ($T_c \sim 3 \times 10^7$ K), it fades away (Figure 2).

Plasma neutrinos and photo-neutrinos (in the first stages) are the dominant energy sinks for WDs. Recombination neutrinos and pair neutrinos are negligible, and bremsstrahlung neutrinos may be relevant in the final stages (see Figure 5 in ref. 105). Until a few years ago, the formulae by De Zotti (37) and Lamb & Van Horn (105) for bremsstrahlung losses and by Beaudet et al (6) for the other neutrinos were usually adopted. Recent work by Munakata et al (140) provides updated and probably definitive or nearly definitive rates (thanks to the final settlement of the theory of the neutral currents) for the main neutrino contributions. It is worth noting that the new rates are on the average slightly lower than the previous ones, at least in the domain of WDs.

The end of neutrino-dominated evolution can be seen in the theoretical luminosity function as a change of slope around $\log L/L_\odot = -1.5$.

3.4 *The Gravitational and Thermal Energy*

When discussing stellar structure, it is useful to define as *gravitational energy* the term

$$\varepsilon_g = -C_p\dot{T} + E_p\dot{P}, \qquad 6.$$

in which both a *thermal* and a *compression* contribution, linked by the virial theorem, are present. In the case of WDs, the external layers are not completely degenerate and settled at constant radius until extremely low luminosities are reached, and thus the contribution of compression to the energy output of the star can be important. In the degenerate core, on the other hand, compression is negligible: In fact, (a) the radius is already relatively close to its final, zero-temperature value at the beginning of the WD phase, and (b) the virial theorem works in such a way that the released compression energy is almost completely absorbed by the consequent necessary increase of the kinetic energies of degenerate electrons (105, 133, 134). Accordingly, then, it is sufficient to discuss (at least for the core) only the variations of thermal energy, including of course the latent heat (if any) liberated during the phase transition (crystallization). Remember, however, that very young WDs can still contract even in the core and slightly increase their internal temperature.

3.4.1 CRYSTALLIZATION This interesting aspect of condensed-matter physics was placed on a firm foundation in the 1960s with works on one-component plasma (OCP) made of nuclei of charge Ze, mass M, number density ρ_n, and temperature T embedded in a neutralizing background of degenerate electrons (e.g. 79).

Coulomb interactions play a most important role here (2, 98, 156); they can be conveniently described by the Coulomb coupling constant Γ:

$$\Gamma = (Ze)^2/akT, \qquad 7.$$

where the radius a of the Wigner-Seitz sphere satisfies the relation $(4/3)\pi a^3 = 1/\rho_n$. High densities and low temperatures bring the nuclei closer and closer together, increasing their electrostatic interaction energies and decreasing their thermal agitation energies. At the same time, the screening effect due to the free electrons decreases, since fewer electrons are left in the energy range in which they can significantly contribute to screening.

For $\Gamma \ll 1$, electrostatic interactions are ineffective in producing correlations among the ions, and the plasma behaves like a gas. As Γ increases, the ions first attain short-range correlations (liquid) without the appearance of any symmetry in their spatial distribution. For even larger values of Γ, the ions eventually arrange themselves into a body-centered-cubic (bcc) (151) periodic lattice structure, which minimizes the total electrostatic interaction energy. This happens through a first-order phase transition as a new symmetry appears in the spatial distribution of ions within the plasma (107, p. 260; 183). Liquefaction does not give rise to a release of latent heat, whereas crystallization does.

A *lower* limit to Γ_{cryst} can be obtained by considering the melting point of a Coulomb solid, for which Mestel & Ruderman (134) found $\Gamma_{cryst} \sim 64$ using the semiempirical Lindemann melting-point formula. Monte Carlo simulations by Brush et al (15) gave $\Gamma_{cryst} \sim 126$. Later on, Hansen (57), from computations with a larger number of particles, obtained $\Gamma_{cryst} \sim 150$, a value that has been used until now in most relevant cooling computations. Recent models by Ogata & Ichimaru (144) give an even larger value ($\Gamma_{cryst} = 180 \pm 3$).

A further complication is the following: Ichimaru et al (80) show that the OCP first attains an amorphous glassy state at $171 \leq \Gamma \leq 210$, before the crystalline state is achieved. If this result is correct, it is not clear if the latent heat is to be released at $\Gamma = 171$, corresponding to the first appearance of microcrystals, or at $\Gamma = 210$, when crystallization becomes macroscopic; similarly, it is unclear at which value of Γ the specific heat per ion must be increased from $(3/2)k$ to $3k$. The possible evolutionary relevance of this transition stage has not yet been included in computations of WD cooling sequences. Further, the thermal conductivity in this stage is *smaller* than in the crystal phase. More recently, Iyetomi & Ichimaru (85) have shown, in the framework of a nonlinear approach to the crystallization of the OCP, that consideration of three-body correlations pushes the value of Γ_{cryst} upward, although their results are only preliminary owing to the instrinsic numerical complexity of the problem.

As an example of a crystallization sequence, Figure 3 schematically shows Γ and the luminosity profile inside several structures of one of the sequences computed in (36). To avoid numerical discontinuities, crystallization is assumed to occur smoothly over the range $145 \leq \Gamma \leq 165$.

The impact of crystallization on the cooling is twofold. First, the crystallization process causes the release of latent heat ($\sim kT$ per ion). While the gravitational energy released by the discontinuous change of density across the phase boundary ($\delta\rho/\rho$) is negligible (105), the latent heat may be more or less significant in the cooling, depending on the luminosity at which crystallization begins and on the rate of energy loss from the star versus its internal temperature.

Lamb & Van Horn (105) estimate that for a pure carbon 1-$M_\odot$ WD, the latent heat release increases the cooling times by $\sim 30\%$ between $L = 10^{-3} L_\odot$ and $L = 10^{-3.5} L_\odot$; Shaviv & Kovetz (169), on the other hand, find only a 5% increase in the cooling times between $L = 10^{-2.8} L_\odot$ (where crystallization sets in for their 0.6-$M_\odot$ star) and $L \sim 10^{-5} L_\odot$ (where the crystallization front reaches the surface). Broadly speaking, the difference between these two results is that crystallization (and release of

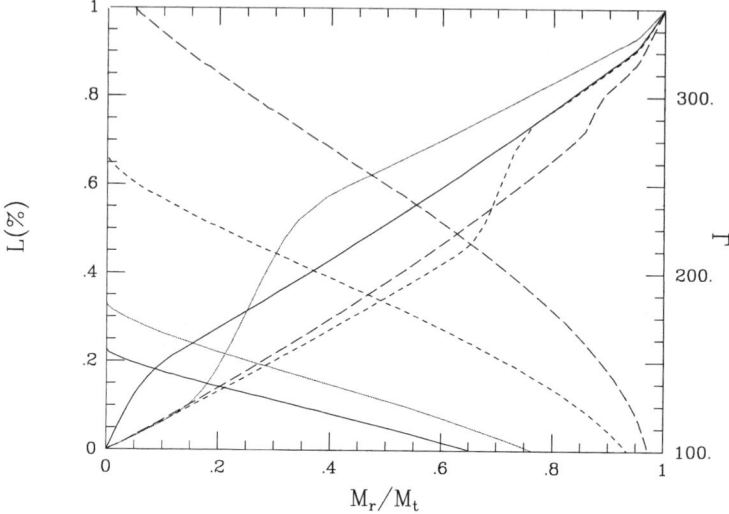

Figure 3 Run of luminosity L and Coulomb coupling constant Γ through four models of a 0.564-$M_\odot$ WD having $M_H = 10^{-8}$ $M_\odot$ (36) during crystallization. Values of (log $L/L_\odot$, log T_{eff}) are as follows: solid curve, (−3.34, 3.87); dotted curve, (−3.53, 3.82); short-dashed curve, (−3.66, 3.79); long-dashed curve, (−3.84, 3.75). The maximum output of luminosity due to latent heat release occurs for the first part of crystallization: In the last two models shown, the already-crystallized part of the structure has a larger specific heat [$3k$ vs $(3/2)k$].

latent heat) is spread in the latter case over a much wider range of luminosities and ages than in the former. This in turn happens because the central temperature in the computations by Shaviv & Kovetz decreases very slowly with respect to luminosity and age, whereas the results of Lamb & Van Horn (as well as more recent computations) show a sharper decrease of T_c with luminosity, so that the crystallization wave moves from the center to the surface in a narrow range of luminosities and ages.

The second effect of crystallization on cooling is that after crystallization occurs, the specific heats of the solid-state phase must be considered. Consequently, we have

$$C_v = 3k \times D(\theta_D/T), \qquad 8.$$

where D is the Debye function, and θ_D is the Debye temperature (107, p. 189), given by (183)

$$\theta_D = 1.74 \times 10^3 \rho_c^{1/2} \left(\frac{2Z}{A}\right). \qquad 9.$$

As the star crystallizes, then, the specific heat first increases up to twice its value in the fluid phase, and cooling tends to slow. [In fact, Figure 3 shows that the maximum percentage contribution of the latent heat release occurs when the inner parts of the structure crystallize, since most of the structure still has $C_v = (3/2)k$.] Then the star enters the Debye phase, the specific heat decreases with a power of the temperature larger than 1, and the thermal content gets smaller and smaller. The function $D(\theta_D/T)$ can be written as a power law:

$$D(\theta_D/T) \propto T^k. \qquad 10.$$

Fast Debye cooling ($k = 3$) is reached only for values of $\theta_D/T \sim 15$.

3.4.2 CHEMICAL SEPARATION OF OXYGEN VERSUS CARBON A further problem with crystallization should be mentioned: Up to this point, only the results from the OCP case have been discussed. Actually, according to Section 2.2, the interior of WDs is a mixture of at least two chemical elements. Apart from the problem of determining Γ_{cryst} for a binary mixture, which has not yet been convincingly solved, this kind of phase transition presents other uncertainties. In fact, while Shaviv & Kovetz (169) assumed that the solid phase would be a disordered alloy of the same composition as the liquid mixture, Stevenson (178) pointed out the possibility of a *eutectic* phase diagram of C-O mixtures, which would give rise to crystallization of oxygen, followed by gravitational separation and sinking of oxygen with respect to carbon, and then by crystallization of carbon. In this case, the gravitational energy release due to sinking of oxygen would slow down the cooling by a nonnegligible amount (136). More recently, careful Monte Carlo simulations seem to have ruled out this possibility (4, 81), restoring the former picture of almost complete miscibility of carbon and oxygen in the alloy phase. This point is discussed further in Section 6.2, focusing on the effects of separation on the luminosity function. Based on the preceding discussion, however, it should be clear that the problem of crystallization in WDs is far from being settled, and that new points of view can still arise.

4. THE TRANSFER OF ENERGY THROUGH THE OUTER ENVELOPE

4.1 *The Fundamental Ingredients*

The problems connected with the evaluation of the internal energy content of WDs, discussed in the previous section, are comparatively well solved compared with those of the energy transfer through the outer nondegenerate or partially degenerate envelope.

The outer structure of WDs is affected by a number of important physical parameters, the first being the *radiative opacity*. If the envelope is convectively unstable, the *equation of state* must also be carefully taken into account, since the temperature gradient in this case depends on the thermodynamic properties of the partially degenerate, partially ionized, strongly coupled gas. When degeneracy is present close to the surface, especially at low luminosities, *electron conduction* can also play a significant role.

This *apparently* simple picture is made complex not only by the intrinsic difficulties in treating a very complicated and not yet well-understood physical scenario, but also by the possible interplay between the three parameters discussed above and by other internal and external agents. Convection, in fact, not only depends on the chemical constitution of the external envelope but also can be responsible for modification of this constitution itself through mixing processes that may produce transitions of spectral type (hydrogen- versus helium-dominated atmospheres) and change the opacity, leading to deep variations in the structure of the envelope and in the cooling rate. Gravitational separation and diffusion are important in the first evolutionary stages in lowering the metal content and determining the atmospheric opacities. In turn, convection in later stages may or may not restore large metal abundances. Accretion of metals and hydrogen from the interstellar medium should be taken into account in the later stages, since it can be by far the most important mechanism in keeping the atmospheric opacities relatively large.

Starting from the center of the star and working toward the surface, we examine in the following sections the problems relative to electron conduction, to convection and the equation of state, and finally to radiative opacity.

4.2 *Electron Conduction*

The high degree of electron degeneracy makes heat transfer in the interior by free electrons highly efficient because the thermal conductivity inside WDs is several orders of magnitude larger than in copper or silver. Actually, when the degeneracy parameter η reaches order ~ 50, one can safely consider the rest of the star isothermal for any purpose. The first treatment of electron conduction appropriate for WDs was given by Mestel (132), who considered only electron-ion collisions. Actually, for dense matter, ion-ion (62) and even electron-electron (106) collisions play an important, and sometimes key, role. Hubbard & Lampe (63) have given a treatment for the nonrelativistic electron conduction, which also includes the effects of ion-ion and electron-electron collisions and of crystallization.

Their results, although formally valid only in the band of the (ρ, T) plane such that

$$10^{4.5}\rho^{0.25} \leq T \leq 10^{7.5}\rho^{0.25}, \qquad \qquad 11.$$

have been largely employed in computations of WD structure even outside their domain of applicability. The treatment of Hubbard & Lampe was extended to the relativistic case by Canuto (18), whose treatment was revised by H. M. Van Horn (see ref. 65).

A more recent treatment of thermal conduction, both in the liquid and solid phases, has been provided by Itoh et al (83, 84) with the aid of more updated input physics and the inclusion of relativistic effects. Computations now make use of the Itoh et al treatment, although its use should be confined to the large-degeneracy case. The new values of conductive opacities are about a factor of 2 larger than the previous values of Hubbard & Lampe (63) in the regions of interest (202, 208).

4.3 Convection and the Equation of State

4.3.1 THE EQUATION OF STATE

As noted in Section 4.1, the study of the external envelopes of WDs is far more complicated than that of the interior properties. In fact, in the outer layers the simplified treatment of matter as fully ionized and strongly degenerate no longer holds: We now must consider a partially ionized, partially degenerate plasma subject to strong Coulomb and short-range interactions. Thermodynamic properties in this complex physical regime were studied in the 1970s by several groups (e.g. 12, 31, 45, 99, 179). The most complete work on the subject, by Fontaine et al (46) and Magni & Mazzitelli (121), is based on the free-energy minimization technique. The main difference between these two studies is in the treatment of the internal partition functions of the bound states. Comparison between the two sets of results shows reasonable agreement, although the availability of better computing facilities in the ensuing decade would make it worthwhile to update these studies. An important by-product of these computations, in fact, could be a better description of the energy levels of bound states, preliminary to a determination of radiative opacity coefficients for a real gas.

As for the effect of a non-ideal-gas treatment of convection in WDs, the ionic and atomic interactions tend to make both molecular dissociation and ionization spread over a wider range of densities and temperatures than for an ideal gas. In turn, the decrease in values of the adiabatic gradient due to dissociation and ionization is less sharp than for an ideal gas. A correct thermodynamic treatment is expected to give larger values of the convective gradients in WDs, leading to larger differences between the central and surface temperatures.

4.3.2 EFFECTS OF CONVECTION Convection affects both the thermal structure and the surface chemistry of WDs, as follows:

1. It keeps at a fixed value the temperature gradient in the outer regions. As long as the radiative layers beneath the thin convection zone are not yet degenerate, local readjustments of the radiative gradients are sufficient for the structure to converge to "zero-radiative" conditions. As soon as degeneracy reaches the bottom of the convective region, however, this readjustment is no longer possible because electron conduction takes over, and the core temperature becomes strongly affected by the temperature profile in the convective region.
2. It may be responsible for the transition from one spectral type to another through the processes of mixing (transition from DA to non-DA), of undershooting (transition from DA to DB), and of dredge-up in competition with diffusion (DQ-type WDs presenting carbon features). Since the surface opacity is important for determining the cooling of WDs (see Section 4.4), this aspect of convection is also crucial for the cooling theories.

The two features discussed above—which cannot be studied separately, since they largely interact with one another—have been explored in papers by Fontaine & Van Horn (43) and D'Antona & Mazzitelli (33). Other envelope convection computations have addressed somewhat different problems, such as the study of acoustic fluxes from WDs (3) and the dependence of convective boundaries on the mixing-length theory adopted. In particular, very efficient convection [the so-called ML3 theory (e.g. 180)] is needed to match the blue edge of the DA and DB instability strips [see Section 5.3 (180, 203)], but the *maximum* extent of convection does not depend on the particular mixing-length theory adopted, so that at least this parameter does not influence the cooling times.

Table 1 shows the thickness in mass fraction of the convective envelopes (q_{ce}), for hydrogen and helium composition, as a function of T_{eff} [from Tassoul et al (180)]. The results presented here employ the ML3 treatment of convection. In addition, the upper boundary of convection is shown, generally coincident with the atmospheric mass fraction (q_{atm}). As for the modifications of chemical stratification due to mixing and the relations of these modifications to variations of spectral type, the complex variety of cases (170, 172) is too large to be even summarized here. We touch on this problem only to consider the implications for cooling, which are twofold: (*a*) Mixing produces variations in the surface opacities, which, as we have discussed, influence the cooling times; and (*b*) mixing gives us hints on the extension of the hydrogen and helium layers, which are not known to the required degree of precision from stellar evolution.

Table 1 Convection boundaries for a 0.6-$M_\odot$ WD[a]

	Helium		Hydrogen	
log $T_{\rm eff}$	log $q_{\rm ce}$	log $q_{\rm atm}$	log $q_{\rm ce}$	log $q_{\rm atm}$
4.82	−15.35	−15.34		
4.800	−15.15	−15.70		
4.775	−15.05	−15.75		
4.750	−15.00	−15.80		
4.700	−14.90	−15.70		
4.650	−14.80	−15.55		
4.600	−14.70	−15.55		
4.550	−14.10	−15.70		
4.500	−13.00	−15.85		
4.450	−12.30	−16.10		
4.400	−11.65	−16.30		
4.350	−11.00	−16.15		
4.300	−7.8	−15.85		
4.250	−6.85	−15.40		
4.225	−6.70	−15.10	−16.80	−16.80
4.200	−6.55	−14.95	−16.65	−17.10
4.175	−6.45	−14.80	−16.50	−17.20
4.150	−6.25	−14.60	−16.30	−17.20
4.125	−6.20	−14.40	−15.40	−17.10
4.100	−6.15	−14.20	−13.00	−16.95
4.075	−6.15	−14.10	−12.40	−16.70
4.050	−6.15	−14.00	−12.25	−16.45
4.025	−6.15	−13.90	−12.10	−16.20
4.000	−6.15	−13.85	−11.10	−16.05
3.975	−6.15	−13.80	−9.40	−15.90
3.950	−6.15	−13.75	−9.25	−15.70
3.925	−6.20	−13.70	−9.10	−15.55
3.900	−6.20	−13.65	−8.50	−15.30
3.850	−6.25	−13.50	−8.15	−15.00
3.800	−6.30	−13.40	−8.00	−14.75

[a] From Tassoul et al (180) and ML3 theory.

4.4 Radiative Opacity

White dwarfs are dense structures even in the atmospheric and subatmospheric layers, and thus the degree of overadiabaticity in convective regions is always negligible, or nearly so. Moreover, owing to the large efficiency of degenerate electrons in transferring heat, radiative opacities play a role only in the thin regions close to the surface where the plasma is neither convective nor largely degenerate. This statement is not meant to underestimate the importance of radiative heat transfer in WDs. To the contrary, owing to the isothermality of the core and to the nearly fixed temperature gradient in convective layers, the energy loss rate of WDs is mainly a function of the opacity of the radiative parts of the envelope and, if degeneracy reaches into the convective region, of the opacity in the optical atmosphere only.

Unfortunately, our present knowledge of the radiative opacities is largely insufficient for the purpose of studying the last cooling phases of WDs; this is probably the main source of uncertainty weighing on the theoretical results. In fact, cool WD atmospheres lie in the (ρ, T) region where both hydrogen and helium are recombined and negligibly contribute to the radiative opacity. Most spectra of cool WDs do not show metal lines. In the few objects for which metals are present (e.g. 100), it is generally believed that the metals have been recently accreted from interstellar clouds and are on the verge of disappearing owing to the action of gravitational settling. The scarcity of metals and the lack of other relevant opacity sources lead to very steep pressure gradients. In turn, this means relatively large values of the density at the base of the optical photosphere. No realistic computations of opacities are yet available in this high-density, low-temperature regime.

This problem was recognized early on by Böhm et al (13) in connection with pure helium envelope models (no metals), but it is also severe if the metals are as scarce as appears from observations ($Z = 10^{-5}$ seems an upper limit in most cases) and if Cox-type opacities (23, 24) give correct indications. Among recent tables collected in (196) computed down to low temperature by means of the Los Alamos opacity library (64), the opacities relative to a hydrogen-free composition of $Z = 10^{-3}$ are much larger than the corresponding Cox opacities. If these larger opacities are also confirmed at lower Z, the following discussion will become much less important for the main phases of cooling.

Figure 4 shows the region of the (ρ, T) plane in which radiative opacities are lacking and where electron conduction is not yet the dominant mechanism in heat transfer. We should mention that even at very low temperatures, electron conduction can take over only after pressure ionization, which sets in at $\log \rho \sim -0.3$ for hydrogen and at $\log \rho \sim 0.3$ for helium. Thus, the low-temperature radiative opacities are presently unknown, *in a crucial region for the determination of the cooling times*, for up to 5 dex in density.

To further clarify this point, we stress that even if cool WDs are convective through most of the region of partial ionization below the photosphere, they are radiative in the atmosphere, and the photospheric pressure at the top of the convective region can differ by *orders of magnitude* according to the different atmospheric opacities. On the other hand, in

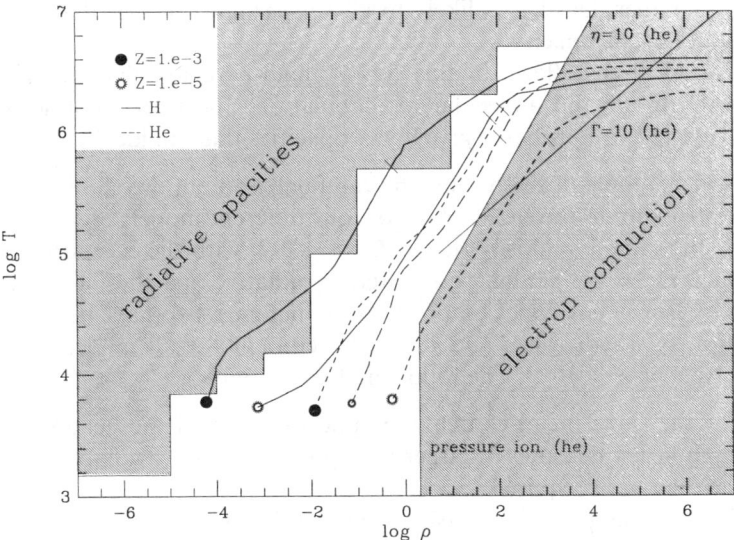

Figure 4 Shown in the (ρ, T) plane are the right boundary of known radiative opacities and the left boundary of the region in which electron conduction dominates for helium, composed by the vertical line of pressure ionization at log $\rho = 0.3$ and by the line at which the degeneracy parameter η is 10. (The corresponding lines for hydrogen are shifted respectively to log $\rho = -0.3$ and by -0.3 dex in ρ.) The line $\Gamma = 10$ for helium is also given, as are five WD models. Envelope models from (33) for a 0.6-$M_\odot$ WD have (log $L/L_\odot$, log T_{eff}) = -3.70, 3.80) and include the $Z = 10^{-3}$ models and the long-dashed model at $Z = 10^{-5}$. The other two models are from (36) for 0.564-$M_\odot$ WD evolution. For these, the He envelope structure is placed at (-3.68, 3.791), whereas the H envelope is much cooler at (-3.87, 3.73). This explains most of the difference in T_c with respect to the $Z = 10^{-3}$ structure.

the coolest convective envelopes, degeneracy reaches into the convective region, so that the gravitational structure—and the pressure—is there dictated by degeneracy. As a result, the lower the surface opacity, the larger is the pressure at the top of the convection region, and consequently the lower is the difference in pressure (and in temperature) between the base of the photosphere and the isothermal core. In other words, the same gravitational structure can correspond to different *thermal* structures as soon as degeneracy reaches into the convective envelope, and the thermal structure will depend on the *atmospheric* opacities.

Given the above warnings, the only way to compute WD cooling times is to extrapolate the values of radiative opacities from the available tables. However, at least two more warnings must be kept in mind:

1. An inspection of the behavior of log κ vs log ρ at constant temperature (see, for instance, 27) suggests that a linear extrapolation for increas-

ingly larger densities is likely to give *upper limits* rather than realistic values of the opacity.
2. Non-ideal-gas effects in this low-T, high-ρ region tend to increase both atomic and molecular recombination, and thus a non-ideal-gas treatment will probably give lower opacities than an ideal-gas one.

As an example of the results one can finally obtain, we show in Figure 4 the structure in the (ρ, T) plane of four different models of WDs, all at about the same luminosity (log $L/L_\odot \approx -3.7$) with the exception of one at log $L/L_\odot = -3.87$, but with different radiative opacities. Two models have hydrogen envelope composition (no helium), two have helium composition (no hydrogen), and the metal content for all models is either $Z = 10^{-3}$ or $Z = 10^{-5}$. The following features emerge:

1. For the same metal content, the photospheres of the helium models reach larger densities (a larger extrapolation of the available opacities has then been required).
2. The H envelope and He envelope models having $Z = 10^{-3}$ converge to central temperatures differing only by 0.04 dex.
3. The same is true for H envelope models of different Z when they have the same luminosity.
4. The two He envelope models *do not* have the same central temperature, and the $Z = 10^{-5}$ model is already degenerate into the convective envelope (e.g. 13).

Hydrogen envelopes with different metals present the same effect at lower T_{eff} (126). As can be seen from Figure 2, the smaller opacities of the He envelope WDs produce a sudden increase of slope in the relation between luminosity and T_c, starting at the point at which the degeneracy boundary reaches the convective layers (see also 180). In turn, this implies that when the core temperature starts to depend on the structure of the *optical* atmosphere, the further *history* of the WD will depend also on *external* parameters, for which it is difficult to say anything other than qualitatively.

We list two of the possible complications produced in the history of cooling:

1. If episodes of accretion of metals from the interstellar medium (such as described, for instance, in ref. 40) occur at low T_{eff}, the increase of opacity in the atmosphere will produce a noticeable variation of the surface parameters of the star. The central temperature of the WD, in fact, is *too low* for its luminosity, so the star will be forced to mimic the structure of a WD with lower luminosity and lower T_{eff}.
2. Another possibility is described in (35, 36). Suppose that the hydrogen

surface layer becomes completely mixed and diluted with helium at low T_{eff}: The core temperature of the helium envelope WD for the same T_{eff} would have been significantly lower. Adjustment of surface parameters will occur, followed by a long phase of cooling at almost constant luminosity, until the core temperature of the helium envelope model is reached.

It is clear, thus, that the solution of these problems requires (*a*) computations of opacities in the unknown region, which must be done by the appropriate equation of state for highly non-ideal-gas conditions; and (*b*) computations of cooling that include the time evolution of accretion of hydrogen and metals from interstellar clouds and their diffusion through the outer envelope.

In addition to the large effect due to the different opacities of helium and hydrogen, which we have described, minor effects on T_c result from the different thicknesses of the helium buffer, as described by Tassoul et al (180). Carbon opacity is larger than helium opacity. Thus, when degeneracy crosses the boundary between helium and carbon, the L-T_c relation will increase its slope. The L-T_c relations for different M_{He} will split each other at the luminosity where this effect occurs. The rule of thumb here is the following: The smaller the helium buffer, the larger will be T_c for a given luminosity, and the longer will be the cooling time.

4.5 *Reliability of Low-Z He-Envelope Models*

The He envelope model with $Z = 10^{-5}$ presented in Figure 4 has very large densities at the photosphere, due to the extrapolation chosen for the opacities in (36). For purposes of comparison, we plot another model—taken from the envelope computations in (33)—having the same "nominal" Z (long-dashed curve in Figure 4). The radiative opacities for this model were extrapolated linearly, and the conductive opacities (63) were also included in the region preceding pressure ionization (where there are essentially no free electrons, and thus where this source of opacity should not have been included). The photospheric density is lower by an order of magnitude, and the central temperature is closer to that of the H envelope model. This comparison stresses our ignorance in this field.

A possible line of attack in determining the reliability of the models predicting the largest densities is to examine whether the spectra of cool WDs give indications about the correct pressure stratification in the atmosphere. Kapranidis (87) and Kapranidis & Böhm (88) computed model atmospheres for pure helium composition, employing an approximate Thomas-Fermi model for the equation of state, and derived densities similar to ours (36). A synthetic spectrum based on these model atmospheres of a WD with one of the lowest known luminosities [LP 701-29 (28)] provides a not unreasonable fit to the spectrum at $T_{\rm eff} = 4500$ K (89), although the presence of both neutral and ionized calcium is not reproduced owing to the difficulties in treating pressure ionization. Still, much work is needed in treating the atmospheres of cool WDs, particularly He-dominated atmospheres, since one needs to know both the equation of state and the self-consistent degrees of ionization that determine the radiative and conductive opacities.

5. THE COOLING TIMES

5.1 General Features

In discussing cooling, we are going to regard as marginal the possible influence of proton-proton nuclear burning on the evolutionary time scales. We make this choice because although this mechanism cannot be ruled out on purely theoretical grounds, it likely does not occur in typical WDs. In fact, we have seen in Section 3.2 that a 50% reduction of the hydrogen envelope with respect to the maximum evolutionary remnant is sufficient to prevent proton-proton burning, and there are several mechanisms operating that can easily provide this reduction, such as diffusion-induced prolongation of CNO burning (70) or just a bit of mass loss [observationally revealed (20, 59)] in the pre-WD.

In Table 2 we summarize several computations of cooling for a $M \simeq 0.6$-$M_\odot$ WD. All of the ages given refer to structures that have emerged from previous evolution, and thus the stratification of carbon and oxygen is that resulting from 3α and $^{12}C+\alpha$ burning. In lieu of the previous statements, the IT84 times of the H-burning sequence given in Table 2 must be considered mainly as upper limits to the age of typical WDs. Also, the DM89 models include p-p burning (see Figure 2) but at a somewhat lower level (despite the larger $M_{\rm H}$) due to the smaller mass.

The above-discussed properties of WD interiors and envelopes provide a natural separation of WD evolution into five main stages:

1. luminous WDs ($L/L_\odot \geq 10^{-1.5}$);
2. fluid cooling ($10^{-1.5} \geq L/L_\odot \geq 10^{-3}$);
3. choice of destiny, depending on external envelope conditions ($10^{-3} \geq L/L_\odot \geq 10^{-4}$);
4. crystallization and possible chemical separation;
5. Debye cooling.

We note, however, that at least stages 3 and 4 can partly overlap. These phases are now summarized based on available model results that have appeared in the literature in recent years.

Table 2 Evolutionary times

log $L/L_\odot$	log (model times, in years)				
	IT84[a]	KS86[b]	DM89[c]	IT84[d]	DM89[e]
2.000	3.873		4.507	4.814	4.423
1.800	4.073		4.819	4.938	4.788
1.600	4.729		5.104	5.068	5.080
1.400	5.149		5.389	5.218	5.376
1.200	5.360		5.642	5.373	5.633
1.000	5.541	5.766	5.852	5.544	5.829
0.800	5.687	5.926	6.019	5.712	6.000
0.600	5.849	6.070	6.170	5.865	6.149
0.400	6.009	6.205	6.309	6.019	6.279
0.200	6.161	6.341	6.433	6.178	6.410
0.000	6.318	6.469	6.556	6.343	6.525
−0.200	6.478	6.598	6.690	6.507	6.641
−0.400	6.656	6.736	6.826	6.660	6.765
−0.600	6.869	6.896	6.982	6.812	6.892
−0.800	7.106	7.080	7.173	6.976	7.035
−1.000	7.422	7.285	7.398	7.144	7.204
−1.200	7.730	7.500	7.602	7.395	7.412
−1.400	7.988	7.705	7.819	7.669	7.648
−1.600	8.179	7.894	7.990	7.901	7.857
−1.800	8.337	8.072	8.143	8.073	8.041
−2.000	8.465	8.231	8.279	8.243	8.202
−2.200	8.596	8.377	8.412	8.411	8.334
−2.400	8.730	8.516	8.536	8.578	8.462
−2.600	8.858	8.645	8.658	8.705	8.585
−2.800	8.978	8.773	8.774	8.832	8.718
−3.000	9.093	8.894	8.891	8.961	8.877
−3.200	9.209	9.015	9.008	9.097	9.050
−3.400	9.352	9.136	9.124	9.266	9.204
−3.600	9.534	9.257	9.279	9.425	9.302
−3.800	9.668	9.377	9.431	9.571	9.370
−4.000	9.756	9.498	9.615	9.705	9.426
−4.200	9.831	9.619	9.716	9.807	9.476
−4.400	9.905		9.774	9.892	9.526
−4.600	9.986		9.821	9.970	9.575
−4.800	10.050		9.857	10.042	9.615
−5.000			9.886		9.652
−5.200			9.911		9.679
−5.400			9.930		9.701
−5.600			9.946		9.719
−5.800			9.961		9.737

[a] From Iben & Tutukov 1984 (75). Parameters: $M = 0.6\ M_\odot$, $M_{He} = 0.026\ M_\odot$, $M_H = 1.5 \times 10^{-4}\ M_\odot$.
[b] From Koester & Schönberner 1986 (102). Parameters: $M = 0.598\ M_\odot$, $M_{He} = 0.014\ M_\odot$, $M_H = 9.4 \times 10^{-5}\ M_\odot$.
[c] From D'Antona & Mazzitelli 1989 (36). Parameters: $M = 0.564\ M_\odot$, $M_{He} = 0.021\ M_\odot$, $M_H = 4 \times 10^{-4}\ M_\odot$.
[d] From Iben & Tutukov 1984 (75). Parameters: $M = 0.6\ M_\odot$, $M_{He} = 0.016\ M_\odot$, $M_H = 0$.
[e] From D'Antona & Mazzitelli 1989 (36). Parameters: $M = 0.564\ M_\odot$, $M_{He} = 0.022\ M_\odot$, $M_H = 0$.

5.2 Cooling During the Five Stages

During stage 1, cooling is a very complicated function of the initial status of the pre-WD, since it depends on the pulse-interpulse phase at which the WD is formed, on how many thermal pulses the star has suffered before the planetary nebula ejection (which determines the gravitational energy available from the helium layer), and on the chemical stratification (how much hydrogen is left). Evolution is also affected by several not well-understood physical features (mass loss, radiative levitation of the elements, gravitational settling and diffusion) acting on this multi-parametric structure. This phase, as we have seen, can also determine the survival of the hydrogen envelope, if diffusion-induced explosive hydrogen burning occurs. Furthermore, the evolution in T_{eff} is sensitive to the *radius* of the star, which depends crucially on the chemical stratification (e.g. 36, 102).

We refer the reader to the many detailed descriptions of the possible evolutionary paths (68, 70, 71, 160, 162, 164, 165). The relevant point is that as soon as neutrino cooling begins to dominate the structure, it also acts to powerfully reset the previous evolution and provides a great uniformity of all the emerging *thermal* structures at $\log L/L_\odot \sim -1.5$. Even the time of evolution down to this luminosity can be considered well known: ~ 6–8×10^7 yr for typical WDs. [Proton-proton burning may increase the age to $\sim 1.2 \times 10^8$ yr (75; see Table 2).] During this stage, a number of different classes of WDs appear, and it is tempting to try to determine their evolution from number ratios in different temperature intervals (e.g. 198). When doing this exercise, we must always keep in mind that this part of the evolution is still much too elusive to allow firm conclusions.

Stage 2 is the best understood evolutionary phase in the WD life: The energy source is mainly the loss of thermal energy from the Coulomb fluid, which is not yet very strongly coupled; in addition, transfer of energy through the envelope is controlled primarily by the relatively large buffer of nondegenerate matter, so that convection (which is already massive in helium envelopes and is just beginning in hydrogen envelopes) does not yet influence the structural properties. Moreover, the order of magnitude of opacities is dominated by helium or hydrogen contributions, and a detailed knowledge of the metal content is not necessary.

The works of different researchers can be compared at this stage in order to avoid the difficulties present at other stages. The cooling time has reached ~ 0.75–1×10^9 yr at $\log L/L_\odot = -3$.

In this phase, we find the ZZ Ceti instability strip, and thus this is the best place to look for confirmation of the cooling rates by the period derivative of these pulsators, as is being done for one of these variables (94, 95). Ultimately, this search may provide information on the core temperature as a function of T_{eff} and, thus, on the opacity (and chemical stratification) of the buffer layer (see Section 4.4).

Stage 3 is characterized by the appearance of new differences in the evolution of WDs, and these begin to affect the relationship between the central temperature and the luminosity as described in Section 4.4. The exact location of the splitting of the L-T_c relations depends on the chemical stratification of the envelope. The most important splitting occurs between the helium envelope and the hydrogen envelope WDs: It appears as a "break" in the relation, and its location depends on the metal content (or, better, on the opacities) assumed in the atmosphere. Figure 2 shows that the break found in the models of D'Antona & Mazzitelli (36), which adopt the *smallest available* opacities ($Z = 10^{-5}$, $Y = 1-Z$), occurs as early as $\log L/L_\odot \sim -3$ for helium envelopes, and at $\log L/L_\odot \sim -3.8$ for hydrogen envelopes. For comparison, we show in Figure 2 the relation from a model by Tassoul et al (180), which refers to a 0.6-$M_\odot$ WD with $M_H/M = 10^{-10}$: It also shows the break occurring at $\log L/L_\odot \sim -3.8$.

The exact location of this break determines the ultimate destiny of the WD: When it appears, the star must lose a large fraction of its thermal content in a narrow luminosity range, and thus it prolongs the evolutionary times at the corresponding luminosity. As a consequence, the WD is left with a lower thermal content to feed the surface luminosity at later stages. Obviously, the larger the luminosity at which the break occurs, the shorter is the corresponding time scale for the drop in T_c, and the shorter also are the cooling times at lower luminosity.

Depending on what happens at stage 3, stage 4 can have either an early or a late onset. An early appearance of the break brings the star into crystallization conditions *earlier* and at a larger luminosity, with *cumulative* "savings" of several billions of years in the total evolutionary times down to Debye cooling. For example, in the case of the H atmosphere model of Figure 2, crystallization is reached *before* the break, whereas for the He atmosphere models with $Z = 10^{-5}$ it occurs *after* the break. Another critical parameter for the occurrence of stage 4 is the core composition ($^{12}C/^{16}O$ ratio), which determines the crystallization temperature. The value of Γ_{cryst} is also important. If $\Gamma_{\text{cryst}} \sim 150$ for a 0.6-$M_\odot$ WD, the crystallization begins at $10^{-3} L_\odot$ for the helium envelope, lowest opacity models (36, 126), which have $^{16}O \sim 80\%$ in the core; at $\log L/L_\odot = -3.3$ for the hydrogen envelope model and the same core composition; but only at $\log L/L_\odot \sim -3.6$ for a 0.6-$M_\odot$ WD with a carbon core (180).

This range of luminosities brackets the main uncertainties, at least for the adopted value of Γ_{cryst}.

All of these effects are cumulative: The total cooling time down to log $L/L_\odot = -4.5$ is $\sim 8.7 \times 10^9$ yr for carbon core, medium-opacity models (205) and $\sim 3.5 \times 10^9$ yr for oxygen-dominated core, lowest opacity models (36). These differences *increase* toward fainter luminosities.

Debye cooling (stage 5) has been studied recently by D'Antona & Mazzitelli (36) for models with very low surface opacities. These models show that the thermal content of the WD is progressively depleted as the central temperature decreases (Figure 5). Interestingly enough, even in this case, it was impossible to reach a "fast" Debye phase ($k = 3$) for the whole star because the most external layers still preserved a θ_D/T ratio that was not large enough for the WD to suddenly cool to "invisibility." Furthermore, residual gravitational contraction in the models at $L \sim 10^{-6} L_\odot$ provides $\sim 30\%$ of the luminosity output. It should also be noted that except for very peculiar and unpredictable events in the shape of the L-T_c relations (30, 36), the Debye phase is not expected to give rise to a sudden disappearance of WDs, but rather to a slow, steady decrease in the theoretical luminosity function (see Section 6.3).

Summing up the uncertainties arising from stages 3–5, it is clear that our present knowledge of the input physics is insufficient to determine

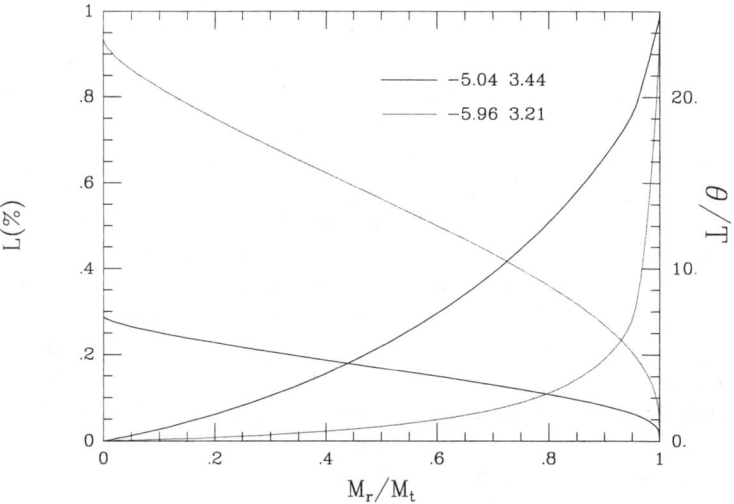

Figure 5 Luminosity (rising curves) and ratio of Debye temperature to temperature (falling curves) are plotted for two models of a 0.564-$M_\odot$ WD after crystallization. Values of (log $L/L_\odot$, log $T_{\rm eff}$) are indicated for each model toward the top of the graph. In the lower luminosity model, the outer 10% of the structure provides more than 80% of the surface luminosity because θ_D/T in this region is still low enough that fast Debye cooling has not been reached, and because gravitational contraction still releases energy at the surface.

theoretically if the cooling times of WDs down to $\log L/L_\odot = -4.5$ (a critical luminosity from the observational point of view) are definitely shorter or definitely longer than the age of the galactic disk.

Of course, orders of magnitude larger low-temperature opacities (196) would in any case give rise to much longer cooling times than presently predicted according to extrapolations of low-temperature Cox opacities.

5.3 Mass Dependence of Cooling

Complete, homogeneous sets of evolutionary computations as a function of the total WD mass, including the correct chemical stratification arising from the evolution of the proper parent mass, do not yet exist in the literature. In addition to the studies relative to a 0.6-$M_\odot$ WD by Iben & Tutukov (75) and Koester & Schönberner (102), the following work has been carried out: Iben & Tutukov (77) have studied the evolution of helium WDs of different-mass ($\leq 0.5\ M_\odot$) remnants of binary evolution. A full evolutionary computation for a 0.84-$M_\odot$ remnant from a 3-$M_\odot$ star, in which the star evolves through a long series of thermal pulses (123), is partially described in D'Antona (29). Furthermore, evolutionary computations have been made for a 0.68-$M_\odot$ WD with three different choices for the envelope composition (126) and for a 0.564-$M_\odot$ remnant of the evolution of a 1-$M_\odot$ star (36). All of these results are so scattered with regard to input physics, chemical composition, and evolutionary history that they do not allow useful comparisons. It is thus necessary to resort to the results of nonevolutionary computations.

In this framework, the most extensive and up-to-date results are those by Winget et al (205) for pure carbon WDs of mass $0.4\ M_\odot \leq M \leq 1.0\ M_\odot$. These models do not take into account that the evolutionary remnants will have chemical stratifications that directly depend on the initial mass of the progenitor through the M_{in}-M_{wd} relation, but they do give some hints of at least the general behavior of the evolution.

As the total mass increases, so too does the evolutionary time at $\log L/L_\odot \geq -3$ owing to the larger thermal content. For large masses, however, crystallization and Debye effects set in earlier due to the larger densities. At low luminosities, then, the larger the mass, the shorter becomes the evolutionary time (up to differences of 4–5 Gyr at $\log L/L_\odot \sim -5$).

The effects due to the presence of oxygen in the core are discussed in (202, 207, 208). In principle, the thermal content decreases and crystallization and Debye cooling commence earlier as the oxygen abundance increases, so that the total cooling times are shorter. In this context, since WDs of larger mass come from parent stars of larger mass too, they should have *lower* oxygen abundances (see 126, and Section 2.2): Taking this effect into account, the differences in ages at low L would be reduced. We warn, however, that qualitative arguments like the ones just discussed, or even quantitative linear analyses like those of Winget & Van Horn (202), have more heuristic value than predictive value, since stars (even WDs) do not behave like linear systems.

5.4 Pulsations and Cooling

By now three main instability strips have been identified in WDs, namely (*a*) the GW Vir or DOV (22) at $T_{\text{eff}} \geq 10^5$ K, (*b*) the DBV (200) at 29,000 K $\geq T_{\text{eff}} \geq$ 24,000 K (119), and (*c*) the DAV or ZZ Ceti class (129, 130) at 13,000 K $\geq T_{\text{eff}} \geq$ 11,400 K (47, 54). The driving mechanism of pulsations in DBVs and DAVs is the κ mechanism, which operates at the inner edge of the partial ionization zone of helium (203) and hydrogen (38, 203), respectively. For DOVs the situation is more complex, and there can again be a κ mechanism operating, owing to the ionization of carbon and oxygen (176; however, see Section 2.3.1 on the expected thickness of the helium envelope), or instead an ε mechanism due to the presence of a still active shell-burning source (93, 171). The latter mechanism, however, predicts periods that are a factor of ~ 3 shorter than the observed ones.

Secular period changes of individual objects are associated with variations of the stellar structure due to evolution and thus could in principle constitute a powerful tool for investigating cooling (e.g. 201).

For DAVs and DBVs, the driving of the instability occurs mainly at the base of the convection zone, exciting pulsations with time scales close to the thermal time scale t_{th} of the envelope (26). As the WD cools, the base of convection reaches deeper into the envelope, which increases t_{th} in the driving region and excites progressively longer period modes. Thus, $P/\dot{P}$ is expected to be positive for these two classes, and it increases as the cooling rate decreases. Kawaler et al (92) expect $P/\dot{P} \sim 1$–6×10^8 yr for DBVs, and Winget gives $P/\dot{P}$ of several gigayears for DAVs (201). As DAVs fall in the best understood cooling region, determination of $P/\dot{P}$ could be a powerful check of the structure in these phases (see also 180). Kepler et al (95) give $P/\dot{P} \geq 8.2 \pm 5.0 \times 10^8$ yr for the DAV G117-B15A, and further observations should provide the value of this derivative with more precision.

In DOVs, the situation is less clear. Winget et al (204) measured $P/\dot{P} = (-1.4 \pm 0.1) \times 10^6$ yr for PG 1159−035. Here models predict a *positive* value of this same order of magnitude (91), which has been attributed to plasma-neutrino cooling of the core, whereas a negative sign would result only for modes trapped near the surface and more sensitive to the residual contraction of the envelope. A positive period derivative is found also from an analysis (92) of the fully evolutionary helium envelope sequences of Iben & Tutukov (75). The models by D'Antona & Mazzitelli (36) were recently analyzed by S. D. Kawaler (private communication), and they also give a positive $P/\dot{P}$. For the latter models, the radius contraction time scale is $R/\dot{R} = -5 \times 10^5\text{--}-2.5 \times 10^6$ yr in the range $5.1 \geq T_{\text{eff}} \geq 4.97$ (see also 102), where PG 1159−035 is found. This residual contraction not only affects the most external layers but also the core, which shows a slight increase in temperature until neutrino cooling begins to dominate, but only at log $L/L_\odot \leq 1.6$ (Figure 2). Thus, the positive $P/\dot{P}$ value obtained from the models cannot be simply attributed to neutrino cooling but instead requires a more complex explanation.

One last warning: We have seen that cooling at several stages depends on very different physical mechanisms. Even if we succeed in tuning the cooling of DAVs at $T_{\text{eff}} \sim 12{,}000$ K by means of stellar seismology, we cannot gain too much insight on the "choice of destiny" of WDs (Section 5.2), since this depends on the outer envelope opacity at lower T_{eff}.

5.5 Some By-Products of Cooling

The problem of determining whether or not there exist WDs at zero temperature not only is of interest in understanding whether WDs can be used as signatures of the age of the galactic disk, but also has several other applications:

1. Do WDs make up any part of the dark matter in the Galaxy? If there are no cool WDs, the answer is obviously no, but if WDs cool in a few billion years, this is possible. In any case, if a constant or slowly changing rate of star formation is assumed, a cooling time of ~ 5 Gyr instead of ~ 10 Gyr can at most double the space density of WDs if we count the possibly invisible population (34). Of course, having invisible WDs becomes more interesting in the scenario of bimodal star formation (108; but see also 209).
2. There is still no completely satisfactory theory of nova outbursts (e.g. 168). If WDs cool very quickly, it is possible that the WD in a nova system becomes very cold ($T_c \leq 10^6$ K) *before* mass transfer from the companion begins. How the nova explosion would be influenced by having accretion on this kind of object remains to be investigated. Qualitatively, it may be inferred that more violent outbursts would be obtained.

6. THE LUMINOSITY FUNCTION

6.1 Observations

The main test of the WD cooling theory lies in its comparison with the observed luminosity function (LF) (192), which in principle could also provide information on the history of star formation in the Galaxy (141, 159) and on the local mass density in the galactic disk.

Since the review on WDs by Liebert (112), and apart from the two small (although statistically significant) samples (39, 82), the current state-of-the-art WD LF has been achieved through work using the Palomar-Green (PG) survey and based on a statistically complete sample of 353 DAs (41) and of DBs and DOs (119, 198). Liebert et al (117) have joined, at low luminosities, the resulting LF—down to $M_v \leq 12.75$—to the one resulting from proper-motion-selected cool WDs from the LHS catalogue (120).

This work (117) also transforms the magnitudes into bolometric luminosities. Uncertainties in the bolometric corrections and different binning of the data provide somewhat different shapes of the LF at the cool end, as shown in Figures 3a–d in (117). Additional uncertainty—no larger than 0.2 dex, according to J. Liebert (private communication, 1989)—also arises from the relative normalization of the PG sample (magnitude limited) with the LHS sample (proper motion limited).

This determination of the LF may not necessarily be the final one: It will be interesting to see whether the proper-motion survey undertaken by Ruiz et al (154, 155), which has a limiting magnitude of $V = 21$ compared with $V = 19$ for the LHS sample, will change the situation at the low-luminosity end. Nevertheless, the theory presently faces the problem of fitting the features of the LF presented in (117). The main features of this LF are the following:

1. A steady increase, more or less following Mestel's cooling-law predictions, is observed, but one that is more articulated. Particular features are (a) a decrease in the slope at $\log L/L_\odot = -1.5$ in the PG sample, and (b) a possible further flattening near the junction between the PG and the LHS samples.
2. The most striking characteristic, however, is the abrupt falloff of the LF around $-4.6 \leq \log L/L_\odot \leq -4.2$, depending on the binning and bolometric corrections adopted.

Although over the last decade a "red deficit" for WDs has been suggested more than once (53, 116), the results in (117) compel us to consider the falloff as an observational point, not just as a lower limit. Thus, the interpretation of this falloff must be found either in the cooling theory itself or in the interplay between cooling and the history of star formation in the galactic disk.

6.2 *The Age of the Disk From the WD Luminosity Function*

D'Antona & Mazzitelli (32) have shown how, for different cooling laws, the finite age of the galactic disk should lead to a more or less marked deficiency of WDs at low luminosity. Winget et al (205), adopting their own updated cooling times, derived an age for the disk of 9.3 ± 2 Gyr. Although Winget et al did not account precisely for the finite lifetime in the main sequence—as was done by Iben & Laughlin (69) and Yuan (209)—this influences only the *morphology* of the LF, whereas the age of the disk is mainly a function of the dropoff luminosity L_{drop} and is the same so long as the cooling times adopted are the same. Iben & Laughlin (69) give the following approximate expression:

$$t_{\text{disk}} \sim 0.9 \times 10^{10} \times (10^{-4.7}/L_{\text{drop}})^{0.28}, \qquad 12.$$

which is based on Winget et al's (205) cooling times. In another work also based on the same cooling times, Noh & Scalo (141) find $t_{\text{disk}} \sim 7$ Gyr: In fact, the observational LF they chose to fit (e.g. Figure 6) presents the dropoff at a larger luminosity (log $L/L_\odot = -4.2$).

While the dropoff location depends on the cooling of average-mass WDs, more massive WDs cool faster, and at log $L <$ log L_{drop} we should

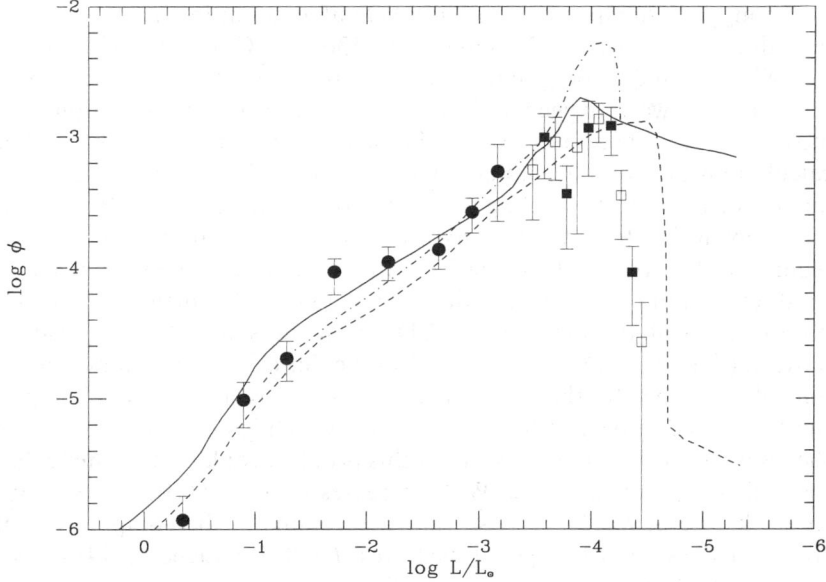

Figure 6 The observational luminosity function is plotted, as are data from Figure 3c (dots and filled squares) and 3d (dots and open squares) of (117). Superimposed are three models that *do not* fit the LF. Solid curve: model from (36), with an H envelope sequence of 0.564 $M_\odot$ and no limitation on disk age; dashed curve: from (205) for $t_{\text{disk}} = 9 \times 10^9$ yr, as given in (117); dash-dotted curve: from (141), with a star formation rate based on the ages of disk stars (182), and with t_{cool} from (205) increased to give the dropoff at log $L/L_\odot = -4.2$.

expect at least a few massive WDs (69, 205). Furthermore, if the age of the halo is substantially greater than that of the disk, a secondary peak due to halo WDs could be present (137).

A relatively low disk age, such as 7–9 Gyr, is perhaps in conflict with other indicators of the disk, namely the isochrone ages (e.g. 56, 182) and the chromospheric age distribution (5). Furthermore (unless there is something fundamentally wrong with our understanding of stellar evolution), galactic globular clusters have ages of $\sim 17 \pm 4$ Gyr according to all evolutionary parameters (16), and current models of the chemical evolution of the Galaxy state that the age gap between the halo and disk cannot be greater than 1–1.5 Gyr (122), although some data seem to imply a large spread of ages [up to 6 Gyr among globular clusters (14, 97, 177)].

The problem is controversial and touches, eventually, on the question of the age of the Universe. It would be interesting to turn the problem around and ask whether we could fit the LF with cooling times that give longer ages at $\log L/L_\odot = -4.2$, so that the estimated age of the disk might become ~ 12–14 Gyr. To do that, we need longer cooling times than those used by Winget and his coworkers.

Among the attempts to extend the WD lifetimes at low L is the line of investigation followed by Mochkovitch (136) and Garcia-Berro et al (51, 52), which includes the gravitational energy release due to the chemical separation between oxygen and carbon at the time of crystallization. The most recent results do not predict this separation (see Section 3.4.2), but calculations are complicated enough that maybe there is still room for a modest increase in the cooling times due to this effect. Garcia-Berro et al have shown that for a binary mixture (half oxygen, half carbon), partial separation fits the falloff of the LF with a disk age of 9.5 Gyr, whereas total separation fits it even at a disk age of 13 Gyr. Unfortunately, a bare prolongation of the lifetimes of WDs from the stage of crystallization onward (that is, at $\log L/L_\odot \leq -3.2$ in the Garcia-Berro et al models) is not sufficient to solve the problem, since in this case a prominent bump in the LF should appear at $\log L/L_\odot \sim -4$, which is not accounted for in the observational data. A way out of this problem could be to assume that the scale height of the oldest WDs increases from ~ 200 pc at $t = 5$ Gyr to ~ 1400 pc at $t = 12$ Gyr, so that the volume in the computation of space densities is much larger at the lowest L (69, 75). Liebert (114) shows, however, that there is no indication of such a large amplification of scale height because the tangential velocities of the dimmest WDs are no different from those of the brightest ones.

Another way of approaching the problem of the scale height is followed by Noh & Scalo (141), who simply take the star formation rate history of the solar neighborhood from the *observed* age distributions derived by

Barry (5) and Twarog (182) for the main sequence. By this assumption, the change of slope at log $L/L_\odot = -1.5$ is attributed to a recent burst of star formation $\sim 3 \times 10^8$ yr ago (see also 69).

Even if the age scales are uncertain by 20–30%, some stars (although not many) have ages definitely larger than the very small disk age required to fit the falloff, so that the corresponding WDs of the same age produce a flat LF. Noh & Scalo (141) then repeat the exercise of increasing the cooling times, as predicted by Garcia-Berro et al (51, 52), and find again an expected bump in the LF at log $L/L_\odot \sim -4$. In this case, the bump *cannot* be lowered by appealing to different space volumes for the oldest WDs because the star formation rate *of the solar neighborhood* has been used, and it implicitly takes account of these possible variations.

Actually, a good fit to the present observational LF is difficult to achieve with standard models because we are taking the observational data too seriously: It is possible that a relative shift is necessary at the junction between the PG sample and the proper-motion-limited sample, which increases by a factor of ~ 2 the low-luminosity space densities and provides a better fit to models of prolonged cooling at log $L/L_\odot \sim -4$. A recent rediscussion of the fit of the LF is given by Yuan (209), who concludes that ages of 6–14 Gyr can fit the data well. The main warning of the preceding discussion is that if we take seriously the dropoff of the LF at log $L/L_\odot \sim -4.2$–-4.6, we must also take seriously the shape of the LF implied by the points at *larger* luminosities, where the observational selection effects are less severe.

6.3 *Fitting the LF by Models That Reach Debye Cooling*

In principle, one can also try to fit the LF with models implying very fast cooling (30, 34, 36). The models by D'Antona & Mazzitelli (36), which show early crystallization and early onset of Debye cooling, provide a very good fit to the observational LF at not-too-low luminosities. In this case, the flattening of the LF at log $L/L_\odot \leq -3$ is not interpreted to be the result of the evolutionary ages prior to WD, but rather as the necessary acceleration of cooling rates in the Debye regime. These models, in any case, predict a flat or slowly decreasing LF in the region in which the dropoff is found, and, as such, they are no better than the previous ones in explaining the dropoff. We conclude this discussion by summarizing the main points:

1. If "relatively long" cooling times [such as those used by Winget et al (205)] apply to WDs, the dropoff of the LF requires a disk age $t_{\text{disk}} \sim 7$–9×10^9 yr according to Equation 12, perhaps in conflict with other evolutionary signatures.

2. If we adopt the hypothesis that *longer* cooling times are correct (this could result either from the separation mechanism in the core or from a new opacity computation), we can fit the LF with $t_{disk} \sim 12 \times 10^9$ yr, but the theory then also predicts a nonobserved excess of WDs at $-3 \leq \log L/L_\odot \leq -4$.
3. If very short cooling times are adopted (36), the general shape of the LF is well reproduced, but the space density expected at the dropoff luminosity is about a factor of 10 larger than the observed one.

7. CONCLUSION

We summarize here our main conclusions:

1. Uncertainties in the stellar evolution theory of the pre-WD phases, particularly in what concerns mass loss and the role of late hydrogen flashes, do not allow a reliable knowledge of cooling times for very young WDs, down to $\log L/L_\odot \sim 1$.
2. At $1 \geq \log L/L_\odot \geq -3.2$ the previous uncertainties are reset by the WD evolution, and the input physics is sufficiently well understood, allowing reliable determinations of the cooling times.
3. At $\log L/L_\odot \leq -3.2$ some physical inputs are still badly missing. We are not yet in a position to claim that WDs cool to $\log L/L_\odot \sim -4.5$ on a time scale either longer or shorter than the age of the galactic disk. Further, we also do not know the age of the disk.

Much work remains to be done, both observationally and theoretically, before the WD luminosity function can be used as a galactic chronometer. Observations and their interpretations through model atmospheres will have the following important roles:

1. For luminous WDs, observations of PNN and hot WDs will continue to provide at least statistical information about the thickness of H and He layers.
2. The definition of the luminosity function at $-3 \geq \log L/L_\odot \geq -5$ must be improved, both by searching for faint WDs in deeper surveys and by studying the correlations between bolometric magnitudes and visual or infrared magnitudes. Model atmospheres and synthetic spectra for cool WDs are needed, along the lines of, for example (89, 138, 190, 191).
3. For faint WDs, more information is needed on the surface chemical abundances, particularly for what concerns metals and trace hydrogen, to help in solving the problem of interstellar accretion. This problem still seems to be far from the grasp of a reliable theory. Observations will also give us valuable information on the *physical* stratifications in

optical atmospheres, which can be compared with the predictions of structural models.

The theory needs improvement in these other fields:

1. We require a better understanding of gravitational sedimentation and (dynamic?) mass loss.
2. We have to determine the internal $^{12}C/^{16}O$ profiles, which depend both on the reaction rates and on the occurrence of mixing processes at the border of convective burning cores.
3. We need further progress in the theory of crystallization of binary mixtures, and we must update the evolutionary computations by adopting the correct value of Γ_{cryst}.
4. It is necessary to extend the computations of radiative opacities in the non-ideal-gas region at $T \leq 3 \times 10^4$ K and $\rho \geq 10^{-5}$ g cm^{-3}; this is necessary both for the interior models and for the model atmospheres.
5. We must establish the role of the interplay between interstellar accretion, gravitational sedimentation, and convective mixing. This interplay, when electron degeneracy reaches into the convective region, can be the most important mechanism in determining the surface chemistry and consequently for the opacity and cooling times.

In our opinion, this last task is the most difficult one. Before a unique, self-consistent theory is able to provide reliable results for the last cooling phases, ample room will exist for parametric theories, tuned in an ad hoc manner to fit the observations.

ACKNOWLEDGMENTS

We thank all our colleagues who kindly sent us reprints and preprints, in particular Gilles Fontaine and G. Vauclair. We also warmly thank Jim Liebert for providing useful comments, Steve Kawaler for assisting us in the process of understanding the derivatives of pulsation periods, Mario Vietri for a careful reading of the manuscript, Raffaele Gratton for helping trace back many references, and Matt Wood for pointing out some misprints in the tables. Finally, we are in debt to Icko Iben, who helped both scientifically and stylistically in the revision of the manuscript.

Literature Cited

1. Aaronson, M., Mould, J. 1985. *Ap. J.* 228: 551
2. Abrikosov, A. A. 1960. *Sov. Phys. JETP* 12: 1254
3. Arcoragi, J.-P., Fontaine, G. 1980. *Ap. J.* 242: 1208
4. Barrat, J. L., Hansen, J. P., Mochkovitch, R. 1988. *Astron. Astrophys.* 199: L15
5. Barry, D. C. 1988. *Ap. J.* 334: 436
6. Beaudet, G., Petrosian, V., Salpeter, E. E. 1967. *Ap. J.* 150: 979

7. Becker, S. A., Iben, I. Jr. 1979. *Ap. J.* 232: 831
8. Becker, S. A., Iben, I. Jr. 1980. *Ap. J.* 237: 111
9. Bertelli, G., Bressan, A. G., Chiosi, C. 1985. *Astron. Astrophys.* 150: 33
10. Boothroyd, A. I., Sackmann, I.-J. 1988. *Ap. J.* 328: 671
11. Bressan, A. G., Bertelli, G., Chiosi, C. 1986. *Mem. Soc. Astron. Ital.* 57: 411
12. Böhm, K. H. 1970. *Ap. J.* 162: 919
13. Böhm, K. H., Carson, T. R., Fontaine, G., Van Horn, H. M. 1977. *Ap. J.* 217: 521
14. Bolte, M. 1989. *Astron. J.* 97: 1688
15. Brush, S. G., Sahlin, H. L., Teller, E. 1966. *J. Chem. Phys.* 45: 2102
16. Buonanno, R., Corsi, C. E., Fusi Pecci, F. 1989. *Astron. Astrophys.* 216: 80
17. Caloi, V. 1990. *Astron. Astrophys.* In press
18. Canuto, V. 1970. *Ap. J.* 159: 641
19. Castellani, V., Chieffi, A., Pulone, L., Tornambè, A. 1985. *Ap. J. Lett.* 294: L31
20. Cerruti Sola, M., Perinotto, M. 1985. *Ap. J.* 291: 237
21. Chin, C. W., Chiu, H.-Y., Stothers, R. 1966. *Ann. Phys.* 39: 280
22. Cox, A. N. 1986. In *Highlights of Astronomy*, ed. J. P. Swings, 7: 229. Dordrecht: Reidel
23. Cox, A. N., Stewart, J. N. 1970. *Ap. J. Suppl.* 19: 243
24. Cox, A. N., Tabor, J. E. 1976. *Ap. J. Suppl.* 31: 271
25. Cox, A. N., Starrfield, S. G., Kidman, R. B., Pesnell, W. D. 1987. *Ap. J.* 317: 303
26. Cox, J. P. 1980. *Theory of Stellar Pulsation*. Princeton, NJ: Princeton Univ. Press
27. Cox, J. P., Giuli, R. T. 1965. *Principles of Stellar Structure*. New York: Gordon & Breach
28. Dahn, C. C., Hintzen, P. M., Liebert, J. W., Stockman, H. S., Spinrad, H. 1978. *Ap. J.* 219: 979
29. D'Antona, F. 1987. *Mem. Soc. Astron. Ital.* 58: 123
30. D'Antona, F. 1989. See Ref. 187a, p. 44
31. D'Antona, F., Mazzitelli, I. 1975. *Astron. Astrophys.* 44: 253
32. D'Antona, F., Mazzitelli, I. 1978. *Astron. Astrophys.* 66: 453
33. D'Antona, F., Mazzitelli, I. 1979. *Astron. Astrophys.* 74: 161
34. D'Antona, F., Mazzitelli, I. 1986. *Astron. Astrophys.* 162: 80
35. D'Antona, F., Mazzitelli, I. 1987. See Ref. 150a, p. 635
36. D'Antona, F., Mazzitelli, I. 1989. *Ap. J.* 347: 934
37. De Zotti, G. 1972. *Mem. Soc. Astron. Ital.* 43: 89
38. Dolez, N., Vauclair, G. 1981. *Astron. Astrophys.* 102: 375
39. Downes, R. A. 1986. *Ap. J. Suppl.* 61: 569
40. Dupuis, J., Pelletier, C., Fontaine, G., Wesemael, F. 1987. See Ref. 150a, p. 657
41. Fleming, T. A., Liebert, J., Green, R. F. 1986. *Ap. J.* 308: 176
42. Fontaine, G., Michaud, G. 1979. *Ap. J.* 231: 826
43. Fontaine, G., Van Horn, H. M. 1976. *Ap. J. Suppl.* 31: 467
44. Fontaine, G., Wesemael, F. 1987. See Ref. 150a, p. 319
45. Fontaine, G., Van Horn, H. M., Böhm, K. H., Grenfell, J. C. 1976. *Ap. J.* 193: 205
46. Fontaine, G., Graboske, H. C., Van Horn, H. M. 1977. *Ap. J. Suppl.* 35: 293
47. Fontaine, G., McGraw, J. T., Dearborn, D. S. P., Gustafson, J. 1982. *Ap. J.* 258: 651
48. Fowler, W. A., Caughlan, G. R., Zimmerman, B. A. 1975. *Annu. Rev. Astron. Astrophys.* 13: 69
49. Fujimoto, M. Y. 1977. *Publ. Astron. Soc. Jpn.* 29: 331
50. Fujimoto, M. Y., Sugimoto, D. 1979. *Publ. Astron. Soc. Jpn.* 31: 1
51. Garcia-Berro, E., Hernanz, M., Isern, J., Mochkovitch, R. 1988. *Astron. Astrophys.* 193: 141
52. Garcia-Berro, E., Hernanz, M., Isern, J., Mochkovitch, R. 1988. *Nature* 333: 644
53. Greenstein, J. L. 1979. *Ap. J.* 227: 244
54. Greenstein, J. L. 1982. *Ap. J.* 258: 661
55. Greenstein, J. L. 1986. *Ap. J.* 304: 334
56. Grenon, M. 1988. In *Evolutionary Phenomena in Galaxies*, ed. J. E. Beckman, B. E. Pagel, p. 29. Dordrecht: Kluwer
57. Hansen, J. P. 1973. *Phys. Rev. A* 8: 3096
58. Harpaz, A., Kovetz, A., Shaviv, G. 1987. *Ap. J.* 323: 154
59. Heap, S. 1979. In *Mass Loss and Evolution of O-Type Stars, IAU Symp. No. 83*, ed. P. S. Conti, C. W. H. de Loore, p. 99. Dordrecht: Reidel
60. Hine, B. P., Nather, E. 1987. See Ref. 150a, p. 619
61. Hollowell, D., Iben, I. Jr. 1989. *Ap. J.* 340: 966
62. Hubbard, W. B. 1966. *Ap. J.* 146: 858
63. Hubbard, W. B., Lampe, M. 1969. *Ap. J. Suppl.* 18: 297

64. Hübner, W. F., Merts, A. L., Magee, N. H. Jr., Argo, M. F. 1977. *Rep. LA-6760-M*, Los Alamos Natl. Lab., N. Mex.
65. Iben, I. Jr. 1975. *Ap. J.* 196: 525
66. Iben, I. Jr. 1981. *Ap. J.* 246: 278
67. Iben, I. Jr. 1982. *Ap. J.* 260: 821
68. Iben, I. Jr. 1984. *Ap. J.* 277: 333
69. Iben, I. Jr., Laughlin, G. 1989. *Ap. J.* 341: 430
70. Iben, I. Jr., McDonald, J. 1985. *Ap. J.* 296: 540
71. Iben, I. Jr., McDonald, J. 1986. *Ap. J.* 301: 164
72. Iben, I. Jr., Renzini, A. 1982. *Ap. J. Lett.* 259: L79
73. Iben, I. Jr., Renzini, A. 1982. *Ap. J. Lett.* 263: L23
74. Iben, I. Jr., Renzini, A. 1983. *Annu. Rev. Astron. Astrophys.* 21: 271
75. Iben, I. Jr., Tutukov, A. N. 1984. *Ap. J.* 282: 615
76. Iben, I. Jr., Tutukov, A. N. 1985. *Ap. J. Suppl.* 58: 661
77. Iben, I. Jr., Tutukov, A. N. 1986. *Ap. J.* 311: 742
78. Iben, I. Jr., Kaler, J. B., Truran, J. W., Renzini, A. 1983. *Ap. J.* 264: 605
79. Ichimaru, S. 1982. *Rev. Mod. Phys.* 54: 1017
80. Ichimaru, S., Iyetomi, H., Mitake, S. 1983. *Ap. J.* 265: L83
81. Ichimaru, S., Iyetomi, H., Ogata, S. 1988. *Ap. J. Lett.* 334: L17
82. Ishida, K., Mikami, T., Noguchi, T., Machara, H. 1982. *Publ. Astron. Soc. Jpn.* 34: 381
83. Itoh, N., Mitake, S., Iyetomi, H., Ichimaru, S. 1983. *Ap. J.* 273: 774
84. Itoh, N., Kohyama, Y., Matsumoto, J., Seki, M. 1984. *Ap. J.* 285: 758
85. Iyetomi, H., Ichimaru, S. 1988. *Phys. Rev. B* 38: 6761
86. Kaplan, S. A. 1950. *Astron. Zh.* 27: 31
87. Kapranidis, S. 1983. *Ap. J.* 275: 342
88. Kapranidis, S., Böhm, K. H. 1982. *Ap. J.* 256: 227
89. Kapranidis, S., Liebert, J. 1986. *Ap. J.* 305: 863
90. Deleted in proof
91. Kawaler, S. D., Hansen, C. J., Winget, D. E. 1985. *Ap. J.* 295: 547
92. Kawaler, S. D., Winget, D. E., Iben, I. Jr., Hansen, C. J. 1986. *Ap. J.* 302: 530
93. Kawaler, S. D., Winget, D. E., Hansen, C. J., Iben, I. Jr. 1986. *Ap. J. Lett.* 306: L41
94. Kepler, S. O., Vauclair, G., Nather, R. E., Winget, D. E., Robinson, E. L. 1989. See Ref. 187a, p. 341
95. Kepler, S. O., Vauclair, G., Dolez, N., Chevreton, M., Nather, R. E., et al. 1990. *Ap. J.* In press
96. Kettner, K. V., Becker, H. W., Buchman, L., Gorres, J., Kravinkel, H., et al. 1982. *Z. Phys.* 308: 73
97. King, C. R., Demarque, P., Green, E. M. 1989. In *Calibration of Stellar Ages*, ed. A. G. D. Philip, p. 211. Schenectady, NY: L. Davis
98. Kirzhnits, D. A. 1960. *Sov. Phys. JETP* 11: 365
99. Koester, D. 1972. *Astron. Astrophys.* 16: 459
100. Koester, D. 1987. See Ref. 150a, p. 329
101. Koester, D. 1989. See Ref. 187a, p. 206
102. Koester, D., Schönberner, D. 1986. *Astron. Astrophys.* 154: 125
103. Koester, D., Weidemann, V., Zeidler, K.-T. 1982. *Astron. Astrophys.* 116: 147
104. Kovetz, A., Shaviv, G. 1970. *Astron. Astrophys.* 8: 398
105. Lamb, D. Q., Van Horn, H. M. 1975. *Ap. J.* 200: 306
106. Lampe, M. 1968. *Phys. Rev.* 170: 306
107. Landau, L. D., Lifshitz, E. M. 1958. *Statistical Physics*. London: Pergamon
108. Larson, R. B. 1986. *MNRAS* 218: 409
109. Lattanzio, J. C. 1987. *Ap. J. Lett.* 313: L15
110. Lattanzio, J. C. 1989. *Ap. J. Lett.* 344: L25
111. Law, W. Y., Ritter, H. 1983. *Astron. Astrophys.* 123: 33
112. Liebert, J. 1980. *Annu. Rev. Astron. Astrophys.* 18: 363
113. Liebert, J. 1986. In *Hydrogen-Deficient Stars and Related Objects, IAU Colloq. No. 87*, ed. K. Hunger, D. Schönberner, N. K. Rao, p. 367. Dordrecht: Reidel
114. Liebert, J. 1989. See Ref. 187a, p. 18
115. Liebert, J. 1989. In *Planetary Nebulae, IAU Symp. No. 131*, ed. S. Torres-Peimbert, p. 545. Dordrecht: Reidel
116. Liebert, J., Dahn, C., Gresham, M., Strittmatter, P. A. 1979. *Ap. J.* 233: 226
117. Liebert, J., Dahn, C., Monet, D. G. 1989. *Ap. J.* 332: 891
118. Liebert, J., Fontaine, G., Wesemael, F. 1987. *Mem. Soc. Astron. Ital.* 58: 17
119. Liebert, J., Wesemael, F., Hansen, C. J., Fontaine, G. 1986. *Ap. J.* 309: 241
120. Luyten, W. J. 1979. *LHS Catalogue*. Minneapolis: Univ. Minn. Press
121. Magni, G., Mazzitelli, I. 1979. *Astron. Astrophys.* 72: 134
122. Matteucci, F., Francois, P. 1989. *MNRAS* 239: 885
123. Mazzitelli, I. 1987. *Mem. Soc. Astron. Ital.* 58: 117
124. Mazzitelli, I. 1989. *Ap. J.* 340: 249
125. Mazzitelli, I. 1989. See Ref. 187a, p. 29
126. Mazzitelli, I., D'Antona, F. 1986. *Ap. J.* 308: 706
127. Mazzitelli, I., D'Antona, F. 1986. *Ap. J.* 311: 762

128. Mazzitelli, I., D'Antona, F. 1987. See Ref. 150a, p. 351
129. McGraw, J. 1979. *Ap. J.* 229: 203
130. McGraw, J. 1980. *Space Sci. Rev.* 27: 601
131. Mendez, R., Miguel, I., Heber, U., Kudritzki, R. P. 1987. In *Hydrogen-Deficient Stars and Related Objects, IAU Colloq. No. 87*, ed. K. Hunger, D. Schönberner, N. K. Rao, p. 323. Dordrecht: Reidel
132. Mestel, L. 1950. *Proc. Cambridge Philos. Soc.* 46: 331
133. Mestel, L. 1952. *MNRAS* 112: 583
134. Mestel, L., Ruderman, M. A. 1967. *MNRAS* 136: 27
135. Michaud, G., Fontaine, G. 1984. *Ap. J.* 283: 787
136. Mochkovitch, R. 1983. *Astron. Astrophys.* 122: 212
137. Mochkovitch, R., Garcia-Berro, E., Hernanz, M., Isern, J., Panis, J. F. 1990. *Astron. Astrophys.* In press
138. Mould, J., Liebert, J. 1978. *Ap. J. Lett.* 226: L29
139. Muchmore, D. 1984. *Ap. J.* 278: 769
140. Munakata, H., Kohyama, Y., Itoh, N. 1985. *Ap. J.* 296: 197
141. Noh, H.-R., Scalo, J. 1990. *Ap. J.* In press
142. Nomoto, K. 1980. In *Type I Supernovae*, ed. J. C. Wheeler, p. 164. Austin: Univ. Tex. Press
143. Nomoto, K. 1984. *Ap. J.* 277: 791
144. Ogata, S., Ichimaru, S. 1987. *Phys. Rev. A* 36: 5451
145. Ostriker, J. P. 1971. *Annu. Rev. Astron. Astrophys.* 9: 353
146. Paczynski, B. 1971. *Acta Astron.* 21: 417
147. Papaloizou, J. C. B., Pringle, J. E., McDonald, J. 1982. *Ap. J.* 198: 215
148. Paquette, C., Pelletier, C., Fontaine, G., Michaud, G. 1986. *Ap. J. Suppl.* 61: 177
149. Paquette, C., Pelletier, C., Fontaine, G., Michaud, G. 1986. *Ap. J. Suppl.* 61: 197
150. Pelletier, C., Fontaine, G., Wesemael, F., Michaud, G., Wegner, G. 1986. *Ap. J.* 307: 242
150a. Philip, A. G. D., Hayes, D. S., Liebert, J. W., eds. 1987. *The Second Conference on Faint Blue Stars. Proc. IAU Colloq. No. 95.* Schenectady, NY: L. Davis
151. Pollock, E. L., Hansen, J. P. 1973. *Phys. Rev. A* 8: 3110
152. Renzini, A. 1983. In *Planetary Nebulae, IAU Symp. No. 103*, ed. D. R. Flower, p. 267. Dordrecht: Reidel
153. Renzini, A. 1989. In *Planetary Nebulae, IAU Symp. No. 131*, ed. S. Torres-Peimbert, p. 391. Dordrecht: Reidel
154. Ruiz, M. T., Maza, J., Wischnjewski, M., Gonzalez, L. E. 1986. *Ap. J. Lett.* 304: L25
155. Ruiz, M. T., Anguita, C., Maza, J. 1989. See Ref. 187a, p. 122
156. Salpeter, E. E. 1961. *Ap. J.* 134: 669
157. Savedoff, M. P., Van Horn, H. M., Vila, S. C. 1969. *Ap. J.* 155: 221
158. Schatzman, E. 1958. *White Dwarfs.* Amsterdam: North-Holland
159. Schmidt, M. 1959. *Ap. J.* 129: 243
160. Schönberner, D. 1979. *Astron. Astrophys.* 79: 108
161. Schönberner, D. 1981. *Astron. Astrophys.* 103: 119
162. Schönberner, D. 1983. *Ap. J.* 272: 708
163. Schönberner, D. 1984. In *Observational Tests of the Stellar Evolution Theory, IAU Symp. No. 105*, ed. A. Maeder, A. Renzini, p. 209. Dordrecht: Reidel
164. Schönberner, D. 1987. *Late Stages of Stellar Evolution. Astrophys. Space Sci. Libr.*, ed. S. Kwok, S. R. Pottasch, 132: 337. Dordrecht: Reidel
165. Schönberner, D. 1986. *Astron. Astrophys.* 169: 189
166. Schönberner, D., Weidemann, V. 1981. In *Planetary Nebulae, IAU Symp. No. 103*, ed. D. R. Flower, p. 359. Dordrecht: Reidel
167. Schwarzschild, M. 1958. *Structure and Evolution of the Stars.* Princeton, NJ: Princeton Univ. Press
168. Shara, M. M. 1989. *Publ. Astron. Soc. Pac.* 101: 5
169. Shaviv, G., Kovetz, A. 1976. *Astron. Astrophys.* 51: 383
170. Shipman, H. L. 1989. See Ref. 187a, p. 220
171. Sienkiewicz, R. 1980. *Astron. Astrophys.* 85: 295
172. Sion, E. M. 1986. *Publ. Astron. Soc. Pac.* 98: 821
173. Sion, E. M., Greenstein, J. L., Landstreet, J. D., Liebert, J., Shipman, H. L., Wegner, G. 1983. *Ap. J.* 269: 253
174. Sion, E. M., Liebert, J., Wesemael, F. 1985. *Ap. J.* 292: 477
175. Deleted in proof
176. Starrfield, S., Cox, A. N., Kidman, R. B., Pesnell, W. D. 1984. *Ap. J.* 281: 800
177. Stetson, P. B., Vanden Berg, D. A., Bolte, M., Hesser, J. E., Smith, G. H. 1989. *Astron. J.* 98: 1360
178. Stevenson, D. J. 1980. *J. Phys. Suppl.* 41: C2
179. Sweeney, M. A. 1976. *Astron. Astrophys.* 49: 375
180. Tassoul, M., Fontaine, G., Winget, D. E. 1990. *Ap. J. Suppl.* In press
181. Truran, J. W., Livio, M. 1989. See Ref. 187a, p. 498
182. Twarog, B. A. 1980. *Ap. J.* 242: 242

183. Van Horn, H. M. 1968. *Ap. J.* 151: 227
184. Vauclair, G. 1989. See Ref. 187a, p. 176
185. Vauclair, G., Fontaine, G. 1979. *Ap. J.* 230: 563
186. Vauclair, G., Liebert, J. 1987. In *Exploring the Universe with the IUE Satellite*, ed. Y. Kondo et al, p. 355. Dordrecht: Reidel
187. Vauclair, G., Vauclair, S., Greenstein, J. L. 1979. *Astron. Astrophys.* 80: 79
187a. Wegner, G., ed. 1989. *White Dwarfs. Proc. IAU Colloq. No. 114. Lect. Notes Phys.*, Vol. 328. Berlin: Springer-Verlag
188. Wegner, G., Yackovich, F. H. 1984. *Ap. J.* 284: 257
189. Wegner, G., Koester, D. 1985. *Ap. J.* 288: 746
190. Wehrse, R. 1977. *Mem. Soc. Astron. Ital.* 48: 13
191. Wehrse, R., Liebert, J. 1980. *Astron. Astrophys.* 83: 184
192. Weidemann, V. 1968. *Annu. Rev. Astron. Astrophys.* 6: 351
193. Weidemann, V. 1984. *Astron. Astrophys.* 134: L1
194. Weidemann, V. 1990. *Annu. Rev. Astron. Astrophys.* 28: 103
195. Weidemann, V., Koester, D. 1983. *Astron. Astrophys.* 121: 77
196. Weiss, A., Keady, J. J., Magee, N. H. Jr. 1990. *At. Data Nucl. Data Tables.* In press
197. Deleted in proof
198. Wesemael, F., Green, R., Liebert, J. 1985. *Ap. J. Suppl.* 58: 379
199. Williams, R. E. 1985. In *Production and Distribution of CNO Elements*, ed. I. J. Danziger, p. 225. Garching: ESO
200. Winget, D. A. 1986. In *Highlights of Astronomy*, ed. J. P. Swings, 7: 221. Dordrecht: Reidel
201. Winget, D. E. 1988. In *Advances in Helio- and Asteroseismology, IAU Symp. No. 123*, ed. J. Christensen-Dalsgaard, S. Frandsen, p. 305. Dordrecht: Reidel
202. Winget, D. E., Van Horn, H. M. 1987. See Ref. 150a, p. 363
203. Winget, D. E., Van Horn, H. M., Tassoul, M., Hansen, C. J., Fontaine, G., Carrol, B. W. 1982. *Ap. J. Lett.* 252: L65
204. Winget, D. E., Kepler, S. O., Robinson, E. L., Nather, R. E., O'Donoghue, D. 1985. *Ap. J.* 292: 606
205. Winget, D. E., Hansen, C. J., Liebert, J., Van Horn, H. M., Fontaine, G., et al. 1987. *Ap. J. Lett.* 315: L77
206. Wood, P. R., Faulkner, D. J. 1986. *Ap. J.* 307: 659
207. Wood, M. A., Winget, D. E. 1989. See Ref. 187a, p. 282
208. Wood, M. A., Winget, D. E., Van Horn, H. M. 1987. See Ref. 150a, p. 637
209. Yuan, J. W. 1990. *Astron. Astrophys.* 226: 108

THE ORIGIN OF NEUTRON STARS IN BINARY SYSTEMS

Ramon Canal

Departament de Fisica de l'Atmosfera, Astronomia i Astrofisica, Universitat de Barcelona, 08028 Barcelona, Spain, Max-Planck-Institut für Physik und Astrophysik, Institut für Astrophysik, D-8046 Garching, Federal Republic of Germany, and Laboratori d'Astrofisica, Societat Catalana de Ciències, I.E.C., 08028 Barcelona, Spain

Jordi Isern

Centre d'Estudis Avançats de Blanes, C.S.I.C., 17300 Blanes (Girona), Spain, and Laboratori d'Astrofisica, Societat Catalana de Ciències, I.E.C., 08028 Barcelona, Spain

Javier Labay

Departament de Fisica de l'Atmosfera, Astronomia i Astrofisica, Universitat de Barcelona, 08028 Barcelona, Spain, and Laboratori d'Astrofisica, Societat Catalana de Ciències, I.E.C., 08028 Barcelona, Spain

KEY WORDS: close binaries, mass accretion, stellar collapse, tidal capture

1. INTRODUCTION

Neutron stars are found both in isolation and as members of binary systems (and perhaps, in some cases, even of triple systems). Single neutron stars were initially detected by their emission of radio pulses, following their discovery by Hewish et al (65). In contrast, binary systems containing neutron stars were first detected as emitters of high-energy (X-ray) photons: binary X-ray pulsars (48, 96, 129, 145) and X-ray burst sources (11,

54). This picture has since become more complex: binary radio pulsars have been found (68), both single and binary millisecond pulsars have been detected (4, 43), and single neutron stars have been observed to emit X rays and γ rays (63, 90), with γ-ray burst sources now mostly thought to be isolated neutron stars (127).

The origin of isolated neutron stars is universally attributed to the collapse of the central, fuel-exhausted cores of massive ($M \gtrsim 8\text{–}12\ M_\odot$) stars, which also produces a Type II supernova explosion (183, and references therein). The discovery of a pulsar in the Crab Nebula, the remnant of a supernova outburst observed in 1056 A.D. (136), first put this picture on firm ground. Recent confirmation of the above mechanism has been provided by SN 1987A in the Large Magellanic Cloud with the detection of neutrinos produced in the collapse (see 3, and references therein). The problem of how a fraction of the gravitational energy liberated in forming the neutron star is transferred to the overlying mantle and envelope to produce the explosion is still unsolved (see the above references), but there is now little doubt that neutron stars can form in such a way. Also, the theory of massive single-star evolution (at least for stars with $M \lesssim 100\ M_\odot$) unequivocally predicts formation of electron-degenerate Fe-Ni cores [in fact with nuclear statistical equilibrium (NSE) composition] that must collapse when their mass grows above the effective Chandrasekhar limit.

The origin of neutron stars in binary systems is more problematic. First, binary star evolution depends on more parameters than that of isolated stars (masses and initial separation of the two components instead of a single mass). Second, when the two stars are sufficiently close (as in most binaries containing neutron stars), mass transfer episodes take place one or more times during the evolution, a common envelope may form at certain stages, and undetermined amounts of mass can be lost by the system. Third, the "standard" core collapse mechanism for formation of neutron stars, which involves ejection of several solar masses at velocities up to tens of thousands of kilometers per second (as in SN 1987A), can be incompatible with the observed characteristics of an appreciable fraction of these binary systems. When this happens, there are only two possible solutions: either the neutron star was formed in its present system by another, less violent mechanism, such as accretion-induced collapse (AIC) of a white dwarf; or the formation of the neutron star preceded that of the system, and some capture mechanism has since created the binary.

In Section 2 we briefly characterize the different types of binaries containing neutron stars. In Section 3 we review the three formation mechanisms ("standard" collapse, diverse types of captures, and accretion-induced collapse). In Section 4 we discuss the possible relevance of each mechanism to the origin of the systems previously considered. Finally, in

Section 5 we summarize our present knowledge on the origin of neutron stars in binaries, pointing out open problems.

2. NEUTRON STARS IN BINARY SYSTEMS

2.1 *X-Ray Binaries*

Accretion-driven stellar X-ray sources can be divided into two major groups: high-mass and low-mass X-ray binaries (165).

High-mass X-ray binaries are systems whose optical counterparts have normal early-type spectra. Most of the known X-ray pulsars belong to this group. One of them, Cen X-3, by nature of its being an eclipsing source, provided the first clear evidence as to the binary nature of the bright galactic X-ray sources (129). The number of massive X-ray binaries presently known is ~ 40. About half of these show X-ray pulsations with periods ranging from fractions of a second (e.g. A0538−66) up to almost 10^3 s (e.g. 4U 0352+30). Strong magnetic fields ($\sim 10^{12}$ G) are inferred from cyclotron lines in the hard X-ray spectrum of some X-ray pulsars (84). The other half (roughly) of the massive X-ray binaries are nonpulsing sources. According to the spectral type of the optical star, most massive X-ray binaries can also be divided into two additional groups (102): In the first, the star is of spectral type earlier than B2 and luminosity class I, II, or III; in the second, it is a Be star. Mass transfer from the primary to the neutron star would occur through a strong stellar wind or incipient Roche lobe overflow in the first group, whereas in the second the compact object would be fed by a circumstellar disk or shell associated with the rapid rotation of the Be star (158, 164, and references therein). Most of the X-ray sources with Be-type companions are highly variable or transient. Recurrent X-ray bursts have been observed in a number of them, and these have been interpreted as resulting from the motion of the neutron star, in an eccentric orbit, through the region of varying density around the Be star (165).

High-mass X-ray binaries are found along the galactic plane, with a narrow latitude distribution. This further confirms their classification as young Population I stars.

Low-mass X-ray binaries have faint stars as optical counterparts. The optical properties are rather uniform (see, for instance, 165). Low-mass X-ray binaries rarely show X-ray pulsations, which indicates that much weaker magnetic fields exist than in high-mass binaries (or, maybe, that the magnetic field is aligned with the rotation axis). Sources in this group emit X-ray bursts [thought to be produced by thermonuclear flashes in accreted matter (95, 166)]. No source shows both pulsations and bursts, which supports the view that the relative weakness of their magnetic field rather than its alignment causes the differences in behavior between low-mass and high-mass X-ray binaries. X-ray spectra of both steady and transient sources are softer than those of high-mass binaries (35). Additionally, the few low-mass X-ray binaries that show pulsations (GX 1+4, Her X-1, 4U 1626−67) have X-ray spectra as hard as those of high-mass sources. The differences in spectrum hardness thus seem to be related to the geometry of the accretion flow, which is in turn determined by the strength of the magnetic field.

In the Galaxy, 11 bright X-ray sources are located in globular clusters, and 10 of these produce bursts (95). Their properties are those of low-mass X-ray binaries, but optical identification is very difficult owing to the central condensation of most clusters and reddening due to interstellar extinction. On the other hand, about 30 low-mass X-ray sources are concentrated toward the galactic center, and these are referred to as "galactic bulge sources." Among them, 8 are strong sources ($L_X \sim 10^{38}$ erg s^{-1}), while the others are weaker sources ($L_X \sim 10^{36}$–10^{37} erg s^{-1}). Similarities between bulge sources and globular cluster sources have led to the whole being considered as a single class of objects. As for stellar population, the classification of these sources would encompass Population II (globular clusters) and old disk population (galactic bulge).

A large fraction of galactic bulge sources show quasi-periodic oscillations (QPO). Their frequency is correlated with position in the X-ray spectral hardness vs intensity diagram (see 60, 61, 63, 91, 98, 163).

2.2 Binary and Millisecond Pulsars

Out of ~ 500 known pulsars, 12 of them [including the recently discovered PSR 1516+02B in the globular cluster M5 (see 180)] are in binary systems. Four systems are in globular clusters. Following van den Heuvel (157), binary pulsars can be divided into two groups. The PSR 1913+16 group (which includes this object plus PSR 2303+46 and PSR 0655+64) is characterized by relatively short orbital periods, relatively high companion masses (~ 1.0–$1.4\ M_\odot$), and, in two cases (PSR 1913+16 and PSR 2303+46), very eccentric orbits. In contrast, those in the PSR 1953+29 group (which, in addition to PSR 1953+29, also includes the other three binary millisecond pulsars PSR 1957+20, PSR 1855+09, and PSR 1620+26) have relatively long orbital periods (except for PSR 1831+00 and PSR 1957+20), low or even very low companion masses (~ 0.02–$0.4\ M_\odot$), and circular orbits.

Four single millisecond radio pulsars are known [including PSR 1516+02A, also in M5 (see 180)]. Three of them are in globular clusters. We mention them here because in the forthcoming discussion we will see that current scenarios for their formation involve a binary stage.

3. FORMATION MECHANISMS

3.1 Core Collapse of Massive Stars

This is the "canonical" formation mechanism for neutron stars. A supernova explosion must take place simultaneously with the formation of the compact object in order to get rid of the mass above the stability limit for a neutron star ($M \simeq 2.0$–$2.5\,M_\odot$, which means that most of the star's mass must be ejected). A stellar black hole would otherwise result. If neutron stars in double-star systems have been formed this way, in the same binary where they are presently observed, the system must have managed to survive the supernova explosion. A basic concept of stellar evolution is that, at constant mass, the larger the initial mass of the star, the faster its evolution. In a binary, the initially most massive star should thus be the first to explode as a supernova and produce the neutron star. This, in turn, would minimize the chances of survival of the system. In a binary system where a member undergoes a supernova explosion and ejects a considerable fraction of its mass in a time shorter than the orbital period, it is easy to derive the condition by which the system will remain bound after the explosion (for circular orbit): $M_{ej}/M_{tot} < 1/2$, where M_{ej} is the ejected mass and M_{tot} the total mass of the system (see 130). This relation was first shown by Blaauw (13). The effects of the supernova blast wave impinging on the companion further increase the chances of disruption (34, 178). Nonetheless, Hills (67) has considered systems with eccentric preexplosive orbits in which a random kick velocity is given to the neutron star in the explosion and concludes that for long-period binaries, the system may still survive a supernova explosion in which more than half of the mass is lost. The conditions for such survival are moderate eccentricity, and for the supernova to take place either near apastron if the kick velocity is small or near periastron if it is large.

In massive binaries, however, the star that explodes may have already become the less massive component of the system owing to a previous stage of large-scale mass transfer (148, 151, 159). Specific scenarios for such an evolution have been developed and reviewed by van den Heuvel (152–154). "Conservative" mass transfer (total mass and angular momentum of the system are conserved) satisfactorily explains the formation of massive X-ray binaries (see ref. 158 for a more recent review). Later common envelope evolution with mass loss by the system as a whole would produce systems containing two neutron stars, such as PSR 1913+16 and PSR 2303+46, and also binaries consisting of a neutron star plus a massive white dwarf, such as PSR 0655+64 (156, 160).

The formation of a binary containing a neutron star through core collapse of a massive star, in systems where the observed companion of the compact object is a low-mass star, might involve triple-star evolution (36). In this scenario, a massive close binary is accompanied at large distance by a late-type dwarf. Formation of a low-mass X-ray binary would then involve three steps. First, the massive pair evolves into a massive X-ray binary through mass transfer and a supernova explosion. Second, the remaining massive star evolves until the Roche lobe overflows and a common envelope forms. The neutron star then spirals into the center of the red giant, and a Thorne-Żytkov object (146) forms—i.e. a red supergiant with a neutron star core. Third, long-term expansion of its envelope should eventually lead to a common envelope phase with the late-type dwarf. The latter object would then spiral in, and the envelope would likely be lost (its mass having been previously reduced by wind emission, and its structure being distended and loosely bound). A close binary consisting of a neutron star plus a low-mass companion might finally result.

3.2 Capture Mechanisms

Another possible way to form a binary system containing a neutron star is to produce the compact object by core collapse of a single massive star and later to couple the neutron star with a companion in a tidal capture process.

Clark (33) first suggested that globular cluster sources are neutron stars that have acquired close stellar companions by capture. As specific mechanisms, Sutantyo (142) considered direct neutron star–red giant collisions, while Fabian et al (39) examined neutron star–main sequence star tidal collisions.

In close encounters between two stars, tidal forces will produce deformations. The energy required for that comes from the relative kinetic energy of the two stars, and if this energy is rapidly dissipated, the stars may become bound. Press & Teukolsky (122) confirmed the viability of this mechanism by detailed calculations. An approximate treatment is made in (170). Following Verbunt (168, 169), the condition for tidal capture is that the distance of closest approach, d, be

$$d \lesssim 3R\left(\frac{k}{0.14} \frac{m}{M} \frac{m+M}{2M_\odot} \frac{R_\odot}{R}\right)^{1/6} \left(\frac{10 \,\mathrm{km\,s^{-1}}}{v}\right)^{1/3}, \qquad 1.$$

where m is the mass of the neutron star, M and R are the mass and radius, respectively, of the colliding star, k is the apsidal motion constant, v is relative velocity, and the normalization constants are, in this case, typical for globular clusters (where such a process has the best chance to work).

Capture will occur if the compact star approaches to within 2 to 3 times the radius of the other star. Direct hits ($d \lesssim R$) would lead to merging, and closest approaches ($R \lesssim d \lesssim 3R$) to binary formation. The cross section for passage of the two stars within distance d is given by

$$\sigma = \pi d^2 \left(1 + \frac{2G(m+M)}{v^2 d}\right) \simeq \pi d \frac{2G(m+M)}{v^2}, \qquad 2.$$

with the last approximation being valid for small relative velocities (such as are found in globular clusters). Thus, the capture rate for unit volume would be

$$\Gamma = n_c n v \sigma \simeq 6 \times 10^{-11} \frac{n_c}{10^2 \, \text{pc}^{-3}} \frac{n}{10^4 \, \text{pc}^{-3}}$$

$$\times \frac{m+M}{M_\odot} \frac{3R}{R_\odot} \frac{10 \, \text{km s}^{-1}}{v} \text{yr}^{-1} \text{pc}^{-3}, \qquad 3.$$

where n_c and n are the number densities of compact stars and of target stars, respectively.

Most frequent encounters of a neutron star will be with a main sequence star. In one third of the cases, the encounter (a direct hit) will destroy the noncompact star. In the other two thirds of the encounters, a binary may form, with orbital period in the range of hours.

Less frequent encounters of a neutron star with a giant or subgiant star should also be, in one third of all cases, direct hits. Verbunt (167) suggests that these encounters would cause the neutron star and the (white dwarf–like) red giant core to move in closed orbits within the giant envelope. Friction should then lead the two compact objects to spiral together and the envelope to be ejected. A binary system (neutron star plus white dwarf) with orbital period ~ 10 min would result.

In the remaining two thirds of neutron star–giant star collisions, a binary of longer period ($\sim$days) may form. Mass transfer should later occur owing to the giant's expansion. Once the giant envelope is entirely expelled, a wide neutron star–white dwarf binary results (see Figure 2 in ref. 168, or Figure 6 in ref. 169).

Another capture mechanism involves exchange encounters in binaries (40). If a neutron star approaches a binary, there are three (or maybe four—see below) possible outcomes (169): disruption of the binary (ionization), change in the binary parameters (excitation), or replacement of one of the binary members (that is ejected) by the neutron star (exchange) (see Figure 2 in ref. 169). Cross sections for the three processes (but only for the case where all three stars have the same mass) have been calculated by Hut & Bahcall (70) and by Hut (69). These interactions will mostly occur through formation of a temporary triple system in which the three stars move in complicated orbits around each other—a process known as resonance scattering. Its rate, for the case of three equal point masses, is (70)

$$\Gamma_{bin} \simeq 5 \times 10^{-10} \frac{n_c}{10^2 \, pc^{-3}} \frac{n_{bin}}{10^2 \, pc^{-3}} \frac{m}{M_\odot} \frac{a}{1 \, AU} \frac{10 \, km \, s^{-1}}{v} \, yr^{-1} \, pc^{-3}, \quad 4.$$

where, in analogy to Equation 3, n_c and n_{bin} are the number densities of compact objects and of binaries, respectively, and a is the binary separation. Again, values are normalized to typical globular cluster values. For equal masses, the incoming star would remain in the binary in two thirds of the encounters. If one compares Equations 3 and 4, it can be seen that resonance scattering has a larger cross section than tidal capture (a rather than $3R$), but that n_{bin} is much smaller than n (the density of single stars).

Finally, Grindlay (51) has suggested that tidal capture may also lead to formation of hierarchical triple systems. Bailyn (5) infers a relatively high efficiency of production of such systems in globular clusters if the third star encounters a preexisting compact binary in a retrograde orbit. The cross section for tidal capture of the third star into stable orbit should be $\sim 25\%$ of that for tidal capture formation of the initial binary (see also 53).

3.3 Accretion-Induced Collapse of White Dwarfs

White dwarfs are the stellar objects whose characteristics most closely resemble those of neutron stars. They also are, in a broad sense, the sole progenitors of neutron stars. The Fe-Ni cores of the massive stars whose collapse produces neutron stars in the standard scenario are, in fact, "white dwarf" structures embedded in a less dense mantle, surrounded, in turn, by a much more tenuous and extended envelope. Their mass growth up to the point of dynamical instability is fed by the silicon-burning shell. White dwarfs, which result from loss of the envelopes of red giant stars of lower masses in less advanced burning stages, are often members of close binary systems. In a fraction of these systems (novae and cataclysmic variables), mass transfer to the white dwarf from its companion is taking place. If the white dwarf could retain this mass and grow up to the Chandrasekhar limit without first exploding, it should collapse and a neutron star would form. Since its mass ($\simeq 1.4 \, M_\odot$) would be below the maximum mass of neutron stars, and thus no mass ejection would be required, the binary system would always survive. This idea of neutron star formation through accretion-induced collapse (AIC) of white dwarfs in close binary systems was first suggested by Schatzman (128). Later, Canal & Schatzman (30–32) and, independently, Ergma & Tutukov (38) further explored this idea. The problems with AIC are twofold:

1. The material transferred by the companion to the white dwarf must be effectively incorporated into the degenerate core instead of being reejected either by explosive events originating at the surface of the compact star or by some less violent mass loss mechanism. This means that appropriate combinations of chemical composition of the infalling material and of accretion rate must occur.
2. Explosive central (or off-center) ignition followed by disruption of the core [as in Type I supernova events (103, 179)] must equally be avoided.

The first problem posed by AIC is basically the same one that has to be solved to find a plausible evolutionary scenario for Type Ia (or "classical" Type I) supernovae (27, 73, 176, 183). The differences are only in (a) the respective frequencies of the events to be explained and (b) the fact that in (assumed) AIC we actually observe in most cases the binary system that should have resulted from it, and thus we know the companion of the neutron star, the hypothetical previous mass donor.

The second problem involves chemical composition, initial mass and temperature of the white dwarf, and, again, accretion rate (within the limits set by the solution to the first problem). It also depends on the physics of phase transition at high densities, on nuclear reaction rates in dense plasmas, and on the dynamics of nuclear-burning fronts.

3.3.1 MASS GROWTH: THE OUTER LAYERS The identified optical counterparts of low-mass X-ray binaries usually show, among other emission lines, Hα and Hβ. The material accreted by the white dwarf thus was, in all probability, hydrogen rich. In order for the star to grow by a few tenths of a solar mass, not only must explosive ignition (as in novae) of this material be avoided, but also the formation of a red giant envelope (eventually engulfing both the white dwarf and its companion) on top of an H-burning shell (but see below), or even the direct formation of a common envelope, must not occur. Usually quoted limits for hydrogen accretion are as follows:

$\dot{M}_H \lesssim 10^{-9} \, M_\odot \, yr^{-1}$ as leading to nova outburst,

$\dot{M}_H \gtrsim 10^{-6} \, M_\odot \, yr^{-1}$ formation of a red giant envelope,

$\dot{M}_H \gtrsim \dot{M}_{Edd} \simeq 10^{-5} \, M_\odot \, yr^{-1}$ direct formation of a common envelope.

We note that the first limit is based on calculations that assume spherically symmetric and "soft" accretion. By "soft," we mean that material deposited at the surface is at rest and has the same specific entropy as the layers into which it is incorporated. It is thus cold and becomes strongly electron degenerate before igniting. Most likely, however, this material will form a disk around the compact star, and the accretion process will thus include angular momentum and kinetic energy dissipation. When these effects are taken into account, the actual range of $\dot{M}_H$ producing nova explosions is ill defined (131, 133). Further research on this point is thus needed. Nova outbursts can hardly be compatible with mass growth toward the Chandrasekhar limit, since they seem in the long run to reduce, rather than to increase, the mass of the white dwarf. Even in the most favorable cases, nova outbursts would limit the actual mass growth to at most 10% of $\dot{M}_H$ (J. W. Truran, private communication).

The development of a common envelope (either due to accretion above the Eddington limit or to formation of a red-giant envelope) should, in general, induce mass loss by the system as a whole and thus inhibit further growth of the white dwarf. Nonetheless, Hachisu et al (58) propose a model whereby stable mass transfer in a common envelope is possible (through steady hydrogen burning) for a range of parameters of the binary system. Predicted mass accretion rates are $\dot{M}_H \gtrsim 10^{-7} M_\odot \, \text{yr}^{-1}$.

Hypothetical close binaries where $10^{-9} M_\odot \, \text{yr}^{-1} \lesssim \dot{M}_H \lesssim 10^{-6} M_\odot \, \text{yr}^{-1}$ may steadily, or in weak flashes, convert the H into He (and thus would not be novae or cataclysmic variables). Such binaries would be very luminous, both during H-burning episodes (when they should appear as bright EUV sources) and between them (as a result of the energy released by accretion). No known objects are clear candidates. [However, they may exist in globular clusters: It has been suggested by Hertz & Grindlay (59) and by Grindlay (52) that low-luminosity X-ray sources in clusters may be such progenitor mass-accreting white dwarfs.] Nonetheless, their frequency might be low enough to minimize the significance of this apparent absence.

A further limitation is that the He layer resulting from the burning of the accreted H will explosively ignite (instead of burning steadily into C+O) if it is accumulated at a rate $10^{-9} M_\odot \, \text{yr}^{-1} \lesssim \dot{M}_{He} \lesssim 5 \times 10^{-8} M_\odot \, \text{yr}^{-1}$. The lower limit, however, is only valid for initial masses of the white dwarf $M_{WD} \lesssim 1.13 M_\odot$ (111). The same criterion applies if He is directly accreted from a companion (either degenerate or nondegenerate) that has previously lost its hydrogen envelope, as in Type Ia supernova scenarios proposed by Iben et al (72). The latter situation would not fit the observed characteristics of binary X-ray sources with identified optical counterparts (see above), but it might be valid for other systems, such as short-period low-mass X-ray binaries and binary millisecond pulsars.

C+O material might also be directly accreted by a C+O white dwarf from a white dwarf companion with the same composition. Such double

C+O white dwarf systems have been proposed as progenitors of Type Ia supernovae (73, 174). Orbital angular momentum loss by emission of gravitational radiation would lead to Roche lobe overflow by the less massive white dwarf. This overflow should generally have a runaway character: The secondary would be disrupted, and its material would likely form a massive disk around the primary [except for cases where mass transfer widens the binary faster than the mass donor expands (see 124)]. Depending on the viscous dissipation in the disk (109), a Type Ia supernova could result. Very fast accretion ($\dot{M}_{C+O} \gtrsim 5 \times 10^{-6}\ M_\odot\ \mathrm{yr}^{-1}$) would instead induce carbon ignition close to the surface and quasi-hydrostatic burning of the C+O white dwarf into a O+Ne+Mg white dwarf (118). Collapse to a (generally single) neutron star might then result (but see below).

3.3.2 MASS GROWTH: THE CORE   Evolution of the primary star in a close binary system can give rise to either a helium white dwarf, a carbon-oxygen white dwarf, or a oxygen-neon-magnesium white dwarf, depending on the initial parameters of the system. However, not all compositions can be involved in AIC. Helium white dwarfs can immediately be discarded. Their mass growth would lead to explosive helium ignition at the center of the star for central density $\rho_c \lesssim 4 \times 10^8\ \mathrm{g\ cm}^{-3}$, no matter how low the temperature (141). The star's mass would thus be at most $M \simeq 1.30\ M_\odot$, well below the Chandrasekhar mass ($M_{Ch} \simeq 1.46\ M_\odot$ for this composition, at zero temperature). After incineration, electron capture would be very slow, and overpressures would be large enough to start a detonation (supersonic burning propagated by a shock wave): The star would thus be completely disrupted (183).

This leaves only carbon-oxygen and oxygen-neon-magnesium white dwarfs as possible candidates. We consider these groups separately, as AIC poses different problems for each: Both cooling in the stage between the formation of the white dwarf and the onset of mass accretion and reheating by mass accretion are especially relevant in the C+O case, whereas semiconvection associated with electron capture plays an important role in the O+Ne+Mg case.

C+O white dwarfs   These white dwarfs can form in binary systems either by Roche lobe overflow just before or just after ignition of He in initially close binaries or by Roche lobe overflow during the early asymptotic giant branch (AGB) phase or the thermally pulsing AGB phase in initially wide binaries. In the latter case, common envelope evolution should allow enough orbital angular momentum to be lost so that the wide binary evolves into a close binary. An important question here is the upper mass limit for C+O white dwarfs formed this way. Observations of cataclysmic variables and of classical novae give average masses of 0.75 $M_\odot$ and 1.23 $M_\odot$, respectively, and a recent model for the recurrent nova U Sco gives a mass $M \simeq 1.38\ M_\odot$ (138). However, the most massive white dwarfs found in these systems may well be O+Ne+Mg, and not C+O, white dwarfs. Theoretically, C+O cores of single stars should ignite C nonexplosively when $M_{core} \gtrsim 1.1$–$1.2\ M_\odot$ (74).

After the formation of the white dwarf and prior to the onset of mass accretion from the secondary, the electron-degenerate core will only cool down. This introduces an important change if the corresponding time interval is long enough to allow crystallization of the star's interior. Cooling times for onset of crystallization are a function of the white dwarf's mass: for example, $t_{cool} = 1.9 \times 10^9$ yr for $M = 0.6\ M_\odot$, 1.3×10^9 yr for $1.0\ M_\odot$, and 1.1×10^9 yr for $1.2\ M_\odot$ (E. García-Berro et al, private communication). Such a first-order phase transition was predicted by Mestel & Ruderman (104). It has more recently been studied using Monte Carlo simulations, based on the work of Brush et al (18). Most of these studies have dealt with a one-component plasma (OCP) only. The ion sites will form a body-centered cubic (bcc) lattice for a critical value of the plasma-coupling constant $\Gamma = Z^2 e^2 / r_s k T$, where Z is atomic number, e the electron charge, k Boltzmann's constant, T the temperature, and r_s the ion-sphere radius (the radius of the sphere containing, on average, one ion). Recently calculated values of Γ_{crit} are $\Gamma_{crit} = 171 \pm 5$ (132) and $\Gamma_{crit} = 180 \pm 1$ (119). The cases of a C+O plasma and of a binary ion mixture (BIM) in general are more complex. At first sight, from the expression for Γ, Γ_{crit} will be reached first for oxygen in the center of a star cooling at constant density (which means constant r_s). The question thus arises as to whether oxygen will then start to crystallize or rather a carbon-oxygen alloy will form at some lower temperature. Direct Monte Carlo simulations (101) seemed to support the idea that carbon and oxygen are miscible in any proportion in the solid phase, and that the crystallization temperature of the mixture (at a fixed density) should be the weighted (by number) mean of the crystallization temperatures of pure carbon and pure oxygen. Based on this, and assuming that carbon and oxygen sites are distributed at random in the crystal, Canal & Isern (22) concluded that gravitational collapse should follow central carbon ignition in a mass-accreting C+O white dwarf whose innermost layers remained solid during the entire mass growth stage. This conclusion was based on two facts:

1. White dwarf cooling in the previous (detached) stage of the system has lowered the temperature and reduced the nuclear fusion rates to their minimum, pycnonuclear values (20, 126). These rates correspond to the regime where the reactions only take place through collisions between

neighboring nuclei oscillating around their equilibrium positions in the lattice (see also 130). Compression then follows a very low adiabat with a flat slope (the adiabatic index is ~ 0.5), and thermonuclear runaway happens at central densities $\rho_c \simeq 10^{10}$ g cm^{-3}. Electron captures on the incinerated (NSE) material are thus very fast (37, 45). The exact value of the explosive ignition density, however, depends on the approximation adopted for the lattice potential (static or relaxed approximations) plus the presence of impurities and/or crystal defects (see 64a).

2. Unless a detonation forms [this cannot be excluded (see 14, 176, and discussion below)], thermonuclear burning should propagate through the solid layers by conduction alone. The corresponding velocities would be low enough (92) to ensure that the Chandrasekhar mass (due to electron captures) becomes lower than the actual mass of the star before the energy released by the thermonuclear reactions can induce appreciable expansion and unbind the core. Further refinements have confirmed this basic conclusion, but they have also introduced new uncertainties (see below).

Phase transition in BIMs was later considered by Stevenson (139). He adopted the random-alloy-mixing (RAM) model for the free energy in the solid phase and deduced an eutectic phase diagram for the carbon-oxygen mixture, which showed that carbon and oxygen would separate when crystallizing. Oxygen should crystallize first and accumulate at the center. A pure oxygen core would thus progressively grow. Given enough time, it would eventually encompass a fraction of the mass of the star equal to the oxygen mass fraction (107). The implications of such a phase diagram for AIC were first explored by Canal et al (24) and were later considered (together with the mixed, random-alloy alternative) in (17, 25–27, 64, 81, 88, 89). Phase separation of carbon and oxygen with growth of an oxygen core of a few tenths of a solar mass would lead (even in the case of later melting of the white dwarf's core upon mass accretion) to central thermonuclear runaway at $\rho_c \simeq 2 \times 10^{10}$ g cm^{-3}. This runaway follows electron captures on oxygen and is triggered by ignition of the ^{16}C formed by those captures.

More recent work on the phase diagram of BIMs now seems to discard the occurrence of large chemical inhomogeneity in a C+O white dwarf resulting from crystallization. Barrat et al (10), using a density-functional approach for calculating the free energies, deduce a diagram of the spindle form: A random C+O alloy, only slightly more oxygen rich than the fluid phase, would result from crystallization. Transition to an ordered alloy of the ClCs type (bcc lattice with carbon and oxygen ion sites forming simple cubic sublattices) is predicted for values of Γ significantly larger than Γ_{crit}. Independently, Ichimaru et al (76) have shown, using Monte Carlo simulations, that the linear-mixing formula is more accurate than the random-mixing one for calculating the free energies of both fluid and solid phases. They then obtain an azeotropic phase diagram. Again, only a moderate oxygen enrichment of the solid alloy is predicted. In their Monte Carlo simulations of the bcc lattice, Ichimaru et al (76) find the equilibrium final states to take random bcc-solid configurations. This is not incompatible with the prediction of Barrat et al (10) that a transition to an ordered alloy occurs at larger Γ, since the simulations discussed above are only for values of Γ close to $\Gamma_{\rm crit}$.

Finally, Godon et al (49) have approached the problem by solid-state physics methods [linear Muffin-Tin orbitals (LMTO)]. They conclude that, within 1% accuracy, microscopic separation requires no energy, and thus that the issue cannot be decided.

The currently most favored hypothesis—that carbon and oxygen form a random alloy with a chemical composition similar to that of the fluid phase when crystallizing—makes AIC strongly dependent on mass accretion history. When mass accretion starts, the degenerate core will grow in mass, and it will be heated by compression and (if either H or He is accreted) by the inward inflow from the He-burning shell (except for the cases $\dot{M}_{\rm H} \lesssim 10^{-9} M_\odot \, {\rm yr}^{-1}$ or $\dot{M}_{\rm He} \lesssim 10^{-9} M_\odot \, {\rm yr}^{-1}$ and $M_{\rm core} \gtrsim 1.13 \, M_\odot$, where there would only be a H-burning shell in the first case and no shell at all in the second case). Compressional heating will also initially affect the outer layers and then diffuse inward (111). Any solid core that would have formed in the previous, purely cooling stage will thus progressively melt (except for low enough accretion rates). If the entire core becomes fluid before pycnonuclear reactions ignite the star's center, thermonuclear runaway can happen at any central density in the interval $2-3 \times 10^9$ g cm$^{-3} \lesssim \rho_{\rm c} \lesssim 9.5 \times 10^9$ g cm^{-3} (64), where in the lower limit the delaying effects of convective Urca neutrino losses (8, 71) have not been taken into account (see below). This will not be the case if either the accretion rate is low ($\dot{M} \lesssim 10^{-10} M_\odot \, {\rm yr}^{-1}$) or both $M_{\rm core}$ and $\dot{M}$ are large (64). In the first case, outward conduction dominates; in the second pycnonuclear carbon ignition precedes the arrival to the center of the heat flow. In both cases the central layers of the white dwarf are compressed quasi-adiabatically (except for thermal neutrino emission and, in the first case, some heat outflow). Variable fractions of the initial solid core thus survive until carbon ignition, which takes place in the density interval 9.5×10^9 g cm$^{-3} \lesssim \rho_{\rm c} \lesssim 1.5 \times 10^{10}$ g cm^{-3} (64). If we considered the (by-now marginal) possibility of an ordered C+O alloy, central ignition would happen at a still higher density, e.g. $\rho_{\rm c} \simeq 2 \times 10^{10}$ g cm^{-3} (unless it were preceded by off-center ignition at some point in the process of melting of the solid core from outside). This is because in the ordered alloy (provided that

$x_C \lesssim x_O$, where x_C and x_O are the number fractions of carbon and oxygen, respectively), only $^{12}C+^{16}O$ (and $^{16}O+^{16}O$) reactions could take place (except for crystal defects). Ignition would then be triggered by electron captures on ^{16}O at the above density: They would melt the center and thus allow $^{12}C+^{12}C$ reactions in the strong-screening regime.

The main problem for having carbon ignition in the solid core of a C+O white dwarf is the rather high initial mass ($M_{core} \gtrsim 1.13\,M_\odot$) that appears to be required for it (unless hydrogen and helium might be burnt steadily or in weak flashes for low accretion rates).

The dynamical evolution of C+O white dwarfs following thermonuclear runaway at $\rho_c \simeq 10^{10}$ g cm^{-3} was first studied by Canal & Isern (22) and by Isern et al (81). Lower ignition densities ($\rho_c \simeq 4 \times 10^9$ g cm^{-3}), but with ignition still taking place in the solid core, were later considered by Isern et al (80): The motivation for this idea was the exceedingly large enhancement factors of the $^{12}C+^{12}C$ reactions by electron screening then proposed by Ichimaru et al (75). The outcome of ignition at densities in the range 9.5×10^9 g cm$^{-3} \lesssim \rho_c \lesssim 1.5 \times 10^{10}$ g cm^{-3} has been studied in (21, 23, 28, 46, 77–79). Given the ignition density, the question of whether the white dwarf collapses into a neutron star or is instead explosively disrupted (during a Type Ia supernova outburst) depends only on the speed of propagation of burning. (Electron capture rates on NSE matter are reasonably well known.) Conductive propagation of the burning front is the only possible regime in solid layers (if detonation is excluded). However, as we have already noted, the assumption that a detonation does not form upon carbon ignition, even at these high densities, might not be true. Depending on the initial temperature distribution, a region may exist where spontaneous burning creates a supersonic front that leads to a detonation (14, 177). This idea has not yet been explored for ignition at densities as high as those presently considered (nor for ignition in a solid), but it is a further uncertainty involved in AIC that must be kept in mind. Conductive propagation confines v_{burn} within a comparatively narrow range. The expression

$$v_{burn} \simeq 25\left(\frac{\rho}{2 \times 10^9 \text{g cm}^{-3}}\right)^{0.8}\left(\frac{X_C}{0.5}\right) \text{km s}^{-1} \qquad 5.$$

where X_C is carbon mass fraction (183; see also 182), checked by numerical simulations using an extremely fine zoning ($\sim 10^{-5}$ cm) to resolve the burning front, can be considered to be accurate within one order of magnitude. In terms of the local sound speed c_s, Equation 5 gives $v_{burn} \simeq \alpha c_s$, with $0.001 \lesssim \alpha \lesssim 0.01$.

The results of parameterized calculations taken from the above ref-

Table 1 Outcome of explosive carbon ignition at high densities for several combinations of ignition density and velocity of the burning front

Model	ρ_{ign} (10^9 g cm^{-3})	v_{burn}/c_s	Outcome	t_{11}[a] (s)
A	9.50	0.005	Collapse	3.08
A	9.50	0.010	Explosion	—
B	10.00	0.010	Collapse	1.95
B	10.00	0.050	Explosion	—
C	15.00	0.010	Collapse	0.70
C	15.00	0.100	Explosion	—

[a] t_{11} is the time elapsed between explosive ignition and when the central density becomes $\simeq 10^{11}$ g cm^{-3}.

erences are summarized in Table 1, where, for the cases of collapse, the elapsed times between runaway and when the central density becomes as high as $\rho_c \simeq 10^{11}$ g cm^{-3} are given. Time evolution of central density is shown in Figure 1 for central ignition at $\rho_c = 10^{10}$ g cm^{-3} and cases

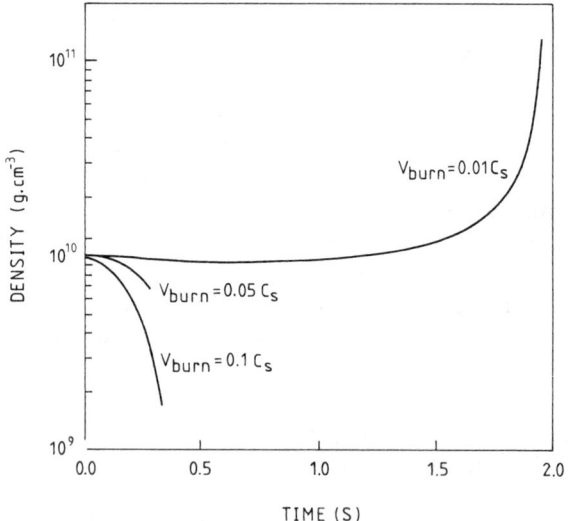

Figure 1 Time evolution of central density following explosive carbon ignition at the center of a C+O white dwarf ($X_C = X_O = 0.50$), for $\rho_c^{ign} = 10^{10}$ g cm^{-3}. The velocity of the burning front is taken to be a constant fraction of the local sound speed: $v_{burn} = \alpha c_s$ (a parameterization that would approximate the behavior of a conductive burning front). Cases $\alpha = 0.01, 0.05,$ and 0.1 are displayed. We see here that bifurcation between collapse and explosion happens for $0.01 \leq \alpha \leq 0.05$.

$\alpha = 0.1$, 0.05, and 0.01. Bifurcation between collapse and explosion happens at $\alpha = 0.02$–0.03. Collapse should thus be a likely outcome of conductive deflagrations starting at such densities.

The preceding results are in contrast with similarly parameterized calculations published by Nomoto (113–116) and by Nomoto & Hashimoto (117), which predict bifurcation at $\alpha \simeq 0.10$–0.15. The latter calculations, however, are clearly wrong, as can be seen, for instance, from the total time they give ($t = 155$ s) for reaching $\rho_c = 10^{11}$ g cm^{-3} with $\alpha = 0.01$ (also starting from $\rho_c = 10^{10}$ g cm^{-3}), plus the total incinerated mass at that time ($M_{\text{inc}} = 0.03\ M_\odot$). Since v_{burn} is $\simeq 10^7$ cm s^{-1}, the result is clearly in error by at least a couple of orders of magnitude. The other two claimed cases of collapse ($\alpha = 0.06$ and $\alpha = 0.1$) are even qualitatively wrong. (They should give explosions.) This apparently resulted from a mistake in trying to simulate a conductive burning front (K. Nomoto, private communication).

A relevant process not previously considered is the interaction of neutrinos emitted by electron captures in the growing central incinerated fireball with the solid layers that still surround this region. Neutrino-electron scattering [47; neutrino-nucleus scattering only plays a minor role (see 19)] heats up these layers and can melt them before the start of dynamical contraction, when the Chandrasekhar mass is still higher than the star's mass (28, 46, 77). The burning front should accelerate in reaching fluid layers, where hydrodynamic instabilities can grow (80). For rather low conductive velocities (within their present range of uncertainty), matter is so neutronized when the burning front emerges into the fluid layers that collapse still ensues for reasonable estimates of hydrodynamic burning velocities. For higher conductive velocities, however, the outcome depends on the poorly known rate of growth of hydrodynamic instabilities (the initial stages of formation of a turbulent burning front). Concerning this, as discussed by Woosley (182) and by Wheeler & Harkness (176), all hydrodynamic deflagration prescriptions (in one dimension) depend on parameters that are equivalent to specifying the propagation speed of the burning front. On the other hand, multidimensional calculations cannot be expected to yield reliable answers in the near future (44). This leaves the issue fairly open.

The acceleration of the burning front due to melting by neutrinos is also important in determining the entropy profile of the collapsing core, and this effect must be included in future collapse calculations. The only calculation of collapse of a C+O white dwarf up to core bounce at nuclear densities has been performed by Baron et al (9). It is based on an initial model that assumes a "convective" deflagration starting very slowly (and thus not propagating from the beginning at $\simeq 0.1\ c_s$, as would be inferred from the paper) at $\rho_c \simeq 10^{10}$ g cm^{-3}. The general conclusion of small mass ejection due to high initial entropy of the core (as compared with the Fe-Ni cores of massive stars) should nonetheless still be valid for more realistic models, since they would have similar characteristics. It must be remembered that ejection of a mass $M_{ej} \leq 0.2\ M_\odot$ would not disrupt the binary system, according to hydrodynamic calculations by Taam & Fryxell (143). Current models for AIC of white dwarfs (either C+O or O+Ne+Mg ones) do not take into account the effects of angular momentum and of magnetic fields. They might nonetheless be very relevant, as has recently been stressed by Narayan & Popham (110) and by Bailyn & Grindlay (7).

Finally, we note that inclusion of convective Urca neutrino losses in the case of carbon ignition at densities $\rho_c \simeq 2$–3×10^9 g cm^{-3} in a completely fluid white dwarf (8, 176, 177) might change the whole picture for AIC of C+O white dwarfs. If burning is actually quenched by neutrino cooling, collapse to a neutron star might be the outcome. Alternatively, ignition in the center of a solid core (where convection can initially develop only to a very limited extent in a small central molten region) or simply in a fluid at higher densities (as can result from complete melting of an initially solid core during the accretion process) eliminates Urca shells as ^{23}Na-^{23}Ne and would thus more easily lead to explosion. Central ignition of solid cores at densities significantly lower than $\rho_c \simeq 10^{10}$ g cm^{-3} might result from the effect of impurities and dislocations (64a), producing explosions such as those already studied by Isern et al (80). The situation would thus be completely reversed.

The case of O+Ne+Mg white dwarfs These white dwarfs would result, in close binary systems, from loss of the helium layer when it expands to red giant size (73, 112). No O+Ne+Mg white dwarfs are expected to form from single-star evolution (56). In contrast, their presence in close binary systems might be indicated by the observation of "Ne novae" (137, 147). They should be, on average, more massive [and less frequent by a factor $\sim 10^4$ (see 73)] than C+O white dwarfs. They would thus have to accrete less material before Ne-O ignition at the center. Miyaji et al (106) calculated that such ignition would occur at $\rho_c \simeq 2 \times 10^{10}$ g cm^{-3}. Actually, this prediction was based on strict adoption of the Schwarzschild criterion for the onset of convective instability. Convection would set in when

$$\nabla > \nabla_{ad}, \qquad\qquad 6.$$

where ∇ is the temperature gradient $d\ln T/d\ln P$ of the mean field, and ∇_{ad} is the adiabatic gradient. A convective core thus starts to grow when heating by electron captures on ^{24}Mg and ^{24}Na (at $\rho_c \simeq 4 \times 10^9$ g cm^{-3})

produces a superadiabatic temperature gradient. But electron captures also create a negative Y_e gradient (Y_e being the electron mole number) that has a stabilizing effect. The situation is such that

$$\nabla_L > \nabla > \nabla_{ad}, \qquad 7.$$

where ∇_L is the Ledoux gradient, defined as

$$\nabla_L \equiv \nabla_{ad} - \left(\frac{\partial \ln T}{\partial \ln Y_e}\right)_{P,\rho} \nabla_{Y_e}, \qquad \nabla_{Y_e} \equiv \frac{d \ln Y_e}{d \ln P}, \qquad 8.$$

where T and P denote temperature and pressure, respectively. This was pointed out by Mochkovitch (108), who suggested that a double diffusive interface forms, analogous to the case of salt-finger instability, at the radius where density equals electron capture threshold density. A detailed numerical analysis would thus be required, since the location of the interface changes with time owing to core compression by accretion.

More recently, Miyaji & Nomoto (105) have followed the evolution of the core of a single star in the mass range $8\ M_\odot \lesssim M \lesssim 10\ M_\odot$ (which should be analogous to that of mass-accreting $O+Ne+Mg$ white dwarfs), now adopting the Ledoux criterion for convective instability. In this case, convection is inhibited up to Ne-O ignition at $\rho_c \simeq 9.5 \times 10^9$ g cm^{-3}. Any form of burning propagation (except dynamical compression) is then (arbitrarily) suppressed, and collapse obviously follows. Note that the above density coincides with the lower limit found in the previous subsection for ignition in a $C+O$ white dwarf with a residual solid core. In the present case, however, no solid layers can remain, even for initially massive and cold white dwarfs, since electron captures on ^{24}Mg and ^{24}Ne at lower densities ($\rho_c \simeq 4 \times 10^9$ g cm^{-3}) already heat the core up to temperatures far above the melting point. Hydrodynamic instability would thus grow from thermonuclear runaway, and a turbulent flame front should propagate burning. A representative parameterization of the velocity of the burning front is given by Woosley & Weaver (183):

$$v_{burn} \simeq F c_s (1 - e^{-r/R_0}) \qquad 9.$$

where r is the distance to the center, and F and R_0 are parameters, for which $F \simeq 0.5$ and $R_0 \simeq 2 \times 10^7$ cm are suggested values for a $C+O$ white dwarf. For velocities of this order, Ne-O ignition would lead to complete disruption of the star (28, 29). It must be noted, however, that the specific energy release when incinerating the $O+Ne+Mg$ mixture is smaller than in the case of a $C+O$ mixture. The buoyancy (to which v_{burn} should be related) is thus also lower. A very slow growth of hydrodynamic instability, as can be derived from Unno's (149) time-dependent theory of convection, for values of the mixing-length parameter $\alpha_{ML} \lesssim 0.7$ ($\alpha_{ML} \equiv l/H_p$, where l is mixing length, and H_p is pressure scale height) would still lead to collapse (K. Nomoto, private communication).

We see that the criterion adopted for the onset of instability and mixing in layers where the temperature gradient takes values between adiabatic and Ledoux's plays a very important role. If semiconvective instability develops and produces mixing (93), explosive ignition might happen for central densities anywhere in the interval $9.5 \times 10^9 \text{ g cm}^{-3} \lesssim \rho_c \lesssim 2 \times 10^{10} \text{ g cm}^{-3}$. A proper analysis should take account of the extra energy needed to bring electrons of lower Fermi level into regions of higher Fermi level, and it should also compare the time scale of mechanical oscillations with that of heat diffusion. But once ignition density has more precisely been fixed, uncertainty as to the rate of growth of the hydrodynamic instability created by explosive burning will still remain (as for C+O ignition in a fluid at similar density).

4. DISCUSSION

We now discuss the possible relevance of the formation mechanisms just described to the origin of the different types of binaries containing neutron stars considered in Section 2.

4.1 *High-Mass X-Ray Binaries*

The origin of high-mass X-ray binary sources can be easily understood (as pointed out in Section 3.1) on the basis of a direct collapse of the nuclearly exhausted core of a massive star (see again ref. 158 for a recent review). Most frequently, mass transfer from the initially more massive component of the binary system will start after the end of core H burning, before He ignition. Mass transfer takes place on a thermal time scale and reduces the primary star to its He core only. (The whole H envelope is accreted by the secondary.) A second stage of mass transfer can take place if $M_{He} < 3.5$ $M_\odot$ owing to expansion of the star during shell C burning (56, 57). However, if $M_{He} > 2.2$ $M_\odot$, core collapse to a neutron star will still eventually occur. This would correspond to initial masses $M > 10$–12 $M_\odot$ of the primary. If mass transfer only starts after core He burning, this lower limit is decreased to $M > 8$–9 $M_\odot$ (see Figure 8.5 in ref. 154).

These conservative evolutionary scenarios satisfactorily account for both early-type and Be-type massive X-ray binaries.

4.2 *Low-Mass X-Ray Binaries*

We first consider globular cluster sources and then discuss galactic bulge sources, since in some scenarios the latter systems originate from the former ones.

4.2.1 GLOBULAR CLUSTER SOURCES Globular clusters contain a relatively large number of low-mass X-ray binaries: Globular clusters make up only $\sim 10^{-4}$ the mass of the Galaxy, but they contain $\sim 10^{-1}$ the number of galactic low-mass X-ray binaries (53, 83, 168, 169). Tidal captures should be favored in globular clusters by nature of both the high densities of stars and the comparatively low relative velocities. In order to evaluate the total number of captures in a globular cluster, Equation 3 in Section 3.2 must be integrated over the cluster volume. In the core region, both n_c and n are much higher than outside this region. Most captures thus occur within a few core radii (172). There is also mass segregation: Stars with higher masses are concentrated toward the core, which enhances n_c there (169).

The number of X-ray sources in globular clusters is compatible (within appreciable uncertainty) with the formation rate expected from Equation 3: in the core of a globular cluster, n can be $\sim 10^5$ pc^{-3}. Compact star density n_c is poorly known. From the models of Verbunt & Meylan (172) for the stellar contents of globular clusters, the total number of neutron stars in cluster cores is $\sim 10^2$, and thus the tidal capture rate should be ~ 1 every 10^9 yr. To compare this with the observed numbers of bright X-ray sources, the typical lifetime of low-mass X-ray binaries must also be known, but there is a large uncertainty on this value. If it were $\sim 10^9$ yr, the expected number would be ~ 1 bright X-ray source per dense cluster, as is observed (see also 154). Combining the two main uncertainties—those on n_c and on the lifetime of the active state of the sources—the final uncertainty is of a factor $\gtrsim 10^2$ (169).

It has been suggested that ultracompact low-mass X-ray sources, such as the 11-min binary 4U 1820−30 in the globular cluster NGC 6624, are in fact hierarchical triples formed by tidal capture of a third star by a compact binary. In this particular case, a 176-day long-term period could be a precession period of the 11-min binary induced by a triple companion in a ~ 2-day orbit (6, 52). According to Bailyn (5) and to Grindlay (53), about 10% of the low-mass X-ray sources in globular clusters might be in hierarchical triples.

Alternatively, neutron stars in globular cluster X-ray sources could also be formed by AIC of white dwarfs (52, 157, and references therein). An argument against the capture mechanism is that neutron stars coming from single (massive) star evolution should be born with kick velocities $\lesssim 30$ km s^{-1} in order to remain in the cluster. From known radio pulsar velocities (2), at most 10% of the neutron stars produced would be retained. More than ~ 1000 neutron stars (and Type II supernovae) should thus typically be produced in a cluster core, instead of the $\sim 10^2$ estimated above. This estimate, in turn, might be in conflict with the observed properties (central condensation and metal enrichment) of the clusters containing high-luminosity X-ray sources (52). This last point, however, is strongly disputed by Verbunt et al (171) on the basis of current models for globular clusters. On the other hand, QPO have been observed in two of the high-luminosity X-ray sources, which would mean that the magnetic fields of the neutron stars exceed $\sim 10^9$ G (160). The neutron stars might thus be relatively young instead of having been formed in the early stages of evolution of the globular cluster. (However, "asymptotic" magnetic field values could well be enough to explain QPO behavior.)

According to Lightman & Grindlay (99), the ratio of the number of white dwarfs to the number of neutron stars in globular cluster cores should be $\gtrsim 100$. This would lead to a high rate of tidal captures of white dwarfs by main sequence stars. Low-luminosity globular cluster X-ray sources form a separate class from the high-luminosity ones (59), and they might be mass-accreting white dwarfs in the low-accretion-rate regime ($\dot{M} \lesssim 10^{-9} M_\odot$ yr^{-1}). Nonetheless, since the companion is a main sequence star, the material should be H rich, and this might be a problem for AIC (see Section 3.3.1).

4.2.2 GALACTIC BULGE SOURCES Given the observed characteristics of galactic bulge sources, it seems unlikely that most of them would have been formed through the collapse and explosion of a massive star (see, however, the discussion in Section 3.1). As for tidal capture mechanisms, one notes that the relative velocities of stars in the galactic bulge are $\gtrsim 100$ km s^{-1}, more than one order of magnitude higher than in globular clusters (154). From Equation 3 in Section 3.2, we see that the corresponding rate of tidal capture should thus be lower by at least the same factor. The star density n is much lower, and so too should be n_c, which also decreases the rate. In addition, the kinetic energy to be dissipated in forming a bound system is larger than the binding energy of a dwarf star. The encounter would thus completely disrupt the latter object (42).

Alternative nonlocal mechanisms involve prior formation of galactic bulge sources (by either tidal capture or AIC) in globular clusters, followed by either escape from the cluster or evaporation or disruption of the cluster itself. Ejection seems unlikely: Even exchange collisions, which may impart a recoil velocity to the newly formed binary, would be unable to significantly alter the original velocity of the neutron star owing to the small mass (a few tenths of a solar mass) of the dwarf star that is exchanged for the neutron star ($\sim 1.5 M_\odot$) (66). It would also be inconsistent with the observed relative populations of cataclysmic variables and of X-ray sources in the bulge (55). Nonetheless, Krolik et al (85, 86) still find that this mechanism might account for a fraction of sources.

Evaporation of globular clusters might be induced by formation of a single, strongly bound central binary (62). Tidal interactions of the clusters with the galactic field can also lead to their disruption (134). The latter possibility was further studied by Bouvier (16), by Ostriker et al (120), and by Spitzer & Chevalier (135). Fall & Rees (41) pointed out that only clusters with masses in the range $\sim 10^4$–10^6 $M_\odot$ would have survived throughout galactic history: Low-mass clusters should lose their stars by evaporation, while high-mass clusters have been tidally disrupted either by other clusters or by repeated passages across the galactic disk. The role of giant molecular clouds in the disruption of globular clusters has been considered by Grindlay (50). This could happen even relatively far from the galactic center (up to galactocentric radii $\lesssim 1$ kpc). Disruption time scales of $\lesssim 3 \times 10^8$ yr are derived for globular clusters orbiting in the ring that extends ~ 4–6 kpc about the galactic center, where giant molecular clouds seem to be concentrated. It is not clear, however, whether the radial distribution of low-mass X-ray binaries in the bulge can be compatible with the cluster disruption hypothesis (36, 94). Also, the luminosity distributions of the low-mass X-ray binaries in the galactic disk and of those in globular clusters are different (97, 173). On the other hand, if the binary 4U 1915−05 were a triple system, it would be a good candidate for a globular cluster origin (53).

Long & van Speybroeck (100) have pointed out that in M31, the 19 sources within 400 pc from the galactic center resemble globular cluster sources in that they are more than twice as bright as the rest of the bulge sources in that galaxy. They might thus have formed in globular clusters that were later dispersed in the bulge. Since the X-ray globular clusters are more strongly concentrated within ~ 400 pc from the center of the Galaxy than the general globular cluster population, this would be a plausible origin for the sources in that region. However, beyond ~ 400 pc from the galactic center, the spatial distribution of globular clusters is the same as that for the rest of the bulge population. This would indicate that hardly any globular clusters have merged with the bulge outside the ~ 400-pc radius. AIC should thus be the preferred formation mechanism beyond that distance.

Scenarios in which neutron star formation by AIC is involved in the production of galactic bulge sources can be found, for instance, in studies by van den Heuvel (154), Taam & van den Heuvel (144), and van den Heuvel (155, 158). In these, a red main sequence star of $\simeq 1$ $M_\odot$ would fill its Roche lobe and transfer mass to a $\simeq 1.4$-$M_\odot$ white dwarf. A shell with mass $\simeq 0.2$ $M_\odot$ would be ejected after the collapse and the bounce of the compact object at nuclear matter densities, with a velocity of $\sim 10^4$ km s^{-1}. This should change the binary into a detached one for a period of $\sim 10^9$ yr, until angular momentum loss by gravitational radiation and/or magnetic braking would again lead to Roche lobe overflow. Less mass ejection and a smaller companion mass might shorten the detached phase down to $\sim 10^8$ yr. It has further been argued (150) that this scenario might explain the observed distribution of novae and bulge X-ray sources in M31. In addition, long detached phases would be consistent with low magnetic field in the neutron star and thus with production of X-ray bursts. Mass transfer driven by gravitational radiation losses and/or magnetic braking should produce accretion rates of $\sim 10^{-10}$–10^{-9} $M_\odot$ yr^{-1} (123). The strong sources in the bulge, on the other hand, should be fed by mass transfer at rates of $\sim 10^{-8}$ $M_\odot$ yr^{-1}. In these cases the companion would be an evolved, low-mass subgiant (175). The systems should resemble Sco X-1, Cyg X-2, and 2S 0921−63 (158).

Triple-star evolution has also been proposed for the origin of low-mass X-ray binaries in the bulge (36; see Section 3.1). From the estimate by van den Heuvel (154) of a formation rate of these sources $\sim 10^{-3}$–10^{-4} times that of massive X-ray binaries, it can be concluded that only one in every 10^3–10^4 massive binaries should undergo triple-star evolution in order to explain the observed numbers of low-mass X-ray binaries. This scenario, however, does not explain the concentration of low-mass X-ray binaries toward the galactic bulge regions.

4.3 Binary and Millisecond Pulsars

Binary pulsars in the PSR 1913+16 group (see Section 2.2) seem to be related to the massive X-ray binaries (157). After formation of such sources, Roche lobe overflow by the massive star should lead to a common envelope phase, with a spiraling in of the neutron star and eventual loss of the envelope. Then, if the companion of the neutron star has a mass $\gtrsim 8$ $M_\odot$, the core will still evolve up to collapse and supernova explosion. This should produce either two neutron stars in a very eccentric orbit (such as PSR 1913+16 and PSR 2303+46) or two runaway pulsars. For lower companion masses (3 $M_\odot \lesssim M \lesssim 8$ $M_\odot$), the core will evolve into a massive white dwarf, and a close system with a circular orbit would form (such as PSR 0655+64).

In contrast, the PSR 1953+29 group of binary pulsars appears to be related to the low-mass X-ray binaries. As was pointed out in the previous section, strong bulge X-ray sources should be fed by mass transfer driven by internal thermonuclear evolution of the companion star. At the end, only the He core of the companion is left, and the neutron star has spiraled out. This scenario would produce systems like PSR 1953+29 (82), which would thus be the descendants of wide low-mass X-ray binaries such as Cyg X-2, 2S 0921−63, Sco X-1, GX1+4, and the bright bulge sources (157). The neutron stars of these systems were likely formed by AIC of a white dwarf. The evolved companion now observed would have sustained mass accretion rates $\dot{M} \gtrsim 10^{-8} M_\odot$ yr^{-1} for periods long enough to allow mass growth of the white dwarf up to the point of collapse. It must be noted, however, that while H would burn steadily or in weak flashes at such rates, explosive ignition of the He layer that grows as a result of H burning would be a problem unless $\dot{M} \gtrsim 5 \times 10^{-8} M_\odot$ yr^{-1} (see discussion in Section 3.3.1). PSR 0820+02, a member of this group, provides strong evidence that a neutron star was recently formed by AIC of a white dwarf. Its magnetic field implies an age $\lesssim 10^7$–10^8 yr, while the companion is a He white dwarf with mass $M = 0.2$–0.4 $M_\odot$. Formation of the latter requires over $\sim 5 \times 10^9$ yr. The paradox of an old binary that contains a young neutron star can be solved by assuming that the neutron star was created in a recent episode of mass accretion, less than $\sim 10^7$ yr ago. That mass transfer terminated recently is inferred from the cooling age ($\lesssim 10^7$ yr) of the companion, a still very hot white dwarf (87).

PSR 1953+29 is one of the four binary millisecond pulsars presently known. The total number of currently active binary millisecond pulsars in the Galaxy should be much larger: up to $\sim 10^4$ if one takes selection effects into account (140). Their spin up, due to mass accretion from a disk, would allow them to reach those very short periods when the magnetic field of the neutron star has reached its "bottom" value of $\sim 10^9$ G (1, 12, 162). Every wide low-mass X-ray binary should evolve into a binary millisecond pulsar to explain the inferred population of the latter.

Single millisecond pulsars also seem to be neutron stars previously spun up by accretion in a close binary system (158). They might result from coalescence of the neutron star with a white dwarf in short-period systems such as PSR 0655+64 (15). Another possible origin for such objects would be "evaporation" of the companion by the large energy flux from the millisecond pulsar itself (121, 125, 161). This seems to be happening in PSR 1957+20 (43).

In globular clusters, single millisecond pulsars might also form either by direct collision of a neutron star with a field star or by the catastrophic encounter of a globular cluster X-ray binary with a field star. In both cases the result would be a single neutron star with a disk around it. Accretion of the disk would spin up the neutron star to millisecond periods. Alternatively, a binary millisecond pulsar can be disrupted by interaction with other cluster stars.

4.4 *Gamma-Ray Burst Sources*

For the sake of completeness, we must also mention that some models for γ-ray burst sources postulate that these sources are made up of a neutron star plus a close companion of low luminosity. The latter would shed the matter that, by accumulating on the neutron star surface, would produce the flashes (181). An origin based on AIC of a white dwarf might thus be invoked. Currently preferred models for these sources do, however, assume that they are single neutron stars (127).

5. SUMMARY

In Table 2 we summarize the present status of our knowledge as to the origin of neutron stars in binaries.

The clearest case is that of Population I X-ray sources (both pulsating and nonpulsing): These sources should be formed by core collapse of a massive star that has previously transferred its envelope to the companion, as discussed in Section 4.1. Binary radio pulsars with companion masses of $\sim 1\ M_\odot$ are also most likely products of massive-star evolution.

Accretion-induced collapse of a white dwarf is a likely formation mechanism of the neutron star in binaries where the companion is a low-mass star and tidal capture of the latter by the neutron star has a very low probability. AIC is especially suggested by old population systems where the neutron star appears to be young, such as the binary X-ray pulsars with low-mass companions. AIC and globular cluster origin (through disruption of the cluster) are still competing hypotheses in the case of the X-ray sources in the galactic bulge. Radial distribution of the sources

Table 2 Possible formation mechanisms of neutron stars in binaries

Type of source	Companion	Number of sources	Origin[a]
Binary X-ray pulsar	Massive star	~ 20	CC
Binary pulsar	Low-mass star	4	AIC
Nonpulsing Pop I X-ray source	Massive star	~ 20	CC
Galactic bulge X-ray source	Low-mass star	~ 30	AIC or GCD
Globular cluster X-ray source	Low-mass star	11	TC or AIC
Binary radio pulsar	Low-mass white dwarf	9	AIC
Binary radio pulsar	Neutron star or massive white dwarf	3	CC
Single millisecond pulsar	None	4	AIC or CL

[a] Abbreviations: CC, core collapse or massive star; AIC, accretion-induced collapse; GCD, globular cluster disruption; TC, tidal capture; CL, collision of single neutron star or binary with another star.

would rather favor AIC origin, whereas confirmed presence of hierarchical triples would strengthen the globular cluster hypothesis (see Section 4.2).

Formation by tidal capture is most likely in globular clusters. AIC might nonetheless account for a fraction of X-ray sources there. Uncertainties as to the number of neutron stars in globular cluster cores, mass segregation, and the lifetimes of low-mass binary X-ray sources make it difficult to ascertain the relative relevance of both mechanisms (TC and AIC) at present.

Binary radio pulsars with low-mass companions (including binary millisecond pulsars) are most easily explained by AIC. Single millisecond radio pulsars in the galactic disk likely started as binaries and then "vaporized" their companions. In globular clusters they might also result from disruption of binary millisecond pulsars by collisions with field stars or from collisions of single neutron stars or of low-mass X-ray binaries with field stars.

The problem of the feasibility of AIC based on the physics of mass-accreting white dwarfs cannot be regarded as solved. Neither $C+O$ nor $O+Ne+Mg$ white dwarfs obviously collapse when they grow in mass and approach the Chandrasekhar limit. As we have seen, their mass growth up to the point of central explosive ignition (which should *always* precede collapse) is already problematic and subjected to severe constraints. Collapse would be facilitated, in $C+O$ white dwarfs, by keeping a central fraction of their cores solid until explosive ignition. This further narrows the range of initial conditions and of mass-accretion rates (the last for a given chemical composition of the inflowing material) that can be allowed. Still, uncertainties in pycnonuclear reaction rates, phase diagrams of $C+O$ mixtures, conductive front velocities, and development of turbulent flame fronts do obscure the final issue. The case is not necessarily better for $O+Ne+Mg$ white dwarfs: Here, the treatment of semiconvection plus, again, the initiation of hydrodynamic burning play the leading roles. At present, AIC might result either from a small fraction of mass-accreting $C+O$ white dwarfs (narrow range of initial conditions), or from most $O+Ne+Mg$ white dwarfs (total numbers much smaller), or from the addition of both types of contributions in some unspecified proportion. It might also finally prove to be physically impossible in all cases. Astronomical evidence, however, points to the contrary. Further work is thus needed on the problems just enumerated. Most of these problems are also relevant to the physics of Type I supernovae (to that of "classical," or Type Ia, events at least). All this also points to the need to still carefully consider other possible origins for neutron stars, even in systems where AIC appears most likely.

ACKNOWLEDGMENTS

We would first like to thank Prof. Evry Schatzman, who introduced us to the problem of the origin of neutron stars in binary systems and whose insights on the physics of white dwarfs and their possible AIC have guided us through much of our work. One of us (RC) would also like to thank Profs. Rudolf Kippenhahn and Wolfgang Hillebrandt and Dr. Ewald Müller for their kind hospitality at the Max-Planck-Institut für Astrophysik, where part of this paper was written. Finally, thanks also to the many authors who sent us preprints and reprints of their papers. The writing of this review was supported in part by CICYT grants PB87-0147 and PB87-0150 at the University of Barcelona and by grant PB034 at the Centre d'Estudis Avançats de Blanes.

Literature Cited

1. Alpar, M. A., Cheng, A. F., Ruderman, M. A., Shaham, J. 1982. *Nature* 300: 728
2. Anderson, B., Lyne, A. 1983. *Nature* 303: 597
3. Arnett, W. D., Bahcall, J. N., Kirshner, R. P., Woosley, S. E. 1989. *Annu. Rev. Astron. Astrophys.* 27: 629
4. Backer, D. C., Kulkarni, S. R., Heiles, C., Davis, M. M., Goss, W. M. 1982. *Nature* 300: 615
5. Bailyn, C. D. 1987. PhD thesis. Harvard Univ., Cambridge, Mass.
6. Bailyn, C. D., Grindlay, J. E. 1987. *Ap. J. Lett.* 316: 625
7. Bailyn, C. D., Grindlay, J. E. 1989. *Ap. J.* In press
8. Barkat, Z., Wheeler, J. C. 1990. Preprint
9. Baron, E., Cooperstein, J., Kahana, S. H., Nomoto, K. 1987. *Ap. J.* 320: 304
10. Barrat, J. L., Hansen, J. P., Mochkovitch, R. 1988. *Astron. Astrophys.* 199: L15
11. Belian, R. D., Conner, J. P., Evans, W. D. 1976. *Ap. J. Lett.* 206: L135
12. Bhattacharya, D., Srinivasan, G. 1986. *Curr. Sci.* 55: 327
13. Blaauw, A. 1961. *Bull. Astron. Inst. Neth.* 15: 265
14. Blinnikov, S. I., Khokhlov, A. M. 1986. *Sov. Astron. Lett.* 12: 131
15. Bonsema, P. F. A., van den Heuvel, E. P. J. 1985. *Astron. Astrophys.* 46: L3
16. Bouvier, P. 1971. *Astron. Astrophys.* 14: 341
17. Bravo, E., Isern, J., Labay, J., Canal, R. 1983. *Astron. Astrophys.* 124: 39
18. Brush, S. G., Sahlin, H. L., Teller, E. 1966. *J. Chem. Phys.* 45: 2102
19. Burrows, A. 1989. In *Supernovae*, ed. A. G. Petschek. Berlin: Springer-Verlag. In press
20. Cameron, A. G. W. 1959. *Ap. J.* 130: 916
21. Canal, R., Hernanz, M., Isern, J., Labay, J., Mochkovitch, R. 1986. In *Accretion Processes in Astrophysics*, ed. J. Audouze, J. Tran Thanh Van, p. 109. Paris: Ed. Front.
22. Canal, R., Isern, J. 1979. In *White Dwarfs and Variable Degenerate Stars*, IAU Colloq. No. 153, ed. H. M. Van Horn, V. Weidemann, p. 52. Rochester, NY: Univ. Rochester Press
23. Canal, R., Isern, J., Hernanz, M., Labay, J., Bravo, E., et al. 1987. In *Advances in Nuclear Astrophysics*, ed. E. Vangioni-Flam, J. Audouze, M. Cassé, J. P. Chièze, J. Tran Thanh Van, p. 251. Paris: Ed. Front.
24. Canal, R., Isern, J., Labay, J. 1980. *Ap. J. Lett.* 241: L33
25. Canal, R., Isern, J., Labay, J. 1982a. In *Supernovae: A Survey of Current Research*, ed. M. J. Rees, R. J. Stoneham, p. 215. Dordrecht: Reidel
26. Canal, R., Isern, J., Labay, J. 1982b. *Nature* 296: 225
27. Canal, R., Isern, J., Labay, J. 1984. In *Problems of Collapse and Numerical Relativity*, ed. D. Banzel, M. Signore, p. 117. Dordrecht: Reidel
28. Canal, R., Isern, J., Labay, J. 1989. In

Timing Neutron Stars, ed. H. Ögelman, E. P. J. van den Heuvel, p. 631. Dordrecht: Kluwer
29. Canal, R., Isern, J., Labay, J., López, R. 1990. In *Supernovae*, ed. S. E. Woosley. Berlin: Springer-Verlag. In press
30. Canal, R., Schatzman, E. 1974a. *C. R. Acad. Sci. Paris* 279: B-681
31. Canal, R., Schatzman, E. 1974b. *Mem. Soc. Astron. Ital.* 45: 763
32. Canal, R., Schatzman, E. 1976. *Astron. Astrophys.* 46: 229
33. Clark, G. W. 1975. *Ap. J. Lett.* 199: L143
34. Colgate, S. A. 1970. *Nature* 225: 247
35. Cominsky, L., Jones, C., Forman, W., Tananbaum, H. 1978. *Ap. J.* 224: 46
36. Eggleton, P. P., Verbunt, F. 1986. *MNRAS* 220: 138
37. Epstein, R. I., Arnett, W. D. 1975. *Ap. J.* 201: 202
38. Ergma, E. V., Tutukov, A. V. 1976. *Acta Astron.* 26: 69
39. Fabian, A. C., Pringle, J. E., Rees, M. J. 1975. *MNRAS* 172: 15P
40. Fall, M., Malkan, M. A. 1978. *MNRAS* 185: 899
41. Fall, M., Rees, M. J. 1977. *MNRAS* 181: 37P
42. Finzi, A. 1978. *Astron. Astrophys.* 62: 149
43. Fruchter, A. S., Stinebring, D. R., Taylor, J. H. 1989. *Nature* 333: 237
44. Fryxell, B. A., Müller, E., Arnett, W. D. 1990. In *Numerical Astrophysics*, ed. P. R. Woodward. New York: Academic. In press
45. Fuller, G. M., Fowler, W. A., Newman, M. J. 1982. *Ap. J. Suppl.* 48: 279
46. García, D., Labay, J., Canal, R., Isern, J. 1990. *Astrophys. Space Sci.* In press
47. Gershtein, S. S., Imshennik, V. S., Nadezhin, D. K., Folomeshkin, V. N., Khlopov, M. Yu., et al. 1976. *Sov. Phys.-JETP* 42: 751
48. Giacconi, R., Gursky, H., Kellogg, E., Schreier, E., Tananbaum, H. 1971. *Ap. J. Lett.* 167: L67
49. Godon, P., Shaviv, G., Ashkenazi, J., Kovetz, A. 1989. In *White Dwarfs, IAU Colloq. No. 114*, ed. G. Wegner, p. 85. Berlin: Springer-Verlag
50. Grindlay, J. E. 1984. *Adv. Space Res.* 3: 19
51. Grindlay, J. E. 1986. In *The Evolution of Galactic X-Ray Binaries*, ed. J. Trümper, W. H. G. Lewin, W. Brinkmann, p. 25. Dordrecht: Reidel
52. Grindlay, J. E. 1987. In *The Origin and Evolution of Neutron Stars, IAU Symp. No. 125*, ed. D. J. Helfand, J.-H. Huang, p. 173. Dordrecht: Reidel
53. Grindlay, J. E. 1988. *Adv. Space Res.* 8(2): 539
54. Grindlay, J., Gursky, H., Schnopper, H., Parsignault, D. R., Heise, J., et al. 1976. *Ap. J. Lett.* 205: L127
55. Grindlay, J. E., Hertz, P. 1985. *Proc. Workshop Cataclysmic Var. and Low-Mass X-Ray Sources, 7th*, ed. D. Q. Lamb, J. Patterson, p. 79. Dordrecht: Reidel
56. Habets, G. H. M. J. 1985. PhD thesis. Univ. Amsterdam, Neth.
57. Habets, G. H. M. J. 1986. *Astron. Astrophys.* 167: 61
58. Hachisu, I., Kato, M., Saio, H. 1989. *Ap. J. Lett.* 342: L19
59. Hertz, P., Grindlay, J. E. 1983. *Ap. J.* 275: 105
60. Hasinger, G. 1988. *Adv. Space Res.* 8: 367
61. Hasinger, G., van der Klis, M. 1989. In *Timing Neutron Stars*, ed. H. Ögelman, E. P. J. van den Heuvel. Dordrecht: Kluwer
62. Heggie, D. C. 1977. *Comments Astrophys. Space Phys.* 7: 43
63. Helfand, D. J., Chanan, G. A., Novick, R. 1980. *Nature* 283: 337
64. Hernanz, M., Isern, J., Canal, R., Labay, J., Mochkovitch, R. 1988. *Ap. J.* 324: 331
64a. Hernanz, M., Isern, J., Canal, R., Labay, J. 1990. *Astrophys. Space Sci.* In press
65. Hewish, A., Bell, S. J., Pilkington, J. D. M., Scott, P. F., Collins, R. A. 1968. *Nature* 217: 709
66. Hills, J. G. 1976. *MNRAS* 175: 1P
67. Hills, J. G. 1983. *Ap. J.* 267: 322
68. Hulse, R. A., Taylor, J. H. 1975. *Ap. J. Lett.* 195: L51
69. Hut, P. 1984. *Ap. J. Suppl.* 55: 301
70. Hut, P., Bahcall, J. N. 1983. *Ap. J.* 268: 319
71. Iben, I. 1982. *Ap. J.* 253: 248
72. Iben, I., Nomoto, K., Tornambè, A., Tutukov, A. V. 1987. *Ap. J.* 317: 717
73. Iben, I., Tutukov, A. V. 1984. *Ap. J. Suppl.* 54: 335
74. Iben, I., Tutukov, A. V. 1986. *Ap. J.* 311: 753
75. Ichimaru, S., Iyetomi, H., Mitake, S., Itoh, N. 1983. *Ap. J. Lett.* 265: L83
76. Ichimaru, S., Iyetomi, H., Ogata, S. 1988. *Ap. J. Lett.* 334: L17
77. Isern, J., Canal, R., García, D., Hernanz, M., Labay, J. 1989. In *White Dwarfs, IAU Colloq. No. 114*, ed. G. Wegner, p. 88. Berlin: Springer-Verlag
78. Isern, J., Canal, R., García, D., García-Berro, E., Hernanz, M., et al. 1988. *Adv. Space Res.* 8(2): 703
79. Isern, J., Canal, R., Labay, J., García,

D. 1990. In *Supernovae*, ed. S. E. Woosley. Berlin: Springer-Verlag. In press
80. Isern, J., Labay, J., Canal, R. 1984. *Nature* 309: 431
81. Isern, J., Labay, J., Hernanz, M., Canal, R. 1983. *Ap. J.* 273: 320
82. Joss, P. C., Rappaport, S. A. 1983. *Nature* 304: 419
83. Katz, J. I. 1975. *Nature* 253: 698
84. Kirk, J., Trümper, J. 1983. In *Accretion-Driven Stellar X-Ray Sources*, ed. W. H. G. Lewin, E. P. J. van den Heuvel, p. 261. Cambridge: Univ. Press
85. Krolik, J., Meiksin, A., Joss, P. C. 1984. *Ap. J.* 282: 466
86. Krolik, J., Meiksin, A., Joss, P. C. 1985. *Proc. Workshop Cataclysmic Var. and Low-Mass X-Ray Binaries, 7th*, ed. D. Q. Lamb, J. Patterson, p. 107. Dordrecht: Reidel
87. Kulkarni, S. 1986. *IAU Circ. No. 4160*
88. Labay, J., Canal, R., García-Berro, E., Hernanz, M., Isern, J. 1985. In *Recent Results on Cataclysmic Variables*, ed. J. Rahe, p. 29. *ESA SP-236*
89. Labay, J., Canal, R., Isern, J. 1983. *Astron. Astrophys.* 117: L1
90. Lamb, D. Q. 1982. In *Gamma-Ray Transients and Related Astrophysical Phenomena*, ed. R. E. Lingenfelter, H. S. Hudson, D. M. Worrall. New York: AIP Press
91. Lamb, F. K. 1988. *Adv. Space Res.* 8(4): 355
92. Landau, L. D., Lifshitz, E. M. 1959. *Fluid Mechanics*. Oxford: Pergamon
93. Langer, N., Sugimoto, D., Fricke, K. 1983. *Astron. Astrophys.* 126: 207
94. Lewin, W. H. G. 1977. *Ann. NY Acad. Sci.* 310: 210
95. Lewin, W. H. G., Joss, P. C. 1983. In *Accretion-Driven Stellar X-Ray Sources*, ed. W. H. G. Lewin, E. P. J. van den Heuvel, p. 41. Cambridge: Univ. Press
96. Lewin, W. H. G., Ricker, G. R., McClintock, J. E. 1971. *Ap. J. Lett.* 169: L17
97. Lewin, W. H. G., van Paradijs, J. 1985. *Astron. Astrophys.* 142: 361
98. Lewin, W. H. G., van Paradijs, J., van der Klis, M. 1988. *Space Sci. Rev.* 46: 273
99. Lightman, A. P., Grindlay, J. E. 1982. *Ap. J.* 262: 145
100. Long, K. S., van Speybroeck, L. P. 1983. In *Accretion-Driven Stellar X-Ray Sources*, ed. W. H. G. Lewin, E. P. J. van den Heuvel, p. 117. Cambridge: Univ. Press
101. Loumos, G. L., Hubbard, W. B. 1973. *Ap. J.* 180: 199
102. Maraschi, L., Treves, A., van den Heuvel, E. P. J. 1976. *Nature* 259: 292
103. Mestel, L. 1952. *MNRAS* 112: 583
104. Mestel, L., Ruderman, M. A. 1967. *MNRAS* 136: 27
105. Miyaji, S., Nomoto, K. 1987. *Ap. J.* 318: 307
106. Miyaji, S., Nomoto, K., Yokoi, K., Sugimoto, D. 1980. *Publ. Astron. Soc. Jpn.* 32: 303
107. Mochkovitch, R. 1983. *Astron. Astrophys.* 122: 212
108. Mochkovitch, R. 1984. In *Problems of Collapse and Numerical Relativity*, ed. D. Banzel, M. Signore, p. 125. Dordrecht: Reidel
109. Mochkovitch, R., Livio, M. 1989. *Astron. Astrophys.* 209: 110
110. Narayan, R., Popham, R. 1989. *Ap. J. Lett.* 346: L25
111. Nomoto, K. 1982. *Ap. J.* 253: 798
112. Nomoto, K. 1984. *Ap. J.* 277: 791
113. Nomoto, K. 1986a. In *Accretion Processes in Astrophysics*, ed. J. Audouze, J. Tran Thanh Van, p. 137. Paris: Ed. Front.
114. Nomoto, K. 1986b. *Prog. Part. Nucl. Phys.* 17: 249
115. Nomoto, K. 1987a. In *The Origin and Evolution of Neutron Stars, IAU Symp. No. 125*, ed. D. J. Helfand, J.-H. Huang, p. 281. Dordrecht: Reidel
116. Nomoto, K. 1987b. *Proc. Tex. Symp. Relativ. Astrophys., 13th*, ed. M. P. Ulmer, p. 519. Singapore: World Scientific
117. Nomoto, K., Hashimoto, M. 1987. *Astrophys. Space Sci.* 130: 395
118. Nomoto, K., Iben, I. 1985. *Ap. J.* 297: 531
119. Ogata, S., Ichimaru, S. 1987. *Phys. Rev. A* 36: 5451
120. Ostriker, J. P., Spitzer, L., Chevalier, R. A. 1972. *Ap. J. Lett.* 176: L51
121. Phinney, E. S., Evans, C. R., Blandford, R. D., Kulkarni, S. R. 1988. *Nature* 333: 832
122. Press, W. H., Teukolsky, S. A. 1977. *Ap. J.* 213: 183
123. Rappaport, S. A., Joss, P. C., Webbink, R. F. 1982. *Ap. J.* 254: 616
124. Rappaport, S. A., Nelson, L. A., Ma, C. P., Joss, P. C. 1987. *Ap. J.* 322: 842
125. Ruderman, M. A., Shaham, J., Tavani, M. 1989. *Ap. J.* 336: 507
126. Salpeter, E. E., van Horn, H. M. 1969. *Ap. J.* 155: 183
127. Schaefer, B. E., Cline, T. L., Desai, U., Teegarden, B. J., Atteia, J.-L., et al. 1987. *Ap. J.* 313: 226
128. Schatzman, E. 1974. Presented at Int. Sch. Cosmol. Gravit., Erice, Italy
129. Schreier, E., Levinson, R., Gursky, H.,

Kellogg, E., Tananbaum, H., Giacconi, R. 1972. *Ap. J. Lett.* 172: L79
130. Shapiro, S. L., Teukolsky, S. S. 1984. *Black Holes, White Dwarfs, and Neutron Stars*. New York: Wiley
131. Shaviv, G., Starrfield, S. 1987. *Ap. J. Lett.* 321: L51
132. Slattery, W. L., Doolen, G. D., DeWitt, H. E. 1982. *Phys. Rev. A* 21: 2087
133. Sparks, W. M., Kutter, G. S. 1987. *Ap. J.* 321: 394
134. Spitzer, L. 1958. *Ap. J.* 127: 17
135. Spitzer, L., Chevalier, R. A. 1973. *Ap. J. Lett.* 176: L51
136. Staelin, D. M., Reifenstein, E. C. 1968. *Science* 162: 1481
137. Starrfield, S. 1990. In *Supernovae*, ed. S. E. Woosley. Berlin: Springer-Verlag. In press
138. Starrfield, S., Sparks, W. M., Shaviv, G. 1989. *Ap. J. Lett.* 325: 235
139. Stevenson, D. J. 1980. *J. Phys. Suppl.* 41(3): C2
140. Stokes, G. H., Taylor, J. H., Dewey, R. J. 1985. *Ap. J. Lett.* 294: L21
141. Sugimoto, D., Nomoto, K. 1980. *Space Sci. Rev.* 25: 155
142. Sutantyo, W. 1975. *Astron. Astrophys.* 44: 227
143. Taam, R. E., Fryxell, B. A. 1984. *Ap. J.* 279: 166
144. Taam, R. E., van den Heuvel, E. P. J. 1986. *Ap. J.* 305: 235
145. Tananbaum, H., Gursky, H., Kellogg, E. M., Levinson, R., Schreier, E., Giacconi, R. 1972. *Ap. J. Lett.* 174: L143
146. Thorne, K. S., Żytkov, A. N. 1975. *Ap. J. Lett.* 199: L19
147. Truran, J. W., Livio, M. 1986. *Ap. J.* 308: 721
148. Tutukov, A. V., Yungelson, L. R. 1973. *Nautsnie Inform.* 27: 58
149. Unno, W. 1976. *Publ. Astron. Soc. Jpn.* 19: 140
150. Vader, J. P., van den Heuvel, E. P. J., Lewin, W. H. G., Takens, R. J. 1982. *Astron. Astrophys.* 113: 328
151. van den Heuvel, E. P. J. 1974. *Proc. Solvay Conf. Phys., 16th*, p. 119. Brussels: Univ. Brussels Press
152. van den Heuvel, E. P. J. 1978. In *Physics and Astrophysics of Neutron Stars and Black Holes*, ed. R. Giacconi, R. Ruffini, p. 828. Amsterdam: North-Holland
153. van den Heuvel, E. P. J. 1981. In *Fundamental Problems in the Theory of Stellar Evolution*, ed. D. Sugimoto, D. Q. Lamb, D. N. Schramm, p. 155. Dordrecht: Reidel
154. van den Heuvel, E. P. J. 1983. In *Accretion-Driven Stellar X-Ray Sources*, ed. W. H. G. Lewin, E. P. J. van den Heuvel, p. 303. Cambridge: Univ. Press
155. van den Heuvel, E. P. J. 1984. *J. Astrophys. Astron.* 5: 209
156. van den Heuvel, E. P. J. 1987. In *The Origin and Evolution of Neutron Stars, IAU Symp. No. 125*, ed. D. J. Helfand, J.-H. Huang, p. 393. Dordrecht: Reidel
157. van den Heuvel, E. P. J. 1988. *Adv. Space Res.* 8(2): 355
158. van den Heuvel, E. P. J. 1989. In *Timing Neutron Stars*, ed. H. Ögelman, E. P. J. van den Heuvel, p. 523. Dordrecht: Kluwer
159. van den Heuvel, E. P. J., Heise, J. 1972. *Nature* 239: 67
160. van den Heuvel, E. P. J., Taam, R. E. 1984. *Nature* 309: 235
161. van den Heuvel, E. P. J., van Paradijs, J. 1988. *Nature* 334: 227
162. van den Heuvel, E. P. J., van Paradijs, J., Taam, R. E. 1986. *Nature* 322: 153
163. van der Klis, M., Jansen, F., van Paradijs, J., Lewin, W. H. G., Sztajno, M., et al. 1987. *Ap. J. Lett.* 313: L19
164. van Paradijs, J. 1983. In *Accretion-Driven Stellar X-Ray Sources*, ed. W. H. G. Lewin, E. P. J. van den Heuvel, p. 185. Cambridge: Univ. Press
165. van Paradijs, J. 1989. In *Timing Neutron Stars*, ed. H. Ögelman, E. P. J. van den Heuvel, p. 191. Dordrecht: Kluwer
166. van Paradijs, J., Lewin, W. H. G. 1989. *Mem. Soc. Astron. Ital.* In press
167. Verbunt, F. 1987. *Ap. J. Lett.* 312: L23
168. Verbunt, F. 1988. *Adv. Space Res.* 8(2): 529
169. Verbunt, F. 1989. In *Timing Neutron Stars*, ed. H. Ögelman, E. P. J. van den Heuvel, p. 593. Dordrecht: Kluwer
170. Verbunt, F., Hut, P. 1987. In *The Origin and Evolution of Neutron Stars, IAU Symp. No. 125*, ed. D. J. Helfand, J.-H. Huang, p. 187. Dordrecht: Reidel
171. Verbunt, F., Lewin, W. H. G., van Paradijs, J. 1989. *MNRAS* 241: 51
172. Verbunt, F., Meylan, G. 1988. *Astron. Astrophys.* 203: 297
173. Verbunt, F., van Paradijs, J., Elson, R. 1984. *MNRAS* 210: 899
174. Webbink, R. F. 1979. In *White Dwarfs and Variable Degenerate Stars, IAU Colloq. No. 53*, ed. H. M. Van Horn, V. Weidemann, p. 426. Rochester, NY: Univ. Rochester Press
175. Webbink, R. F., Rappaport, S. A., Savonije, G. J. 1983. *Ap. J.* 270: 678
176. Wheeler, J. C., Harkness, R. P. 1990. *Fundam. Cosmic Phys.* In press
177. Wheeler, J. C., Harkness, R. P., Barkat, Z., Swartz, D. 1986. *Publ. Astron. Soc. Pac.* 98: 1018

178. Wheeler, J. C., Lecar, M., McKee, C. F. 1975. *Ap. J.* 200: 145
179. Whelan, J., Iben, I. 1973. *Ap. J.* 186: 1007
180. Wolszczan, A., Anderson, S., Kulkarni, S. R., Prince, T. 1989. *IAU Circ. No. 4880*
181. Woosley, S. E. 1984. In *High Energy Transients in Astrophysics*, ed. S. E. Woosley, p. 485. New York: AIP Press
182. Woosley, S. E. 1989. In *Supernovae*, ed. A. G. Petschek. Berlin: Springer-Verlag. In press
183. Woosley, S. E., Weaver, T. A. 1986. *Annu. Rev. Astron. Astrophys.* 24: 205

H I IN THE GALAXY[1]

John M. Dickey

Department of Astronomy, University of Minnesota, Minneapolis, Minnesota 55455

Felix J. Lockman

National Radio Astronomy Observatory, Charlottesville, Virginia 22903

KEY WORDS: galactic structure, interstellar medium, 21-cm radiation, ultraviolet spectra

1. PREFACE

This is an observational review. The topic of H I in the Galaxy is so vast that it cannot be covered in any single article. Recent comprehensive reviews by Kulkarni & Heiles emphasizing the thermodynamics and astrophysics of the atomic phase (152, 153), and by Burton on the galactic distribution and morphology of H I (34), together run to more than 120 pages and free us from the burden of discussing those topics here. Instead, we focus on observations of interstellar atomic hydrogen, from Lyman-α and the 21-cm line through indirect tracers like dust and gamma rays; on what one can and cannot learn from these observations; on what the H I sky looks like; and, finally, on the state of our knowledge and prospects for the future. We also discuss results from recent H I studies that bear on two important areas: the spatial organization of interstellar H I, and its vertical distribution in the Galaxy. There is no mention of the galactic nucleus (see 28, 163), and we are unable to summarize or even list the many observations of H I associated with objects such as individual H II regions, supernova remnants, high-velocity clouds, or specific interstellar clouds (see, for example, 56).

The sheer abundance of galactic H I data drives us to speak of it in terms derived from theoretical models that we do not have space to discuss. The temperature variations of the interstellar medium are so dramatic, its

[1] The US Government has the right to retain a nonexclusive royalty-free license in and to any copyright covering this paper.

pressure structure is so problematic, and its dynamics, hydrodynamics, and/or magnetohydrodynamics are so intricate that it has commanded the attention of several generations of the finest theoretical astrophysicists. We assume in this review that the reader is familiar enough with terms like "cloud" and "the two-phase model" to get the gist of what we say, while being thoughtful enough to recognize that, whatever else is true about galactic H I, the last word on its properties has yet to be spoken.

2. 21-cm OBSERVATIONS

2.1 Critical Review of the Observational Tools

2.1.1 GENERAL COMMENTS The only radio transition of ground-state neutral hydrogen is the 21-cm hyperfine line at 1420.4058 MHz. In contrast to the typically highly saturated ultraviolet lines of H I, a Gaussian 21-cm line has an opacity at line center of

$$\tau = 5.2 \times 10^{-19} \frac{N_\text{H}}{\Delta v T_\text{s}}, \qquad 1.$$

where the spin temperature T_s is the kinetic temperature in most cases, and Δv is the line's full width at half-maximum, in kilometers per second (93, 153).

Since typically $T_\text{s} \geq 50$ K and $\Delta v \sim 10$ km s^{-1}, the column density N_H must be greater than 10^{21} cm^{-2} before the line center is opaque. This is about 10 times the total column density in the direction of the galactic poles (168), but values of 10^{21} cm^{-2} in a narrow velocity interval are common at low latitudes or toward dense clouds. So, unlike Lα, the 21-cm line often has a moderate or even low opacity, and a single spectrum may have a number of line components whose opacities range from $< 10^{-4}$ to ≥ 4. The spectra usually contain a bewildering amount of kinematic detail.

2.1.2 OBSERVING EMISSION Emission at 21 cm was first detected nearly simultaneously by three groups (89, 187), and it was studied intensely in the 1950s by astronomers in the Netherlands and Australia using telescopes of the 25-m class (see ref. 248 for a historical review). The ubiquity of atomic hydrogen in the Galaxy, the low energy of the 21-cm transition, its long lifetime, and its relatively high excitation temperature (compared with the radio continuum brightness temperature in most directions) make it easily detectable in emission in every direction despite its generally low opacity. It is the strongest thermal spectral line in radio astronomy. Profiles have a peak brightness temperature never less than about 0.5 K and as high as 125 K at low galactic latitude.

2.1.2.1 *Sensitivity* The accuracy of galactic 21-cm spectra is not always determined by noise from the receiving equipment. A filled-aperture (single-dish) radio telescope with a state-of-the-art receiver can take spectra at 1 km s^{-1} resolution with a noise level of only $\Delta T_b \sim 0.5\ t^{-1/2}$ K (where t is the integration time, in seconds), independent of the telescope size, structure, or angular resolution. Thus, a minute's integration on a modern single dish gives an H I spectrum with a signal-to-noise ratio of about 15 at the galactic poles and more than 1500 in the plane. Receiver noise is no longer the limiting factor in filled-aperture 21-cm emission studies.

The same is not true for observations using a radio interferometer or "synthetic aperture." Here the noise, expressed as a brightness temperature ΔT_b, depends on the telescope configuration because in these systems it is the noise in units of flux density per beam, i.e. $\Delta T_b \times \text{HPBW}^2$, that is a constant. (HPBW is the full width at half-power of the antenna beam.) In general, a synthesis telescope produces a map that has a brightness noise that goes as the inverse square of the beam size, so increasing the angular resolution of a synthesis telescope by moving the individual elements further apart reduces the telescope's sensitivity to surface brightness. The trade-off is extremely advantageous when observing an object of small angular size and high surface brightness, but it has drawbacks for galactic H I emission work. As an order-of-magnitude example, consider the Very Large Array (VLA), which currently obtains 21-cm spectra with a brightness temperature noise in a 1 km s^{-1} channel of $\Delta T_b \sim 100 \times [\text{HPBW (arcmin)}]^{-2}(t^{-1/2})$ K. In its most compact configuration, the VLA has a HPBW at 21 cm of $\sim 1'$. It would thus take more than 6 hr to get a 3σ detection on a 1-K galactic H I line from a cloud 1' in diameter. At the higher angular resolutions of which the VLA is capable (e.g. 1"), the integration time required to detect even the bright H I in the galactic plane is prohibitively large.

2.1.2.2 *Stray radiation* Radio telescopes have some response (sidelobes) in all directions, not just in the vicinity of the main beam, and this can compromise H I emission measurements. Sidelobes arise principally from Fraunhofer diffraction at the dish edge and from scattering off the feed-support legs or any other obstacle in the aperture. The aperture-blockage sidelobes, which can lie many tens of degrees off the telescope axis, cause the most problems for galactic H I studies. Although the sidelobes have small amplitude, they cover a large solid angle and can contain a significant fraction of the telescope's total response. When pointed toward the galactic poles, many single dishes receive about equal amounts of radiation from the galactic plane via a far sidelobe as from the pole itself (168). This so-called stray radiation has the effect of reducing the dynamic range of 21-

cm observations and may even create spurious and time-varying spectral features (135).

Stray radiation is not a significant problem for telescopes that have an unblocked aperture, e.g. horns or horn-reflectors, and these have been used to measure highly accurate H I profiles (120, 245, 282). Unfortunately, all existing unblocked telescopes are relatively small and thus have poor angular resolution (HPBW of a few degrees). Larger telescopes can be used for accurate H I studies if their spectra are corrected for stray radiation, either by modeling the sidelobes and compensating for the emission expected in each (135, 267) or by "bootstrapping" observations to those made with unblocked apertures (168). Both methods require substantial off-line computer processing of the data and have been used in only a few directions. The new 100-m class telescope under construction at Green Bank, West Virginia, has been designed to minimize stray radiation and should be a superb instrument for measuring galactic H I.

2.1.2.3 *Aperture synthesis telescopes* Aperture synthesis telescopes have a different sidelobe problem that complicates their use for galactic H I emission studies—namely, the sidelobes are both positive and negative. Because of the very high degree of angular correlation of galactic H I emission, almost all of the signal entering the main beam will be canceled by that which enters nearby negative sidelobes (e.g. 45a, p. 197). This severely reduces the usefulness of most aperture synthesis telescopes for the study of most galactic H I emission unless their data can be combined with accurate single-dish observations to restore the missing signal.

In more formal terms, a multiplying interferometer artificially sets the map average of each spectral channel to zero because it cannot measure the "zero-spacing" flux, i.e. the flux in the u-v plane pixel centered on $(0,0)$. Most galactic H I emission, however, contributes *only* to the $(0,0)$ pixel and so is absent from the interferometer measurement. Even with perfect knowledge of the sidelobes, the zero-spacing flux still cannot be recovered.

Aperture synthesis radio telescopes are very useful for studying H I emission features of moderate angular size that are isolated, either because they are much brighter than their surroundings or they are at a different velocity than most galactic gas. Examples of these are H I clouds near H II regions, the brighter high-velocity clouds, and, of course, most other galaxies. The unique wide-field aperture synthesis telescope at the Dominion Radio Astrophysical Observatory (Penticton, Canada) is particularly well designed for measuring galactic H I emission. The array consists of small-diameter dishes that allow very short u-v spacings to be sampled and also produce maps that cover a wide area ($1°.7$) at $1'$ resolution (e.g. 157, 217a).

2.1.2.4 *State of 21-cm emission surveys* A useful table of the available H I surveys is given by Burton (34). Although every part of the sky has been measured at least once, it is only near the galactic plane that sensitive surveys are complete with ~10′ angular resolution. Thus, while we know the galactic H I sky well enough to characterize its main features, a researcher looking for a high-quality H I emission spectrum in any given direction may be disappointed with what is available. One problem is that large single-dish telescopes measure 21-cm emission lines no faster than do small telescopes, merely with better angular resolution. Large telescopes have such small beams that it is not feasible to map large areas completely. Also, many existing surveys are compromised by stray radiation at the ≲1-K level. Much work needs to be done before there is a satisfactory account of the 21-cm sky.

2.1.3 OBSERVING 21-cm ABSORPTION H I viewed against a bright continuum source may appear in absorption. In the simplest general case, the brightness temperature T_b of the 21-cm signal at a velocity v, relative to the signal level in the continuum, is a mix of emission and absorption defined by

$$\Delta T_b(v) = [T_s - T_0][1 - e^{-\tau(v)}]. \qquad 2.$$

Absorption dominates emission at velocity v when the brightness temperature of the continuum source averaged over the antenna beam, T_0, is greater than the H I spin temperature averaged over the antenna beam, T_s, and the optical depth τ is high. Typically, the emission brightness at a given velocity does not vary strongly with beam size, whereas the observed brightness of a small background continuum source goes as the inverse of the antenna-beam solid angle. In practice, most telescopes measure an average temperature for a point continuum source at 21 cm of $T_0 \sim 100\ S(\mathrm{Jy})\ [\mathrm{HPBW\ (arcmin)}]^{-2}$. Typical bright extragalactic radio continuum sources are several janskys, and the smallest beam that one gets from a single dish is about 3′.2 at Arecibo, so except in a few special directions T_0 is at most several tens of kelvins. This is smaller than the typical values of T_s. Single dishes, then, generally measure a net H I emission profile even when pointed at a relatively bright continuum source, albeit an emission profile that has been reduced somewhat by absorption.

To untangle $\tau(v)$ from Equation 2 and thus obtain the average kinetic temperature and column density as a function of velocity, it is necessary to determine what the emission would be in the absence of a continuum source by one of two techniques: (*a*) use a variable continuum source, such as a pulsar, as a target and measure the pure emission term when the source is "off"; or (*b*) observe the H I emission in directions near the continuum source and interpolate an "expected" emission profile toward the source.

Observation of 21-cm absorption against pulsars is quite difficult because the typical pulsar is weak at 21 cm ($S \ll 1$ Jy). As a result, most spectra are noisy and only a small fraction of all pulsars have been observed (49). Nonetheless, the results have special value because of a pulsar's extremely small angular size, which probes the narrowest possible line of sight. The observations have also helped establish distances to pulsars and thus map the galactic electron distribution using the pulsars' dispersion measures (278).

2.1.3.1 *Single-dish absorption studies* Even the largest single dish rarely measures a net absorption spectrum [i.e. $\Delta T_b(v)$ is usually positive], so the accuracy of the observations depends on the accuracy with which the expected emission profile can be determined. Natural emission fluctuations across the sky *always* dominate the error budget of the resulting absorption spectrum. The rms uncertainty of spectra from the Arecibo telescope is typically ~ 0.5 K at moderate galactic latitude, which is greater than the noise in a spectrum for integration times longer than a few seconds; it may rise to $\gtrsim 2$ K at low latitudes. This spectral-line confusion appears only at velocities where real emission is present, so the reliability of an absorption spectrum cannot be estimated from the spectral baseline away from the line. The accuracy of the absorption spectrum is a strong function of telescope size. The smaller the telescope, the larger is its beamwidth, so not only is the absorbing component of Equation 2 weaker, but the reference spectrum must be taken farther from the continuum source, where it is less likely to be representative of the "on" position. Liszt (162) has argued from simulations that a telescope must be > 100 m in diameter to produce a trustworthy absorption spectrum, except for the strongest continuum sources (> 10 Jy). Ironically, the emission fluctuations, or noise of any sort, limit the measurement of high-opacity lines more than that of optically thin lines because the "optical depth noise" goes as $\Delta \tau = \sigma_\tau \times e^\tau$. Additional discussion of the observational techniques is given by Colgan et al (50).

2.1.3.2 *Absorption studies using interferometers* Interferometers can have higher angular resolution than any single dish, and most background continuum sources are well matched to their angular response. For example, the VLA in the "A" configuration has a beam of $\sim 1''$ at 21 cm, so a 1-Jy point source has an average brightness over the beam of $\sim 5 \times 10^5$ K, whereas the receiver noise in the spectrum is two orders of magnitude smaller after only an hour's integration. Absorption lines with $\tau \gtrsim 0.1$ are detected easily, but the emission, even if it has a $T_b \sim 100$ K, is below the noise level. In addition, the negative sidelobes of an interferometer, which interfere with its ability to measure H I in emission, are an asset for absorption measurements because they further diminish the smooth emission component that varies slowly across the sky. Recent studies have used aperture synthesis instruments to generate a very small, clean beam and measure absorption toward compact sources (78) or map the distribution of absorption across the face of an extended background source (60, 105).

Although interferometers reject the bulk of the 21-cm emission, there is still the potential for confusion if the emission has some structure on angles as small as the fringe separation for some or all of the baselines used in the aperture synthesis. At low latitudes, even baselines as long as 1 km (fringe separation 44") can show traces of galactic emission. Since the amplitude of the emission fringes decreases roughly as the angle to the power two to three (59), and the brightness of the absorption goes as the inverse beamwidth squared, the ultimate absorption sensitivity that a telescope can achieve is proportional to its angular resolution to about the fourth power.

The telescope dynamic range sets a limit on the maximum absorption sensitivity also, since at some level the absorption spectra toward nearby continuum sources blend with the spectrum toward the source of interest. If the absorption is different toward different continuum sources in the field, there will be errors at the level of $\Delta\tau/R$, where $\Delta\tau$ is the variation in optical depth over the field and R is the dynamic range (typically 100 to 1000 or more for well-calibrated observations).

Observational techniques for synthesis telescope observations of 21-cm absorption are discussed by van Gorkom & Ekers (266). The experimental results can be particularly sensitive to details of these techniques, especially if the background continuum source is extended. For example, if individual channels are "cleaned" separately before subtracting the continuum, spurious variations in optical depth may be generated. Also, variation in the surface brightness of the background continuum source causes even a constant level of noise in channel maps to generate a wildly varying noise level in $e^{-\tau}$ and in τ.

As an example of what is currently possible, it takes about 4 hr at the VLA to obtain a spectrum with optical depth noise $\sigma_\tau \sim 0.1$ at a velocity resolution of 1 km s^{-1} toward a 50-mJy source. It is impractical to go much deeper in optical depth, or to get a spectrum with even such meager sensitivity toward a much fainter background source, because the telescope time required soon becomes prohibitively large. But since there is typically one extragalactic background source per square degree with $S \gtrsim 50$ mJy, it should be possible to obtain an absorption spectrum within a degree or so of any direction of interest. This will be adequate for many purposes because there is usually a fair correlation between absorption spectra separated by a degree or less at high and intermediate latitudes.

2.1.3.3 *Absorption surveys* The first observations of 21-cm absorption were made by Hagen et al (107) toward very strong galactic continuum sources using a single-dish telescope, followed by Muller (186) and Shuter & Verschuur (235), but after that until the mid-1970s most advances in absorption studies were made by interferometers at Caltech (46, 47, 127) and at Parkes (207). Further single-dish absorption studies became possible when instruments of the 100-m class became available (69, 180, 281). In particular, the Nançay radio telescope, because of its special geometry, is particularly suited for absorption studies (61, 137, 158). The upgraded Arecibo telescope has been used for these observations since the late 1970s (50, 62, 79, 80, 148, 197–199, 237). In addition, combined observations using interferometers to measure the absorption and single dishes to measure the emission have been done at the Parkes Radio Observatory and at Green Bank (44, 75, 100, 181, 209, 210, 216). More recently, there have been several absorption studies using the VLA (78, 97), the Arecibo interferometer (156), and the Ratan 600 telescope (2). About 500 absorption spectra are now available with sensitivity to $\tau \gtrsim 0.1$; some 50 to 100 have sensitivity to lines a factor of 3 to 10 weaker.

2.1.3.4 *VLBI* Absorption by the 21-cm line can be measured and even mapped at an angular resolution of $\leq 0\rlap{.}''1$ using VLBI techniques if the background continuum source is suitable. Dieter et al (81) did the first successful experiment of this kind and detected a surprisingly strong variation in the optical depth of absorbing gas in front of 3C 147 over about $0\rlap{.}''2$. On this line of sight this corresponds to a linear size of 70 AU or less. This result has recently been confirmed by Diamond et al (73) using the European VLBI Network; similar small-scale structure is suggested in two other directions.

These results are stunning, particularly since the background sources used are extragalactic, so the directions studied should give random samples of low-latitude interstellar gas. The column density needed for the absorption is a few times 10^{19} cm^{-2}, which implies a volume density of nearly 10^5 cm^{-3} if the optical depth variation is caused by density change alone. Whether these are distinct structures or some sort of discontinuous internal variations in the density and/or temperature inside larger clouds is not yet known. If they are discrete structures, they need contain only 10^{-8} $M_\odot$. The highest resolution VLA maps of absorption toward Orion (262) also show very small concentrations of gas. However, with sizes of a few tenths of a parsec and masses typically of 0.01 $M_\odot$, these subparsec clouds are vastly larger and more massive than those found by the VLBI experiments. Both may be the result of the passage of a shock front through a much larger cloud. The recently discovered structure in the ionized medium on scales of ~10 AU (92c, 281a) might be related to the small-scale structure in the cool neutral medium if, for example, both are caused by instabilities behind supernova shocks, but this connection is only speculative.

2.1.4 OBSERVING SELF-ABSORPTION Just as interstellar Lα absorption can be detected against the broad Lα emission lines from late-type stars, so too can a cold H I cloud appear in absorption against H I emission from warmer gas behind it. Two circumstances are illustrated in Figure 1, along with the 21-cm profiles that result: The telescope on the left measures the sum of the two profiles because the hot cloud is optically thin, and the telescope on the right sees a self-absorbed profile.

Self-absorption was first detected against the low-latitude, bright H I toward the galactic center and anticenter (109, 205). Its occurrence is direct evidence that the kinetic temperature of galactic H I has a large range.

The phenomenon can be difficult to recognize in a complex spectrum, and Knapp (144) has proposed that self-absorbed features be identified by

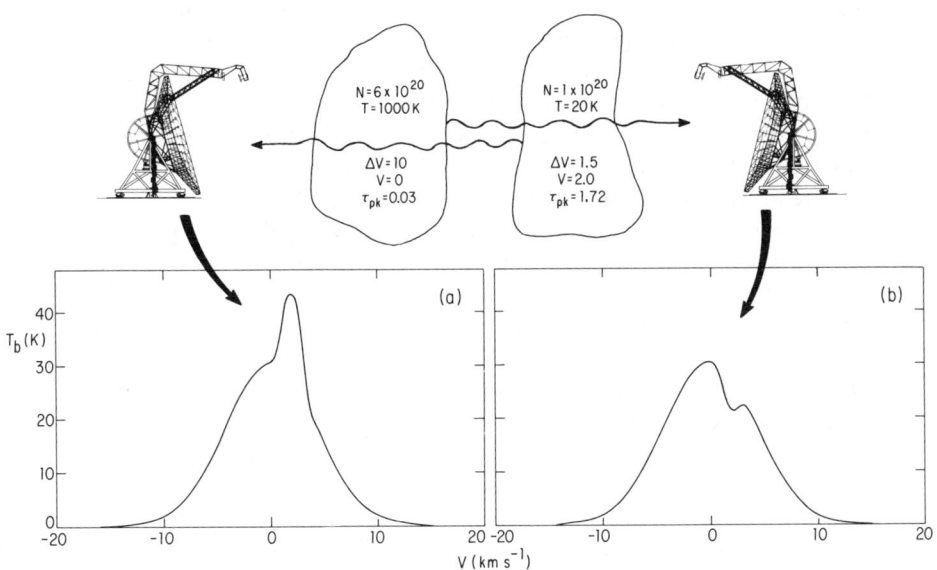

Figure 1 Schematic of the geometry of 21-cm self-absorption. The structure of an emission profile depends on the relative location of hot and cold clouds as viewed by the observer. Superpositions like this are very common at low and intermediate latitudes. The profile on the right (*b*) is self-absorbed.

the following characteristics: (*a*) They should have a relatively narrow line width of only a few kilometers per second, implying that the kinetic temperature is low; (*b*) they should have steep sides, so that they are unlikely to be just the chance sum of emission components; and (*c*) they should be relatively localized in space. A self-absorption feature is also suspected where a sharp minimum is seen in an H I emission profile at the velocity of a molecular cloud in the same direction. Conditions for observable self-absorption are most favorable at low galactic latitudes, where the H I is fairly bright, or in the direction of local gas clouds that subtend many arcminutes. The technique is useful for determining cloud properties in directions lacking suitable background radio continuum sources (5, 8, 37, 178, 220, 222, 234).

In analyzing a self-absorbed profile, such as Figure 1*b*, one usually estimates the background emission by drawing a smooth curve connecting the apparently unabsorbed portions of the spectrum. This step introduces considerable uncertainty, especially for complex spectra (159), and still leaves two unknowns (the spin temperature and the opacity of the cloud) and only one measured quantity (ΔT_b). The ambiguity can be resolved if the cloud covers so large an angle that one can literally move off the background H I, as one moves off a background continuum source, and obtain a pure emission profile for the cloud. Circumstances like this are more common than one might expect (206). At least one cloud goes from emission to self-absorption as it crosses the bright H I in the galactic plane (9), and van der Werf et al (261) have observed a cloud that appears in self-absorption at its center but in emission at its edges. Where the cloud disappears, its spin temperature must be close to the brightness temperature of the background H I, in this case about 25 K.

Most self-absorption studies, however, can give only a range of temperature and density for the absorbing cloud, usually with significant uncertainty. Although the range can be restricted in some cases (219), there is little or no way to determine a confidence level for an individual measurement of τ.

It is possible that self-absorption causes much of the structure in low-latitude galactic emission profiles. Prominent examples of self-absorption remain a sobering reminder that there are places in the interstellar medium where the 21-cm brightness decreases as the H I column density goes up.

2.1.5 COLUMN DENSITY MEASUREMENT   One of the major deductions from 21-cm observations is the total column density of H I, or the column density at a particular velocity or in a particular spectral component. If the 21-cm line is optically thin in the direction of interest, then the column density is given by the velocity integral (in K km s^{-1}) under the profile:

$$N_H(\tau \ll 1) = 1.823 \times 10^{18} \int T_b \, dv \, \text{cm}^{-2}. \qquad 3.$$

Given only an emission spectrum and not a measurement of τ itself, a reasonable rule of thumb is that if the peak T_b is only a few tens of kelvins or less, the optically thin assumption is probably not too bad. Most H I is hotter than 20 K, so a peak T_b less than that suggests a modest opacity if self-absorption and other evidence of cool gas are absent.

A first-order opacity correction can be made by adopting a constant T_s for a spectrum. Most large-scale emission surveys have been analyzed this way using values of T_s in the range 120–150 K (e.g. 35, 143, 194).

If the 21-cm line has significant opacity, a unique interpretation of the data is usually impossible because a cloud with an optical depth τ greater than a few tenths partially obscures the emission from gas behind it. At the velocity of this cloud, the emission brightness temperature is a function of the spin temperature and optical depth of the cloud, plus the column density of any foreground optically thin gas and the attenuated column density of the background gas. Even with both emission and absorption observations, and thus a measurement of $\langle T_s \rangle$ and $\langle \tau \rangle$ at each velocity, there is no unique $N_H(v)$, since correction for opacity depends on the detailed juxtaposition of warm and cool gas. The algebra of this radiative transfer problem is discussed by many authors (50, 75, 153). As an example, consider the spectra in Figure 1. The optically thin assumption gives an N_H that is 10% too low for spectrum (*a*) and 20% too low for spectrum (*b*). The assumption that the gas is isothermal at 125 K gives an N_H that is only 5% too high for (*a*) but 10% too low for (*b*).

The need to correct for opacity in 21-cm profiles ultimately limits the accuracy of many N_H determinations (88). In most directions, the total column density derived from an optically thin analysis of an emission spectrum should be multiplied by factors of 1.1–1.3 to account for self-absorption. There are, however, a few places where the factor gets large as cold clouds shield a region of velocity crowding. Observation of 21-cm absorption toward extragalactic continuum sources and other evidence (200) shows that considerably less than half of all velocities in low-latitude spectra are covered by H I with $\tau > 1$. While opacity effects make it difficult to derive a very accurate N_H at low galactic latitude ($|b| \lesssim 10°$), it is unlikely that 21-cm emission surveys have missed more than about 20–30% of the mass of galactic H I, on average (80a).

2.1.6 THE ZEEMAN EFFECT AT 21-cm The line-of-sight component of the interstellar magnetic field shifts the frequency of the two circular polarizations of the 21-cm line by a small amount. This effect was first detected by Verschuur (269). There have been several recent comprehensive reviews of Zeeman observations along with other methods for measuring the interstellar magnetic field (115, 152, 254, 256).

2.1.6.1 *Observational considerations* The Zeeman effect is extremely difficult to measure at 21 cm, not only because very high sensitivity is needed to achieve even a significant upper limit, but because of the subtle calibration required and the many sources of spurious polarized signals that mask and even mimic the effect. High sensitivity is necessary because the frequency difference between the two polarizations is very small, only 2.8 Hz $(\mu G)^{-1}$, so that for the typical interstellar field strength of a few microgauss the splitting is more than two orders of magnitude smaller than the H I line width. Thus, rather than seeing separate lines in the two polarizations, one sees at best an almost imperceptible shift in the line center. This is generally measured differentially—for example, using a switch that synchronously subtracts the signals from two circularly polarized feeds (254).

Zeeman splitting can be observed in emission or in absorption. The former technique benefits from the strength of the lines on even a small telescope, whereas the latter has the advantage of narrower lines and a sampling of a more restricted volume in space that reduces the possibility that the effect will be canceled by a reversal of the magnetic field along the line of sight. But even with a relatively strong $B_\parallel$ of 10 μG, the amplitude of the circular polarization differential signal is only 10^{-3} of the total brightness in emission, and perhaps 3 times that in absorption. Thus, to detect the Zeeman splitting takes about 10^6 times longer than the time required to detect the line itself.

2.1.6.2 *General results* Most recent observations of the Zeeman effect in 21-cm emission have been made with the 25-m telescope at the Hat Creek Observatory (reviewed by Heiles in ref. 115; see also 116a, 117, 271). This telescope has been carefully calibrated in circular polarization, and integration times of many days per spectrum have been achieved. Because of the enormous time and effort needed to obtain a good measurement, the body of existing data is neither large nor systematic. It consists mainly of observations in "interesting" directions.

Measured field strengths typically range from ~ 10 to ~ 100 μG and vary considerably and to some extent unpredictably from point to point. The magnetic field seems to be concentrated in filamentary structures that lie on the surfaces of H I shells and supershells. This concentration may be the result of the interstellar shocks associated with the shells. The magnetic field pressure P/k is typically several times 10^4 cm^{-3} K, which is much larger than the thermal pressure in the H I gas (typically $2-3 \times 10^3$

cm^{-3} K) and larger than the equivalent pressure in macroscopic and turbulent motions (typically 2–5 times the thermal pressure) indicated by the 21-cm line width. It is tempting to associate these line widths with Alfvén waves (92) and consider the macroscopic motion of the gas to be a response to magnetic pressure gradients resulting from the field configuration (117). The magnetic field evidently dominates the small- and intermediate-scale kinematics of interstellar H I in many regions. In one interpretation (56), the magnetic field can also moderate the response to thermal instabilities in the neutral gas and so determine the thermodynamic phase balance in interstellar H I. On the other hand, many directions have upper limits to the magnetic field of 3–5 μG or less. These areas, where the magnetic pressure is less than or comparable to the pressure due to macroscopic motions, are probably more typical of interstellar H I as a whole than those exceptional regions of strong fields that give the positive detections of Zeeman splitting.

In 21-cm absorption, few regions have been well observed because of the long integration times needed even to place interesting limits on the field strength. Detailed maps of the field distribution toward the very bright supernova remnant Cas A have been made with the Westerbork synthesis telescope (24, 229). In one spectral feature the magnetic field strength more than doubles, from 20 to 50 μG, over a distance of 4 pc. The regions of higher magnetic field strength correspond with the position of molecular clumps embedded in the cool H I cloud (253). Recent VLA studies of the 21-cm Zeeman effect in Orion (257, 262) and W3 (263) show again that the magnetic pressure is sufficient to overcome gas pressure, bulk motions, and gravitational forces in large areas around these H II regions. Small-scale structure is seen in the field configuration with reversals over lengths less than 1 pc, suggesting a tangled field. This may be the result of the recent nearby star formation.

On larger scales the existence of a correlation between field strength and gas density in the atomic phase is tentative (254–256), although strong fields are seen in the 18-cm lines of OH, originating where the density is several hundred atoms per cubic centimeter or higher (64, 191, 253).

Although interest in measuring 21-cm Zeeman splitting is increasing and more telescopes are being used to measure the effect, we still know less about the strength and direction of the magnetic field than about any other major aspect of galactic H I. This area of research is ripe for observational advances.

2.2 *General Characteristics of 21-cm Observations*

All 21-cm spectra, despite their diversity, have a number of common attributes, which are illustrated in Figures 2 and 3. The data in Figure

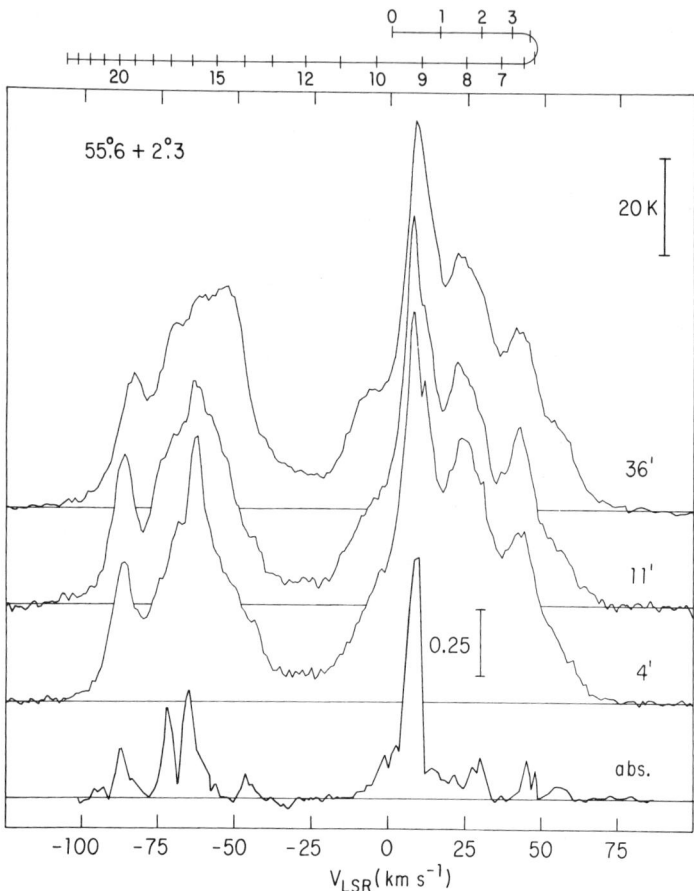

Figure 2 Representative 21-cm emission and absorption profiles at low galactic latitude. All are in the same direction $(l, b) = (55°.6, +2°.3)$. The emission profiles were taken with angular resolutions of 36′, 11′, and 4′, while the absorption is measured against an extragalactic source that subtends $<0″.1$ (9, 78, 277). The emission profiles are in brightness temperature on the scale given by the bar labeled "20 K." The absorption profile (inverted here) gives the value of $1 - e^{-\tau} \sim \tau$ at each velocity, and its reference bar has $\tau = 0.25$. At the top of the figure is a curve that shows the relation between velocity and distance from the Sun (in kiloparsecs) for an assumed galactic rotation law (36). While there are only minor differences among the emission profiles despite the differing angular resolution, the absorption profile looks quite different. This is not an effect of angular resolution but of the bias of absorption spectra to the coldest H I.

2 were taken near $(l, b) = (55°\!.6, +2°\!.3)$ in emission at several angular resolutions and in absorption against an extragalactic continuum source that has an angular size of $<0''\!.1$. The curve at the top of the figure shows how the velocity, if interpreted entirely as galactic rotation, corresponds to distance from the Sun [using a rotation curve (36) scaled to $R_0 = 8.5$ kpc]. Positive velocities imply galactocentric radii <8.5 kpc, i.e. within the solar circle. This particular line of sight intersects the galactic "disk" twice: first near the Sun, then again as it passes through the galactic warp more than 15 kpc from the Sun and >500 pc above the nominal galactic plane. The relatively bright and complex emission at negative velocities comes from the galactic warp. Emission spectra with this sensitivity can be obtained in about 1 min with current receivers. The absorption spectrum was obtained in 1 hr using the VLA.

Low-latitude emission spectra, such as those shown here, are always very complex; the signal commonly extends over several hundred kilometers per second. The shape of the emission spectra does not change greatly with angular resolution in the inner Galaxy (6), which here lies at positive velocities. Except for self-absorbed features, high-angular-resolution observations near the plane have revealed surprisingly little not already seen in lower resolution observations. This relative absence of angular structure is caused in part by "velocity crowding," a term first used by Burton (32) to emphasize the fact that galactic rotation, projected on the line of sight, often does not impart very much of a velocity difference to spatially separated objects. A long distance can be "crowded" into a small velocity interval. On Figure 2, for example, the velocities from 20 to 30 km s^{-1} cover >1.5 kpc. When a lot of space is squashed into a few kilometers per second, individual interstellar components blend completely, and little structure is revealed by increasing the angular resolution except, perhaps, for an occasional cold foreground cloud that can appear in self-absorption. The absence of significant small-scale angular structure at positive velocities is not the consequence of opacity of the line: The absorption spectrum clearly shows that the gas is optically thin at most velocities.

Small-scale angular structure in the negative velocity emission of Figure 2, e.g. near -65 km s^{-1}, arises because this gas is so distant that the 36' beam covers 160 pc, which is comparable to the size of major structures in the interstellar medium (ISM), and even to the thickness of the H I layer.

The overall shape of the emission profiles in Figure 2 is mainly determined by the large-scale distribution and kinematics of H I in the Galaxy. The bright, low-velocity gas is local, and the emission diminishes with distance as the line of sight gets farther from the galactic plane. H I profiles at low galactic latitude always have several distinct peaks. The origin of these is somewhat controversial; they are manifestations of variations in velocity crowding as well as genuine density enhancements (33, 34). Large features in "longitude-velocity" space are seen in both 21-cm and CO surveys (33), but their identification with spiral arms or "superclouds" is ambiguous (87).

The picture in H I absorption is substantially different from that in emission. The quantity plotted in Figure 2 is the fractional absorption $(1-e^{-\tau})$, which (unlike the emission) varies strongly with the gas temperature. Absorption lines are always narrower than emission lines. Absorption near the plane is common but highly variable. Some velocities have H I emission but no detectable H I absorption. For example, there is often little or no absorption between major emission peaks, whereas all allowed velocities always show significant H I emission. There was dispute for some time over whether this effect could be caused by the different solid angles of the absorption background source and the telescope beam used to measure emission (e.g. 104). However, with the introduction of smaller beams and large absorption surveys the explanation has been convincingly shown to be that suggested originally by Clark (46)—i.e. that the difference between emission and absorption results mainly from variation in the spin temperature of the H I along the line of sight.

The absorption spectrum of Figure 2 is not particularly rich in individual features. To some extent it suffers from the same latitude influence as the emission spectra: H I far from the galactic plane usually has a smaller opacity than that close to the plane (154). But once the sight line pierces the galactic warp, at negative velocities, absorption is again seen. It is interesting to note that cool H I clouds are found even 15–20 kpc from the galactic center.

Unlike emission spectra, which do not look like the superposition of a few Gaussians with unambiguous line centers and widths, absorption spectra can usually be decomposed into Gaussian components, although it may be difficult to extract the weakest lines from a blend. In comparison with 21-cm emission, it is surprising how nearly Gaussian most absorption features are. This is because the absorption lines arise only in regions of cool gas, which are more distinct along the line of sight, and which have narrower intrinsic line widths, than the gas that contributes to H I emission.

Emission at high latitudes is typified by the spectrum in Figure 3, which is in the direction of the radio galaxy 3C 287. Sensitive observations have detected no 21-cm absorption in this direction, implying that the peak 21-cm opacity is less than 1% (80). It is common for there to be little or no H I absorption in directions with $|b| \gtrsim 45°$.

H I emission profiles are never simple, single Gaussians; virtually all

H I IN THE GALAXY   231

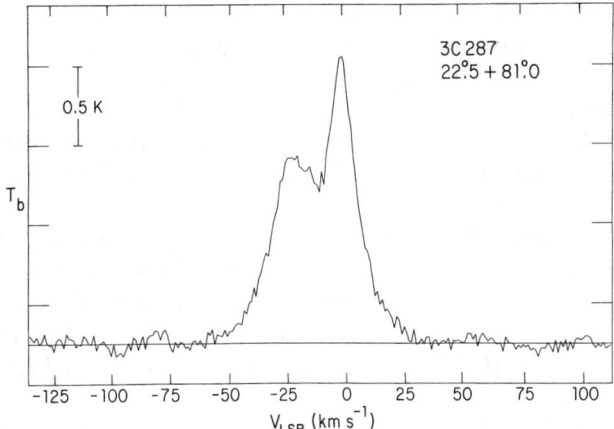

Figure 3 A representative 21-cm emission profile at high galactic latitude. This profile is toward the radio source 3C 287. No H I absorption has been detected in this direction, which implies that $\tau < 0.01$. Multiple spectral components, often with different velocities, are common in galactic H I profiles, even at high latitude. The component at -25 km s^{-1} is an example of the apparently infalling gas that is widespread at northern latitudes (80b, 279; see also Section 6.1).

show multiple components, line asymmetries, or broad wings. Sometimes, as in Figure 3, they contain components whose velocity cannot arise from galactic rotation. It is not generally useful to try to decompose emission spectra into Gaussian components because a profile's shape is determined as much by the vagaries of galactic rotation, velocity crowding, and streaming, modified by opacity and variations in temperature, as it is by the superposition of discrete elements. A Gaussian analysis provides unambiguous information only when an emission feature has a very odd velocity (such as high-velocity clouds) or is very much brighter than its surroundings, so that it is not blended with most other emission.

2.3 *Integral Properties of the H I Sky*

Some fundamental aspects of the H I sky can be seen in the distribution of total column densities; thanks to a number of recent surveys (38, 48, 130, 141, 245, 277), reasonably accurate data are available. The following discussion is based on these surveys merged and averaged into $1° \times 1°$ bins in l and b; this gridding is adequate for most purposes, except near the galactic plane. Column densities were calculated for $T_s = 200$ K and include all emission within ± 250 km s^{-1} of zero velocity (LSR), except for that associated with the Magellanic Clouds.

The total H I column density varies by a factor of at least 500 across the sky. The location of the highest N_H is not well known because of opacity effects; the highest value in our data set, 2.6×10^{22} cm^{-2} at $(l, b) = (339°, 0°)$, is certainly less than the true maximum. The lowest column density, however, has been fairly well established as $4.4 \pm 0.5 \times 10^{19}$ cm^{-2} at $(l, b) = (152°, +62°)$ in Ursa Major (130, 168). Toward the galactic poles, the column density is $\sim 10^{20}$ cm^{-2}.

The frequency of occurrence of a given column density is shown in Figure 4. The general form of this figure, with its steep decrease at the low-N_H end, and the long tail to high N_H, would be seen in any uniform H I disk observed from within, but the large breadth of the distribution indicates that there are substantial variations in N_H at all latitudes. The cutoff below $N_H \sim 10^{20}$ is real and not an artifact of limited sensitivity or stray radiation.

The distribution of N_H with latitude shown in Figure 5 is dominated by the galactic plane, which has a FWHM of about 4° on average. As mentioned above, the values of N_H in our data set span a range of > 500, but most of this is a csc $|b|$ effect. The actual deviations from a simple plane-parallel layer are shown in Figure 6, where $N_H \sin |b|$ is plotted against $\sin |b|$. These column densities vary by a factor of ~ 10 owing in part to several large H I concentrations (e.g. in Orion and Taurus) that lie far from the plane. The fact that $N_H \sin |b| \to 0$ as $\sin |b| \to 0$ is mainly a

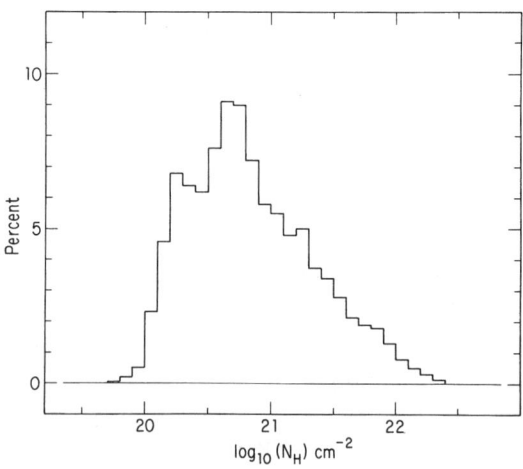

Figure 4 The percentage of the sky covered by H I at a given N_H. Each column density has been averaged over $1° \times 1°$. This figure is somewhat incomplete at the high-N_H end but should otherwise be representative. It was derived using data and assumptions described in Section 2.2. The same data set is used for Figures 5–8.

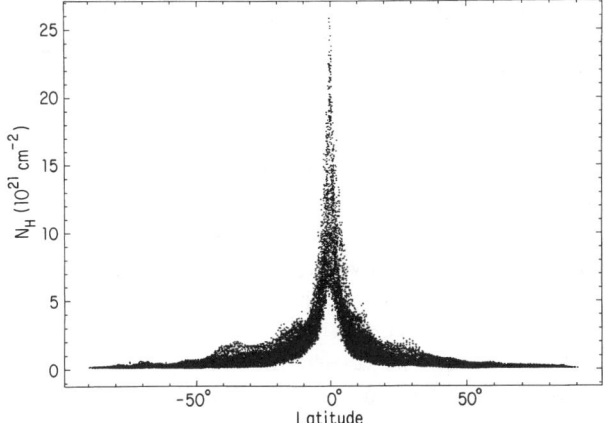

Figure 5 H I column densities, each averaged over $1° \times 1°$, versus galactic latitude for the entire sky.

consequence of the finite size of the galactic disk, although opacity effects and the grid size also contribute somewhat.

The data in Figure 6 (for $|b| > 2°.5$) have a mean of 2.9×10^{20} csc $|b|$ and a dispersion of 1.4×10^{20} cm^{-2}. These values are consistent with other determinations of $\langle N_H \rangle$ (145, 162). The mean, however, changes with latitude in a systematic deviation from a uniform plane-parallel layer (145).

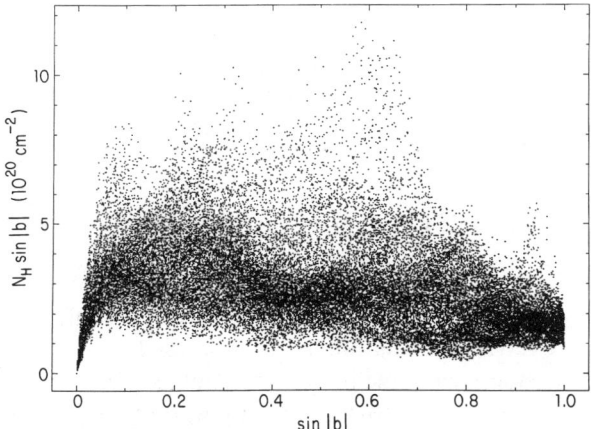

Figure 6 H I column densities multiplied by sin $|b|$ and plotted versus sin $|b|$ for the entire sky. They go to zero as the latitude goes to zero, mainly because the Galaxy has a finite size. The decrease in $\langle N_H \sin |b| \rangle$ at high latitude is an effect of the hot cavity in the interstellar medium that extends some 100 pc around the Sun.

A fit to the binned median of the distribution gives $N_H = 3.84\,\text{csc}\,|b| - 2.11$, in units of 10^{20} cm^{-2}. More than two thirds of the points lie within 1.0×10^{20} csc $|b|$ of this line. The minimum N_H does not lie at the poles but near $b = +55°$.

The decrease in $N_H \sin |b|$ at high latitudes is a consequence of the Sun's location in a very low density region of the interstellar medium (reviewed in ref. 57). A number of supernovae have apparently excavated an irregular cavity 100–200 pc in diameter in which the Sun resides. The cavity contains little H I but is filled with low-density, fully ionized gas that radiates in the soft X-ray bands. The vertical extent of the excavated region is comparable to the usual vertical thickness of the H I layer, so the high-latitude H I sky has been perturbed substantially by this event. This is discussed in detail by McCammon & Sanders (177a) elsewhere in this volume.

The H I sky is shown in Figure 7. The top panel gives the total N_H, while the bottom panel gives $N_H \sin |b|$ and therefore the deviations of the sky from a uniform plane-parallel system. The regions of excess $N_H \sin |b|$ often lie in arching filaments, some associated with well-known structures like the North Polar Spur and the shell around the North Celestial Pole (135°, +35°). These are no doubt regions like our own local cavity, though in many cases more vast, that have been arranged by a series of supernovae. Pictures like this have led to suggestions that the organization of interstellar H I is better described with words like "sheet" or "filament" than the more usual "cloud" (111). We discuss this topic in Section 5.

The southern galactic sky ($b < 0°$) has about 20% more H I than the northern sky. This is partly due to a large tongue of H I that extends south in the anticenter, but the phenomenon is more general: There is more H I in the south than in the north at all latitudes. This may be another effect of the local cavity (57), for high-latitude molecular clouds have a similar asymmetry (174).

3. ULTRAVIOLET OBSERVATIONS

The Lyman series of H I transitions lies in the ultraviolet between 1216 and 912 Å and cannot be observed from the ground. Diffuse Lα emission from backscattered solar radiation has been studied to obtain information on the neutral interstellar medium surrounding the Sun (12), but the biggest contribution of ultraviolet observations to the study of galactic H I has come from the Lyman series of lines measured in absorption against background targets, mainly stars (see the reviews in refs. 54, 243). The first measurements of interstellar Lα absorption were made from sounding rockets (43, 185). Observations have now been made toward about 300

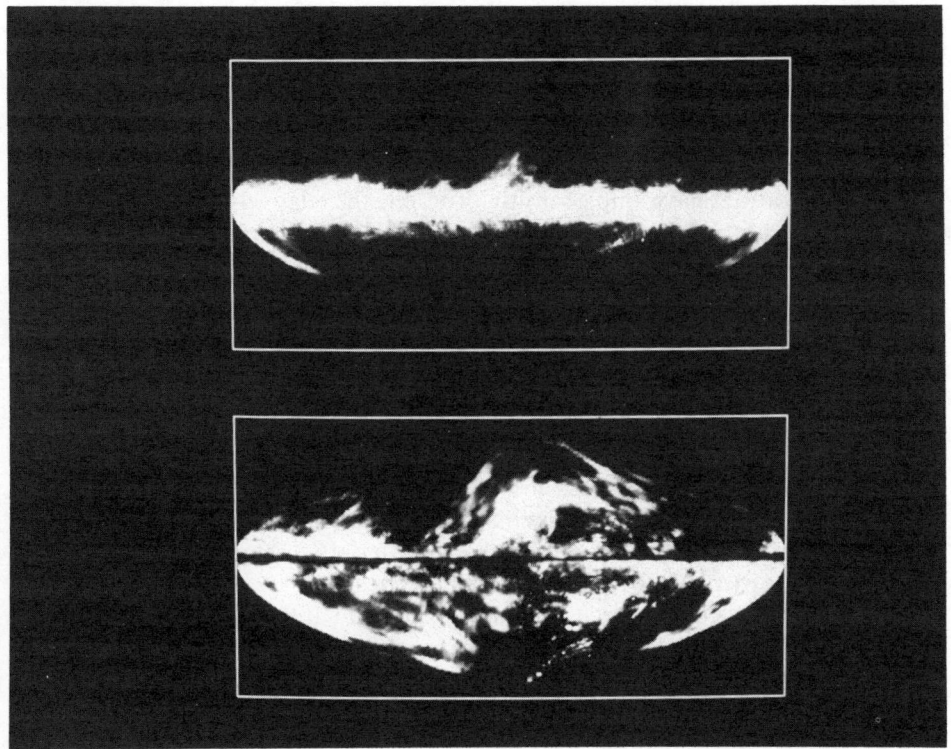

Figure 7 Two views of the H I sky. Both maps are centered on the galactic center with longitude increasing to the left. (*Top panel*) The distribution of total H I column densities. (*Bottom panel*) The distribution of $N_H \sin |b|$. This shows the deviation of the sky from a uniform, infinite, plane-parallel layer. The loops and arching filaments have probably been swept up by supernovae.

stars from the space observatories *OAO-2* (224), *Copernicus* (20), and *IUE* (233). The Hubble Space Telescope (HST) is intended to be the next in this series of observatories.

Lyman-α at 1216 Å is the strongest interstellar absorption line. It is saturated in almost all directions and thus it does not yield radial velocities, but conversely it does give a fairly accurate column density. Equivalent widths are derived by fitting model profiles to the observed spectra (226, 233); column densities have uncertainties in the range 10–40%. Currently the major source of uncertainty in a derived N_H comes from errors in the estimated stellar continuum; new instruments to be launched in the next decade should reduce the total error to $<10\%$.

Most targets for Lα absorption measurements are early-type stars of spectral class no later than B2. Stars of later spectral class have strong stellar Lα absorption lines that contaminate the interstellar lines. The dividing line between useful and suspect types, although generally placed at B2, is not always precise, for it depends somewhat on the interstellar H I column density (226). Early-type stars are so intrinsically luminous that Lα observations have ample sensitivity despite the relatively small aperture of the telescopes that have flown thus far. For example, the *IUE* has detected interstellar Lα absorption toward unreddened B0V stars as faint as $m_V = 12–13$ (223), which corresponds to distances of up to ~ 20 kpc. This sensitivity is sufficient to detect Lα absorption toward nearly every early-type star that is at moderate and high galactic latitude. The increased sensitivity of future instruments like the HST will greatly expand H I studies of stars at low latitude and those behind dense clouds, and it will also increase the number of other objects that can be used for Lα studies. For example, only a few Lα absorption measurements have been made against the chromospheric emission from very nearby, late-type stars (189) because with the *IUE* they can be observed to only ~ 30 pc from the Sun.

Higher order members of the Lyman series (Lβ, Lγ, Lδ, ...) are in principle accessible to check column densities and, if the higher order lines are not saturated, to obtain velocity information (e.g. 272). Only the *Copernicus* satellite, however, was able to observe substantially shortward of Lα, e.g. to Lβ at 1026 Å, and it will be some time before there is general coverage of these wavelengths again, for the HST is limited to $\lambda > 1060$ Å.

For a number of years there was a discrepancy between the values of N_H derived from Lα absorption and from 21-cm emission measurements made in the same direction at moderate and high galactic latitude (42, 131). In all cases $N_H(21\ cm)$ was greater than $N_H(L\alpha)$, sometimes by an order of magnitude. Plots of $N_H(L\alpha)$ vs $N_H(21\ cm)$ showed little if any correlation, even on a log-log scale (224). Blame for the disagreement was variously laid to the poor angular resolution of radio telescopes or to some radiative transfer phenomenon.

The structure of the local ISM and of the Galaxy caused most of the early problems [as suggested by Spitzer & Jenkins (243)]. Stray radiation in the 21-cm observations contributed to the confusion, but it was not the main effect. Lα observations of nearby stars show less H I than 21-cm observations because the Sun is located in a low-density cavity (reviewed in ref. 57); nearly all galactic H I lies behind the nearby stars. It was an unfortunate coincidence that in the region of the Orion Nebula, which was the target of many early studies, most of the gas is also behind the bright stars, so again $N_H(L\alpha) \ll N_H(21\ cm)$. A bias of this sort is expected in most ultraviolet/radio comparisons because ultraviolet studies must use sightlines with low reddening, so if there is a substantial amount of H I in a given direction, it must lie behind the stars observed in the UV. Finally, the general thickness and irregularity of the galactic H I layer were not appreciated. Significant amounts of H I can occur many hundreds of parsecs from the galactic plane, so very high $|z|$ stars are needed for an accurate $L\alpha$/21-cm comparison.

The current situation is summarized in Figure 8, where $N_H(L\alpha)$ is plotted against $N_H(21\text{ cm})$ for a selected set of stars. Figure 8 contains all available observations that meet the following criteria (from ref. 123): (a) The target stars are of spectral type B2 or earlier, (b) the stars are $\gtrsim 30°$ from the galactic plane, (c) they are at least 1500 pc above or below the galactic plane, and (d) the 21-cm observations are not contaminated by stray radiation. The $L\alpha$ data are from the *Copernicus* and *IUE* satellites; the 21-cm data are from the Green Bank 140-ft telescope (67, 167, 225). The error bars are $\pm 1\sigma$ experimental errors. The agreement between ultraviolet and radio estimates of N_H is excellent and somewhat astonishing, considering that the angular resolutions, experimental techniques, and telescopes differ in all respects. Even the rather coarse 21' angular resolution of the radio data apparently does not introduce a large error. Within the experimental uncertainties, and given the rather select nature of the target directions, the $L\alpha$ and 21-cm techniques give consistent results.

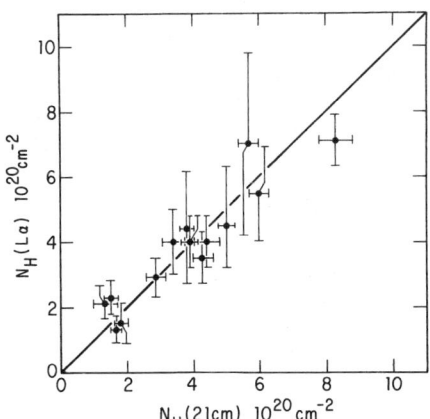

Figure 8 High-latitude H I column densities derived from radio (21-cm) observations over a 21' beam compared with column densities derived from UV ($L\alpha$) observations toward very distant stars in the same direction. Error bars are experimental $\pm 1\sigma$ values. The excellent agreement between the two vastly different ways of measuring H I is reassuring and also indicates that there is not very much angular structure in N_H at high latitudes (see also Section 5.2).

4. INDIRECT TRACERS OF H I

Alternatives to Lα and 21-cm observations give independent estimates of the properties of galactic H I, often over paths that cannot otherwise be studied. This is not a comprehensive account of the relationship between H I and other interstellar species, but rather a summary of some of those that can be used as predictors of H I.

4.1 *Spectral Lines*

Interstellar absorption lines from species like Ca, Na, K, and Ti can be used to infer the distribution of H I when variations in excitation and depletion by adsorption on the grains are small. [This topic was recently reviewed by Cowie & Songaila (54; see also 242, 243).] The optical absorption line of Ti II at $\lambda 3384$ Å is perhaps the best *qualitative* tracer of H I because it is the dominant ionization stage in H I regions and has an ionization potential almost exactly that of hydrogen. This line has been used for several studies of neutral gas (1, 246). Unfortunately, the depletion of Ti is variable and the line cannot be used for quantitative estimates of N_H. The gas-phase abundance of Ti II depends so sensitively on physical conditions (because of depletion, the abundance is inversely proportional to the ambient density) that the scale height of Ti II is substantially larger than that of H I (85). Where one sees Ti II there must be neutral hydrogen, but the amount is uncertain. Optical lines of Na I and K I correlate well with N_H, with perhaps a factor of 2 scatter from star to star, but the slope of the relationship between N(Na I) and N_H may well be a function of N_H (91, 122). The H and K lines of Ca II correlate poorly with hydrogen and, according to Hobbs (122), are the least suitable for most purposes.

A review of techniques of optical and UV spectroscopy can be found in (54). Modern detectors have high-velocity resolution (~ 1 km s^{-1}) and good sensitivity to lines from clouds with an equivalent H I column density $\gtrsim 10^{19}$ cm^{-2}. Detailed component-by-component comparisons between 21-cm, optical, and UV spectra are now being made in a few directions (e.g. 67). This multiwavelength approach will be the pattern of future studies of the ISM.

4.2 *Dust*

4.2.1 REDDENING AND EXTINCTION The connection between H I and dust has been studied by making comparisons between N_H and reddening $E(B-V)$, or between N_H and extinction A_V, ever since the detection of the 21-cm line (30, 31, 68, 95, 131, 147, 160, 177, 258). Because of the obvious advantage in measuring the H I and reddening over an identical path, the early radio work was superseded by ultraviolet/optical comparisons (20,

132, 243). These have established that $\langle N_H/E(B-V)\rangle = 6\pm2 \times 10^{21}\,\text{cm}^{-2}$ mag^{-1} for sight lines dominated by neutral atomic gas, in agreement with determinations from other methods (20). Indeed, the small but select group of stars [all of which have known $E(B-V)$ values] used for Figure 8 give this value, including the dispersion, provided that the two stars with a reported $E(B-V) = 0$ are omitted. The existence of anomalies like these two stars has provoked a small but important debate on whether there might not be sight lines that have H I but virtually no reddening (e.g. 119). We sidestep this issue by noting that to an H I observer the scatter in $\langle N_H/E(B-V)\rangle$ is so large, no matter what zero point for reddening is adopted or what correction is made for H_2, that reddening must be considered a relatively poor estimator of N_H, uncertain by a factor of 2 for any star. It is, however, the only way to get N_H in the direction of the large number of stars that are too cool to be good Lα targets (e.g. spectral type A), yet are bright enough to be used for study of Ca, Ti, Na, and other interstellar species. An excess of $E(B-V)$ along a given path is also a signal to check for ionized H II or molecular gas, since a significant amount of dust may reside in these phases.

4.2.2 INFRARED EMISSION The *Infrared Astronomical Satellite* (*IRAS*) detected far-IR emission from all directions of the sky. This satellite was so sensitive that it could measure the dust in an equivalent H I column density of order a few times $10^{19}\,\text{cm}^{-2}$ at its longer wave bands of 100 and 60 μm (reviewed in ref. 11). While the *IRAS* results have been very important for interstellar studies, it seems that interstellar conditions vary so widely that far-IR emission is only a qualitative predictor of galactic H I; it also follows H_2, dust associated with H II regions, and the interstellar radiation field (21, 118).

The galactic, diffuse IR emission that *IRAS* measured can be separated into two general components: The dominant one is a relatively smooth background that varies as csc $|b|$. Superimposed on that are large, filamentary "cirrus" features (170). Both components have exact analogs in H I, and some cirrus are seen optically as reflection nebulae (221a). At high galactic latitudes and in the outer Galaxy where molecules are relatively rare, there is a detailed point-by-point correlation between $I(100\,\mu\text{m})$ and N_H, but the IR emissivity per H atom varies by a factor of 3 from place to place (21, 251). This is because the far-IR brightness and color depend on the mixture of small and large grains and on their temperature. The grain-size distribution may well vary from place to place depending on the grain creation and destruction processes at work in the local environment (21a), while the interstellar radiation field is a strong function of position (83, 201).

Other anomalies in the H I–IR correlation can be attributed to the presence of significant amounts of molecular hydrogen (often traced by emission in the radio lines of CO) or ionized gas. This has stimulated attempts to compare all convenient dust and gas tracers, an idea that of course predates the *IRAS* survey (e.g. 65). The cirrus clouds appear to be at the transition between primarily molecular and primarily atomic gas; with column densities of $3–5 \times 10^{20}$ cm^{-2} they can barely shield their CO and H$_2$ from photodissociation (see Section 5.3). So neither 21-cm nor CO emission studies give the total gas column density directly, and even with both in hand there are uncertainties in the conversion factor between CO emission intensity and H$_2$ column density. Finally, the gas-to-dust ratio is probably not constant. Variations in the Milky Way have been found from pre-*IRAS* studies (31, 51, 120a), and Schwering (231, pp. 195–252) has shown that it has a different value in the Small Magellanic Cloud.

One of the more exciting questions about the relationship between *IRAS* cirrus and ordinary clouds traced by 21-cm emission is the frequent correspondence between far-IR emission and multiple velocity components in the H I profiles. Often the H I shows highly supersonic velocity discontinuities in the vicinity of cirrus, and anomalous intermediate velocity material is often seen associated with far-IR bright clouds (70, 118). The prototype for this effect is the "Dracula" cloud (in Draco), which has been studied for many years by Mebold and others (182, and references therein; 219a). All these cases suggest enhancement of the far-IR emission behind shocks such as those caused by cloud collisions. On the other hand, the good overall correlation between far-IR and 21-cm emission means that it cannot exclusively be anomalous velocity gas that corresponds to the cirrus, since this gas makes up a fairly small fraction of the total H I on most lines of sight. No 100-μm emission has been detected from high-velocity H I clouds (274).

4.3 *Absorption in the EUV and Soft X-Ray Bands*

Any measurement of the opacity of the ISM at wavelengths between the Lyman limit and the C-band edge (0.28 keV; $\lambda = 44$ Å) can be used to determine the column density of neutral hydrogen, because in this energy range only hydrogen and helium contribute significantly to the absorption cross section (184). At the lower energies, the high H I cross section limits the observations to very near (more precisely, very low N_H) objects, but in the soft X-ray bands (for example, near 0.2 keV), the total opacity through the galactic disk at high latitudes may be only ~ 2. To derive an accurate N_H from these observations corrections must be made for molecular gas along the line of sight and for the neutral helium in any low-excitation H II regions that might be encountered. The background source must also have a strong EUV or soft X-ray flux and, most importantly, an intrinsic spectrum that is well known, so that the photoelectric absorption can be measured precisely.

The H I density near the Sun is so low that the spectra of a number of white dwarfs have been observed from the near-ultraviolet right through the Lyman limit (124). This is possible only because the stars are nearby, are bright, and have total column densities in their direction $\ll 10^{19}$ cm^{-2}. Such targets are rare. Of more general use are estimates of the total N_H from soft X-ray observations of extragalactic sources like active galaxies and quasars. But here the limit is the uncertainty in the "expected" X-ray spectrum. Indeed, the first attempts to solve simultaneously for the X-ray spectral index and the column density of intervening H I for a number of quasars yielded values for N_H that were much too small. It is now recognized that QSOs have complex X-ray spectra, and this makes them of little use as tools for deriving an accurate galactic N_H (see 280). Even for sources that have a simple power-law spectrum, the uncertainty in the N_H derived from current data is relatively large (e.g. 173).

The next generation of X-ray observatories, beginning with *ROSAT*, will measure spectra of many tens of thousands of objects. These will be useful for determining N_H in some situations and will help to identify regions of the sky that have excess X-ray opacity caused by the presence of molecular gas or neutral He in a low-ionization-state H II region. Attenuation of soft X rays by He in the diffuse ionized component, in particular, may be substantial (215). The extragalactic X-ray observations will be an independent complement to the IR observations in disentangling the state of the ISM at high galactic latitudes.

4.4 *Gamma Rays*

Surveys of diffuse gamma-ray emission by orbiting observatories such as *SAS-2* and *COS*-B have provided another way of measuring the abundance of interstellar matter, including H I (92a; recently reviewed in ref. 15). Gamma rays with energies in the range of a few hundred to a few thousand MeV are produced mainly by bremsstrahlung of cosmic-ray electrons with energies ~1 GeV and by decay of π^0 particles produced by collisions of cosmic rays with the ISM. By observing the gamma-ray surface brightness and comparing it with the distribution of interstellar gas, one can deduce the density of cosmic rays as a function of position in the Galaxy (12a, 245a). Conversely, since the galactic nonthermal radio emission gives a good estimate of the cosmic-ray density in this energy range (17, 214), one could derive the total amount of interstellar matter from the gamma-ray brightness. In practice, the most useful approach has been to start from the surveys of 21-cm and CO emission as tracers of the H I and H_2, respectively, and to try to reproduce the observed gamma-ray brightness using various galactic models (247, and references therein).

There are observational and theoretical limitations to this technique. The gamma-ray observations have low angular resolution ($\geq 1°\!.5$ for *COS-B*), so the survey of diffuse brightness is confusion limited by small sources. Also, large-angle, nearby interstellar structures, like radio Loop I, must be removed (15). In addition, there are uncertainties in interpreting the gamma-ray flux due to our limited understanding of cosmic-ray propagation in the Galaxy, both on small scales (e.g. magnetic reflection from clouds) and large scales (e.g. leakage from the disk). When higher resolution data become available they will give a more detailed picture of both the cosmic-ray distribution in the Galaxy and its interaction with interstellar clouds (13, 16, 17). This should occur this year (1990) with the launch of the *Gamma Ray Observatory* (92b).

It is unlikely that the techniques described in this section will supplant the more direct methods of measuring interstellar H I, but they do serve as an independent check that is especially important in tracing gas not easily measured otherwise, such as molecular hydrogen and the low-density, ionized component.

5. THE ORGANIZATION OF INTERSTELLAR H I

There has been a significant change in our understanding of the morphology of galactic H I as new observations have revealed structures on scales ranging from 1 kpc to $\lesssim 100$ AU. With this has also come a breakdown in the rigid distinction between atomic and molecular clouds. The main message of these new data may be that galactic H I cannot be understood unless viewed in conjunction with all other phases of the ISM, both locally and globally.

5.1 *Bubbles, Shells, and Supershells*

It has been evident since the early emission studies of Heiles (110) and Verschuur (270) that the most prominent interstellar atomic hydrogen structures are shaped like sheets and filaments, often with velocity gradients and often arranged in shells. The most convincing shells show an empty cavity centered on some velocity v_0, two bright "polar caps" at the central position but with velocity offsets $v_0 \pm v_{\rm exp}$, and in between a symmetric change of shell size with velocity, with the largest size at v_0 (150). While few shells appear so cleanly in the data, shells or shell fragments are seen throughout the Galaxy with a very broad range of size (for catalogs, see 112, 114, 244). This diversity probably reflects the variety of physical processes that generate the shells. It is possible that the expanding ring of gas and stars that we call Gould's Belt is a shell viewed from within (40, 161, 193, 222).

Supershells, with diameters of more than a kiloparsec, are often larger than the disk thickness and may open into the halo. Some of these (e.g. the anticenter shell) have so much kinetic energy (some 10^{53} ergs) that they have been attributed to the impact of infalling high-velocity clouds on the disk (155, 183, 249, 250). Other supershells can be explained by the combined effect of multiple supernovae in an association (116).

There are a great many shells with diameters of a few hundred parsecs; most of these are probably the old remnants of one or more supernovae that are a few times 10^6 or even 10^7 years old. Often these have associated shells of molecular gas and radio continuum emission (108, and references therein; see also 18, 113, 133, 239). Expansion velocities tend to decrease with increasing size, more or less as expected for old supernova remnants, with typical values for v_{exp} being 25 km s^{-1} or less. The oldest shells have v_{exp} as low as 3 km s^{-1}, sizes of a few hundred parsecs, and ages of several million years. Younger supernova remnants also show distinct H I shells; the interaction of these remnants with the surrounding medium has been the topic of numerous studies (e.g. 51, 84, 102, 103, 157, 212, 213).

The energy and, more importantly, the accumulated momentum available in stellar winds are sufficient to generate shells with properties nearly identical to those caused by old supernova remnants (276, and references therein). Many H I shells (these in particular are often called "bubbles") have OB associations near their center (e.g. 39, 41, 260). Typically these bubbles are 50 to 150 pc in diameter, and their range of sizes overlaps that of supernova remnants. Many H II regions show prominent H I velocity discontinuities and cavities, which suggest that H I shells are in the process of forming (22, 134, 212, 213, 217, 268). Striking examples are seen in the VLA maps of Orion A, W3, and BG 2107+49 (262–264).

On the smallest scales (1 pc or less), H I is associated with molecular outflows ($v_{exp} \lesssim 120$ km s^{-1}) centered on infrared stars in star-forming regions such as NGC 2071 (7, 71, 72, 218). This atomic gas is likely the dissociation product of molecular gas, either behind the shock formed by the outflow on the surrounding molecular cloud or on the inner edge of the photodissociation front around the young stars. Such high-velocity outflows may represent the earliest stages of bubble formation.

High-resolution maps of H I in other galaxies show disks pockmarked with shells, as well as evidence of the systematic infall or outflow of large neutral clouds (26, 27, 221, 259). These, combined with galactic data, have at times led many H I observers to mentally picture most galactic H I as residing in giant structures shaped by a history of local overpressure from stellar winds or supernovae, rather than in a hierarchy of equilibrium clouds. Evidence for this extreme view, however, is still circumstantial. There is no solid estimate of the fraction of galactic H I that is tied up in bubbles, shells, and supershells or in their old fragments.

5.2 Small-Scale Structure in Atomic Clouds

Aperture synthesis observations of 21-cm absorption in front of extended background sources almost always detect some structure in the absorption on angular scales as small as the beam size (see, for example, the discussion of VLBI results in Section 2.1.3). These data are clearly telling us something about the internal composition of interstellar clouds. While it may seem that the telescopes must be resolving ever smaller, independent "cloudlets," this model is not satisfactory. For one thing, the covering factor of the absorbing gas parcels must be large, not small, because the angular scale over which absorption is measured does not seem to affect 21-cm absorption line widths or line shapes or derived spin temperatures (199). That is, the statistical properties of H I absorption lines are identical whether the absorption is measured against point-source pulsars or is averaged over several arcminutes (74). Furthermore, H I emission lines do not show the large point-to-point fluctuations that would occur if a substantial fraction of the ISM were concentrated into small cloudlets (128, 129, 158, 167). There are directions where small-scale density fluctuations can be large, such as the vicinity of H II regions and supernova remnants, but this is not the case in more typical regions. In general, no more than 10% of all H I can be in these small-scale structures, and the fraction is usually much smaller on the subparsec scale. Skeptics on this point should study Figure 8 and note the high correlation between N_H measured over a 21′ beam and N_H measured over the solid angle of a star. Also, the suggestion that broad, 21-cm emission lines are composed of many beam-smeared, narrow-velocity components (271a) is inconsistent with the observation of absorption spectra toward distant stars (e.g. Ti II; 1), which often show broad lines identical to those in 21 cm. Cloudlets cannot be the fundamental form of a significant fraction of interstellar H I.

Nevertheless, aperture synthesis observations do show some variation in 21-cm opacity on all angular scales. One of the most detailed 21-cm observational studies ever done on a single field (136) found H I clouds in a complex of clumps with sizes of 0.5–1 pc and masses of 0.03–0.5 $M_\odot$. The solid-angle covering factor of the clumps is large, so they overlap in front of the background source, although Gaussian decomposition of the absorption spectrum shows these features to be distinct in velocity. About three quarters of the H I, however, is not in these clouds but rather is spread so smoothly across the field on scales $> 10′$ that the interferometer could not detect it. Structure on linear scales of 0.1 pc and smaller has been probed by only a few experiments (60, and references therein; 73, 105), but these also show variations in optical depth of perhaps 20% at a given velocity across a background radio source (see Section 2.1.3).

We believe that the evidence is against large variations in column density on small scales, and that the concept of "cloudlets" is not a useful one in describing most H I. The fluctuations in optical depth that are observed on scales smaller than 1 pc or so most probably arise from turbulence and temperature fluctuations within clouds. This implies that H I clouds are both far from quiescence and far from thermal and dynamic equilibrium.

5.3 The Relation Between Atomic and Molecular Gas

Molecules require shielding against UV photodissociating radiation; this requires a layer of partially atomic gas whose thickness (column density) depends on the rate of molecule formation, and hence on the space density (265; reviewed in refs. 257b,c). Molecular hydrogen is a significant component of those interstellar clouds that have $N_H \gtrsim 5 \times 10^{20}$ cm^{-2} (224a). In low-density environments it is hard to trace the molecular hydrogen column density using the most common tool (millimeter-wave CO emission lines), in part because the shielding of CO is rather different from that of H_2 and in part because a total density of about 10^3 cm^{-3} is needed to keep the collision rate high enough to maintain the 3-mm transition in local thermodynamic equilibrium. Even so, CO emission is detected from some primarily atomic clouds (63, 175, and references therein). Also, OH absorption lines show that molecular cores are present inside clouds that show H I absorption (76, 132a), and larger, primarily molecular clouds have atomic halos (275) or residual cool H I (37, 101).

It is at high latitudes that the various types of atomic and low-mass molecular clouds can best be distinguished. Beginning with the pioneering work of L. Magnani and L. Blitz, high-latitude molecular clouds have been studied in surveys of CO, OH, C_3H_2, and other molecules (57a, 173a, 175, 175b, 257a, 281b). The nearby high-latitude clouds have masses of only a few hundred solar masses, much lower than that of the giant molecular clouds in the disk (which can reach $10^6 M_\odot$ or more) and lower than such nearby large molecular complexes as Orion and Taurus. Often they are found in the catalog by Lynds (171b) of dust clouds selected as discrete, intermediate-A_V regions on the Palomar Sky Survey.

Observations of CO, CH, and OH can also be used to trace molecules in much lower A_V regions that have been called "translucent molecular clouds" (132a, 175a, 281b), although they are predominantly atomic and have masses of only a few hundred solar masses and $A_V < 1$. Typically, the 21-cm line is in self-absorption in such clouds. Some of the far-IR cirrus clouds correspond to this translucent molecular cloud population, but most cirrus clouds show no trace of CO or any other radio molecular species, which is not surprising since the sky covering factor of CO-bearing clouds is only about 0.5% (174), whereas cirrus covers almost 10% of the high-latitude sky. Thus, most cirrus clouds are probably diffuse atomic clouds.

The physical conditions in the translucent molecular clouds and the diffuse atomic clouds are very similar; they have densities of 50–100 cm^{-3} and temperatures of 50–100 K. The molecules simply reside in the higher column density regions, where they are shielded from photodissociating UV. Thus, the translucent molecular clouds, and even the Lynds-type clouds at high latitude, may be simply the cores of the larger atomic diffuse clouds. "Core," however, is not the best description, since in fact these shielded regions often have an octopus- or spongelike morphology.

Most of the cool atomic hydrogen at the solar circle and beyond, where the surface density of atomic gas is greater than that of molecular gas (34, 231a), is probably in clouds that are primarily atomic rather than molecular. In the inner Galaxy, where the molecular gas surface density exceeds that of H I, much of the H I may be associated with molecular clouds, either as halos, as a photodissociation product near young stellar associations [as is the case in M51 (252)], or as the low-density substrate for the molecular "mist" suggested by Solomon et al (240). Warm H I, at least half of which is not associated with clouds (50, 162), remains about as abundant in the inner Galaxy as it is in the solar neighborhood.

5.4 *Cloud Mass Spectrum*

The range of cloud masses spans more than a factor of 10^6, from the giant molecular clouds to the diffuse atomic clouds. To measure the interstellar cloud mass spectrum, i.e. the distribution function of cloud masses, different tracers are needed for different types of clouds, and results obtained from these may be difficult to put on the same scale. The problem is that absorption surveys, whether in the 21-cm, optical, or UV lines, give the number of clouds per unit line-of-sight distance (the mean free path) as a function of optical depth, whereas emission surveys (e.g. in CO) give the number of clouds per unit volume (cloud abundance or density) as a function of flux or brightness temperature (86, and references therein; 230a, and references therein). The latter can be converted, at least approximately, into cloud abundance versus mass, and the former into mean free path versus column density, but the issues of cloud cross section and internal density structure still need to be addressed in order to put these kinds of measurements together.

Dickey & Garwood (77) made rough guesses for these quantities to produce the abundance function of Figure 9, which shows the number of

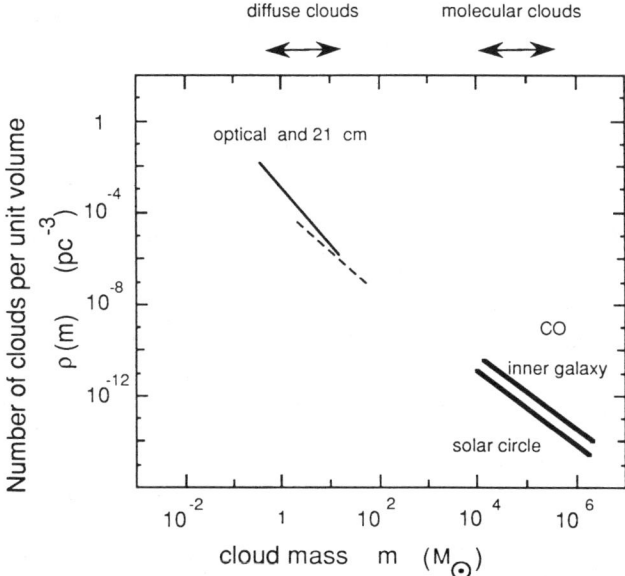

Figure 9 The "mass spectrum" of interstellar clouds. This figure plots an estimate of the distribution function by mass of clouds, i.e. the number of clouds per unit interstellar volume as a function of cloud mass. Relatively complete inventories of the galactic cloud population are available only over limited ranges of mass. The two solid lines on the right show results from molecular cloud surveys, normalized to give the molecular gas surface density at the solar circle and at the inner Galaxy peak. The solid and dashed lines on the left represent estimates for atomic clouds based on 21-cm absorption surveys and two assumptions for cloud temperatures (77).

clouds per unit volume $\rho(m)$ as a function of cloud mass m. Cloud mass is defined so that $2 \times \langle|z|\rangle \times \int \rho(m)\,dm$ gives the surface density of gas in the cloud phase. The mass ranges that are directly observed are indicated; the center of this figure is blank, not because it is difficult to see such clouds (e.g. with masses $10^3\ M_\odot$), but because they are difficult to inventory over a large area owing to incompleteness in the surveys. Also, we do not know the cutoff at the low-mass end, which determines the total mass in clouds if the slope is as steep as it appears among the low-mass clouds detected in optical and 21-cm absorption. In computing this cloud mass spectrum it is important to avoid using cloud sizes, if possible, because observationally, cloud size has only a weak dependence on mass. It must be that cloud sizes are determined by a variety of mechanisms (77, 190).

Our observational knowledge of the cloud mass spectrum is still so rudimentary that it is difficult to know how much theoretical interpretation of Figure 9 is justified. It is reassuring that the observed slopes for atomic clouds and for molecular clouds are about the same; the slope measured from molecular surveys for clouds in the mass range of a few times 10^4–10^6 $M_\odot$ is about -1.6, whereas the slope measured for diffuse clouds with masses 1–10 $M_\odot$ is about -2.1. Given that there are roughly equal gas surface densities in the molecular and cool atomic phases, the slope is necessarily near -2, which is the critical slope giving an equal-mass contribution from each decade mass range. A simple theory of collisional agglomeration (the "Oort model") predicts a slope of -1.5 (94), but the theory has many possible variations depending on the physics of cloud collisions and the velocity distribution of clouds (e.g. 52, 53, 227, and references therein; 236, 283). A slope near -1.5 is not necessarily ruled out by the data, at least over mass ranges of several orders of magnitude.

On a more fundamental level, the interpretation of the survey results implicit in Figure 9 does not take account of the hierarchical clustering of interstellar clouds (e.g. 126, 226a). Aggregates of smaller clouds make up larger clouds, all arranged typically on large shells or at least sheets, sometimes punctured by smaller, younger shells or bubbles. In the inner Galaxy, a large fraction of the cool atomic gas may be associated with molecular clouds. This association raises the question of which structures are distinct clouds and which are mere halos, cores, or the substrata of a larger complex. The term cloud, a metaphor drawn from the terrestrial atmosphere, is fraught with such imponderables, which are reflected in the difficulty of objectively delimiting discrete features in emission surveys (176).

5.5 *Random Motions*

To measure random velocities (as distinct from the large-scale galactic rotation), either the atomic gas must be clearly associated with an object that has a known distance [e.g. an H II region (232)] or a particular region of l-v space must be studied in which the effect of galactic rotation is somehow isolated. Examples of the latter are the lines of sight to the galactic center and the anticenter (151, 211, 230), velocities near the terminator (subcentral point) in the inner Galaxy (151), and high-latitude clouds (58). Results from the various studies have generated some recent controversy and have revealed several surprising properties of interstellar H I.

There is evidence for several kinematic populations of interstellar H I, not counting the high- and intermediate-velocity clouds that are defined empirically as having a velocity well outside the range permitted by galactic rotation (267a). The cloud population studied in absorption has a velocity dispersion of ~ 7 km s^{-1} (11a, 58, 230). In emission the dispersion is somewhat higher for the bulk of the H I, but it depends on the method of analysis and the specific region of the Galaxy being studied (151). The dispersion may increase toward the inner Galaxy. In addition, a component of H I seen in both emission and absorption has a velocity dispersion in the range 15–35 km s^{-1} (3, 232, and references therein), although the amount of gas in this phase was overestimated at first (see 151). Whether this component is distinct from the intermediate-velocity gas, or even from some of the high-velocity gas, is not clear (273). At rms velocities of 15–35 km s^{-1} the velocity distribution is quite symmetric, whereas the higher velocities have a clear preference for infall in a class of high-velocity clouds. There is a predominance of infalling gas at high positive galactic latitudes even at velocities of -20 to -50 km s^{-1}, but this seems to be local to the solar neighborhood (279).

A very interesting result from a study by Kulkarni & Fich (151) is that at high latitudes the kinetic energy $[v^2 N(v)]$ per unit velocity in H I is approximately constant from 10 to 70 km s^{-1}, i.e. the amount of gas decreases as velocity to the power -2 (see also 1). The physical cause of this equipartition of kinetic energy among equal-velocity intervals is not clear.

In addition to these truly random motions, there are systematic departures from galactic rotation associated with galactic structure. A galactic bar or density wave will induce large-scale H I streaming, and a spiral shock pattern can explain many of the velocity anomalies observed in absorption toward H II regions (10, 33, 104a, 232).

6. VERTICAL STRUCTURE

6.1 *Thickness of the H I Layer*

In recent years there has been renewed interest in measuring and understanding the vertical distribution of H I and other components of the ISM. The concentration of H I to the galactic plane was evident in the earliest observations (195); in the inner Galaxy most of it lies in a layer that has a thickness between half-density points of only about 220 pc. The thickness is approximately constant from $0.4R_0$ to R_0, but it decreases to <100 pc near the nucleus and increases enormously at $R > R_0$ (171). The vertical density profile $n(z)$ is approximately Gaussian near the plane with a low-density tail to high z (165). The gas lies in a flattened system interior to the solar orbit but is systematically warped in its outer parts (29, 139, 142). A cross section of the Galaxy that illustrates most of these features is given by Burton (34). Recent research has focused on topics like the detailed shape of $n(z)$ and its variation in the Galaxy, the maximum extent of the neutral layer, and the perennial, provocative question of what holds the whole thing up.

It is necessary to consider UV, optical, and radio data together because each has revealed a different aspect of the vertical structure of H I. The classic Ca II observations of Münch & Zirin (188) showed that there were neutral interstellar clouds up to $z \sim 1$ kpc, a discovery that has been confirmed and broadened through optical and UV measurements of many species (1, 66, 137a, 185a, 225). The 21-cm emission data provide most of the information on vertical structure but are usually dominated by low-z gas; Lα studies show a component that is 3 times wider than the main layer deduced from 21-cm data (20). The cumulative effect of these observations has led to a new appreciation of the vertical stratification of the ISM. One can lay out a hierarchy of increasing physical temperature (or ionization state) with increasing scale height that goes from molecular clouds (the species most confined to the plane) through H I and on into the gas traced in highly excited ions like C IV, Si IV, and N V, which extends several kiloparsecs above the disk (225). It is natural to ask if there might also be separate stratified components within the atomic hydrogen itself, and this question has led to attempts to decompose the 21-cm data into populations with separate scale heights, usually in the spirit of the two-phase model (5, 90, 154, 208). Typically, the hotter H I is found to have a scale height of order 1.5 times that of the colder clouds.

The vertical stratification shows dramatically in Lα/21-cm comparisons toward high-latitude stars, where it is observed that the scale height of H I is usually inversely proportional to the mean density along the line of sight (167). This implies that dense clouds are in the thinnest layer and probably accounts for the general difference between Lα and 21-cm scale heights. A recent Lα study (225) toward stars chosen to have "the lowest attainable color excess and H I column densities per kiloparsec" measured values of N_H that are consistent with a single exponential distribution with $n(0) = 0.1$ cm^{-3} and a scale height of 800 pc. This particular result is a product of the target selection and may not pertain to any interstellar H I component, but it again illustrates the composite nature of the vertical structure and reminds us that along some sight lines, often spanning many kiloparsecs, integrated properties vary by a factor of 5 from global averages. Indeed, density fluctuations are often many times larger than the mean at all altitudes, as one might guess from Figure 7. Moderate-density H I clouds are found many hundreds of parsecs to several kiloparsecs from the plane, and parts of H I supershells extend to $z > 600$ pc (112, 114, 165, 238). In all, the layer is not very tidy, and it looks like a smooth function of z only when averaged over large volumes of the Galaxy (see, for example, 45).

In recent years there has been special interest in the highest z H I, which lies in the lower galactic halo at $|z| > 500$ pc. This medium extends far beyond the H I disk. It may regulate the evolution of supernovae and supershells (172), may contain much of the kinetic energy of the ISM (151), and may be connected with events high in the galactic halo via a galactic fountain (23, 125, 192). Halo H I is difficult to measure in the 21-cm line because of blending with the much brighter disk gas and contamination of profiles by stray radiation. The most conspicuous neutral components are clouds whose velocities are peculiar enough to make them stand out against the rest of the ISM.

Such objects are common locally, and a prime example is the intermediate-velocity H I that covers much of the northern galactic polar regions at column densities occasionally as high as a few times 10^{20} cm^{-2} (80b, 279). It is falling toward the galactic plane at about -50 km s^{-1}. Absorption-line studies toward halo stars have bracketed its vertical distance between 340 and 760 pc, which implies that it has a mass of $10^5 \, M_\odot$ (66). This cloud (or, more likely, "sheet") contains $\sim 10\%$ of the H I mass in the solar neighborhood and is probably connected with the evolution of the local bubble (57). It is a nearby example of a general galactic process that may produce most high-velocity clouds (23).

This leads to the question of whether halo H I is all in structures either rising or falling, or whether there is also a truly diffuse, static medium. Studies of general halo H I are not yet able to give a definitive answer, but we feel that the observations point toward a halo in which the neutral gas is mainly composed of discrete structures, such as intermediate-velocity clouds or the fragments of supershells that are so evident in Figure 7.

6.2 *Average* n(z)

Recent analyses have suggested the following general form for the vertical density function $n(z)$: In the innermost few kiloparsecs, inside the radius of the galactic bulge but outside of the galactic nucleus, the layer is extremely thin and is approximated by a single Gaussian with a $\sigma \lesssim 70$ pc; in the outermost regions, beyond the solar circle at $R > R_0$, the layer flares dramatically and nearly linearly, reaching $\langle z^2 \rangle^{1/2} \gtrsim 3$ kpc (34, 82, 121, 149, 165). Gas in the outer Galaxy is distributed in a complicated multicomponent layer. Faint emission extends very far from the main concentrations, reminiscent of the "thick envelopes" of H I that are seen around some galaxies (e.g. 221).

From radii of perhaps 4–8 kpc, the layer has approximately constant average properties (although, as noted above, with very large fluctuations). This is the most studied part of the galactic disk. Our "best estimate" of $n(z)$ over this region is given in Figure 10. It is a combination of two

Gaussians, of central densities 0.395 and 0.107 cm^{-3} and FWHMs of 212 and 530 pc, and an exponential with $n(0) = 0.064$ cm^{-3} and a scale height of 403 pc. The distribution has a FWHM of 230 pc, a central density of 0.57 cm^{-3}, and $N_H = 6.2 \times 10^{20}$ cm^{-2} through a full disk. This function was derived from the data given by Lockman (165) scaled to $R_0 = 8.5$ kpc and corrected for the undercounted H I at low z in that study by increasing the central density of the narrowest component.

For comparison, Figure 10 also shows other density profiles that have been derived for galactic H I. The UV data (Savage & Massa) were obtained along very low N_H lines of sight, which accounts for the low $n(0)$ although not necessarily for the large scale height, while the earlier radio data did not have sufficient sensitivity or vertical coverage to catch the faint tail. Descriptions of the central layer, however, have been remarkably consistent among radio observers (see also 45, 140), and in all, the various differences displayed in Figure 10 probably reflect the diversity of the neutral ISM.

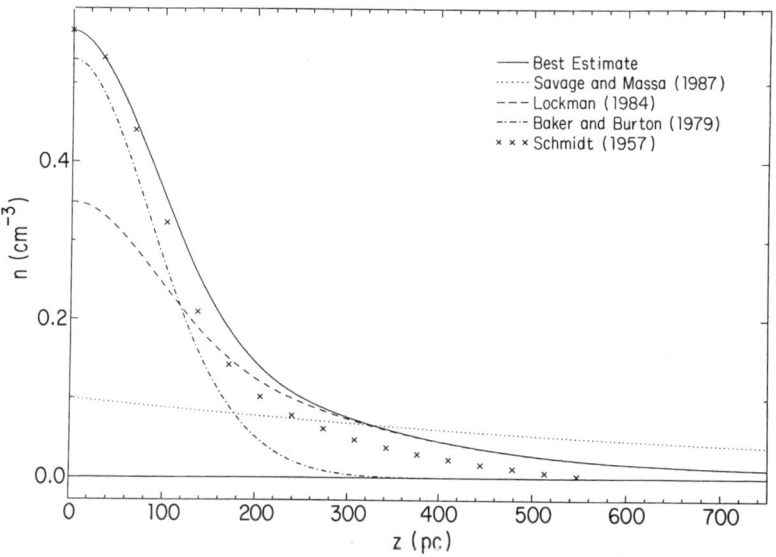

Figure 10 Recent estimates of the vertical density distribution of galactic H I in the range $0.4 R_0 \lesssim R \lesssim R_0$. All are based on 21-cm data (5, 165), except for the Savage & Massa (225) curve, which comes from Lα observations of a sample of exceptionally low N_H lines of sight. The curve marked "Best Estimate" is what we believe to be the most likely distribution and is discussed in the text. For comparison, the crosses show the values $n(z)$ derived by Schmidt (228) in 1957 after they are adjusted to a central density of 0.57 cm^{-3}. All curves have been scaled to $R_0 = 8.5$ kpc.

6.3 Support of the Layer

The kinetic temperature of H I is far too low to support it to its observed height against the galactic potential. It must be held up by turbulent pressure (random motion of H I clouds) or by nonthermal pressure from cosmic rays and magnetic fields. Models of global equilibrium have been suggested that incorporate all three forms of support (4, 14, 55, 56, 96, 138, 196; reviewed by 284); most of the models are dynamically unstable. It is not clear that cosmic rays or magnetic fields have a vertical pressure gradient large enough to provide the needed support, but we do not discuss the possibilities here [see Spitzer's article in this volume (242a) for more on this topic]. Instead, it is interesting to consider the vertical distribution that would result just from the H I motions that are measured in 21-cm emission profiles. Examination of high-latitude 21-cm spectra in the solar neighborhood shows that there is enough kinetic energy in H I motion to maintain a layer to a FWHM of 350 pc, and also that about 15% of the H I could be in a component with a scale height of ~ 500 pc (151, 169). This distribution is reasonably like the measured one (Figure 10) given the large uncertainties, so we consider it possible that cloud-cloud or bulk motion rather than nonthermal pressure supports the neutral ISM, at least in the solar neighborhood.

An item that puzzled Oort in 1962 (195), and which is still unresolved, is the apparent constancy of the layer's thickness from $0.4R_0$ to $\sim R_0$, despite the change in the gravitational potential by a factor of several over this range (19, 152; but see ref. 146 for a different opinion). Beyond the solar circle the layer thickens to an extent consistent with a decrease in the gravitational potential, but there is no corresponding narrowing interior to the solar circle. Suggestions that increased cosmic-ray and magnetic pressure exactly compensate the change in gravity seem artificial. No other solution to the dilemma is apparent unless one is prepared to consider the entire ISM, rather than just the atomic component, as one entity that is in vertical balance. In that case, the overall layer does narrow substantially in the inner Galaxy because molecular gas, which increasingly dominates the total surface density at small R, is much more confined to the plane than the H I.

6.4 Deviations From a Plane

6.4.1 CORRUGATIONS   When H I observations were used to redefine the galactic coordinate system it was noticed that there was a residual "waviness" of the inner-Galaxy H I layer about the new galactic plane (106). Since then, there have been several attempts to determine if the waviness has some large-scale pattern, for in some areas the deviations of the average

z from zero seem to be tantalizingly regular, as if they are manifestations of ordered "corrugations" that have an amplitude of a few tens of parsecs. A similar, though perhaps not identical, effect appears in H II regions and other constituents of the ISM (164, 202–204, 241). The form of these distortions is not at all certain, for it is difficult to disentangle galactic structures except in special directions, and it is even harder to follow individual H I features for any distance in the Galaxy. An unambiguous observational picture of the inner Galaxy corrugations might best be done in species that are less abundant and therefore less confused than H I.

6.4.2 THE WARP The H I warp is one of the most striking but least understood features of the Galaxy. Recent research has been reviewed by Burton (34). The warp has an integral-sign shape; its line of nodes is close to the Sun-center line and does not change significantly with R. The warp begins just outside of the solar circle and extends to the Galaxy's edge; it reaches an altitude of 4 kpc in the north, whereas in the south it seems to turn back toward the plane after reaching only ~ -1.5 kpc. But these numbers do not do justice to the extent of the warp, for with the combined flaring of the layer and its vertical displacement it is easily seen to latitudes of $\sim 30°$ in some directions (82). At its outermost edge the warp is fluted as if by a higher frequency vertical distortion (249). Many other spiral galaxies show even more extreme large-scale warps (25).

Accurate measurement of H I in the warp is an observational challenge because of the high angular resolution, sensitivity, and freedom from stray radiation that are required (166). Our knowledge of the warp, and thus of the ultimate extent and size of the Galaxy, is certain to increase in the coming years as better observational and theoretical tools are used to study it.

6.4.3 CYLINDRICAL ROTATION Is it possible that when the gas layer is significantly and systematically displaced from a simple plane, the rotational motion of the H I deviates from strict corotation with the motion at $z = 0$ and has full three-dimensional structure? Is gas in the galactic warp, for example, rotating as if it were firmly fixed to the extension of the galactic plane that lies a scale height below it? As of now, these questions can be answered only for special subsets of galactic H I. It seems likely that H I in the inner Galaxy, for example, is corotating to at least 1 kpc, while high-velocity clouds, though they have some component of galactic rotation, are following paths of their own (98, 99, 165). Also, there are hints that some of the highest scale-height ions (C IV) are lagging behind galactic rotation (225). Determination of the rotation curve away from the galactic plane is critical in establishing a distance to gas in the outermost parts of the Galaxy and the galactic halo (179, 223), but it will probably be done with indirect tracers of H I, such as optical and UV absorption lines, rather than with measurements of H I itself. The importance of the assumption of corotation should not be underestimated when analyzing velocities in the halo and outer Galaxy.

ACKNOWLEDGMENTS

We are grateful to numerous colleagues for useful conversations and advice. In particular, we thank Tom Bania, Hans Bloemen, Leo Blitz, Don Cox, Carl Heiles, George Helou, Lew Hobbs, Harvey Liszt, Phil Maloney, Dan McCammon, Ed Salpeter, Blair Savage, John Scalo, Lyman Spitzer, and Barry Turner for their helpful suggestions. We are also very grateful for the expert assistance of Terry Thibeault, Executive Secretary in the Astronomy Department at the University of Minnesota. JMD wishes to thank the Nederlandse Wetenschappelijk Onderzoek (ZWO) and the hospitality of the Sterrewacht Leiden, where much of this work was done. This work was supported in part by the National Science Foundation under grant 87-22990 to the University of Minnesota. The National Radio Astronomy Observatory is operated by Associated Universities, Inc., under agreement with the National Science Foundation.

Literature Cited

1. Albert, C. E. 1983. *Ap. J.* 272: 509–39
2. Alferova, Z. A., Venger, A. P., Gosachinskij, I. V., Grashev, V. G., Egorova, T. M., et al. 1987. *Astrofiz. Issled. Izv. Spets. Astrofiz. Obs.* 24: 93–107
3. Anantharamaiah, K. R., Radhakrishnan, V., Shaver, P. A. 1984. *Astron. Astrophys.* 138: 131–39
4. Badhwar, G. D., Stephens, S. A. 1977. *Ap. J.* 212: 494–506
5. Baker, P. L., Burton, W. B. 1975. *Ap. J.* 198: 281–97
6. Baker, P. L., Burton, W. B. 1979. *Astron. Astrophys. Suppl.* 35: 129–52
7. Bally, J., Stark, A. A. 1983. *Ap. J. Lett.* 266: L61–64
8. Bania, T. M. 1983. *Astron. J.* 88: 1222–27
9. Bania, T. M., Lockman, F. J. 1984. *Ap. J. Suppl.* 54: 513–45
10. Bash, F. N., Leisawitz, D. 1985. *Astron. Astrophys.* 145: 127–34
11. Beichman, C. A. 1987. *Annu. Rev. Astron. Astrophys.* 25: 521–63
11a. Belfort, P., Crovisier, J. 1984. *Astron. Astrophys.* 136: 368–70
12. Bertaux, J. 1984. In *The Local Interstellar Medium, IAU Colloq. No. 81*. *NASA Conf. Publ. 2345*, ed. Y. Kondo, F. C. Bruhweiler, B. D. Savage, pp. 3–23. Greenbelt, Md: NASA
12a. Bignami, G. F., Fichtel, C. E. 1974. *Ap. J. Lett.* 189: L65–67
13. Bloemen, J. B. G. M., Strong, A. W., Blitz, L., Cohen, R. S., Dame, T. M., et al. 1986. *Astron. Astrophys.* 154: 25–41
14. Bloemen, J. B. G. M. 1987. *Ap. J.* 322: 694–705
15. Bloemen, H. (J. B. G. M.) 1989. *Annu. Rev. Astron. Astrophys.* 27: 469–516
16. Bloemen, J. B. G. M., Bennett, K., Bignami, G. F., Blitz, L., Caraveo, P. A., et al. 1984. *Astron. Astrophys.* 135: 12–22
17. Bloemen, J. B. G. M., Reich, P., Reich, W., Schlickeiser, R. 1988. *Astron. Astrophys.* 204: 88–100
18. Bochkarev, N. G. 1984. *Sov. Astron. Lett.* 10: 76–78
19. Bohigas, J. 1988. *Astron. Astrophys.* 205: 257–66
20. Bohlin, R. C., Savage, B. D., Drake, J. F. 1978. *Ap. J.* 224: 132–42
21. Boulanger, F., Perault, M. 1988. *Ap. J.* 330: 964–85
21a. Boulanger, F., Falgarone, E., Puget, J. L., Helou, G. 1990. *Ap. J.* In press

22. Braunsfurth, E. 1983. *Astron. Astrophys.* 117: 297–304
23. Bregman, J. N. 1980. *Ap. J.* 236: 587–91
24. Bregman, J. D., Troland, T. H., Forster, J. R., Schwarz, U. J., Goss, W. M., Heiles, C. 1983. *Astron. Astrophys.* 118: 157–62
25. Briggs, F. H. 1990. *Ap. J.* 352: 15–29
26. Brinks, E., Braun, R., Unger, S. W. 1989. In *Structure and Dynamics of the Interstellar Medium, IAU Colloq. No. 120*, ed. G. Tenorio-Tagle, M. Moles, J. Melnick, pp. 524–29. Berlin: Springer-Verlag
27. Brinks, E., Bajaja, E. 1986. *Astron. Astrophys.* 169: 14–42
28. Brown, R. L., Liszt, H. S. 1984. *Annu. Rev. Astron. Astrophys.* 22: 223–65
29. Burke, B. F. 1967. *Astron. J.* 63: 90
30. Burstein, D., Heiles, C. 1978. *Ap. J.* 225: 40–55
31. Burstein, D., Heiles, C. 1984. *Ap. J. Suppl.* 54: 33–79
32. Burton, W. B. 1966. *Bull. Astron. Inst. Neth.* 18: 247–55
33. Burton, W. B. 1976. *Annu. Rev. Astron. Astrophys.* 14: 275–306
34. Burton, W. B. 1988. In *Galactic and Extragalactic Radio Astronomy*, ed. K. Kellermann, G. L. Verschuur, pp. 295–358. New York: Springer-Verlag
35. Burton, W. B., Deul, E. R. 1987. In *The Galaxy*, ed. G. Gilmore, B. Carswell, pp. 141–72. Dordrecht: Reidel
36. Burton, W. B., Gordon, M. A. 1978. *Astron. Astrophys.* 63: 7–27
37. Burton, W. B., Liszt, H. S., Baker, P. L. 1978. *Ap. J. Lett.* 219: L67–72
38. Burton, W. B., te Lintel Hekkert, P. 1986. *Astron. Astrophys. Suppl.* 65: 427–63
39. Cappa de Nicolau, C. E., Niemela, V. S. 1984. *Astron. J.* 89: 1398–1403
40. Cappa de Nicolau, C. E., Niemela, V. S., Arnal, E. M. 1986. *Astron. J.* 92: 1414–19
41. Cappa de Nicolau, C. E., Poppel, W. G. L. 1986. *Astron. Astrophys.* 164: 274–99
42. Carruthers, G. R. 1969. *Ap. J. Lett.* 156: L97–100
43. Carruthers, G. R. 1970. *Space Sci. Rev.* 10: 459–82
44. Caswell, J. L., Murray, J. D., Roger, R. S., Cole, D. J., Cooke, D. J. 1975. *Astron. Astrophys.* 45: 239–58
45. Celnik, W., Rohlfs, K., Braunsfurth, E. 1979. *Astron. Astrophys.* 76: 24–34
45a. Christiansen, W. N., Hogbom, J. A. 1985. *Radio Telescopes.* Cambridge: Univ. Press. 2nd ed.
46. Clark, B. G. 1965. *Ap. J.* 142: 1398–1422
47. Clark, B. G., Radhakrishnan, V., Wilson, R. W. 1962. *Ap. J.* 135: 151–74
48. Cleary, M. N., Heiles, C., Haslam, C. G. T. 1979. *Astron. Astrophys. Suppl.* 36: 95–127
49. Clifton, T. R., Frail, D. A., Kulkarni, S. R., Weisberg, J. M. 1988. *Ap. J.* 333: 332–40
50. Colgan, S. W. J., Salpeter, E. E., Terzian, Y. 1988. *Ap. J.* 328: 275–98
51. Colomb, F. R., Dubner, G. M., Giacani, E. B. 1984. *Astron. Astrophys.* 130: 294–300
52. Cowie, L. L. 1980. *Ap. J.* 236: 868–79
53. Cowie, L. L. 1981. *Ap. J.* 245: 66–71
54. Cowie, L. L., Songaila, A. 1986. *Annu. Rev. Astron. Astrophys.* 24: 499–535
55. Cox, D. P. 1986. In *Model Nebulae*, ed. D. P. Pequignot, pp. 11–22. Meudon, Fr: Obs. Paris
56. Cox, D. P. 1987. See Roger & Landecker 1987, pp. 73–90
57. Cox, D. P., Reynolds, R. J. 1987. *Annu. Rev. Astron. Astrophys.* 25: 303–44
57a. Cox, P., Gusten, R., Henkel, C. 1988. *Astron. Astrophys.* 206: 108–16
58. Crovisier, J. 1978. *Astron. Astrophys.* 70: 43–50
59. Crovisier, J., Dickey, J. M. 1983. *Astron. Astrophys.* 122: 282–96
60. Crovisier, J., Dickey, J. M., Kazes, I. 1985. *Astron. Astrophys.* 146: 223–34
61. Crovisier, J., Kazes, I., Aubry, D. 1978. *Astron. Astrophys. Suppl.* 32: 205–82
62. Crovisier, J., Kazes, I., Aubry, D. 1980. *Astron. Astrophys. Suppl.* 41: 229–44
63. Crovisier, J., Kazes, I., Brillet, J. 1984. *Astron. Astrophys.* 138: 237–45
64. Crutcher, R. M., Kazes, I., Troland, T. H. 1987. *Astron. Astrophys.* 181: 119–26
65. Dall'Oglio, G., de Bernardis, P., Masi, S., Melchiorri, F., Moreno, G., Trabalza, R. 1985. *Ap. J.* 289: 609–12
66. Danly, L. 1989. *Ap. J.* 342: 785–806
67. Danly, L., Savage, B. D., Lockman, F. J., Meade, M. R. 1990. *Ap. J.* In press
68. Davies, R. D. 1956. *MNRAS* 116: 443–52
69. Davies, R. D., Cummings, E. R. 1975. *MNRAS* 170: 95–113
70. Deul, E., Burton, W. B. 1990. *Astron. Astrophys.* In press
71. Dewdney, P. E., Roger, R. S. 1982. *Ap. J.* 255: 564–76
72. Dewdney, P. E., Roger, R. S. 1986. *Ap. J.* 307: 275–85
73. Diamond, P. J., Goss, W. M., Romney, J. D., Booth, R. S., Kalberla, P. M. W., Mebold, U. 1989. *Ap. J.* 347: 302–6
74. Dickey, J. M., Weisberg, J. M., Rankin, J. M., Boriakoff, V. 1981. *Astron. Astrophys.* 101: 332–41

75. Dickey, J. M., Benson, J. M. 1982. *Astron. J.* 87: 278–305
76. Dickey, J. M., Crovisier, J., Kazes, I. 1981. *Astron. Astrophys.* 98: 271–85
77. Dickey, J. M., Garwood, R. W. 1989. *Ap. J.* 341: 201–7
78. Dickey, J. M., Kulkarni, S. R., van Gorkom, J. H., Heiles, C. E. 1983. *Ap. J. Suppl.* 53: 591–621
79. Dickey, J. M., Salpeter, E. E., Terzian, Y. 1977. *Ap. J. Lett.* 211: L77–81
80. Dickey, J. M., Salpeter, E. E., Terzian, Y. 1978. *Ap. J. Suppl.* 36: 77–114
80a. Dickey, J. M. 1990. In *The Interstellar Medium in External Galaxies*, ed. J. M. Shull, H. A. Thronson. Dordrecht: Kluwer. In press
80b. Dieter, N. H. 1964. *Astron. J.* 69: 288–93
81. Dieter, N. H., Welch, W. J., Romney, J. D. 1976. *Ap. J. Lett.* 206: L113–15
82. Diplas, A., Savage, B. D. 1990. *Ap. J.* In press
83. Draine, B. T., Anderson, N. 1985. *Ap. J.* 292: 494–99
84. Dubner, G. M., Colomb, F. R., Giacani, E. B. 1986. *Astron. J.* 91: 343–53
85. Edgar, R. J., Savage, B. D. 1989. *Ap. J.* 340: 762–74
86. Elmegreen, B. G. 1987. In *Interstellar Processes*, ed. D. J. Hollenbach, H. A. Thronson, Jr., pp. 259–80. Dordrecht: Reidel
87. Elmegreen, D. M., Elmegreen, B. G. 1987. *Ap. J.* 314: 3–9
88. Elvis, M., Lockman, F. J., Wilkes, B. J. 1989. *Astron. J.* 97: 777–82
89. Ewen, H. I., Purcell, E. M. 1951. *Nature* 168: 350–56
90. Falgarone, E., Lequeux, J. 1973. *Astron. Astrophys.* 25: 253–60
91. Ferlet, R., Vidal-Majar, A., Gry, C. 1985. *Ap. J.* 298: 838–43
92. Ferriere, K. M., Zweibel, E. G., Shull, J. M. 1988. *Ap. J.* 332: 984–94
92a. Fichtel, C. E., Hartman, R. C., Kniffen, D. A., Thompson, D. J., Bignami, G. F., et al. 1975. *Ap. J.* 198: 163–82
92b. Fichtel, C. E. 1989. In *Large Scale Surveys of the Sky. Proc. NRAO Workshop No. 20*, ed. J. J. Condon, F. J. Lockman, pp. 143–60. Green Bank, W. Va: NRAO
92c. Fiedler, R. L., Dennison, B., Johnston, K. J., Hewish, A. 1987. *Nature* 326: 675–78
93. Field, G. B. 1958. *Proc. Inst. Radio Eng.* 46: 240–50
94. Field, G. B., Saslaw, W. C. 1965. *Ap. J.* 142: 568–83
95. Fong, R., Jones, L. R., Shanks, T., Stevenson, P. R. F., Strong, A. W., et al. 1987. *MNRAS* 224: 1059–72
96. Fuchs, B., Thielheim, K. O. 1979. *Ap. J.* 227: 801–7
97. Garwood, R. W., Dickey, J. M. 1989. *Ap. J.* 338: 841–61
98. Giovanelli, R. 1980. *Astron. J.* 85: 1155–81
99. Giovanelli, R. 1986. In *Gaseous Halos of Galaxies. Proc. NRAO Workshop No. 12*, ed. J. N. Bregman, F. J. Lockman, pp. 99–114. Green Bank, W. Va: NRAO
100. Goss, W. M., Radhakrishnan, V., Brooks, J. W., Murray, J. D. 1972. *Ap. J. Suppl.* 24: 123–59
101. Goss, W. M., Retallack, D. S., Felli, M., Shaver, P. A. 1983. *Astron. Astrophys.* 117: 115–26
102. Gosachinskij, I. V., Khersonskij, V. K. 1987. *Astrophysics* 26: 40–44
103. Gosachinskij, I. V., Khersonskij, V. K. 1987. *Sov. Astron. AJ* 31: 621–23
104. Greisen, E. W. 1973. *Ap. J.* 184: 363–90
104a. Greisen, E. W., Lockman, F. J. 1979. *Ap. J.* 228: 740–47
105. Greisen, E. W., Liszt, H. S. 1986. *Ap. J.* 303: 702–17
106. Gum, C. S., Kerr, F. J., Westerhout, G. 1960. *MNRAS* 121: 132–49
107. Hagen, J. P., Lilley, A. E., McClain, E. F. 1955. *Ap. J.* 122: 361–75
108. Handa, T., Sofue, Y., Reich, W., Furst, E., Suwa, I., Fukui, Y. 1986. *Publ. Astron. Soc. Jpn.* 38: 361–78
109. Heeschen, D. J. 1955. *Ap. J.* 121: 569–84
110. Heiles, C. 1967. *Ap. J. Suppl.* 15: 97–130
111. Heiles, C. 1974. In *Galactic Radio Astronomy, IAU Symp. No. 60*, ed. F. J. Kerr, S. C. Simonson, III, pp. 13–44. Dordrecht: Reidel
112. Heiles, C. 1979. *Ap. J.* 229: 533–44
113. Heiles, C. 1980. *Ap. J.* 235: 833–39
114. Heiles, C. 1984. *Ap. J. Suppl.* 55: 585–95
115. Heiles, C. 1987. In *Interstellar Processes*, ed. D. J. Hollenbach, H. A. Thronson, Jr., pp. 171–92. Dordrecht: Reidel
116. Heiles, C. 1987. *Ap. J.* 315: 555–66
116a. Heiles, C. 1988. *Ap. J.* 324: 321–30
117. Heiles, C. 1989. *Ap. J.* 336: 808–21
118. Heiles, C., Reach, W. T., Koo, B.-C. 1988. *Ap. J.* 332: 313–27
119. Heiles, C., Stark, A. A., Kulkarni, S. 1981. *Ap. J. Lett.* 247: L73–76
120. Heiles, C., Wrixon, G. T. 1976. In *Methods of Experimental Physics*, ed. M. L. Meeks, 12C: 58–77. New York: Academic

120a. Helou, G. 1989. In *Interstellar Dust, IAU Symp. No. 135*, ed. L. J. Allamandola, A. G. G. M. Tielens, pp. 285–301. Dordrecht: Kluwer
121. Henderson, A. P., Jackson, P. D., Kerr, F. J. 1982. *Ap. J.* 263: 116–22
122. Hobbs, L. M. 1974. *Ap. J.* 191: 381–93
123. Hobbs, L. M., Morgan, W. W., Albert, C. E., Lockman, F. J. 1985. *Ap. J.* 263: 690–95
124. Holberg, J. B. 1984. In *The Local Interstellar Medium, IAU Colloq. No. 81. NASA Conf. Publ. No. 2345*, ed. Y. Kondo, A. C. Bruhweiler, B. D. Savage, pp. 41–94. Greenbelt, Md: NASA
125. Houck, J. C., Bregman, J. D. 1990. *Ap. J.* 352: 506–21
126. Houlahan, P., Scalo, J. 1990. *Ap. J. Suppl.* 72: 133–52
127. Hughes, M. P., Thompson, A. R., Colvin, R. S. 1971. *Ap. J. Suppl.* 23: 323–70
128. Jahoda, K., McCammon, D., Dickey, J. M., Lockman, F. J. 1985. *Ap. J.* 290: 229–37
129. Jahoda, K., McCammon, D., Lockman, F. J. 1986. *Ap. J. Lett.* 311: L57–61
130. Jahoda, K., Lockman, F. J., McCammon, D. 1990. *Ap. J.* 354: In press
131. Jenkins, E. B. 1970. In *Ultraviolet Stellar Spectra and Related Ground-Based Observations, IAU Symp. No. 36*, ed. L. Houziaux, H. E. Butler, pp. 281–301. Dordrecht: Reidel
132. Jenkins, E. B., Savage, B. D. 1974. *Ap. J.* 187: 243–55
132a. Jenniskens, P. M. M., Wouterloot, J. G. A. 1990. *Astron. Astrophys.* 227: 553–62
133. Jonas, J. L. 1986. *MNRAS* 219: 1–12
134. Joncas, G., Dewdney, P. E., Higgs, L. A., Roy, J. R. 1985. *Ap. J.* 298: 596–613
135. Kalberla, P. M. W., Mebold, U., Reich, W. 1980. *Astron. Astrophys.* 82: 275–86
136. Kalberla, P. M. W., Schwarz, U. J., Goss, W. M. 1985. *Astron. Astrophys.* 144: 27–36
137. Kazes, I., Aubry, D. 1973. *Astron. Astrophys.* 22: 413–20
137a. Keenan, F. P., Dufton, P. L., McKeith, C. D., Blades, J. C. 1983. *MNRAS* 203: 963–75
138. Kellman, S. A. 1972. *Ap. J.* 175: 353–62
139. Kerr, F. J. 1958. *Rev. Mod. Phys.* 30: 924–25
140. Kerr, F. J. 1969. *Annu. Rev. Astron. Astrophys.* 7: 39–66
141. Kerr, F. J., Bowers, P. F., Jackson, P. D., Kerr, M. 1986. *Astron. Astrophys. Suppl.* 66: 373–504
142. Kerr, F. J., Hindman, J. V., Carpenter, M. S. 1957. *Nature* 180: 677–79
143. Kerr, F. J., Westerhout, G. 1965. In *Galactic Structure (Stars and Stellar Systems*, Vol. 5), ed. A. Blaauw, M. Schmidt, pp. 167–202. Chicago: Univ. Chicago Press
144. Knapp, G. R. 1974. *Astron. J.* 79: 527–40
145. Knapp, G. R. 1975. *Astron. J.* 80: 111–16
146. Knapp, G. R. 1987. *Publ. Astron. Soc. Pac.* 99: 1134–43
147. Knude, J. 1981. *Astron. Astrophys.* 98: 74–80
148. Kuchar, T. A., Bania, T. M. 1990. *Ap. J.* 352: 192–206
149. Kulkarni, S. R., Blitz, L., Heiles, C. 1982. *Ap. J. Lett.* 259: L63–66
150. Kulkarni, S. R., Dickey, J. M., Heiles, C. 1985. *Ap. J.* 291: 716–21
151. Kulkarni, S. R., Fich, M. 1985. *Ap. J.* 289: 792–802
152. Kulkarni, S. R., Heiles, C. 1987. In *Interstellar Processes*, ed. D. J. Hollenbach, H. A. Thronson, Jr., pp. 87–122. Dordrecht: Reidel
153. Kulkarni, S. R., Heiles, C. 1988. In *Galactic and Extragalactic Radio Astronomy*, ed. K. I. Kellerman, G. L. Verschuur, pp. 95–153. New York: Springer-Verlag
154. Kulkarni, S. R., Heiles, C., van Gorkom, J., Dickey, J. M. 1984. In *The Local Interstellar Medium, IAU Colloq. No. 81. NASA Conf. Publ. No. 2345*, ed. Y. Kondo, F. C. Bruhweiler, B. D. Savage, pp. 269–73. Greenbelt, Md: NASA
155. Kulkarni, S. R., Mathieu, R. 1986. *Astrophys. Space Sci.* 118: 531–33
156. Kulkarni, S. R., Turner, K., Heiles, C., Dickey, J. M. 1985. *Ap. J. Suppl.* 57: 631–42
157. Landecker, T. L., Roger, R. S., Dewdney, P. E. 1982. *Astron. J.* 87: 1379–89
158. Lazareff, B. 1975. *Astron. Astrophys.* 42: 25–35
159. Levinson, F. H., Brown, R. L. 1980. *Ap. J.* 242: 416–23
160. Lilley, A. E. 1955. *Ap. J.* 121: 559–68
161. Lindblad, P. O., Grape, K., Sandqvist, Aa., Schober, J. 1973. *Astron. Astrophys.* 24: 309–12
162. Liszt, H. S. 1983. *Ap. J.* 275: 163–74
163. Liszt, H. S. 1988. In *Galactic and Extragalactic Radio Astronomy*, ed. K. Kellermann, G. L. Verschuur, pp. 359–80. New York: Springer-Verlag

164. Lockman, F. J. 1977. *Astron. J.* 82: 408–13
165. Lockman, F. J. 1984. *Ap. J.* 283: 90–97
166. Lockman, F. J. 1988. In *The Outer Galaxy*, ed. L. Blitz, F. J. Lockman, pp. 79–88. New York: Springer-Verlag
167. Lockman, F. J., Hobbs, L. M., Shull, J. M. 1986. *Ap. J.* 301: 380–94
168. Lockman, F. J., Jahoda, K., McCammon, D. 1986. *Ap. J.* 302: 432–49
169. Lockman, F. J., Gehman, C. 1990. *Ap. J.* In press
170. Low, F. J., Beintema, D. A., Gautier, T. N., Gillett, F. C., Beichman, C. A., et al. 1984. *Ap. J. Lett.* 278: L19–22
171. Lozinskaya, T. A., Kardashev, N. S. 1963. *Sov. Astron. AJ* 7: 161–66
171b. Lynds, B. T. 1962. *Ap. J. Suppl.* 7: 1–52
172. MacLow, M. H., McCray, R. 1988. *Ap. J.* 324: 776–85
173. Madejski, G. M., Schwartz, D. A. 1989. In *BL Lac Objects: 10 Years After*, ed. L. Maraschi, T. Maccaro, M.-H. Ulrich, pp. 267–80. Heidelberg: Springer-Verlag
173a. Magnani, L., Blitz, L., Mundi, L. 1985. *Ap. J.* 295: 402–21
174. Magnani, L., Lada, E. A., Blitz, L. 1986. *Ap. J.* 301: 395–97
175. Magnani, L., Blitz, L., Wouterloot, J. G. A. 1988. *Ap. J.* 326: 909–23
175a. Magnani, L., Lada, E. A., Sandell, G., Blitz, L. 1989. *Ap. J.* 339: 244–57
175b. Magnani, L., Siskind, L. 1990. *Ap. J.* In press
176. Maloney, P. 1990. *Ap. J.* In press
177. Massa, D., Savage, B. D., Fitzpatrick, E. L. 1983. *Ap. J.* 266: 662–83
177a. McCammon, D., Sanders, W. T. 1990. *Annu. Rev. Astron. Astrophys.* 28: 657–88
178. McCutcheon, W. H., Shuter, W. L. H., Booth, R. S. 1978. *MNRAS* 185: 755–69
179. McGee, R. X., Newton, L. M., Morton, D. C. 1983. *MNRAS* 205: 1191–1205
180. Mebold, U., Hills, D. L. 1975. *Astron. Astrophys.* 42: 187–94
181. Mebold, U., Winnberg, A., Kalberla, P. M. W., Goss, W. M. 1982. *Astron. Astrophys.* 115: 223–41
182. Mebold, U., Cernicharo, J., Velden, L., Reif, K., Crezelius, C., Goerigk, W. 1985. *Astron. Astrophys.* 151: 427–34
183. Mirabel, I. F. 1982. *Ap. J.* 246: 112–19
184. Morrison, R., McCammon, D. 1983. *Ap. J.* 270: 119–22
185. Morton, D. C. 1967. *Ap. J.* 147: 1017–24
185a. Morton, D. C., Blades, J. C. 1986. *MNRAS* 220: 927–48
186. Muller, C. A. 1959. In *Paris Symp. Radio Astron.*, ed. R. N. Bracewell, pp. 360–65. Stanford, Calif: Stanford Univ. Press
187. Muller, C. A., Oort, J. H. 1951. *Nature* 168: 357–58
188. Münch, G., Zirin, H. 1961. *Ap. J.* 133: 11–28
189. Murthy, J., Henry, R. C., Moos, H. W., Landsman, W. B., Linsky, J. L., et al. 1987. *Ap. J.* 315: 675–86
190. Myers, P. C. 1987. In *Interstellar Processes*, ed. D. J. Hollenbach, H. A. Thronson, Jr., pp. 71–86. Dordrecht: Reidel
191. Myers, P. C., Goodman, A. A. 1988. *Ap. J.* 329: 392–405
192. Norman, C. A., Ikeuchi, S. 1989. *Ap. J.* 345: 372–83
193. Olano, C. A., Poppel, W. G. L. 1987. *Astron. Astrophys.* 179: 202–18
194. Oort, J. H. 1955. In *Vistas in Astronomy*, ed. A. Beer, 1: 607–16. London: Pergamon
195. Oort, J. H. 1962. In *The Distribution and Motion of Interstellar Matter in Galaxies*, ed. L. Woltjer, pp. 3–22, 71–77. New York: Benjamin
196. Parker, E. N. 1966. *Ap. J.* 145: 811–33
197. Payne, H. E., Dickey, J. M., Salpeter, E. E., Terzian, Y. 1978. *Ap. J. Lett.* 221: L95–98
198. Payne, H. E., Salpeter, E. E., Terzian, Y. 1982. *Ap. J. Suppl.* 48: 199–218
199. Payne, H. E., Salpeter, E. E., Terzian, Y. 1983. *Ap. J.* 272: 540–50
200. Peters, W. C., Bash, F. N. 1987. *Ap. J.* 317: 646–52
201. Puget, J. L., Legerl, A., Boulanger, F. 1985. *Astron. Astrophys. Lett.* 142: L19–22
202. Quiroga, R. J. 1974. *Astrophys. Space Sci.* 27: 323–42
203. Quiroga, R. J. 1977. *Astrophys. Space Sci.* 50: 281–310
204. Quiroga, R. J., Schlosser, W. 1977. *Astron. Astrophys.* 57: 455–59
205. Radhakrishnan, V. 1960. *Publ. Astron. Soc. Pac.* 72: 296–302
206. Radhakrishnan, V. 1974. In *Galactic Radio Astronomy, IAU Symp. No. 60*, ed. F. J. Kerr, S. C. Simonson, III, pp. 3–12. Dordrecht: Reidel
207. Radhakrishnan, V., Brooks, J. W., Goss, W. M., Murray, J. D., Schwarz, U. J. 1972. *Ap. J. Suppl.* 24: 1–14
208. Radhakrishnan, V., Goss, W. M. 1972. *Ap. J. Suppl.* 24: 161–66
209. Radhakrishnan, V., Goss, W. M., Murray, J. D., Brooks, J. W. 1972. *Ap. J. Suppl.* 24: 49–121
210. Radhakrishnan, V., Murray, J. D., Lockhart, P., Whittle, R. P. J. 1972. *Ap. J. Suppl.* 24: 15–47

211. Radhakrishnan, V., Sarma, N. V. G. 1980. *Astron. Astrophys.* 85: 249–51
212. Read, P. L. 1980. *MNRAS* 192: 11–32
213. Read, P. L. 1980. *MNRAS* 193: 487–94
214. Reich, P., Reich, W. 1988. *Ap. J. Suppl.* 74: 7–23
215. Reynolds, R. J. 1989. *Ap. J. Lett.* 339: L29–32
216. Roger, R. S., Caswell, J. L., Murray, J. D., Cole, D. J., Cooke, D. J. 1978. *MNRAS* 182: 209–18
217. Roger, R. S., Irwin, J. A. 1982. *Ap. J.* 256: 127–38
217a. Roger, R. S., Landecker, T. L., eds. 1987. *Supernova Remnants and the Interstellar Medium. Proc. IAU Colloq. No. 101.* Cambridge: Univ. Press
218. Roger, R. S., Pedlar, A. 1981. *Astron. Astrophys.* 94: 238–50
219. Rohlfs, K., Braunsfurth, E., Mebold, U. 1972. *Astron. J.* 77: 711–17
219a. Rohlfs, R., Herbstmeier, U., Mebold, U., Fink, U. 1990. *Astron. Astrophys.* In press
220. Sancisi, R. 1971. *Astron. Astrophys.* 12: 323–31
221. Sancisi, R. 1988. In *QSO Absorption Lines*, ed. J. C. Blades, D. Turnsheck, C. A. Norman, pp. 241–55. Cambridge: Univ. Press
221a. Sandage, A. 1976. *Astron. J.* 81: 954–57
222. Sandquist, Aa., Tomboulides, H., Lindblad, P. O. 1988. *Astron. Astrophys.* 205: 225–34
223. Savage, B. D., de Boer, K. S. 1979. *Ap. J. Lett.* 230: L77–82
224. Savage, B. D., Jenkins, E. B. 1972. *Ap. J.* 172: 491–522
224a. Savage, B. D., Bohlin, R. C., Drake, J. F., Budich, W. 1977. *Ap. J.* 216: 291–307
225. Savage, B. D., Massa, D. 1987. *Ap. J.* 314: 380–96
226. Savage, B. D., Panek, R. J. 1974. *Ap. J.* 191: 659–74
226a. Scalo, J. 1990. In *Physical Processes in Fragmentation and Star Formation*, ed. R. Capuzzo-Dolcetta, C. Chiosi, A. DiFazio. Dordrecht: Kluwer. In press
227. Scalo, J. M., Struck-Marcel, C. 1986. *Ap. J.* 301: 77–82
228. Schmidt, M. 1957. *Bull. Astron. Inst. Neth.* 13: 247–68
229. Schwarz, U. J., Troland, T. H., Albinson, J. S., Bregman, J. D., Goss, W. M., Heiles, C. 1986. *Ap. J.* 301: 320–30
230. Schwarz, U. J., Ekers, R. D., Goss, W. M. 1982. *Astron. Astrophys.* 110: 100–4
231. Schwering, P. B. W. 1989. PhD thesis. Rijksuniv. Leiden, Neth.
231a. Scoville, N. Z., Saunders, D. B. 1987. In *Interstellar Processes*, ed. D. J. Hollenbach, H. A. Thronson, Jr., pp. 21–50. Dordrecht: Reidel
232. Shaver, P. A., Radhakrishnan, V., Anantharamaiah, K. R., Retallack, D. S., Wamsteker, W., Danks, A. C. 1982. *Astron. Astrophys.* 106: 105–8
233. Shull, J. M., Van Steenberg, M. E. 1985. *Ap. J.* 294: 599–614
234. Shuter, W. L. H., Dickman, R. L., Klatt, C. 1987. *Ap. J. Lett.* 322: L103–8
235. Shuter, W. L. H., Verschuur, G. L. 1964. *MNRAS* 127: 387–404
236. Shaya, E. J., Federman, S. R. 1987. *Ap. J.* 319: 76–83
237. Silverglate, P., Terzian, Y. 1978. *Astron. J.* 83: 1412–16
238. Smith, G. P. 1963. *Bull. Astron. Inst. Neth.* 17: 203–8
239. Sofue, Y., Nakai, N. 1983. *Astron. Astrophys. Suppl.* 53: 57–70
240. Solomon, P. M., Rivolo, A. R., Barrett, J., Yahil, A. 1987. *Ap. J.* 319: 730–41
241. Spicker, J., Feitzinger, J. V. 1986. *Astron. Astrophys.* 163: 43–55
242. Spitzer, L. 1985. *Ap. J. Lett.* 290: L21–24
242a. Spitzer, L. Jr. 1990. *Annu. Rev. Astron. Astrophys.* 28: 71–101
243. Spitzer, L. Jr., Jenkins, E. B. 1975. *Annu. Rev. Astron. Astrophys.* 13: 133–64
244. Stacy, J. G., Jackson, P. D. 1982. *Astron. Astrophys. Suppl.* 50: 377–422
245. Stark, A. A., Bally, J., Linke, R. A., Heiles, C. 1990. In press
245a. Stecker, F. W., Solomon, P. M., Scoville, N. Z., Ryter, C. E. 1975. *Ap. J.* 201: 90–97
246. Stokes, G. M. 1978. *Ap. J. Suppl.* 36: 115–41
247. Strong, A. W., Bloemen, J. B. G. M., Dame, T. M., Grenier, I. A., Hermsen, W. 1988. *Astron. Astrophys.* 207: 1–15
248. Sullivan, W. T. 1984. *The Early Years of Radio Astronomy.* Cambridge: Univ. Press
249. Tenorio-Tagle, G., Franco, J., Bodenheimer, P., Różyczka, M. 1987. *Astron. Astrophys.* 179: 219–30
250. Tenorio-Tagle, G. 1980. *Astron. Astrophys.* 88: 61–65
251. Terebey, S., Fich, M. 1986. *Ap. J. Lett.* 309: L73–77
252. Tilanus, R. P. J., Allen, R. J. 1989. *Ap. J. Lett.* 339: L57–62
253. Troland, T. H., Crutcher, R. M., Heiles, C. 1985. *Ap. J.* 298: 808–17
254. Troland, T. H., Heiles, C. 1982. *Ap. J.* 252: 179–92

255. Troland, T. H., Heiles, C. 1982. *Ap. J. Lett.* 260: L19–22
256. Troland, T. H., Heiles, C. 1986. *Ap. J.* 301: 339–45
257. Troland, T. H., Heiles, C., Goss, W. M. 1989. *Ap. J.* 337: 342–54
257a. Turner, B. E., Rickard, L. J, Xu, L.-P. 1989. *Ap. J.* 344: 292–305
257b. Turner, B. E. 1988. In *Galactic and Extragalactic Radio Astronomy*, ed. K. Kellermann, G. L. Verschuur, pp. 154–99. New York: Springer-Verlag
257c. Turner, B. E., Ziurys, L. M. 1988. In *Galactic and Extragalactic Radio Astronomy*, ed. K. Kellermann, G. L. Verschuur, pp. 200–54. New York: Springer-Verlag
258. van de Hulst, H. C., Muller, C. A., Oort, J. H. 1954. *Bull. Astron. Inst. Neth.* 12: 117–49
259. van der Hulst, T., Sancisi, R. 1988. *Astron. J.* 95: 1354–59
260. van der Bij, M. D. P., Arnal, E. M. 1986. *Astrophys. Lett.* 25: 119–25
261. van der Werf, P. P., Goss, W. M., Vanden Bout, P. A. 1988. *Astron. Astrophys.* 201: 311–26
262. van der Werf, P. P., Goss, W. M. 1989. *Astron. Astrophys.* 224: 209–24
263. van der Werf, P. P., Goss, W. M. 1990. *Astron. Astrophys.* In press
264. van der Werf, P. P., Dewdney, P. E., Goss, W. M., Vanden Bout, P. A. 1989. *Astron. Astrophys.* 216: 215–29
265. van Dishoeck, E. F., Black, J. H. 1988. *Ap. J.* 334: 771–802
266. van Gorkom, J. H., Ekers, R. D. 1985. In *Synthesis Imaging*, ed. R. A. Perley, F. R. Schwab, A. H. Bridle, pp. 177–87. Green Bank, W. Va: NRAO
267. van Woerden, H., Takakubo, K., Braes, L. L. 1962. *Bull. Astron. Inst. Neth.* 16: 321–60
267a. van Woerden, H., Schwarz, U. J., Hulsbosch, A. N. M. 1985. In *The Milky Way Galaxy, IAU Symp. No. 106*, ed. H. van Woerden, R. J. Allen, W. B. Burton, pp. 387–408. Dordrecht: Reidel
268. Venger, A. P., Gosachinskij, I. V., Grachev, V. G., Egorova, T. M., Ryzhkov, N. F., et al. 1984. *Astrophys. Space Sci.* 107: 271–87
269. Verschuur, G. L. 1969. *Ap. J.* 156: 861–74
270. Verschuur, G. L. 1970. *Astron. J.* 75: 687–94
271. Verschuur, G. L. 1989. *Ap. J.* 339: 163–70
271a. Verschuur, G. L., Schmelz, J. T. 1989. *Astron. J.* 98: 267–78
272. Vidal-Majar, A., Ferlet, R., Laurent, C., York, D. G. 1982. *Ap. J.* 260: 128–40
273. Wakker, B. P. 1990. *Astron. Astrophys.* In press
274. Wakker, B. P., Boulanger, F. 1986. *Astron. Astrophys.* 170: 84–90
275. Wannier, P. G., Lichten, S. M., Morris, M. 1983. *Ap. J.* 268: 727–38
276. Weaver, R., McCray, R., Castor, J. 1977. *Ap. J.* 218: 377–95
277. Weaver, R., Williams, D. R. W. 1973. *Astron. Astrophys. Suppl.* 8: 1–503
278. Weisberg, J. M., Boriakoff, V., Rankin, J. M. 1987. *Astron. Astrophys.* 186: 307–11
279. Wesselius, P. R., Fejes, I. 1973. *Astron. Astrophys.* 24: 15–34
280. Wilkes, B. J., Elvis, M. 1987. *Ap. J.* 323: 243–62
281. Williams, D. R. W. 1973. *Astron. Astrophys.* 28: 309–11
281a. Wolszczan, A., Cordes, J. M. 1989. *Ap. J. Lett.* 320: L35–39
281b. Wouterloot, J. G. A. 1981. PhD thesis. Rijksuniv, Leiden, Neth.
282. Wrixon, G. T., Heiles, C. 1972. *Astron. Astrophys.* 15: 444–49
283. Wyse, R. F. G. 1986. *Ap. J. Lett.* 311: L41–45
284. Zweibel, E. G. 1987. In *Interstellar Processes*, ed. D. J. Hollenbach, H. A. Thronson, Jr., pp. 195–221. Dordrecht: Reidel

SOLAR CONVECTION

H. C. Spruit

Max-Planck-Institut für Physik und Astrophysik,
Karl-Schwarzschildstrasse 1, D-8046 Garching,
Federal Republic of Germany

Å. Nordlund

University Observatory, Øster Voldgade 3, DK-1350 Copenhagen K,
Denmark

A. M. Title

Lockheed Palo Alto Research Laboratories, 3251 Hanover Street,
Palo Alto, California 94304

KEY WORDS: Sun, granulation, solar interior

1. INTRODUCTION

Our understanding of convection in the Sun derives from optical observations of the surface, from helioseismological observations of the interior, and from theories and simulations of compressible convection. In all three of these fields, significant progress has been made in recent years. This progress has been documented in recent workshops on solar granulation (144), on the solar photosphere (77) and on helioseismology (32, 141).

This review does not include the extensive subject of convection interacting with magnetic fields. Observational data can be found in various conference proceedings (5, 148, 151), and theoretical reviews are given in (73a, 126, 137).

1.1   *Granulation Observations*

Before the development of modern telescopes and detectors, solar observations were visual. Most early observers were not at excellent solar seeing sites, and great patience was required in order to glimpse the rare instants of good seeing. The fleeting nature of the excellent seeing prevented direct recording of the solar images. Therefore, communication of the nature of the surface depended on descriptions and drawings from the observer's memory.

In 1801 Sir William Herschel (71) observed that the surface of the quiet Sun was nonuniform in intensity and consisted of small "rice grains," which he thought were hot clouds floating over a cooler surface. Nasmyth (111) later confirmed Herschel's observation of elongated structures and claimed that the surface features had the shape of "willow leaves." This description was strongly contested by Dawes (36), who saw no evidence for a particular shape. He described the surface as granulated—covered with merging and separating clouds. He coined the word "granule" to describe the elements of the surface texture and estimated that they had a typical scale of 1 to 1.5 arcsec.

Much of the controversy based on visual observations ended in 1896 when Janssen (76) produced good photographic images of the solar surface. These photos clearly demonstrated that granules were not typically elongated and had a size of 1 to 2 arcsec, thus confirming the reports of Dawes. Observers of the time agreed that Janssen's images were nearly as good as could be seen in good seeing conditions. Janssen's research did not focus on the nature of the granules because he was more interested in his discovery of larger scale features, which he named *"réseau photospherique"*; he was unaware that these were due to seeing-induced distortions.

In 1908 Chevalier (31a), describing his own work and reporting on the results of Hansky (62), showed how atmospheric and particularly telescope seeing could cause both the large-scale structures seen in the solar surface and the rapid horizontal motions of granules. Time sequences of images were used to separate seeing and solar changes and to measure a granule lifetime of 5 min.

With the report of Hansky, most of the observable features of granulation were well described in the sense that they did not differ significantly from those discussed in de Jager's 1959 *Handbuch der Physik* article (38) on the structure of the solar atmosphere. During the intervening 50 years, Krat and his colleagues at the Pulkovo Observatory (82, 83) and Rösch at the Observatoire de Pic du Midi (139, 140) obtained excellent images from the ground, while Blackwell et al (14) obtained similar quality images from low-altitude (6.5 km) manned balloons. Finally, in 1958 Schwarzschild and the Princeton group successfully flew a high-altitude (30 km) solar balloon telescope and obtained some diffraction-limited (30-cm aperture) images of the solar surface (153).

Although the observational picture did not change much during the first half of the twentieth century, the theoretical picture changed completely. The physical processes responsible for energy generation and transport in stars were appreciated. Granules were no longer hot clouds covering a cool surface; instead the mixing-length formalism used for stellar convection zones contributed to the idea that granules were the tops of convection cells. However, the very high Reynolds number of the solar atmosphere made the cellular convection suggested by apparent sharp edges and the irregular polygonal shapes observed in the best images hard to understand.

In the late 1940s and early 1950s turbulence was a popular topic in astronomy. From measurements and theory, Richardson & Schwarzschild (138) and, in particular, Uberoi (180) argued that the true size of granules was much less than an arcsecond. Their interpretations were based on the assumption that the surface was an example of a random turbulent flow field, and that the measurements of the solar surface were dominated by properties of the Earth's turbulent atmosphere. A consequence of Uberoi's arguments was that a positive and negative granulation photograph (bright and dark areas reversed) would look the same. The primary goal of the Princeton balloon program was to establish the expected Kolmogoroff spectrum of granulation sizes.

During the 1960s and 1970s there were high-altitude balloon flights of telescopes and telescope-spectrograph systems by the Schwarzschild, the Krat, and the Kiepenheuer groups. These flights yielded some outstanding solar photographs but did not provide convincing evidence for a turbulent origin to the solar granulation. The best of the balloon images achieved the diffraction limit of a 50-cm aperture (0.″25). On the ground, Rösch's group at Pic-du-Midi, by using high-speed bursts of exposures and postexposure image selection, also produced images that on occasion achieved 0.25-arcsec resolution.

In an *Annual Reviews* article in 1963 Robert Leighton (89) argued, based upon the quasi-polygonal bright pattern of granules separated by narrow lanes present in the balloon images, that granules were indeed cells as claimed by the visual observers rather than homogeneous isotropic turbulence. The Stratoscope pictures were not reversible: The dark lanes formed a largely connected network, while the bright granules were isolated peaks. Leighton et al (88) observed that the entire photosphere was oscillating with a period of approximately 5 min. They concluded that the fine-scale photospheric velocity field was driven by the buffeting of the atmosphere by hot rising granules. Their Dopplergrams (vertical velocity maps) also clearly showed a larger scale (30 Mm) supergranulation pattern. The monograph *Solar Granulation* [(21), revised edition (22)] amplified on the theme that granulation consists of a "cellular" pattern.

By the beginning of 1980, most solar astronomers thought that the granulation was convective overshoot rather than turbulence. Measurable properties of granulation were well summarized by Wittmann (196), and the physical processes in the photosphere and the techniques of their observation were reviewed by Beckers (10). At that point, the lifetime, scale, and size of the thermal plumes were considered well measured, although the nature of the convective processes and how the cells penetrated into the photosphere were not well understood. The role of the 5-min oscillation in complicating the measurement of the penetration of the granular flow into the photosphere was appreciated, but good techniques for separating flows and oscillations had not yet been developed.

Because of pointing jitter, focus shifts, and thermal bending of the optics of the balloon telescopes, excellent time sequences were not produced in the balloon era. Movies of good quality were produced on the ground by H. Ramsey at Rye Canyon, J. Rösch at Pic du Midi, and R. B. Dunn at Sacramento Peak, but these were disappointing to the observers because of the apparent large seeing waves that rapidly crossed the field of view. As a result they were not widely shown, and efforts to produce granulation movies languished.

In 1985, the Solar Optical Universal Polarimeter (SOUP) on the *Spacelab 2* Space Shuttle flight, developed by the Lockheed Solar Physics Group, produced diffraction-limited, 30-cm aperture image sequences of 20–40 min duration that were completely free of atmospheric and telescope seeing and from blurring produced by pointing errors (175). Quite surprisingly, waves similar to those seen in the ground-based movies were also visible in the SOUP data. They were soon demonstrated to be caused by the solar "5-min" oscillations (176). Since 1987, the Swedish Vacuum Solar Telescope at the Observatorio del Roque de los Muchachos on La Palma in the Canary Islands, using real-time image selection techniques, has produced time sequences with durations of hours and an image quality of 0.3 arcsec in a high percentage of selected frames (145). Similar high-quality image sequences have been produced by digitizing and aligning the best images from sets of bursts obtained at Pic du Midi.

1.2 Stellar Convection Theories

From the theoretical point of view, the greatest practical advance was made by the introduction of the mixing-length model (12, 15, 194). Although this model was also inspired by theories for laboratory and geophysical convection, it addressed the astrophysical situation directly, making the strong stratification of stellar envelopes its central ingredient. Its main free parameter, the mixing-length to scale-height ratio, is determined by requiring that a global model for the Sun has the correct radius of

6.96×10^{10} cm. Current understanding of convective zones in stars is based almost completely on this model.

Numerous efforts were made in the 1960s and 1970s to improve upon the simplistic ideas behind the mixing-length model (reviewed in refs. 163, 164). This resulted in a wealth of information on the behavior of convection in various model problems, mostly centering around the Boussinesq (incompressible) model. Because these were on the wrong side of the gap between the laboratory and astrophysical situations, however, they attracted little interest in astrophysics. In parallel, more elaborate schemes were devised for quantitative numerical calculations in stellar envelopes. Three lines of approach are discernible. In the linear superposition approach, one attempts to construct convection zones by superposition of linear convective modes in an assumed background stratification (17, 67, 97, 110). Since convection is very much a nonlinear process, it is uncertain how relevant such schemes are. In the "modal" approach (52, 85, 86, 113, 114) the horizontal structure is reduced to just one (or a few) mode(s) of given horizontal scales. This results in a one-dimensional problem that can be solved with detailed physics (stratification, ionization, radiation). In stars, the relevant horizontal scales vary with depth by orders of magnitude, however, so that these calculations do not give meaningful results unless they are restricted to a narrow depth range [such as the solar surface (113, 114)].

Most effort has gone into a "turbulent" approach, in which the mixing-length concept is extended in various ways to include more of the physics of turbulence (4, 108a, 118, 133, 162, 178, 179, 181–184, 186, 187, 199). In contrast to the mixing-length model, these are nonlocal theories, so that they can deal with phenomena such as overshooting into stable layers. All contain at least one essential free parameter, and since until recently only one parameter could be determined from observations, the predictive power of these theories is not much better than that of the mixing-length formalism. Application of these theories has been limited (179, 187). Some theories also address time-dependent convection, which is of special importance for theories of the excitation and damping of p-mode oscillations. [For a critical review, see (9).] In those theories that also address the surface layers of a star, the predicted amount of convective flux penetrating into the atmosphere differs greatly. This has helped focus attention on the role of radiative energy transfer in the surface layers (118).

Several numerical simulations not intended directly for solar applications have been made (27, 30, 31, 60, 198). All of these extend over several scale heights but do not include radiative transfer at the top boundary. The most extensive are the three-dimensional calculations of Chan & Sofia (30, 31). These authors also express their results in mixing-length terms by listing numerical values for the unknown coefficients that go into the formalism (such as the mixing-length to scale-height ratio) as functions of the ratio of specific heats of the gas.

Theories for laboratory convection, based on the "direct interaction" approach (25, 63, 87) or on the mixing length concept (178), have been developed. These are fairly successful in reproducing the observed (26, 58) relation between heat flux and temperature gradient in laboratory convection. Theories for laboratory convection do not address the problems of stratification or the radiative boundary at a stellar surface, so their application is to the deeper layers of convection zones. In these layers, the flux-gradient relation is not the primary astrophysical concern. (The stratification is very close to adiabatic anyway.) For stellar applications the laboratory convection theories, like their more astrophysical counterparts, must be judged on factors such as the realism with which they represent the flow field (needed for dynamo and differential rotation theories). For more thoughts on the application of these and related theories to stars, see (70).

The most important part of the convection zone is a boundary layer of a few hundred kilometers just below the surface, where the mean stratification deviates significantly from an isentropic one. The ingredients needed for a direct numerical simulation of this layer—namely, radiative transport theory, thermodynamics of partially ionized gas, opacities, and numerical methods for the hydrodynamic equations—have been available since the 1950s. Nelson (113, 114) and Nordlund (119) were the first to realize that quantitative numerical models for the surface region were within reach. These models, though simple from the hydrodynamic point of view, correctly identified the vertical stratification, the thermodynamics, and the radiative transport as the key ingredients. A little later, full three-dimensional time-dependent simulations were started by Nordlund. Their goal was a simulation without tunable parameters, using only the hydrodynamic equations and the known thermodynamic properties of the solar plasma. To the extent that such ab initio calculations agree quantitatively with the observed phenomena, one may claim that they represent the physics of solar convection and can use them as a basis for developing our physical understanding of stellar convection. The success of the present generation of simulations is encouraging in this respect.

2. OBSERVATIONS

In parallel with the development of the theory, observational data have greatly improved over the last decade. New methods for analyzing time sequences of data have been made possible by digital image processing and large on-line memory systems. Among these methods are (*a*) *digital registration and destretching*, which removes small telescope pointing errors and some of the distortions due to seeing; (*b*) *space-time filtering*, by which convective processes and oscillations can be separated; and (*c*) *local correlation tracking* for determining horizontal flows ("proper motions"). In the following, these methods are described along with recent observational results.

The new techniques have mostly been used to describe properties of the surface as measured in the continuum. Work is now in progress at several places in which digital processing of images taken in narrow spectral windows is being used to establish how the upper photosphere responds to the flows in the surface. Very high resolution spectra enable detailed testing of the response of the atmosphere to convective processes predicted by models (42, 73, 94, 100).

High-resolution observations are not the only way to characterize the convective processes. Observations of average line profile shapes and shifts as a function of position on the disk play an important role in the definition of convective processes (42, 79, 96). They are also a key indicator of long-term variations in convection (29, 80, 96, 117).

Measurements of solar rotation and of large-scale surface flows provide the main tests for theories of the deeper regions of the convection zone. The status of these measurements has been thoroughly reviewed by Schröter (150) and Bogart (16). Differential rotation can be measured to an accuracy of 20 m s^{-1}. For large-scale flows (sizes greater than 0.5 $R_\odot$) only upper limits have been found; these are on the order of 10 m s^{-1} for meridional flows, 10 m s^{-1} for large-scale flows near the pole (35), 20 m s^{-1} on the disk (150), with some authors claiming even lower upper limits (28). Recently, the techniques of helioseismology have been refined to the point that rotation below the surface to at least a depth of $0.7R$ can be measured as a function of solar latitude and depth (23, 34, 65, 92).

2.1 *Interpretation of the Surface Structures*

Movies (e.g. 176, 177) show that in addition to the granulation, there are global oscillations, wave phenomena, and cellular flows of different scales occurring at the solar surface. For qualitative descriptions or quantitative measurements of the convective components, the different dynamical processes must be identified and then isolated to the degree possible. Besides the complication introduced by the waves, data obtained with SOUP, with the Swedish Vacuum Solar Telescope at La Palma, and at Pic du Midi show that the solar granulation varies with the amount of magnetic field present, and that even in the quiet Sun the evolution of the small-scale pattern varies on larger scales.

The relationship between spatial ($\mathbf{k}$) and temporal (ω) frequencies is well known for both acoustic and free surface waves, and the domain of the oscillations in $\mathbf{k}$-ω space is well separated from most of the other, much slower and smaller scale surface processes. Title et al (175) showed that it is possible to remove the 5-min oscillations from a movie by Fourier transforming the sequence of images into $\mathbf{k}$-ω space, applying a filter that removes the oscillations, and then transforming back to the spatial-temporal domain to produce a time sequence free of oscillations. We call this process "space-time filtering."

Besides the 5-min oscillations, there are other wave phenomena in the surface. These are visible as internal disturbances in individual granules and as waves propagating across limited regions of the image. As these waves move across the surface, granules appear to divide and often, but not always, reform. Unlike the 5-min oscillations, these waves are not well isolated in $\mathbf{k}$-ω space and cannot easily be removed by space-time filtering.

Because the human perceptual system is much more sensitive to rapid motions than to slow motions, removal of the oscillations makes the slower evolution of granules much clearer. For example, "exploding" granules, which develop dark centers and then expand as bright annuli, were thought to be relatively rare by earlier observers (24, 101, 109). However, the space-time-filtered SOUP images showed them to be quite frequent, with a birth rate of 7.7×10^{-11} km^{-2} s^{-1} (177). When combined with the average diameter at maximum expansion of 7.2×10^{6} km^{2}, this means that the expansion fronts of exploding granules could cover the entire solar surface in less than 1.8×10^{3} s (30 min) if they were distributed uniformly.

The frequency of exploding granules, combined with their lifetime of about 20 min, gives space-time-filtered movies a clear "arrow of time." In addition, exploding granules are not uniformly distributed but rather tend to occur in cells that cover about half the nonmagnetic solar surface. As exploding granules expand at the rate of 1–2 km s^{-1}, the granules in their immediate vicinity are either eliminated or displaced by the "exploder." The prominence of exploding granules means that there are at least three major patterns of granule evolution—the exploders themselves, neighbors of exploders, and unaffected granules. As an exploder evolves, it forms a bright annulus that becomes thinner and finally breaks into a number of fragments that do not always immediately disappear. Because of the number of exploding granules, there are always many remnants on the surface.

It has long been known that granulation appears to be different in magnetic regions (46). Without the removal of the 5-min oscillations, it is difficult to see the difference via statistical techniques, such as spatial autocorrelation functions (49, 50, 95, 155). Figure 1 shows that the difference in texture between magnetic and nonmagnetic regions is quite striking.

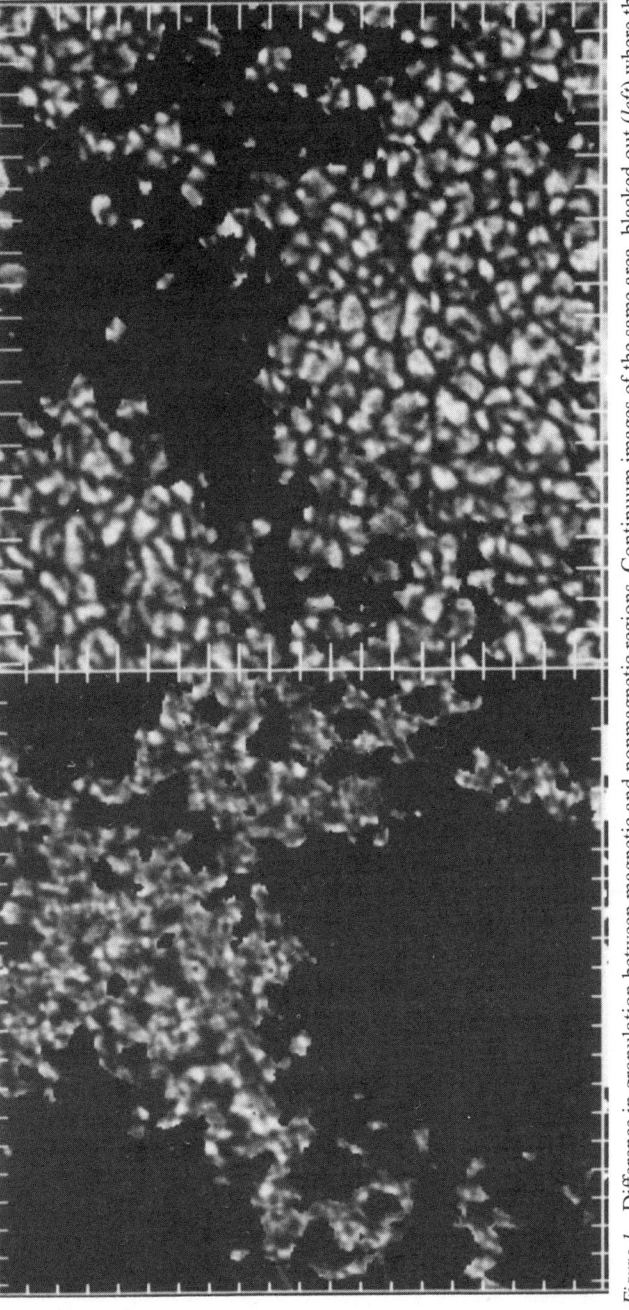

Figure 1 Difference in granulation between magnetic and nonmagnetic regions. Continuum images of the same area, blacked out (*left*) where the average field strength is less than 75 G, and (*right*) where the field strength is larger than 75 G. Data taken with the Swedish Vacuum Solar Telescope.

Furthermore, movies of the magnetic and nonmagnetic regions show very different evolution. For example, the rms intensity fluctuation is a factor of 2 smaller at points where the magnetic field is greater than 75 G. When the image sequences are space-time filtered to separately isolate the convective, f, and p modes before computing the (rms) time-averaged intensity fluctuations, it is found that all components show drops of about a factor of 2 in magnetic regions (173). The same reduction is seen in the rms vertical velocities from, for example, Dopplergrams in the nonmagnetic iron line λ 5576 Å. Small-scale magnetic fields are bright and are preferentially located in the intergranular lanes. This explains some of the reduction of the rms intensity contrast in magnetic regions, but it does not account for the longer lifetimes (see below) of granules or for the different vertical velocity amplitudes.

2.2 Intensity Fluctuations

An important test for any theoretical model of the surface is the observed temperature variation. Traditionally this is measured by the rms intensity fluctuation in the continuum. The highest contrast uncorrected images have been obtained with the 50-cm Swedish Vacuum Solar Telescope refractor using a real-time image-selection system developed by Scharmer (146). These images have an rms contrast of 10–11% in a 25-Å band centered at 4696 Å. If one neglects atmospheric degradation and assumes that the telescope is perfect, a lower limit of 14% rms is obtained for the intrinsic contrast, including the 5-min oscillations. Previous observers (22, pp. 69–75) have measured values of 6.6–8.8% and, after correction, obtained values between 7.2 and 17.9% rms. Karpinsky (77) has reported 22% rms at 5000 Å based on the *Soviet Stratospheric Solar Observatory* observations. The correction factors of often 3 or more that are applied to the measurements to obtain the actual surface intensity variations depend sensitively on the treatment of the wings of the atmospheric and instrumental point-spread function (41, 121).

The appropriate theory for inferring and correcting for the instantaneous atmospheric transfer function, required for the proper correction of the best data, has not yet been developed. However, because of the great sensitivity of the correction value to the details of the atmospheric and telescope transfer function, it would not be surprising to eventually find an rms intensity variation in the region of 25% at 4600 Å. Part of the intensity variation is due to f- and p-mode oscillations. Removal of the oscillations reduces the rms intensity fluctuations by 35% in SOUP data (177). The relative contamination with 5-min oscillation signal is resolution dependent, since the granulation has smaller scale details than the 5-min oscillations.

2.3 Lifetime and Size of Granules

In addition to the value of the rms intensity fluctuations, measurements of the size, the size distribution, and the lifetime of the granules have been the main parameters used to characterize the surface patterns. Quantification of these parameters has been a major problem because of the difficulty of identifying granules. Two different approaches have been used. One is based on the spatial autocorrelation (AC) function, the product of the spatial intensity and the same function displaced in space. The major advantage of this technique is that it avoids the issue of identifying individual granules, but a disadvantage is that it mixes all of the processes occurring in the surface layers. The results of earlier AC measurements are well summarized by Wittmann (196). The second approach is direct identification of granules, which can be done visually or via a computer algorithm. Here, individual granules are identified and their areas measured (101). Lifetimes are found by following individual granules forward and backward in time to their birth and death. Alissandrakis et al (1) have discussed counting methods and compared these methods with the AC techniques.

The autocorrelation size of granules is determined from the full width at half-maximum (FWHM) of the spatial AC function. The FWHM yields a diameter of about 1000 km (1.4 arcsec) and a spacing of about 1400 km (1.9 arcsec) for the nearest neighbor distance (89). The spatial AC function yields an area- and intensity-weighted estimate of size and spacing. Small, dim granules have relatively little effect on either the FWHM, the size, or the position of secondary peaks (which yield the spacing).

Autocorrelation lifetimes are determined from the temporal AC function, i.e. the normalized product of the intensity at an initial time with the intensity at a later time. The standard interpretation of the e-folding time of the AC function as a lifetime carries the implied assumptions that (*a*) the granules have uniform evolutionary properties in the area measured, (*b*) the granules do not move a significant fraction of their diameter, and (*c*) granule evolution is the dominant cause of the decorrelation. All three assumptions are now known to be invalid to a significant degree.

Before the development of space-time filtering, the lifetime estimates from AC measurements were significantly contaminated by the 5-min oscillations (177). When these are removed, the AC lifetime in the quiet Sun increases from about 5 to about 8 min. The effect is even greater in magnetic regions, where it is increased from 7 to 15 min by removal of the oscillations. Thus, the filtered AC measurements efficiently discriminate between magnetic and nonmagnetic regions. However, granules move a significant fraction of their size during their lifetime, so that even the oscillation-filtered autocorrelation lifetimes are only lower limits to granulation lifetimes (in magnetic as well as nonmagnetic regions).

Direct measurements of granule sizes and lifetimes depend on the techniques used to identify and track individual granules. Until recently, most authors used visual identification of granules, a necessarily subjective process. Waves crossing granules, fragmentation of granules near the end of their lives, and image-quality variations may easily be interpreted differently by different individuals. In attempts to remove this "observer dependence," numerical identification algorithms have been developed (19, 142, 157, 177).

Several single-image granulation-recognition procedures have been attempted, including peak finding, valley finding, and contour selection. The simplest conceptually of these is *peak finding*, whereby granule centers are identified by testing every point in the image to see if it is a local intensity maximum. The weakness of this procedure is that it is sensitive to the noise in the image, since the difference between a local peak and its surroundings must be greater than the noise in the image. *Valley finding* is the logical opposite of peak finding. The algorithm examines every image point to test for local minima in at least one direction. This process is much more robust than peak finding because a granule is surrounded by a boundary containing many points but has only one center. A completion algorithm can fill in a few missing points in an incomplete boundary, further reducing the sensitivity to noise. The effectiveness of valley finding depends fundamentally on the fact that intergranular lanes form a connected network. When valley- and center-found granules are compared on the same image, virtually all of the center-found granules are also identified by the valley finder. But the valley-finder method also discovers nearly as many additional granules.

Global oscillations can be expected to have a significant effect on any contour-selection procedure. For this reason images must be filtered to remove low-spatial-frequency structures. Short-period waves and noise also must be filtered before a contouring algorithm yields satisfyingly smooth structures (142).

As the number of granules detected increases, the relative number of small granules rapidly increases on the high-resolution data. Define the number density f of granules of diameter d as the number per unit of surface area per unit of d. Procedures that do not reject small granules as "fragments" yield an f that monotonically increases toward smaller sizes to the resolution limit of the observations—0.25 arcsec (105, 106, 177). The lack of a peak in f may indicate that a cutoff in the number density occurs at a scale much smaller than can presently be resolved. Probably the most important feature of the granule size distribution is that there are virtually no granules with sizes greater than 3 arcsec (106, 177). This may seem to contradict observations of exploding granules, which can be followed out to much larger sizes. Their bright annuli, however, are already well fragmented by the time they expand to 3-arcsec diameter.

Regardless of the sensitivity of the identification process used, the area distribution, defined here as the *area* covered by granules per unit of d, has a peak between $d = 0.9$ and $d = 1.3$ arcsec. This occurs because while different identification processes may disagree about the number of small granules, they substantially agree on the area occupied by granules larger than 0.7 arcsec. For the same reason there is a well-defined peak in the distribution of nearest neighbor distances. Lifetimes based on visual inspection are generally longer (about 15 min) than those measured by computer algorithms. Granules identified by lane finding have lifetimes of less than 8 min in space-time-filtered image sets, whereas center-identified granules yield lifetimes slightly less than 10 min.

Granule sizes, lifetimes, and their distribution functions are of limited use for directly quantifying physical processes. However, they are useful for differential measurements (for example, for use in quantitatively establishing that there exists a difference between convection in magnetic and nonmagnetic regions) and for quantitatively comparing simulations with observations.

2.4 *Horizontal Flows and Patterns*

The SOUP data allowed measurement of velocities perpendicular to the line of sight by local correlation tracking (LCT) (131, 158, 160, 176, 177). In LCT, the local displacement between pairs of images is measured. A window function (usually a truncated Gaussian) is applied in turn to each mesh point of a regular grid, and the relative displacement that maximizes the correlation between the small-scale patterns visible in the windows is sought. The displacement of the small-scale patterns is assumed to be due to advection by a larger scale horizontal flow. The process averages over the measuring window, so that the flow amplitude and the scale of the measured horizontal flow depend on the size of the windows used. Because LCT tracks intensity patterns, it will track brightness waves in addition to mass motions. Bogart et al (18) have discussed the effect of resolution on tracking accuracy, and November & Simon (129) have shown that LCT can also be reliably applied to ground-based data when the seeing is sufficiently good.

Data from SOUP, the Swedish Vacuum Solar Telescope, and Pic du Midi have been successfully used to measure the horizontal flow field in quiet and magnetic regions of the solar surface by means of this technique. The average (10–30 min) flow field in the quiet Sun exhibits a cellular structure 6–8 arcsec in diameter (19, 131, 177). As the size of the correlation window is increased from 0.7 to 1.5 arcsec, using a fixed 0.4-arcsec grid spacing, the 20-min average flow velocity decreases from 800 to 400 m s^{-1}, but the flow pattern remains the same. Figure 2 (color insert) shows both a flow map superimposed on a continuum image and a map formed by taking the horizontal divergence of the flow field. The latter map shows that virtually all exploding granules occur in regions of positive divergence.

The rate of expansion of exploding granules is sufficient to cover the positive-divergence regions in about 10 min. Thus, it is not surprising that the properties of granules—size, brightness, lifetime, and expansion rate—are different in regions of positive and negative divergence (19).

Owing to mass conservation, the horizontal divergence is closely related to the vertical component of the velocity field. November (130) has shown that divergence maps made from continuum images correlate well with Dopplergrams observed in solar lines formed in the mid- and upper photosphere. Thus, the horizontal flow fields measured by local correlation tracking, the mesogranulation discovered earlier (39, 128) in Doppler measurements, and the families of granules reported by Kawaguchi (78) and Oda (132) may be closely related.

The measurement of horizontal velocities has led to the development of a number of new techniques to characterize the surface evolution. One of these methods is to add markers on the frames of a movie and to update the positions of the markers from frame to frame according to the horizontal flow speed. We call these "cork" movies. Corks do not disturb the flow, they do not feel other corks, and they are unaffected by the vertical component of the flow. By overlaying the vector flow field and the cork patterns on the original images and viewing these overlays as movies, it is possible to obtain a qualitative impression of the time evolution of granulation, mesoscale flows, and supergranulation. Some of these impressions can be quantified. For example, by calculating the mean displacement and mean-square displacement of the corks as a function of time, it is possible to measure flow rates and a diffusion coefficient of the evolving flow pattern.

Starting from an original square grid with 0″.4 spacing, almost all of the corks are contained in the dark boundaries of the granules after about 10 min. In 15 min the granulation scale network has begun to disappear, and in 30 min a larger, 6–8 arcsec diameter mesoscale pattern emerges. Corks then flow along the mesogranulation boundaries and collect into points. This evolution is illustrated in Figure 3.

The rate of formation of the mesoscale pattern may be described by the cumulative distribution function of the separation of pairs of corks. If $N(S)$ is the number of pairs with separation less than S, its slope

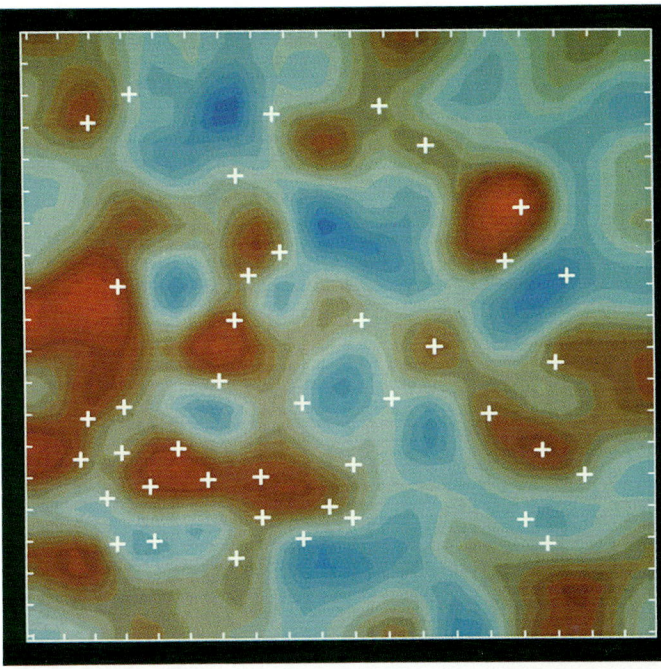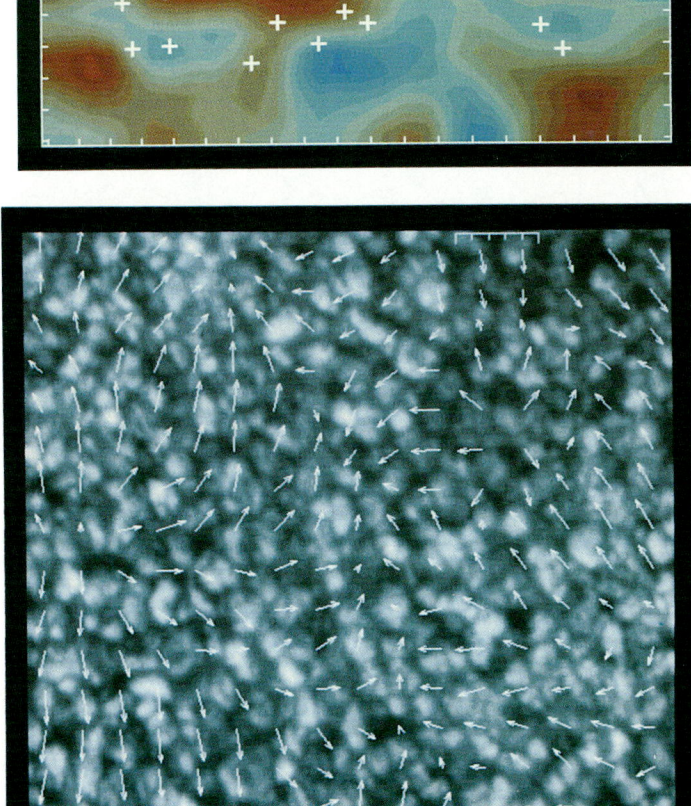

Figure 2 Horizontal flows derived by local correlation tracking. (*left*) Flow field superimposed on a continuum image. (*right*) The horizontal divergence of the same flow field. Crosses mark the positions of exploding granules.

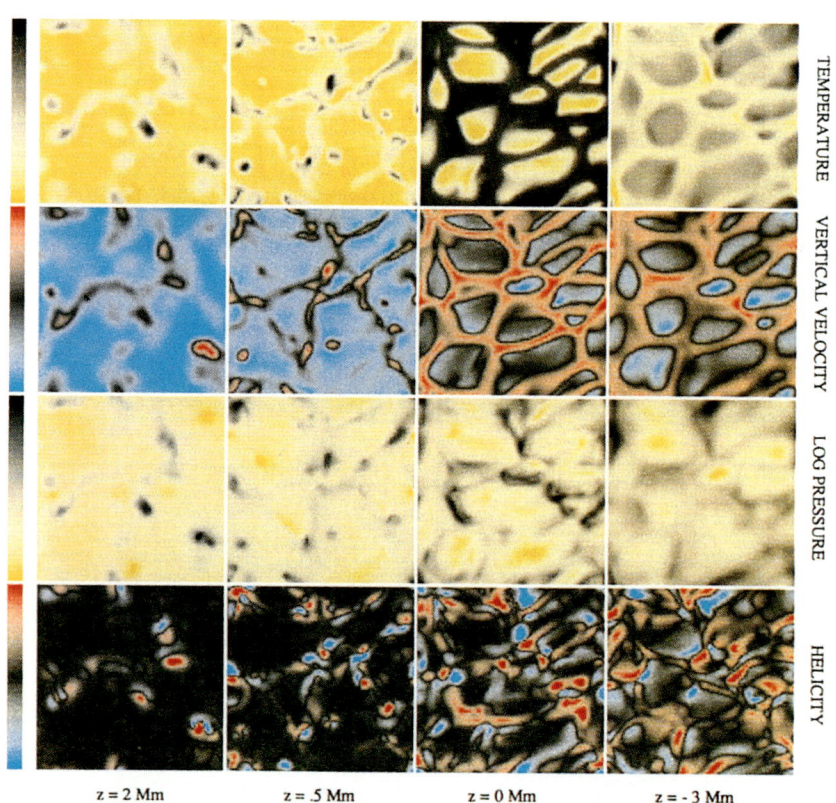

Figure 5 Change of flow topology with depth z (positive into the Sun). The surface pattern consisting of lanes surrounding granules changes into a pattern of disconnected downdrafts. From (166).

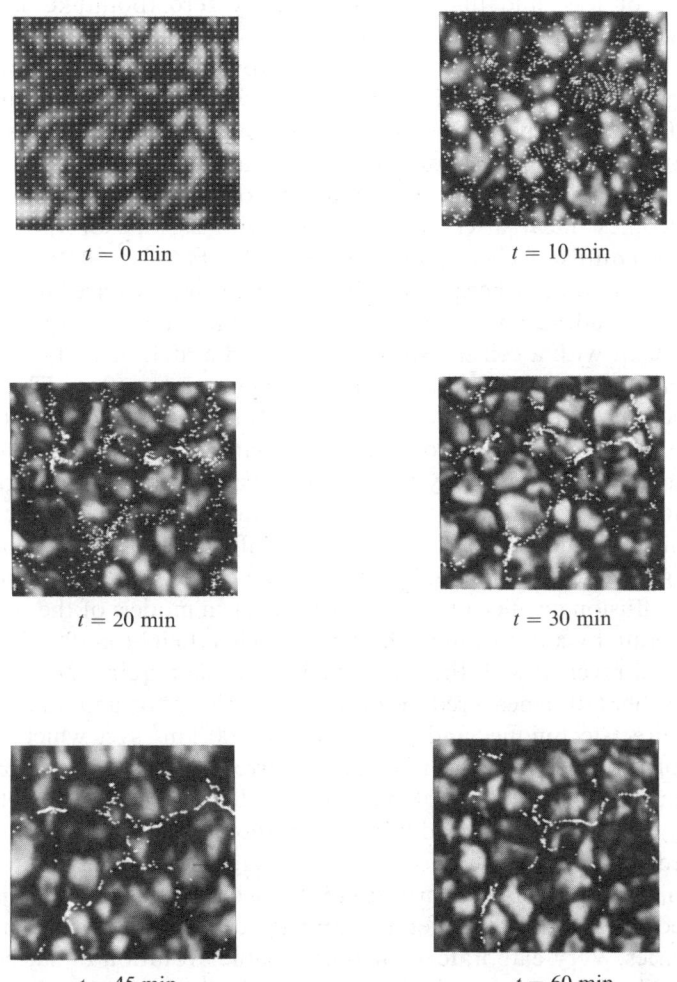

Figure 3 "Cork" evolution. The frames show the positions of markers (corks) floating horizontally with the flows measured with LCT, starting with a regular grid of particles at $t = 0$ min. Data taken with the Swedish Vacuum Solar Telescope.

$K = d (\log N)/dS$ is related to the fractal dimension of the pattern at that scale. Figure 4 shows that K drops to one (corresponding to one-dimensional structures) in 25–30 min for pairs with separations of 3 arcsec or less, and that it falls to nearly zero (pointlike structures) in about 70 min.

The velocity of the corks as they move toward the granule boundaries (during the initial 10 min) is about 1.6 km s^{-1}, is 1.0 km s^{-1} (during the next 20 min) as they move toward the mesoscale boundaries, and is 0.5 km s^{-1} as they move along the boundaries.

The mean-square displacement as a function of time is initially parabolic and becomes linear after about 20 min with a slope of 950 km^2 s^{-1}, implying a diffusion coefficient D of 240 km^2 s^{-1} ($\langle x^2 \rangle = 4Dt$). The initial behavior is parabolic because the displacements initially are linear in time, causing a quadratic mean-square displacement. The diffusion coefficient is consistent with a cell size of 6–8 arcsec and a lifetime of 1–2 hr, and it is a factor of 2 larger than that due to the granulation flow. Thus, mesoscales are more important for the passive diffusion of tracers on the solar surface than granulation. Cork-based estimates of the diffusion coefficient thus far have been made only on fields small compared with a supergranule, and thus they underestimate the solar diffusion coefficient. From these initial measurements it is clear that cork diffusion is a strong function of the magnetic field, dropping by a factor of 3 or more in plage regions.

The diffusion coefficient is a key parameter in models of the solar cycle that operate by a diffusion of the surface fields. Leighton (90) found that polar field reversal with the observed 11-yr cycle requires $D = 1100$ km^2 s^{-1}. Mosher (104) measured the diffusion coefficient by a number of direct and indirect techniques and obtained $D = 200$ km^2 s^{-1}, which is of the order of the mesoscale diffusion rate referred to above. He argued that this value is significantly less than that needed in dynamo models of the activity cycle, and that an additional meridional flow of 10–20 m s^{-1} was required. A lower diffusion coefficient in plage is also consistent with the semiempirical models of Schrijver (154) in which diffusion coefficients are adjusted to reproduce the lifetime of plage fields and their sharp magnetic boundaries. Very elaborate simulations of the evolution of surface magnetic fields during the cycle have been carried out by Sheeley and his colleagues. The most modern of these (195) require a diffusion coefficient of 600 km^2 s^{-1} and a meridional flow of 10–20 m s^{-1}. Not only is this diffusion coefficient higher than the measured value determined from horizontal flows, it is also not clear that magnetic fields actually diffuse at this rate. For example, the size of supergranulation cells (see ref. 202) in active regions and the arguments of Schrijver (154) on the strength and lifetime of plage suggest that there may be a basic problem with diffusion models.

SOLAR CONVECTION 279

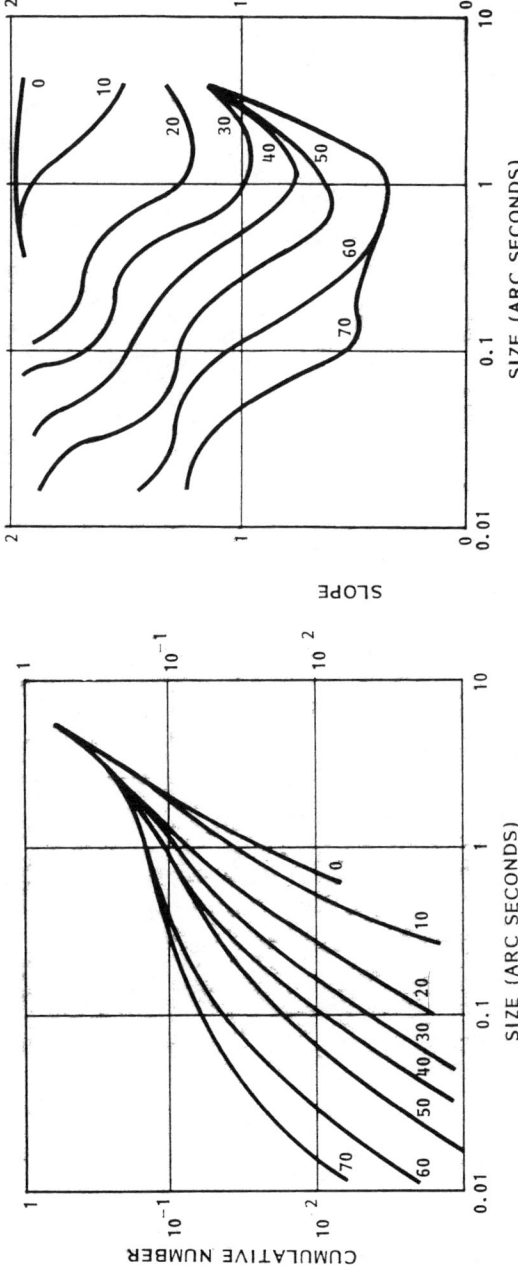

Figure 4 Cumulative distribution $N(S)$ of pairs of corks separated by a distance S or less as a function of time for the pictures shown in Figure 3. (*Left*) $N(S)$. (*Right*) $d \log N / d \log S$.

Although there is a characteristic size for granules as seen in brightness, there is increasing evidence that in the *flow field*, distinct scales of convection do not exist. If we assume that magnetic fields are carried passively by the horizontal flows (which probably is not quite correct), they can then serve as a convenient proxy for the flow field itself, which is hard to measure at large scales and low amplitudes. Measurements of the surface distribution of the magnetic field yield a fractal dimension between 1.4 and 1.6 (177). The variation in the dimension depends on the details of defining the location of clumps of field. For a given definition the same dimension is valid for scales from 0.5 to 1000 arcsec. Measurements of the Fourier transform of the field distribution also show no peaks (81a, 177), which is consistent with the self-similarity of fractal patterns. In spite of these measurements, visual inspection of magnetograms gives the immediate impression that a special scale (the supergranulation) exists. This apparent discrepancy has not yet been resolved.

3. THEORY

For the purpose of this section we divide the convection zone into two parts: a top layer extending from the atmosphere down to a few thousand kilometers below the surface, and the deeper layers. In the deeper layers convection is so efficient that the mean stratification is nearly isentropic irrespective of the details of convective transport. Models for this region aim mainly at calculating the flow fields that determine observables such as the solar dynamo and the differential rotation.

The top layer is not isentropically stratified. Calculation of the entropy difference between the surface and the deeper layers is a prime goal of stellar convection theory, since in a stellar-structure calculation it is this difference that determines the radius of stars with convective envelopes. A successful ab initio calculation of this quantity for the Sun is therefore also of practical importance for stellar evolution theory.

Simulation of the entire convection zone is not possible at present because of the large range of time and length scales present.

3.1 *The Top Layers*

Accurate modeling of the surface region must include the following:

1. *The density stratification.* The density scale height at the surface is a factor of 5–10 times smaller than the horizontal size of a granule. Such a strong stratification makes the flow very asymmetric, upward flows behaving quite differently from downflows.
2. *Radiative transport.* Because of the rapid increase of opacity with depth,

the radiative exchange of heat in the flow is unimportant except in a thin layer (about 100 km thick) near the observed surface. The losses from this thin layer to the outside world, however, determine the temperature difference between upward and downward flows and hence drive the entire flow.
3. *Opacities and equation of state.* Opacities and the thermodynamic properties of the (partially ionized) gas should be realistic enough for a proper treatment of the energy exchange between the gas and the radiation field.
4. *Three-dimensional geometry.* Real solar phenomena (with the notable exception of solar oscillations) have no spatial symmetries or temporal periodicities. Spatially complex topologies (and apparently chaotic temporal behavior) are the rule. Ideally, the computational domain should be large compared with a granule, and the grid spacing should be small enough to study the influence of small-scale processes. These are conflicting requirements, and compromises between size and resolution must be made.

Along with the parameters of the constitutive equations (such as abundances and opacities), the parameters for the models are the known values of the acceleration of gravity at the solar surface and the effective temperature.

Early calculations by Nelson (107, 113–116) and by Nordlund (119) assumed a given horizontal dependence of all variables, or a two-component structure, and therefore were essentially one dimensional. Yet these models satisfied the first two requirements listed above and were fairly successful in reproducing classical observables, including the center-to-limb variation of spectral-line shifts (11). Nordlund (120, 122, 124) developed three-dimensional simulations using the anelastic approximation to the continuity equation, with special care taken in treating the radiative transport. Fully compressible calculations are presented in (166). Simulations of a similar scope have been reported by Steffen et al (169, 172), Gadun (6, 54), and Uus (188–190). The models that presently come closest to fulfilling the criteria listed above are those of Stein & Nordlund (166).

In several of the calculations mentioned, the radiative losses at the surface and other effects of radiative energy transfer are calculated by solving the radiative transfer equation along a large number of rays through the model. Some use a gray opacity (equated with the Rosseland mean) (172); others take into account that in the upper observable layers the energetically significant radiative transfer takes place mainly in the spectral lines (120, 166, 189). A single gray opacity produces hardly any interaction between the radiation field and the gas in the upper layers and therefore cannot be expected to produce an accurate temperature structure. A much more accurate way to include the effects of the millions of spectral lines in the solar spectrum is by sorting the monochromatic opacities into a few bins, each covering one order of magnitude in the value of the absorption coefficient (120, 127). The detailed frequency dependence of the opacity is represented by these binned opacity distribution functions, and these functions are used routinely in static atmosphere models (61, 84).

Critical to the computations is the treatment of the boundaries of the computational domain. Periodic conditions may be used in the horizontal directions. This avoids the introduction of actual boundaries but still limits the maximum size of the structures that can be represented. To prevent the artificial turnover of gas that is forced by closed boundaries, the upper and lower boundaries should be open. Open boundaries allow for inflow and outflow of mass, and ideally also allow waves to be transmitted, but are difficult to stabilize.

In Stein & Nordlund's (166) calculations, the top boundary (typically at a height of 500 km above the surface) is open and is stabilized by a "fiducial" layer about one scale height (150 km) above, which is assumed to be stress free and locally in hydrostatic equilibrium with respect to the boundary layer. At the lower boundary, the inflowing mass is given a uniform entropy and the pressure is kept constant, so as to keep the mass in the computational domain constant. The choice of a constant entropy at inflow is justified by the nature of the flow (see Section 3.1.2).

In the Sun, where all diffusivities are very small, dissipative effects are significant only in a very small fraction of the volume. In numerical simulations, artificial dissipation must be introduced to prevent dissipative effects from taking place on scales smaller than can be numerically resolved. Through the introduction of molecular viscosity and resistivity terms, the results can be related more easily to laboratory situations. If such diffusion coefficients are used, it is important that they do not scale with the local density. The use of a constant dynamic viscosity, for example, implies that it must be tuned to the densest regions of the model, which inevitably makes the least dense regions of the model extremely viscous. For solar simulations, molecular-type viscosity terms waste computer resources, since they also influence the well-resolved motions significantly. In the calculations of Nordlund and of Uus, more subtle damping processes were used—broadly speaking, of the "sub-grid-scale eddy diffusivity" type (37, 120, 161, 168). The method used by Steffen et al (172) is based on the characteristics of the hydrodynamic equations and is expected to be intrinsically relatively free of numerical viscosity.

3.1.1 GRANULATION One of the main results of the numerical simulations is that the pattern seen as granulation at the surface is a very shallow phenomenon (122, 123, 166). The conversion of the solar luminosity from a dominantly convective flux beneath the surface to a dominantly radiative flux above the surface takes place in a very thin (50–100 km) cooling layer at the surface. This transition is due to the enormous temperature (T) sensitivity of the opacity (κ) of the photospheric plasma ($\partial \ln \kappa / \partial \ln T \approx 10$). In the ascending plasma at the center of granules, the temperature drops from some 11,000 K just below the surface to photospheric temperatures (5000–6000 K) in less than 100 km. The peak-to-peak temperature difference just below the visible surface is about 5000 K. The observed intensity contrast in the granulation corresponds to much smaller differences because the optical depth unity surface is not flat owing to the strong temperature dependence of the opacity [see (109, 113, 114) and Figure 12 in (8)].

The granulation is characterized by cells of hot, ascending plasma surrounded by lanes of cool, descending plasma. The lanes form an interconnected network, with the vertices of the network being the coolest points. This topology changes qualitatively below the surface. Figure 5 (see color insert) shows the topology in horizontal planes at several depths and illustrates how the pattern of connected lanes in the surface layer changes into a horizontally intermittent one at a depth of only about 500 km (a fraction of a cell size). The topology below the surface is one with slow, expanding upflows intermixed and nearly isentropic; the cool downdrafts are filamentary and twisting. The entropy fluctuations are all associated with downdrafts that started as intergranular lanes at the surface.

3.1.2 FLOWS ON LARGER SCALES The 6 Mm × 6 Mm × 3 Mm simulations of Stein & Nordlund (166) show flows on a scale somewhat larger than granulation. These mesoscale flows are similar to the observed ones discussed in Section 2.4. In the simulations, as in the observations, these larger scales are hard to detect on single snapshots of surface features but show up, for example, in time sequences as proper motion of granules.

Filamentary downdrafts of individual granules merge below the surface into fewer, more widely separated downdrafts [cf. Figure 5 (color insert)]. These larger scale downdrafts have longer lifetimes than do individual granules, and at the surface their locations are the points where the horizontal motion of the granules converges.

The large-scale flow topology can be displayed with *trace particle plots*. One selects a set of test particles at a certain time and traces their position forward and backward in time, using the stored velocity field from the simulations. Figure 6 illustrates the history of all surface particles that are ascending at a certain time. The two midpanels show the locations of the trace particles at the reference time. Trace particles were placed at all grid points with upward velocity in the plane $z = 0$ at $t = 0$; consequently, these particles indicate the surface granulation pattern. Nine solar minutes earlier, these particles were ascending, in a narrow region at a depth of about 1 Mm (*bottom left panel*). Note that the horizontal distribution (*upper left panel*) is very uneven, and groups of granules at the surface originate from isolated regions below the surface. The source regions expand as parts of larger scale updrafts. As in the observations, granulation properties (size, lifetime) are modulated by these mesoscale flows in the simulations. The two panels on the right show the location of the trace particles 9 solar min after the reference time. The majority of the (originally ascending) particles are found *below* the surface (*bottom right panel*). In sharp contrast to the upflow regions, there is a large variation in the depth of the particles in the downflows.

These properties of the flow are the consequence of the stratification and the radiating surface. The density changes by about a factor of 100 in the first 3 Mm below the surface. Plasma that ascends over that interval must expand about 2 orders of magnitude in volume, or a factor of 5 in linear size, if the expansion is isotropic. Consequently, only a small fraction of the ascending plasma at depth ever reaches the surface. Conversely, a

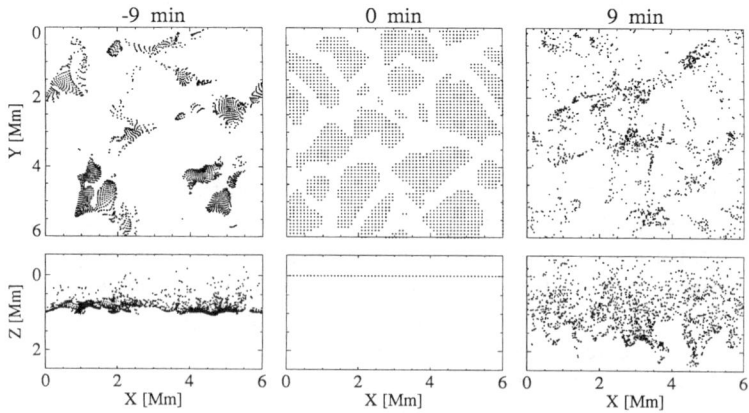

Figure 6 The three-dimensional flow illustrated with trace particles. Particles are released at the surface $z = 0$ and $t = 0$ in the upward-moving parts of the flow (*middle panel*). The *left panel* shows where these particles were 9 min earlier, and the *right panel* shows their position 9 min later.

descending parcel of plasma contracts to a small fraction of its surface volume and becomes only a small fraction of the descending plasma at depth.

Only plasma that reaches the surface loses entropy directly by radiation. Plasma that turns over below the surface moves *adiabatically*, except for turbulent mixing with cool plasma from the surface. Such turbulent mixing affects only part of the flow—namely, the entrainment regions adjacent to the filamentary downdrafts; the expanding upflows are expected to be largely free of turbulence (cf. discussion below). Since all ascending plasma at a certain level originates from a tiny fraction of the ascending plasma at a deeper level, the ascending plasma (and all plasma that overturns below the surface) is also nearly *isentropic*.

Plasma that reaches the surface cools very efficiently as it flows into intergranular lanes and cell vertices. Plasma at the cell vertices is coolest because it has been exposed to radiative cooling for the longest time. The cooler plasma at the cell vertices therefore obtains larger descent velocities than the plasma in the intergranular lanes. As a consequence, filamentary downdraft columns are formed below the surface under the vertices. Surrounding plasma is "sucked into" these columns. Because all the cool plasma from the surface eventually ends up near the cell vertices, the pattern of connected intergranular lanes at the surface is replaced below the surface by isolated downflow regions.

3.1.3 VORTICITY AND TURBULENCE The plasma that overturns into the filamentary downdrafts in general has nonvanishing angular momentum with respect to the centers of the downdrafts. This sets up circular (cyclonic) motions around the centers of downdrafts, and the circular velocity is amplified as the plasma contracts (the "bathtub effect"). The simulations often show such vortices (30; Figure 2 in ref. 123; Figure 2 in ref. 166). Because of their small size, such effects are not now directly observable at the granulation scale on the Sun, but similar effects have been observed at mesoscales by Brandt et al (20).

The boundaries between downflowing filaments and the surrounding upflows are regions of strong shear, and one would expect shear turbulence to be produced there. Because of the small size of the filaments in the lower part of the computational volume, such turbulence is only partially resolved in the simulations. The effect of turbulence is expected to be *entrainment* of surrounding fluid, which causes the filament to grow in width and to decrease in mean density excess and downward acceleration.

As a filament descends and is compressed, energy is fed into the turbulence. This is understood by considering the turbulence as a collection of rotating eddies; the conservation of angular momentum of these eddies implies an increase of kinetic energy per unit mass upon compression. For this reason, the boundaries of the filaments should be long-lasting sources of turbulence.

The same argument implies that any turbulence in an *upflow* decreases in strength: Irregularities in the upflows are smoothed out by the expansion of the fluid. Also, since the upflows are very nearly isentropic, the upflows internally are convectively neutral (no local convective instability). Thus, we expect the upflows to be relatively smooth and weak in turbulence. These upflows dominate the visual appearance of the solar surface in the form of bright granules.

The preceding discussion explains why granulation looks so much more like cellular convection than like turbulence, in spite of the enormous Reynolds numbers of the flows. It also demonstrates how the density stratification can qualitatively change the nature of a flow.

3.1.4 GRANULE DYNAMICS, EXPLODING GRANULES The chaotic evolution of the granulation pattern is caused by the unending competition between individually expanding granules. Expanding plasma from neighboring granules meets along common borders, and mass conservation implies that the expanding plasma must be deflected either down or along the borders. Dynamically, the horizontal expansion is driven by excess pressure in the granules, and the deflection at the intergranular lanes is effected by a pressure excess that builds up over the intergranular lanes (74). The plasma in the intergranular lanes is dense because of the excess pressure and because it has been radiatively cooled, and it descends because of this excess density.

In order to accommodate the increasing total ascending mass, the amplitude of the horizontal velocity must increase as the size of a granule increases. Thus, granules that become large, either because they are particularly powerful or happen to be surrounded by relatively weak granules, develop a substantial central pressure excess. The resulting density excess eventually reduces the upward flow velocity in the center of large cells (108, 112, p. 128), an effect called "buoyancy braking" (85, 99). As the ascent velocity falls below the minimum value necessary to sustain the surface energy losses, the center of the granule cools, which leads to further density excess and a rapid darkening and reversal of the vertical velocity (109, 122, 123). The cold and dark center, surrounded by a ring of expanding, hot plasma, is the observational characteristic of "exploding" granules (Section 2.37 in ref. 21; 109). Exploding granules are often seen in the simulations and are quite common in time-lapse video recordings of solar granulation.

3.1.5 GRANULE SIZE Assuming that buoyancy braking is the limiting factor that determines the maximum size of granules, an upper limit to the size of granules may be obtained by estimating at what size the pressure excess becomes large enough to brake the ascent (166).

Since the peak-to-peak temperature fluctuations in granules are comparable to the temperature itself (see Section 3.1.1), the relative pressure fluctuations must be of order unity before buoyancy braking becomes effective. According to the Bernoulli principle, the pressure fluctuations that drive the horizontal flow increase with the square of the horizontal velocity and become comparable with the total pressure for sonic velocities.

For anelastic flows, the ratio of horizontal to vertical velocity amplitude is given by $u_{\text{hor}}/u_{\text{vert}} = 1/kH$, where k is the horizontal wave number, and H is the scale height of the vertical mass flux. This means that granules cannot be larger than some $2\pi H u_{\text{hor}}/u_{\text{vert}}$. Even though the thermal energy excess in the plasma ascending to the surface in a granule is several times kT per particle (because of the hydrogen ionization), the plasma still must ascend with at least ≈ 2 km s^{-1} to supply enough energy to compensate for the surface radiative losses (120). Using the sound speed as an upper limit for u_{hor}, 2 km s^{-1} as a lower limit for u_{vert}, and 150 km as a typical value for the vertical mass flux scale height gives an estimate of the upper limit of the size of granules equal to ≈ 4 Mm. This line of reasoning was first advanced by Nelson & Musman (116).

Even though buoyancy braking is most likely the main factor determining the maximum size of granules, it is not the whole story. In Stein & Nordlund's (166) simulations, which were large enough to accommodate scales larger than granulation, the sizes of even the largest granules were sensitive to the amount of numerical diffusion in the simulations. This shows that turbulence aids in the breaking up of granules, and it also illustrates the importance of introducing only the minimum amount of diffusion in numerical simulations.

Whether a *minimum* granule size exists is less clear from the simulations because of the limitations on spatial resolution. Nelson & Musman (116), however, show that at lengths smaller than 200 km, horizontal exchange of radiation across the granule produces a powerful damping of temperature fluctuations. For this reason, we do not expect to see *granules*, i.e. overturning cells, much below the present observational limit. This agrees with estimates based on interferometric measurements (40, 66, 193). This argument does not exclude the existence of small scales in the flow field. The downdrafts are sources of both small-scale horizontal (by their tendency to be vortical) and vertical motions (through the shear with the upflows).

3.1.6 ATMOSPHERIC COOLING Since the radiative processes can be included fairly realistically in the present generation of models, the effect of the overshooting flow on the energy balance in the atmosphere can be assessed realistically. The overshooting plasma, which ascends to the temperature minimum in typically a few hundred seconds, experiences a pressure drop of 3 orders of magnitude. If the plasma were not heated, the temperature would drop adiabatically (i.e. by a factor of $10^{0.4 \times 3} \approx 15$ for an ideal gas). In fact, the plasma *is* heated quite efficiently, by the radiation field, to such an extent that the temperature only drops by about 40% and approaches that of radiative equilibrium. The temperature does not quite reach that of radiative equilibrium though, and this is probably the main reason why attempts to directly estimate the mechanical heating of the chromosphere and upper photosphere by means of calculating the radiative losses from empirical models of the solar atmosphere invariably have indicated mechanical *cooling* of the upper photosphere and the temperature minimum region (e.g. 2, 7; Figure 5 of ref. 8).

In the simulations by Nordlund (120) and Stein & Nordlund (166), most of the radiative heating is by reabsorption of photospheric radiation in a large number of weak spectral lines [represented by a flux-weighted absorption in one bin in the four-bin representation of the spectrum (cf. 127)]. Pure continuum absorption would result in a significantly lower heating by reabsorption in the upper photosphere. In the calculations by Steffen et al (172), who use a gray Rosseland opacity, the weaker coupling to the radiation field is more than compensated for by the substantially lower photospheric velocities (and hence a weaker convective cooling) found in these models. The resulting temperature structures in the upper photosphere differ significantly between the two sets of simulations. Qualitatively, however, the results are similar, with a negative correlation of temperature and ascent velocity in the overshoot layer (cf. Figure 5 and refs. 171, 172). Due to the combined effect of granule evolution and the expansion cooling, temperature inhomogeneities in the photosphere are often inclined. The characteristic signatures of such inclined inhomogeneities were observed in photospheric spectral lines and correctly interpreted about 25 years ago (51).

3.1.7 SHOCKS Because of the small density scale heights in the photosphere (100–200 km) as compared with the typical size of granules (1–2 Mm), the horizontal velocities in the photosphere are substantially higher than the vertical ones. This is evidenced observationally by the increase in width of photospheric spectral lines toward the limb; this increase is also reproduced by synthetic spectral-line calculations based on the numerical simulations (11, 122). The horizontal expansion velocities in compressible simulations occasionally become supersonic, an effect that is assisted by the drop of temperature (and hence of sound speed) with height. The expanding flows from neighboring granules collide along common boundaries, and the shocks that form when the velocities are supersonic have the character of "stand-off shocks"—i.e. the shocks are nearly stationary, with supersonic plasma from the interior of a granule being slowed down and compressed in the shock (168). Such shocks have not yet been identified in the observations.

3.1.8 COMPARISON WITH OBSERVATIONS For an example of the simulated surface topology and its time evolution, see Figure 1 of (166). Wöhl & Nordlund (197) compared geometrical properties of observed solar granulation and synthetic granulation images from anelastic simulations covering a 3 Mm × 3 Mm × 1.5 Mm volume (120, 122) and found quantitative agreement.

Lites et al (94) compared high-resolution slit-jaw observations from La Palma with synthetic granulation images from fully compressible simulations covering 6 Mm × 6 Mm × 3 Mm (129, 166), and found that the numerical resolution was not sufficient to reproduce all of the small-scale structure of the observations, but that the topology and time evolution were consistent with the observations. The rms intensity variations and velocities cannot be compared directly with the simulations because of the unknown atmospheric transfer function, to which they are very sensitive (121). However, Lites et al (94) found that the *ratio* of intensity and velocity fluctuations is insensitive to the atmospheric transfer function, and that this ratio was consistent with the observations. Thus, when degrading the simulated images with a model transfer function such that the rms velocities agreed (at 0.52 km s^{-1}), the simulations gave an rms contrast of 8.1%, compared with 8.0% in the observations.

A further sensitive test is the comparison of mean spectral-line shapes averaged over many granules because these can be measured with high precision. The width of spectral lines is a measure of the rms velocity amplitudes. Their blueshift is a measure of the product of velocity and intensity fluctuations, and the spectral-line asymmetry ("C-shape") is a characteristic fingerprint that is a complicated functional of the combined velocity, density, and temperature fluctuations (45). Comparison with observations (Figures 3–4 of ref. 120; see also 123, 170) showed close agreement in the line shift, line width, and C-shape. No adjustments in the line position or model parameters were made in making these comparisons. Discrepancies in the cores of strong lines may be due to non-LTE effects (125).

Anelastic calculations, similar to the solar ones, have been carried out for stars with parameters in the vicinity of the Sun in the Hertzsprung-Russell diagram (127), and synthetic spectral lines from these simulations have been compared with observed spectral lines for α Cen A, α Cen B, β Hyi, and Procyon (43, 44), with quite satisfactory agreement.

3.1.9 CONNECTIONS WITH MIXING-LENGTH MODELS Mixing-length models give a mean stratification for the thermodynamic variables. Comparing this mean stratification with the horizontal mean of the simulations gives an impression of the accuracy of the mixing-length model. The main features of the mean stratification are two nested boundary layers near the surface (194). On top is the radiative layer where the transport changes from convective to radiative; below it is the nonisentropic layer, where individual fluid elements move adiabatically but the mean stratification is not isentropic. Together, these are called the superadiabatic (or nonadiabatic) layer in the stellar-structure literature. Finally, we have the main body of the convection zone, where the flow is nearly adiabatic and the stratification nearly isentropic.

In spite of the significant differences between the mixing-length view of the nonisentropic layer and the picture shown by the simulations, the entropy profiles in this layer are qualitatively similar. Mixing-length estimates of vertical velocity amplitudes as a function of depth may also be qualitatively correct.

In the classical mixing-length picture, the velocity amplitude and entropy fluctuation are the result of a balance between convective driving and turbulent mixing between downward-moving and upward-moving fluid elements. In reality, the mean entropy profiles may be the result of a much simpler averaging effect. Since the downward mass flux in the lower entropy filaments is roughly constant with depth, their cross section decreases with depth (inversely with the density, for a constant flow speed). The weight of the cool downdrafts in horizontally averaged quantities decreases in proportion. The resulting depth dependence of the mean entropy would be similar to that in mixing-length models, even if no turbulent mixing were to take place between the downward-moving and upward-moving fluid elements. In practice, mixing and entrainment do take place, but their effect is not as essential for the mean stratification as it is in the mixing-length view. It is interesting to note that Böhm-Vitense (15) actually based the astrophysical application of mixing-length theory on arguments about mixing being forced by overturning due to the stratification.

An important result of the numerical simulations of convection in the solar surface layers is a *calibration of the mixing length* value to be used in stellar evolution calculations. The main effect of the mixing-length parameter in these calculations is to determine the jump in entropy across the superadiabatic layer (59, 152, Ch. 16; 194). For a small range of models in the vicinity of the Sun in the Hertzsprung-Russell diagram, Nordlund

& Dravins (127) found that the deep layers corresponded well with mixing-length models calculated with a mixing length equal to about 1.6 times the local pressure scale height [using the Henyey et al (69) formulas]. With helioseismology, the depth of the solar convection zone has recently been determined with great precision (33) to be 200 Mm ($\pm 2\%$). The depth that corresponds to the entropy measured at the base of the simulations of Stein & Nordlund (166) is consistent with this value.

3.2 *The Deeper Layers*

3.2.1 SUBSURFACE TOPOLOGY Based on the behavior of larger scale flows near the surface seen in the simulations, Stein & Nordlund (166) propose a hierarchical scenario for flows at larger depths and on larger scales. In the deeper layers, a more gradual change of topology is expected to take place. The filamentary downdrafts that originate from the cell vertices and intergranular lanes at the surface merge into fewer, more widely separated downdrafts. These are the downdrafts corresponding to larger scale "cells" in the horizontal velocity field, and the merging is effected by the advection of filaments by the horizontal velocity components of these larger scale flows. The *driving* of the larger scale cells is provided by the entropy deficiency of the plasma advected into the cell vertices and, hence, ultimately by the surface radiative cooling. The argument is self-consistent and also recursive; downdrafts are expected to merge on successively larger scales at successively larger depths. The resulting structure of the downflows is sketched in Figure 7 and is approximately a self-similar

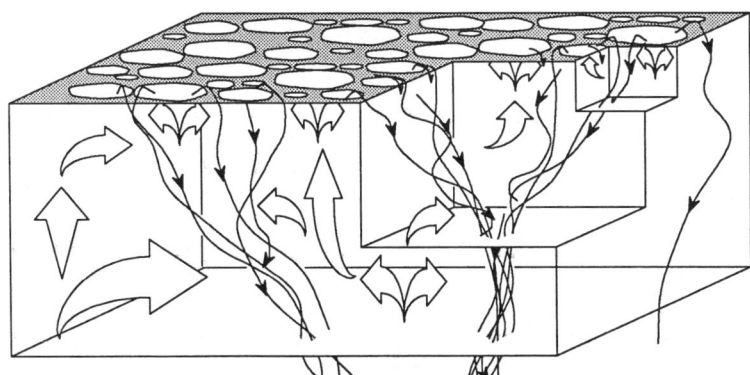

Figure 7 Flow lines showing the merging of the downdrafts on successively larger scales (schematic). The boxes cut out illustrate how the same process occurs on (in this illustration) three different scales.

fractal "tree." Note that the free structure has its origin entirely in the density stratification of the envelope, and that its length scales have nothing to do with turbulence [which does not mean that turbulence does not also play a role (see Section 3.1.3)].

The density scale height increases rapidly below the surface because of the increase in temperature and the decrease in molecular weight. Thus, increasingly larger scale flows are allowed at increasing depth below the surface. The allowed size also depends on the ratio of horizontal to vertical velocities (cf. the discussion above). The numerical simulations show ratios of order unity that rarely exceed a factor of 2–3, which in turn imply ratios of horizontal size to vertical scale height of the order of 5–20.

The flow time scales below the surface are long, and thus the flow is nearly *anelastic*. In this approximation, it can be shown (cf. Section 3.6 in ref. 120) that pressure fluctuations generated at depth extend over many scale heights. Thus, the pressure fluctuations of larger scale cells extend up to the surface, and so do the horizontal velocities, which are driven by these fluctuations. These horizontal velocities advect smaller scale downdrafts into the vertices of the larger scale cells. The resulting picture is that granular downdrafts merge into larger scale downdrafts, which again merge into even larger ones.

Presumably, flows on global scales are also driven; however, because of the rapid decrease of convection amplitudes with depth, flows on scales larger than supergranulation are hard to detect. Mixing-length models of the solar convection zone indicate smoothly decreasing velocity amplitudes with depth. According to the arguments above, we should expect these vertical velocities at successively larger depths to map onto surface velocities at successively larger horizontal scales. There is no obvious reason why, in this process, certain distinct scales should be favored. The often-mentioned helium ionization zone, for example (159), which extends over a large range of depths, has no particular significance in this scenario, except for reducing the adiabatic temperature gradient and hence for reducing the density scale height somewhat.

The scenario suggested by Stein & Nordlund (166) is, in a sense, an inverse cascade, since the flows on larger scales are driven by flows on smaller scales. The inverse cascade is a consequence of specific circumstances—namely, a vertically stratified, nearly adiabatic convection zone driven by the localized cooling at the upper surface. This picture is just the opposite of the one proposed by Zahn (201), who suggests that granulation is caused by the cascading of energy from larger scale, supergranular flows into smaller scales.

3.2.2 GLOBAL CONVECTION MODELS The deeper layers have also been investigated as a problem of global convection in a spherical shell. These investigations primarily aim at finding the large-scale flows, such as giant cells and the differential rotation. In addition to being observable at the surface, these flows also are key ingredients in theories of the solar dynamo. Simulations using the anelastic approximation (55–57, 68) should be very good in the deeper layers. At the present state of modeling, it is not possible to include the top layers because the range of length and time scales that can be included is limited. These simulations are thus complementary to those of the surface layers discussed above. The neglect of the top layers, however, introduces uncertainties when interpreting surface observations with these models. Also, the top boundary condition in these calculations replaces the solar surface as the ultimate source of driving for the deeper, larger scale flows. It is therefore important that this boundary condition somehow incorporates the filamentary character of the downdrafts and their typical horizontal separation at the depth where the boundary condition is applied.

Current simulations of the deeper layers show a strong tendency for the rotation to be constant on cylinders coaxial with the rotation axis, especially in the lower half of the convection zone. This behavior is what one expects for an incompressible, inviscid fluid (Proudman-Taylor constraint), and it should manifest itself also in a compressible fluid at small Rossby numbers (i.e. small ratio of rotation period to convective overturning time scale). Since the Rossby number is of order unity or less near the base of the solar convection zone, rotation on cylinders was thought to be understandable. Glatzmaier's (56, 57) results were rather successful at reproducing the observed differential rotation at the surface. Helioseismology measurements of the internal differential rotation of the convection zone, however, indicate that, to a first approximation, the rotation is constant with spherical radius rather than constant with coaxial cylinders (23, 48, 72). That is, below the surface, the latitude dependence of the rotation rate is the same as at the surface. The global convection models are not presently consistent with these observations (23).

The approximately constant rotation rate on radii indicates that the solar convection zone is "vertically stiff"—i.e. that ascending and descending plasma together tend to conserve angular speed, rather than angular momentum.

Glatzmaier's (56, 57) models show giant cells (the dominant convective scale in these simulations) with horizontal scales of the order of 300 Mm and surface velocities of 200 m s^{-1}. These velocities are 10 times higher than the observational upper limits (150). For reasons of numerical stability, artificial viscosity and heat transport coefficients were used in the models. Since these coefficients were not tailored to minimize their influence on the large scales, the effective Reynolds and Rayleigh numbers of the simulations are not very large. Simulation of a full spherical shell also limits the presently attainable resolution to something of the order of 100 Mm. This is insufficient to represent the highly intermittent flows found in the higher resolution (but nonglobal) simulations discussed above. This is one possible reason for their failure; another might be their neglect of the upper layers. Van Ballegooijen (192) argues that these layers will actually "shield" the large-scale flows to a great extent.

Theories explaining the differential rotation as a result of anisotropy or inhomogeneity of small-scale motions have a long history (see 13, 47, 81, 103, 143, and references therein). The simulations so far may have missed such effects because of the isotropy and homogeneity of the viscosity used in representing the small scales.

3.2.3 LABORATORY RESULTS AT HIGH RAYLEIGH NUMBERS Convection at Rayleigh numbers up to 10^{13} has been studied in the laboratory (26, 58, 91), but because these experiments lack stratification, they can be compared only with the situation near the base of the convection zone. At a Rayleigh number of 2.5×10^5, these experiments show a transition from temporal chaos (with spatially ordered flow) to turbulence (spatial as well as temporal stochasticity). From Ra = 2×10^7 to 10^{13}, turbulence with a different spatial structure is found ("hard turbulence"). This is interpreted as indicating a highly intermittent flow, consisting of thin sheets of fluid deviating in temperature from the bulk. These sheets are thought to be shed by turbulent boundary layers at the top and bottom of the container.

A similar phenomenon may occur at the base of the convection zone. We would expect that thin rising plumes would be generated there rather than broad, scale-height-sized upflows. This might have implications for the way active regions form from a sheet of magnetic field at the base. There would be a broad similarity between these intermittent upflows and the intermittent downflows in the rest of the convection zone. However, the causes are quite different in the two cases: The downflows are intermittent mainly because of *stratification*. This same stratification will prevent the upflows from remaining narrow very long, since their size increases by a factor of order e for every scale height traveled upward.

The experiments show a dependence of heat flux on temperature difference that is well represented by a power law with index 0.282 (26). Theories reproducing this dependence have been made using both a direct interaction formalism (25, 63) and nonlocal mixing-length ideas (178).

3.2.4 OVERSHOOT LAYER AT THE BASE Below the depth where the entropy gradient just vanishes (the formal base of the convection zone), there is a layer where the gradient is subadiabatic, but only weakly so because of the mixing action of overshooting flow. For most applications in stellar structure, it is the depth of the convection zone including this layer that matters. A simple estimate based on the kinetic energy of an eddy hitting the base of the convection zone yields a penetration into the interior of only a few hundred kilometers. This eddy also mixes, however, so that one is interested in the steady state that takes these mixing processes into account. Most authors (99a, 134, 149, 156, 191) assume a more or less steady flow pattern, driven by a large-scale eddy in the convection zone proper, and its depth dependence is found by solving hydrodynamic equations. In another view (98), it is argued that if a Kolmogoroff spectrum of turbulence exists near the base, it will be the *smallest* scales of motion that are most effective in mixing into the stable interior (see also 200). In both kinds of models, the overshoot layer has a thickness that is some fraction of a scale height. In the first kind, however, there is a thin boundary layer below the overshoot layer in which the temperature increases rapidly. In this layer the flow is rapidly braked, and the convective energy flux is high but of the wrong sign and is compensated by a high radiative flux. In the second view, a jump is not envisaged, but detailed calculations for such a model are not available.

Because of the uncertainties involved, a jump such as that in the first kind of overshoot model is usually not included in solar models that are compared with the p-mode oscillation frequencies, and it is not clear how much effect such a jump has. The frequencies of global g-modes are sensitive to the details of the overshoot layer (53, 174). If such modes are detected reliably, we may expect to learn more about this region.

3.2.5 GENERATION OF INTERNAL GRAVITY WAVES Unsteady convective flow near the base of the convection zone causes time-dependent displacements of the interface with the stable interior. Seen from the interior, such displacements correspond to internal gravity waves. These propagate inward and dissipate their energy in deeper layers. The waves may lead to mixing in the interior (135, 136) by processes comparable with those in the Earth's oceans. If so, this would be relevant for the lithium depletion problem (193a). The waves might also transport angular momentum. A naive estimate using a mixing-length model for the kinetic energy, length, and time scale of the dominant mode of convection near the base shows that a significant flux of energy may result. The waves generated by this dominant mode have a low frequency, however, and as a result their vertical wavelength is so small that they are damped by radiative exchange within a very short distance. The waves one is interested in are generated by the high-frequency, long-wavelength corner in the spectrum of convective motions (135). To estimate this, a more sophisticated model than the mixing-length theory is needed. It is conceivable that the fast, narrow downdrafts expected (see above) would be efficient in generating higher frequencies. More may be learned from numerical simulations of a convective layer above a stable one, such as those in (75) and (102).

4. CONCLUDING REMARKS

The last decade has shown great advances in the observation of time-dependent phenomena on the solar surface. In part this is due to better observations and instrumentation, in part to the greater ease with which large quantities of data can now be handled, and in part to the development of new algorithms for extracting and quantifying the physical information from these data.

At the same time, ab initio numerical simulations of the photosphere have produced intensity and velocity patterns that are remarkably similar to the observations. The calculated average spectral-line profiles agree closely with the observed ones, showing that the flows contributing to the main observables (including "microturbulent" and "macroturbulent" velocities) are reproduced in the simulations. The simulations predict the properties of the flows below and above the surface. Below the surface, the flows are organized into broad isentropic upflows mixed with narrow downdrafts originating at the surface; the observed granulation pattern of cells and lines is a shallow surface effect. The flows cannot be described by a "turbulent cascade" from large scales to granular scales. Helioseismology offers the opportunity to directly verify the predicted properties of the flows below the surface, while high-resolution one- and two-dimensional spectroscopy can be used for sensitive quantitative tests of the predicted intensity and flow fields in the atmosphere.

Improved simulations in the future will cover larger horizontal and vertical scales and will resolve small scales better. Simulations have already been made for other stars near the Sun in the Hertzsprung-Russell diagram, allowing accurate diagnostics by using the line shapes directly without resorting to empirical turbulence parameters. Such simulations also yield mean entropy profiles for convective envelopes that are accurate enough for use in stellar evolution calculations.

Over the next decade, helioseismology using ground-based networks like GONG and space instrumentation on *SOHO* is expected to be a major new source of information on solar convection. In addition, with local correlation tracking using the MDI instrument on *SOHO* (147), it should be possible to accurately measure horizontal flows on scales ranging from mesoscales to global scales.

A number of convection-related phenomena have not been covered, or have only been touched upon, in this review; for example, the interaction of convection with the solar p-mode oscillations. The excitation of p-mode oscillations is observed in the numerical simulations of the top layers of the convection zone (3, 168, 169). There is reasonable hope that such results can be used to obtain an understanding of the driving and damping of solar p-modes and to investigate the effects of the inhomogeneous surface layers on the exact frequencies of solar p-modes. Another issue is that of radius and luminosity variations of the Sun due to, for example, magnetic fields varying with the cycle. For a review of this topic, see (165).

Literature Cited

1. Alissandrakis, C. E., Dialetis, D., Tsiropoula, G. 1987. *Astron. Astrophys.* 174: 275–80
2. Anderson, L. S. 1989. *Ap. J.* 339: 558–78
3. Antia, H. M., Chitre, S. M., Gough, D. O. 1988. See Ref. 32, pp. 371–74
4. Antia, H. M., Chitre, S. M., Narasimha, D. 1982. *Sol. Phys.* 77: 303–27
5. Athay, R. G., Spicer, D., eds. 1987. *Theoretical Problems in High Resolution Solar Physics II. NASA CP-2483.* Washington, DC: NASA
6. Atroschenko, I. N., Gadun, A. S., Kostik, R. I. 1989. See Ref. 144, pp. 135–43
7. Avrett, E. H. 1985. In *Chromospheric Diagnostics and Modeling*, ed. B. W. Lites, pp. 67–127. Sunspot, N. Mex: Sacramento Peak Obs.
8. Avrett, E. H. 1990. See Ref. 172a, pp. 3–22
9. Baker, N. 1987. In *Physical Processes in Comets, Stars and Active Galaxies*, ed. W. Hillebrandt, E. Meyer-Hofmeister, H.-C. Thomas, pp. 105–24. Heidelberg: Springer-Verlag
10. Beckers, J. M. 1981. In *The Sun as a Star. NASA-SP-450*, ed. S. Jordan, pp. 11–64. Washington, DC: NASA
11. Beckers, J. M., Nelson, G. D. 1978. *Sol. Phys.* 55: 243–61
12. Biermann, L. 1948. *Z. Astrophys.* 25: 135–44
13. Biermann, L. 1951. *Z. Astrophys.* 28: 304–9
14. Blackwell, D. E., Dewhirst, D. W., Dolfus, A. 1957. *Nature* 180: 211–13
15. Böhm-Vitense, E. 1958. *Z. Astrophys.* 46: 108–43
16. Bogart, R. S. 1987. *Sol. Phys.* 110: 23–34
17. Bogart, R. S., Gierasch, P. J., MacAuslan, J. M. 1980. *Ap. J.* 236: 285–93
18. Bogart, R. S., Ferguson, S. H., Scherrer, P. H., Tarbell, T. D., Title, A. M. 1988. *Sol. Phys.* 116: 205–14
19. Brandt, P. N., Ferguson, S., Scharmer, G. B., Shine, R. A., Tarbell, T. D., et al. 1989. In *High Spatial Resolution Solar Observations*, ed. O. von der Lühe, pp. 473–88. Sunspot, N. Mex: Natl. Sol. Obs.
20. Brandt, P. N., Scharmer, G. B., Ferguson, S., Shine, R. A., Tarbell, T. D., Title, A. M. 1988. *Nature* 335: 238–40
21. Bray, R. J., Loughhead, R. E. 1967. *The Solar Granulation*. London: Chapman & Hall
22. Bray, R. J., Loughhead, R. E., Durrant, C. J. 1984. *The Solar Granulation*. London: Chapman & Hall. 2nd ed.
23. Brown, T. M., Christensen-Dalsgaard, J., Dziembowski, W. A., Goode, P., Gough, D. O., Morrow, C. A. 1989. *Ap. J.* 343: 526–46
24. Callier, A., Chauveau, F., Hugon, M., Rösch, J. 1968. *C. R. Acad. Sci. Paris* B 266: 199–201
25. Canuto, V. M., Goldman, I., Chasnov, J. 1987. *Phys. Fluids* 30: 3391–3417
26. Castaing, B., Gunaratne, G., Heslot, F., Kadanoff, L., Libchaber, A., et al. 1989. *J. Fluid Mech.* 204: 1–30
27. Cattaneo, F., Hurlburt, N. E., Toomre, J. 1990. *Ap. J. Lett.* 349: L63–66
28. Cavallini, F., Ceppatelli, G., Righini, A. 1989. See Ref. 151, pp. 21–23
29. Cavallini, F., Ceppatelli, G., Righini, A. 1989. See Ref. 144, pp. 283–87
30. Chan, K. L., Sofia, S. 1986. *Ap. J.* 307: 222–41

31. Chan, K. L., Sofia, S. 1989. *Ap. J.* 336: 1022–40
31a. Chevalier, J. 1908. *Ap. J.* 27: 12–24
32. Christensen-Dalsgaard, J., Frandsen, S., eds. 1988. *Advances in Helio- and Asteroseismology. Proc. IAU Symp. No. 123*. Dordrecht: Kluwer
33. Christensen-Dalsgaard, J., Gough, D. O., Thompson, M. J. 1989. Preprint
34. Christensen-Dalsgaard, J., Schou, J. 1988. See Ref. 141, pp. 149–56
35. Cram, L. E., Durney, B. R., Guenther, D. B. 1983. *Ap. J.* 267: 442–51
36. Dawes, W. R. 1864. *MNRAS* 24: 161–65
37. Deardorff, J. W. 1973. *J. Fluids Eng.* 95: 429–38
38. de Jager, C. 1959. In *Handbuch der Physik*, ed. S. Flügge, 52: 81–362. Berlin: Springer-Verlag
39. Deubner, F.-L. 1974. *Sol. Phys.* 36: 299–301
40. Deubner, F.-L. 1988. *Astron. Astrophys.* 204: 301–5
41. Deubner, F.-L., Mattig, W. 1975. *Astron. Astrophys.* 45: 167–71
42. Dravins, D. 1989. See Ref. 144, pp. 153–60
43. Dravins, D., Nordlund, Å. 1990. *Astron. Astrophys.* 228: 184–202
44. Dravins, D., Nordlund, Å. 1990. *Astron. Astrophys.* 228: 203–17
45. Dravins, D., Lindegren, L., Nordlund, Å. 1981. *Astron. Astrophys.* 96: 345–64
46. Dunn, R. B., Zirker, J. B. 1973. *Sol. Phys.* 33: 281–304
47. Durney, B. R. 1989. *Sol. Phys.* 123: 197–216
48. Duvall, T. L. Jr., Harvey, J. W. 1984. *Nature* 310: 19–22
49. Edmonds, F. N. Jr. 1960. *Ap. J.* 131: 57–60
50. Edmonds, F. N. Jr. 1962. *Ap. J. Suppl.* 6: 357–406
51. Evans, J. W. 1964. *Astrophys. Norv.* 9: 33–54
52. Fox, P., Van der Borght, R. 1985. *Proc. Astron. Soc. Aust.* 6: 60–62
53. Gabriel, M. 1986. In *Seismology of the Sun and Distant Stars*, ed. D. O. Gough, pp. 177–86. Dordrecht: Reidel
54. Gadun, A. S. 1986. Preprint [Inst. Theor. Phys., Kiev, USSR (in Russian)]
55. Gilman, P. A., Miller, J. 1986. *Ap. J. Suppl.* 61: 585–608
56. Glatzmaier, G. A. 1985. *Geophys. Astrophys. Fluid Dyn.* 31: 137–50
57. Glatzmaier, G. A. 1985. *Ap. J.* 291: 300–7
58. Goldstein, R. J., Tokuda, S. 1980. *Int. J. Heat Mass Transfer* 23: 738–50
59. Gough, D. O., Weiss, N. O. 1976. *MNRAS* 176: 589–607
60. Graham, E. 1969. In *Problems of Stellar Convection, IAU Colloq. No. 38. Lect. Notes Phys.*, ed. E. A. Spiegel, J.-P. Zahn, 71: 151–55. Berlin: Springer-Verlag
61. Gustafsson, B., Bell, R. A., Eriksson, K., Nordlund, Å. 1975. *Astron. Astrophys.* 42: 407–32
62. Hansky, A. 1906. *Bull. Soc. Astron. Fr.* 1906(64): 178–81
63. Hartke, G. J., Canuto, V. M., Dannevik, W. P. 1988. *Phys. Fluids* 31: 256–62
64. Deleted in proof
65. Harvey, J. W. 1988. See Ref. 141, pp. 55–66
66. Harvey, J. W. 1985. See Ref. 148, p. 183
67. Hart, M. 1973. *Ap. J.* 184: 587–603
68. Hathaway, D. 1982. *Sol. Phys.* 77: 341–56
69. Henyey, L., Vardya, M. S., Bodenheimer, P. 1965. *Ap. J.* 142: 841–54
70. Herring, J. R. 1987. In *The Internal Solar Angular Velocity*, ed. B. Durney, S. Sofia, pp. 725–88. Dordrecht: Reidel
71. Herschel, W. 1801. *Philos. Trans. R. Soc. Part 1*, p. 265
72. Hill, F., Rust, D. M., Appourchaux, T. 1988. See Ref. 32, pp. 49–52
73. Holweger, H., Kneer, F. 1989. See Ref. 144, pp. 173–86
73a. Hughes, D. W., Proctor, M. R. E. 1988. *Annu. Rev. Fluid Mech.* 20: 187–223
74. Hurlburt, N. E., Toomre, J., Massaguer, J. M. 1984. *Ap. J.* 282: 557–73
75. Hurlburt, N. E., Toomre, J., Massaguer, J. M. 1986. *Ap. J.* 311: 563–77
76. Janssen, J. 1896. *Ann. Obs. Astrophys. Paris Meudon* 1: 91
77. Karpinsky, V. N. 1990. See Ref. 172a. In press
78. Kawaguchi, I. 1980. *Sol. Phys.* 65: 207–20
79. Keil, S. L. 1980. *Astron. Astrophys.* 82: 144–51
80. Keil, S. L. 1984. In *Small-Scale Dynamical Processes in Quiet Stellar Atmospheres*, ed. S. L. Keil, pp. 148–56. Dordrecht: Reidel
81. Kichatinov, L. L. 1987. *Geophys. Astrophys. Fluid Dyn.* 38: 273–92
81a. Knobloch, E., Rosner, R. 1981. *Ap. J.* 247: 300–10
82. Krat, V. A. 1954. *Izv. Gl. Astron. Obs. Pulkove* 152: 1
83. Krat, V. A. 1956. *Izv. Gl. Astron. Obs. Pulkove* 155: 17
84. Kurucz, R. L. 1979. *Ap. J. Suppl.* 40: 1–340
85. Latour, J., Toomre, J., Zahn, J.-P. 1983. *Sol. Phys.* 82: 387–400

86. Legait, A. 1986. *Astron. Astrophys.* 168: 173–83
87. Leith, C. E., Kraichnan, R. H. 1972. *J. Atmos. Sci.* 29: 1041–58
88. Leighton, R. B., Noyes, R. W., Simon, G. W. 1962. *Ap. J.* 135: 474–94
89. Leighton, R. B. 1963. *Annu. Rev. Astron. Astrophys.* 1: 19–40
90. Leighton, R. B. 1964. *Ap. J.* 140: 1547–62
91. Libchaber, A. 1987. *Proc. R. Soc. London Ser. A* 413: 63–69
92. Libbrecht, K. G. 1988. See Ref. 141, pp. 131–36
93. Deleted in proof
94. Lites, B. W., Nordlund, Å., Scharmer, G. B. 1989. See Ref. 144, pp. 349–57
95. Livingston, W. C. 1968. *Ap. J.* 153: 929–42
96. Livingston, W. C. 1987. See Ref. 151, pp. 14–20
97. MacAuslan, J. 1985. *Sol. Phys.* 99: 55–78
98. Marcus, P. S., Press, W. H., Teukolsky, S. A. 1983. *Ap. J.* 267: 795–821
99. Massaguer, J. M., Zahn, J.-P. 1980. *Astron. Astrophys.* 87: 315–27
99a. Massaguer, J. M., Latour, J., Toomre, J., Zahn, J.-P. 1984. *Astron. Astrophys.* 140: 1–16
100. Mattig, W., Hanslmeier, A., Nesis, A. 1989. See Ref. 144, pp. 187–93
101. Mehltretter, J. P. *Astron. Astrophys.* 62: 311–16
102. Métais, O., Herring, J. R. 1989. *J. Fluid Mech.* 202: 117–48
103. Monin, A. S., Rakhmanova, N. K., Ruzmaikin, A. A. 1985. *Astrophys. Space Sci.* 114: 157–63
104. Mosher, J. M. 1977. PhD thesis. Calif. Inst. Technol., Pasadena
105. Muller, R. 1984. *Sol. Phys.* 100: 237–56
106. Muller, R. 1989. See Ref. 144, pp. 101–22
107. Musman, S., Nelson, G. D. 1976. *Ap. J.* 207: 981–88
108. Musman, S., Nelson, G. D. 1976. *Ap. J.* 214: 912–17
108a. Nakano, T., Fukushima, T., Unno, W., Kondo, M. 1979. *Publ. Astron. Soc. Jpn.* 31: 713–35
109. Namba, O., van Rijsbergen, R. 1969. In *Problems of Stellar Convection, IAU Colloq. No. 38. Lect. Notes Phys.*, ed. E. A. Spiegel, J.-P. Zahn, 71: 119–25. Berlin: Springer-Verlag
110. Narasimha, D., Antia, H. M. 1982. *Ap. J.* 262: 358–68
111. Nasmyth, J. 1862. *Mem. Lit. Philos. Soc. Manchester, Ser. 3* 1: 407
112. Nelson, G. D. 1978. PhD thesis. Univ. Wash., Seattle
113. Nelson, G. D. 1978. *Ap. J.* 238: 659–66
114. Nelson, G. D. 1978. *Sol. Phys.* 60: 5–18
115. Nelson, G. D., Musman, S. 1977. *Ap. J.* 214: 912–16
116. Nelson, G. D., Musman, S. 1978. *Ap. J. Lett.* 222: L69–72
117. Nesis, A., Fleig, K. H., Mattig, W. 1989. See Ref. 144, pp. 289–94
118. Nordlund, Å. 1974. *Astron. Astrophys.* 32: 407–22
119. Nordlund, Å. 1976. *Astron. Astrophys.* 50: 23–39
120. Nordlund, Å. 1982. *Astron. Astrophys.* 107: 1–10
121. Nordlund, Å. 1984. In *Small-Scale Dynamical Processes in Quiet Stellar Atmospheres*, ed. S. L. Keil, pp. 174–80. Sunspot, N. Mex: Natl. Sol. Obs.
122. Nordlund, Å. 1984. In *Small-Scale Dynamical Processes in Quiet Stellar Atmospheres*, ed. S. L. Keil, pp. 181–221. Sunspot, N. Mex: Natl. Sol. Obs.
123. Nordlund, Å. 1985. *Sol. Phys.* 100: 209–35
124. Nordlund, Å. 1985. See Ref. 148, pp. 101–23
125. Nordlund, Å. 1985. In *Problems in Stellar Spectral Line Formation Theory*, ed. J. O. Beckman, L. Crivellari, pp. 215–24. Dordrecht: Reidel
126. Nordlund, Å. 1990. See Ref. 172a, pp. 191–212
127. Nordlund, Å., Dravins, D. 1990. *Astron. Astrophys.* 228: 155–83
128. November, L. J. 1989. In *High Spatial Resolution Observations*, ed. O. von der Lühe. Sunspot, N. Mex: Natl. Sol. Obs. In press
129. November, L. J., Simon, G. W. 1988. *Ap. J.* 333: 427–42
130. November, L. J. 1989. *Ap. J.* 344: 494–503
131. November, L. J., Simon, G. W., Tarbell, T. D., Title, A. M., Ferguson, S. H. 1987. See Ref. 5, pp. 121–28
132. Oda, N. 1984. *Sol. Phys.* 93: 243–55
133. Parsons, S. B. 1969. *Ap. J. Suppl.* 18: 127–65
134. Pidatella, R. M., Stix, M. 1986. *Astron. Astrophys.* 157: 338–40
135. Press, W. H. 1981. *Ap. J.* 245: 286–303
136. Press, W. H., Rybicki, G. B. 1981. *Ap. J.* 248: 751–66
137. Proctor, M. R. E., Weiss, N. O. 1982. *Rep. Prog. Phys.* 45: 1317–79
138. Richardson, R. S., Schwarzschild, M. 1950. *Ap. J.* 111: 351–61
139. Rösch, J. 1955. *C. R. Acad. Sci. Paris* 240: 1630
140. Rösch, J. 1956. *C. R. Acad. Sci. Paris* 243: 478

141. Rolfe, E. J., ed. 1988. *Seismology of the Sun and Sunlike Stars, ESA SP-286*. Paris: ESA
142. Roudier, T., Muller, R. 1986. *Sol. Phys.* 107: 11–26
143. Rüdiger, G., Tuominen, I. 1987. In *The Internal Solar Angular Velocity*, ed. B. Durney, S. Sofia, pp. 361–70. Dordrecht: Reidel
144. Rutten, R. J., Severino, G., ed. 1989. *Solar and Stellar Granulation*. Dordrecht: Kluwer
145. Scharmer, G. B. 1987. See Ref. 151, pp. 349–53
146. Scharmer, G. B. 1989. See Ref. 144, pp. 161–67
147. Scherrer, P. H., Hoeksema, J. T., Bogart, R. S. 1988. See Ref. 141, pp. 375–79
148. Schmidt, H. U., ed. 1985. *Theoretical Problems in High Resolution Solar Physics*. Garching bei München: Max-Planck-Inst. Phys. Astrophys.
149. Schmitt, J. H. M. M., Rosner, R., Bohm, H. U. 1984. *Ap. J.* 282: 316–29
150. Schröter, E.-H. 1984. *Sol. Phys.* 100: 141–69
151. Schröter, E.-H., Vázquez, M., Wyller, A. A., eds. 1987. *The Role of Fine-Scale Magnetic Fields on the Structure of the Solar Atmosphere*. Cambridge: Univ. Press
152. Schwarzschild, M. 1958. *Structure and Evolution of Stars*. Princeton, NJ: Princeton Univ. Press
153. Schwarzschild, M., Rogerson, J. B., Evans, J. E. 1958. *Astron. J.* 63: 313 (Abstr.)
154. Schrijver, C. J. 1988. *Sol. Phys.* 122: 193–208
155. Semel, M. 1962. *C. R. Acad. Sci. Paris* 254: 3978–80
156. Shaviv, G., Salpeter, E. E. 1973. *Ap. J.* 184: 191–200
157. Simon, G. W. 1967. *Z. Astrophys.* 65: 345–63
158. Simon, G. W., Weiss, N. O. 1989. *Ap. J.* 345: 1060–78
159. Simon, G. W., Leighton, R. B. 1964. *Ap. J.* 140: 1120–47
160. Simon, G. W., Title, A. M., Topka, K. P., Tarbell, T. D., Shine, R. A., et al. 1988. *Ap. J.* 327: 964–67
161. Smagorinsky, J. 1963. *Mon. Weather Rev.* 91: 99–164
162. Spiegel, E. A. 1963. *Ap. J.* 138: 216–25
163. Spiegel, E. A. 1971. *Annu. Rev. Astron. Astrophys.* 9: 323–52
164. Spiegel, E. A. 1972. *Annu. Rev. Astron. Astrophys.* 10: 261–304
165. Spruit, H. C. 1990. In *The Sun in Time*, ed. C. Sonett, M. Magisos. Tucson: Univ. Ariz. Press. In press
166. Stein, R. F., Nordlund, Å. 1989. *Ap. J. Lett.* 342: L95–98
167. Deleted in proof
168. Stein, R. F., Nordlund, Å., Kuhn, J. R. 1989. See Ref. 144, pp. 381–99
169. Steffen, M. 1988. See Ref. 32, pp. 379–82
170. Steffen, M. 1987. See Ref. 151, pp. 47–52
171. Steffen, M. 1989. See Ref. 144, pp. 425–39
172. Steffen, M., Ludwig, H.-G., Krüss, A. 1989. *Astron. Astrophys.* 213: 371–82
172a. Stenflo, J. O., ed. 1990. *Solar Photosphere: Structure, Convection, and Magnetic Fields. Proc. IAU Colloq. No. 138*. Dordrecht: Kluwer
173. Tarbell, T. D., Peri, M., Frank, Z., Shine, R., Title, A. 1988. See Ref. 141, pp. 315–19
174. Tassoul, M. 1980. *Ap. J. Suppl.* 43: 469–90
175. Title, A. M., Tarbell, T. D., Simon, G. W., and the SOUP Team. 1986. *Adv. Space Res.* 6(8): 253–62
176. Title, A. M., Tarbell, T. D., Acton, L., Duncan, D., Ferguson, S., et al. 1987. See Ref. 5, pp. 55–77
177. Title, A. M., Tarbell, T. D., Topka, K. P., Ferguson, S. H., Shine, R. A., and the SOUP Team. 1989. *Ap. J.* 336: 475–94
178. Tooth, P. D., Gough, D. O. 1988. See Ref. 141, pp. 463–69
179. Travis, L. D., Matsushima, S. 1971. *Ap. J.* 180: 975–85
180. Uberoi, M. S. 1955. *Ap. J.* 122: 466–76
181. Ulrich, R. K. 1970. *Astrophys. Space Sci.* 7: 71–86
182. Ulrich, R. K. 1970. *Astrophys. Space Sci.* 7: 183–200
183. Ulrich, R. K. 1970. *Astrophys. Space Sci.* 9: 80–96
184. Ulrich, R. K. 1976. *Ap. J.* 207: 564–73
185. Deleted in proof
186. Unno, W., Kondo, M.-A. 1986. *Astrophys. Space Sci.* 118: 223–25
187. Unno, W., Kondo, M.-A., Xiong, D. R. 1985. *Publ. Astron. Soc. Jpn.* 37: 235–44
188. Uus, U. 1986. *Tartu Astrofüüs. Obs. Publ.* 51: 20–34
189. Uus, U. 1987. *Tartu Astrofüüs. Obs. Publ.* 52: 33–47
190. Uus, U., Poljakov, E. V., Karpinsky, V. N. 1990. See Ref. 172a. In press
191. van Ballegooijen, A. A. 1982. *Astron. Astrophys.* 113: 99–122
192. van Ballegooijen, A. A. 1986. *Ap. J.* 304: 828–37

193. Von der Lühe, O., Dunn, R. B. 1987. *Astron. Astrophys.* 177: 265–76
193a. Vauclair, S. 1988. *Ap. J.* 335: 971–75
194. Vitense, E. 1953. *Z. Astrophys.* 32: 135–64
195. Wang, Y.-M., Nash, A. G., Sheeley, N. R. Jr. 1989. *Science* 245: 712–17
196. Wittmann, A. 1979. In *Small-Scale Motions on the Sun. Mitt. Kiepenheuer Inst.*, 179: 29–53. Freiburg: Kiepenheuer Inst.
197. Wöhl, H., Nordlund, Å. 1985. *Sol. Phys.* 97: 213–21
198. Woodward, P., Porter, D. H. 1987. *Bull. Am. Astron. Soc.* 19: 1023 (Abstr.)
199. Xiong, D. R. 1979. *Acta Astron. Sin.* 20: 338 (Engl. transl. in *Chin. Astron.* 2: 118)
200. Xiong, D. R. 1985. *Astron. Astrophys.* 150: 133–38
201. Zahn, J.-P. 1988. In *Solar and Stellar Physics. Lect. Notes Phys.*, ed. E. H. Schröter, M. Schüssler, 292: 55–71. Heidelberg: Springer-Verlag
202. Zwaan, C. 1978. *Sol. Phys.* 60: 213–29

Annu. Rev. Astron. Astrophys. 1990. 28: 303–45

QUANTITATIVE SPECTROSCOPY OF HOT STARS[1]

R. P. Kudritzki

Institut für Astronomie und Astrophysik, Scheinerstrasse 1, 8000 München 80, Federal Republic of Germany

D. G. Hummer

Institut für Astronomie und Astrophysik, Scheinerstrasse 1, 8000 München 80, Federal Republic of Germany, and Joint Institute for Laboratory Astrophysics, University of Colorado and National Institute of Standards and Technology, Boulder, Colorado 80309-0440[2]

KEY WORDS: stellar atmospheres, stellar winds, stellar evolution, photospheric parameters, elemental abundances

1. INTRODUCTION

The expression "quantitative spectroscopy" implies the systematic acquisition and analysis of accurate spectroscopic (and, to a lesser extent, photometric) data by means of detailed quantitative modeling of the objects concerned in order to determine accurate values of the stellar parameters, including chemical abundances. The term "hot stars" is used because many sufficiently hot objects of different stellar types can be understood in terms of the same group of physical processes. As the atmospheres of these objects are relatively (but certainly not absolutely) simple compared with those of cooler stars, it appears that our understanding of the relevant physics, while not in any way complete, is adequate for the task at hand. This area of study has recently experienced great progress, thanks to the dramatic advances in observational and com-

---

[1] The US Government has the right to retain a nonexclusive royalty-free license in and to any copyright covering this paper.
[2] Staff Member, Quantum Physics Division, National Institute of Standards and Technology.

putational facilities of the past decade, as well as to a new generation of modeling techniques and calculations of accurate atomic data.

The assumption of a relatively simple atmospheric structure and the strategy of solving accurately the obvious conservation equations as the basis for investigating the atmospheres of hot stars have been criticized strongly by certain individuals (e.g. 144a, 146a), who hold up the complexities of the Sun or manifestly nonstationary objects as examples. Among the ignored phenomena are magnetic fields, known or unknown sources of mechanical energy, and gas flows in directions other than radial. But in hot stars there is little or no evidence that these phenomena are important for quantitative spectroscopy of hot stars. For example, numerous attempts to graft coronal structures onto hot stars have found no observational support (21b). In view of (a) the large amount of accurate spectral data that can be understood in terms of the obvious (and accurately treated) physics and (b) the self-consistent picture that is emerging, we believe our approach to be valid. Consistency might be the criterion of fools, but for these objects the contributions of the "wise" have been negative.

Nearly all hot stars display a systematic, relatively constant outflow of matter—the so-called stellar wind. Although these winds were first detected in 1967 by Morton (115a) and have subsequently been the object of intensive study, in most of the discussions the winds and photospheres are treated as quite different entities. Only recently has the intrinsic unity of these two components of the stellar atmosphere become fully apparent. Nevertheless, the approximate division of the atmosphere in this way is in many situations necessary, for until recently the driving mechanism for hot-star winds was only poorly understood, whereas the theory of stellar photospheres was in much better condition. The latter theory is also sufficient for certain purposes, as is discussed below.

The motivation for the accurate determination of stellar parameters is manyfold. (a) Much more precise constraints on the theory of evolution of hot stars can be obtained than hitherto possible from the stellar parameters of a large group of hot stars, especially if the sample includes objects from galaxies with very different compositions. This is especially valuable information in view of the uncertainties arising in evolutionary calculations from unresolved problems in the treatment of mass loss, convection and convective overshooting, and even opacities. (b) Hot, massive stars play important roles in the chemical and dynamical evolution of galaxies as sources of kinetic energy, ionizing radiation, gas, and processed elements. (c) At the same time, these stars provide a uniquely reliable source of information concerning the physical conditions of the gas in the host galaxies.

In organizing this review, we use the division of the stellar atmosphere into a photosphere and a wind as an expository device. The validity of this separation is related to that of the so-called core-halo picture in which the radiation field in the relatively low-density halo is entirely determined by the underlying core. This picture has proved extremely valuable for developing techniques and has led to substantial understanding of many kinds of stars. The subsequent sections are concerned with progressive stages in the breakdown of this picture: photospheres, optically thin stellar winds, unified wind-photosphere models, and opaque winds.

As we are concerned here as much with methods as with results, it is impossible in the space available to review all of the recent analyses of hot-star spectra, much less to summarize our current understanding of all aspects of the atmospheres of hot stars. Work based on largely "uncontrolled" approximations is not discussed. As we are reviewing recent developments, we do not follow every trail to its origin, but instead give references to earlier reviews where possible.

A number of recent reviews cover many aspects of the present one in more detail. A very extensive presentation at a more pedagogical level has been given by Kudritzki (89). The role of massive stars in galaxies is discussed by Kudritzki & Hummer (85). Chemical abundances in hot stars are discussed by Kudritzki (84) and for all classes are completely reviewed by Gehren (43). The quantitative spectroscopy of hot stars in the EUV is discussed by Kudritzki et al (94).

2. PHOTOSPHERES

In view of the ubiquity of stellar winds, it is useful to define the photosphere as that part of the star in which radiative equilibrium and hydrostatic equilibrium hold but the Saha-Boltzmann and Planck distributions cannot be assumed. This region, for which we take the sonic point (where the flow speed equals the sonic speed) as the outer limit, includes the region in which the optical continuum in "normal" stars is formed, i.e. where the optical depth at some reference wavelength (such as 5000 Å) is unity. In many O stars and central stars of planetary nebulae (CSPN), the He II ground-state continuum is formed above the sonic point, and in most Wolf-Rayet (WR) stars the entire continuum is formed in the wind.

In addition to radiative equilibrium and hydrostatic equilibrium the structure of the photosphere is determined by the radiative transfer equation and by the condition of statistical equilibrium, which in a steady state requires the rate of all transitions into and out of each atomic state to be equal. An equation of state (usually the ideal gas law), the elemental abundances, and the geometry of the atmosphere must be specified, as must the collisional and radiative cross sections and rate constants. The mathematical statement of these conservation laws yields a nonlinear set of integrodifferential equations that determine the temperature, the densities of each species making up the gas, the radiation field at all frequencies, and the atomic-level populations. To make this general problem tractable, a number of simplifying assumptions are made.

The most general assumption is that of plane-parallel stratification—i.e. that the physical conditions depend on only one spatial coordinate, which is specified as the mass in a column of unit-cross-section measure from some point well outside the star, or the optical depth in an appropriate mean opacity. Plane-parallel stratification is valid when the mean distance between the emission and next scattering or absorption of the photons is very small compared with the radius at that point. Only rarely has plane-parallel stratification been replaced by spherical symmetry, in which the depth variable is the radius. Spherical symmetry is necessary for stellar winds and certain very luminous stars. Nearly all atmospheric models contain only hydrogen and helium in the atmospheric structure calculation. The statistical equilibrium and radiative transfer equations for other ions are solved by assuming that the atmospheric structure is independent of these "trace ions," in view of their low abundance relative to H and He. The effects of other elements are discussed below.

Much more controversial was the replacement of statistical equilibrium by local thermodynamic equilibrium (LTE), in which the atomic-level populations were specified by the Saha-Boltzmann law at the local electron temperature. A sufficient condition for LTE is that the collisional rate between each atomic state is much larger than the corresponding radiative rate. Local thermodynamic equilibrium and plane-parallel stratification, together with parameterized opacities, formed the basis for the first model photospheres calculated in the 1930s.

The modern age of model atmospheres and quantitative spectroscopy of hot stars (QSHS) began in the 1960s, when C. H. Auer and D. Mihalas developed computational techniques that made it possible to abandon LTE in favor of statistical equilibrium and then demonstrated brilliantly that the statistical equilibrium models produced spectra, especially in strong lines, that agreed substantially with the observations, whereas LTE models always failed to do so. There is no longer any serious question that the lines and strong continua of hot stars must be calculated with statistical equilibrium (often called non-LTE, a clumsy and inaccurate term that should be abandoned). The early history of model stellar atmospheres is sketched in the textbook by Mihalas (111), who also summarizes in detail the technical basis of the statistical equilibrium calculations and the resulting insight into the nature of hot-star atmospheres.

As all published work in QSHS is based on the work by Auer & Mihalas (10a), either through the application of codes based on theirs (113) or on the independent codes of Kudritzki (80) and of D. Husfeld & H.-G. Groth in Munich (private communication), we summarize their techniques very briefly, along with subsequent important technical developments. We then consider the recent progress in understanding the relevant physical processes, which has led to the dramatic development of QSHS. Although remarkable progress has been made very recently in solving very complex radiative transfer and statistical equilibrium problems—Anderson's (7) brilliant treatment of the effect of many metal lines, and the development and first applications of the "approximate lambda iteration" method—these techniques, with very few exceptions (see 157), have not yet been used in QSHS. In view of their enormous potential for the very near future, it is unfortunate that we cannot summarize these techniques here.

2.1 Complete Linearization

The crucial element introduced by Auer & Mihalas was the concept of complete linearization, in which the nonlinear and strongly coupled equations describing the atmosphere are discretized in depth and frequency and then solved iteratively by utilizing a set of linear equations for corrections to the values of all quantities—densities, atomic-level populations, temperature, and the intensity of radiation at each frequency—from the previous iteration. Thus, complete linearization is a generalization of the familiar Newton-Raphson method for one dimension and shares the advantage of quadratic convergence near the solution point. Complete linearization has been the workhorse of stellar atmosphere computations of the past two decades; only now are more effective methods being developed.

The latest and most powerful program using complete linearization is that of Hubeny (68), which by allowing linearization with respect to only selected quantities makes possible the calculation of models with a larger number of frequency points and atomic species. It also offers improved convergence, in the hands of an experienced operator.

Despite the strong convergence of complete linearization near the solution point, difficulty can be experienced in the early stages of iteration starting from an LTE model, which is calculated with relative ease. Hummer & Voels (71) have introduced a technique of artificially enhancing the ratio of collisional to radiative rates and then, as the iteration proceeds, gradually modifying the rates to their correct values, thus smoothing the way between LTE and statistical equilibrium. This so-called collisional-radiative switching has been quite useful in treating models with enhanced helium abundances (149). An even more powerful rescaling technique for improving the convergence in difficult cases has been developed recently by Hubeny (68a).

2.1.1 AUER-HEASLEY METHOD   An especially fast procedure for treating combined radiative transfer and statistical equilibrium is Auer & Heasley's (9) reformulation of Rybicki's (133) method. This has been utilized by Giddings (45) in a set of programs for solving radiative transfer and statistical equilibrium in a predetermined atmospheric structure (DETAIL), and also for calculating the emergent line profiles (SURFACE). These codes were further developed by Butler (20) and have been used in much of the work from Munich discussed below, but despite their great power and generality, they have never been published or even widely circulated. They are located in the computer library of the Co-ordinated Computational Project No. 7 of the UK Science and Research Council, but the documentation (21) is minimal.

2.2 Atomic Data and Atomic Models

To determine elemental abundances using quantitative spectroscopy, one needs not only the capability of solving radiative transfer and statistical equilibrium in model atmospheres but also adequate atomic models. Recent development of computational methods in atomic physics and of the very powerful programs written by the atomic physics groups at University College London and Queens University Belfast [especially their joint efforts with collaborators from three continents within the Opacity Project (142)] has led to a qualitative increase in the amount and quality of atomic radiative data. Individual workers in these schools have also produced many accurate electron collision data. This material has been utilized by K. Butler, S. R. Becker & F. Eber to construct very elaborate model atoms and to calculate equivalent widths using a statistical equilibrium model for O II (11, 12), N II (13, 14), N III (20), C II (36), and C III (35). So-far unpublished work with models of comparable completeness includes ions of Ne II and Si II, III, and IV (K. Butler & S. R. Becker, in preparation). Equivalent widths have also been calculated from statistical equilibrium models using less elaborate model atoms by Lennon et al (101) for Si III and Si IV and by Dufton et al (33) for Al III.

The value of the very elaborate atomic models is shown in (36), where the discrepancy with the observed C II 4277-Å line found by Lennon (100) using a less complete model atom is resolved. Becker & Butler (11) found for stars like τ Sco and 10 Lac that no microturbulence parameter was required to fit the equivalent widths of weak and strong lines of O II with the same abundance value, whereas for LTE models a carefully chosen value is necessary.

Accurate abundance determinations also require reliable line-broadening profiles. Schöning & Butler (139, 140) have calculated new broadening functions for He II, using the unified theory of Vidal et al (147), but now with hyperbolic electron trajectories. Their results agree better with observations than do those obtained with earlier broadening data. Improved broadening functions for a large number of metal lines have been recently presented by Dimitrijević (31) and Seaton (141).

2.3 Metal Opacities and Line Blanketing

As mentioned, statistical equilibrium atmospheric models of hot stars have until recently contained only H and He, apart from those of Mihalas (112) and Mihalas & Hummer (114), which included a "mean light ion" to roughly represent the ionization of carbon, nitrogen, and oxygen (CNO). Early attempts by Husfeld (72) to include these ions using complete linearization encountered severe convergence problems, but to the extent that his models converged, the effects were important only in the EUV. Later, Husfeld et al (73) published a flux distribution for a converged model with $T_{eff} = 100,000$ K, $\log g = 5$, and solar CNO abundances, which accounted for the bound-free transition of the metal ions. For this model, a substantial effect arising from CNO was found only shortward of the O VI edge.

Of much more importance than the photoionization of CNO is line blanketing—the combined effect of an enormous number of weak lines in absorbing radiation flowing up from deep layers, which leads to lower surface temperatures and "backwarming" (an increase in the temperature gradient). Although this effect has been treated in the LTE approximation by a number of workers, most extensively in the heroic calculations of Kurucz (98), it is essential to provide a statistical equilibrium treatment of line blanketing, the lack of which is probably the weakest point of hot-star atmospheric modeling at present. Although progress in including both bound-free metal opacities (155, 156) and line blanketing (7, 8, 98, 137) has been considerable, the number of available models is still both extremely small and not really in the parameter range of most interest here; unfortunately, we do not have the space to discuss this important work.

2.4 Wind Blanketing

If photons streaming from the photosphere drive a stellar wind, then some of them must be scattered back into the photosphere. This effect was first investigated by Hummer (69), using a gray atmosphere and a representation of the wind backscattering by a frequency-dependent albedo function, which was parameterized as a step function. This simple model showed that significant heating of the outer layers of the atmosphere could result from this effect, although the temperature gradient at great depth was not changed, in contrast to line blanketing, which causes lower surface temperatures and an increased temperature gradient. Independently, Husfeld (72) found similar results using albedos that were constant and different from zero either over the whole spectrum or over one of the ranges 228–504 Å or 228–912 Å, and a statistical equilibrium atmosphere code with H and He. He found no observable change in the shape of the observable continuum, but for models with the nonzero albedo redward of 228 Å, the He II continuum became significantly brighter.

Abbott & Hummer (5) then used a Monte Carlo method developed by Abbott & Lucy (4) to calculate more realistic albedo functions from models of radiation-driven winds, which provided the outer boundary conditions for a statistical equilibrium atmosphere calculation with H and He. This work showed that for O stars the continuum is not significantly changed, but that the strengths of photospheric lines are very sensitive to this effect. For stars with effective temperatures on the order of 40,000 K and strong winds, wind blanketing reduced the effective temperature by as much as 10%. However, recently D. Husfeld & J. Puls (private communication; see also 90), using improved wind models, have obtained significantly smaller temperature changes. An additional source of uncertainty comes from the breakdown of the core-halo picture in the region of the He II continuum. Abbott & Hummer arbitrarily set the albedo there to unity, but since this continuum is formed in the wind, this value of the albedo is clearly too high. On the other hand, if the effective temperature is less than roughly 30,000 K, the amount of radiation in the region of the He II continuum is sufficiently small that this uncertainty is then inconsequential. For WR winds, which are optically thick at all wavelengths, the wind-blanketing mechanism must have a major effect on the surface temperature, although one can no longer clearly define an albedo.

2.5 *Determination of Photospheric Parameters and Abundances*

Quantitative analysis begins with the determination of $T_{\rm eff}$, log g, and the relative helium abundance by number of atoms, i.e. $Y = N({\rm He})/[N({\rm H}) + N({\rm He})]$. These parameters specify the structure of the atmosphere (unless the metals are extremely overabundant or the atmosphere is very extended). The metal abundances are determined subsequently. The determination of $T_{\rm eff}$ and log g in hot stars has been the subject of considerable controversy.

2.5.1 CONTINUUM VS. LINE METHODS   The most obvious method of finding the temperature from the relative energy distribution (color temperature) immediately runs afoul of the fundamental limitation arising from the linear temperature dependence of the Rayleigh-Jeans function in the part of the spectrum at present observable for hot stars ($\lambda > 1100$ Å), quite apart from dereddening problems. The direct determination of $T_{\rm eff}$ by measuring the total flux and the angular diameter requires knowledge of the bolometric correction and is limited to stars that are bright enough for interferometry. This method has been shown by Abbott & Hummer (5) to be extremely unreliable for stars with $T_{\rm eff} > \sim 30,000$ K, for two reasons: (a) the integrated flux redward of 1100 Å is only weakly dependent on temperature, i.e. the Rayleigh-Jeans phenomenon; and (b) the flux depends strongly on the gravity in stars approaching the Eddington limit, a point stressed earlier by Kudritzki et al (83).

As angular diameters are known for only a small number of stars, Blackwell & Shallis (17) estimated this quantity by comparing the flux observed at Earth with the model flux at one or more wavelengths in order to find $T_{\rm eff}$ for cool stars. Abbott & Hummer showed that with hot stars this method is even less reliable than that described above, for it involves ratios of model fluxes that are largely independent of $T_{\rm eff}$. Tobin (145) reached the same conclusion independently. Moreover, not one of these methods determines the gravity or helium abundance. As this conclusion was strongly challenged by Underhill (146), Hummer et al (70) showed that the value of $T_{\rm eff}$ derived from the continuum of ζ Puppis by Underhill led to a line spectrum very different from that observed, and that the observed continuum could be fit to within observational errors by models with a large range of temperatures.

2.5.1.1 *Measured flux distributions* It is interesting to inquire if precise continuous flux distributions observed down to at least 1215 Å, or even to the Lyman limit, can be compared with accurate model fluxes in order to determine $T_{\rm eff}$ and log g. A partial answer to this question comes from the work of Massa & Savage (107), who deredden *IUE* spectra of 26 lightly and moderately reddened O stars in open clusters using UV extinction curves derived from main sequence B stars in the same cluster and shown to be uniform across the cluster. They find that the observed continua are too cool shortward of ~ 1800 Å compared with models by Mihalas (112), where $T_{\rm eff}$ and log g were estimated from the spectral type. This conclusion had also been reached earlier by Brune et al (19) from a smaller sample and less precise dereddening and is supported by the calibration of Woods et al (160). Massa & Savage show that the one star (15 Mon) for which they carry out a detailed comparison can be fit with an extended atmosphere from Kunasz et al (97), but as these models have a parameterized radiation force, this finding is only suggestive. The work of Massa &

Savage is extremely valuable, not as a method of finding stellar parameters, but for checking the physics of the models. The UV model fluxes are uncertain for several reasons, including line blanketing and atmospheric extension; both of these are subjects of current investigations.

2.5.2 LINE-FITTING TECHNIQUES Fitting a number of observed line profiles to the same model not only gives reliable values of the stellar parameters and an indication of the internal consistency of the model, but also reduces the problems of nonuniqueness. Fits made with equivalent widths are too insensitive to give reliable parameters, but they can be used for abundance determinations from stars with established structures.

Models are fit to spectra by two methods. Kudritzki (82) constructs curves in the ($T_{\rm eff}$, log g) plane, along which each model profile fits the observed one, and takes the intersection of the two as the locus of the solution and the intersection area as an indication of the error; this can be done for a number of values of the assumed He abundance. Bohannan et al (18) and Voels et al (149) use an interactive computer system, modifying the models as the process proceeds. The first method has the advantage that many stars can be analyzed from the same set of models, whereas the second is useful if no model grid exists, as is necessarily the case if wind blanketing is included. The second method also allows one to use the systematics of the profile differences at every stage as an indication of the changes to be made to the provisional parameters. All of the profiles in both methods are fit to one final model and in principle should lead to the same results. In cases for which the spectrophotometric precision was comparable, the same results are obtained (for example, see the results for HD 46223 in Table 1).

2.6 Derived Atmospheric Parameters

We have collected in Tables 1 and 2 the atmospheric parameters determined by QSHS for galactic and extragalactic O stars. Because of space limitations and the existence of satisfactory data collections (159), we have not done the same for B stars.

2.6.1 GALACTIC O STARS In Table 1 we have summarized data for 38 objects, including spectral types from various papers of Walborn, temperatures and gravities obtained without and with wind blanketing (WB), and the helium abundance Y by number. The errors quoted are those of the original authors and are, of course, purely internal uncertainties. The spectra used by the Boulder group (ref. B in Table 1) all have a signal-to-noise ratio of $S/N = 300$, whereas the spectra of the Munich-Kiel group (62, 76, 91, 138), except for the older photographic work (82, 83, 143), are in the range $S/N = 50$–100, with the lower values occurring for fainter objects.

Some comments are in order. The discrepancies between the early and recent temperatures for stars 1–4 and 8 reflect improvements in observational material and helium line-broadening data. The recent results are more reliable. The importance of wind blanketing (WB) for the early O stars is shown by stars 5, 6, and 8, for which T_{eff} from WB is about 3000 to 5000 K lower, i.e. from 10 to 15%. For later types, the differences are small or nonexistent. As discussed above, the WB temperatures may be somewhat too low, but those without WB are certainly too high. The gravity inferred for Of and supergiant stars has been shown by Pauldrach et al (122) to be slightly too small because the line radiation force in the photosphere is not included in photospheric models; the emission in Hγ by the wind also plays a role. These effects are treated by the "unified models" discussed in Section 4.

The validity of plane-parallel models for the analysis of line spectra of giants and supergiants may be questioned, in view of the extension of the outer atmosphere arising from the low gravities. In an analysis of four O9.5-type stars of types Ia, Ib, II, and V, using spectra with $S/N = 300$, Voels et al (149) obtained excellent fits to six He I lines (2P-3, 4, 5D, singlets and triplets) that were consistent with the He II and H lines for the main sequence object but found that as the gravities became lower only progressively weaker lines could be fit. This implies that the plane-parallel models represent the deeper layers well but not the outer part of the extended atmospheres. That the observed He I lines corresponding to the lower series members were too strong had been recognized for years and discussed under the label of the "dilution effect" because of its origin in the severe overpopulation of the metastable $N = 2$ levels in dilute radiation fields (cf. 44). The shift of helium ionization toward neutrality caused by the lower temperatures in the outer regions could also play a role; Voels et al (149) introduced the term "generalized dilution effect" for this phenomenon. Heasley et al (55) found that even for B0–5 V stars with the lowest gravities, the He I 5876 Å and 6678 Å lines were too strong relative to plane-parallel models. This problem is being addressed in Munich with the "unified model" calculations discussed in Section 4.

For the earliest spectral types the range of gravities between classes I and V is numerically small because the main sequence lies close to the Eddington limit; however, the spectra are very sensitive to the gravity, so using models with incorrect gravities leads to large errors in T_{eff}. For later types, the relative decrease in T_{eff} between classes I and V is substantially larger than in hotter stars. Attempts to calibrate spectral types uniquely with temperature are therefore wrong in principle, gravity being the second *required* parameter.

Table 1 Stellar parameters of galactic O and early-B stars

Star name (assoc.)	Spectral type	T_{eff}	$\log g$	T_{eff} WB[b]	$\log g$ WB[b]	Y	Ref.
1. HD 93250	O3 V((f))	52.5 ± 2.5	3.95 ± 0.15	—	—	0.09	82
2. HDE 303308	O3 V((f))	51.0 ± 1.5	3.90 ± 0.10	—	—	0.09	76
	O3 V((f))	45.5 ± 3.0	3.90 ± 0.15	—	—	0.09	143
3. HD 93128	O3 V((f))	48.0 ± 1.5	3.90 ± 0.01	—	—	0.09	76
	O3 V((f))	48.0 ± 3.0	3.85 ± 0.15	—	—	0.09	143
4. HD 93129A	O3 If*[c]	52.0 ± 1.5	3.90 ± 0.10	—	—	0.09	82
	O3 If*[c]	45.0 ± 2.0	3.60 ± 0.15	—	—	0.09	143
5. HD 164794 9 Sgr	O4 V((f))	$50.5^{+1.0}_{-1.5}$	$3.75^{+0.01}_{-0.05}$	—	—	0.09	76
	O4 V((f))	47.0 ± 1.0	3.90 ± 0.15	43.0 ± 1.0	3.80 ± 0.15	0.10	B[a]
6. HD 46223	O4 V((f))	49.0 ± 1.0	3.90 ± 0.15	46.0 ± 1.0	3.80 ± 0.15	0.10	B
	O4 V((f))	50.0 ± 1.5	3.95 ± 0.10	—	—	0.09	76
7. NGC 6611 #205	O4 V((f))	50.5 ± 1.5	3.95 ± 0.10	—	—	0.09	76
8. HD 66811 ζ Pup	O4 I(n)f	42.0 ± 2.5	3.50 ± 0.15	—	—	0.14 ± 0.03	83
	O4 I(n)f	47.0 ± 1.0	3.60 ± 0.10	42.0 ± 1.0	3.50 ± 0.10	0.20 ± 0.03	B
9. HD 15570 (Cas OB6)	O4 If^{+c}	$48.0^{+4.0}_{-2.0}$	$3.60^{+0.2}_{-0.1}$	—	—	0.12 ± 0.03	62
10. HD 193682 (Cyg OB1)	O5 V	47.0 ± 1.5	3.80 ± 0.10	—	—	0.15 ± 0.03	62
11. HD 46150	O5 V((f))	48.0 ± 1.5	3.90 ± 0.10	—	—	0.09	76
12. HD 15629 (Cas OB6)	O5 V((f))	48.0 ± 1.0	3.80 ± 0.10	—	—	0.09 ± 0.03	62
13. HD 15558 (Cas OB6)	O5 III(f)	4.70 ± 1.0	3.80 ± 0.10	—	—	0.09 ± 0.02	62
14. NGC 6611 #197	O6 V((f))	49.0 ± 1.5	3.95 ± 0.10	—	—	0.09	76
15. HD 210839 λ Cep	O6 I(n)fp	40.0 ± 1.0	3.50 ± 0.10	37.0 ± 1.0	3.40 ± 0.10	0.17 ± 0.03	B
16. BD +60°513	O7 Vn	40.0 ± 1.5	3.70 ± 0.15	—	—	0.09 ± 0.03	62
17. HD 93222	O7 III((f))	38.5 ± 1.5	3.65 ± 0.10	—	—	0.09	76
18. HD 34656 (Aur OB1)	O7 II(f)	38.0 ± 1.0	3.50 ± 0.10	—	—	0.20 ± 0.03	62

19.	HD 192639 (Cyg OB1)	O7 Ib(f)	36.0±1.0	3.30±0.10	—	—	0.18±0.03	62
20.	HD 193514 (Cyg OB1)	O7 Ib(f)	38.0±1.0	3.50±0.10	—	—	0.12±0.03	62
21.	HD 48279	O8 V	37.5±1.5	4.00±0.20	—	—	0.15±0.04	138
22.	HD 46966	O8 V	37.5±1.5	3.80±0.10	—	—	0.09	76
23.	NGC 6611 #401	O8 V	40.0±1.5	4.00±0.10	—	—	0.09	76
24.	HD 14633	ON8 V	35.5±1.5	3.70±0.20	—	—	0.15±0.04	138
25.	NGC 6611 #166	O9 V	37.5±1.5	3.75±0.10	—	—	0.09	76
26.	HD 214680 10 Lac	O9 V	38.0±1.0	4.25±0.20	—	—	0.09±0.02	138
27.	NGC 6611 #367	O9.5 V	35.0±1.5	4.15±0.10	—	—	0.09	76
28.	HD 34078 AE Aur	O9.5 V	35.5±1.0	3.95±0.15	35.5±1.0	3.95±0.15	0.10	B
29.	HD 36486 δ Ori A	O9.5 II	33.0±1.0	3.45±0.15	33.0±1.0	3.45±0.15	0.10	B
30.	HD 89137	O9.5 III(n)p	30.0±1.5	3.25±0.20	—	—	0.26±0.04	138
31.	HD 16429 (Cas OB6)	O9.5 II((n))	31.5±1.0	3.15±0.10	—	—	0.12±0.03	62
32.	HD 30614 α Cam	O9.5 Ia	32.0±1.0	3.00±0.10	30.0±1.0	2.90±0.10	0.18±0.03	B
33.	HD 37742 ζ Ori A	O9.7 Ib	32.0±1.0	3.20±0.10	32.0±1.0	3.20±0.10	0.10	B
34.	HD 18409	O9.7 Ib	31.0±1.5	3.10±0.10	—	—	0.15±0.02	62
35.	HD 37128 ε Ori	B0 Ia	26.0±1.0	2.75±0.10	—	—	0.20±0.07	91
36.	HD 149438 τ Sco	B0.2 V	33.0±1.0	4.15±0.20	—	—	0.10±0.02	138
37.	HD 93030 Θ Car	B0.2 Vp?	32.5±1.5	4.15±0.20	—	—	0.17±0.04	138
38.	HD 38771 κ Ori	B0.5 Ia	25.0±1.0	2 70±0.10	—	—	0.20±0.07	91

[a] B = ((148), which includes values from (18), (149), and unpublished work of S. A. Voels, B. Bohannan, D. G. Hummer, and D. C. Abbott.
[b] Value obtained by assuming wind blanketing.
[c] The superscripts * and + indicate spectral subtypes in Walborn's notation.

Table 2 Stellar parameters of O and early-B stars in the Magellanic Clouds

Galaxy[a]	Star	Spectral type	T_{eff}	log g	Y	Ref.
LMC	1. Sk $-66°172$	O3–4	50.0 ± 3.0	4.20 ± 0.2	0.09 ± 0.04	42
	2. Sk $-70°69$	O3–4	46.5 ± 3.0	4.10 ± 0.2	0.09 ± 0.04	42
	3. LH $-43°81$	O7	45.0 ± 3.0	4.05 ± 0.2	0.09 ± 0.04	42
	4. Sk $-65°21$	B0 Ia	27.0 ± 1.0	2.70 ± 0.1	0.35 ± 0.07	91
	5. Sk $-68°41$	B0.5 Ia	25.0 ± 1.0	2.60 ± 0.1	0.23 ± 0.07	91
	6. HDE 268685	B1 Ia	22.0 ± 1.0	2.35 ± 0.1	0.23 ± 0.07	91
	7. Sk $-69°213$	B1 Ia	22.0 ± 1.0	2.40 ± 0.1	0.15 ± 0.07	91
	8. HDE 269504	B2 Ia	20.0 ± 1.0	2.25 ± 0.1	0.23 ± 0.07	91
SMC	9. NGC 346#3	O3 III(f)	$55.0 \pm^{5.0}_{2.5}$	3.9 ± 0.1	0.09 ± 0.02	90
	10. AzV 388	O4 V	47.0 ± 3.0	4.0 ± 0.2	0.09 ± 0.04	42
	11. NGC 346#6	O4–5 V	40.0 ± 2.0	3.7 ± 0.15	0.09 ± 0.02	90
	12. NGC 346#1	O4 III(f)	43.5 ± 1.5	3.6 ± 0.1	0.09 ± 0.02	90
	13. NGC 346#4	O5–6 V	42.0 ± 2.0	3.8 ± 0.1	0.09 ± 0.02	90
	14. AzV 243	O5–6 V	45.0 ± 3.0	4.0 ± 2.0	0.09 ± 0.04	42
	15. AzV 238	O8.5	35.0 ± 3.0	3.75 ± 0.2	0.09 ± 0.04	42
	16. Sk 159	B0.5 Ia	25.0 ± 1.0	2.55 ± 1.0	0.35 ± 0.07	91
	17. Sk 119	B2 Ia	20.0 ± 1.0	2.30 ± 0.2	0.15 ± 0.07	91

[a] LMC = Large Magellanic Cloud, SMC = Small Magellanic Cloud.
[b] Sk = Sanduleak, LH = Lucke-Hodge, AzV = Azzopardi & Vigneau.

2.6.2 MAGELLANIC CLOUD (MC) O STARS Results for 17 objects are collected in Table 2, illustrating the use of QSHS for stars down to 13th magnitude. The values given by Gehren et al (42) are only preliminary. As the metallicities are much smaller than for galactic stars, wind blanketing is negligible, as has been confirmed by Kudritzki et al (90). Although the helium abundances are normal for the O stars, they are as much as 3 times larger for B-type supergiants.

2.6.3 CENTRAL STARS OF PLANETARY NEBULAE One of the big successes of QSHS is the determination of the masses and luminosities of the central stars of planetary nebulae (CSPN). For decades, the physical parameters of these important objects near the final stage of stellar evolution were heavily debated. The major reason was that their stellar spectra were not used directly for spectral analysis; instead, the parameters of the CSPN were inferred indirectly from the recombination lines of the surrounding planetary nebulae (PN), together with recombination theory. This situation has now changed drastically. Both photospheric theory and observational techniques have now reached a standard that allows the photospheric spectra of rather faint ($11^m \leq m_V \leq 14^m$) CSPN to be analyzed quantitatively with high precision. Roberto Méndez and the Munich group plus collaborators in Kiel, Pasadena (Caltech), and Tenerife have used spectra with a resolution of 0.2 Å and a signal-to-noise ratio of $S/N = 50$–100 for this purpose (59, 92, 106, 108–110), and they have determined effective temperatures (accurate to $\pm 10\%$), gravities (± 0.2 dex), and helium abundances (± 0.1 dex) for a large sample (≈ 40) of CSPN. These results enabled them to locate the objects in the ($\log g$, $\log T_{\text{eff}}$) diagram and to compare them with post–asymptotic giant branch (AGB) evolutionary tracks that were transformed into this diagram (see Figure 1), which allows a direct determination of stellar masses with high precision ($\pm 10\%$ or better). Using these masses with the measured gravities, they obtained the radii and distances, the latter with 25% accuracy.

These results could be used to discuss critically the alternative methods of distance and luminosity determination by means of the nebular spectra (108), and they have made it possible to investigate in detail the connection between the nebular expansion time scale and the evolutionary time scale

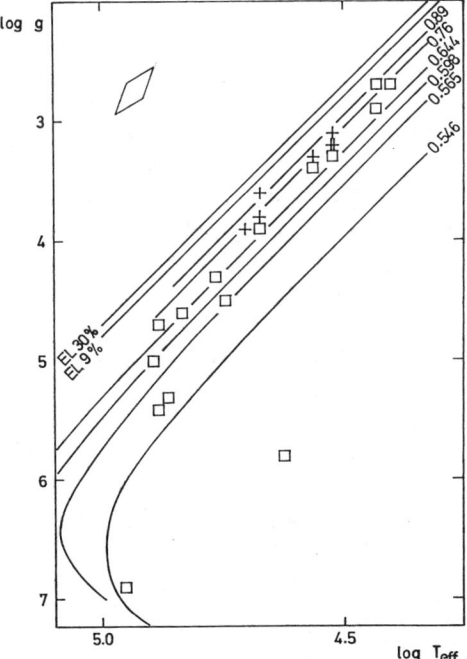

Figure 1 The ($\log g$, $\log T_{\text{eff}}$) diagram for central stars of planetary nebulae compared with the evolutionary tracks of post-AGB stars (labeled by their mass, in solar units). The Eddington limits (EL) for normal (9%) and enhanced (30%) helium abundance and a typical error box are also shown (from 108).

(106). Thus, based on these results, the future use of PN as extragalactic standard candles appears to be very promising.

2.6.4 HOT SUBDWARFS IN GLOBULAR CLUSTERS The reliability of the spectroscopic mass and luminosity determinations using the (log g, log T_{eff}) diagram can be tested by means of faint blue stars in globular clusters, where the distances are known from main sequence fitting. A very convincing example is the hot subdwarf ROB 162 ($m_V = 13.3$) in NGC 6397, analyzed by Heber & Kudritzki (56), where the spectroscopically derived parameters agree very well with the values obtained from the known distance (see Table 3, and refs. 92, 108).

Similar results were obtained for the hot subdwarfs and horizontal branch stars in NGC 6752 by Heber et al (57), Crocker et al (26), and Glaspey et al (46), which again demonstrate the potential of the spectroscopic method. Very recently, M3, M15, M92 and NGC 288 were also studied with similar techniques by Crocker et al (27).

2.6.5 SUBLUMINOUS OB STARS IN THE GALACTIC FIELD Quantitative spectroscopy of subluminous OB stars in the galactic field based on detailed statistical equilibrium models was first introduced by Kudritzki & Simon (81). Since then, a large number of stars have been studied, mainly by the Kiel group in a worldwide collaboration with other groups. The results are summarized in several review papers published in the proceedings of the Second Conference on Faint Blue Stars (58, 74, 88). Very recently, additional progress has been made in the analyses of extremely hot, extremely helium-rich subdwarfs (75, 157), giving more insight into the nature of these still enigmatic objects.

2.7 Chemical Abundances

Extensive programs of QSHS are under way for both galactic and MC objects, but only the initial results have appeared. Schönberner et al (138) made statistical equilibrium analyses of four OBN stars (Table 1, Nos. 21, 24, 30, 37) differentially with two O stars. These objects have helium

Table 3 Mass, radius, and luminosity of the hot subdwarf ROB 162 in the globular cluster NGC 6397

Property	Spectroscopic method	Cluster distance method
$R/R_\odot$	$0.70 \pm ^{0.18}_{0.14}$	0.66 ± 0.04
$M/M_\odot$	$0.56 \pm ^{0.04}_{0.02}$	$0.50 \pm ^{0.3}_{0.2}$
$\log L/L_\odot$	3.5 ± 0.2	3.4 ± 0.1

abundances in the range $0.15 \leq Y \leq 0.26$. The results show that the total abundance of CNO is about normal, but that N is enriched, C is depleted, and O is about normal. This result indicates an incomplete CNO cycle, with only the CN equilibrium well established. The authors suggest that homogeneous evolution with very fast evolution was terminated by rotational spindown caused by angular momentum loss through the stellar wind; subsequently, inhomogeneous evolution sets in with an increased helium abundance.

Quite complete analyses of two B supergiants in the LMC (Table 2, Nos. 4, 5) have been made by Kudritzki et al (87), who find substantial enhancements of helium and nitrogen, indicating that material from the CN bi-cycle of CNO burning was brought to the surface by strong mass loss. The metallicity lies well below the solar value, in accord with the generally assumed value for the LMC.

3. OPTICALLY THIN WINDS

3.1 Introduction

Rapidly expanding stellar winds around hot stars are ubiquitous, a conclusion that is now well established by observations ranging from the radio domain to the soft X-ray part of the observed spectra. Numerous conference proceedings and review papers (cf. 21a, 86a) have been published that discuss the results of wind diagnostics with regard to velocities, mass loss rates, wind driving mechanisms, etc, and that demonstrate the overall importance of winds for the evolution of hot stars, the energy and momentum balance of the interstellar medium, and the evolution of galaxies. We do not repeat these discussions here. Instead, we concentrate on reviewing the recent progress in using the spectral characteristics of stellar winds as tools for quantitative spectroscopy. The various improvements over the past few years in calculating self-consistent models of hot star winds have led to a situation in which—at least for the case of optically thin winds—the variations of strength and shape of specific ionic lines formed in the wind can be used to determine stellar parameters with a precision comparable with, or even better than, that obtained from the analysis of photospheric lines. Excitingly, the information that can be obtained from the wind lines—stellar luminosity and mass—is complementary to, but also partially redundant with, the information available from the photospheric lines—effective temperature and gravity. Thus, the radius (and therefore the stellar distance) can in principle for the first time be obtained from the stellar wind lines by pure quantitative spectroscopy, whereas the mass is determined independently from both photospheric and wind lines, which provides an additional check of the consistency of both procedures.

The basis for this approach is the concept of radiation-driven winds. It has been well known since the early 1960s (1, 3, 22, 102, 144) that the absorption by thousands of metal lines of intense continuous radiation field of hot stars leads to an overwhelming outward force that is sufficient to initiate and to maintain supersonic stellar winds. Thus, any theoretical attempt to model the observed winds must include this physical mechanism. Although it cannot be excluded that additional mechanisms might also be important, the simplest strategy of considering only radiative driving has proven to be surprisingly successful in explaining both the major spectroscopic wind features and the sources of variability in the wind flow.

In this section we discuss the case of optically thin winds. For the dynamics of radiation-driven winds, this is an important simplification. Thus, the continuous radiation providing most of the momentum (in practice, the Lyman and/or Balmer continuum), after being formed in the quasi-hydrostatic photosphere, encounters no significant continuous opacity in the atmospheric layers where the *line absorption* leads to a transfer of photon momentum to the wind flow. Such a situation appears in the winds of O stars, early-B main sequence stars and not too extreme supergiants, CSPN less massive than 0.75 $M_\odot$, and O subdwarfs.

Optically thick winds, where the continuum photons are emitted by layers that are already accelerated to velocities much higher than the sound speed, are treated in Section 5. Typical examples for this case are Wolf-Rayet stars.

3.2 *Stationary Models for Optically Thin Winds*

The basic framework for computing stationary line-driven winds in the case of optically thin continua was developed in a pioneering paper by Castor, Abbott & Klein (22; hereinafter CAK). The essential concept is that of representing the contribution of the thousands of spectral lines to the radiative acceleration in the hydrodynamic calculations by a power-law distribution function $N(l \geq l_0)$ of line strengths l: $N(l \geq l_0) \sim k l_0^{\alpha - 1}$.

The steepness parameter α and the constant k—proportional to the effective number of lines—are not free but are fitted to the distribution of line strengths calculated for all the individual lines that are taken into account (cf. 2, 22, 89, 123, 130). This power-law distribution function is then used to approximate the radiative line acceleration, $g_{\text{rad}}^{\text{line}}$, which depends on the luminosity L, the density ρ, and the velocity gradient dv/dr:

$$g_{\text{grad}}^{\text{line}} \sim k \frac{L}{r^2} \left(\rho^{-1} \frac{dv}{dr} \right)^\alpha. \qquad 1.$$

Several comments on this procedure are necessary. First, the parameters k and α of the distribution function depend on the selection of the line sample and on the calculation of occupation numbers in the wind plasma. Originally, CAK used a sample of C III lines only and adopted LTE. Abbott (2) finally introduced a realistic sample of 250,000 lines essentially complete for all ions from H to Zn in the ionization stages I to VI. This sample remains the basis of all recent work on radiation-driven wind dynamics (see below), at least for objects with $T_{\mathrm{eff}} \leq 50{,}000$ K; for hotter stars (e.g. CSPN, O subdwarfs), higher ionization stages will have to be included. A simplified statistical equilibrium for the ionization was adopted by Abbott (2) that considered only ionization by geometrically diluted radiation from the ground state and direct recombination back to it. All other radiative transitions as well as electron collisions were neglected. These approximations can lead to significant errors. An improved treatment was made by Pauldrach (123), who performed for the first time full multilevel statistical equilibrium calculations for all relevant ions—133 in total—driving the wind. In deeper layers, electron collisions exciting the lower levels and ionization from excited levels cause an important shift to higher ionization stages. Moreover, the ionizing radiation field for bound-free transitions shortward of the He II edge at 228 Å has to be treated by the full solution of the transfer equation in spherical symmetry, since this part of the spectrum remains optically thick in the wind, even in the case of low-density winds that is discussed here. In this way, an additional shift to higher ionization states is obtained.

The line acceleration is obtained from the line-strength distribution function with the help of two approximations: (a) The wind is optically thin for continuous radiation, and (b) the Sobolev approximation for the line transfer is valid. Since most of the continuum photon momentum is absorbed by lines longward of 228 Å, the first approximation is justified for O stars and not too extreme B supergiants. The validity of the Sobolev approximation for the calculation of the radiative line acceleration and wind dynamics has been investigated by Pauldrach, Puls & Kudritzki (122; hereinafter PPK) and Puls & Hummer (131) and also found to be sufficiently accurate.

Two further effects are important for the line acceleration: the radial change of ionization in the wind, and the "finite cone angle effect." As shown by Abbott (2) and Pauldrach (124), the first effect can be taken into account by multiplying $g_{\mathrm{rad}}^{\mathrm{line}}$ with the factor $(n_{\mathrm{E}}/W)^{\delta}$, where n_{E} is the electron density (in 10^{11} cm^{-3}), W is the dilution factor, and δ is another fitting parameter obtained from a comparison with the contributions of all individual lines. The second effect arises from the fact that a volume element in the wind is accelerated not simply by radially streaming photospheric photons, but rather by a radiation field within a finite cone angle formed by the photospheric disk. This leads to the "finite cone angle correction factor"

$$\mathrm{CF} = \frac{1}{\alpha+1}\frac{x}{1-h}\left[1-\left(1-\frac{1}{x^2}+\frac{h}{x^2}\right)^{\alpha+1}\right],\qquad 2.$$

where $h^{-1} = d \ln v/d \ln x$, and x is the radial coordinate in units of the photospheric radius. As this correction factor complicates the dynamic equations significantly, it was neglected by CAK and Abbott (2). However, as shown independently by PPK and by Friend & Abbott (38), this factor is crucial for the wind dynamics and leads to much better agreement with the observed velocity fields. A detailed discussion of the finite cone angle effect, together with analytical solutions for wind models, is given by Kudritzki et al (93).

One of the crucial problems in the theoretical description of radiation-driven winds is caused by line overlaps. The UV and EUV contain thousands of important lines, which overlap because of the velocity-induced Doppler shifts. Thus, a photon emitted by one line can be absorbed by another redshifted line, and so on. This affects the wind dynamics in a complicated way. One would at first expect the repeated momentum transfer resulting from the multiple absorption in different line transitions to lead to stronger winds. However, multiple absorption and line overlap also mean that the photons from the photosphere are partially blocked before reaching the outer wind layers, since part of the previously absorbed photons are reemitted backward to the photosphere. Thus, although photons that reach the outer wind layers can transfer momentum several times, the number of these photons becomes smaller because of absorption in the intervening low-velocity regions. A priori, it is not clear which of the two competing processes (multiple momentum transfer or line blocking) will dominate. The situation is further complicated by the fact that owing to the multiple absorption and reemission, a diffuse radiation field is created that not only contributes to the line acceleration but also changes the radiative transition probabilities in the statistical equilibrium conditions and therefore the occupation numbers.

The first self-consistent treatment of this complex problem using the full sample of 250,000 lines and the full system of multilevel rate equations of 133 ions was presented by Puls (130), who also discusses the earlier studies (4, 37, 120). For the case of the O4f star ζ Puppis, he found that both mass loss rate and terminal velocity are mildly reduced relative to the standard calculations (123), which neglect line blocking and multiple momentum transfer. Although this means that the effects of line overlaps are much smaller than previously thought, it remains important to consider these effects in the detailed quantitative analysis of individual stars (see below).

To summarize the situation of stationary models for optically thin line-driven winds: The state of the art is given by the combined code of Pauldrach (123) and Puls (130), which solves the hydrodynamic equation of radiation-driven winds simultaneously with the full multilevel statistical equilibrium equations for all relevant elements and ions. In total, the calculations comprise 133 ionization stages of 26 elements (H to Zn), with altogether 4000 levels and 10,000 bound-bound transitions. Electron collisions and the correct radiative transfer for the ionizing radiation field are included. The effects of line overlaps (i.e. multiple momentum transfer and blocking) are also taken into account. The occupation numbers obtained in this way are used to calculate the contributions of more than 100,000 lines to the line acceleration using the updated and improved line list of Abbott (2).

Figure 2 demonstrates what can be achieved within this theoretical framework. Note that this almost perfect agreement between observations and theory is not an empirical fit, where mass loss rate and velocity field have been adjusted, but is the result of a consistent calculation in which only stellar effective temperature, gravity, and radius are used as free parameters. The various applications of this work for the purposes of quantitative spectroscopy are discussed in the next section.

Finally, we must mention the deficiencies of the stationary optically thin models in their present stage. Since there are several problems, we start

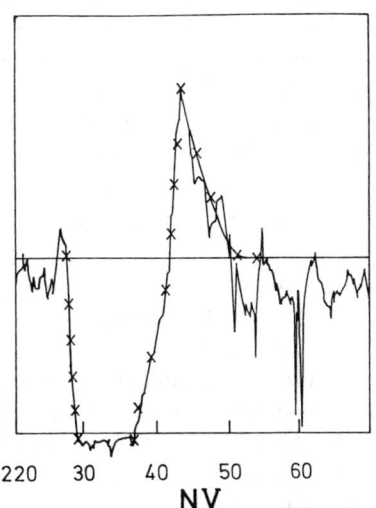

Figure 2 The observed N V resonance line of the O4f star ζ Puppis (*IUE* high resolution) compared with the theoretical profile (*crosses*) calculated for a consistent line-driven wind model as described in the text. The residual intensity is zero below the abscissa because of overestimated background subtraction in this part of the spectrum (from 130).

with those for which the improvements appear to be easiest to implement. The first is the inclusion in the wind dynamics of the contribution to radiative acceleration from photospheric radiation absorbed and scattered by continuum processes. At present, only Thomson scattering is taken into account, and bound-free absorption and free-free absorption are neglected in the equation of motion. Since, for the statistical equilibrium equations, the transfer of the continuum radiation is correctly included, this modification appears to be straightforward. For optically thin winds we do not expect strong effects but rather modifications that are possibly important for quantitative work. The same is true for the absorption in the collisionally and radiatively broadened wings of strong lines, which has been neglected so far. An extension of the work by Puls & Hummer (131) would allow this additional accelerating mechanism to be included. A necessary improvement would be the general inclusion of dielectronic recombination into the statistical equilibrium equations. This process is taken into account in the present codes only for all neutral and singly ionized species and for N III and C III. Again we expect only small, but perhaps quantitatively important, modifications for the other ions. Advection terms due to the wind flow are also neglected in statistical equilibrium. As rediscussed recently (89), these terms might be important for trace ions with very low recombination rates or in situations with strong ionization or excitation gradients. Including these terms would also be straightforward.

More complex is the question of the photospheric input radiation field in bound-free continua at wavelengths $\lambda \geq 228$ Å, which strongly influences the statistical equilibrium. At the moment two options exist: LTE line-blanketed or statistical equilibrium unblanketed photospheric model fluxes. While the use of statistical equilibrium models is clearly indicated (see Section 2), their neglect of photospheric line blanketing is crucial, since this omission means that the ionizing flux [or "radiation temperature" $T_{\rm rad}(\lambda)$] is overestimated for ions like N III, C III, and Si IV that are important for the quantitative wind diagnostics. Work is now under way to extend the line-overlap calculation by Puls (130) to the deeper photospheric layers and to obtain in this way a more reasonable continuous radiation field.

The most important, but also most difficult, point is the energy balance [or temperature stratification $T(r)$] in the wind. The dynamical models of Pauldrach (123) and Puls (130) adopt the ad hoc relation $T(r) \approx T_{\rm eff}$ or a stratification obtained from simplified radiative equilibrium models neglecting metal-line opacity. Although it was an enormous success of these models that the observed lines of N V, O VI and C IV could be reproduced without requiring temperatures of 10^5 or 10^6 K, it is still unsatisfactory that the local wind temperature is not calculated in a strictly self-consistent way. A first reasonable procedure would be to adopt full radiative equilibrium (including all the lines considered in the treatment of the radiative acceleration in the equation of motion), including an energy sink term for the transfer of photon energy into kinetic energy of the wind. Since it is by far more difficult technically to treat the energy equation for all the lines than the equation of motion, the models are still far from this stage of completeness. The importance of metal-line opacity for radiative equilibrium, at least in the outer wind layers, has been demonstrated by Drew (32), who, however, used a very approximate statistical equilibrium treatment in which the collisional excitation of excited levels is neglected and the transfer of continuous radiation is treated by a method subject to large errors. Moreover, the transfer of photon energy into kinetic wind energy via line absorption was also neglected. Despite these simplifications, Drew's conclusion that strict radiative equilibrium in the outer layers will yield a significant temperature drop below the effective temperature is convincing. The question, however, is whether radiative equilibrium is really fulfilled in these layers. The observation of soft X rays emerging from the outer wind layers plus our knowledge of the instability of line-driven winds (see Section 3.4) suggest that additional energy dissipation might occur.

A careful discussion of this problem has been given by Pauldrach (127), who argues that the ad hoc assumption of $T(r) \approx T_{\text{eff}}$ is in good agreement with the observations, and that a detailed, self-consistent calculation of the energy balance including a correct evaluation of the line opacity and emissivity from the equations of statistical equilibrium is the only way to determine empirically the amount of energy dissipated in the outer layers. Work in this direction is under way in Munich.

3.3 Applications of Stationary Wind Models in Quantitative Spectroscopy

In the past, stationary wind models were used mainly to demonstrate that the theory of line-driven winds is able to reproduce the observed wind properties. As a result of the progress achieved during recent years (and described in the previous section), a further step has become possible: The theory can now be applied to determine stellar masses, radii, and luminosities using the stellar wind properties determined by spectroscopic diagnostics. In this section, we review the progress in this new field.

3.3.1 THE $(v_\infty, \log T_{\text{eff}})$ DIAGRAM The classical approach in discussing stellar evolution is to place stars in the HR diagram and to read off stellar masses from the evolutionary tracks. Since this procedure requires a priori knowledge of stellar distances, it is affected by unnecessary uncertainties in many cases. An alternative approach is to use the (log g, log T_{eff}) diagram, since gravity and effective temperature are directly obtained by photospheric spectroscopy independent of any assumption concerning the distance (see Section 2). Interestingly, stellar winds provide an additional possibility: the (v_∞, log T_{eff}) diagram, where the terminal velocity v_∞ of the stellar wind is used as the crucial observational quantity.

The theory of line-driven winds predicts a relation between v_∞ and v_{esc} (the escape velocity from the photospheric surface) of the form $v_\infty = v_{\text{esc}} f(v_s/v_{\text{esc}}, \alpha, \delta)$, where v_s is the sound velocity at the critical point. [The critical point is a singular point of the gas-dynamical equations, slightly larger than r_c, introduced in (22); for a discussion of $f(v_s/v_{\text{esc}}, \alpha, \delta)$, see (93).] This means that v_∞ varies primarily with $v_{\text{esc}} \sim (M/R)^{1/2}$ during stellar evolution, a result that was shown by Kudritzki et al (86) (see Figure 3). Consequently, the (v_∞, log T_{eff}) diagram can be used to determine stellar parameters by comparison with evolutionary tracks in a way similar to the more conventional (log g, T_{eff}) diagram or the HR

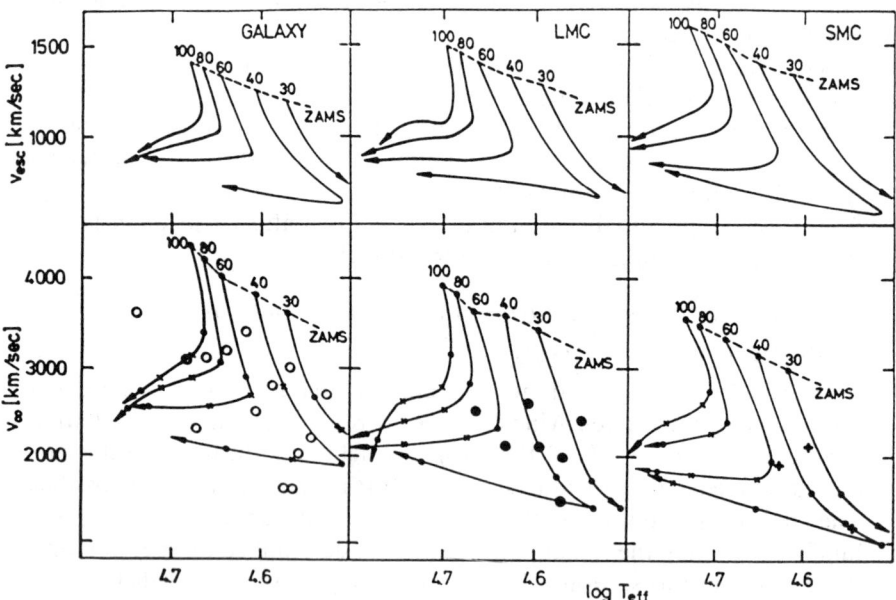

Figure 3 The "alternative" HR diagram of massive star evolution. (*Upper frames*) Surface escape velocity v_{esc} versus log T_{eff} for different evolutionary tracks (labeled by mass, in solar units). (*Lower frames*) Terminal velocity v_∞ versus log T_{eff} for wind calculations along the same tracks. The positions of O stars observed in the Galaxy, the LMC, and the SMC are also shown. For further discussion, see the text (from 86).

diagram. Since v_∞ can in principle be measured with high precision ($\pm 10\%$) from the shapes of the blue edges of the UV P Cygni profiles (47, 60, 67, 129), this independent method has the potential to become an interesting alternative for massive stars.

The (v_∞, log T_{eff}) method has already been successfully applied (125) to CSPN. Since these objects evolve with almost constant luminosity from the AGB toward the white dwarf region, their v_{esc} and consequently v_∞ increase continuously with increasing T_{eff}. Because of the core mass-luminosity relation of post-AGB stars, this allows a direct mass determination, as demonstrated by Figure 4. The masses derived in this way agree well with the masses found for white dwarfs. A direct comparison with masses of CSPN determined from the (log g, log T_{eff}) diagram indicates tentatively that "wind masses" appear to be systematically lower by 0.1 $M_\odot$. For a discussion of this small discrepancy, we refer the reader to (125).

3.3.2 MORPHOLOGY OF STELLAR WIND SPECTRA AS A FUNCTION OF TEMPERATURE AND LUMINOSITY From the work of Walborn and collaborators (150–154) it became evident that UV stellar wind spectra show a systematic dependence on luminosity, temperature, and abundance. In particular, the Si IV resonance line shows a strong luminosity dependence. Using detailed stationary wind models calculated along evolutionary tracks, Pauldrach et al (126) were able to reproduce this effect. Figure 5 gives an example from their paper. The quantitative use of Si IV for direct determinations

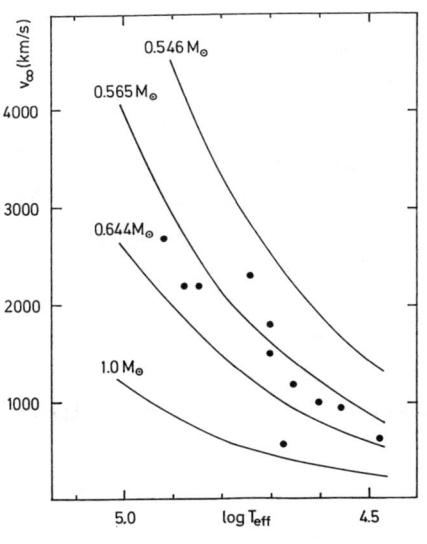

Figure 4 Terminal velocity v_∞ versus log T_{eff} calculated along the evolutionary tracks of post-AGB stars. Observed values for CSPN are also shown (from 125).

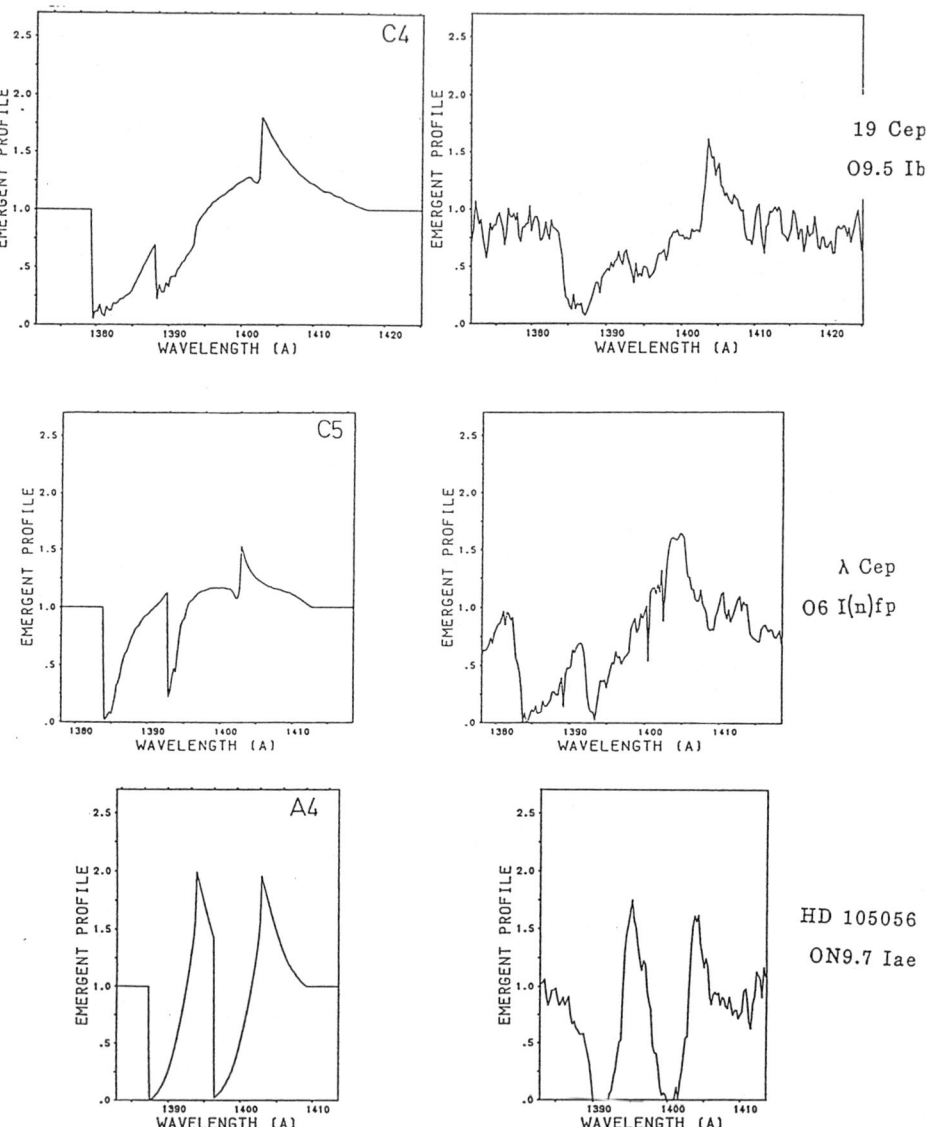

Figure 5 The strong Si IV stellar wind resonance line for Of supergiants: theoretical profiles (*left*) versus observations (*right*) (from 126).

of luminosity, therefore, appears to be feasible. In view of the UV capacity of the Hubble Space Telescope, this is an important tool for future quantitative spectroscopy of massive hot stars.

3.3.3 THE WIND OF P CYGNI The prototype of a mass-losing hot star is the luminous blue variable P Cygni. Its wind is characterized by an extremely high mass loss rate ($\approx 10^{-5} M_\odot$ yr^{-1}) and a low outflow velocity (≈ 200 km s^{-1}). Very recently, Pauldrach & Puls (128) were able to show that radiation-driven wind theory is able to reproduce the observed wind properties of P Cygni in great detail. The reason for the high-density, low-velocity wind is the proximity of the star to the Eddington limit. The theory can be used to determine the stellar parameters with high precision. In addition, it can be shown that the radiation-driven wind of P Cygni is bistable, which might explain the observed ejection of shells.

3.3.4 DIRECT DETERMINATION OF MASSES AND RADII FROM TERMINAL VELOCITIES AND MASS LOSS RATES Since the terminal velocity of radiation driven winds depends on the escape velocity v_{esc} [$\sim (M/R)^{1/2}$] and gravity g ($\sim M/R^2$), the values of the latter quantity obtained from photospheric spectroscopy can be used to determine directly mass and radius. This concept was outlined and applied first by Ebbetts & Savage (34) for a subluminous O star using the simple empirical relation $v_\infty = 3v_{esc}$. Kudritzki & Hummer (85) used the rate v_∞/v_{esc} from detailed wind model calculations and the results of quantitative statistical equilibrium analyses to obtain in this way independent distances of the most luminous O stars in the Carina cluster.

The disadvantage of this method is that for objects close to the Eddington limit, the usual uncertainties of ± 0.1 dex in log g lead to large errors. Thus, just for those objects, the most luminous stars, which are observable at the largest distances, the method fails. Kudritzki (95, 96) has therefore suggested that the use of log g be avoided in favor of the wind parameters, v_∞ and $\dot{M}$. Two possibilities exist. First, if the stellar distances (and therefore radii) are known by other methods, masses can be determined directly from v_∞ using the relation $v_\infty = v_{esc} f(v_s/v_{esc}, \alpha, \delta)$ (see Section 3.3.1 and ref. 93). Table 4 summarizes the results and compares them with stellar masses obtained directly from the photospheric log g.

The masses obtained by the two independent methods agree surprisingly well. However, the uncertainties of the v_∞ masses are significantly smaller, since they reflect mainly the uncertainty of the radius, whereas the log g masses are also affected by the uncertainty in the log g determination.

On the other hand, when the distances are not well determined, even for stars close to the Eddington limit, the mass loss rate and terminal velocity can be used together to determine radii (distances) and gravities (masses)

Table 4 Stellar masses from $\log g$ and v_∞ for objects with known distances (and radii)

Star	Log $M/M_\odot$		Ref.
	From v_∞	From $\log g$	
HD 93129A	2.08 ± 0.10	2.07 ± 0.17	95
HD 93250	2.01 ± 0.11	2.06 ± 0.18	95
HD 303308	1.61 ± 0.12	1.75 ± 0.21	95
ζ Puppis	1.68 ± 0.10	1.62 ± 0.20	95
HD 15629	1.40 ± 0.16	1.36 ± 0.26	61
HD 15558	1.70 ± 0.15	1.70 ± 0.26	61
HD 34656	1.42 ± 0.28	1.42 ± 0.40	61
HD 193514	1.46 ± 0.16	1.44 ± 0.26	61
HD 192639	1.37 ± 0.15	1.26 ± 0.26	61

directly from the wind properties by means of radiation-driven wind theory. Figures 6 and 7 are typical examples. Figure 6 yields $R/R_\odot = 17 \pm 5$ and $\log M/M_\odot = 1.64^{+0.18}_{-0.25}$ for ζ Puppis. While the uncertainties appear to be rather large at first glance, it is important to note that they reflect only ±0.5 mag in the distance modulus. An extension to much fainter objects using the Hubble Space Telescope in the UV for v_∞ and optical spectroscopy for Hα or He II 4686 Å emission to obtain information on $\dot{M}$ without losing accuracy in the distance modulus appears to

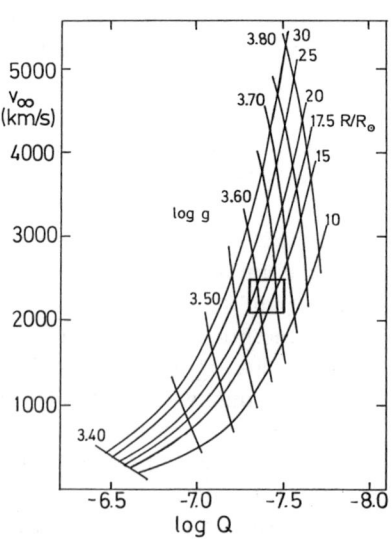

Figure 6 Terminal velocity v_∞ and $\log Q = \log \dot{M} (M_\odot \text{ yr}^{-1}) - 3/2 \log R/R_\odot$ of the O4f star ζ Puppis calculated for wind models with different $\log g$ and $R/R_\odot$. The observed values are indicated by the rectangular box. (Note that $\log Q$ is obtained directly either from radio observations or from Hα or He II 4686 Å emission.) The diagram allows one to read off directly the stellar gravity and radius (from 95, 96).

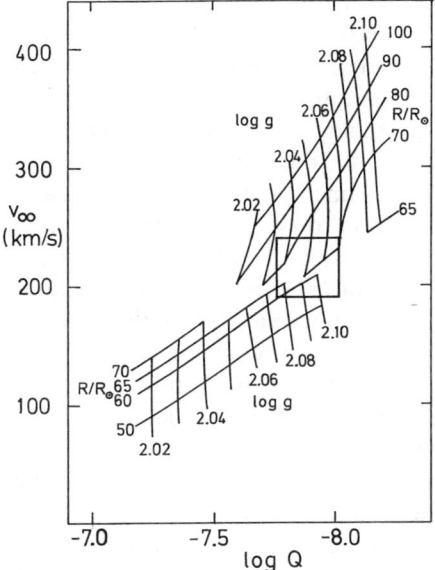

Figure 7 Same as Figure 6 for the case of P Cygni. Note the precision of the log g determination in this case.

be straightforward. Figure 7 leads to $R/R_\odot = 70 \pm 15$ and log $M/M_\odot = 1.30 \pm 0.15$ for P Cygni, or an uncertainty of ± 0.4 mag in the distance modulus. With refinements now being developed in Munich, this method appears to be extremely promising for the determination of distances of luminous blue stars. Systematic investigations of the possibility of using luminous blue stars in this way as extragalactic distance indicators are under way in both Boulder and Munich.

3.4 The Imperfections of Stationary Models: Flow Instabilities, Variabilities, Shocks, and X Rays

Despite the success of the stationary monotonic flow wind models, the observation of nonstationary components in the UV spectral lines formed in the wind (60, 67, 129) and the detection of X rays by the Einstein Observatory indicate that significant additional improvements are needed. We summarize here briefly the important steps in this direction that have been made during recent years.

From the work of Owocki & Rybicki (116–118) it is clear that line-driven winds are unstable against radial velocity perturbations. A recent extension to three-dimensional stability analysis (134), moreover, indicates that horizontal perturbations are damped, and that the conclusions obtained from the one-dimensional radial analysis are realistic. Therefore, the next step would be to take a detailed stationary model as described in

Section 3.2, to superimpose perturbations, and to follow their development by time-dependent one-dimensional hydrodynamic calculations. Because of (a) the complexity of calculating the radiative line acceleration and (b) the stringent limitations on time and spatial steps required to ensure numerical accuracy and stability, such calculations are not yet available. However, an impressive first step was presented recently by Owocki, Castor & Rybicki (119; hereinafter OCR), who assumed pure line absorption and a constant temperature. Dramatic fluctuations in density and velocity were obtained that showed the existence of both forward and reverse shocks. It is at present not clear how seriously the pure absorption approximation affects the results. The assumption of an isothermal flow with the same temperature across a shock certainly favors the tendency toward very high shock amplitudes. We will therefore have to wait until these two approximations have been dropped before it can be concluded how much of the stationary flow model must be modified.

An alternative concept is to avoid the detailed computation of the temporal development of line acceleration instabilities (because it is numerically and physically extremely complicated) by accepting that shocks exist and studying their dynamic development with a simplified treatment of the line acceleration. This idea was introduced by Lucy (103, 104) as a means to interpret qualitatively the X-ray emission of O stars. Krolik & Raymond (78) studied the behavior of individual shocks with regard to their thermodynamics and radiation characteristics. Recently, MacFarlane & Cassinelli (105) presented a very promising new approach. They adopt a phenomenological, temperature-dependent line force that vanishes for high temperatures of the shocked gas because of ionization. In this way, a two-component shock zone with forward and reverse shocks of different temperatures is obtained, together with a dense region between the shocks that could account for the observed narrow components. The disadvantage of this approach relative to that of OCR is that the velocity field and mass loss rate are prespecified and not calculated consistently with a realistic line force. The advantage is that the energy constraint and the effects of increased shock temperature are—at least in principle—taken into account.

4. UNIFIED MODEL ATMOSPHERES

Hydrostatic photospheric models neglect the emission of the surrounding stellar wind envelope. They are therefore unable to produce the wind-induced strong emission features in the optical spectra of hot stars (Hα, He II 4686 Å, IR excess) or the distortion of photospheric absorption profiles (Hβ, He II 5876 Å, He II 5412 Å) that are observed in many cases.

The normal procedure to account for these effects is to put a wind on top of a hydrostatic photosphere and then to study its influence on the spectrum. This method has crucial limitations because it makes an artificial division between photospheres and winds and introduces unnecessary free parameters that strongly influence the calculations at the boundary between the two regions (for a discussion, see 40, 89). The alternative is the concept of "unified model atmospheres" introduced recently by the Munich group (40). This new model atmosphere code takes the mass loss rate, density, and velocity structure from the stationary wind code (see Section 3) and calculates the temperature structure and hydrogen and helium occupation numbers by a detailed statistical equilibrium treatment using radiative equilibrium and correct spherical radiative transfer. These models need only $T_{\rm eff}$, log g, and $R/R_\odot$ at an inner boundary far below the sonic point as the natural free parameters and are in principle self-consistent. They are spherically extended and yield the entire sub- and supersonic atmospheric structure, the stellar energy distribution, and the hydrogen and helium line spectra (including the effects of the outflow velocity field). The weakness of these models in their present stage is that the transfer of photon energy from the continuum into kinetic wind energy due to metal-line absorption is not yet considered in the energy equation. On the other hand, these models are as good as the standard hydrostatic photospheric statistical equilibrium models as far as opacities and radiative equilibrium are concerned, and they are therefore ideal for studying differentially the influence of sphericity and winds on the observed energy distribution and on hydrogen and helium lines.

These effects are quite spectacular. First, the energy distribution is changed dramatically in the far-IR and in the EUV. The unified models reproduce the observed IR excess (see Figure 8) and yield a factor of 1000 more photons in the EUV for wavelengths shorter than the He II edge at 228 Å. This latter effect is important for the ionization of H II regions and planetary nebulae.

Second, the observed emission lines, such as Bα, Pα, Hα, He II 10124 Å, and He II 4686 Å, are also reproduced by the unified models. Even more interesting is that the calculations show a very pronounced luminosity dependence of these lines (see ref. 91 and Figures 9 and 10). Thus, the strength of these lines can be used as a direct luminosity indicator.

Third, the strategic lines normally used for quantitative spectral analysis by means of the photospheric models (Hγ, Hδ, He II 4542 Å, He II 4200 Å) are partially affected by sphericity and winds. This has systematic consequences for the gravity determination of O stars with dense winds, such as Of stars, and results in gravities that are higher by roughly log $g \approx +0.1$ dex.

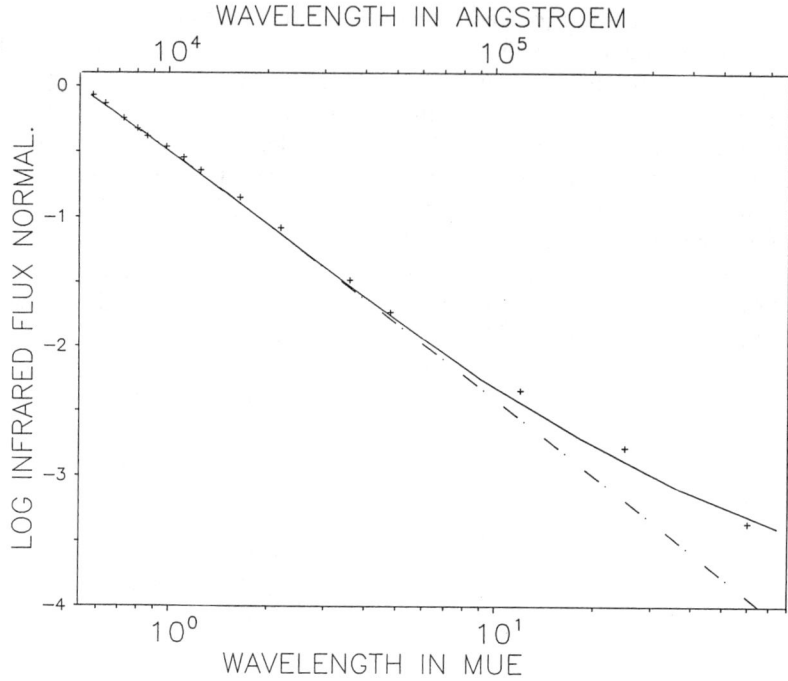

Figure 8 IR energy distribution for ζ Puppis calculated with the unified model atmosphere code (*solid curve*) and with a standard hydrostatic statistical equilibrium code (*dash-dotted*). The crosses represent observations (from 40).

A modification of the unified model technique was introduced by Gabler et al (39) to determine accurate mass loss rates from the He II 4686 Å and Hα emission. In this case, it is assumed that the stellar parameters are known so that they are kept fixed in the analysis, but line acceleration parameters k, α (see Section 3.2) are varied to produce a sequence of models with the same v_∞ but wth different mass loss rates (see Figure 11). A comparison of observed and calculated profiles then yields mass loss rates with a precision of about ±0.2 dex. This approach is superior to the more common procedure in which the wind emission is simply added on top of a "photospheric" Hα profile obtained from hydrostatic calculations (see ref. 39 and references therein, and compare with discussion in ref. 99). Since He II 4686 Å and Hα emission profiles with intermediate resolution ($\approx$ 1–2 Å) can be obtained for objects as faint as 19th magnitude with present-day techniques, this appears to be an ideal way to obtain mass loss rates of hot luminous stars in distant galaxies.

QUANTITATIVE SPECTROSCOPY 335

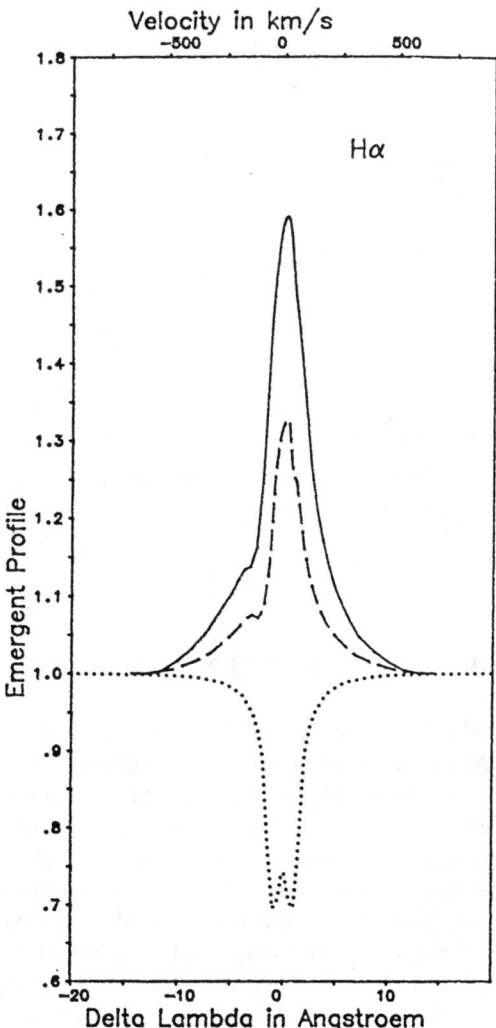

Figure 9 Hα line profiles for a hydrostatic model (*dotted*) with $T_{\rm eff} = 32{,}000$ K, $\log g = 2.9$ and for two unified models with the same temperature and gravity but different luminosity ($\log L/L_\odot = 5.8$ and 6.1) (from 91).

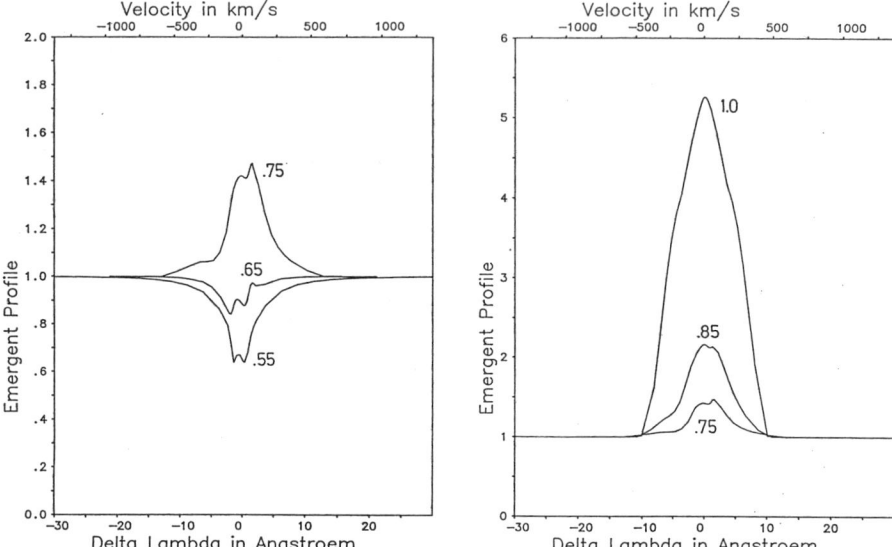

Figure 10 Hα line profiles of unified models calculated for CSPN with T_{eff} = 50,000 K. Different profiles belong to models with different stellar masses (in solar units). Because of the core mass-luminosity relation for post-AGB stars, models with higher masses have higher luminosities (from 40).

5. STARS WITH OPTICALLY THICK WINDS

We are concerned here primarily with Wolf-Rayet stars. The formation of both lines and continua in the winds of such objects and the consequent breakdown of the core-halo picture not only prevent the determination of stellar parameters from a straightforward analysis, even to a low level of accuracy, but also make it difficult to identify the relevant physical mechanisms and lead to enormous technical problems in model calculations. Hence, the quantitative spectroscopy of such objects is still in its infancy. Much of the older published data, obtained by methods used (mostly incorrectly) for O stars, are totally unreliable. Because the H I and He II continua are formed in the winds, their intensities cannot be related to T_{eff} in any known way, and hence Zanstra temperatures are of unknown significance.

Recently, estimates of various stellar parameters have been obtained from empirical models of WR stars, in which the run of temperature and flow speed are represented by ad hoc forms containing as parameters a characteristic temperature, radius, and mass loss rate that are then varied

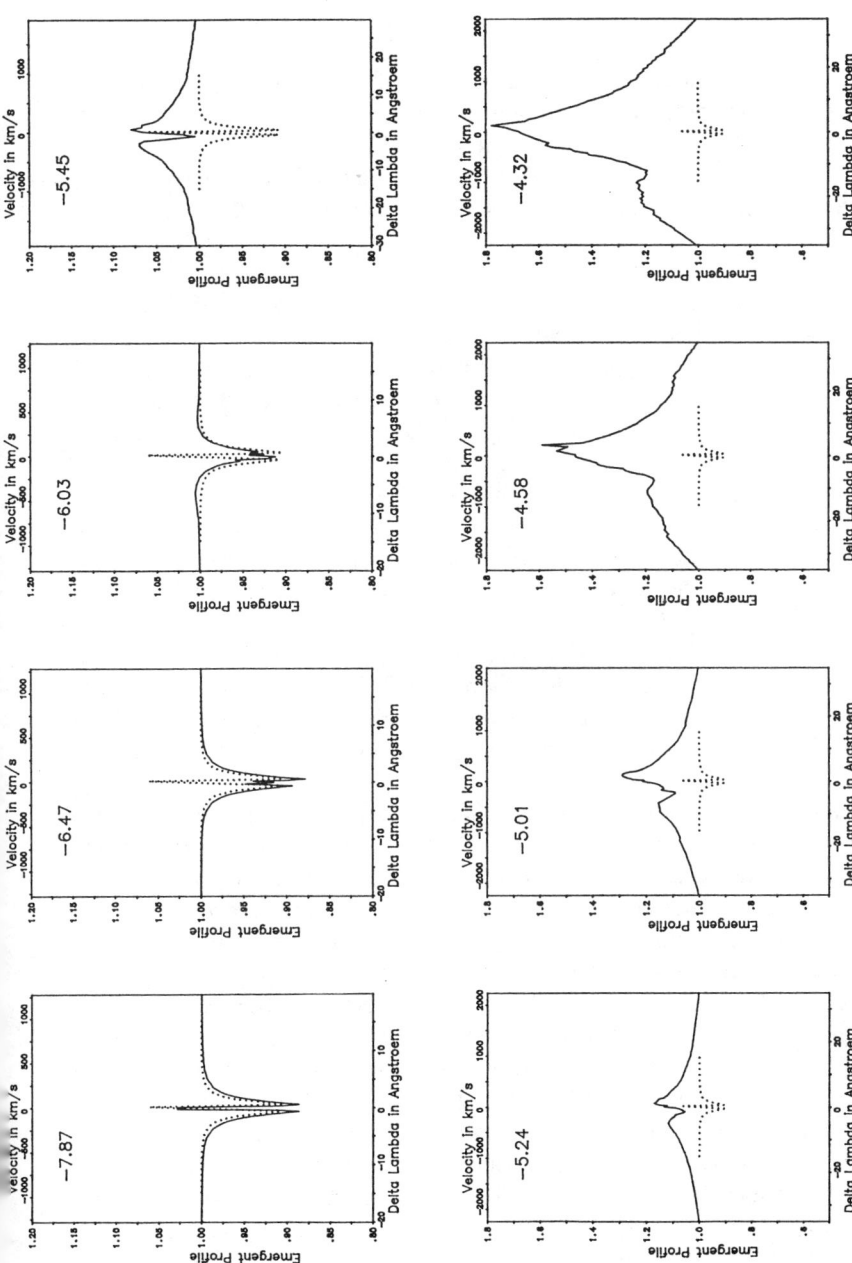

Figure 11 He II 4686-Å profiles calculated for a sequence of models with the same stellar parameters, with the same v_∞, but with different mass loss rates. [The individual profiles are labeled by log $(\dot{M}/M_\odot \text{ yr}^{-1})$.] The corresponding profile of a plane-parallel, hydrostatic model without wind is also shown (*dotted*).

to reproduce observed line and continuous spectra. Nevertheless, many questions remain open, including the fundamental one of the driving mechanism for the wind. At present, the insight given into the structure and physical mechanisms of the stars is probably more valuable than the numerical values of the derived parameters; we stress this below. Excellent recent reviews of WR stars include those of Abbott & Conti (6) and Conti (24). The discussions by Underhill (146a,b) are marred by numerous errors and expound a personal point of view that we do not find convincing.

5.1 Helium Models of WR Stars

Such models have been calculated by D. J. Hillier and by W.-R. Hamann, W. Schmutz and their collaborators. In a series of papers the latter group has produced pure helium models of increasing refinement and made fits to a large number of stars; nearly all of this work was done with the accelerated lambda iteration technique in the form developed by Hamann (49–51). In the early papers (48, 52, 158) they investigated the effect of various parameters on the observable features of the He spectrum. They established the importance of modeling neutral helium lines, which (because He I is present only as a trace ion) are very sensitive to the velocity and temperature distributions. In particular, Hamann, Schmutz, and collaborators showed the correct temperature distribution in the outer layers to be much closer to that given by a gray spherical LTE approximation than by the corresponding plane-parallel case, which points to the validity of radiative equilibrium. Although quite good fits to the He II lines and the continua could be obtained, minor inconsistencies among the He I lines remain.

Hillier (64) solved the radiative transfer and statistical equilibrium problems for He II alone and then calculated the statistical equilibrium for He I (and H I), since these impurity ions have little effect on the ionization of the He II–III balance; he used both the comoving frame procedure and the Sobolev approximation. Hillier assumed radiative equilibrium where the UV continuum was formed, and at larger radii he used a free parameterization of the temperature. The effect of electron scattering on the line profiles was included as described in (63), and it was shown to provide a useful upper limit on $\dot{M}$. Hillier found both He I and He II lines, as well as the continua, and that (at least for stars of roughly type WN5) helium went from mostly twice ionized to singly ionized in the outer region of the wind, in accord with the difference between IR and radio spectral indices, and the line spectra. Consequently, the mass loss rate inferred from radio observations (cf. ref. 161 for methodology) referred to a region in which the helium was singly, and not doubly, ionized, a result that increased the radio mass loss estimates by a factor of 2.7; this conclusion had also been reached by Schmutz & Hamann (135) for all but the hottest WR stars. Hillier made the additional point that although optical and IR fluxes are controlled by $\dot{M}$, the UV fluxes are much less sensitive to this parameter, which is a valuable diagnostic tool.

The WN5 star HD 50896 plays the same role in the study of WR objects as ζ Puppis does for O stars. Hillier (65) fit his models, assuming a wind temperature of 30,000 K, to about 20 lines of He I and He II in the UV, optical, and IR regions of the spectrum and to the observed flux distribution from 1300 Å to 5 GHz. The model equivalent widths for the IR lines are 20–40% too small, and the goodness of the other fits varies from excellent to moderate, but the overall representation is reasonably good, considering that no attempt was made to optimize the fit in view of both the observational uncertainties and the simplified physical picture inherent in the model. Hillier found that the spectrum could be reproduced by the set of parameters $M/\dot{M}_\odot = 5 \times 10^{-5}$, $R_{\text{core}}/R_\odot = 2.5$, $L/L_\odot = 5\text{–}10 \times 10^4$, and $v_\infty = 1600$ km s^{-1}, although this set is not unique in view of the very uncertain distance. The wind velocity is considerably smaller than $v_\infty = 2500$ km s^{-1}, as found from the UV lines, which indicates that the wind has reached only $\sim 70\%$ of its terminal speed at $R = 30 R_{\text{core}}$; Hillier concludes that there must be a significant input of momentum for $R \geq 50 R_{\text{core}}$. It is interesting that Puls & Pauldrach (132) find that the multiline effects in stellar wind theory predict an acceleration in the outer regions of the WR wind, and that such an effect has been inferred from eclipse observations of HD 90657 by Auer & Koenigsberger (10). Hillier also confirmed that $N(\text{H}^+)/N(\text{He}^{++}) < 0.3$.

Hamann et al (53) analyzed HD 50896 using the technique of fit curves with the equivalent widths and peak fluxes of four lines. The solutions for three values of the mass loss rate appear to be of equal quality and produced line profiles that fit the observations quite well. Only on the basis of a highly approximate distance of 2 kpc could they conclude that $\dot{M}/M_\odot = 10^{-4}$ yr^{-1}, $L/L_\odot = 10^{5.3}$, and $R_*/R_\odot = 4.7$, where R_* is a radius with negligible flow speed. The effective temperature at $\tau_{\text{Ross}} = 2/3$ was 32,000 K. The authors indicate that their models are in substantial agreement with those of Hillier.

Subsequently, Schmutz et al (136) observed spectra of 30 WR stars, which were fit to models using only the equivalent widths of He I 5876 Å and He II 5412 Å and the absolute visual magnitude. As the spectrum is determined largely by the wind density, which was also stressed by Hillier (64), the analysis could be made in terms of a temperature parameter T_* and a scaled radius $R_t \sim R_*(v_\infty/\dot{M})^{2/3}$, where T_* is the effective temperature at R_*. Schmutz et al determined the stellar parameters for these 30 stars, including the effective temperatures at $\tau_{\text{Ross}} = 2/3$, which were surprisingly low and did not correlate well with spectral types. Their mass loss rates agreed to within 25% for 11 of these stars for which radio determinations had previously been made. The derived luminosities and temperatures are not in accord with the picture of post-red-giant evolution.

Hamann et al (54) have used their models to synthesize the light curve of the eclipsing binary V444 Cygni (WN5+O6?). The WR parameters were determined at phase 0.24, where both components are at maximum separation. A variety of assumptions were made concerning the O star. The light curve was then synthesized by accounting for the absorption of O star radiation as it passed through the WR atmosphere, which provides a sensitive measure of the WR mass loss rate. Although good fits to the light curve are found, they are not unique. The derived parameters disagree rather strongly with those derived by Cherepashchuk et al (23) from light curves.

5.2 WR Models With Carbon and Oxygen

Hillier (66) has investigated extensively the effects of C and N on the He models for HD 50896 discussed above, which he has modified to find the temperature distribution on the basis of radiative equilibrium by including approximately the cooling by collisional excitation of nitrogen. The main effect is a reduction of the temperature in the outer parts of the wind from around 30,000 K to 10,000–15,000 K, which should lead to an improved fit of the IR spectra. The resulting model, which still contains several drastic simplifications, reproduces the lines of C IV to within a factor of 2, those of N IV to within a factor of 3, and the N V 4609 Å line to within a factor of 4. However, the stellar parameters obtained by pure He models are essentially unchanged. Hillier also reports C/N ratios by number of 0.07, while the N/He ratio is uncertain but could be as large as 4×10^{-3}. Hamann et al (54) have reached much the same conclusions from their He models by including nitrogen. Both groups are actively extending their treatment of metals. In our opinion, the work of Bhatia & Underhill (15, 16), despite the complexity of their atomic models, has no relevance to WR stars because of the extreme crudity of their "one-point" atmosphere and the large number of uncontrolled parameters. Very recently, Lamers et al (99a) have summarized a number of lines of evidence showing Underhill's picture to be untenable.

5.3 WR Wind Models

Using the stellar parameters of V444 Cygni determined by Cherepashchuk et al (23) from an analysis of multicolor light curves, Pauldrach et al (121) used a radiation-driven wind model based on the core-halo picture to obtain the photospheric radius, mass loss rate, and terminal velocity, as well as the run of density and velocity. The derived quantities were in remarkably good agreement with those found from the light curves, despite the failure of the core-halo picture. The resulting effective temperature of 90,000 K and luminosity of roughly 5.5 dex are in disagreement with the models of Schmutz et al (136), and the ratio of L/M is somewhat larger than that given from evolutionary calculations. A correct treatment of radiation driving in WR stars, accounting for the interaction of the direct and diffuse radiation fields with both lines and continua, is badly needed.

5.4 Comments

We believe that the diagnostics of those regions of the WR stars where the observed spectra are formed in the models discussed in Sections 5.1 and 5.2 are essentially correct, but we suggest that perhaps the values of T_{eff} and luminosity may be incorrect. There seem to be two questionable aspects of the present models. First, the temperature and radius parameters in the empirical temperature and velocity formulas cannot be related uniquely to a definite radius in view of their ad hoc nature and their application to parts of the atmosphere that are too deep to be represented by the observed spectra. Thus, an unknown mass of low-velocity gas could be hidden under the parts of the atmosphere to which the models are fit. Second, this gas could form a quasi-static extended atmosphere that would force some, and perhaps a large part, of the energy to escape far into the EUV, as suggested by the wind-blanketing calculations of Hummer (69) and Husfeld (72). The analysis of WR eclipse data, such as that of Cherepashchuk and his collaborators and of Auer, Koenigsberger, and Moffat (10, 79, 115), should be very useful in correctly locating the radiating gas in the system. EUV flux measurements could also give useful insights in this matter.

6. CONCLUSION

Despite spectral details that cannot yet be explained and an incomplete knowledge of the physical mechanisms at work in hot stars, we believe that the results presented in this review demonstrate the reliability of quantitative spectroscopy as a unique source of accurate values of the global parameters—effective temperature, surface gravity, and elemental abundances—of these objects.

ACKNOWLEDGMENTS

We thank all of our colleagues, primarily in Boulder, Kiel, and Munich, for discussions and for preprints and private communications. The preparation of this review was supported in part by a NATO travel grant, the A. von-Humboldt Stiftung, the Max-Planck-Institut für Extraterrestrische Physik, NSF grant AST 88-02937, and NASA grant NAGW-766 through the University of Colorado, and by various grants of the Deutsche Forschungsgemeinschaft (Ku 474/7–13).

Literature Cited

1. Abbott, D. C. 1979. In *Mass Loss and Evolution of O-Type Stars, IAU Symp. No. 83*, ed. P. S. Conti, C. W. H. de Loore, pp. 237–39. Dordrecht: Reidel
2. Abbott, D. C. 1982. *Ap. J.* 259: 282–301
3. Abbott, D. C. 1985. In *Relations Between Chromospheric-Coronal Heating and Mass Loss in Stars. Workshop, Sacramento Peak Obs., Sunspot, N. Mex., 1984*, ed. R. Stalio, J. B. Zirker. 265 pp.
4. Abbott, D. C., Lucy, L. B. 1985. *Ap. J.* 288: 679–93
5. Abbott, D. C., Hummer, D. G. 1985. *Ap. J.* 294: 286–302
6. Abbott, D. C., Conti, P. S. 1987. *Annu. Rev. Astron. Astrophys.* 25: 113–50
7. Anderson, L. S. 1985. *Ap. J.* 298: 848–58
8. Anderson, L. S. 1987. See Ref. 77, pp. 163–90
9. Auer, L. H., Heasley, J. N. 1976. *Ap. J.* 205: 165–71
10. Auer, L. H., Koenigsberger, G. 1990. See Ref. 41, pp. 291–93
10a. Auer, L. H., Mihalas, D. 1969. *Ap. J.* 158: 641–55
11. Becker, S. R., Butler, K. 1988. *Astron. Astrophys.* 201: 232–46
12. Becker, S. R., Butler, K. 1988. *Astron. Astrophys. Suppl.* 74: 211–25
13. Becker, S. R., Butler, K. 1989. *Astron. Astrophys.* 209: 244–54
14. Becker, S. R., Butler, K. 1988. *Astron. Astrophys. Suppl.* 76: 331–38
15. Bhatia, A. K., Underhill, A. B. 1986. *Ap. J. Suppl.* 60: 323–56
16. Bhatia, A. K., Underhill, A. B. 1988. *Ap. J. Suppl.* 67: 187–223
17. Blackwell, D. E., Shallis, M. I. 1977. *MNRAS* 180: 177–91
18. Bohannan, B., Abbott, D. C., Voels, S. A., Hummer, D. G. 1986. *Ap. J.* 308: 728–35
19. Brune, W. H., Mount, G. H., Feldman, P. D. 1979. *Ap. J.* 227: 884–99
20. Butler, K. 1984. PhD thesis. Univ. London, Engl.
21. Butler, K., Giddings, J. 1985. *Newsletter* 9: 7–30. London: Collab. Comput. Proj. No. 7
21a. Cassinelli, J. P. 1979. *Annu. Rev. Astron. Astrophys.* 17: 275–308
21b. Cassinelli, J. P. 1985. In *The Origin of Nonradiative Heating/Momentum in Hot Stars. NASA Conf. Publ. 2358*, ed. A. B. Underhill, A. G. Michalitsianos, pp. 2–23. Greenbelt, Md: NASA
22. Castor, J. I., Abbott, D. C., Klein, R. I. 1976. *Ap. J.* 195: 157–74 (CAK)
23. Cherepashchuk, A. M., Eaton, J. A., Khaliullin, Kh. F. 1984. *Ap. J.* 281: 774–88
24. Conti, P. S. 1988. See Ref. 25, pp. 81–156
25. Conti, P. S., Underhill, A. B., eds. 1988. *O Stars and Wolf-Rayet Stars. NASA SP-497*. Washington, DC: NASA. 428 pp.
26. Crocker, D. A., Rood, R. T., O'Connell, R. W. 1986. *Ap. J. Lett.* 309: L23–26
27. Crocker, D. A., Rood, R. T., O'Connell, R. W. 1988. *Ap. J.* 332: 236–46
28. Danziger, I. J., Matteucci, F., Kjär, K., eds. 1985. *Proc. ESO Workshop on Production and Distribution of C, N, O Elements*. Garching bei München: ESO. 429 pp.
29. Davis Phillip, A. G., Hayes, D. S., Liebert, J. W., eds. 1987. *The Second Conference on Faint Blue Stars. Proc. IAU Colloq. No. 95*. Schenectady, NY: L. Davis. 778 pp.
30. de Loore, C. W. H., Willis, A. J., Laskarides, D., eds. 1986. *Luminous Stars and Associations in Galaxies. Proc. IAU Symp. No. 116*. Dordrecht: Reidel. 533 pp.
31. Dimitrijević, M. S. 1988. *Astron. Astrophys. Suppl.* 76: 53–59
32. Drew, J. 1989. *Ap. J. Suppl.* 71: 267–91
33. Dufton, P. L., Brown, P. J. F., Lennon, D. J., Lynas-Gray, A. E. 1986. *MNRAS* 222: 713–18
34. Ebbetts, D. C., Savage, B. D. 1982. *Ap. J.* 262: 234–43
35. Eber, F. 1987. Diplomarbeit. Univ. Munich, Fed. Rep. Germ. 182 pp.

36. Eber, F., Butler, K. 1988. *Astron. Astrophys.* 202: 153–58
37. Friend, D. B., Castor, J. I. 1983. *Ap. J.* 272: 259–72
38. Friend, D. B., Abbott, D. C. 1986. *Ap. J.* 311: 701–7
39. Gabler, A., Gabler, R., Pauldrach, A. W. A., Puls, J., Kudritzki, R. P. 1990. See Ref. 41, pp. 218–29
40. Gabler, R., Gabler, A., Kudritzki, R. P., Puls, J., Pauldrach, A. W. A. 1989. *Astron. Astrophys.* 226: 162–82
41. Garmany, C. D., ed. 1990. *Hot Star Workshop: Intrinsic Properties of Hot Luminous Stars*. ASP Conf. Ser., Vol. 7. San Francisco: Astron. Soc. Pac.
42. Gehren, T., Husfeld, D., Kudritzki, R. P., Conti, P. S., Hummer, D. G. 1985. See Ref. 30, pp. 413–14
43. Gehren, T. 1989. *Rev. Mod. Astron.* 1: 52–101
44. Ghobros, R. A. 1962. *Z. Astrophys.* 56: 113–26
45. Giddings, J. 1981. PhD thesis. Univ. London, Engl.
46. Glaspey, L. W., Michard, G., Moffat, A. F. J., Demers, S. 1989. *Ap. J.* 339: 929–32
47. Groenewegen, M. A. T., Lamers, H. J. G. L. M., Pauldrach, A. W. A. 1989. *Astron. Astrophys.* 221: 78–88
48. Hamann, W.-R. 1985. *Astron. Astrophys.* 145: 443–48
49. Hamann, W.-R. 1985. *Astron. Astrophys.* 148: 364–68
50. Hamann, W.-R. 1986. *Astron. Astrophys.* 160: 347–51
51. Hamann, W.-R. 1987. See Ref. 77, pp. 35–65
52. Hamann, W.-R., Schmutz, W. 1987. *Astron. Astrophys.* 174: 173–82
53. Hamann, W.-R., Schmutz, W., Wessolowski, U. 1988. *Astron. Astrophys.* 194: 190–96
54. Hamann, W.-R., Wessolowski, U., Schwarz, E., Dünnebeil, G., Schmutz, W. 1990. See Ref. 41, pp. 259–70
55. Heasley, J. N., Wolff, S. C., Timothy, J. G. 1982. *Ap. J.* 262: 663–74
56. Heber, U., Kudritzki, R. P. 1986. *Astron. Astrophys.* 169: 244–50
57. Heber, U., Kudritzki, R. P., Caloi, V., Castellani, V., Danziger, J., Gilmozzi, R. 1986. *Astron. Astrophys.* 162: 171–79 (Erratum: *Astron. Astrophys.* 166: 396)
58. Heber, U. 1987. See Ref. 29, pp. 79–88
59. Heber, U., Werner, K., Drilling, J. S. 1988. *Astron. Astrophys.* 194: 223–29
60. Henrichs, H. 1988. See Ref. 25, pp. 199–235
61. Herrero, A., Kudritzki, R. P., Vilchez, J. M. 1990. *Proc. Eur. IAU Conf.* In press
62. Herrero, A., Kudritzki, R. P., Vilchez, J. M. 1990. See Ref. 41, pp. 50–52
63. Hillier, D. J. 1984. *Ap. J.* 280: 744–48
64. Hillier, D. J. 1987. *Ap. J. Suppl.* 63: 947–64
65. Hillier, D. J. 1987. *Ap. J. Suppl.* 63: 965–81
66. Hillier, D. J. 1988. *Ap. J.* 327: 822–39
67. Howarth, I. D., Prinja, R. K. 1989. *Ap. J. Suppl.* 69: 527–92
68. Hubeny, I. 1988. *Comput. Phys. Commun.* 52: 103–32
68a. Hubeny, I. 1990. See Ref. 41, pp. 93–96
69. Hummer, D. G. 1982. *Ap. J.* 257: 724–32
70. Hummer, D. G., Abbott, D. C., Voels, S. A., Bohannan, B. 1988. *Ap. J.* 328: 704–8
71. Hummer, D. G., Voels, S. A. 1988. *Astron. Astrophys.* 192: 279–80
72. Husfeld, D. 1982. Diplomarbeit. Univ. Kiel, Fed. Rep. Germ.
73. Husfeld, D., Kudritzki, R. P., Simon, K. P., Clegg, R. E. S. 1984. *Astron. Astrophys.* 134: 139–46
74. Husfeld, D. 1987. See Ref. 29, pp. 237–46
75. Husfeld, D., Butler, K., Heber, U., Drilling, J. S. 1989. *Astron. Astrophys.* 222: 150–70
76. Imhoff, J. 1989. Diplomarbeit. Univ. Munich, Fed. Rep. Germ.
77. Kalkofen, W., ed. 1987. *Numerical Radiative Transfer*. Cambridge: Univ. Press
78. Krolik, J. H., Raymond, J. C. 1985. *Ap. J.* 298: 660–75
79. Koenigsberger, G., Auer, L. H. 1985. *Ap. J.* 297: 255–65
80. Kudritzki, R. P. 1973. *Astron. Astrophys.* 28: 103–7
81. Kudritzki, R. P., Simon, K. P. 1978. *Astron. Astrophys.* 70: 653–63
82. Kudritzki, R. P. 1980. *Astron. Astrophys.* 85: 174–83
83. Kudritzki, R. P., Simon, K. P., Hamann, W.-R. 1983. *Astron. Astrophys.* 118: 245–54
84. Kudritzki, R. P. 1985. See Ref. 28, pp. 277–301
85. Kudritzki, R. P., Hummer, D. G. 1986. See Ref. 30, pp. 3–18
86. Kudritzki, R. P., Pauldrach, A. W. A., Puls, J. 1987. *Astron. Astrophys.* 173: 293–98
86a. Kudritzki, R. P., Pauldrach, A. W. A., Puls, J. 1988. See Ref. 25, pp. 173–99
87. Kudritzki, R. P., Groth, H. G., Butler, K., Husfeld, D., Becker, S., et al. 1987. In *ESO Workshop on the SN 1987A*, ed. I. J. Danziger, pp. 39–51. Garching bei München: ESO. 688 pp.

88. Kudritzki, R. P. 1987. See Ref. 29, pp. 177–89
89. Kudritzki, R. P., Yorke, H. W., Frisch, H. 1988. In *Radiation in Moving Gaseous Media. 18th Adv. Saas-Fee Course*, ed. Y. Chmielewski, T. Lanz, pp. 3–192. Sauverny-Versoix: Geneva Obs. 448 pp.
90. Kudritzki, R. P., Cabanne, M. L., Husfeld, D., Niemela, V. S., Groth, H. G., et al. 1989. *Astron. Astrophys.* 226: 235–48
91. Kudritzki, R. P., Gabler, A., Gabler, R., Groth, H. G., Pauldrach, A., Puls, J. 1989. In *Physics of Luminous Blue Variables, IAU Colloq. No. 113*, ed. K. Davidson, A. F. J. Moffat, H. J. G. L. M. Lamers, pp. 67–80. Dordrecht: Kluwer. 328 pp.
92. Kudritzki, R. P., Mendez, R. H. 1989. In *Planetary Nebulae, IAU Symp. No. 131*, ed. S. Torres-Peimbert, pp. 273–92. Dordrecht: Kluwer
93. Kudritzki, R. P., Pauldrach, A., Puls, J., Abbott, D. C. 1989. *Astron. Astrophys.* 219: 205–18
94. Kudritzki, R. P., Puls, J., Gabler, R., Schmitt, J. H. M. M. 1990. In *Extreme Ultraviolet Astronomy*, ed. R. F. Malin, S. Beyer. In press
95. Kudritzki, R. P. 1990. In *Frontiers of Stellar Evolution*, ed. D. L. Lambert. In press
96. Kudritzki, R. P. 1990. See Ref. 41, p. 63
97. Kunasz, P. B., Hummer, D. G., Mihalas, D. 1975. *Ap. J.* 202: 92–113
98. Kurucz, R. L. 1979. *Ap. J. Suppl.* 40: 1–340
99. Leitherer, C. 1988. *Ap. J.* 326: 356–67
99a. Lamers, H. J. G. L. M., Maeder, A., Schmutz, W., Cassinelli, J. P. 1990. *Ap. J.* In press
100. Lennon, D. J. 1983. *MNRAS* 205: 829–38
101. Lennon, D. J., Lynas-Gray, A. E., Brown, P. J. F., Dufton, P. L. 1986. *MNRAS* 222: 713–18
102. Lucy, L. B., Solomon, P. 1970. *Ap. J.* 159: 879–93
103. Lucy, L. B. 1982. *Ap. J.* 255: 278–85
104. Lucy, L. B. 1982. *Ap. J.* 255: 286–92
105. MacFarlane, J. J., Cassinelli, J. P. 1990. *Ap. J.* 347: 1090–99
106. McCarthy, J. K., Mould, J. R., Méndez, R. H., Kudritzki, R. P., Husfeld, D., et al. 1990. *Ap. J.* 351: 230–44
107. Massa, D., Savage, B. D. 1985. *Ap. J.* 299: 905–16
108. Méndez, R. H., Kudritzki, R. P., Herrero, A., Husfeld, D., Groth, H. G. 1988. *Astron. Astrophys.* 190: 113–36
109. Méndez, R. H., Groth, H. G., Husfeld, D., Kudritzki, R. P., Herrero, A. 1988. *Astron. Astrophys.* 197: L25–28
110. Méndez, R. H., Herrero, A., Manchado, A. 1990. *Astron. Astrophys.* In press
111. Mihalas, D. 1987. *Stellar Atmospheres*. San Francisco: Freeman. 632 pp. 2nd ed.
112. Mihalas, D. 1972. *Non-LTE Model Atmospheres for B and O-Stars. NCAR-TN/STR-76*. Boulder, Colo: NCAR
113. Mihalas, D., Heasly, J. N., Auer, L. H. 1975. *Non-LTE Model Atmospheres for B and O-Stars. NCAR-TN/STR-104*. Boulder, Colo: NCAR
114. Mihalas, D., Hummer, D. G. 1974. *Ap. J. Suppl.* 28: 343–72
115. Moffat, A. F. J., Koenigsberger, G., Auer, L. H. 1989. *Ap. J.* 344: 734–46
115a. Morton, D. C. 1967. *Ap. J.* 147: 1017–24
116. Owocki, S. P., Rybicki, G. B. 1984. *Ap. J.* 284: 337–50
117. Owocki, S. P., Rybicki, G. B. 1985. *Ap. J.* 299: 265–76
118. Owocki, S. P., Rybicki, G. B. 1986. *Ap. J.* 309: 127–40
119. Owocki, S. P., Castor, J. I., Rybicki, G. B. 1988. *Ap. J.* 335: 914–30 (OCR)
120. Panagia, N., Macchetto, F. 1982. *Astron. Astrophys.* 106: 266–73
121. Pauldrach, A., Puls, J., Hummer, D. G., Kudritzki, R. P. 1985. *Astron. Astrophys.* 148: L1–4
122. Pauldrach, A. W. A., Puls, J., Kudritzki, R. P. 1986. *Astron. Astrophys.* 164: 86–100 (PPK)
123. Pauldrach, A. W. A. 1987. *Astron. Astrophys.* 183: 295–313
124. Pauldrach, A. W. A. 1987. PhD thesis. Univ. Munich, Fed. Rep. Germ.
125. Pauldrach, A. W. A., Puls, J., Kudritzki, R. P., Méndez, R. H., Heap, S. R. 1988. *Astron. Astrophys.* 207: 123–31
126. Pauldrach, A. W. A., Kudritzki, R. P., Puls, J., Butler, K. 1990. *Astron. Astrophys.* 228: 125–54
127. Pauldrach, A. W. A. 1990. See Ref. 41, pp. 171–88
128. Pauldrach, A. W. A., Puls, J. 1990. *Astron. Astrophys.* In press
129. Prinja, R. K., Howarth, I. D. 1986. *Ap. J. Suppl.* 61: 357–418
130. Puls, J. 1987. *Astron. Astrophys.* 184: 227–48
131. Puls, J., Hummer, D. G. 1988. *Astron. Astrophys.* 191: 87–98
132. Puls, J., Pauldrach, A. W. A. 1990. See Ref. 41, pp. 203–17
133. Rybicki, G. B. 1971. *J. Quant. Spectros. Radiat. Transfer* 11: 589–95

134. Rybicki, G. B., Owocki, S. P., Castor, J. I. 1990. *Ap. J.* 349: 274–85
135. Schmutz, W., Hamann, W.-R. 1986. *Astron. Astrophys.* 166: L11–14
136. Schmutz, W., Hamann, W.-R., Wessolowski, U. 1989. *Astron. Astrophys.* 210: 236–48
137. Schmutz, W., Abbott, D. C., Russell, R. S., Hamann, W.-R., Wessolowski, U. 1990. *Ap. J.* In press
138. Schönberner, D., Herrero, A., Butler, K., Becker, S., Eber, F., et al. 1988. *Astron. Astrophys.* 197: 209–22
139. Schöning, T., Butler, K. 1989. *Astron. Astrophys.* 219: 326–33
140. Schöning, T., Butler, K. 1989. *Astron. Astrophys. Suppl.* 78: 51–87
141. Seaton, M. J. 1988. *J. Phys. B* 21: 3033–53
142. Seaton, M. J. 1990. In *Atomic Spectra and Oscillator Strengths for Astrophysics and Fusion Research*, Amsterdam: K. Ned. Akad. Wet. In press
143. Simon, K. P., Jonas, G., Kudritzki, R. P., Rahe, J. 1983. *Astron. Astrophys.* 125: 34–44
144. Snow, T. P., Morton, D. C. 1976. *Ap. J. Suppl.* 32: 429–65
144a. Thomas, R. N. 1988. See Ref. 25, pp. IX–XLIX
145. Tobin, W. 1983. *Astron. Astrophys.* 125: 168–71
146. Underhill, A. B. 1988. *MNRAS* 230: 55–68
146a. Underhill, A. B. 1986. *Publ. Astron. Soc. Pac.* 98: 897–913
146b. Underhill, A. B. 1988. See Ref. 25, pp. 273–419
147. Vidal, C. R., Cooper, J., Smith, E. W. 1970. *J. Quant. Spectros. Radiat. Transfer* 10: 1011–63
148. Voels, S. A. 1989. PhD thesis. Univ. Colo., Boulder. 70 pp.
149. Voels, S. A., Bohannan, B., Abbott, D. C., Hummer, D. G. 1989. *Ap. J.* 340: 1073–90
150. Walborn, N. R., Panek, R. J. 1984. *Ap. J. Lett.* 280: L27–30
151. Walborn, N. R., Panek, R. J. 1984. *Ap. J.* 286: 718–24
152. Walborn, N. R., Panek, R. J. 1985. *Ap. J.* 291: 806–11
153. Walborn, N. R., Nichols-Bohlin, J., Panek, R. J. 1985. *IUE Atlas of O-Type Spectra From 1200 to 1900 Å*. NASA Ref. Publ. 1155. Washington, DC: NASA
154. Walborn, N. R., Nichols-Bohlin, J. 1987. *Publ. Astron. Soc. Pac.* 99: 40–53
155. Werner, K. 1988. *Astron. Astrophys.* 204: 159–76
156. Werner, K. 1989. *Astron. Astrophys.* 226: 265–69
157. Werner, K., Heber, U., Hunger, K. 1990. See Ref. 41, pp. 86–89
158. Wessolowski, U., Schmutz, W., Hamann, W.-R. 1988. *Astron. Astrophys.* 194: 160–66
159. Wolff, S. C., Heasley, J. N. 1985. *Ap. J.* 292: 589–600
160. Woods, T. N., Feldman, P. D., Bruner, G. H. 1985. *Ap. J.* 292: 676–86
161. Wright, A. E., Barlow, M. J. 1975. *MNRAS* 170: 41–51

RADIO IMAGES OF THE PLANETS

Imke de Pater[1]

Astronomy Department, University of California, Berkeley, California 94720

KEY WORDS: atmospheres, magnetospheres, subsurface layers

INTRODUCTION

At radio wavelengths in the millimeter–centimeter regime one typically probes regions in planetary atmospheres, magnetospheres, and surfaces that are inaccessible at optical or infrared wavelengths. At centimeter wavelengths, one typically examines regions in an atmosphere well below the visible cloud layers. If the planet has a solid crust, the observations pertain to depths ~ 10 wavelengths into the surface. In addition to the thermal emission from the planet, one may receive nonthermal emission from energetic electrons in a planet's magnetosphere. Hence, observations at radio wavelengths, combined with optical and infrared data, allow us to make detailed studies of planetary atmospheres, magnetospheres and surface layers. In this paper I address in particular the question of what we can learn from interferometric radio images of planets. However, since the interpretation of such images builds upon knowledge from single-dish data, I also discuss results from single-element radio observations.

The planets can roughly be divided into two classes: terrestrial (Mercury, Venus, Earth, Mars, and Pluto) and giant (Jupiter, Saturn, Uranus, and Neptune). However, even within these classes, large differences exist. For example, Venus has a very hot and dense CO_2 atmosphere, whereas Mercury has virtually no atmosphere. Models of interior structures of the giant planets show large differences between the two outermost planets (Uranus and Neptune) and the inner ones (Jupiter and Saturn). While the bulk of the atmospheric mass in Jupiter and Saturn consists primarily of

---

[1] Alfred P. Sloan Foundation Fellow.

H_2, Uranus and Neptune contain a large amount of "ices." In addition, Uranus is quite different from the other three giant planets in that its internal heat source is small [$\leqslant 13\%$ of the received sunlight (Conrath et al 1988)], if there is any at all. The other three planets emit about twice as much energy as they receive from the Sun.

Giant Planets

ATMOSPHERES Observations at microwave wavelengths typically probe pressure levels of ~ 0.5–10 bars in the atmospheres of Jupiter and Saturn, and down to 50–100 bars in the Uranus and Neptune atmospheres. Much information is contained in the planet's radio spectrum: a graph of the disk-averaged brightness temperature of the planet as a function of wavelength. These spectra generally show an increase in brightness temperature with increasing wavelength beyond 1.3 cm due to the combined effects of a decrease in opacity at longer wavelengths and an increase in temperature at increasing depth in the planet. At millimeter–centimeter wavelengths the main source of opacity is ammonia gas, which has a broad absorption band at 1.3 cm. At longer wavelengths (typically >10 cm), absorption by water vapor and droplets becomes important, whereas at short millimeter wavelengths the contribution of collision-induced absorption by hydrogen gas becomes noticeable.

Radio spectra of the planets are usually analyzed by comparing the observed spectra with synthetic spectra obtained by integrating the equation of radiative transfer through a model atmosphere (for details, see, e.g., de Pater & Massie 1985, Briggs & Sackett 1989, de Pater et al 1989). The disk-averaged brightness $B_\nu(T_D)$ at frequency ν is given by

$$B_\nu(T_D) = 2 \int_0^1 \int_0^\infty B_\nu(T) \exp(-\tau/\mu) \, d(\tau/\mu) \, d\mu. \qquad 1.$$

The brightness $B_\nu(T)$ is given by the Planck function, and optical depth $\tau_\nu(z)$ is the integral of the total absorption coefficient over the altitude range z at frequency ν. The parameter μ is the cosine of the angle between the line of sight and local vertical. As mentioned above, the main source of opacity at radio wavelengths is ammonia gas. The absorption profile is very broad owing to the high pressures in the atmospheres. The line profile is usually modeled with a Ben Reuven line shape at centimeter wavelengths; at millimeter wavelengths a Van Vleck–Weisskopf profile might be more appropriate. In this paper a modified Van Vleck–Weisskopf line profile is used at millimeter wavelengths, as suggested by de Pater & Massie (1985).

Before the integration in Equation 1 can be carried out, the atmospheric structure, as defined by the composition and temperature-pressure profiles, needs to be defined. Since the temperature, pressure, and composition of an atmosphere are all related, they should be defined at some deep level in the atmosphere (well below the formation of clouds). As we step up in altitude from this base level, the new temperature can be calculated assuming a dry adiabatic lapse rate (or appropriate wet lapse rate in the cloud formation regions), and the new pressure can be obtained by using hydrostatic equilibrium. The partial pressures of the trace gases in the atmosphere are compared with the saturation vapor pressures, and if the partial pressure exceeds the latter value, a cloud of that condensate forms. The following cloud layers are expected to form in the giant planet atmospheres (moving from deep levels upward in attitude): aqueous ammonia solution ($H_2O-NH_3-H_2S$), water ice, ammonium hydrosulfide solid, ammonia ice, hydrogen sulfide ice, and (on Uranus and Neptune) methane ice.

To a first approximation the spectra of both Jupiter and Saturn resemble those expected for an atmosphere of solar composition, although some variation with depth is present. On the other hand, the spectra of Uranus and Neptune indicate a depletion of ammonia gas compared with the solar value by ~ 2 orders of magnitude (Gulkis et al 1978, de Pater & Massie 1985). Resolved images of both Jupiter and Saturn show bands of enhanced brightness temperature on the planets' disks (de Pater & Dickel 1986, and in preparation; Grossman et al 1989), implying latitudinal variations in the precise ammonia abundance. Uranus shows an enhanced brightness temperature toward the visible (south) pole (e.g. Jaffe et al 1984, de Pater & Gulkis 1988); images of Neptune have too low a resolution at this point to infer latitudinal variations in brightness temperature.

SYNCHROTRON RADIATION At centimeter wavelengths we may receive an additional component of emission: nonthermal or synchrotron radiation S from relativistic electrons gyrating in the planet's magnetic field (Legg & Westfold 1968):

$$S_v dv = \int CB \sin \alpha N(E, \alpha, r) F(v/v_c) dE dv \qquad 2.$$

Here, the emissivity polarization function $F(x)$ is given by

$$F(x) = x \int_0^\infty K_{5/3}(\eta) d\eta,$$

where K is a modified Bessel function. The pitch angle α is the angle between the magnetic field direction and the particle's motion. The number density N is a function of energy E, pitch angle α and position r. B is the local magnetic field strength, C a constant, and v_c the critical frequency (see Equation 3); (for further details, see Legg & Westfold 1968, de Pater 1981a). Since the strength of the synchrotron radiation is proportional to the product NB, we only receive emission if the inner radiation belts of the planet are populated with energetic electrons (~ 1–100 MeV). Synchrotron radiation dominates Jupiter's emission at wavelengths longward of 6 cm. No synchrotron radiation has been detected from Saturn or Uranus. In Saturn's magnetosphere, all the energetic particles are absorbed by Saturn's rings (McDonald et al 1980), since the rotation and magnetic axes are closely aligned. Uranus' magnetic field strength is too weak for electrons to produce a measurable amount of synchrotron emission (de Pater & Gulkis 1988). A small fraction of Neptune's radio emission at 20 cm has been attributed in the past to synchrotron radiation (de Pater & Goertz 1989); at present, this is not supported by data from the recent *Voyager 2* flyby.

Terrestrial Planets

Microwave observations of Mercury, Venus, and Mars probe the atmosphere as well as the (sub)surface layers of the planets. Both Venus and Mars have atmospheres that consist primarily of CO_2 gas, which is the primary source of atmospheric microwave opacity. The photolysis product of CO_2, CO, has strong rotational transitions at millimeter wavelengths, which can be utilized to determine the atmospheric temperature profile and the CO abundance in the altitude regions probed.

The temperature structure of the (sub)surface layers of airless bodies depends upon a balance between solar insolation, heat transport within the crust, and reradiation outward. The conductive heat transport into the subsurface depends upon the thermal conductivity and the heat capacity of the material; reradiation from the surface depends upon the emissivity of the material. The emissivity is described by the dielectric constant ϵ and the loss tangent of the material tan Δ, which are determined by the bulk density or composition of the material and its state of compaction (e.g. Campbell & Ulrichs 1969). A typical radio skin depth is ~ 10 wavelengths. Hence, by obtaining a spectrum of the objects, properties regarding its (sub)surface layers can be derived. However, since most constants entering the equations for heat conduction and radiative transfer vary with wavelength as well as with composition of the material, interpretation of such spectra remains difficult.

If an atmosphere overlies the crust of the planet, the crust can be dramatically heated by the greenhouse effect in the atmosphere. Venus' surface, for example, is warmer by a factor of ~ 3 as compared with the temperature expected from solar insolation alone. Thus, models including both the atmosphere and crust layers need to be developed (e.g. Muhleman et al 1979).

Because the planets are so different, I discuss each of them individually in what follows. For historical reasons, I start with Jupiter, the largest planet in our solar system and the first from which radio emission was detected. I then work my way out from and subsequently in toward, the Sun. Before discussing the planets, however a short tutorial is given on the observing and reduction techniques used to obtain high-quality images of planets.

2. OBSERVING AND REDUCTION TECHNIQUES

The basic principles of radio astronomy and interferometry have been described by, for example, Kraus (1986), Rohlfs (1986), Thompson et al (1986), and Perley et al (1989). In this section I concentrate on the construction of radio images of planets, a topic that has not been widely discussed in the existing literature. Obtaining high-quality radio images of planets is not always a straightforward process. The best results are obtained by using a radio interferometer that consists of many elements, like the Very Large Array (VLA) in New Mexico. The resolution for an interferometer pair is roughly λ/D, with λ the observing wavelength and D the projected antenna-pair separation. A radio image can be constructed from observations obtained with many interferometer pairs at different spacings. To image a planet, one typically needs data at both short and long spacings: At short baselines the entire object can be "seen," but details on the planet are washed out due to the low resolution of such baselines; longer baselines details on the planet can be distinguished, but the large-scale structure of the object gets resolved out and hence would be invisible on the image unless short spacing data are included as well. Therefore, before observing time is requested, one has to define the goals of the observations: If an accurate total flux density is desired, short-spacing data are a must; if the object is large and a high resolution is desired, one needs short- as well as long-spacing data. With the VLA, one might consider the combination of data from different array configurations, under the assumption that the radio emission is stable. The 2- and 6-cm images of Jupiter discussed in Section 3 and the highest resolution images of Saturn (Section 4) were obtained from data taken with a few different array configurations. Obviously, when data from different array configurations are used, they need to be scaled and rotated such that the orientation and size of the planet is the same in all data sets.

In general, to construct an image of an object, the response of all the individual interferometer pairs, the visibility data, are gridded into cells having uniform intervals in projected antenna-pair separation on the sky. These spacings are measured in the north-south (called u) and east-west (called v) directions; this is commonly referred to as the uv plane. This grid of data is then Fourier transformed to give a map of the brightness distribution. In this process one needs to choose a weighting function for the uv data [see Sramek (1982) and Sramek & Schwab (1989) for specific details]. Three basic options exist: natural, uniform, and superuniform weighting. In the first option, each cell is given a weight proportional to the number of data points inside that cell. Since the cells near the uv origin contain relatively more data points than the cells elsewhere in the uv plane, the short-spacing data are weighted quite heavily, which produces a rather broad beam with large sidelobes. It gives the best signal-to-noise ratio and hence is best for detection experiments. This option also appears to give the best results when mapping a weak, extended (relative to the beam size) planet. For example, Uranus (Section 5) could only be imaged well with the natural weight option. Uniform weighting gives equal weight to each cell, so the size of the beam is smaller, and the sidelobes are reduced. Many of the high-resolution radio images of planets were constructed using this option. However, in so-called snapshot observations (few minutes to hour-long integrations), this weighting is often not uniform enough, since, for the VLA, the cells that contain any data are mainly concentrated toward the center and along the arms of the array. By using a so-called super-uniform weighting function, the lightly sampled isolated cells get a relatively larger weight compared with the better sampled parts of the uv plane. This effect is somewhat similar to an inverse taper. The sidelobe level is considerably reduced, while the beam is slightly narrower than for a uniform map. This option gave the best images of Jupiter's nonthermal radiation at 20 cm (Section 3), where the integration time was 15–30 mins.

Once the image is obtained, one needs to remove the response of the antenna beam from the images (see, e.g., Cornwell & Braun 1989). The most widely used technique is CLEAN (Högbom 1974, Clark 1980), which deconvolves an image by an iterative procedure in which a fraction of the interferometer response to the brightest point remaining on the map is successively subtracted. This process is repeated until the entire source is removed from the map. The CLEAN components are then convolved with a Gaussian beam that best fits the central part of the antenna pattern and are restored to the map. One may decide to choose a larger or smaller beam size for restoration, depending on the quality of the image. However, based upon the sampling theory, the beam size cannot be smaller than $\sim 0.5\lambda/D_{max}$, where D_{max} is the maximum baseline length in wavelengths λ.

The CLEAN procedure obviously will not work on a large plateau of emission, since subtraction of the antenna beam creates positive and negative intensity fluctuations on the disk, which CLEAN interprets as real intensity fluctuations during its next cycle. The final map then consists of a stripy or mottled pattern on the planetary surface, the exact appearance of which depends upon the interference pattern of the telescope array. Good results can sometimes be obtained by first subtracting a set of CLEAN components that best model the large plateau of emission. For planets, the model usually consists of a set of equal-brightness points uniformly filling an oblate disk. Their summed flux density is equal to the integrated flux density of the planet, or that which would be measured by a single antenna whose beam covers the entire source (the so-called zero-spacing flux density). Once this large plateau has been removed from the map, normal CLEANing is continued, allowing both positive and negative components to be found. This will correct for deviations from the subtraction of the uniformly bright disk. After CLEANing, all components are convolved with a Gaussian beam that best fits the inner, elliptically shaped part of the antenna beam and are then restored to the map. This deconvolution technique appears to work well on planets that have a dynamic range in brightness (ratio of the peak brightness to the level of one standard deviation per beam) of the order of 20 or better.

If a spectrum of a planet is obtained, one has to be careful not to distort the line profile by the CLEANing technique. Since the brightness distribution on the planet may vary from channel to channel, the antenna response should be removed from all the individual channel maps. Best results are obtained by subtracting a uniform oblate disk with a total flux density scaled according to the disk-averaged line profile, as obtained from the "dirty" images (before CLEANing). After further CLEANing, the disk is restored to the map after convolution with the appropriate beam.

Instead of the CLEAN deconvolution technique, one can also make use of a maximum entropy method (MEM) to deconvolve the image (e.g. Gull & Daniell 1978, Cornwell & Braun 1989). While CLEAN works directly on an image, MEM selects an image that has maximum smoothness and fits the visibility data. One drawback is that the resolution in an MEM map depends upon the signal-to-nose ratio, which results in a resolution that varies across the image. However, one can convolve the image with a Gaussian beam to get a more uniform resolution in the image. In practice, it appears that for planetary observations MEM images are at best similar to CLEANed images.

A third deconvolution technique has been developed by R. W. Gerchberg & W. O. Saxton (Gerchberg 1974). In this technique, the map is constrained to be nonzero and positive only in a region defined by the observer. Data and image-plane constraints are imposed alternately while Fourier transforming to and from the image plane, until convergence has been reached. This algorithm is very simple, and not widely used. However, for noisy planetary data, the results are far superior than those obtained from either CLEAN or MEM. Several of the images of Uranus (Section 5) have been made with the Gerchberg-Saxton technique.

None of these procedures corrects for artifacts or "noise" in the image created by random phase and amplitude fluctuations in the receivers or the Earth's atmosphere during the observations. These fluctuations limit the reliability and final dynamic range of the maps. Generally one can remove most of these fluctuations by making use of a "self-calibration" technique, which is essentially the same as the phase closure technique applied to many VLBI maps (e.g. Schwab 1980, Bridle et al 1981). It consists of adjusting the complex gains of the individual antennas such that the response of each interferometer pair most closely resembles the response expected for a certain model of the source.

The source model generally consists of a point for compact sources or the CLEAN components of an extended source after the image has been cleaned. For planets we usually use the brightness points that represent the oblate model disk, sometimes supplemented by additional CLEAN components. By applying several self-calibration passes to the data, very high dynamic range maps can be obtained. The updated model for additional passes is obtained by CLEANing a new map, obtained from the adjusted data base. If all N antennas are correlated with one another, there are $N(N-1)/2$ baselines, and the N complex gains can be solved for by a least-squares fitting procedure. If the number is sufficiently large (at the VLA, $N = 27$), the antenna-based errors can indeed be determined without destroying the true response to the source brightness distribution on the individual baselines.

Although in principle the self-calibration method ought to work on planets, in practice convergence is not met if the source is weak (less than a few tens of millijanskys) and/or if there are not enough data at short spacings. If the flux density is $\gtrsim 1$ Jy and there are data at short spacings, a solution can generally be found by choosing the right averaging time per solution interval and the right uv range for the data allowed to be used in the solution. Typically, VLA radio images of Jupiter, Saturn, and the terrestrial planets can be self-calibrated satisfactorily, whereas those of Uranus and Neptune cannot.

3. GIANT PLANETS

Jupiter

Radio signals from Jupiter were first detected in 1955 at a frequency of 22.2 MHz (Burke & Franklin 1955). This emission was sporadic in character and was confined to frequencies less than 40 MHz. It is likely due to cyclotron radiation from electrons with mirror points close to Jupiter's ionosphere. Excellent reviews on this topic are those by, for example, Carr & Desch (1976) and Carr et al (1983).

A detailed historic overview on radio observations of Jupiter at microwavelengths was given by Berge & Gulkis (1976). They noted that the first detection of microwave radiation from the planet was made in 1956 at 3-cm wavelength by Mayer et al (1958). The measured flux density corresponded to a blackbody temperature of ~ 140 K. In subsequent years, observations at longer wavelengths were conducted, which yielded temperatures of a few thousand degrees at wavelengths $\lambda \gtrsim 10$ cm. Even though one expects a temperature gradient in an atmosphere, the measured spectrum was too steep to be caused by a reasonable atmospheric gradient (e.g. an adiabatic gradient). Interferometric observations by Radhakrishnan & Roberts (1960) in 1960 at a wavelength of 31 cm, and a year later by Morris & Berge (1962) at 31 and 22 cm, showed that the radiation was $\sim 30\%$ linearly polarized at both wavelengths and had a linear extent roughly 3 times the planet's diameter in the equatorial direction, while the north-south extent agreed with the planetary diameter. This led to the suggestion that Jupiter's radio emission at wavelengths $\gtrsim 6$ cm was dominated by synchrotron radiation, emitted by high-energy electrons in a Jovian Van Allen belt. Emission at shorter wavelengths is dominated by the thermal radiation from the planet's atmosphere.

SYNCHROTRON RADIATION   Synchrotron radiation is emitted by relativistic electrons gyrating around magnetic field lines. The radiation is beamed in the forward direction within a cone $1/\gamma$, with $\gamma = 2E$ (where E is the energy of the radiating electron, in MeV). The radiation is emitted over a wide range of frequencies but shows a maximum at $0.29\nu_c$, with ν_c (the critical frequency, in MHz) equal to

$$\nu_c = 16.08 E^2 B. \qquad 3.$$

In Equation 3 the energy E is in MeV, and the field strength B is in gauss. For a peak emission at 20 cm, we require $E^2 B = 320$. If B is 0.5 G, the typical energy of electrons emitting at 20 cm is close to 25 MeV. At lower field strengths and/or higher observing frequencies, the typical energy increases. Hence, we probe a different electron population when observing at different frequencies. Further, since the magnetic field strength decreases with planetary distance r approximately as r^3 for a dipole field, we also observe different electron populations at different distances from the planet.

Synchrotron emission is polarized, and thus we express the observed quantities in terms of the Stokes parameters I, Q, U, and V. The degree of linear polarization is given by $(Q^2+U^2)^{1/2}/I$, with the position angle of the electric vector $PA = \frac{1}{2}\arctan(U/Q)$. In the absence of Faraday rotation, which is a reasonable assumption for Jupiter (e.g. de Pater 1980), the projection of the magnetic field can be found by rotating PA over 90°. Note, however, that the emission is integrated along the entire line of sight (Equation 2), weighted most heavily by the regions that emit the most radiation. The degree of circular polarization V/I is a measure of the strength of the component of the magnetic field directed along the line of sight. In general, one expects zero circular polarization for a dipole field if the observer is in the magnetic equatorial plane, and maxima (with opposite sign) when the magnetic poles are facing the observer.

Since the "discovery" of Jupiter's synchrotron radiation, this component of the planet's microwave emission has been studied in detail. The variation of the total nonthermal intensity and polarization characteristics during one Jovian rotation (so-called beaming curves) is indicated in Figure 1 for the total intensity S, the position angle PA of the electric vector, the linearly and circularly polarized flux densities P_L and P_C and the magnetic latitude of the Earth ϕ_m. The orientation of Jupiter's magnetosphere is indicated at the top. The maxima and minima in S and P_L occur approximately at $\phi_m = 0$ and $|\phi_m|$ = maximum, respectively; whereas the circularly polarized flux density P_C is zero where S and P_L show maxima, and P_C shows a positive or negative maximum where S and P_L show minima. These curves indicate that Jupiter's magnetic field is approximately dipolar in shape, offset from the center of the planet by $\sim 0.1\ R_J$ ($1 R_J = 1$ Jovian radius) toward a longitude of 140°, and inclined $\sim 10°$ from the rotation axis (see, e.g., the review by de Pater & Klein 1990). Most electrons are confined to the equatorial plane. The magnetic north pole is in the northern hemisphere, tipped toward a longitude of 200°. The 40-MHz cutoff at decametric wavelengths implies a strength of ~ 10 G at the surface. When the spacecraft *Pioneer 10* and, in particular *Pioneer 11* flew by Jupiter, all findings reported above were confirmed. However, none of the spacecraft to date have passed close enough to Jupiter for detailed observations of the radiation belts. *Pioneer 11* came within 1.6 R_J but along a fast north-south track. As a result of this relatively close approach, the quadruple and octupole terms of the magnetic field could be measured with decent accuracy. This, together with the information on particles at larger distances from the planet, can be used as boundary conditions to model radio data of Jupiter's synchrotron radiation (e.g. de Pater 1981a,b).

The first radio image of Jupiter was constructed by Berge (1966) from model fits to visibility data obtained with the Owens Valley interferometer at 10.8 cm. It shows a peak in emission at approximately 1.6 R_J at each

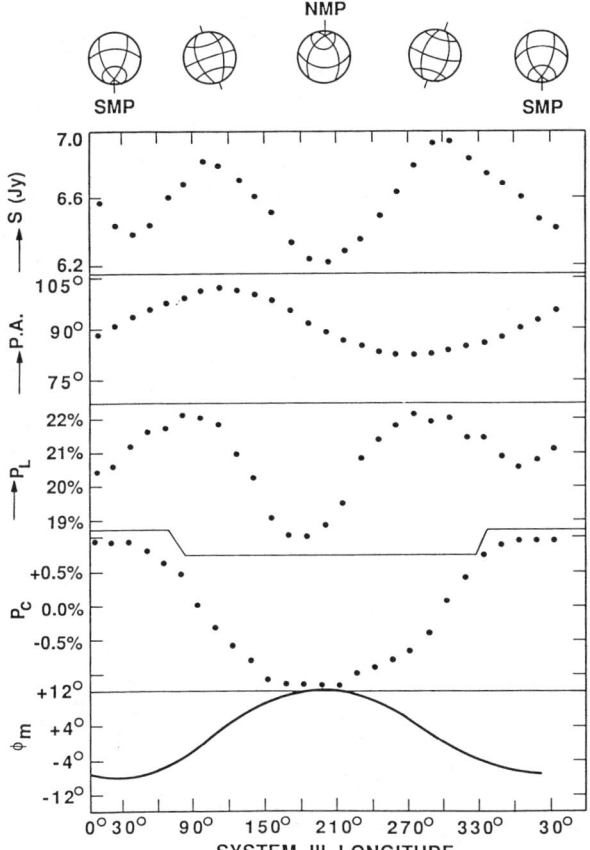

Figure 1 An example of the modulation of Jupiter's synchrotron radiation due to Jupiter's rotation (from de Pater & Klein 1989; after de Pater 1980). The orientation of the planet is indicated at the top; the different panels show subsequently the total flux density S, the position angle PA of the electric vector, the degree of linear and circular polarization P_L and P_C, and the magnetic latitude of the Earth ϕ_m. This latitude can be calculated using the following relation: $\phi_m = D_E + \beta \cos(\lambda - \lambda_0)$, where D_E is the declination of the Earth, β the angle between Jupiter's magnetic and rotational axes, λ the longitude, and λ_0 the longitude of the magnetic north pole.

side of the planet. (All distance scales are from the planet's center.) When telescope arrays were built (e.g. the 1-mile east-west telescopes in Cambridge and Westerbork, the 5-km east-west array in Cambridge, and the Y-shaped VLA), the visibility data could be Fourier transformed directly to yield an image of the planet. The first direct image (Branson 1968) showed a clear asymmetry between the radiation peaks—one of the peaks appeared stronger, while the ratio between the peaks changed with the planet's rotation. This was interpreted to be due to a "hot spot" in Jupiter's radiation belts, near the longitude of the magnetic north pole. Data taken 6 years later (de Pater & Dames 1979) showed the "hot spot" at a longitude of 255°; the 60° migration in longitude of this region is still not understood (de Pater & Klein 1989). The first snapshot images, which showed a rotational smearing of only 15° as opposed to the 120° in Branson's (1968) maps, were obtained with the Westerbork telescope (de Pater 1980), an east-west array of 12 (presently 14) dishes. She obtained excellent images in all four Stokes parameters. The asymmetry between the radiation peaks was clearly visible and changed from one rotational aspect to the next. A sample of the images is shown in Figure 2. Images in the circularly polarized flux component show to first approximation the changing polarity and flux density as expected for a rotating dipole field; the radiation is severely modified, however, by the nondipole character of the field, causing large east-west asymmetries in these images.

De Pater (1981a,b) developed an elaborate model to simulate the radio images, using results from the Pioneer spacecraft to constrain the model. She used the multipole magnetic field configuration as determined by *Pioneer 11* and calculated the electron distribution using adiabatic theory and a diffusion model consistent with the Pioneer results. The O4 octupole magnetic field model, as derived from the Pioneer data by Acuna & Ness (1976), gave a good fit to the radio data. De Pater noticed that the electron spectrum had to be flatter than that measured by the spacecraft; this observation was later confirmed with calculations by de Pater & Goertz (1990), who attributed the increasing flatness of the spectrum with decreasing planetary distance to scattering by the dust particles in the planet's rings and general environment. Absorption effects by the satellites Thebe and Amalthea as well as the rings, cause the electron distribution to be confined to the equatorial plane. The "hot spot" could be explained partly by the multipole character of the field, together with a dusk-dawn electric field over the magnetosphere. An excess of electrons, however, is still needed near longitudes of 240–360°.

More recently, high-resolution images of Jupiter's synchrotron radiation were obtained by Roberts et al (1984) and de Pater & Jaffe (1984) using the VLA at 20 cm. One of the images is shown in Figure 3. It reveals more details of the emission, and visually confirms the confinement of the radiation to the magnetic equatorial plane out to a distance of approximately 3 R_J, the orbital distance of the satellite Thebe. The most intriguing new features are the secondary emission peaks just north and south of the main peaks. They must be produced by electrons at their mirror points,

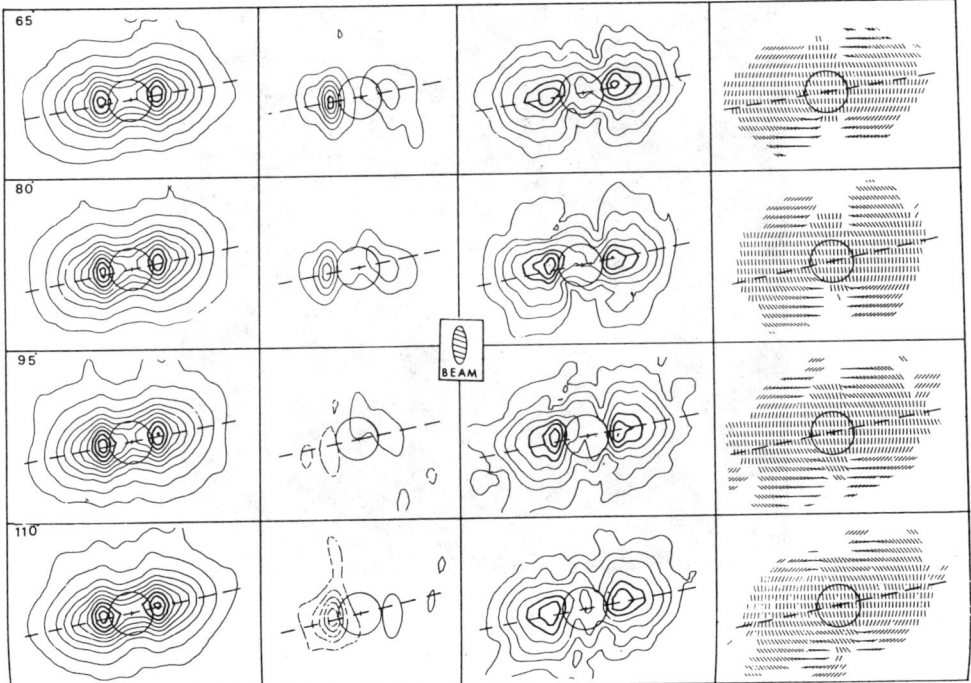

Figure 2 Images of Jupiter's flux density as obtained with the Westerbork Synthesis Radio Telescope (from de Pater 1980). Indicated are (from left to right) maps of the total intensity *I*, the circularly and linearly polarized flux densities *V* and *P*, and a vector diagram of the magnetic field of the planet. Dashed contours indicate left-handed circular polarization and solid contours right-handed circular polarization. The contours belonging to the three highest values in all three maps are drawn with heavy lines. The central meridian longitude is indicated in the top left corner. Contour values are *I*: 9.5 K, 65–1065 K in steps of 125 K; *V*: 1.9–21.5 K in steps of 2.8 K; *P*: 9.5 K, 32 K, 65–325 K in steps of 65 K.

implying a rather large number of particles between 2.5 and 3 R_J with small pitch angles (de Pater 1983); these particles were neither seen by spacecraft nor predicted from de Pater's (1981a,b) model calculations.

THERMAL EMISSION   In order to compile a thermal spectrum for Jupiter, one needs to separate the thermal from the nonthermal radiation. This topic has been addressed in detail by Berge & Gulkis (1976), and was extended by de Pater et al (1982). The most widely used technique is based upon the polarization properties of the two emission components. One assumes that the thermal emission is essentially unpolarized, and that the

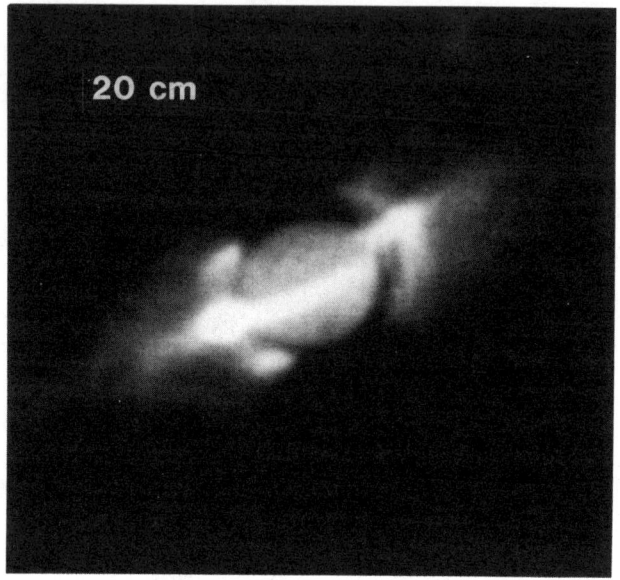

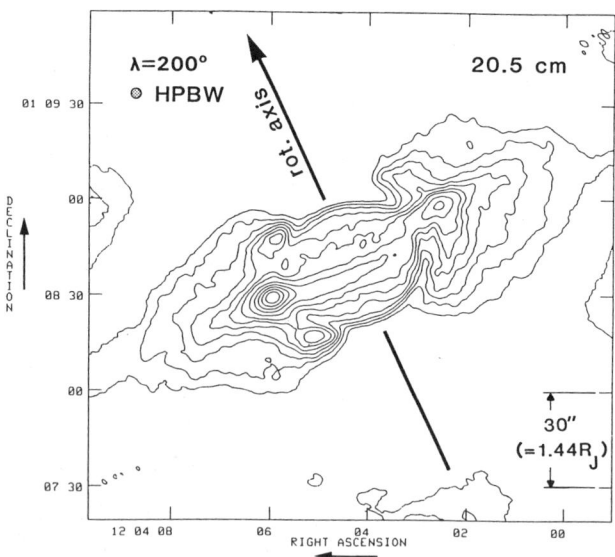

Figure 3 Radio photo (*top*) and contour map (*bottom*) of Jupiter at wavelength of 20 cm (from de Pater & Dickel 1986). The radio photo is at a central meridian longitude of 117°, and the contour map is at 200°. Contour values (in kelvins) are 119, 178, 237, 297, 356, 416, 475, 594, 712, 831, 950, and 1070.

degree of linear polarization in the synchrotron radiation is 22% at all wavelengths. This number was defined from the degree of polarization at long wavelengths, averaged over Jupiter's rotation; at long wavelengths the thermal contribution is negligible. It was assumed (Berge & Gulkis 1976) that the degree of linear polarization remains constant at shorter wavelengths; this is a fairly good assumption, since de Pater (1981a) showed later that it decreases only by $\sim 2\%$ between 20 and 6 cm. When high-resolution images were obtained, one could in principle separate the thermal and nonthermal contributions visually; however since the region subtended by the disk is also partly influenced by synchrotron radiation, more refined models were needed. De Pater et al (1982) used de Pater's (1981a,b) model calculations and high-resolution images of the planet to determine more accurate values for the thermal flux density at wavelengths of 6, 11, and 21 cm.

Figure 4 shows a disk-averaged spectrum of Jupiter, with data taken from de Pater & Massie (1985, and references therein) and Klein & Gulkis

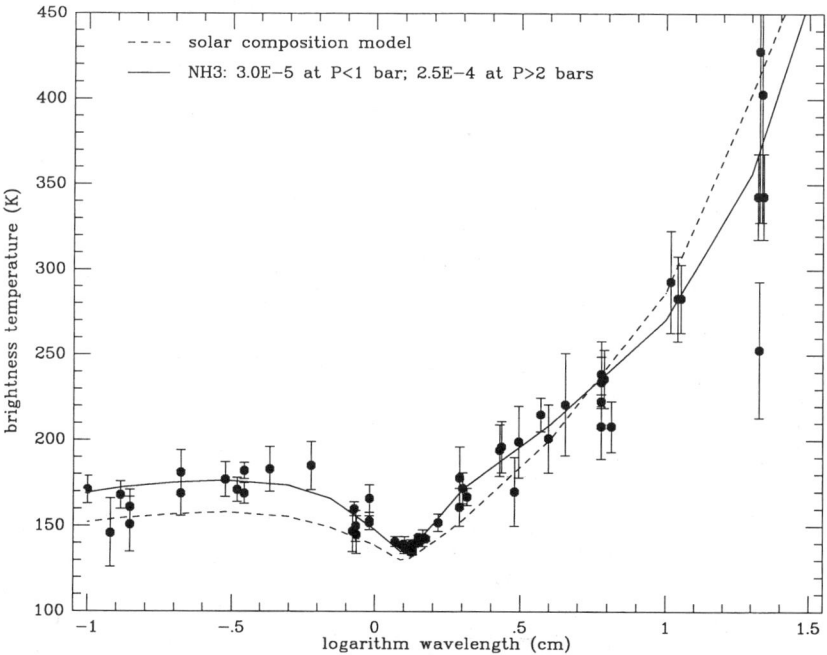

Figure 4 Jupiter's radio spectrum. Superimposed are various model atmosphere calculations. Dashed curve—solar composition atmosphere. Solid curve—NH_3: 3×10^{-5} at $P < 1$ bar and 2.5×10^{-4} at $P > 2$ bars. The gas is subsaturated at $P < 0.6$ bars.

(1978). Superimposed are model atmosphere calculations (after de Pater & Massie 1985). The dashed curve is for a solar composition model[2], and the solid curve is for a model atmosphere in which ammonia gas is depleted compared with the solar nitrogen value by a factor of ~ 5 at $P < 1$ bar and is enhanced by a factor of 1.5 at $P > 2$ bars. In addition, NH_3 gas is subsaturated at $P \lesssim 0.6$ bars to fit the radio spectrum near 1.3 cm. The latter model provides a good fit to the data. The loss in NH_3 gas at $1 < P < 2$ bars is likely due to the formation on an NH_4SH cloud. The existence of such a cloud was first postulated by Lewis (1969) to explain the apparent nondetection of H_2S gas in Jupiter's atmosphere. If indeed the NH_3 mixing ratio drops by a factor of ~ 8 owing to the formation of an NH_4SH cloud, H_2S needs to be enhanced by a factor of 6–7 compared with the solar value (de Pater 1986). Such an enhancement is compatible with theories on the formation and evolution of the giant planets (Hubbard 1984, Pollack & Bodenheimer 1989). The subsaturation of ammonia gas at $P \lesssim 0.6$ bars is likely due to photolysis.

High-resolution radio images of the planet at 2- and 6-cm wavelength are displayed in Figure 5 (from de Pater & Dickel 1986). The resolution is 1.2" and 2", respectively. The images show bright horizontal bands across the disk, which coincide with the brown belts seen at visible and infrared (IR) wavelengths. These bands have a higher brightness temperature, which most likely is due to a depletion in ammonia gas relative to the zonal regions. Figure 6 shows a summary of the atmospheric structure in Jupiter's North Equatorial Belt (NEB) and Equatorial Zone (EZ) (from de Pater 1986). The NH_3 gas is enhanced by a factor of 1.5 at $P > 2.2$ bars. We see a gradual decrease in NH_3 gas between 2 and 1 bars, where in the NEB the gas is depleted over a small altitude range by a factor of ~ 10, and in the EZ by a factor of ~ 5 over a larger altitude range. The depletion in both regions is likely caused by the formation of an NH_4SH cloud layer, which extends over a larger altitude range in the EZ than in the NEB. At $P < 0.6$ bars, NH_3 is condensed out and partially destroyed by photodissociation effects, as mentioned above.

De Pater (1986) shows that the difference in latent heat release upon formation of an extensive NH_4SH cloud in the EZ, versus the latent heat release from a 3 times smaller cloud in the NEB, is 3–4 K, enough to drive the zonal winds observed on the planet. This theory predicts the winds to extend down to a depth of ~ 2 bars. It further confirms the historical picture of rising gas in the warmer zones, with subsidence in the belts. With the smaller NH_3 gas reservoir above the belts, the NH_3 ice clouds

---

[2] The solar mixing ratios are as follows: CH_4/H_2, 8.35×10^{-4}; NH_3/H_2, 1.74×10^{-4}; H_2O/H_2, 1.38×10^{-3}; H_2S/H_2, 3.76×10^{-5}.

are expected to be thinner in belts than in zones. This general cloud picture is in agreement with the structure suggested by West et al (1986) based upon ground-based and spacecraft data at visible and IR wavelengths. Note, however, the difference between the two observing techniques: The radio data probe directly the gas from which the clouds are formed, whereas IR and visible data are sensitive to "a" cloud layer. Hence, together, the data contain a full picture of Jupiter's cloud structure.

Saturn

ATMOSPHERE The results of microwave observations of Saturn have been compiled by Klein et al (1978). Since most observations have been obtained with single-dish telescopes, which have a very low spatial resolution, the flux density of the entire Saturnian system was recorded. Measurements obtained with radio interferometers provided some strong constraints on the microwave properties of Saturn's ring system. Klein et al (1978) used this information to develop a simple model for the influence of the planet's rings on its microwave spectrum and corrected the radio data for it. The resulting thermal spectrum of Saturn is shown in Figure 7. [The data are complemented with data by Dowling et al (1987), Grossman et al (1989), Briggs & Sackett (1989), and I. de Pater & J. R. Dickel (in preparation).] Superimposed is a model calculation after Briggs & Sackett (1989) for a solar composition atmosphere (dashed curve) and an atmosphere in which H_2O and CH_4 are enhanced by a factor of 5, NH_3 gas is enhanced by a factor of 3, and H_2S gas is enhanced by a factor of 11 in the planet's deep atmosphere (solid line). In contrast to de Pater & Massie's (1985) work, these model calculations include the cloud condensation effects of NH_4SH and the solution cloud. The NH_3 mixing ratio decreases with altitude primarily as a result of the formation of an NH_4SH cloud at the 3–5 bars level. The data at short centimeter wavelengths imply a larger decrease in NH_3 gas than is suggested by Briggs & Sackett, which can be obtained by increasing the H_2S abundance to 12 or 13 times the solar value. The enhancement of NH_3 gas in the deep atmosphere is necessary to reproduce the long-wavelength range of Saturn's spectrum. In the past, Klein et al (1978) suggested that the low brightness temperature at wavelengths longward of ~ 6 cm was due to the presence of a water cloud at levels in the atmosphere below ~ 270 K. However, an enhancement in the water mixing ratio changes the spectrum at 6–20 cm only slightly if the atmosphere is in thermochemical equilibrium.

The first radio image of the planet was obtained by Schloerb et al (1979) at a wavelength of 3.7 cm. The data were obtained with the interferometer of the Owens Valley Radio Observatory using 13 different baselines. The resolution was $8'' \times 15''$. After subtraction of a uniform disk from the map,

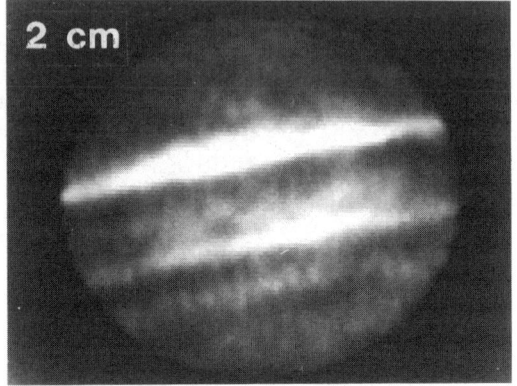

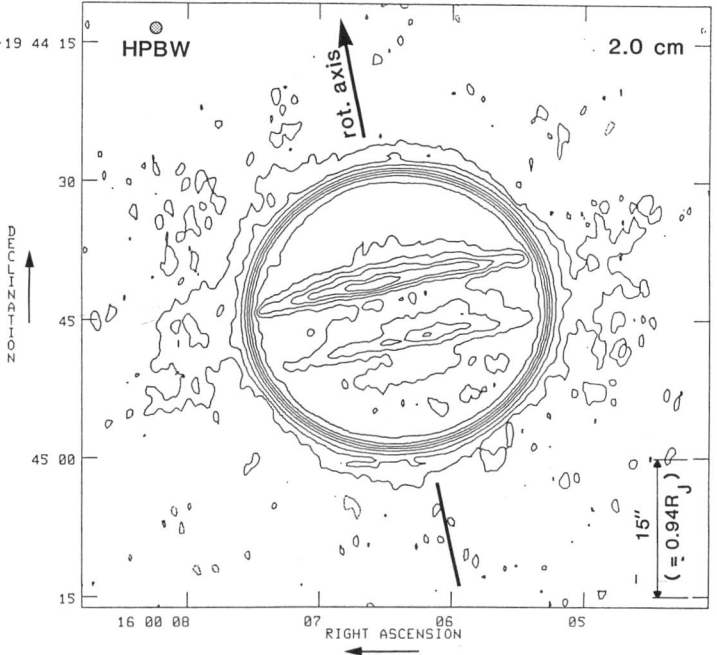

Figure 5 Radio photo (*above, top*) and contour map (*above, bottom*) of Jupiter at a wavelength of 2 cm. Contour values (in kelvins) are 1.8, 5, 9, 18, 44, 71, 98, 124, 151, 160, 168, and 174 K: (*facing page*) Contour map of Jupiter's total intensity at 6 cm. Contour levels (in kelvins) are 8.5, 14, 20, 28, 43, 57, 71, 114, 157, 200, 242, 256, 270, and 279. On all contour maps, negative values, with the same absolute levels, are indicated by dashed contours (from de Pater & Dickel 1986).

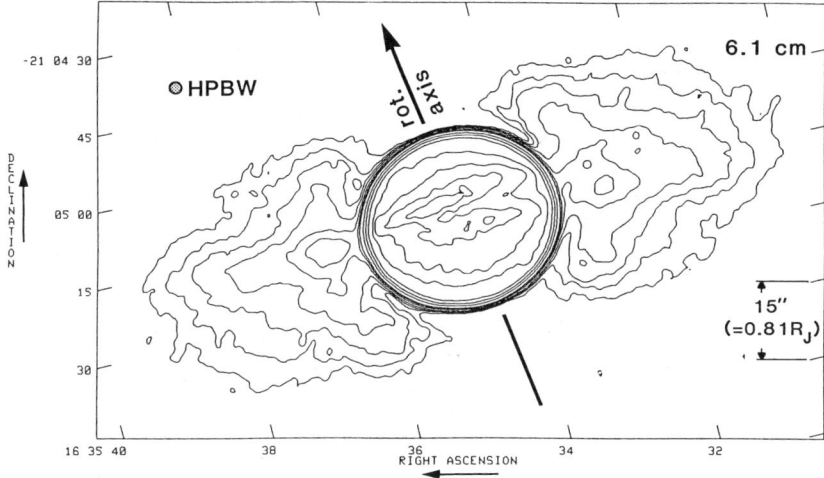

Figure 5 (continued)

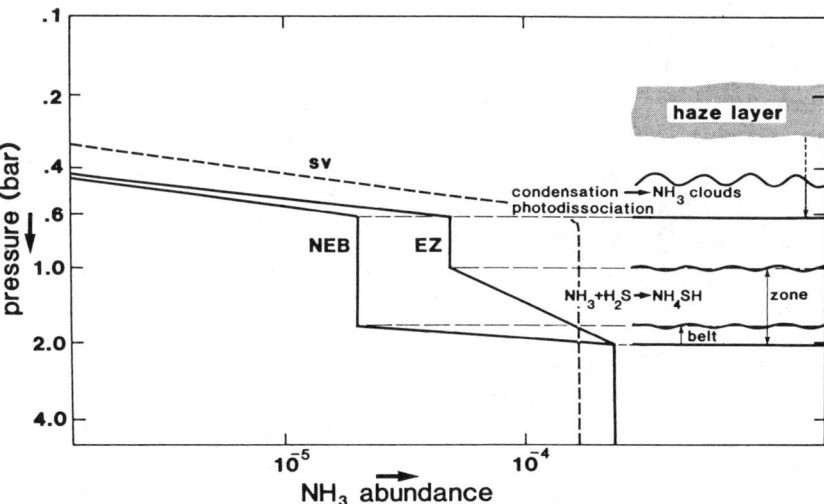

Figure 6 The altitude distribution of ammonia gas in Jupiter's atmosphere, in the NEB and EZ (from de Pater 1986). The various cloud layers are sketched on the right side; the saturated vapor curve for ammonia gas of solar concentration is indicated by the line labeled "sv".

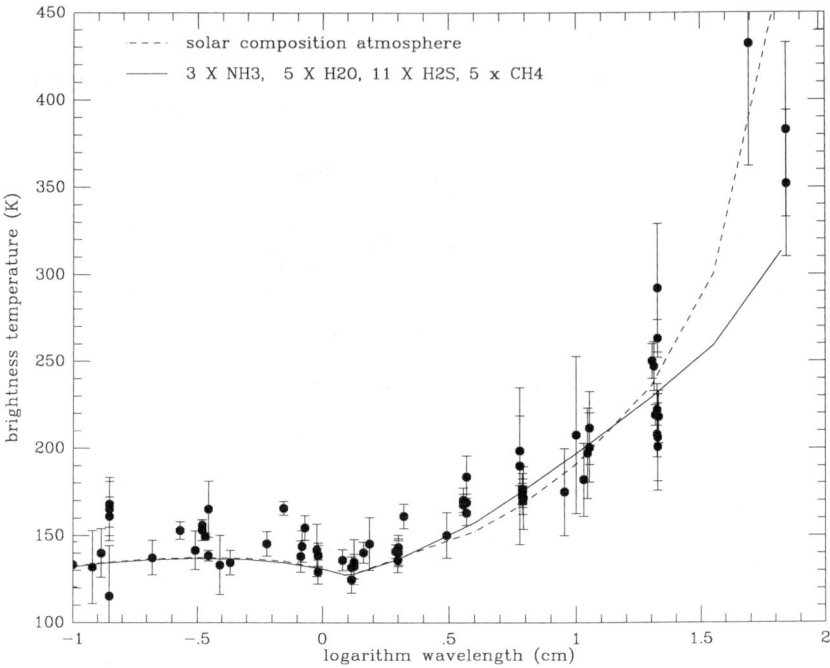

Figure 7 Saturn's radio spectrum. Superimposed are model atmosphere calculations after Briggs & Sackett (1989). See text for further explanation.

the contribution from the rings was visible as a positive signature at either side of the planet, and as a negative signature where the rings obscured part of the planet's radio emission. The first VLA images were published by de Pater & Dickel (1982). Subsequently, images with a better quality have been obtained (e.g. de Pater & Dickel 1983, de Pater 1985, Grossman et al 1989, I. de Pater & J. R. Dickel, in preparation). Figure 8 shows a few of the VLA images (I. de Pater & J. R. Dickel, in preparation) at 2 and 6 cm. The resolution in these images is 1.5". Figures 8 (*top, left*) and 8 (*top, right*) show radio photos at 2 and 6 cm, at ring inclination angles $B = 26°$ and 20°, respectively. Figure 8c shows a contour map of an image at 6 cm when $B = 25°$. The images clearly show the A and B rings, separated by the Cassini Division. At 2 cm, the planetary disk shows no structure, other than that the planet seems less limb darkened in the north-south direction than expected for a uniform atmosphere. At 6 cm, however, there is a clear bright band across the planet at approximately 30° latitude. Figure 9 shows meridional scans through several images at 6 cm, taken in different years (August 1981, January 1982, January 1984, and June 1986).

RADIO IMAGES OF THE PLANETS 367

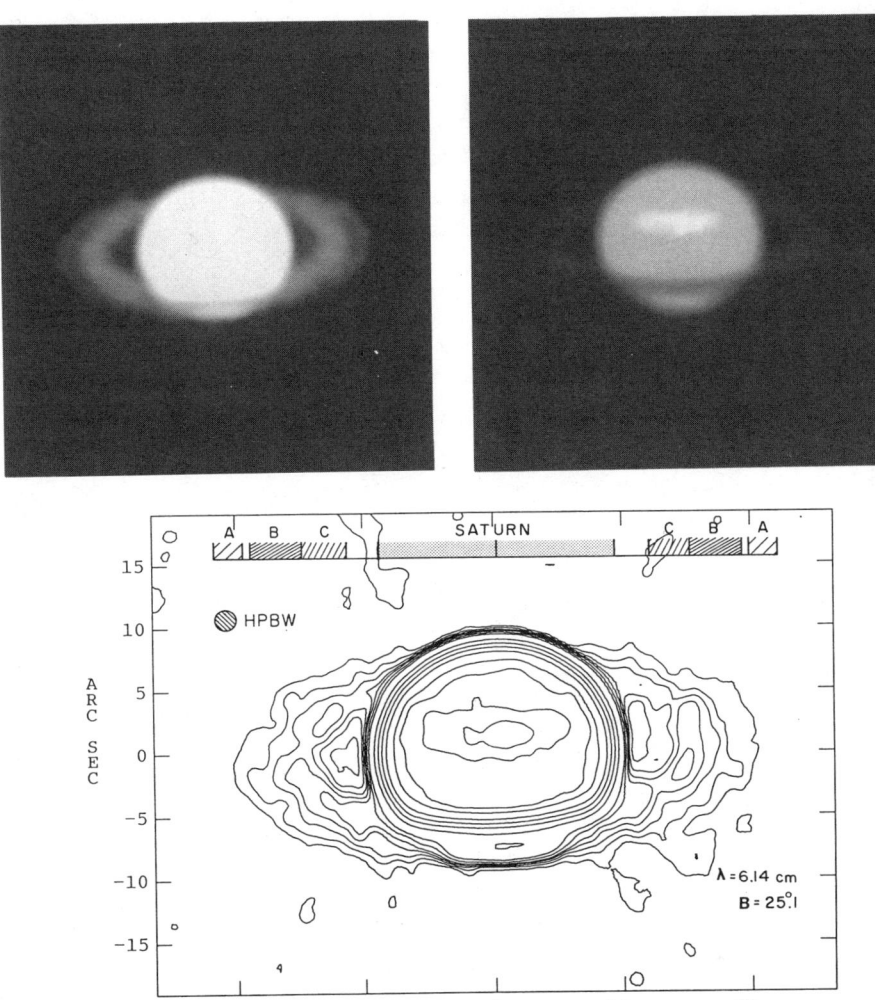

Figure 8 Radio images of Saturn (from I. de Pater & J. R. Dickel, in preparation), at 2 and 6 cm and at different ring inclination angles B. The resolution in all images is 1.5″. (*top, left*) Radio photo at $\lambda = 2$ cm, $B = 26°$; (*top, right*) radio photo at $\lambda = 6$ cm, $B = 20°$; (*bottom*) contour map at $\lambda = 6$ cm, $B = 25°$. Contour values (in kelvins) are 1.9, 3.7, 5.6, 7.4, 9.3, 13.0, 18.6, 46.5, 74.3, 102.2, 130.0, 158.0, 167.0, 176.5, and 182.1. Negative contour values, with the same absolute levels, are indicated by dashed contours. The images were taken in December 1986, January 1984, and June 1986, respectively. The bar at the top shows schematically the A, B, and C rings at either side of Saturn.

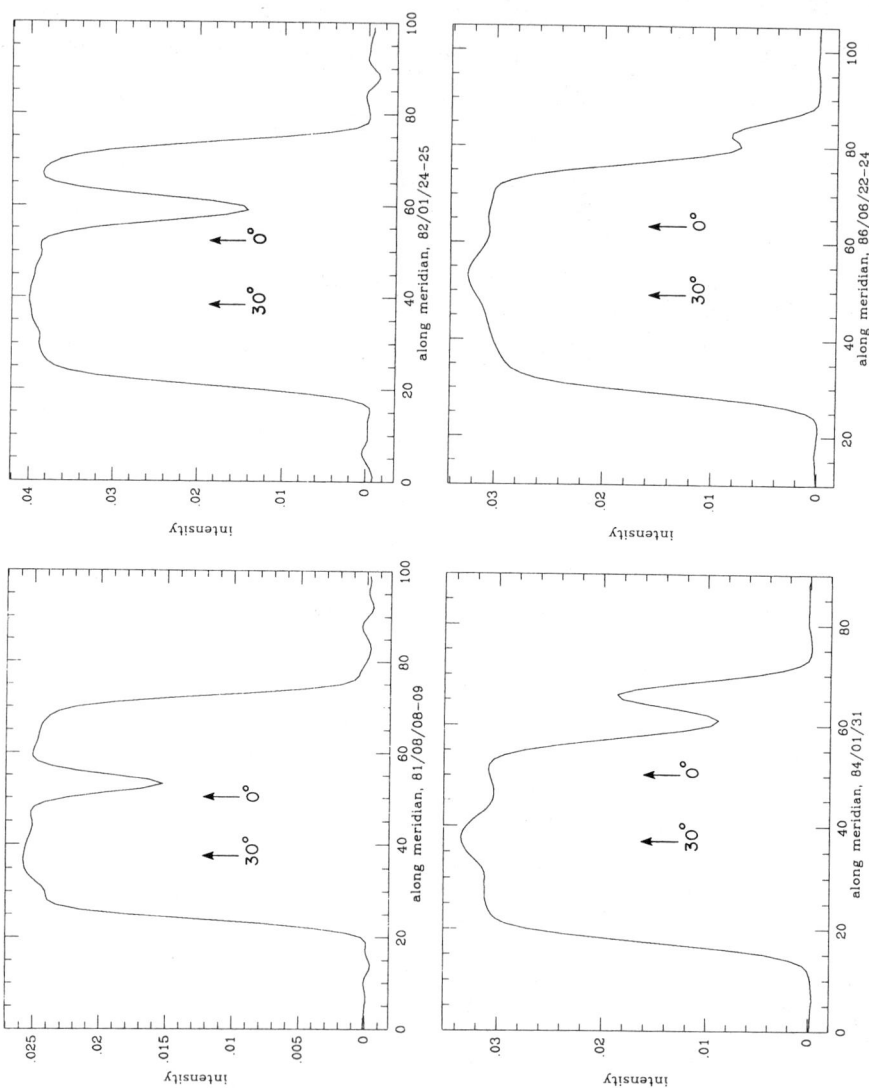

Figure 9 Meridional scans through 6-cm images of Saturn. The data were taken in August 1981, January 1982, January 1984, and June 1986. The planetocentric latitudes of 0° and 30° are indicated by arrows.

The absorption effects by the rings differ from year to year owing to the varying ring inclination angle. The planetocentric latitudes of 0° and of 30° are indicated by arrows. Note that the bright band in the atmosphere appears to have moved southward over the years. Also, the intensity of this band seems to have changed, as has the amount of limb darkening toward the pole. Higher resolution images suggest the presence of a second, much weaker bright band near the equator, as well as another one in the southern hemisphere.

Model atmosphere calculations show that the difference in brightness temperature between the bright band and the rest of the planet can be explained by a difference in the ammonia mixing ratio at levels in the atmosphere where $P \sim 1$–5 bars. Grossman et al (1989) suggest a 30% decrease in the NH_3 mixing ratio in the bright band. However, with a 3 times solar mixing ratio of NH_3 gas at $P \gtrsim 5$ bars, the ammonia abundance in the bright band must be decreased by nearly 50% at $P < 5$ bars. Such an effect would also be visible at 2 cm wavelength. I. De Pater & J. R. Dickel (in preparation) suggest that the global NH_3 abundance in the upper atmosphere is about 5×10^{-5}. The low abundance extends to deeper levels in Saturn's atmosphere in the bright band than at other positions (down to 4–4.5 bars in the bright band, and to 2.5–3 bars everywhere else). This implies that the NH_4SH cloud layer forms at a deeper level in the atmosphere in the bright band than at other latitudes. In addition, the vertical extent of the cloud layer is about 0.5 bar in both regions. Note that this is opposite to what we saw on Jupiter.

From limb-darkening curves it appears that in the atmospheres of both Jupiter and Saturn the ammonia abundance above the NH_4SH cloud deck decreases by approximately a factor of 2 toward the poles, while the underlying higher ammonia abundance starts at higher levels in the atmosphere near the polar regions. Since the ammonia abundance above the NH_4SH cloud layer is largely determined by the abundance of H_2S, the latter may increase somewhat from the equator to the pole. Also, the altitude at which the NH_4SH cloud forms apparently varies with latitude. The reaction $NH_3 + H_2S \rightarrow NH_4SH$ is heterogeneous, and so it requires the presence of solid surfaces as, for example, aerosols. Hence, the variation in the NH_3 abundance with latitude and altitude depends upon the H_2S abundance as well as on the aerosol distribution, which is likely tied in with the dynamics on the planet.

Far-infrared (IRIS) observations obtained with the Voyager spacecraft probe pressure levels of ~ 0.3–0.7 bars. A warm band is visible at mid-latitudes, similar to the band seen at radio wavelengths (Bézard et al 1984). However, if, at these pressure levels, this region is hot owing to an enhancement in the physical temperature of this band, the hot band should show up in the radio images at 2 cm. Likely, as suggested by the authors, the feature seen at infrared wavelengths is due to a latitudinal variation in cloud opacity. A thinner NH_3-ice cloud at midlatitudes would imply a region of downwelling, which is consistent with the interpretation of an altitude variation with latitude of the NH_4SH clouds derived from the radio data.

RINGS   As mentioned above, radio interferometric observations were used to extract information on the microwave properties of Saturn's rings. Observations at different wavelengths and polarizations can be used to determine the compositions and sizes of the ring particles through their scattering characteristics. Cuzzi et al (1980) present detailed theoretical models of the brightness of Saturn's rings at microwave wavelengths, including both intrinsic ring emission and diffuse scattering of the planetary emission by the rings.

Schloerb et al (1980) show that the effective normal optical depth of the A and B rings decreases with decreasing ring inclination angle (such as is expected for the classical A and B rings), with a clear open gap (the Cassini Division) in between. Figure 10 shows a graph of the data points of the

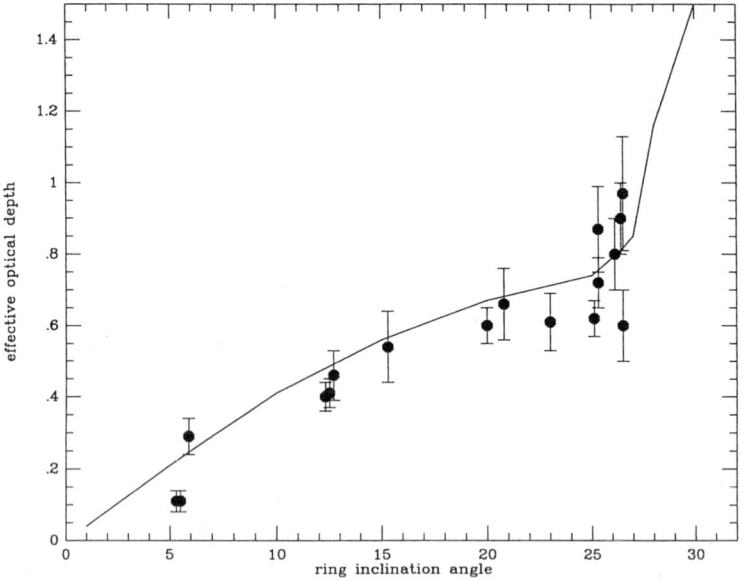

Figure 10   A graph of the effective optical depth of the combined A and B rings as a function of ring inclination angle (after Schloerb et al 1980). The solid line is a prediction for the effective optical depth if $\tau_A = 0.7$ and $\tau_B = 1.5$

effective optical depth τ_{eff}. Superimposed on the graph is a theoretical calculation of τ_{eff} (after Schloerb et al 1980; not a fit to the data!):

$$\exp^{-\tau_{\text{eff}}/\sin|B|} = f_A \exp^{-\tau_A/\sin|B|} + f_B \exp^{-\tau_B/\sin|B|} + f_{CD}, \qquad 4.$$

where f_A, f_B, and f_{CD} are the fractions of the obscured region of the planet blocked by the A and B rings and the Cassini Division, respectively; B is the inclination angle of the ring plane. The Cassini Division was assumed to be entirely transparant at radio wavelengths. The optical depth in the A ring (τ_A) was taken to be 0.7, and in the B ring (τ_B) as 1.5. It is clear that the trend in the observed effective optical depth follows the theoretical prediction. In this figure no differentiation is made between points at different wavelengths, although Cuzzi et al (1980) showed that the optical depth of the rings may decrease with wavelength if the rings are wide open. They used this fact to constrain the size of the ring particles to be primarily less than a meter.

Schloerb et al (1980) point out that at small ring inclination angles, the effective optical depth approaches the optical depth of the least optically thick region. With a transparent Cassini Division, or other clear open gaps in the rings, this will be zero, as shown on the graph. Cuzzi et al (1980) show that the ring particles scatter preferentially in the forward direction. High-resolution VLA observations of the ring system, such as those presented in Figure 8 confirm this suggestion, since the far side of the ring system is usually weaker than the near side. The fact that the particles scatter preferentially in the forward direction will also cause the effective ring optical depth to decrease with decreasing ring inclination angle.

Thermal radiation from the ring particles has been observed at wavelengths between ~ 10 μm and ~ 1 cm. Somewhere between 100 μm and 1 mm wavelength, the ring brightness temperature drops below the blackbody behavior, and longward of 1-cm wavelength the rings behave nearly like a perfect reflecting surface. Esposito et al (1984) show that the brightness temperature rises approximately as $\sim 1/\lambda$ shortward of 1 cm, as expected from an optically thick slab of particles that are nearly conservative scatterers (Cuzzi et al 1980). The low intrinsic brightness temperature of the rings at centimeter wavelengths, together with the observed perfect reflector behavior, hints at an icy composition for the particles. In addition, the ring particles cannot be primarily smaller than ~ 1 cm, independent of their precise composition (Cuzzi et al 1980).

The high-resolution images better constrain the optical depth and ring brightness temperature of the individual A, B, and C rings. Rather than making model fits to the ultraviolet (UV) data, one can use the images directly to determine the ring properties. Grossman et al (1989) show the optical depth and ring brightness temperature for the three rings and

Cassini Division separately, as determined directly from the images. They conclude from their measurements that the A and C rings contain many particles in the size range 0.6–2.0 cm, whereas the B ring contains a greater population of large particles.

Uranus

The angular diameter of Uranus subtends an angle of less than 4" as seen from the Earth; therefore, until the VLA could be used to image the planet, most radio observations were of the unresolved disk. The Planetary Radio Astronomy group at the Jet Propulsion Laboratory (M. J. Klein, personal communication) assembled a catalogue of all radio data of Uranus and recalibrated older data points so that all temperatures are on the same flux density scale. Their catalogue was published in a review paper on Uranus by Gulkis & de Pater (1984); since this publication, more observations have been made (see de Pater & Gulkis, 1988, Berge et al 1988), which are included in the spectrum shown below.

SPECTRUM A spectrum of disk-averaged brightness temperatures of Uranus is shown in Figure 11. Since Uranus' brightness temperature

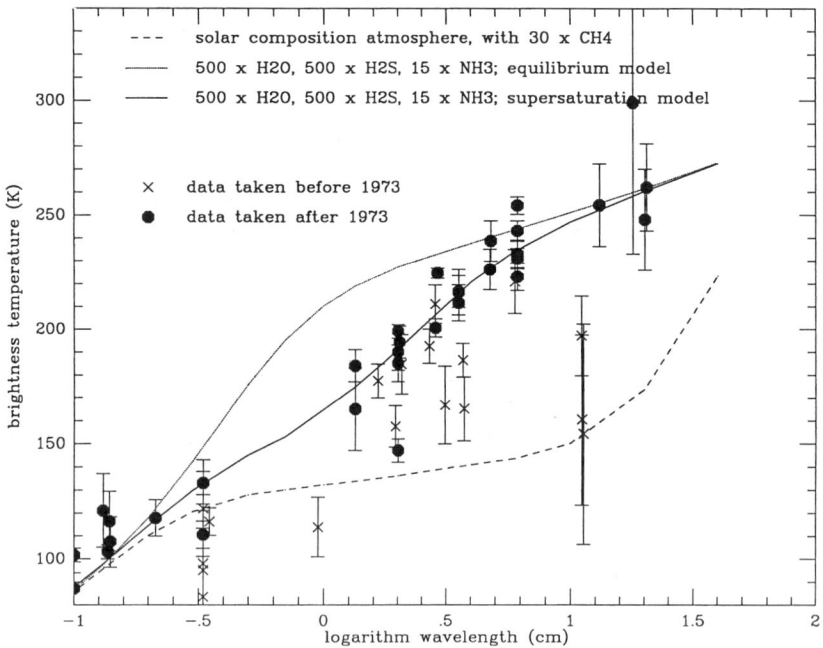

Figure 11 Radio spectrum of Uranus. Superimposed are various model atmosphere calculations from de Pater et al (1989), as indicated in the figure. See text for further explanations.

steadily increased until 1973 (e.g. Klein & Turegano 1978, Gulkis et al 1983, Gulkis & de Pater 1984), the data points taken before and after 1973 are distinguished by the use of different symbols (crosses for data taken before 1973; filled circles for data taken after 1973). This increase in brightness temperature is due to a large temperature gradient between a warm pole and a cold equator. Such a temperature profile results in an increase in the disk-averaged brightness temperature when Uranus' pole comes into view (Gulkis & de Pater 1984).

As was first pointed out by Gulkis et al (1978), Uranus' brightness temperature is too warm to be matched by a solar composition atmosphere, as indicated by the dashed curve on Figure 11. Atmospheric models by de Pater et al (1989) include the formation of the NH_4SH and solution cloud. Their results are also shown in Figure 11: The dotted curve is a model for an atmosphere in thermochemical equilibrium in which both the H_2S and H_2O abundances are enhanced by a factor of 500 compared with the solar S and O values, respectively, and in which NH_3 is enhanced by a factor of 15 compared with the solar N value. [Note that CH_4 was always assumed to be 30 times enriched above the solar C value to match the 2% mixing ratio in the upper troposphere, as observed by Lindal et al (1987).] The solid curve is for the same model but with NH_4SH "supersaturated" at $P < 30$ bars ($T < 240$ K, where the NH_3 abundance is 3.5×10^{-7}). In the latter model no condensation NH_4SH occurs, even though the product of the partial pressures of H_2S and NH_3 exceeds the equilibrium constant at these pressure levels. This effectively "forced" the ammonia abundance to be constant over a large range in altitude, a feature that is necessary to match the steep part of the planet's radio spectrum at wavelengths between 0.3 and 6 cm. A similar result can be obtained by a rapid vertical mixing in the atmosphere. However, since this required the eddy diffusion coefficient to be unrealistically large for this process to be effective, de Pater et al dismissed the mixing model. In addition to the supersaturation model, the H_2S concentration needs to be large (>100 times the solar S value) to force the formation of the NH_4SH cloud deep in the atmosphere, so ammonia gas gets depleted significantly at large depths in the atmosphere. De Pater et al also considered the possibility of a subsolar NH_3 abundance throughout the atmosphere but dismissed this case: The synthetic spectra did not fit the data as well as those in which ammonia gas was equal to or enhanced above the solar value. In addition, if we assume that the N/C ratio is ~ 0.5, as is expected from planet formation theories (e.g. Pollack & Bodenheimer 1989), nitrogen must be present in the form of N_2 if the NH_3 mixing ratio is subsolar. However, the upper limit to the detection of N_2 by the *Voyager 1* UVS experiment is on the order of 1 ppb at the 1 mbar level (de Pater et al 1989). This implies that most of the nitrogen in Uranus' atmosphere must be in the form of NH_3, not N_2. Hence, the authors concluded the abundances of the various constituents to be as follows: $S > 100$ times solar; $N > 1-10$ times solar; $O >$ solar; and $S/N > 3$ if $H_2O \gtrsim 100$ times solar, or $S/N > 5$ if $H_2O \lesssim 100$ times solar. Unfortunately, they could not place a tighter constraint on the water abundance.

LATITUDINAL BRIGHTNESS DISTRIBUTION Briggs & Andrew (1980) were the first to directly observe a large temperature gradient on Uranus between the equator and the pole, at a wavelength of 6 cm. Their conclusions were based upon visibility data obtained with an interferometer. The first radio images were published by Jaffe et al (1984). At 2 cm the image showed a symmetric disk, as expected for a uniform gaseous planet, with the brightest point at the sub-Earth point and limb darkening occurring toward the limb. At a wavelength of 6 cm, however, the planet appeared asymmetric in that it showed the brightest point on the planet to be near the pole rather than the subsolar point. Since these initial observations, Uranus has been imaged regularly at 2 and 6 cm (de Pater & Gulkis 1988, Berge et al 1988). Examples of the images are shown in Figure 12, which displays a 6-cm image from 1982, and a 2-cm image from 1984 (from de Pater & Gulkis 1988). The asymmetry at 6 cm is always present, although the detailed latitudinal distribution of the brightness temperature varied significantly over the years (de Pater et al 1989). At 2 cm, the position of the brightest point on the disk appears to have moved from the subsolar point to a point closer to the pole between the years 1980 and 1984–85.

Even though the detailed latitudinal structure changes drastically on time scales of about a year (or shorter), the general zonal distribution always seems present: a hot polar region at latitudes $>70°$ (270–280 K at 6 cm, and 240–250 K at 2 cm, a cold equatorial band below roughly 30–40° (220 K at 6 cm, and 160–170 K at 2 cm), and up to two bands at midlatitudes (~ 250 K at 6 cm, and 200–220 K at 2 cm). These zonal variations can be explained in terms of a latitudinal variation in the ammonia abundance: Using de Pater et al's (1989) "supersaturation" model, the NH_3 abundance is 2×10^{-6} at $T < 240$ K in the equatorial region, $1-2 \times 10^{-7}$ at $T < 220$ K at midlatitudes, and 10^{-7} down to 280 K in the polar region. The gaseous ammonia abundance will follow the saturated vapor curve at $T \lesssim 145$ K when NH_3 gas condenses out. These abundances hint at a general upwelling of gas at midlatitudes, with subsidence in the equatorial region (and at higher latitudes). This atmospheric circulation supports the meridional flow as derived from Voyager observations by Flasar et al (1987). It further causes condensation nuclei to be present at higher altitudes at midlatitudes than in the equatorial region.

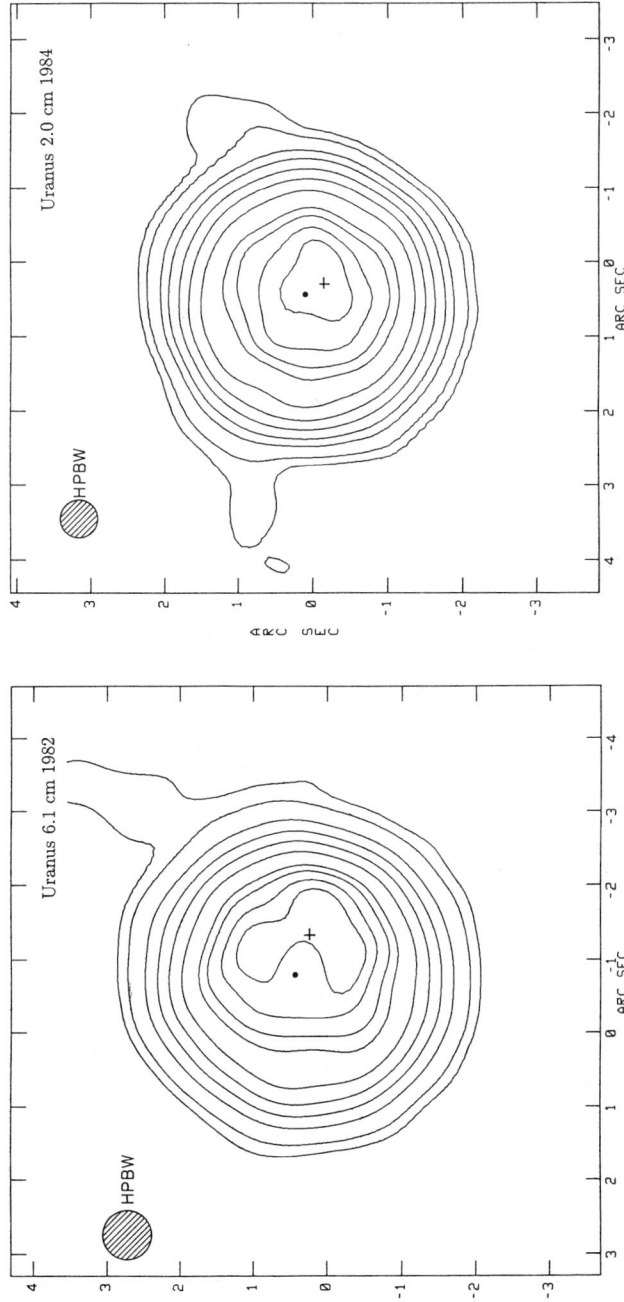

Figure 12 Radio images of Uranus at (*left*) 6 cm and (*right*) 2 cm (from de Pater & Gulkis 1988). The cross indicates the position of the pole, and the dot of the position of the subsolar point. The resolution of the images is 0.65" for the 6-cm image and 0.5" for the 2-cm image. The contour values (in kelvins) are (*a*) 14, 28, 69, 110, 151, 193, 234, 248, 261, and 270; and (*b*) 11, 22, 55, 88, 121, 154, 187, 198, 209, and 216.

Since these nuclei are necessary for the reaction $NH_3 + H_2S \rightarrow NH_4SH$, to occur, it supports the idea that the formation of NH_4SH is confined to deeper levels in the atmosphere in the equatorial region than at higher latitudes. The low ammonia abundance in the polar regions, down to levels as deep as 280 K, cannot be explained in terms of the "supersaturation" model; de Pater et al suggest the existence of strong downdrafts of dry air (from which the NH_3 has been removed by condensation) to a depth of at least 280 K.

Neptune

The first radio astronomical detection of Neptune was made in 1966 by Kellermann & Pauliny-Toth (1966) at a wavelength of 1.9 cm. Despite the lapse of time since this measurement, the spectrum of the planet is still poorly defined. Obviously, this situation is mainly due to the faintness of this distant object, which causes many problems due to confusion in the signals received by single antennas. Such confusion is minimized when the source is observed with an interferometer. De Pater & Richmond (1989) recently published a paper on VLA observations of Neptune and refined the planet's spectrum. Their spectrum is shown in Figure 13, with the VLA data points indicated by filled circles, and other data indicated by crosses. It is clear that Neptune, like Uranus, is too warm at centimeter wavelengths for a solar composition atmosphere (dashed curve). A straight model atmosphere calculation (after de Pater & Massie 1985) gives a best fit to the data if the NH_3 abundance is equal to 3×10^{-6} throughout the atmosphere (dot-dash curve). Model atmosphere calculations after Romani et al (1989) are presented by the dotted and solid curves: The dotted curve is a calculation in thermochemical equilibrium for a planet with the same composition as Uranus (i.e. $30 \times CH_4$, $500 \times H_2O$, $500 \times H_2S$, and $15 \times NH_3$ compared with the solar C, O, S, and N values). The solid curve is the same calculation for an atmosphere in which the NH_4SH cloud was assumed to be supersaturated at layers where $T < 240$ K. Although the latter curve fits the data shortward of ~ 10 cm reasonably well, neither the solid nor the dotted curve can match the rather high brightness temperature of 318 ± 16 K at 20 cm. These data can only be matched if NH_3 gas is less than solar throughout the atmosphere by a factor of ~ 50 compared with the solar N value, as shown by the dot-dash curve. Based upon planetary formation theories and a comparison with nearby Uranus, a depletion of NH_3 gas by nearly two orders of magnitude compared with the solar value seems highly unlikely.

SYNCHROTRON RADIATION? In analogy with Jupiter, de Pater & Richmond (1989) suggested that the excess emission at 20 cm (0.4 ± 0.13 mJy) would

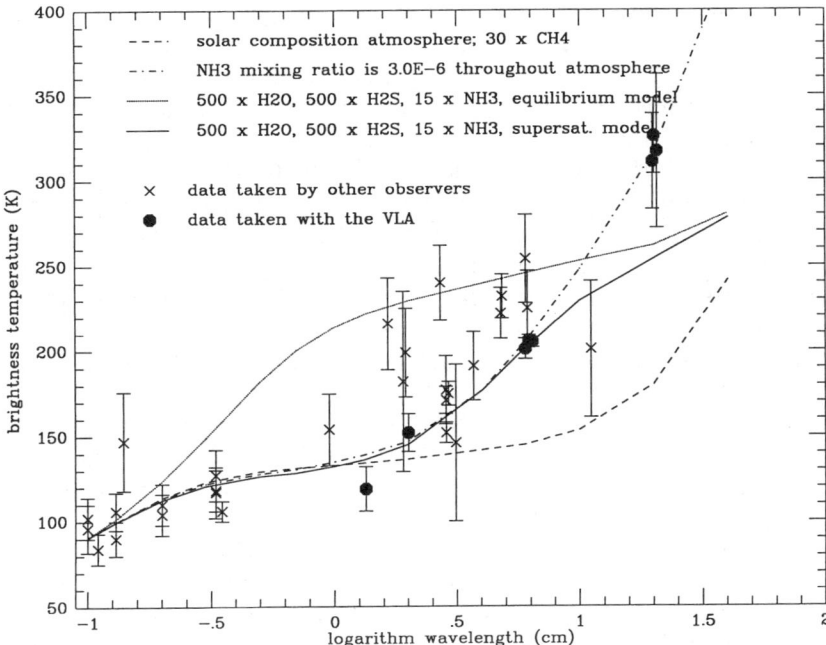

Figure 13 Radio spectrum of Neptune (after Romani et al 1989). Superimposed are various model atmosphere calculations, as explained in the text.

be due to synchrotron radiation emitted by energetic electrons in a Neptunian magnetic field. De Pater & Goertz (1989) performed some calculations on the inward diffusion of energetic electrons and suggested that Neptune's surface magnetic field strength at the equator would be ~ 0.5 G if the particle distribution is similar to that in Jupiter's radiation belts. The field strength has been confirmed by the recent *Voyager 2* flyby, but no energetic particles were observed.

Comparison of the Giant Planets and Implications on Formation Theories

The radio spectra and resolved images of the four giant planets all show a subsolar ammonia mixing ratio in the upper atmosphere and an enhanced mixing ratio in the lower atmosphere. On Jupiter and Saturn, NH_3 gas is depleted by a factor of $\sim 3-5$ at $P \lesssim 1$ (Jupiter)–3.5 (Saturn) bars and is enhanced by a factor of 1.5 on Jupiter and by a factor of 3–4 on Saturn at deeper levels in their atmospheres. Bright bands across the two planetary disks imply a latitudinal variation in the precise ammonia abundance.

Uranus and Neptune show a depletion in NH_3 gas of nearly two orders of magnitude over a large altitude range in the atmosphere. The gas is likely enhanced by an order of magnitude or more at deeper levels. In addition, Uranus shows a large pole-to-equator gradient in the ammonia abundance.

The loss of NH_3 gas in the atmospheres of all four giant planets is most likely due to the formation of NH_4SH, which in thermochemical equilibrium calculations is expected to form at the pressure levels where the decrease in the ammonia gas abundance is observed. To obtain a large enough loss in ammonia gas at the right pressure levels, the H_2S abundance in Jupiter and Saturn needs to be enhanced by factors of 6–7 and 10–15, respectively, compared with the solar S value, and by factors $\geqslant 100$ on Uranus and Neptune. In addition, the S/N ratio on the outer two planets must exceed 3 if the H_2O abundance is larger than a few hundred times the solar O value, or exceed 5 if the H_2O abundance is less than this value.

Due to the variations in enrichment factors for the heavy elements in all four giant planets, the cloud structures of the planets are rather different. On Jupiter we do not expect to find a solution cloud, on Saturn it is small (base at $\sim$ 20-bars level), but on Uranus and Neptune it is very extensive (base near 2000-bars level). Water ice will form on all four planets at a temperature of about 270 K. At higher altitudes, the NH_4SH cloud layer will form. The base level of the cloud is roughly near 210–230 K (2–5 bars level) on Jupiter and Saturn, and near 280 K ($\sim$ 100-bars level) on Uranus and Neptune. On Jupiter and Saturn, no H_2S gas will be left above this cloud layer. At temperatures of 140–150 K, NH_3 gas will freeze out and form ammonia ice. This is the cloud layer "visible" at optical wavelengths. On Uranus and Neptune there is much H_2S gas present above the NH_4SH cloud layer. This gas will condense out at a temperature of about 170–180 K ($\sim$ 10–15 bars). Since NH_3 and H_2S gas is supersaturated above the NH_4SH cloud layer, we find a small ammonia-ice cloud near 120 K (3–4) bars level). On the latter two planets, the temperature also gets cold enough for CH_4 gas to freeze out, at about 80 K ($\sim$ 1-bar level). The latter cloud deck is "seen" at visible wavelengths.

Table 1 summarizes the abundances of various heavy elements in the giant planets (after Pollack & Bodenheimer 1989; supplemented with the values presented in this paper). These abundances should be compared with current models of planetary formation. Pollack & Bodenheimer (1989) favor the "core-instability" model for planetary formation, in which the core of the giant planets is formed first by solid-body accretion, similar to the formation of the terrestrial planets. When the mass reaches a critical value, gas accretion from the surrounding protoplanetary nebula becomes very rapid. This model accounts for the fact that the total mass made up

Table 1 Composition of the atmospheres of the giant planets (after Pollack & Bodenheimer 1989; supplemented with values presented in this paper)

	Abundance with respect to the solar volume mixing ratio				
Element	Jupiter	Saturn	Uranus	Neptune	References
C	2.3 ± 0.2	5.1 ± 2.3	35 ± 15	40 ± 20	Courtin et al 1984, Lindal et al 1987, Orton et al 1987
P	$> 1.4 \pm 0.4$	$> 2.8 \pm 1.6$			Courtin et al 1984
S	6–7	10–15	> 100	> 100	de Pater 1986, I. de Pater & J. R. Dickel, in preparation, Briggs & Sackett 1989, de Pater et al 1989, Romani et al 1989
O	> 1	> 1	> 1	> 1	Carlson et al 1988, de Pater & Massie 1985, de Pater et al 1989, Romani et al 1989
N	1.5 ± 0.2	3.5 ± 1	\multicolumn{2}{c}{$1 < N < 100$}	de Pater 1986, I. de Pater & J. R. Dickel, in preparation, Briggs & Sackett 1989, de Pater et al 1989, Romani et al 1989
He	0.65 ± 0.15	0.2 ± 0.15	1 ± 0.15		Conrath et al 1987

of "heavy" elements (those with atomic masses larger than H_2 and He) is similar for all four giant planets. For Jupiter, Saturn, and Uranus/Neptune this mass is roughly 5, 25, and 300 times larger, respectively, than would be expected from solar elemental abundances. Furthermore, models of the interior structure of the planets show that the envelopes of the planets also contain large amounts of heavy elements. This can be accounted for in the "core-instability" hypothesis by the fact that late accreting planetesimals find it increasingly difficult to penetrate through the denser and denser envelope.

Based upon the observed methane abundances in the atmosphere of the giant planets (2.3, 5.1, and 30–40 times the solar value on Jupiter, Saturn, and Uranus/Neptune, respectively), Pollack & Bodenheimer (1989, and references therein) estimate that about 10% of the carbon in the outer solar nebula was in the condensed phase during the epoch of planetary formation. Water and rocky and refractory elements, like Mg, Si, P, and S, were entirely in the solid phase, although part of the O was present in the form of mineral oxides and CO gas. The regular satellites of the giant planets are all made of rock and ice, with a mean rock/ice ratio of 55/45 by mass. However, since water is somewhat easier to dissolve in the envelopes of the forming giant planets, Pollack & Bodenheimer predict that the envelopes of the planets contain somewhat more water than rock by mass. Thus, they expect the O/S ratio to be slightly larger than the solar value. Like carbon, nitrogen is expected to have been present in both the gas (e.g. N_2) and solid phases. Since the N/C ratio was found to be subsolar in meteorites and comets, Pollack & Bodenheimer predict the N/C ratio in the envelopes of the giant planets to also be subsolar by a factor of ~ 2. Hence, the abundances for the various elements as given in Table 1 tend to support Pollack & Bodenheimer's theory of planetary formation. The N/C ratio, wherever determined, is typically of the order of 0.5, and the S/C ratio is of the order of 5 or larger; unfortunately, no constraints on the oxygen abundance are known, other than that H_2O is at least equal to the solar O value on all planets. Hence, the O/S ratio can indeed exceed unity, as the formation theories predict.

The latitudinal variation in the brightness temperature on the various giant planets is most likely due to a latitudinal variation in the precise ammonia abundance, caused by latitudinal variations in the location and thickness of the ammonium hydrosulfide cloud in the atmosphere. Since this cloud only forms in the presence of solid surfaces, the location likely coincides with layers of aerosols. The location and abundance of these particles depend upon the origin of the aerosols, as well as upon the dynamics in the atmosphere. From the images at centimeter wavelengths as presented in this paper, it appears that the conditions for the NH_4SH cloud differ from one planet to the next. On Jupiter, the vertical extent of the NH_4SH cloud is larger in zones than in the belts. The difference in latent heat release between the clouds in the zones and in the belts may drive the zonal winds, and it causes upwelling of gas in the zones and subsidence in the belts. On Saturn, the extent of the cloud in the bright band is similar to that in other regions, but it is confined to a deeper part of the atmosphere. This will cause latitudinal variations in the temperature-pressure profile between 2 and 5 bars, the region where the NH_4SH cloud forms. No clear correlation is seen with the wind profile measured with the Voyager spacecraft. The Voyager infrared data show enhanced temperatures at 730 mbar near $30°$ latitude which hints at a thinner NH_3-ice cloud deck above the bright band than in neighboring regions. Both the IR and radio data imply downwelling in the bright band and upwelling in the neighboring regions. On Uranus, we need a general increase in the NH_4SH cloud extent or in the aerosol distribution toward the pole. The distribution hints at a general upwelling of air at midlatitudes, with subsidence in the equatorial region and at higher latitudes. In addition, there must be strong downdrafts of dry air in the polar region. Unfortunately, we do not yet have high-resolution images of Neptune on which we can resolve individual zonal bands. We expect the planet to have a zonal distribution of gases/winds like those of the other three giant planets.

4. TERRESTRIAL PLANETS

Mars

SURFACE The temperature structure of the (sub)surface layers of Mars depends upon a balance between the solar illumination cycle, the heat transport within the crustal layers, and reradiation outward. As a result of the variation in heliocentric distance r_M, the surface brightness temperature will vary as $r_M^{-0.5}$. Scattering in the subsurface layers will change the radiative flux, a phenomenon that is hard to take into account in calculations. The temperature of the crust may be slightly increased above that expected from solar illumination by the atmospheric "greenhouse effect." In addition, although atmospheric dust storms are transparent at radio wavelengths, these storms may have a direct influence on the Martian flux: The dust particles contribute to the opacity at wavelengths shorter than ~ 40 μm, and therefore they affect the temperature gradient in the atmosphere and reduce the amount of solar radiation to the surface, thus causing a decrease in the surface brightness temperature. The latter effects have not yet been taken into account in models of Mars' surface.

Sunlight will heat Mars' surface during the day; the heat will be transported downward mainly by conduction. The amplitude and phase of the diurnal temperature variations and the temperature gradient with depth in the crust are largely determined by the thermal inertia $\gamma = (K\rho C)^{1/2}$ and the thermal skin depth of the material $L_t = (2K/\Omega\rho C)^{1/2}$. In these expressions, K is the thermal conductivity, ρ the density of the material, C the heat capacity, and Ω the angular velocity. Temperature variations are largest at the surface, where the temperature is determined directly by the solar radiation. Obviously, it is hottest near noon and coldest just before sunrise. Since the heat is transported downward by conduction, it takes time for the subsurface layers to heat up. Thus we see a phase lag in the diurnal temperature variation, and the amplitude of the variation will be diminished. When the thermal inertia, or more precisely the thermal conductivity, is low, the amplitude of the temperature wave is large and does not penetrate deeply into the crust. If the inertia, or thermal conductivity, is high, temperature variations are smaller but penetrate to greater depths in the subsurface layers.

Disk-averaged brightness temperatures of Mars at infrared as well as radio wavelengths show a variation with central meridian, or sub-Earth, longitude (CML; Andrew et al 1977, Epstein et al 1983); data at 20 μm, 3.5 mm, and 2.8 cm during the 1978 Mars opposition period are shown in Figure 14 (from Epstein et al 1983). Note that all observations are of Mars' day side; we cannot observe the night side from the ground. Since the disk-averaged surface temperature of the planet will be warmer than the subsurface layers, the 20-μm brightness temperature is higher than the radio brightness temperature. The temperature variations at infrared wavelengths are out of phase with those at radio wavelengths, a situation that can be understood in terms of the thermal inertia or conductivity discussed above. Thus, regions with a low inertia will have a high brightness temperature at infrared wavelengths and a low temperature at radio wavelengths. Epstein et al (1983) show that the reversal between the curves should occur at a wavelength between 0.2 and 3 mm. One can see in Figure 14 that the rotation curve amplitude at 3.5 mm is larger than that seen at 2.8 cm and 20 μm; Epstein et al attribute this to subsurface scattering from rocks or roughness on a scale less than 1.5 cm.

Rudy et al (1987; Rudy 1987) imaged Mars at 2 and 6 cm using the VLA in the most extended, or A, configuration. They obtained data in both the total intensity and polarized flux density. The latter is useful in determining the dielectric constant of the surface. Two seasons were observed: late spring in the northern hemisphere [at a planetocentric orbital longitude $L_s \sim 60°$ (the beginning of spring is at $L_s = 0°$)] and early summer in the southern hemisphere ($L_s \sim 305°$). Rudy et al determined the vari-

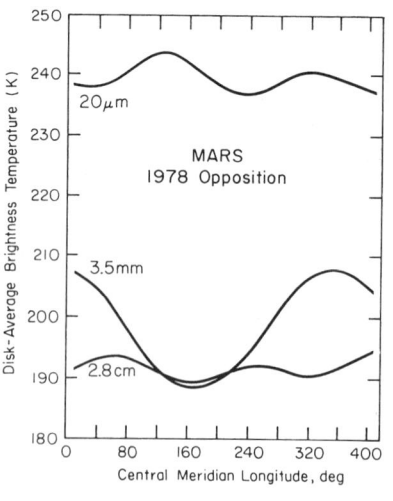

Figure 14 Best fits to the disk-averaged Martian brightness temperature at 20 μm, 3.5 mm, and 2.8 cm as a function of central meridian longitude. The data were taken during the 1978 Mars opposition period (from Epstein et al 1983).

ation in brightness temperature with latitude on the planet; since the data were smeared in longitude, no longitudinal information could be obtained. In both the northern and southern hemispheres there is a region near the polar caps that has a lower brightness temperature at 2- and 6-cm wavelength. Table 2 summarizes disk-averaged brightness temperatures, dielectric constants, subsurface densities, and the brightness temperatures of the cold poles for both sets of observations.

Rudy et al (1987) point out that the cold polar regions are likely the result of the presence of CO_2 frost on the surface. The thickness of the frost, including its distribution with space as well as depth, modifies the microwave brightness temperatures. Note, further, that the surface of Mars never gets much colder than the sublimation temperature of CO_2, whereas the temperature of layers underneath the CO_2 frost can be warmer than the sublimation temperature.

ATMOSPHERE Mars' atmosphere can be probed at millimeter wavelengths in the $J = 1$–0 and $J = 2$–1 rotational transitions of CO. CO is produced by photodissociation of CO_2, the primary constituent of Mars' atmosphere. Whether the line is seen in emission, absorption, or a combination thereof depends mainly on the temperature-pressure profile in the atmosphere and how this profile compares with the brightness temperature of the surface. The Viking temperature profiles (Seiff & Kirk 1977) show a roughly constant temperature of 140 K down to an altitude of about 60 km, below which it increases to ~ 210–220 K at the surface. The 115- and 230-GHz ^{12}CO lines are optically thick; thus, the core of the line is formed high up in the atmosphere where it is cold and hence is seen in absorption against the continuum background from the planet's surface. The wings of the line are formed in the lower atmosphere, just above the surface. As a consequence of the surface emissivity, the brightness temperature of the surface is somewhat less than the kinetic temperature in the atmosphere

Table 2 Properties of Mars' surface as derived by Rudy (1987) from VLA observations at 2 and 6 cm

Hemisphere	Wavelength (cm)	Whole-disk brightness temperature[a]	Whole-disk dielectric constant	Whole-disk subsurface density	Polar cold region
North	2	189 ± 7	2.34 ± 0.05	1.24 ± 0.16	126 ± 7
South	2	198 ± 7	2.02 ± 0.03	1.02 ± 0.16	183 ± 7
North	6	187 ± 5	2.70 ± 0.09	1.45 ± 0.18	150 ± 5
South	6	192 ± 5	2.47 ± 0.06	1.31 ± 0.16	148 ± 5

[a] Whole-disk brightness temperatures are normalized to a solar distance of 1.524 AU.

just above it. The wings of the line are therefore seen in emission against the continuum background.

Photolysis of CO_2 proceeds at a rate of $\sim 10^{12}$ cm^{-2} s^{-1} (Hunten 1974), while recombination of CO and O (CO + O + M → CO_2 + M) is very slow. Therefore the photochemical lifetime of CO is approximately 3 yr. Yet, the mixing ratio of CO is only $\sim 10^{-3}$. This suggests a catalytic recombination. Hunten (1974) suggested catalysis by odd hydrogen (H, OH, HO_2), aided by very rapid downward mixing and photolysis of H_2O_2. As a result of sublimation and freezing the abundance of odd hydrogen is unstable, which can lead to changes in the CO abundance by a factor of ~ 2 on a time scale of ~ 1 yr. Based upon the Phobos/UV, visible, and infrared data, Atreya & Blamont (1990) proposed that heterogeneous chemistry between CO and O involving aerosols may play a major role in the recombination process. Owing to large spatial as well as time variations in the abundance of such particles, the CO abundance may locally change over time scales as short as 10 days.

Time variability in CO over time scales between months and years has been reported more than once (Good & Schloerb 1981, Clancy et al 1983, Lellouch et al 1989). Since the determination of the CO abundance and the choice of temperature-pressure profile both influence the line profiles, a unique solution of the CO abundance can only be found if measurements of both the optically thick ^{12}CO and optically thin ^{13}CO lines are made. R. T. Clancy (personal communication) analyzed CO microwave spectra and suggests that the CO mixing ratio (averaged over the disk!) is constant with time between 1967 and 1988, but that a distinct change takes place in the temperature-pressure profile. Images of Mars in the CO lines might shed some light on possible spatial variations in the CO abundance with latitude and local time, and these variations will likely help us better understand the production and destruction of CO in Mars' atmosphere. Such data have recently been obtained at 3 mm with the Hat Creek Radio Interferometer[3] by D. L. Mitchell during Mars' opposition in 1988. In addition to adding to our knowledge regarding the atmosphere, the 3-mm continuum images of Mars' surface will also contribute to our understanding of the the surface characteristics of the planet.

Venus

SPECTRUM Microwave observations of Venus probe the atmosphere and, at wavelengths longward of ~ 4 cm, the surface. A spectrum of the disk-

---

[3] Operated by the University of California at Berkeley, the University of Illinois, and the University of Maryland, with support from the National Science Foundation.

averaged brightness temperature is shown in Figure 15 (from Muhleman et al 1979). Between a few millimeters and 7 cm the brightness temperature increases from ~ 300 K to ~ 650 K. At the longer wavelengths one probes deeper levels in the atmosphere, which (owing to the adiabatic temperature structure in the Venus' atmosphere) results in the steep temperature increase in the spectrum. There are no diurnal temperature variations, owing to the large heat capacity of the atmosphere. At wavelengths longward of 7 cm the subsurface layers of the planet are probed. In order to interpret Venus' radio spectrum, a model needs to be developed that includes both the atmospheric and surface emission/absorption characteristics. Such a model was developed by Muhleman et al (1979), and their results are superimposed on the data in Figure 15.

SURFACE Chapman (1986) and Pettengill et al (1988) imaged Venus at 6- and 20-cm wavelength with the VLA. Figure 16(*left*) shows an image at 20 cm, taken on 12 November 1983. The Aphrodite region is near the center

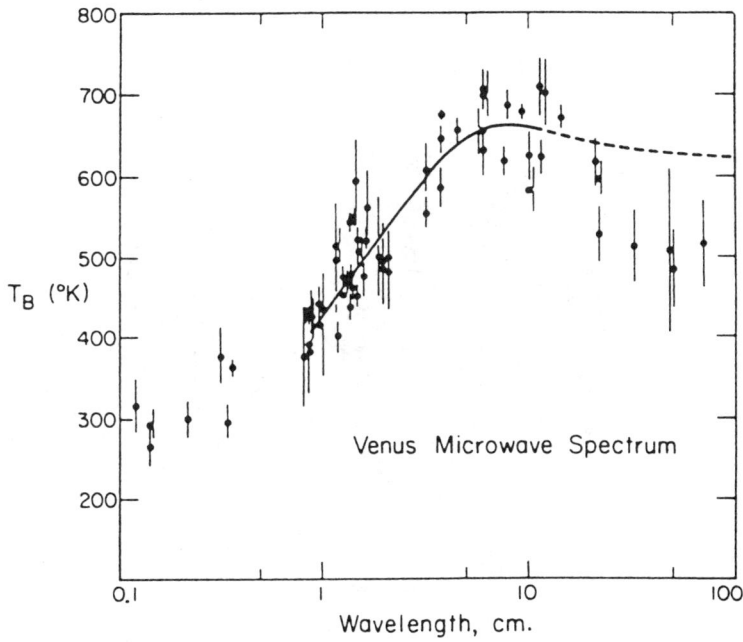

Figure 15 Venus microwave spectrum. Superimposed is Muhleman et al's model (from Muhleman et al 1979).

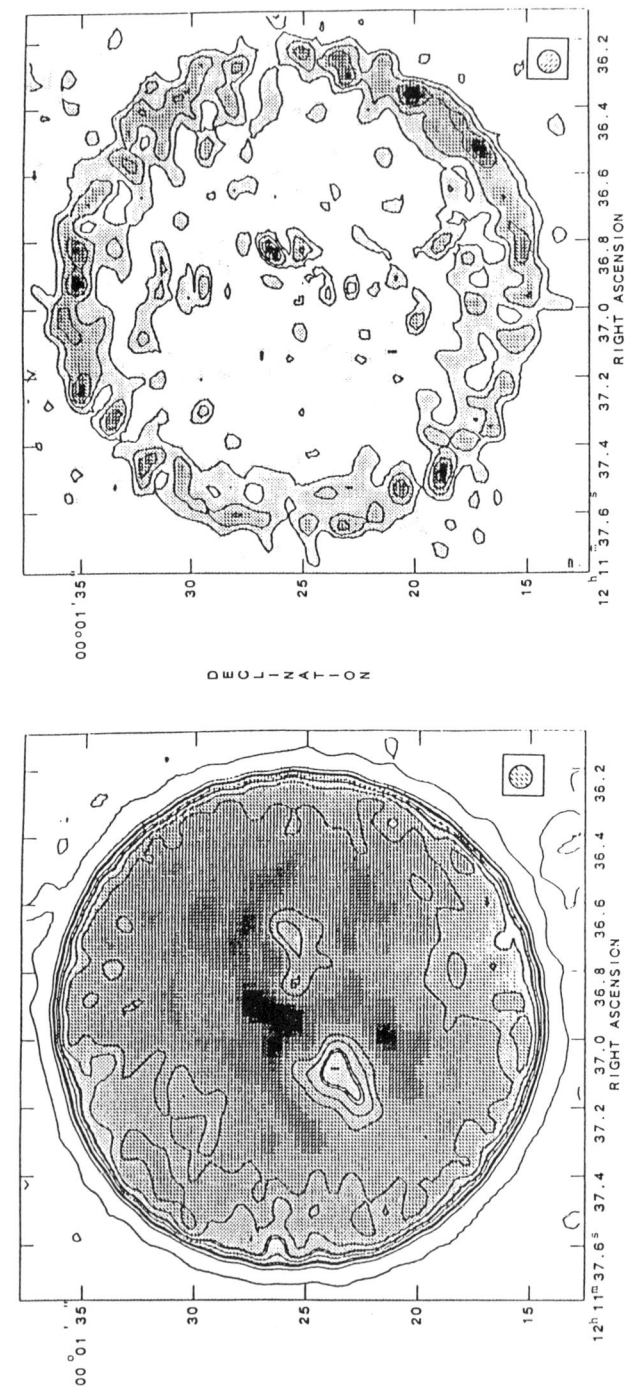

Figure 16 Radio images of Venus (from Chapman 1986). (*left*) The total intensity—contour levels (in kelvins) are 90, 315, 371, 382, 405, 450, 483, 494, 539, and 584; (*right*) linearly polarized flux density—contour values (in millijanskys per beam) are 0.1, 0.15, 0.2, and 0.25.

of the planet, and the two cool regions coincide with the locations of Thetis and Ovda. The brightness temperatures of the latter regions are 515 and 450 K, respectively, compared with the disk-averaged temperature of 636 ± 28 K. The high brightness temperature to the north of the Aphrodite region has a peak value of 772 K. From the radio brightness temperatures one can obtain the emissivity of the material if the physical temperature of the observed region is known. With a surface temperature of 735 K and a vertical adiabatic lapse rate of 8 K km^{-1} (e.g. Seiff 1983), the temperature of the mountains can be calculated if the altitude of the area is known (from, for example, radar observations). The Thetis and Ovda regions appear to have emissivities of 0.71 and 0.62, respectively, whereas the disk-averaged value is 0.86. The latter value implies a dielectric constant for the material of 5, in close agreement with the radar results. The low emissivities of the Thetis and Ovda regions imply dielectric constants as high as 20–30.

Information on the dielectric constant can also be obtained from images of the polarized intensity (Chapman 1986). Figure 16(*right*) shows the intensity of the linear polarization for the image shown in Figure 16(*left*). Note that the polarized intensity increases as one approaches the limb. At transmission angles less than 30–40° the signal is buried in the noise. At larger angles the signal-to-noise ratio improves, and the dielectric constant can be determined. The data suggest that the dielectric constant decreases with increasing angle of transmission, from a value close to 4 at transmission angles near 30° down to 2.5 near the limb. These values are lower than those obtained from the radio emissivities and the radar reflectivity data. This fact, as well as the apparent decrease in the dielectric constant when approaching the limb, may be caused by surface roughness, which will reduce the degree of polarization. An alternative explanation would be the existence of a vertical gradient in the density of the surface material.

Venus' surface properties can only be determined from the VLA images when combined with the radar results. The radar scattering and reflectivity characteristics contain additional information on the (sub)surface properties of the planet. Pettengill et al (1988) analyzed results from the *Pioneer Venus* radar altimetry experiment in combination with the VLA measurements. The radar data show that the reflection coefficient varies over the disk from 0.14 ± 0.03 in the lowlands to roughly 0.40 in the highlands. Regions with a high reflectivity show low brightness temperatures (both on VLA images and in measurements by the *Pioneer Venus* radiometer) and thus low emissivities. Table 3 gives details on the parameters for the regions that can be distinguished on the different VLA images. The emissivities of the regions were obtained from the brightness temperatures measured in the VLA images, combined with those obtained from the *Pioneer Venus* radiometer.

Table 3 Properties of Venus' surface (after Pettengill et al 1988)

Feature	Reflection coefficient	Brightness temperature	Estimated physical temperature	Emissivity	Dielectric constant
Lowlands or whole-disk average	0.14±0.03	636±28	735	0.86±0.04	5.0±0.9
Maxwell Montes	0.40±0.05	420±25	663	0.50±0.07	38.7±15.0
Ovda region	0.39±0.05	450±35	695	0.55±0.06	24.3±8.8
Thetis region	0.37±0.08	515±35	703	0.60±0.07	20.5±9.0

The results of the planet's emissivity (and hence its dielectric constant) can be interpreted in terms of its surface composition (Chapman 1986, Pettengill et al 1988). Dielectric constants of $\varepsilon \sim 2$ imply porous surface materials; $\varepsilon \sim 5$–9 is typical for solid rocks (granite-basalt). Much higher dielectric constants can be caused by the inclusion of metallic and/or sulfide material. Hence, Venus' surface is overlain, at most, by only a few centimeters of soil or dust and likely consists of dry solid rock. The highlands probably contain substantial amounts of minerals and sulfides close to the surface. On Earth, such material is often formed in volcanic areas.

ATMOSPHERE As mentioned above, Muhleman et al (1979) modeled Venus' spectrum and concluded that the atmospheric opacity is due to CO_2 gas and some unknown absorber. Steffes (1985, 1986) suggests that the extra absorptivity in Venus' atmosphere at microwave wavelengths longward of 1.8 cm is due to H_2SO_4 gas, and at $1.2 < \lambda < 1.8$ cm it is due to SO_2 gas. Both the CO_2 and SO_2 absorptivity vary as λ^{-2}, but the H_2SO_4 opacity shows a pronounced peak at 2.2 cm. Steffes et al (1990) obtained a good fit to Venus' radio spectrum using a model atmosphere with the microwave absorbers CO_2, H_2SO_4, and SO_2. They assumed an abundance of 96% for CO_2. The SO_2 abundance was taken to be zero above the cloud layer (> 48–50 km) and 40 ppm below the main cloud layer down to the surface. Gaseous H_2SO_4 was assumed to be uniformly distributed at 5 ppm between 38 and 48 km; it is fully dissociated at lower levels, and above 48 km condensation starts, and the gas follows the saturated vapor curve.

Unfortunately, images of Venus' atmosphere at short centimeter wavelengths have not yet been published. Such images contain information on the spatial distribution of the absorbing gases and would be extremely helpful for unraveling possible time variations in Venus' microwave brightness temperature (Steffes 1986).

At millimeter wavelengths, in the $J = 1$–0 and $J = 2$–1 transitions of the CO line, one probes altitudes between 70 and 120 km (the mesosphere of Venus), a region not well studied at other wavelengths. CO is produced upon photodissociation of CO_2 by solar UV radiation in the 70–120 km altitude region. This is a transition region between the massive lower atmosphere (altitudes $\lesssim 70$ km), in which the radiative time constant is much greater than a solar day, and the upper atmosphere (altitudes $\gtrsim 120$ km), which has a low heat capacity. As mentioned earlier, the temperature structure in the lower atmosphere follows an adiabatic curve; in the upper atmosphere a strong day-to-night gradient in the temperature is expected and observed (e.g. Seiff 1983). The difference in temperature structure between the two atmospheric regions will cause a very different wind pattern to exist as well. A strong retrograde zonal wind is observed in the visible cloud layers ($\sim$48–63 km altitude), with velocities of the order of ~ 100 m s^{-1}; in the upper atmosphere the day-to-night temperature gradient should drive strong day-to-night winds, which have been recently observed by Goldstein et al (1988). Since CO is formed in the transition region, CO line spectra can be used to derive the thermal structure in the mesosphere, the wind speeds and directions, and the CO abundance as a function of altitude.

Because the continuum emission at millimeter wavelengths originates in the low, warm part of the atmosphere, the CO lines are seen in absorption against the continuum background. The lines are pressure broadened by Venus' dense atmosphere. The $J = 1$–0 transition of this line was first observed by Kakar et al (1976). Soon after its discovery it was found that the line-to-continuum ratio of CO on the day and night side of the planet was quite different: the night-side line was approximately 3 times deeper than the day-side line (e.g. Schloerb 1985, and references therein). The core of the line seemed to vary symmetrically from the day to the night side, and the wings of the line appeared to have their peak absorption just after local midnight. These results were interpreted with a CO abundance that is larger on the night side of the planet than the day side. This is just the opposite of what one would expect if CO is formed by photodissociation of CO_2. Clancy (1983; Clancy & Muhleman 1985) suggested that the CO is likely carried from its place of formation at the day side to the night side by strong day-to-night winds such as those expected to exist in Venus' mesosphere. According to Clancy (1983; Clancy & Muhleman 1985), a small influence of the retrograde cloud circulation on the CO abundance at the lower levels of the CO line formation [where the line wings originate (~ 70–90 km)] may explain the few-hour shift away from midnight in the maximum absorption of the line wings.

Recently, I. de Pater, F. P. Schloerb & A. Rudolph (in preparation) imaged Venus at a wavelength of 3 mm (112 GHz) using the Hat Creek millimeter array while Venus was at western elongation (January 1987). The spatial resolution was 3.5″. Figure 17 shows the image of the 3-mm continuum emission. The night side of the planet is warmer than the day side, by approximately 30 K. In addition, the tropical regions on the night side (in particular, in the northern hemisphere) are brighter than the equatorial band. Since there is no day-to-night variation in the thermal structure in this part of the atmosphere, the difference in brightness temperature is due to a variation in opacity over the disk. The main sources of opacity at radio wavelengths are CO_2, SO_2, and gaseous H_2SO_4. The absorptivity of CO_2 accounts for roughly half the total opacity (e.g.

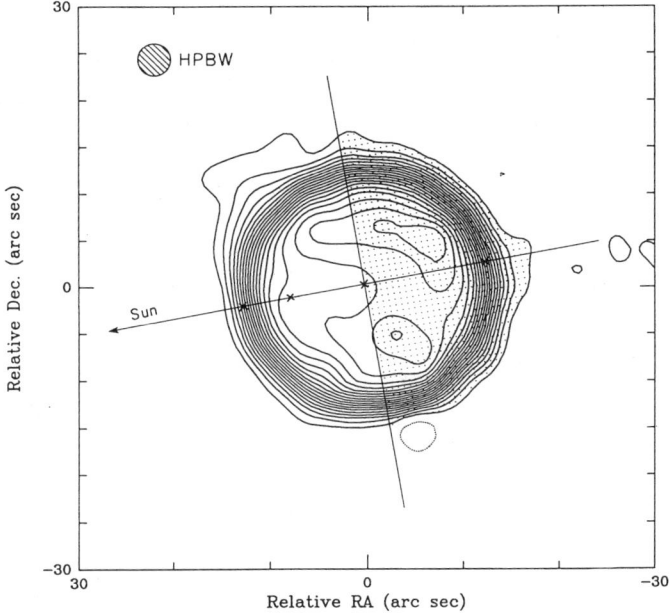

Figure 17 Radio image of Venus at 3 mm (from I. de Pater, F. P. Schloerb & A. Rudolph, in preparation). The direction to the Sun and the terminator are indicated. The contour levels are in 5% intervals, with a maximum of ∼365 K. The crosses along the Sun-Venus line indicate the locations of the spectra displayed in Figure 18.

Muhleman et al 1979). The 3-mm continuum measurements of Venus probe the altitude region in which the clouds form. Although the clouds are transparent at centimeter wavelengths, we expect some absorptivity from the cloud particles at millimeter wavelengths. Unfortunately, the opacity due to sulfuric acid droplets has never been measured. The dielectric constant may be similar to that of water, in which case the optical depth is roughly 0.1–0.2 km^{-1} in the main cloud deck (P. G. Steffes, personal communication). For comparison, the optical depth at 3 mm in the main cloud deck due to gaseous H_2SO_4 (abundance 5–10 ppm) and SO_2 (abundance 100–180 ppm) is roughly 0.2–0.5 km^{-1} for each gas (P. G. Steffes, personal communication). Hence, we expect a similar amount of opacity due to gaseous and liquid H_2SO_4 as well as SO_2.

The clouds on Venus form most likely from photochemically produced sulfuric acid (e.g. Esposito et al 1983, and references therein). This process must be more effective on the day side than the night side. The day-to-night asymmetry in the 3-mm radio image of the planet may be due to small day-to-night differences in the sulfuric acid abundance. Since the gaseous H_2SO_4 likely follows the saturated vapor curve in the region of cloud formation, the larger opacity at the day side might be due to a larger concentration of cloud particles rather than the gas at the day side.

Together with the 3-mm continuum image of Venus, I. de Pater, F. P. Schloerb & A. Rudolph (in preparation) also obtained images in the $J = 1-0$ CO line (115 GHz), with a frequency resolution of 125 MHz. In this transition, altitudes well above the cloud layers are probed (90–120 km). Figure 18 shows spectra at four different places along Venus' equator: at noon (the limb at the day side), at 8:30 A.M. (midmorning), at 6 A.M. (terminator), and at midnight (limb at the night side). All of the spectra are taken from images convolved with a beam with FWHM of $3'' \times 10''$, with the long axis parallel to the terminator. As expected, the line is deep and narrow at noon, which implies that there is much CO relatively high up in the atmosphere. Since CO is photochemically produced from CO_2, one indeed expects the line to be narrow and deep. At the morning side, however, the line is broader and less deep. When crossing the meridian, the line gets narrower and deeper again. At midnight, the line is narrow and rather deep. The spectra agree qualitatively with single-dish spectra of the planet. No detailed anaylsis of the Hat Creek spectra have yet been made. I. De Pater, F. P. Schloerb & A. Rudolph (in preparation) hope to derive more detailed information on the CO formation, transportation, and destruction processes.

Mercury

Radio observations of Mercury probe the planet's surface and subsurface layers. As on Mars, the temperature structure of these layers depends upon a balance between solar insolation, heat transport within the crust, and reradiation outward. As a result of the 3/2 resonance between Mercury's rotational and orbital periods in combination with Mercury's large orbital eccentricity, the average diurnal insolation varies significantly. Regions along Mercury's equator near longitudes $\phi = 0°$ and $180°$ (the subsolar longitudes when the planet is at perihelion) receive roughly 2.5 times as much sunlight on the average than longitudes $\phi = 90°$ and $270°$. As a result of this nonuniform heating, the diurnal temperature variation depends upon Hermographic longitude. The nighttime surface temperature is approximately 100 K, independent of longitude, but the peak (noon) surface temperature varies between 700 K at $\phi = 0°$ and $180°$ to 570 K for $\phi = 90°$ and $270°$ (Soter & Ulrichs 1967). While the subsurface temperature is below that of the surface during the day, it is above the surface temperature at night. Thus, heat is transported downward during the day and upward at night. In addition, owing to the phase lag between solar

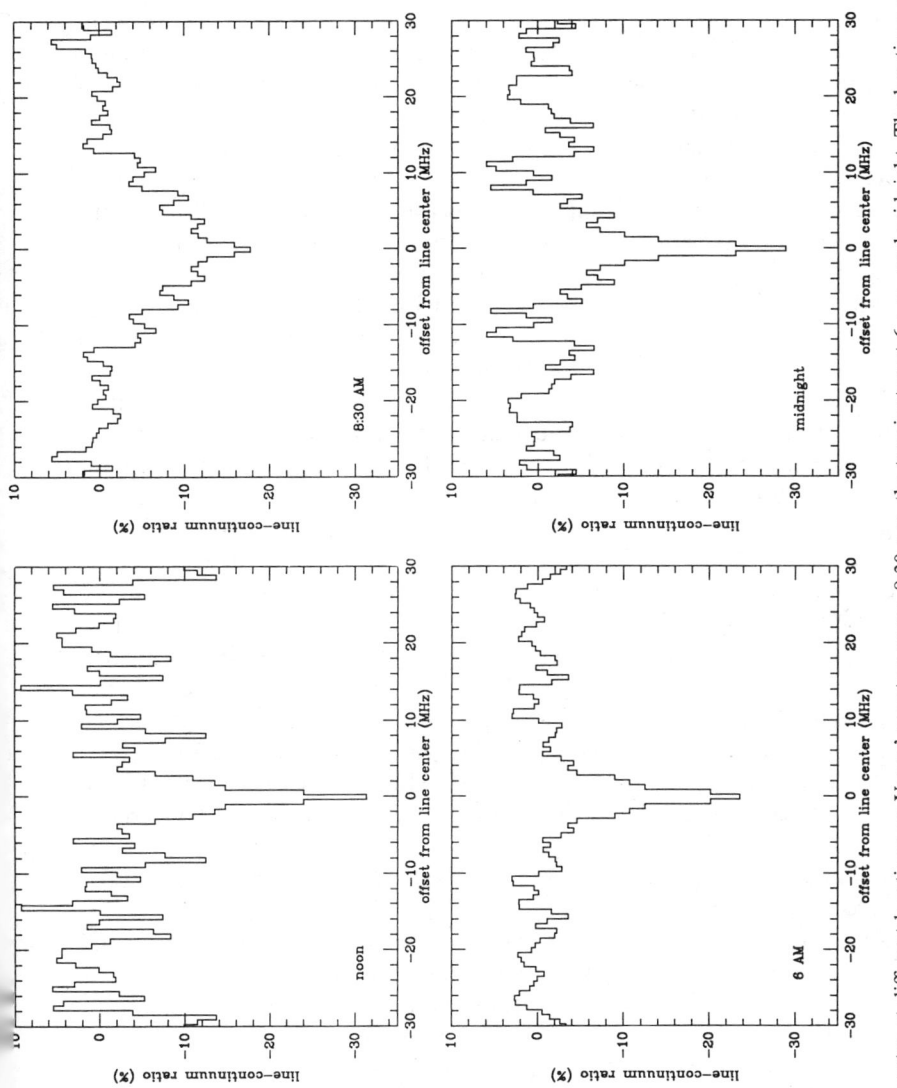

Figure 18 Spectra at different locations on Venus' equator: noon, 8:30 A.M. the terminator at 6 A.M., and midnight. The locations are indicated by the crosses in Figure 17. The spectra are expressed in line-to-continuum ratios, and they are averaged over a Gaussian beam with FWHP 3" × 10", with the long axis parallel to the terminator.

insolation and heat transport downward, the highest subsurface temperature is reached in the afternoon, rather than at noon.

Morrison (1970) summarized the microwave observations of Mercury obtained prior to 1970. These are disk-averaged brightness temperatures at wavelengths between 0.3 and 11 cm. The most extensive data set is from Klein (1970). He shows that the diurnal temperature variations indeed depend upon Hermographic longitude, with a minimum disk-averaged brightness temperature of ~ 300 K independent of longitude, and a peak temperature between ~ 380 and ~ 440 K, depending on longitude.

The best model to date has been developed by Cuzzi (1974). He derived subsurface properties by comparison with disk-averaged brightness temperatures at wavelengths between 0.3 and 18 cm. Mercury's surface appears similar to that of the Moon. The top few centimeters consist of a low-density ($\rho \sim 0.6$–1.0 g cm^{-3}) powder with dielectric constant $\varepsilon \sim 1.5$–2. The density and dielectric constant increase with depth in the crust to $\rho \sim 1.5$–2 g cm^{-3} and $\varepsilon \sim 3$ at a depth of about 2.5 m.

Radio images of the planet show a brightness variation across the disk, which displays the history of the solar insolation. The first VLA images were obtained by Chapman (1986) in April 1985 at 2- and 6-cm wavelength. In July 1986 Burns et al (1987) observed the planet at 6-cm wavelength. Figure 19 shows their 6-cm image. The planet had an eastern elongation of 22° and phase of 0.22. The viewing geometry of the planet, the Hermographic longitude of 0°, and the equator are indicated in the insert. Clearly, the disk shows an asymmetric brightness temperature distribution, with the hottest area centered near 0° longitude. In contrast to the radio images of Venus and Mars, no correlations between spatial brightness variations and known optical/infrared features on the surface of Mercury have been distinguished (Chapman 1986). Figure 20 shows a preliminary image at 3-mm wavelength obtained with the Hat Creek Radio Interferometer (D. L. Mitchell, I. de Pater & M. C. H. Wright, in preparation) while the planet was at perihelion and at greatest eastern elongation. A large day-to-night difference in temperature is readily visible.

At 6 cm we probe 0.5–1 m down into the surface, where the diurnal temperature variation at any longitude is minimal. If the thermal properties such as albedo, emissivity, thermal inertia, and loss tangent do not vary across the disk, any spatial variations in the brightness temperature across the disk must be due to variations in the physical temperature at these depths. Burns et al (1987) show that the observed brightness temperature variations at 6 cm indeed more or less mimic the spatial variations expected from the nonuniform heating of the planet. Chapman's images show the night side of Mercury, at longitudes $\sim 180°$ away from Burns et al's picture.

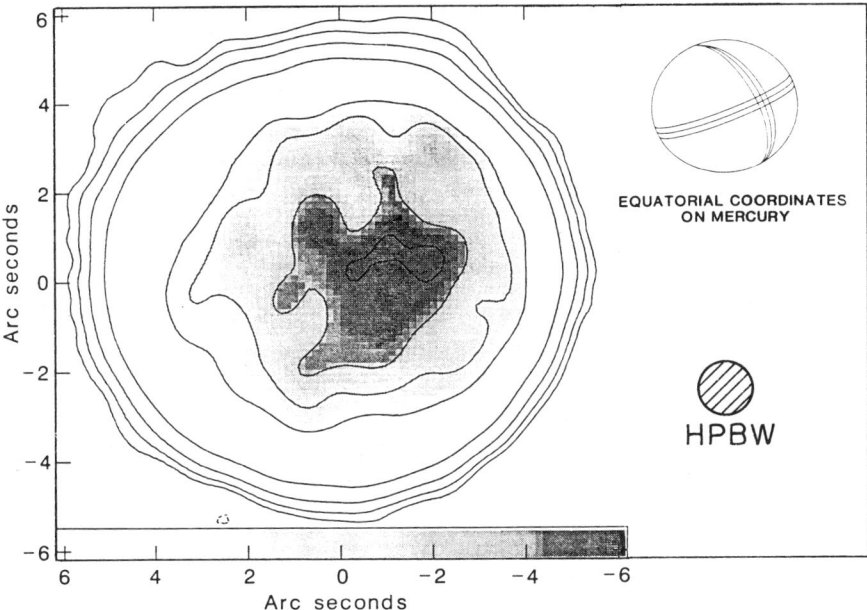

Figure 19 Radio image of Mercury at 6-cm wavelength (from Burns et al 1987). Contour levels (in kelvins) are 20, 40, 80, 160, 310, 330, 350, and 370. In the upper right corner the geometry of the planet is shown, with the equator, circles of latitude ±5°, and meridians of longitude 0° and ±5° shown. HPBW of the beam is shown in the lower right corner.

Also, in Chapman's 6-cm data the region at a longitude near ~180° is clearly warmer than the surroundings.

At 3-mm wavelength one probes only a few centimeters into the crust, where the diurnal temperature variation is relatively large. Since in Figure 20 the Hermographic longitude 0° coincides with local noon at the planet, the brightness temperature should be maximal at this longitude (~500–600 K; see D. L. Mitchell, I. de Pater & M. C. H. Wright, in preparation). The temperature at midnight should be much smaller (~200 K). The coldest area on the disk coincides with the sub-Earth point and the 6 A.M. meridian. The brightness temperature of this region will likely be slightly less than that at $\phi = 180°$. The 3-mm image indeed shows these temperature variations; detailed simulations are still in progress, however.

5. FUTURE RESEARCH

With the radio detection of Pluto (Altenhoff et al 1988) all planets have now been detected at radio wavelengths, and most of them have been

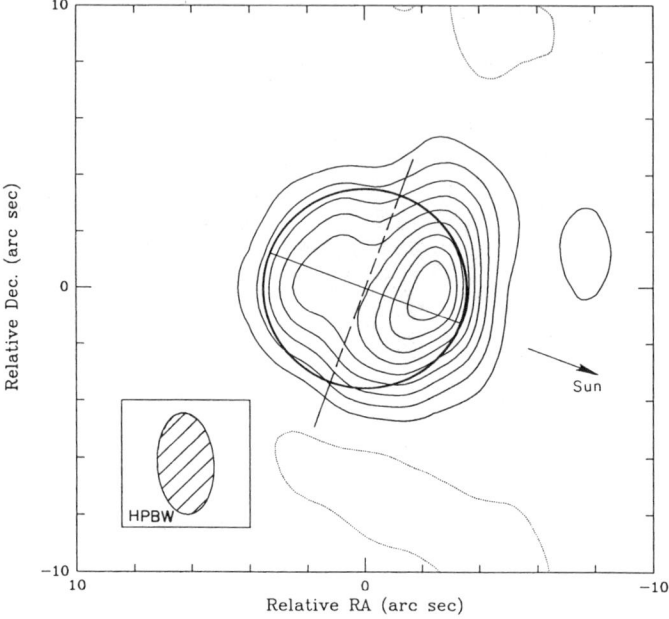

Figure 20 Preliminary radio image of Mercury at 3 mm (from D. L. Mitchell, I. de Pater & M. C. H. Wright, in preparation). Contour levels are at 10% intervals. The direction to the Sun, the size of the disk, the terminator, and the HPBW are indicated. The subsolar longitude is at 0°.

imaged as well, as described in this review. By no means does this imply that this type of research has been completed; rather it serves as a first step to identify the important issues that can be addressed through radio imaging of planets. In the future the following topics should be studied in detail: (*a*) a physical explanation for the detailed structure seen in images of Jupiter's synchrotron radiation, and the time variations therein; (*b*) detailed dynamical modeling of the planetary atmospheres and comparison with multiwavelength radio images obtained over a significant time interval; (*c*) extraction of longitudinal information from images of the atmospheres of the giant planets using improved deconvolution techniques; (*d*) a search for minor constituents such as HCN, SO_2, CO, and PH_3, in the atmospheres of the giant planets; (*e*) examination of Saturn's rings and comparison with Voyager data; (*f*) detailed imaging and modeling of the atmospheres and surfaces of the terrestrial planets.

Of particular interest would be the simultaneous imaging of the $J = 1–0$, and $J = 2–1$ rotational transitions of both ^{12}CO and ^{13}CO on Venus and

Mars at different locations in their orbits. Furthermore, to best constrain thermophysical models of Mercury's subsurface layers, images at infrared-to-centimeter wavelengths are required at different orbital positions and sub-Earth longitudes. Observations at millimeter wavelengths are possible with the Owens Valley and Hat Creek arrays. The latter is particularly suitable for planetary research because of its large field of view and good instantaneous UV coverage in its future expanded (six or more dishes) configuration.

ACKNOWLEDGMENTS

Over the years I have worked with a number of collaborators on various aspects of the research described in this review, including S. K. Atreya, J. R. Dickel, C. K. Goertz, S. Gulkis, M. J. Klein, S. T. Massie, D. L. Mitchell, M. Richmond, P. N. Romani, and F. P. Schloerb. I would like to thank them all for their contributions to this work. I further appreciate comments on this manuscript by S. K. Atreya, B. D. Chapman, J. R. Dickel, D. Mitchell, F. P. Schloerb, and P. G. Steffes. The research described in this paper is supported by NSF grants AST-8514896 and AST-8900156 and NASA grant NAGW 1805 to the University of California at Berkeley, and by the Alfred P. Sloan Foundation.

Literature Cited

Acuna, M. H., Ness, N. F. 1976. In *Jupiter*, ed. T. Gehrels, pp. 830–47 Tucson: Univ. Ariz. Press
Altenhoff, W. J., Chini, R., Hein, H., Kreysa, E., Mezger, P. G., et al., 1988. *Astron. Astrophys.* 190: L15–17
Andrew, B. H., Harvey, G. A. Briggs, F. H. 1977. *Ap. J. Lett.* 213; L131–34 (Erratum: 1978. *Ap. J. Lett.* 220: L61
Atreya, S. K., Blamont, J. E. 1990. *Geophys. Res. Lett.* 17: 287–91
Berge, G. L. 1966. *Ap. J.* 146: 767–98
Berge, G. L., Gulkis, S. 1976. In *Jupiter*, ed. T. Gehrels, pp. 621–92. Tucson: Univ. Ariz. Press
Berge, G. L., Muhleman, D. O., Linfield, R. 1988. *Astron. J.* 96: 388–95
Bézard, B., Gautier, D., Conrath, B. 1984. *Icarus* 60: 274–88
Branson, N. J. B. A. 1968. *MNRAS* 139: 155–62
Bridle, A. H., Fomalont, E. B., Cornwell, T. J. 1981. *Astron. J.* 86: 1294–1305
Briggs, F. H., Andrew, B. H. 1980. *Icarus* 41: 269–77
Briggs, F. H., Sackett, P. D. 1989. *Icarus* 80: 77–103

Burke, B. F., Franklin, K. L. 1955. *J. Geophys. Res.* 60: 213–17
Burns, J. O., Gisler, G. R., Borovsky, J. E., Baker, D. N., Zeilik, M. 1987. *Nature* 329: 234–36
Carlson, B. E., Lacis, A. A., Rossow, W. B. 1988. *Bull. Am. Astron. Soc.* 20: 869
Campbell, M. J., Ulrichs, J. 1969. *J. Geophys. Res.* 74: 5867–81
Carr, T. D., Desch, M. D. 1976. In *Jupiter*, ed. T. Gehrels, pp. 693–737. Tucson: Univ. Ariz. Press
Carr, T. D., Desch, M. D. Alexander, J. K. 1983. In *Physics of the Jovian Magnetosphere*, ed. A. J. Dessler, pp. 226–84. Cambridge: Univ. Press
Chapman, B. D. 1986. PhD thesis, Mass. Inst. Technol., Cambridge
Clancy, R. T. 1983. PhD thesis, Calif. Inst. Technol., Pasadena
Clancy, R. T., Muhleman, D. O. 1985: *Icarus* 64: 183–204
Clancy, R. T., Muhleman, D. O., Jakosky, B. M. 1983. *Icarus* 55: 282–301
Clark, B. G. 1980. *Astron Astrophys.* 89: 377–78

Conrath, B., Gautier, D., Hanel, R., Lindal, G., Marten, A. 1987. *J. Geophys. Res.* 92: 15,003–10

Conrath, B. J., Pearl, J. C., Appleby, J. F., Lindal, G. F., Orton, G. S., Bézard, B. 1988. Paper presented at Uranus Colloq., Pasadena, Calif.

Cornwell, T., Braun, R. 1989. See Perley et al 1989, pp. 167–83

Courtin, R., Gautier, D., Marten, A., Bézard, B., Hanel, R. 1984. *Ap. J.* 287: 899–916

Cuzzi, J. N. 1974. *Ap. J.* 189: 577–86

Cuzzi, J. N., Pollack, J. B., Summers, A. L. 1980. *Icarus* 44: 683–705

de Pater, I. 1980. *Astron. Astrophys.* 88: 175–83

de Pater, I. 1981a. *J. Geophys. Res.* 86: 3397–3422

de Pater, I. 1981b. *J. Geophys. Res.* 86: 3423–29

de Pater, I. 1983. *Adv. Space Res.* 3(3): 31–37

de Pater, I. 1985. *Proc. Int. Workshop, Alpbach, Austria. ESA SP*-241, pp. 203–8. Paris: ESA

de Pater, I. 1986: *Icarus* 68: 344–65

de Pater, I., Dames, H. A. C. 1979. *Astron. Astrophys.* 72: 148–60

de Pater, I., Dickel, J. R. 1982. *Icarus* 50: 88–102

de Pater, I., Dickel, J. R. 1983. *Adv. Space Res.* 3(3): 39–41

de Pater, I., Dickel, J. R. 1986. *Ap. J.* 308: 459–71

de Pater, I., Goertz, C. K. 1989. *Geophys. Res. Lett.* 16: 97–100

de Pater, I., Goertz, C. K. 1990. *J. Geophys. Res.* 95: 39–50

de Pater, I., Gulkis, S. 1988. *Icarus* 75: 306–23

de Pater, I., Jaffe, W. J. 1984. *Ap. J. Suppl.* 54: 405–19

de Pater, I., Klein, M. J. 1989. In *Time Variable Phenomena in the Jovian System*. *NASA SP*-494, ed. M. J. S. Belton, R. A. West, J. Rahe, pp. 139–50. Washington, DC NASA

de Pater, I., Massie, S. T. 1985. *Icarus* 62: 143–71

de Pater, I., Richmond, M. 1989. *Icarus* 80: 1–13

de Pater, I., Kenderdine, S., Dickel, J. R. 1982. *Icarus* 51: 25–38

de Pater, I., Romani, P. N., Atreya, S. K. 1989. *Icarus* 82: 288–313.

Dowling, T. E., Muhleman, D. O., Berge, G. L. 1987. *Icarus* 70: 506–16

Epstein, E. E., Andrew, B. H., Briggs, F. H., Jakosky, B. M., Palluconi, F. D. 1983. *Icarus* 56: 465–75

Esposito, L. W., Knollenberg, R. G., Marov, M. Ya., Toon, O. B., Turco, R. P. 1983. In *Venus*, ed. D. M. Hunten, L. Colin, T. M. Donahue, V. I. Moroz, pp. 484–564. Tucson: Univ. Ariz. Press

Esposito, L. W., Cuzzi, J. N., Holberg, J. B., Marouf, E. A., Tyler, G. L., Porco, C. C. 1984. In *Saturn*, ed. T. Gehrels, pp. 463–545. Tucson: Univ. Ariz. Press

Flasar, F. M., Conrath, B. J., Gierasch, P. J., Piraglia, J. A. 1987. *J. Geophys. Res.* 92: 15,011–18

Gerchberg, R. W. 1974. *Opt. Acta* 21: 709–20

Goldstein, J. J., Mumma, M. J., Kostiuk, T., Deming, D., Espenak, F., Zipoy, D. 1988. *Bull. Am. Astron. Soc.* 20: 833

Good, J. C., Schloerb, F. P. 1981. *Icarus* 47: 166–72

Grossman, A. W., Muhleman, D. O., Berge, G. L. 1989. *Science* 245: 1211–15

Gulkis, S., de Pater, I. 1984. In *Uranus and Neptune*. *NASA Conf. Publ. 2330*, ed. J. T. Bergstralh, pp. 225–62. Washington, DC: NASA

Gulkis, S., Janssen, M. J., Olsen, E. T. 1978. *Icarus* 34: 10–19

Gulkis, S., Olsen, E. T., Klein, M. J., Thompson, T. J. 1983. *Science* 221: 453–55

Gull, S. F., Daniell, G. J. 1978. *Nature* 272: 686–90

Högbom, J. A. 1974. *Astron. Astrophys. Suppl.* 15: 417–26

Hubbard, W. B. 1984. *Planetary Interiors*. New York: Van Nostrand Reinhold

Hunten, D. M. 1974. *Rev. Geophys. Space Phys.* 12: 529–35

Jaffe, W. J., Berge, G. L., Owen, T., Caldwell, J. 1984. *Science* 225: 619–21

Kakar, R. K., Waters, J. W., Wilson, W. J. 1976. *Science* 191: 379–80

Kellermann, K. I., Pauliny-Toth, I. I. K. 1966. *Ap. J.* 145: 954–57

Klein, M. J. 1970. *Radio Sci.* 5: 397

Klein, M. J., Gulkis, S. 1978. *Icarus* 35: 44–60

Klein, M. J., Turegano, J. A. 1978. *Ap. J. Lett.* 224: L31–34

Klein, M. J., Janssen, M. A., Gulkis, S., Olsen, E. T. 1978. In *The Saturn System*. *NASA Conf. Publ. 2068*, ed. D. M. Hunten, D. Morrison, pp. 195–216. Washington, DC: NASA

Kraus, J. D. 1986. *Radio Astronomy*. Powell, Ohio: Cygnus Quasar Books

Legg, M. P., Westfold, K. C. 1968. *Ap. J.* 154: 499–514

Lellouch, E., Gerin, M., Combes, F., Atreya, S. K., Encrenaz, T. 1989. *Icarus* 77: 414–38

Lewis, J. S. 1969. *Icarus* 10: 365–78

Lindal, G. F., Lyons, J. R., Sweetnam, D. N., Eshleman, V. R., Hinson, D. P., Tyler, G. L. 1987. *J. Geophys. Res.* 92: 14,987–15,002

Mayer, C. H., McCullough, T. P., Sloanaker, R. M. 1958. *Ap. J.* 127: 11–16
McDonald, F. B., Schardt, A. W., Trainor, J. H. T. 1980. *J. Geophys. Res.* 85: 5813–30
Morris, D., Berge, G. L. 1962. *Ap. J.* 136: 276–82
Morrison, D. 1970. *Space Sci Rev.* 11: 271–307
Muhleman, D. O., Orton, G. S., Berge, G. L. 1979. *Ap. J.* 234: 733–45
Orton, G. S., Baines, K. H., Bergstrahl, J. T., Brown, R. H., Caldwell, J., Tokunaga, A. T. 1987. *Icarus* 69: 230–38
Perley, R. A., Schwab, F. R., Bridle, A. H., eds. 1989. *Synthesis Imaging in Radio Astronomy. Proc. NRAO Workshop. No. 21.* San Francisco: Astron. Soc. Pac.
Pettengill, G. H., Ford, P. G., Chapman, B. D. 1988. *J. Geophys. Res.* 93(B12): 14,881–92
Pollack, J. B., Bodenheimer, P. 1989. In *Origin and Evolution of Planetary and Satellite Atmospheres*, ed. S. K. Atreya, J. B. Pollack, M. S. Matthews, pp. 564–602. Tucson: Univ. Ariz. Press
Radhakrishnan, V., Roberts, J. A. 1960. *Phys. Rev. Lett.* 4: 493–94
Roberts, J. A., Berge, G. L., Bignell, R. C. 1984. *Ap. J.* 282: 345–58
Rolhfs, K. 1986. *Tools of Radio Astronomy.* Berlin/Heidelberg: Springer-Verlag
Romani, P. N., de Pater, I., Atreya, S. K. 1989. *Geophys. Res. Lett.* 16: 933–36
Rudy, D. J. 1987. PhD thesis. Calif. Inst. Technol., Pasadena

Rudy, D. J., Muhleman, D. O. Berge, G. L., Jakoksky, B. M., Christensen, P. R. 1987. *Icarus* 71: 159–77
Schloerb, F. P. 1985. *Proc. ESO-IRAM-Onsala Workshop (Sub)millimeter Astron. Aspenas. Swed.*, pp. 603–16
Schloerb, F. P., Muhleman, D. O., Berge, G. L. 1979. *Icarus* 39: 214–31
Schloerb, F. P., Muhleman, D. O., Berge, G. L. 1980. *Icarus* 42: 125–35
Schwab, F. R. 1980. *Proc. Soc. Photo-Opt. Instrum. Eng.* 18–25
Seiff, A. 1983. In *Venus*, ed. D. M. Hunten, L. Colin, T. M. Donahue, V. I. Moroz, pp. 215–79. Tucson: Univ. Ariz. Press
Seiff, A., Kirk, D. B. 1977. *J. Geophys. Res.* 82: 4364–78
Soter, S., Ulrichs, J. 1967. *Nature* 214: 1315–16
Sramek, R. A. 1982. In *Synthesis Mapping. Proc. NRAO-VLA Workshop, Socorro, N. Mex.*, ed. A. R. Thompson, L. R. D'Addrico, pp. 2.1–2.13. Green Bank, W. Va: NRAO
Sramek, R. A., Schwab, F. R. 1989. See Perley et al 1989, pp. 117-38
Steffes, P. G. 1985. *Icarus* 64: 576–85
Steffes, P. G. 1986. *Ap. J.* 310: 482–89
Steffes, P. G., Klein, M. J., Jenkins, J. M. 1990. *Icarus* 84: 83–92
Thompson, A. R., Moran, J. M., Swenson, G. W. Jr. 1986. *Interferometry and Synthesis in Radio Astronomy.* New York: Wiley
West, R. A., Strobel, D. F., Tomasko, M. G. 1986. *Icarus* 65: 161–217

GAMMA-RAY BURSTS

James C. Higdon

Joint Science Department, The Claremont Colleges, Claremont, California 91711

Richard E. Lingenfelter

Center for Astrophysics and Space Sciences, University of California, San Diego, La Jolla, California 92093

KEY WORDS: gamma-ray sources, X-ray sources, neutron stars

1. INTRODUCTION

Of the more than 500 sources that have been detected in gamma rays, most are highly variable—seen only in their transient emission and collectively known as gamma-ray bursts. These bursts, discovered by Klebesadel et al (1973), are intense, gamma-ray transients with fluxes at energies >30 keV as large as 10^{-2} erg cm^{-2} s^{-1}, orders of magnitude greater than that of the most intense, steady-state X-ray and gamma-ray sources. Their emission has been measured over a wide range of photon energies from ~ 1 keV to ~ 100 MeV, but the bulk of their emission is at gamma-ray energies >0.1 MeV. Both absorption and emission features have been observed in their spectra. Their observed durations range from about 10^{-2} s to at least 10^3 s, and their rise times can be as short as $\sim 10^{-4}$ s. Despite the large number of bursts observed, only three of the gamma-ray burst sources are known to repeat, and no quiescent counterpart in the optical, infrared, radio, or X-ray range has been identified for any of the gamma-ray bursts. Only the 5 March 1979 gamma-ray burst position has been associated with any previously identified object—N49, a diffuse supernova remnant in the Large Magellanic Cloud.

A wide variety of sources and processes have been suggested for these bursts, ranging from stellar flares to cosmic strings (see reviews by Ruderman 1975, Lamb 1984, Taam 1987). As we shall see, however, the observational evidence strongly suggests that the sources of most bursts are nearby, magnetic neutron stars, where episodic accretion, thermonuclear runaway, and core quakes might all produce gamma-ray bursts. More than one type of source, however, may be needed to account for the great diversity of bursts.

Although gamma-ray bursts have been observed for 20 years, they have not previously been treated in the *Annual Review of Astronomy and Astrophysics*. Various observational and theoretical aspects of the bursts, however, have been discussed in the proceedings of three conferences devoted to them (Lingenfelter et al 1982, Woosley 1984a, Liang & Petrosian 1986); more recently, the observations have been reviewed by Hartmann & Woosley (1988), Klebesadel (1988), and Hurley (1989a,b), and the theory has been discussed by Taam (1987), Zdziarski (1987), and Lamb (1988). Here we review the observations to define the properties of the bursts and examine what constraints the observations can place on the nature of the burst emission processes and their sources. We consider first their nomenclature and classification; then review their temporal behavior, their spatial distribution and distances, and their energy spectra and emission processes; and finally summarize the general nature of their sources and energetics.

1.1 *Nomenclature and Catalogs*

Because of the lack of accurate source locations for most gamma-ray bursts, they are referred to by their date of occurrence. If more than one burst occurs during a day, they are indicated as a, b, etc. Thus the burst of 5 March 1979, mentioned above, is known as GB790305b. When source locations are known to at least degree accuracy, they are named by their position coordinates. The source of the 5 March 1979 burst, located at right ascension $5^h\ 26^m$ and declination $-66°$, is therefore designated GBS 0526−66.

The most extensive published set of gamma-ray burst observations is that of the Konus experiment on the *Venera 11* and *12* spacecraft (Mazets et al 1981a). Many more observations, however, have been made by experiments on the *Vela, Pioneer Venus Orbiter (PVO), High Energy Astronomical Observatory (HEAO) 1* and *3, Solar Maximum Mission (SMM), International Cometary Explorer (ICE), Venera 13* and *14, Prognoz, Helios 2, Hakucho, Hinotori, Ginga, Phobos*, and other spacecraft. The Konus burst detectors on the Venera spacecraft are described by Andreev et al (1983), the detectors on *PVO* by Klebesadel et al (1980), *SMM* by Matz (1986), and *Ginga* by Murakami et al (1989). The Burst and Transient Source Experiment (BATSE) to be launched on the *Gamma Ray Observatory* (GRO) is described by Fishman et al (1989), and the planned High Energy Transient Experiment (HETE) by Ricker et al (1988). All of the gamma-ray bursts reported prior to 1984 are summarized in a catalog by Baity et al (1984), and catalogs of bursts reported subsequently have been prepared by Yoshimori et al (1984), Kuznetsov et al (1986b), Matz (1986), Atteia et al (1987b), Estulin et al (1987), Golenetskii et al (1987a), Hueter (1987), Murakami (1988), and Mitrofanov et al (1989).

1.2 Classification

There appear to be at least two distinct classes of gamma-ray transients, as roughly outlined by Klebesadel et al (1982). These are as follows:

1. *Classical bursts* account for the majority of the observed bursts and are characterized most often by multipeaked emission lasting a few seconds or longer, hard, time-varying spectra with both absorption and emission features, and lack of observed repetition on time scales less than several years.
2. *Soft repeaters* presently consist of three sources, which have produced over 100 bursts, characterized (Mazets et al 1982a, Atteia et al 1987a, Laros et al 1987, Cline et al 1988) by their repetition, short duration (generally < 1 s), simple time profiles, and soft, time-independent energy spectra.

A variety of other classification schemes have been proposed (e.g. Mazets & Golenetskii 1981, 1987, Fishman et al 1986), based primarily on temporal characteristics, that would differentiate the very diverse classical bursts, for example, into singled-peaked, double-peaked, and complex, multipeaked bursts. However, there is no agreement on either the number of such possible classes or on their precise definitions.

2. TEMPORAL BEHAVIOR

2.1 Durations

Gamma-ray burst durations vary greatly, spanning at least five orders of magnitude, from the shortest of < 12 ms (FWHM) for GB820405 (Mazets et al 1983) to the longest of ~ 1000 s for GB840304 (Klebesadel et al 1984b). The distribution of burst durations, however, is not well defined. The closest approximation is probably that determined (Mazets et al 1981a, Golenetskii et al 1987a) from a sample of 216 classical gamma-ray bursts detected by the Konus experiment on *Venera 11–14* (shown in Figure 1). These durations represent the total time from the onset of the burst until the latest time that a significant flux was detected in the energy band from 50 to 200 keV. As Mazets & Golenetskii (1987) have pointed out, the apparent burst duration is a detector-dependent parameter because intense nearby events, which are well resolved above background, can appear to be of longer duration than identical but more distant events,

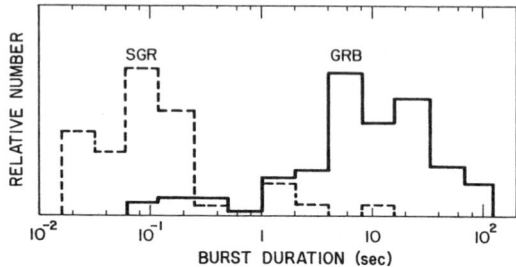

Figure 1 Distribution of burst durations for 216 classical bursts (solid line) from the Konus experiment on *Venera 11–14* (Mazets et al 1981a, Golenetskii et al 1987a), and a sample of repeating bursts (dashed line) summarized by Hurley (1989b).

which appear as less intense bursts only weakly resolved above background.

The sample of bursts (Hurley 1989b) that are identified with one of the three soft repeaters are plotted separately from those that have not been observed to repeat. As can be seen, the distributions of these two populations are quite different. The median duration of the classical, non-repeating bursts is ~ 10 s, whereas that of the repeating bursts is ~ 0.1 s, nearly two orders of magnitude shorter. The apparent duration of a burst can also depend strongly on the observed energy band because of spectral changes during the burst, as we discuss later.

The relative number of very long duration bursts is underrepresented in this sample because data were saved for only the first 66 s of each burst on *Venera 11* and *12*, and for the first 141 s on *Venera 13* and *14*. Unfortunately, most of the other data cut off even sooner, so there is very little information on the longer duration bursts. There are also other biases resulting from the differing burst identification schemes used in the various detector systems. These can result in rather large differences between the observed ratios of short and long bursts. In the Konus sample, the fraction of the detected bursts shorter than 1 s is roughly 16%, whereas the same fraction reported from the *ICE* gamma-ray burst detector is 37% (Norris et al 1984). The *ICE* detector triggered when the time to accumulate a preset number of photons was less than a fixed time (Norris 1983, Norris et al 1984). This strategy is best for detecting bursts with very short rise times (≤ 100 ms), but it is not well suited for detecting more slowly changing, longer bursts.

2.2 Temporal Structure

Gamma-ray bursts display very diverse temporal structures, so there is no typical burst light curve. Although some bursts, particularly the soft repeaters, appear to have relatively simple time profiles, most of the classical bursts show complex time profiles with multiple peaks, or spikes, and statistically significant fine structure on time scales (FWHM) as short as a few milliseconds. Many bursts are also preceded by weak precursor events that may trigger the detectors.

Some of the diversity in the classical burst time profiles can be seen in Figure 2. The long, complex, multipeaked burst (Figure 2a) GB820331 (Mazets & Golenetskii 1987) has a small precursor that triggered the detector ~ 60 s before the main burst and at least a score of peaks of differing intensity and duration. Similar complexity is also found in much shorter intense bursts, such as (Figure 2b) GB841215 (Laros et al 1985a), which within a span of only 0.3 s has several peaks with widths as short as 5 ms. Such rapid flux variations suggest a source size of $< 1.5 \times 10^8$ cm. Many of the seemingly simple time profiles of short, less intense bursts may also actually have such fine structure that was not resolved.

Nonetheless, there are classical bursts with relatively simple time profiles, such as (Figure 2c) GB830801b (Kuznetsov et al 1986a), which has no measurable fine structure, except for a short (0.1 s), weak precursor ~ 0.15 s prior to the onset of the main burst and some broader structure at late (> 10 s) times.

Because of spectral evolution during a burst, however, the time profiles are also energy dependent. Even the relatively simple burst profile of GB830801b begins to show more structure at lower and perhaps higher energies, as can be seen in Figure 3. Moreover, the width (FWHM) of the peak changes from ~ 5 s in the 39-68 keV band to ~ 1 s in the 2.5-4.5 MeV band, and the time of peak flux shifts by ~ 3 s between the same energy bands.

The short, soft repeaters nearly all appear to have simple, single-peaked time profiles. The most conspicuous exception is the extraordinary gamma-ray burst GB790305b, shown in Figure 4, which had additional, but much weaker, periodic emission (to be discussed below). The duration of the initial burst (~ 120 ms) was nonetheless typical of soft repeaters. However, its intensity was $\sim 10^3$ times that of the 16 bursts subsequently detected (Golenetskii et al 1987b) from that source.

The rise time of this burst is also the shortest discovered so far. Its intensity increased by two orders of magnitude, from background level to nearly peak intensity, in less than 1 ms, corresponding to an e-folding rise time of ~ 0.2 ms (Cline et al 1980, Cline 1982). This requires a source size of < 60 km, only somewhat larger than the diameter of a neutron star, if bulk motions in the emitting region are nonrelativistic. The mean rise time, however, of other short bursts is ~ 20 ms, 10^2 times greater than that of GB790305b (Barat et al 1984a).

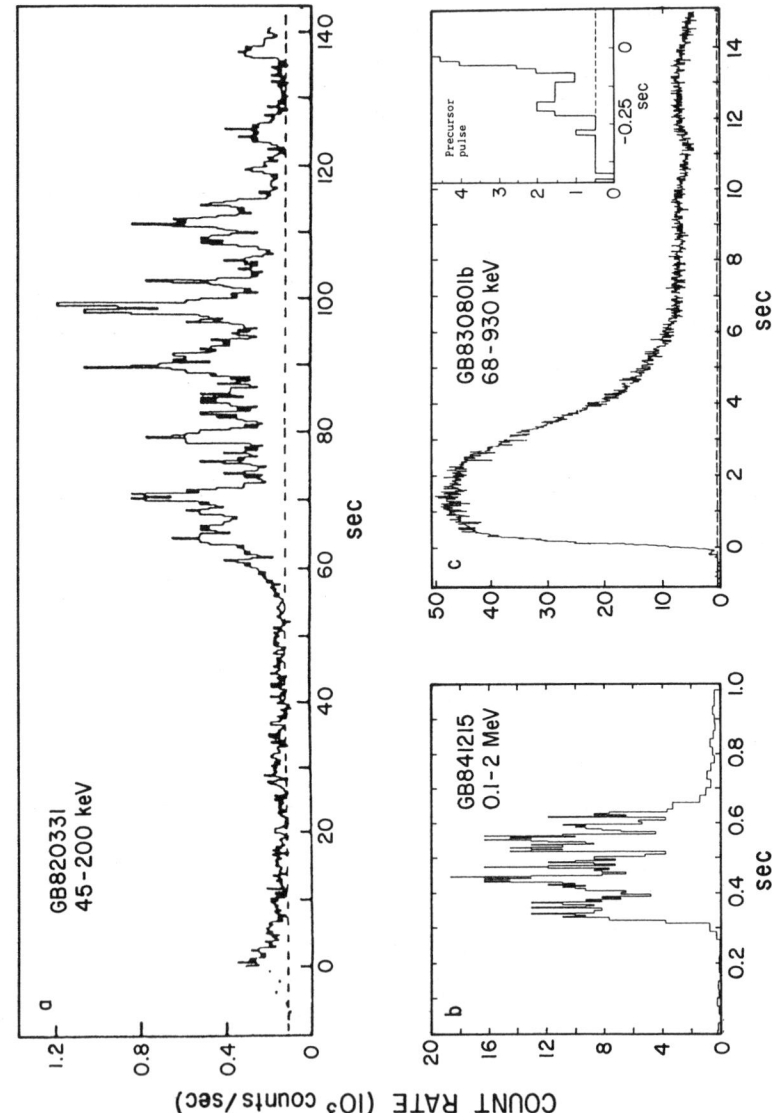

Figure 2 Diverse time profiles of classical bursts, showing examples of multipeaked structure on (*a*) long (GB820331; Mazets & Golenetskii 1987) and (*b*) short (GB841215; Laros et al 1985a) time scales and (*c*) a simple single peak (GB830801b; Kuznetsov et al 1986a). The dashed lines indicate the detector background prior to the burst.

GAMMA-RAY BURSTS 407

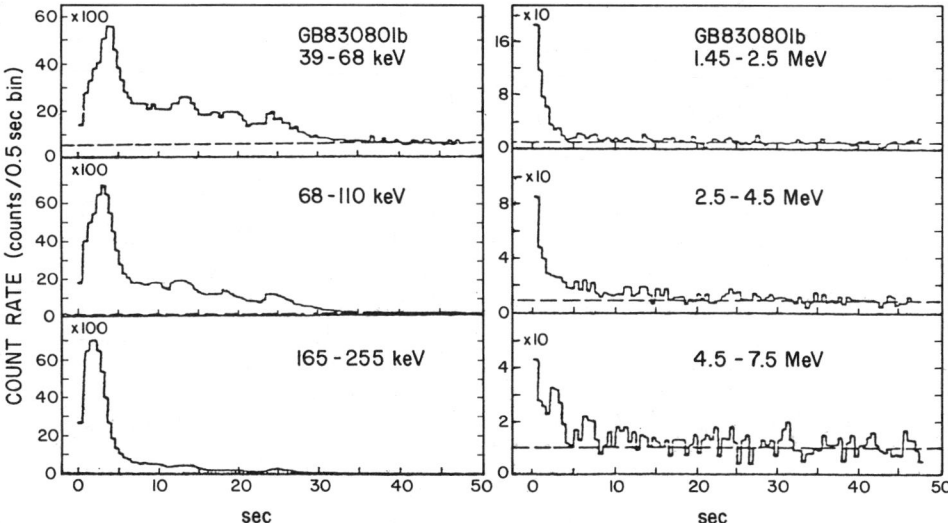

Figure 3 Energy dependence of time profiles of the classical burst GB830801b (Kuznetsov et al 1986a), showing a 3-s shift in the peak between the lowest and highest energy bands and more complex structure in some bands. The dashed lines indicate the detector background prior to the burst.

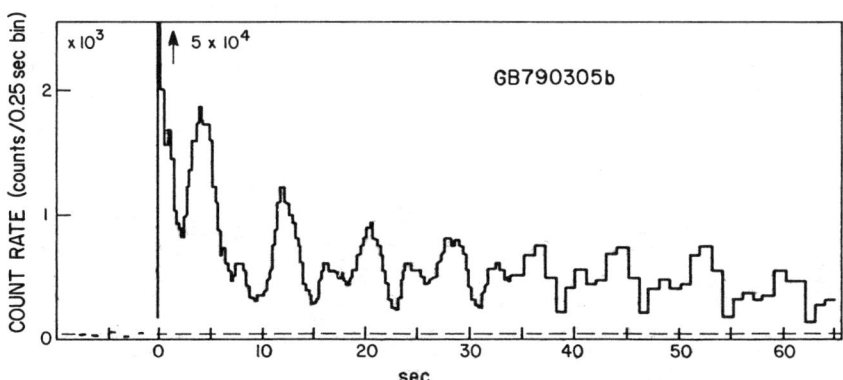

Figure 4 Time profile of the extraordinary burst GB790305b (Mazets et al 1979a) from a soft repeater, showing the short burst followed by much less intense periodic emission. The dashed line indicates the detector background prior to the burst.

2.3 Periodicities

Although the majority of gamma-ray bursts have multiple peaks, searches for periodicities in their light curves have generally been unsuccessful (e.g. Wood et al 1986). Only one "indisputable" period has been found (Mazets et al 1979a, Barat et al 1979, Cline et al 1980, Terrell et al 1980), and it is in the postburst emission of the unique event GB790305b from a soft repeater. This 8.00 ± 0.05-s periodicity is clearly seen for 8 cycles in the truncated Konus data (Figure 4) and was detected for 22 cycles in the *Helios 2* data. The double-pulsed form of this emission is very similar to that of pulsars, and the period has likewise been generally interpreted (Mazets et al 1979a, Barat et al 1979, Cline et al 1980, Terrell et al 1980) as a neutron star rotation period, although precession (Brecher 1982) and plasma oscillations (Woosley & Wallace 1982) have also been suggested. In order that the surface velocity does not exceed that for centrifugal breakup, such a spin period P would set a nominal limit on the mean density $\rho > 3\pi/GP^2$ of 2×10^6 g cm^{-3}. This exceeds typical white dwarf densities ($\sim 6\times 10^5$ g cm^{-3}) by more than a factor of 3, suggesting that the source could only be a neutron star. During an earlier observation of this general region in 1976, a ~ 90-s X-ray transient emission was reported (Loznikov et al 1980) with an 8.6 ± 0.6-s period. It has been suggested (Ramaty et al 1980, Kazanas 1988) that the extraordinary burst GB790305b resulted from a core quake that may have spun up the star. The 120-ms duration of the burst is also within the expected range of the gravitational-wave damping times of neutron star vibrations (Ramaty et al 1980, Lindblom & Detweiler 1983).

A ~ 23-ms quasi-periodic oscillation of much lower statistical significance was also reported (Barat et al 1983) in the time profile of GB790305b during the burst peak. This periodicity is evident for seven cycles in the *Venera 12* Signe observations, but it was not apparent in the *Venera 11* Signe or *Helios 2* observations, which were at slightly higher energies and lower counting rates. Periods of this order could be expected (Van Horn 1980) from torsional vibrations of a neutron star.

Suggestions of periodicities ranging from 2 to 18 s have been reported in the time profiles of more than a dozen classical, multipeaked gamma-ray bursts (Desai 1981, Wood et al 1981, 1986, Norris 1983, Barat et al 1984b, Kouveliotou et al 1988). However, these suggested periodicities exhibit complex temporal behavior, whose significance is difficult to quantify. Using Monte Carlo simulations, Schaefer & Desai (1988) calculated the probability that random events could mimic such periodicities; applying their model results to these bursts, they concluded that only GB790305b shows an "indisputable" periodicity at 8.0 s.

2.4 Recurrence

Only three repeating sources of gamma-ray bursts have been found, and their recurrence, plus similar temporal and spectral characteristics, have distinguished them as a distinct class—the "soft repeaters." No recurrent bursts have been identified from the classical gamma-ray bursts, and only rough upper and lower limits can be set on their recurrence times.

2.4.1 SOFT REPEATERS The three repeating sources—GBS 0526−66 with 17 bursts (Golenetskii et al 1979, 1987b), GBS 1806−20 with >100 bursts (Atteia et al 1987a, Laros et al 1987), and GBS 1900+14 with 3 bursts (Mazets et al 1979b)—have been identified from the overlap of their burst positions on the sky. Most of the bursts attributed to the prolific burster GBS 1806−20, however, lack good source localizations and have been selected on the basis of similarities in both their time profiles and their energy spectra.

These sources, as mentioned above, differ from other gamma-ray bursters not just in repetition alone (Kouveliotou et al 1987, Cline et al 1988). The repeating bursts have simple time profiles of typically short (~ 0.1 s) duration, unlike the classical bursts that commonly have multipeaked time profiles of longer (~ 10 s) duration. These repeating bursts also have soft energy spectra, which are roughly constant in time, unlike the bulk of the classical bursts, which have much harder energy spectra that soften with time. Thus, the repeaters clearly represent a separate class from the classical bursters.

GBS 0526−66 Although the intense burst GB790305b was detected by 10 different experiments, the 16 much weaker recurring bursts from this source were observed only by the Konus experiments (Golenetskii et al 1984, 1987b). Thirteen of these bursts were located at the same sky position to within 0.01–1° uncertainties, and three to within 15° uncertainties; all were consistent with the previously determined site of GB790305b. This position, which lies within the diffuse supernova remnant N49 in the Large Magellanic Cloud (Evans et al 1980), also points to a neutron star as the source.

The timing of these recurrent bursts does not show any simple periodicity. However, Rothschild & Lingenfelter (1984) have found that the times between these recurrences show a marginally significant (2.6σ) degree of ordering, implying a 164-day period, which they interpreted as evidence for periodic accretion onto a neutron star from a binary companion in a highly elliptical orbit.

These repeating bursts differ from those observed from the other two repeating sources, GBS 1806−20 and GBS 1900+14, in one important respect. The peak intensities at energies >30 keV for 13 out of the 16 recurrent bursts following GB790305b do not vary significantly (Golenetskii et al 1984, 1987b), i.e. within 1σ, from a mean value of $2\pm0.4\times10^{-6}$ erg cm^{-2} s^{-1}. At the distance of the Large Magellanic Cloud (55 kpc), such an intensity would imply a peak luminosity of $\sim 6\times 10^{41}$ ergs s^{-1}. The two exceptions were only ~ 2–3 times more luminous, whereas the intense pulse of GB790305b was 10^3 times brighter. The durations of the repeaters, on the other hand, ranged over two orders of magnitude, from ~ 0.1 to 9 s, but the median was ~ 0.2 s.

GBS 1806−20 This prolific source is located near the direction of the galactic nucleus at a galactic longitude $l\sim 9.8°$ and latitude $b\sim -0.24°$. The first burst identified from this source was the soft, short burst GB790107 (Mazets et al 1981a, Laros et al 1986), and 110 repeating bursts have since been observed (Laros et al 1987) through June 1986 with an experiment on *ICE* in the 5–30 keV energy band. Eighteen of these bursts were also seen by other spacecraft, and overlapping positions were determined. Laros et al (1987) identified the remaining bursts as repeaters from the similarity of their energy spectra and of their temporal properties.

The occurrence of these repeating bursts is highly nonuniform in time. The time between bursts has ranged from 1 s to 5×10^7 s, and more than half of all the bursts observed over the ~ 8 yr of *ICE* observations occurred within a 2-week period in early November 1983. Thus, the mean time between bursts ($\sim 10^6$ s) differs from the median time ($\sim 10^4$ s) by a factor of 10^2. There is no periodicity in the occurrence times, and there appears to be no correlation between burst intensity and the time between occurrences. Unlike the bursts from GBS 0526−66, the peak intensities of these bursts varied by a factor of 30 and were as high as 4×10^{-5} erg cm^{-2} s^{-1} at energies >30 keV. At the nominal distance of the galactic center, such intensities would correspond to a peak luminosity of $\sim 4\times 10^{41}$ ergs s^{-1}. This value is quite comparable to that of the repeaters from GBS 0526−66, if it is in the Large Magellanic Cloud. Both source luminosities are $>10^3$ times higher than the Eddington luminosity for a neutron star. The durations of these bursts range from <16 to 200 ms (Atteia et al 1987a).

Kouveliotou et al (1987) have suggested that these soft burst repeaters resemble Type II X-ray bursts in both the shape of their time profiles and their constant energy spectra. However, there are many more differences: The Type II X-ray bursters have much longer mean rise times (~ 1 s), blackbody spectra with a temperature of only ~ 1.5 keV, regular ~ 6-month recurrence periods, and burst intensities correlated with the time between occurrences (Lewin & Joss 1983).

GBS 1900+14 This source position corresponds to the galactic coordinates $l \sim 47°$, $b \sim 4°$ (Mazets & Golenetskii 1981). The three weak repeating bursts GB790324, GB790325a, and GB790327a, which occurred within 3 days of one another, were detected solely by the Konus detectors on *Venera 11* and *12*. Based on their common sky position and similarities in their time profiles and energy spectra, Mazets et al (1979a) concluded that these bursts were produced by the same source. All three also had very similar peak intensities of $\sim 2 \times 10^{-5}$ erg cm^{-2} s^{-1}, with durations of 50–200 ms and rise times of <10 ms (Mazets et al 1981a, 1982a).

2.4.2 CLASSICAL BURSTS The most comprehensive search to date for repetition of the classical bursts is that made by Atteia et al (1987b). They looked for coincidences in the positions in a sample of 84 bursts, which occurred over a 17-month period and were accurately localized using arrival-time analyses of multiple detections by widely separated spacecraft. They found that the number of overlapping burst locations was consistent with random sky positions, and thus that there was no evidence for repetitions. From these data they placed a 3σ lower limit on the mean recurrence time of ~ 8 yr for any repeating bursts above the detector thresholds. They also found, however, that if the repeating bursts had a power-law distribution in luminosity spanning three or more decades, so that the bulk of the hypothetical repeaters occurred below the instrumental thresholds, then the limit on the mean recurrence time could be <1 yr. Very similar limits were set by Schaefer & Cline (1985) from a different sample of bursts.

Model-dependent upper bounds on the mean recurrence time can be estimated (e.g. Jennings & White 1980, Shklovskii & Mitrofanov 1985, Hartmann et al 1990) from the observed frequency of bursts and the possible number of sources. For example, if the burst sources are old, nearby galactic neutron stars with a number density of $\sim 10^{-3}$ pc^{-3}, then there could be $\sim 10^4$ potentially observable sources within a scale height of about 0.2 kpc required by their isotropy (discussed below), and the observed rate of $\sim 10^2$ yr^{-1} would imply a recurrence time of no more than $\sim 10^2$ yr, even if all are burst sources.

3. POSITIONS AND SPATIAL DISTRIBUTION

Burst source positions have been measured by two different means. Positional error boxes of several square degrees have been determined from the relative intensities seen by orthogonally oriented detectors on single spacecraft (e.g. Mazets et al 1981a). Much higher precision error boxes, which can be as small as a fraction of a square arcminute, were determined by arrival-time differences between detections by widely separated spacecraft (e.g. Atteia et al 1987b). More precise burst positions have been searched for quiescent counterparts, while the more numerous, less precise burst positions show the distribution of the sources on the sky.

3.1 Search for Quiescent Counterparts

Quiescent counterparts of gamma-ray bursts have been sought both in survey catalogs of unusual objects that might lie in any of the burst error boxes and in deep observations of the more precise error boxes in various wave bands. Although close to 200 burst positions have been determined to accuracies of a few degrees or better, there are only 10 burst sources that have been localized to error boxes of ≤ 20 arcmin^2, which are small enough to permit deep searches (Atteia et al 1987b). Such precise positions, determined from arrival-time differences, are rare because they require measurements of statistically significant temporal variations over times of <100 ms (e.g. Klebesadel 1988).

So far no quiescent counterpart in the optical, infrared, radio, or X-ray ranges has been identified for any of the gamma-ray bursts. Only the soft repeater source of the extraordinary GB790305b gamma-ray burst has been associated with any previously identified object. Nonetheless, the lack of candidates in other error boxes sets constraints on the nature of the burst sources.

The smallest error box of 0.05 arcmin^2 is that of GB790305b from the soft repeater GBS 0526−66. This error box lies within the ~ 2-arcmin^2 X-ray contours of the diffuse supernova remnant N49 in the Large Magellanic Cloud (Cline et al 1982). Such an association strongly suggests (e.g. Cline 1980) a neutron star source for this burst, as do the temporal and spectral observations. Although the identification of this source location in the supernova remnant has been questioned (e.g. Golenetskii et al 1984), the random probability of an association with any of the $\sim 6 \times 10^4$ known sources that were searched is estimated (Felten 1981) to be quite small ($<10^{-3}$).

The diffuse emission from N49 limits the depth of optical and other searches. An optical search (Pedersen et al 1986) of the GBS 0526−66 error box down to $m_V = 21.5$ has failed to reveal either a candidate source or a likely binary companion, which might be expected as a source of accreting material needed to power its repeating bursts. At the distance of the Large Magellanic Cloud, this only eliminates sources or companions with an absolute magnitude $M_V < 2.8$, corresponding to main sequence F stars.

Several *Einstein* X-ray observations in the energy range 0.1–4.5 keV also show (Helfand & Long 1979, Pizzichini et al 1986) no discrete source in the GBS 0526−66 error box. These observations limit the quiescent source flux to $<10^{-13}$ erg cm^{-2} s^{-1} if it is blackbody emission from the polar cap of a neutron star (Pizzichini et al 1986).

Deep optical searches (Pedersen et al 1983, Barat et al 1984c, Cline et al 1984, Laros et al 1985b, Motch et al 1985, Hartmann & Pogge 1987) of the smallest error boxes for classical bursts also show no likely quiescent source candidates. These searches go down to objects as faint as $m_V \sim 25$ (Motch et al 1985), requiring sources and possible companions with $M_V > 15$ (fainter than main sequence M7 stars at galactic distances of ≤ 1 kpc). Archival plate searches for optical transients in these error boxes are discussed below.

X-ray limits (Pizzichini et al 1986, Boer et al 1988) for several of these error boxes have been used (Hurley 1987) to constrain the maximum temperatures of possible neutron star sources to $< 10^6$ K and their minimum ages to $> 10^5$ yr, assuming current cooling models and galactic distances of ≤ 1 kpc. Constraints have also been placed (Pizzichini et al 1986, Boer et al 1988) on the accretion rate of possible neutron star sources from these X-ray limits, which would appear to exclude steady accretion rates as high as those required by some thermonuclear flash models (e.g. Hameury et al 1983). These constraints, however, are very model dependent.

3.2 Optical Transients

No simultaneous optical and gamma-ray burst emission has yet been observed. Still, several optical transients have been reported at other times from sites located within high-precision error boxes (Schaefer 1981, Schaefer et al 1984, Pedersen et al 1984). Repeated optical flashes have also been reported (Hudec et al 1988) from a position 5 arcmin, and 15σ (Laros 1988), outside of the error box of GB790325b. These identifications have not yet been confirmed, however, and those found on archival plates have been questioned (Zytkow 1990).

Schaefer (1981) and Schaefer et al (1984) searched archival photographic plates with a total cumulative exposure of 2.7 yr for a record of transient images from seven error boxes. They reported that three optical transients (OTs) occurred on the nights of 17 November 1928, 4 October 1901, and 19 February 1944 in the sky fields of GB781119, GB791105, and GB790113, respectively. However, from an independent examination of these plates, Zytkow (1990) concludes that at least one of the reported transients is a plate defect, although Schaefer (1990) disputes this conclusion. If they are not artifacts, these archival images, designated OT1928, OT1901, and OT1944, would have m_B of 10.2, 13.7, and 12.1, respectively (Schaefer et al 1984). The implied ratio of energy emitted at optical wavelengths in these archival images to energy released in gamma rays (>30 keV) in the present-day bursts is $\sim 10^{-3}$. The quiescent emission from the sites of these possible optical transient sources is very faint. Their sky fields contain weak stellar sources with $m \sim 23$ (Schaefer et al 1983, Pedersen et al 1983).

Pedersen et al (1984) detected an optical transient with a maximum m_V of 8.7 on 8 February 1984 from the field of N49, associated with the location of GBS 0526−66. Its light curve resembles that of the initial ∼120-ms spike observed (Mazets et al 1981a) in GB790305b, but its duration was 1.7 s. Also, it occurred at zero phase of the 164-day period determined by Rothschild & Lingenfelter (1984) from gamma-ray burst recurrences. The lack of detectable accompanying gamma-ray emission places a lower limit on the ratio of the optical to gamma-ray (>30 keV) luminosity of 10^{-2}, much more than that suggested for archival flashes. This flash differs significantly from optical transients generated by satellite glints and meteor trails, the primary sources of the background rate of such events (Schaefer et al 1987).

A variety of models have been proposed to explain the generation of such optical transients in gamma-ray burst events. These include the reprocessing of burst gamma rays in either the surface layers of a low-mass ($<0.06\ M_\odot$) stellar companion (Rappaport & Joss 1985, Melia et al 1986, Cominsky et al 1987) or in a cold, degenerate accretion disk surrounding a neutron star (Epstein 1985a, Melia 1988); cyclotron reprocessing in the magnetosphere of a neutron star (Woosley 1984b, Hartmann et al 1988); and plasma processes in neutron star magnetospheres (Liang 1985, Sturrock 1986, Ruderman 1987).

3.3 Distribution on the Sky

The most extensive sample of burst source positions is that compiled by Golenetskii (1988), primarily from the Konus experiments aboard *Venera 11–14*. This sample of 175 burst directions localized to single error boxes is shown in Figure 5 in galactic coordinates. These burst positions are not concentrated toward either the galactic plane or the galactic center, as would be expected of more distant members of a disk or spheroid population. Nor are these positions correlated with known galaxies or galactic clusters, as might be expected of extragalactic sources. The distribution of positions is instead random. However, quantitative analyses of these sky positions are complicated by an instrumental artifact: a galactic latitude asymmetry in the ratio of bursts in the Northern to Southern Hemispheres caused (Mazets & Golenetskii 1982) by the shadowing of the Konus detectors by the radio antenna of the Venera spacecraft.

Fifty-four bursts have been localized to single error boxes with greater accuracy than that of the Konus experiment by multispacecraft detections of the second interplanetary network (Atteia et al 1987b). The distribution of these error boxes is also isotropic. Analyses of these positions (Hartmann & Epstein 1989) show no evidence for either a dipole or a quadrupole component. Angular correlation studies (Hartmann & Blumenthal 1989)

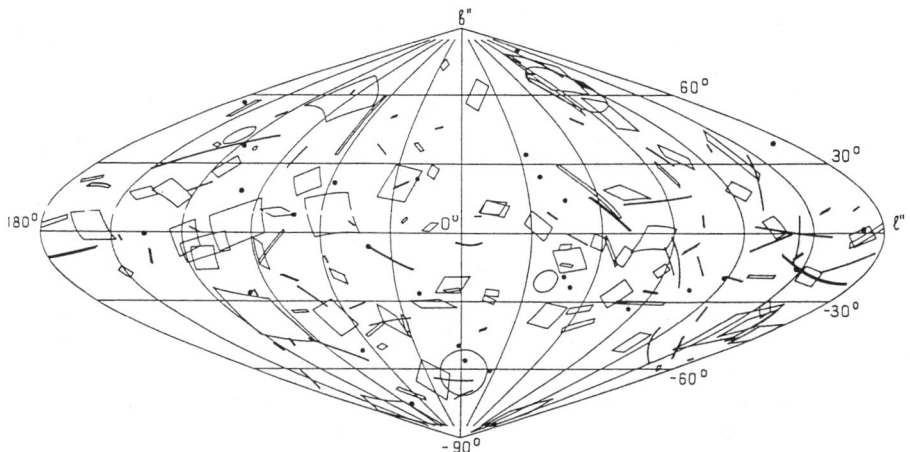

Figure 5 Distribution of 175 burst positions in galactic coordinates from the Konus experiment (Golenetskii 1988). The isotropy of these positions is consistent with nearby galactic sources.

also show no clustering in the burst source positions. Together these studies indicate that the burst sources must be either nearby members of a galactic disk population lying well within a scale height or members of a very distant extragalactic population lying beyond ~ 30 Mpc.

Since the typical energy flux in this burst sample is of the order of 10^{-5} erg cm^{-2} s^{-1}, typical burst luminosities of $< 10^{38}$ ergs s^{-1} are indicated if the burst sources are nearby members of a disk population lying within half a scale height (probably < 0.3 kpc). Much larger luminosities ($> 10^{49}$ ergs s^{-1}) would be required if the brighter ($\sim 10^{-4}$ erg cm^{-2} s^{-1}) burst sources were members of a very distant (> 30 Mpc), nearly cosmological population, which would also be consistent with the observed isotropy.

3.4 *Size-Frequency Distributions*

Because of the lack of identifiable counterparts in any of the burst source error boxes, special attention has been given to studies of the size-frequency distribution of the classical bursts in an attempt to learn something about their spatial distribution and typical distances. In these studies the size of the bursts is usually measured either by their peak energy flux or by their total energy fluence, which is the energy flux integrated over the duration of the burst; the frequency is the cumulative discovery rate per year of bursts larger than a given size. Since the observed peak flux or fluence of a burst is proportional to the inverse square of its distance and the total number of bursters is proportional to the cube of the distance (if they are uniformly distributed in space), the expected frequency of observation of bursts with fluxes or fluences greater than some value should be proportional to the $-3/2$ power of that flux or fluence. This slope is independent of either beaming of the emission or variations in the luminosity. A deviation of the observed size-frequency distribution from a $-3/2$ power law should thus indicate a nonuniformity in the spatial distribution. However, the size-frequency distributions of gamma-ray bursts as functions of either energy flux or fluence are subject to severe selection biases.

The most widely studied size-frequency distribution of gamma-ray bursts has been that in the time-integrated energy flux (or fluence) $N(>S)$, the number of bursts detected per year at a fluence greater than S. Some recent measurements of this distribution are shown in Figure 6 as a function of fluence (ergs per square centimeter) at photon energies greater

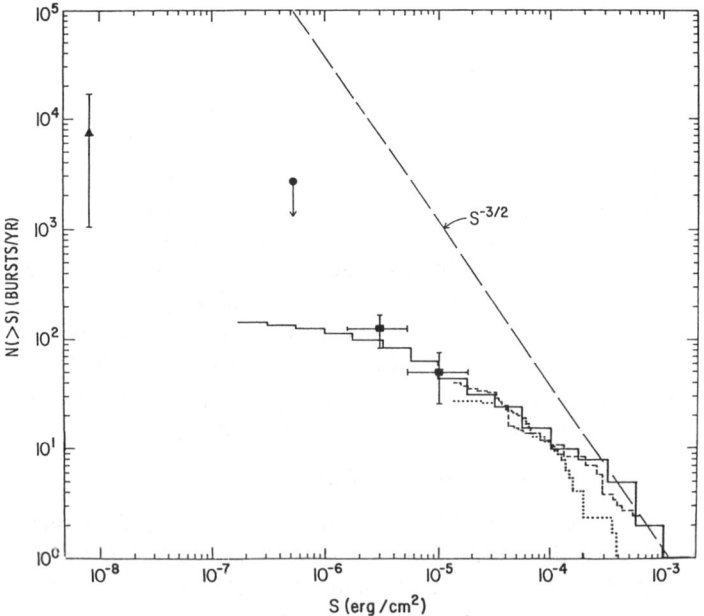

Figure 6 Recent cumulative size-frequency distributions of gamma-ray bursts as a function of fluence S at photon energies >30 keV from observations of Mazets et al (1981a; solid curve), Barat et al (1982; dashed curve), Klebesadel et al (1984a; dotted curve), Meegan et al (1984; circle), and Beurle et al (1981; triangle). Also shown for comparison is the size-frequency distribution $S^{-3/2}$ that would be expected for a uniform distribution of sources.

than 30 keV. These observations include the spacecraft surveys of Konus (Mazets et al 1981a), Signe (Barat et al 1982), and *PVO* (Klebesadel et al 1984a), which reported fluences greater than $\sim 10^{-6}$ erg cm^{-2}, and the much briefer, but more sensitive, balloon measurements (Beurle et al 1981, Meegan et al 1984, 1985), which set limits on the detections of low-fluence bursts. Shown for comparison is the $-3/2$ power-law distribution expected for a complete sample of bursts from a uniform distribution of sources in Euclidean space.

Although there are systematic differences among these measurements resulting from differing instrumental responses, burst identification procedures, and sky coverage, the measurements nonetheless all show that the cumulative distributions deviate significantly from the $-3/2$ power law at fluences less than $\sim 10^{-4}$ erg cm^{-2}. This flattening at low fluences was previously taken as proof of a galactic origin of the bursts, reflecting a nonuniform spatial distribution of the more distant sources (Yoshimori 1978, Fishman 1979, Jennings & White 1980, Puget 1981, Jennings 1982, 1984).

However, such nonuniform spatial distributions are not consistent with the isotropic distribution of burst positions on the sky (Mazets et al 1981a, Mazets & Golenetskii 1987, Atteia et al 1987b) if the burst sources are members of a galactic disk or halo population. This apparent contradiction between isotropic sky positions and the size-frequency distributions at low fluences led to suggestions that the gamma-ray burst sources belonged to exotic populations in an extended galactic corona at distances approaching ~ 0.25 Mpc (Jennings 1984, Shklovskii & Mitrofanov 1985), or at even greater burst distances at cosmological redshifts of 1 or 2 (Paczynski 1986). However, Higdon & Lingenfelter (1984) pointed out that most of the flattening of the cumulative distributions at low fluences resulted from incomplete sampling due to instrumental selection biases, not spatial non-uniformity, and thus these distributions and the isotropic sky positions were both consistent with a nearby ($\leq 10^2$ pc) portion of a disk population of gamma-ray burst sources.

Such selection biases arise because typical burst searches do not trigger on fluence or flux but on the detector counting rate in some energy band and integration-time interval. Thus, searches are complete only above the threshold counting rate for burst detection, which corresponds to a very wide range of fluxes and fluences, because of the extremely large variation in the energy spectra and durations of the classical bursts (Higdon & Lingenfelter 1985, 1986, Mazets 1985).

A clear illustration of the dominance of selection biases on the cumulative burst distributions is shown in Figure 7 (Mazets & Golenetskii 1987). Here the same long-duration (≥ 1 s) classical bursts detected by the Konus experiments aboard *Venera 11–14* are displayed as separate cumulative

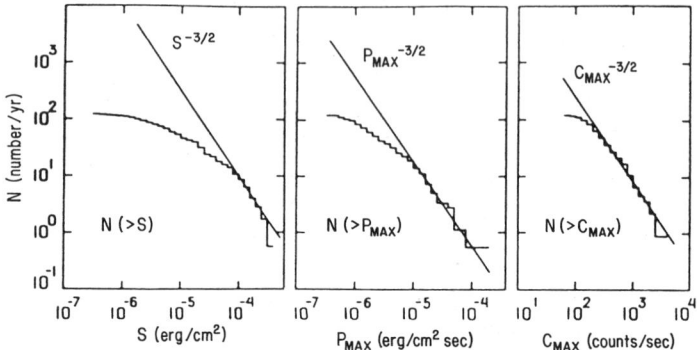

Figure 7 Cumulative size-frequency distribution of the Konus bursts (>1 s) vs. fluence, peak energy flux, and maximum counting rate (Mazets & Golenetskii 1987), showing differences due to instrumental selection biases resulting from variations in the burst durations and energy spectra.

distributions in fluence S, peak energy flux P_{max}, and maximum count rate C_{max}. The deviations of $N(>S)$ and $N(>P_{max})$ from the $-3/2$ power law are caused solely by variations in the burst durations and energy spectra, since even the small deviation of $N(>C_{max})$ from the $-3/2$ power law at low C_{max} is an instrumental bias (Mazets & Golenetskii 1987). Note that the observed range of C_{max} from $\sim 10^2$ to 4×10^3 s^{-1} corresponds to a relatively small variation in source distances of only a factor of 6 between the nearest and farthest sources observed in this sample.

Some recent studies (Yamagami & Nishimura 1986, Jennings 1988, Paczynski & Long 1988) of the Konus count-rate distribution reached contradictory conclusions about the spatial uniformity of the burst sample, resulting in part from different treatments of the instrumental biases. Higdon & Schmidt (1989) have employed an alternative technique, the V/V_{max} test (Schmidt 1968, Schmidt et al 1988), to test the spatial uniformity of the Konus distribution. Unlike cumulative distributions as functions of C_{max}, the V/V_{max} test is independent of threshold variations and instrumental sensitivities. The value of V/V_{max} for each burst is the ratio of the spherical volume defined by the distance of the burst and the maximum volume from which the burst could have been detected. If the burst population has a uniform spatial distribution, then values of V/V_{max} will be uniformly distributed between 0 and 1, and the mean value would be 0.5 with a root-mean-square error of $(12N)^{-1/2}$, where N is the number of bursts in the sample. Values <0.5 would predominate if the bursts belong to a galactic disk population, sampled to distances larger than their scale height, or if cosmological sources were evaluated using Euclidean geometry. For the complete sample of 123 classical bursts detected by Konus on *Venera 11* and *12*, Higdon & Schmidt (1990) found $V/V_{\max} = 0.45 \pm 0.03$, consistent at the 2σ level with the spatial uniformity and at the 1σ level with values expected (Hartmann et al 1990, Paczynski 1990) from a nearby disk population of neutron stars.

Nonetheless, spatial nonuniformities doubtless account for the major portion of the deviations from a $-3/2$ slope observed at the very low fluences of the cumulative distributions, obtained from high-sensitivity balloon measurements (Beurle et al 1981, Meegan et al 1984, 1985) that are shown in Figure 6. In the most sensitive of these measurements, having a threshold *count rate* ~ 10 times smaller than that of the Konus experiment, Meegan et al (1985) detected only a single burst at a sensitivity where they would have expected to have found 43 if the burst population were still spatially uniform. Thus, bursts at such low fluence levels most likely come from a disk population at distances much greater than the scale height of the population. If so, the distribution of low-fluence bursts on the sky should be anisotropic, concentrated at low galactic latitudes and longitudes. The BATSE experiment (Fishman et al 1989) on the *Gamma Ray Observatory* (*GRO*) will be able to test this.

4. ENERGY SPECTRA

Gamma-ray burst emission has been measured over a broad band of energies from ~ 1 keV to ~ 100 MeV. No single measured spectrum spans this range, however; the Konus spectral observations and most others cover only the range from 30 keV to about 1 MeV, while the higher energy spectra measured by *SMM* are incomplete below 300 keV. Only a small fraction of the measured spectra fully cover the important range from a few tens of keV to 10 MeV, which would allow distinct high- and low-energy components of the emission to be identified.

The determination of gamma-ray energy spectra from the count-rate data, moreover, is not unique. Because the photons generally do not deposit all of their energy in the detector, there is a probability distribution for the energy of each photon, and some assumptions must be made about the form of the incident spectrum (e.g. Fenimore et al 1983, Teegarden 1984, Loredo & Epstein 1989). For this reason, the shapes of gamma-ray continuum spectra and even the existence of line features have been contentious issues (e.g. Zdziarski 1987, Lamb 1988).

4.1 Continuum

The continuum spectra of the soft repeaters and the classical bursts are quite different, although the same emission processes have been suggested for both types of bursts. Thus we first discuss the spectral observations of these two classes separately, and then examine the observations of spectral features, before we consider the overall constraints on the emission processes collectively.

4.1.1 SOFT REPEATERS With the exception of the unique GB790305b event, the energy spectra of bursts from the three soft repeaters GBS 0526−66, GBS 1806−20, and GBS 1900+14 are relatively simple, featureless continua that vary little with time (Golenetskii et al 1984, 1987b, Atteia et al 1987a, Laros et al 1987). Their energy spectra can all be approximated quite well by a simple exponential with an e-folding energy of about 25–40 keV, consistent with either thermal bremsstrahlung or synchrotron emission. Such a spectrum with a characteristic energy of 35 keV also fits the soft (<150 keV) emission of both the GB790305b burst and its subsequent 8-s oscillatory emission (Mazets et al 1982a). In that burst, however, there is an additional, much harder component with annihilation radiation, as is discussed below. Representative spectra of the soft burster GBS 1806−20 (Atteia et al 1987a) clearly show (Figure 8a) that the spectra are very similar, and that the bulk of the power in these bursts is emitted at comparatively low photon energies (<100 keV).

4.1.2 CLASSICAL BURSTS Unlike the soft repeaters, the classical burst spectra have their peak power at gamma-ray energies >100 keV. The continuum spectra can be highly variable, and some appear to have separate components, above and below a few hundred keV [as can be seen (Figure 9a) in the spectrum of GB780325 (Hueter 1987)], while others are quite monotonic over this range.

Spectral variability is, in fact, a characteristic feature of classical bursts (Mazets et al 1982b). Significant variations have been observed down to the shortest times (0.25 s) over which spectra can be resolved (Mazets et al 1983, Mitrofanov et al 1984). Such variability is complex, and no single pattern describes the bulk of the bursts. However, the burst continua are often hardest in the initial burst phase and soften with time (Mazets et al 1982b, Teegarden 1982, Kuznetsov et al 1986a), as can be seen (Figure 8b) in the rapidly evolving spectrum of the burst GB830801b, whose time-dependent fluxes are shown in Figure 3. Similar hard-to-soft spectral evolution has also been observed within individual pulses of complex, multipeaked bursts (Norris 1983, Norris et al 1986).

There are significant differences in the shape of the continua above and below a few hundred keV; there often appears to be a power-law component at the higher energies and a quasi-exponential component at the lower energies, with possibly a third, blackbody component at keV energies at late times.

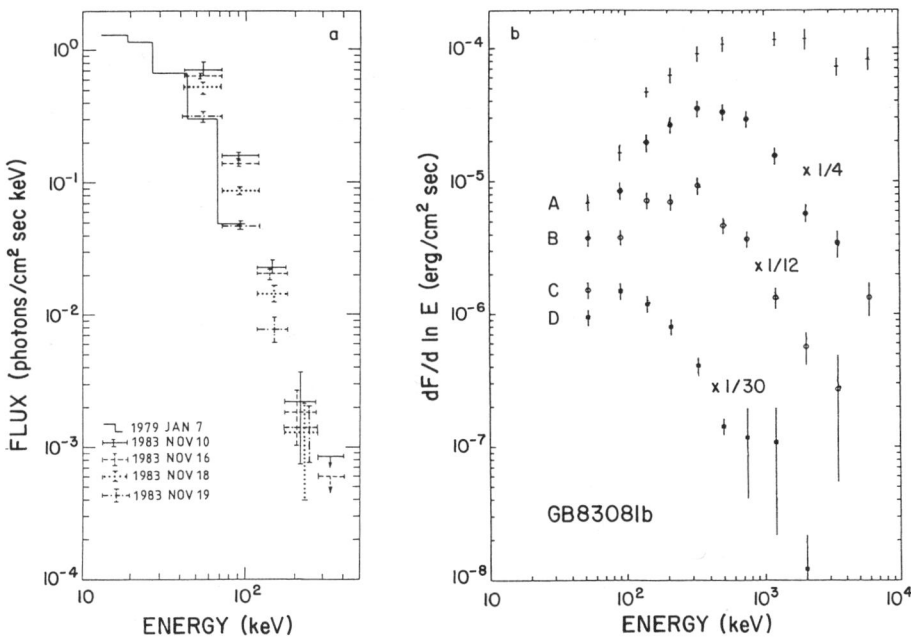

Figure 8 Examples of continuum spectra of bursts: (*a*) the very similar quasi-exponential spectra of bursts from the repeater GBS 1806−20 (Atteia et al 1987a); (*b*) the rapidly evolving classical burst spectra in successive 0.5-s intervals in GB830801b (Kuznetsov et al 1986a, Zdziarski 1987).

At energies less than a few hundred keV, the classical burst photon spectra have been approximated (e.g. Mazets et al 1982b, Laros & Nishimura 1986) by an exponentially truncated inverse power law, $E^{-n} \times \exp(-E/kT)$, where n lies between about 0.8 and 1.5, and the effective "temperature" T varies greatly with time. These spectra are consistent with either thermal bremsstrahlung or synchrotron. Such a shape can fit the observed spectra down to energies as low as ~ 1 keV (Wheaton et al 1973, Gilman et al 1980, Laros et al 1984, Murakami 1988). The maximum observed temperatures for some 70 Konus bursts were broadly scattered around a mean of ~ 200 keV, or $\sim 2 \times 10^9$ K (Mazets et al 1982b). Because the Konus detectors triggered on the flux at lower energies, there was a strong selection bias against detection of weak, high-temperature bursts (Higdon & Lingenfelter 1986).

The soft X-ray emission often decays much more slowly than the gamma-ray emission, lasting as much as $\sim 10^2$ s longer (Laros & Nishimura 1986, Murakami 1988). Analyses of the *Ginga* X-ray spectra show (Mura-

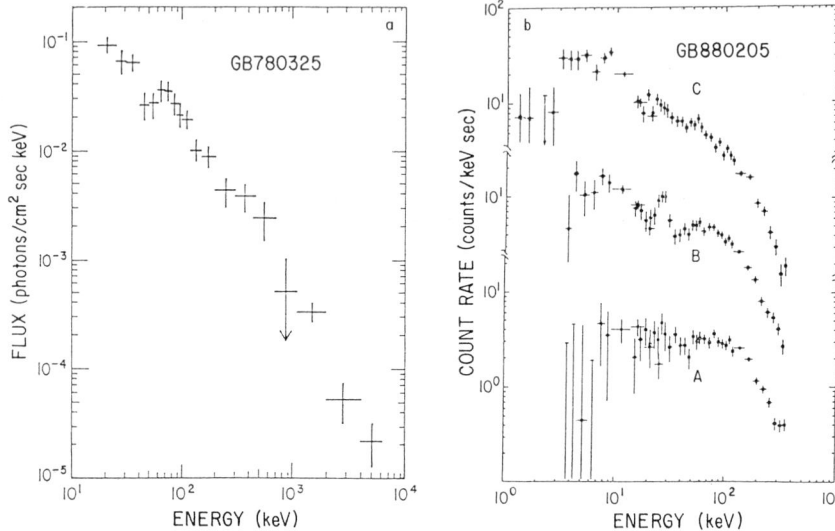

Figure 9 Examples of absorption features in the spectra of the bursts GB780325 (panel *a*) measured by Hueter (1987) and GB880205 (panel *b*, spectrum *B*) measured by Murakami et al (1988).

kami 1988) that the burst spectra are monotonically cooling, and that at late times the soft X-ray emission from at least one burst (GB870303) could be fitted by a blackbody spectrum with a temperature of ~ 1 keV, although thermal bremsstrahlung and thermal cyclotron emission could also give an acceptable fit. With the blackbody emissivity as an upper bound, the detected flux of 0.9×10^{-8} erg cm^{-2} s^{-1} clearly requires (Murakami 1988) a galactic origin for this gamma-ray burst if the X rays come from the surface of a cooling neutron star, since the source distance D (in units of kiloparsecs) must be less than the radius of the emission region R (in units of kilometers), i.e. $D(\text{kpc}) \leq R(\text{km})$.

Although less than a few percent of the gamma-ray burst power is emitted in soft X rays (Helfand & Vrtilek 1983, Murakami 1988), this lack of significant X-ray emission also constrains the geometry of the gamma-ray emission region. In order that the neutron star surface not absorb a larger fraction of the burst energy, which would be reemitted as blackbody radiation, it has been suggested (Epstein 1986, Lamb 1988) that the gamma rays must be emitted anisotropically away from the stellar surface, and that any isotropic emission must come from several stellar radii above the surface, so that the star subtends only a small solid angle. However, Compton scattering can also suppress (Brainerd 1989, Ho & Epstein 1989, Dermer 1989) the X-ray emission.

At energies greater than a few hundred keV, the photon flux spectra of the classical bursts can usually be fitted (Matz 1986) up to energies as high as ~ 100 MeV by power laws of the form E^{-n} with a time-dependent n ranging anywhere from about 1 to 2.5. This emission could arise in a number of ways, as is discussed below. The *SMM* observations have shown (Matz et al 1985, Matz 1986) that such high-energy gamma-ray emission is a common feature of the classical bursts; out of 71 bursts, 60% had significant emission above 1 MeV, and 25% had emission above 4 MeV.

There is no indication (Matz 1986) of a sharp cutoff in most of these spectra at high photon energies (e.g. Figure 8*b*), as might be expected from e^+e^- pair production. This lack of a high-energy cutoff places constraints (e.g. Harding et al 1986, Meszaros et al 1989) on both the transverse component of the magnetic field seen by the high-energy photons and the density of the high-energy photons. Gamma rays of energy greater than the single-photon pair-production threshold of 1.022 MeV and moving at an angle θ with respect to an intense magnetic field can interact with the field to produce e^+e^- pairs (e.g. Daugherty & Harding 1983). Thus, if gamma rays of energy E are emitted at the surface of a neutron star, the transverse component of the magnetic field seen by them, $B \sin \theta$, must be less than $\sim 2E(\mathrm{MeV})^{-0.9}\, 10^{12}$ G for them to escape without significant attenuation from pair production (Harding et al 1986) because the pair-production mean free path would be less than 1 km.

However, such power-law components without a cutoff up to at least 6 MeV, requiring $B \sin \theta \leq 0.4 \times 10^{12}$ G, have been observed (Hueter 1984, 1987) simultaneously with a ~ 50-keV cyclotron absorption feature in the lower energy component, requiring $B \sin \theta \approx 4 \times 10^{12}$ G, as is discussed below (see GB780325 in Figure 9*a*). This implies that the lower energy component (less than a few hundred keV) is produced near the surface of a neutron star, while the higher energy power-law component must have been produced in a different region, probably well above the surface, where the magnetic field was at least an order of magnitude weaker (Brainerd & Lamb 1987). A similar conclusion was previously reached (Hueter & Lingenfelter 1983) from consideration of the photon-photon pair-production opacity.

Even without such intense magnetic fields, high-energy gamma rays can be attenuated by e^+e^- pair production in photon-photon collisions (e.g. Schmidt 1978, Cavallo & Rees 1978). In a field-free vacuum, e^+e^- pair production by the interaction of two photons of energies E_1 and E_2 moving at an angle θ with respect to one another can occur when $E_1 E_2 \times (1 - \cos \theta) \geq 2m_e^2 c^4$. In a uniform emission region of radius R, the e^+e^- pair-production opacity exceeds unity above the threshold at $m_e c^2$, and the burst spectrum should steepen when the "compactness parameter"

$\sigma_T L/(Rm_e c^3) > 1$, where L is the burst luminosity and σ_T is the Thomson cross section (e.g. Zdziarski & Lamb 1986).

If one assumes an emission region size, the lack of a high-energy cutoff and the observed energy flux can place a limit on the distance of the burst source from the "compactness parameter," $d(\text{pc}) < 2(R_6/P_{-4})^{1/2}$, where the emission region radius R is in units of 10^6 cm and the energy flux P is in units of 10^{-4} erg cm^{-2} s^{-1}. Thus, in the burst GB780325 mentioned above, if the emission greater than 1 MeV came from a neutron star polar cap with a radius of $\sim 10^5$ cm, as the absorption feature implies for the lower energy emission, then this source with an energy flux P of only $\sim 10^{-6}$ erg cm^{-2} s^{-1} would have to lie within a distance $d < 6$ pc. This is uncomfortably close, since this burst was relatively weak and there are several bursts observed each year with more than 100 times that energy flux, many of which would then presumably have to be within 1 pc! Thus, again, the hard component of the emission (>1 MeV) must come from a separate, larger region well above the surface of the neutron star.

Limits on the burst distances have also been obtained from this opacity constraint by taking an upper limit to the emission region radius equal to the light travel distance for the smallest observed time scale of the flux variations. These radius limits also imply (Schmidt 1978, Epstein 1985a, Zdziarski 1987) a galactic origin for the bursts with distances ≤ 20 kpc. These galactic distance constraints, however, are not as compelling as the blackbody and spatial distribution arguments, discussed above, since these constraints implicitly assume isotropic emission, and much greater distance limits could be found by assuming relativistic bulk motions (Guilbert et al 1985) or beaming. For example, pair production does not constrain burst distances in the relativistic fireball model (Paczynski 1986, Goodman 1986), where photon energies are less than $m_e c^2$ in the reference frame of the pair outflow. The absorption features, discussed below, however, do exclude such large distances.

4.2 Spectral Features

Superimposed on the continuum spectra of many bursts are absorption and emission features (Mazets et al 1979a, 1981b, 1982b, Teegarden & Cline 1980, Dennis et al 1982, Barat et al 1984d, Hueter 1984, 1987, Golenetskii et al 1986, Murakami et al 1988), which we now discuss.

4.2.1 ABSORPTION LINES Absorption lines in the energy range from 20 to 70 keV have been observed in the spectra of the classical bursts. Although there was initially some question (Fenimore et al 1982a) about whether the features first observed (Mazets et al 1981b, 1982b) by the Konus experiment were caused by absorption or by time-varying continuum emission, detailed temporal analysis of such features observed by *SMM* (Dennis et al 1982), *HEAO-1* (Hueter 1987), and *Ginga* (Murakami et al 1988) have clearly demonstrated that they are, in fact, absorption lines. The lines are, however, a variable feature of the spectra and are mostly commonly observed only in the initial phase of a burst (Mazets et al 1981b, 1982b).

An example of such lines in the spectrum of the burst GB780325 measured by *HEAO-1* (Hueter 1987) is shown in Figure 9a, and the best resolved lines measured by *Ginga* (Murakami et al 1988) in the spectrum of the burst GB880205 are shown in Figure 9b. The spectrum in a 5-s interval (spectrum B in Figure 9b) just prior to the peak of this burst shows two absorption features, centered at 19.3 ± 0.7 and 38.6 ± 1.6 keV, differing by a factor of 2 in energy. The lack of such lines in the two spectra from 5-s intervals (spectra A and C) before and after this spectrum shows the transient nature of these features.

Mazets et al (1981b) first suggested that such features resulted from absorption of continuum emission at the cyclotron frequency by surrounding colder plasma in a very strong magnetic field. The absorption-line energies, corresponding to transitions between Landau levels n, are $E_n = n\hbar eB/m_e c = 11.6 n B_{12}$ keV, where B_{12} is the magnetic field in units of 10^{12} G, not including relativistic effects. The observed lines at 20–70 keV thus require intense magnetic fields, comparable to those deduced from pulsar spindown (Manchester & Taylor 1981), which strongly supports a neutron star origin for these bursts.

The lines at ~ 20 and 40 keV seen (Murakami et al 1988, Fenimore et al 1988) in the spectrum of GB880205 have thus been interpreted (Fenimore et al 1988, Lamb et al 1989, Wang et al 1989) as photon scattering leading to transitions from ground level to the first and second Landau levels in a magnetic field of $\sim 1.7 \times 10^{12}$ G. The ~ 20-keV line results from cyclotron resonant scattering of photons with energies near the first harmonic, and the ~ 40-keV line results from Raman scattering of photons with energies near the second harmonic. Most of the electrons excited to the second Landau level undergo radiative transitions to the first Landau level, since the branching ratio for radiative transitions from the second level directly to ground state is $\sim B/B_c$, where the critical field $B_c = 4.4 \times 10^{13}$ G (Daugherty & Ventura 1981). Photons scattered near the second and higher harmonics thus act as they would in an absorption line, as long as $B \ll B_c$. The radiative deexcitation of these higher Landau levels through intermediate levels generates new photons with energies near the first harmonic that fill in for many of the photons scattered out at those energies.

Because of such filling in at the first harmonic, the relative depth of the first and second harmonic lines can be comparable (Bussard & Lamb 1982), as is seen in Figure 9b, even though the probability of scattering at the second harmonic is small compared with that at the first. Absorption at the third harmonic is expected to be only a fraction of that at the second, which is also consistent with the observed spectrum. This would also imply that the absorption lines at < 60 keV observed by the Konus experiment (Mazets et al 1981b) could all be second-harmonic transitions, since those detectors had an energy threshold of 30 keV and would have missed the first line. Wang et al (1989) have shown that these scattering lines could be formed in a thin ($\sim 2 \times 10^{-3}$ g cm^{-2}), relatively cool (~ 5 keV) plasma sheet. The thickness of the sheet is determined by the column depth within which first-harmonic resonant scattering dominates the heating and cooling rates, and the temperature is determined by the balance between these rates (Lamb et al 1990).

These lines are a common feature of the spectra of the classical gamma-ray bursts. Mazets et al (1981b) have found absorption lines in the spectra of 19 of the ~ 60 bursts observed by Konus on *Venera 11* and *12* that had fluences $> 6 \times 10^{-6}$ erg cm^{-2}, or about one third of those bursts with enough photons to resolve such lines. Moreover, the actual fraction of burst spectra with such lines could be significantly higher, since the Konus detectors would have missed lines at energies below their 30-keV threshold, as well as higher energy lines that lasted significantly less than their 4-s spectral integration time. Thus, the observation of these lines also suggests that a major fraction, if not all, of the classical gamma-ray bursts have a neutron star origin.

4.2.2 EMISSION LINES   Gamma-ray emission lines have been observed in the spectra of many of the classical bursts and the soft repeater burst GB790305b. These features have been observed (Mazets et al 1979a, 1981b, 1982b, Teegarden & Cline 1980, Barat et al 1984d, Hueter 1984, 1987, Golenetskii et al 1986) simultaneously by detectors on more than one spacecraft: *Venera 11–14*, *ISEE*, and *PVO*. Although one feature at ~ 470 keV, which was observed (Mazets et al 1983) by the Konus experiment from the burst GB811231a, was not seen (Nolan et al 1984, Nolan 1986) by *SMM* detectors with a 300-keV threshold, this nondetection may have resulted from contamination of the *SMM* spectrum by burst photons Compton scattered from the Earth's atmosphere.

Nearly all of these emission features have been observed at energies of ~ 350 to 500 keV, and they have generally been interpreted (e.g. Liang 1986) as hot electron-positron annihilation radiation, gravitationally redshifted to $[511 + (3/4)kT]/(1+z)$ by neutron stars with a redshift z of 0.2 or more. The annihilation line is not only thermally broadened, it is also thermally blueshifted by $\sim(3/4)kT$ because both the rest mass and the kinetic energy in the center of mass are annihilated (Zdziarski 1980, Ramaty & Meszaros 1981). Such lines are a relatively common feature, occurring in the spectra of roughly one fifth of the classical bursts detected (Golenetskii et al 1986) by the Konus experiment on *Venera 13* and *14*.

The best resolved and narrowest of these lines was that observed (Mazets et al 1982a) during the impulsive burst GB790305b from the soft repeater GBS 0526−66, shown in Figure 10*a*. This line, centered at 430 keV, was attributed to redshifted 511-keV emission at $z \sim 0.2$, which is consistent with values of the gravitational redshift expected at the surface of a neutron star. The line width (FWHM) of ~ 100 keV constrains the temperature of the annihilating pairs, $kT(\text{eV}) \sim 0.7(1+z)^2[\text{FWHM}(\text{keV})]^2$, to be ~ 10 keV, at least an order of magnitude less than that implied by the continuum emission. This could result from cyclotron cooling of the pairs on a time scale shorter than their annihilation time (Ramaty et al 1981) if the magnetic field was $> 10^{10}$ G for the expected surface densities of neutron stars. The accompanying continuum is composed of two exponential components with e-folding energies of 33 and 500 keV. These two components have been attributed to gyrosynchrotron emission by MeV electrons in 10^{11}–10^{12} G fields (Ramaty et al 1981) and to the further Compton scattering of that emission by the same electrons (Liang 1981). This is the only

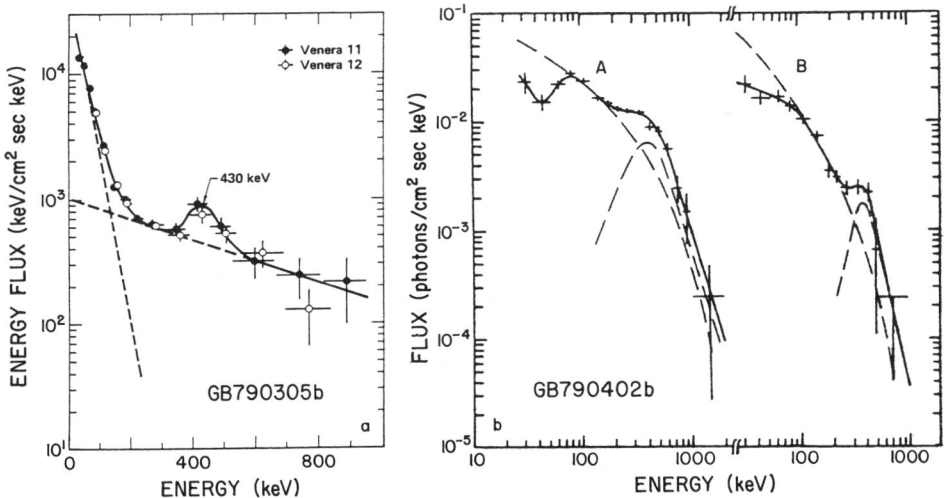

Figure 10 Examples of (*a*) a narrow emission feature in the spectrum of the extraordinary burst GB790305b from the soft repeater GBS 0526−66 (Mazets et al 1982a) and (*b*) a broader feature in two spectra of the classical burst GB790402b separated by 4 s (Golenetskii et al 1986).

line emission observed in the spectra of any repeating bursters, but it clearly suggests that such sources are neutron stars.

Time-varying emission lines are also seen in the classical bursts, as can be seen in the evolving spectrum of the burst GB790402b (Figure 10b) and in GB780325 (Figure 9a). The narrower (FWHM $\sim$ 100 keV), well-defined lines, similar to that from GB790305b (Figure 10a), are quite consistent with positron annihilation in hot ($kT \sim 10$ keV) plasmas on neutron stars with surface redshifts $z \sim 0.2$, which are compatible with current neutron star models. However, as Kluzniak (1989) has shown, if the broader (FWHM >300 keV) emission lines, such as those seen in the early phase of GB790402b (Figure 10b) and in GB780325 (Figure 9a), come from single-temperature annihilation regions, then the temperatures ($kT > 100$ keV) that would be implied by the width would give significant thermal blueshifts ($z \sim -3kT/4m_e c^2$). This would increase the required redshifts to values of $z > 0.5$, which are not consistent with those expected in current models.

However, Golenetskii et al (1986) have suggested that these broader emission lines are instead the superposition of annihilation lines coming from an extended region having a broad range of temperatures, so that the coolest determines the peak energy of the line and the hottest determines the line width. The required redshifts for the broader lines would then be consistent with current neutron star models. The smooth continuation of many of these broader lines into power-law spectra at higher energies further suggests (Golenetskii et al 1986) that this higher energy continuum emission may all result from electron-positron pair annihilation in such extended multitemperature regions. Opacity arguments, discussed above, also suggest that the higher energy emission does not come from the surface of a neutron star.

A narrow, higher energy emission line was also reported (Teegarden & Cline 1980) with a high-resolution Ge detector on *ICE* at 738 ± 10 keV in the spectrum of the burst GB781119. This narrow line, containing $\sim 9\%$ of the total luminosity of the burst, was attributed to ^{56}Fe nuclear deexcitation at 847 keV that was gravitationally redshifted with a $z \sim 0.10$–0.18, possibly at or near the surface of a neutron star. However, a weaker line, containing $\sim 1.5\%$ of the total luminosity, was also observed (Teegarden & Cline 1980, Mazets et al 1979c) at 420 ± 20 keV in the same burst and was attributed to redshifted annihilation radiation with a $z \sim 0.16$–0.28.

4.3 Emission Processes

A variety of thermal emission processes have been suggested for the burst continuum emission: optically thin bremsstrahlung (Gilman et al 1980,

Mazets et al 1981b), gyrosynchrotron (Ramaty et al 1981, Lamb 1982, Liang 1983, Hameury et al 1985, Imamura et al 1985, Pavlov & Golenetskii 1986), Compton scattering (Fenimore et al 1982b), and blackbody (Goodman 1986, Paczynski 1986). For a review of these emission processes, see Petrosian & Harding (1986) and Lamb (1988).

For the quasi-exponential, low-energy component below a few hundred keV, which is common to both the classical bursts and the soft repeaters, the first three mechanisms (bremsstrahlung, gyrosynchrotron, and Compton scattering) all seem capable of reproducing the observed spectra, given the compliant nature of the spectral fitting procedures (Liang et al 1983, Teegarden 1984). However, the common appearance of cyclotron absorption features in this component of the classical burst spectra clearly requires that the emission come from a region of intense ($\sim 10^{12}$ G) magnetic fields. In such an environment, synchrotron must be the dominant emission mechanism, since it becomes more efficient than bremsstrahlung when the magnetic field is $> 10^{10}$ G at the surface densities of neutron stars (Lamb 1982, 1984), and it dominates Compton scattering if the field is $> 10^9$ G (Fenimore et al 1984). This may also be the case for the soft repeaters if their magnetic fields are $> 10^{10}$ G, as was suggested (Ramaty et al 1981) by the annihilation line width in the burst GB790305b.

Blackbody spectra of relativistic fireballs (Goodman 1986) are inconsistent with both the observed burst spectra at energies < 100 keV and the absorption- and emission-line features observed in the spectra. As discussed above, however, blackbody emission at a temperature of around 1 keV could be responsible for the soft X-ray component observed at late times in the GB870303 burst by *Ginga* (Murakami 1988).

For the higher energy component of the emission extending far beyond 1 MeV, which is frequently observed (Matz 1986) in the classical bursts, single-temperature thermal models cannot explain the power-law spectra, and thus multitemperature nonthermal processes have also been proposed. These include e^+e^- pair annihilation (Lingenfelter & Hueter 1984, Golenetskii et al 1986) in an extended multitemperature (10^8–10^{10} K) electron plasma, or a nonthermal distribution; synchrotron emission (Sturrock 1986, Brainerd & Lamb 1987, Ruderman & Cheng 1988, Preece & Harding 1989) from nonthermal electron distributions; and Compton scattering (Zdziarski & Lamb 1986, Dermer 1989, Ho & Epstein 1989, Zdziarski et al 1990) by nonthermal distributions. Unlike the lower energy component, there is no indication as to which process or processes are the dominant source of the higher energy component of the emission. The current status of the problem has been reviewed by Petrosian & Harding (1986), Zdziarski (1987), and Lamb (1988).

5. SUMMARY

As we have seen, even though none of the more than 500 bursts have been identified with any known source, there is nonetheless a considerable body of circumstantial evidence suggesting that most of the common classical burst sources are single, intensely magnetic neutron stars in the nearby galactic disk, and that the few soft repeater sources are more distant neutron stars with slightly weaker magnetic fields and possible binary companions or surrounding cometary clouds.

The neutron star origin of most of the classical bursts and the soft repeaters is strongly suggested by various lines of evidence: (a) by the observation of absorption lines at 20–70 keV in a large ($\sim 1/3$) fraction of the resolved spectra of classical bursts, which are attributed to cyclotron absorption in 10^{12}-G magnetic fields; (b) by the emission lines seen at 350–500 keV in the spectra of both classes of bursts, which are attributed to gravitationally redshifted electron-positron annihilation radiation with neutron star redshifts of 0.2 or more; (c) by the very small source size limits of <1500 km implied by rapid flux variations in classical bursts on time scales as short as 5 ms, and of <60 km implied by soft repeater rise times as short as 0.2 ms; (d) by the very high mean density of $>2 \times 10^6$ g cm^{-3} implied for solid-body rotation of the source if the 8-s period in the late time emission of the soft repeater burst GB790305b is rotational; (e) by the location of the latter source position within the supernova remnant N49 in the Large Magellanic Cloud; and (f) by the failure to detect any likely quiescent counterparts at any wavelength in the source positional error boxes.

The local galactic origin of the classical burst sources lying within about half the scale height of the disk population is indicated by the uniform distribution implied by the V/V_{max} test, together with the isotropic distribution of burst source positions on the sky; by the galactic distances implied by the blackbody limit on the recently detected soft X-ray emission from the burst GB870303; and by the pair-production opacity limits on the MeV emission. The peak luminosities of these bursts is thus probably within an order of magnitude either way of 10^{38} ergs s^{-1}, which is roughly the Eddington luminosity of a neutron star. Greater distances are implied for the three soft repeater sources by their source directions pointing to the inner part of the Galaxy and the Large Magellanic Cloud. The implied peak luminosities of the repeating bursts from these sources would be a few times 10^{41} ergs s^{-1}, which is about 10^3 times that of the classical bursts, and the luminosity of the extraordinary burst GB790305b would be another factor of 10^3 brighter than that if its source is in fact in the Large Magellanic Cloud.

The solitary nature of the classical burst sources is suggested by the lack of any detectable quiescent counterparts or identifiable companion objects in any of the source error boxes, while binary companions or cometary clouds have been suggested around the soft repeater sources to supply episodically accreting matter needed to feed the recurrent bursts.

The burst energy spectra cannot be generally interpreted in terms of a single emission process in a single source region. The spectra often appear to have at least two components: a quasi-exponential spectrum below a few hundred keV, and a power-law spectrum above this limit. The detection of cyclotron absorption features in the low-energy emission due to $\sim 10^{12}$-G magnetic fields implies a gyrosynchrotron origin for the low-energy component, whereas the simultaneous detection of power-law continuum emission extending up to several MeV without a cutoff expected from pair production implies that this higher energy component must be produced well above the stellar surface, where the magnetic field is much weaker. The lack of significant soft X-ray emission expected from reprocessing also implies that gamma rays must be radiated anisotropically from the surface of the source, and that any isotropic emission must come from several radii above the surface.

The origin of these bursts still remains a mystery, although a variety of possibilities have been suggested [see Lamb (1984) and Taam (1987) for recent reviews]. Within the context of a neutron star origin, however, the most likely sources of the burst energy are gravitational and nuclear. The impulsive release of gravitational energy could take place with the accretion of a comet, asteroid, or planetesimal onto the surface of a neutron star, yielding ~ 100 MeV per nucleon of accreted matter (Harwit & Salpeter 1972, Colgate & Petschek 1981, Joss & Rappaport 1984, Epstein 1985b, Tremaine & Zytkow 1986, Livio & Taam 1987), or by disk instabilities in gas accreted from a stellar companion (Taam & Lin 1984). Gravitational energy could also be released impulsively by a neutron star quake, which could dissipate a fraction of its energy in magnetoacoustic wave acceleration of particles in the stellar magnetosphere (Tsygan 1975, Ramaty et al 1980, Kazanas 1988, Blaes et al 1989). Nuclear energy could be impulsively released by the runaway thermonuclear burning of matter accreted onto the surface of a neutron star, yielding several MeV per nucleon (Woosley & Taam 1976, Woosley & Wallace 1982, Hameury et al 1982, Woosley 1984b, Bonazzola et al 1984). All three of these processes would appear to be able to supply the total energies of 10^{37}–10^{41} ergs inferred for all of the classical and soft repeating bursts, except for the extraordinary burst GB790305b. Only a star quake releasing at least 10^{-9} of the neutron star gravitational binding energy of $\sim 10^{53}$ ergs would appear to be capable of supplying the 10^{44} ergs that would be required for the latter burst, if it is in the Large Magellanic Cloud.

Moreover, because of the temporal and spectral diversity of the bursts collectively known at present as classical bursts, it is quite possible that these bursts may arise from more than one type of process in more than one population of sources.

More detailed calculations are now being made of the energy released and of the emission spectra expected from some models that can be more directly tested against observations. With the launch of the *Gamma Ray Observatory* and the planned High Energy Transient Experiment we can look forward to more definitive tests of these models based on a much broader data base of burst observations with improved temporal and spectral resolution.

ACKNOWLEDGMENTS

We would like to thank Jerry Fishman, Dieter Hartmann, Kevin Hurley, Don Lamb, Tom Loredo, Bill Mahoney, Pat Nolan, Allan Sandage, Gerry Share, and John Wang for valuable comments on the manuscript, and NASA for support of this work through grants NAG 5-1011 (JCH), NAGW 1970 (REL), and NAS 8-36081 (REL).

Literature Cited

Andreev, O. N., Aptekar, A. L., Golenetskii, S. V., Guryan, Yu. A., Dyachov, A. V., et al. 1983. *Kosm. Issled.* 21: 480–88

Atteia, J.-L., Barat, C., Hurley, K., Niel, M., Vedrenne, G., et al. 1987b. *Ap. J. Suppl.* 61: 305–82

Atteia, J.-L., Boer, M., Hurley, K., Niel, M., Vedrenne, G., et al. 1987a. *Ap. J. Lett.* 320: L105–10

Baity, W. A., Hueter, G. J., Lingenfelter, R. E. 1984. See Woosley 1984a, pp. 434–84

Barat, C., Chambon, G., Hurley, K., Niel, M., Vedrenne, G., et al. 1979. *Astron. Astrophys.* 79: L24–25

Barat, C., Chambon, G., Hurley, K., Niel, M., Vedrenne, G. 1982. *Astron. Astrophys.* 109: L9–11

Barat, C., Hayles, R. I., Hurley, K., Niel, M., Vedrenne, G., et al. 1983. *Astron. Astrophys.* 126: 400–2

Barat, C., Hayles, R. I., Hurley, K., Niel, M., Vedrenne, G. 1984a. *Ap. J.* 285: 791–800

Barat, C., Hurley, K., Niel, M., Vedrenne, G., Cline, T. L., et al. 1984b. *Ap. J. Lett.* 286: L5–9

Barat, C., Hurley, K., Niel, M., Vedrenne, G., Evans, W. D., et al. 1984c. *Ap. J.* 280: 150–53

Barat, C., Hurley, K., Niel, M., Vedrenne, G., Mitrofanov, I. G., et al. 1984d. *Ap. J. Lett.* 286: L11–13

Beurle, K., Bewick, A., Mills, J. S., Quenby, J. J. 1981. *Astrophys. Space Sci.* 77: 201–14

Blaes, O., Blandford, R., Goldreich, P., Madau, P. 1989. *Ap. J.* 343: 839–48

Boer, M., Atteia, J.-L., Gottardi, M., Hurley, K., Niel, M., et al. 1988. *Astron. Astrophys.* 202: 117–23

Bonazzola, S., Hameury, J. M., Heyvaerts, J., Lasota, J. P. 1984. *Astron. Astrophys.* 136: 89–97

Brainerd, J. J. 1989. *Ap. J. Lett.* 341: L67–70

Brainerd, J. J., Lamb, D. Q. 1987. *Ap. J.* 313: 231–62

Brecher, K. 1982. See Lingenfelter et al 1982, pp. 293–97

Bussard, R. W., Lamb, F. K. 1982. See Lingenfelter et al 1982, pp. 189–200

Cavallo, G., Rees, M. J. 1978. *MNRAS* 183: 359–65

Cline, T. L. 1980. *Comments Astrophys.* 9: 13–22

Cline, T. L. 1982. See Lingenfelter et al 1982, pp. 17–33
Cline, T. L., Desai, U. D., Pizzichini, G., Teegarden, B. J., Evans, W. D., et al. 1980. *Ap. J. Lett.* 237: L1–5
Cline, T. L., Desai, U. D., Teegarden, B. J., Barat, C., Hurley, K., et al. 1984. *Ap. J. Lett.* 286: L15–18
Cline, T. L., Desai, U. D., Teegarden, B. J., Evans, W. D., Klebesadel, R. W., et al. 1982. *Ap. J. Lett.* 255: L45–48
Cline, T. L., Kouveliotou, C., Norris, J. 1988. *Proc. Int. Cosmic Ray Conf., 20th, Moscow* 9: 121–24
Colgate, S. A., Petschek, A. G. 1981. *Ap. J.* 248: 771–82
Cominsky, L., London, R. A., Klein, R. I. 1987. *Ap. J.* 315: 162–79
Daugherty, J. K., Harding, A. K. 1983. *Ap. J.* 273: 761–73
Daugherty, J. K., Ventura, J. 1981. *Astron. Astrophys.* 61: 723–27
Dennis, B. R., Frost, K. J., Kiplinger, A. L., Orwig, L. E., Desai, U., et al. 1982. See Lingenfelter et al 1982, pp. 153–62
Dermer, C. D. 1989. *Ap. J. Lett.* 347: L13–16
Desai, U. D. 1981. *Astrophys. Space Sci.* 75: 15–20
Epstein, R. I. 1985b. *Ap. J.* 291: 822–33
Epstein, R. I. 1985a. *Ap. J.* 297: 555–63
Epstein, R. I. 1986. In *Radiation Hydrodynamics in Stars and Compact Objects*, ed. D. Mihalas, K. Winkler, pp. 305–21. Berlin: Springer-Verlag
Estulin, I. V., Mitrofanov, I. G., Vilchinaskaya, A. S., Dolidze, V. Sh., Dyachkov, A. V., et al. 1987. Preprint 1218, Inst. Kosm. Issled. Moscow: Akad. Nauk SSSR. 58 pp.
Evans, W. D., Klebesadel, R. W., Laros, J. G., Cline, T. L., Desai, U. D., et al. 1980. *Ap. J. Lett.* 237: L7–9
Felten, J. E. 1981. *Proc. Int. Cosmic Ray Conf., 17th, Paris* 9: 52–55
Fenimore, E. E., Conner, J. P., Epstein, R. I., Klebesadel, R. W., Laros, J. G., et al. 1988. *Ap. J. Lett.* 335: L71–74
Fenimore, E. E., Klebesadel, R. W., Laros, J. G. 1983. *Adv. Space Res.* 3(4): 207–10
Fenimore, E. E., Klebesadel, R. W., Laros, J. G. 1984. See Woosley 1984a, pp. 590–96
Fenimore, E. E., Klebesadel, R. W., Laros, J. G., Stockdale, R. E., Kane, S. R. 1982b. *Nature* 297: 665–67
Fenimore, E. E., Laros, J. G., Klebesadel, R. W., Stockdale, R. E., Kane, S. R. 1982a. See Lingenfelter et al 1982, pp. 201–9
Fishman, G. J. 1979. *Ap. J.* 233: 851–56
Fishman, G. J., Meegan, C. A., Wilson, R. B., Paciesas, W. S., Parnell, T. A., et al. 1989. *Proc. Gamma-Ray Obs. Sci. Workshop*, ed. W. N. Johnson, 2: 39–50. Greenbelt, Md: NASA
Fishman, G. J., Paciesas, W. S., Meegan, C. A., Wilson, R. B. 1986. *Adv. Space Res.* 6(4): 23–26
Gilman, D. A., Metzger, E., Parker, R. H., Evans, L., Trombka, J. I. 1980. *Ap. J.* 236: 951–57
Golenetskii, S. V. 1988. *Adv. Space Res.* 8(2–3): 653–57
Golenetskii, S. V., Aptekar, R. L., Guryan, Yu. A., Ilyinskii, V. N., Mazets, E. P. 1987b. *Sov. Astron. Lett.* 13: 166–68
Golenetskii, S. V., Guryan, Yu. A., Dumov, G. B., Sheshin, L. O. 1987a. Preprint 1119, Fiz. Tekh. Inst. Ioffe. Leningrad: Akad. Nauk SSSR. 9 pp.
Golenetskii, S. V., Ilyinskii, V. N., Mazets, E. P. 1984. *Nature* 307: 41–43
Golenetskii, S. V., Mazets, E. P., Aptekar, R. L., Guryan, Yu. A., Ilyinskii, V. N. 1986. *Astrophys. Space Sci.* 124: 243–78
Golenetskii, S. V., Mazets, E. P., Ilyinskii, V. N., Guryan, Yu. A. 1979. *Sov. Astron. Lett.* 5: 340–42
Goodman, J. 1986. *Ap. J. Lett* 308: L47–50
Guilbert, P. W., Fabian, A. C., Rees, M. J. 1985. *MNRAS* 205: 593–603
Hameury, J. M., Bonazzola, S., Heyvaerts, J., Ventura, J. 1982. *Astron. Astrophys.* 111: 242–51
Hameury, J. M., Heyvaerts, J., Bonazzola, S. 1983. *Astron. Astrophys.* 121: 259–64
Hameury, J. M., Lasota, J. P., Bonazzola, S., Heyvaerts, J. 1985. *Ap. J.* 293: 56–68
Harding, A. K., Petrosian, V., Bussard, R. 1986. See Liang & Petrosian 1986, pp. 127–35
Hartmann, D. H., Blumenthal, G. R. 1989. *Ap. J.* 342: 521–26
Hartmann, D. H., Epstein, R. I. 1989. *Ap. J.* 346: 960–66
Hartmann, D. H., Epstein, R. I., Woosley, S. E. 1990. *Ap. J.* 348: 625–33
Hartmann, D. H., Pogge, R. W. 1987. *Ap. J.* 318: 363–69
Hartmann, D. H., Woosley, S. E. 1988. In *Multiwavelength Astrophysics*, ed. F. A. Cordova, pp. 189–233. Cambridge: Univ. Press
Hartmann, D. H., Woosley, S. E., Arons, J. 1988. *Ap. J.* 332: 777–803
Harwit, M., Salpeter, E. E. 1973. *Ap. J. Lett.* 186: L37–39
Helfand, D. J., Long, K. S. 1979. *Nature* 282: 589–91
Helfand, D. J., Vrtilek, S. D. 1983. *Nature* 304: 41–43
Higdon, J. C., Lingenfelter, R. E. 1984. See Woosley 1984a, pp. 568–77
Higdon, J. C., Lingenfelter, R. E. 1985. *Proc.*

Int. Cosmic Ray Conf., 19th, La Jolla 1: 37–40
Higdon, J. C., Lingenfelter, R. E. 1986. *Ap. J.* 307: 197–204
Higdon, J. C., Schmidt, M. 1990. *Ap. J.* In press
Ho, C., Epstein, R. 1989. *Ap. J.* 343: 277–91
Hudec, R., Borovicka, J., Danis, S., Franc, V., Peresty, R., Valnicek, B. 1988. *Adv. Space Res.* 8: 665–68
Hueter, G. J. 1984. See Woosley 1984a, pp. 373–77
Hueter, G. J. 1987. PhD thesis. Univ. Calif., San Diego, La Jolla. 304 pp.
Hueter, G. J., Lingenfelter, R. 1983. In *Positron-Electron Pairs in Astrophysics*, ed. M. L. Burns, A. K. Harding, R. Ramaty, pp. 89–93. New York: Am. Inst. Phys.
Hurley, K. 1987. In *The Origin and Evolution of Neutron Stars, IAU Symp. No. 125*, ed. D. J. Helfand, J.-H. Huang, pp. 489–500. Dordrecht: Reidel
Hurley, K. 1989a. In *Cosmic Gamma Rays, Neutrinos, and Related Astrophysics*, ed. M. Shapiro, E. Wefel, pp. 337–80. Boston: Kluwer
Hurley, K. 1989b. *Ann. NY Acad. Sci.* 571: 442–59
Imamura, J. N., Epstein, R. I., Petrosian, V. 1985. *Ap. J.* 296: 65–68
Jennings, M. C. 1982. *Ap. J.* 258: 110–20
Jennings, M. C. 1984. See Woosley 1984a, pp. 412–21
Jennings, M. C. 1988. *Ap. J.* 333: 700–18
Jennings, M. C., White, R. S. 1980. *Ap. J.* 238: 110–21
Joss, P. C., Rappaport, S. 1984. See Woosley 1984a, pp. 555–57
Kazanas, D. 1988. *Nature* 331: 320–21
Klebesadel, R. W. 1988. In *Physics of Neutron Stars and Black Holes*, ed. Y. Tanaka, pp. 387–404. Tokyo: Univers. Acad. Press
Klebesadel, R. W., Evans, W. D., Glore, J. P., Spalding, R. E., Wymner, F. J. 1980. *IEEE Trans. Nucl. Sci.* GE-18: 76–80
Klebesadel, R. W., Fenimore, E. E., Laros, J. G. 1984a. See Woosley 1984a, pp. 429–33
Klebesadel, R. W., Fenimore, E. E., Laros, J. G., Terrell, J. 1982. See Lingenfelter et al 1982, pp. 1–15
Klebesadel, R. W., Laros, J. G., Fenimore, E. E. 1984b. *Bull. Am. Astron. Soc.* 16: 1016 (Abstr.)
Klebesadel, R. W., Strong, I. B., Olson, R. A. 1973. *Ap. J. Lett* 182: L85–88
Kluzniak, W. 1989. *Ap. J.* 336: 367–75
Kouveliotou, C., Desai, U. D., Cline, T. L., Dennis, B. R., Fenimore, E. E., et al. 1988. *Ap. J. Lett.* 330: L101–5
Kouveliotou, C., Norris, J. P., CLine, T. L., Dennis, B. R., Desai, U. D., et al. 1987. *Ap. J. Lett.* 322: L21–25
Kuznetsov, A. V., Sunyaev, R. A., Terekhov, O. V., Barat, C., Boer, B., et al. 1986a. *Sov. Astron. Lett.* 12: 315–18
Kuznetsov, A. V., Sunyaev, R. A., Terekhov, O. V., Yakubtsev, L. A., Barat, C., et al. 1986b. *Sov. Astron. Lett.* 12: 311–15
Lamb, D. Q. 1982. See Lingenfelter et al 1982, pp. 249–72
Lamb, D. Q. 1984. *Ann. NY Acad. Sci.* 422: 237–48
Lamb, D. Q. 1988. In *Nuclear Spectroscopy of Astrophysical Sources*, ed. N. Gehrels, G. H. Share, pp. 265–84. New York: Am. Inst. Phys.
Lamb, D. Q., Wang, J. C. L., Wasserman, I. M. 1990. Submitted for publication
Lamb, D. Q., Wang, J. C. L., Loredo, T. J., Wasserman, I. M., Fenimore, E. E. 1989. *Ann. NY Acad. Sci.* 571: 460–81
Laros, J. G. 1988. *Nature* 333: 124
Laros, J. G., Evans, W. D., Fenimore, E. E., Klebesadel, R. W., Middleditch, J., et al. 1985b. *Ap. J.* 290: 728–34
Laros, J. G., Evans, W. D., Fenimore, E. E., Klebesadel, R. W., Shulman, S., Fritz, G. 1984. See Woosley 1984a, pp. 378–89
Laros, J. G., Fenimore, E. E., Fikani, M. M., Klebesadel, R. W., Barat, C., et al. 1986. *Nature* 322: 152–53
Laros, J. G., Fenimore, E. E., Fikani, M. M., Klebesadel, R. W., van der Klis, M., et al. 1985a. *Nature* 318: 448–49
Laros, J. G., Fenimore, E. E., Klebesadel, R. W., Atteia, J.-L., Boer, M., et al. 1987. *Ap. J. Lett.* 320: L111–15
Laros, J. G., Nishimura, J. 1986. See Liang & Petrosian 1986, pp. 79–84
Lewin, W. G. H., Joss, P. C. 1983. In *Accretion Driven Stellar X-Ray Sources*, ed. G. H. Lewin, P. J. van den Heuvel, pp. 41–115. Cambridge: Univ. Press
Liang, E. P. 1981. *Nature* 292: 319–21
Liang, E. P. 1983. *Ap. J. Lett.* 268: L89–92
Liang, E. P. 1985. *Nature* 313: 202–4
Liang, E. P. 1986. *Ap. J.* 304: 682–87
Liang, E. P., Jernigan, T., Rodriques, R. 1983. *Ap. J.* 271: 766–77
Liang, E. P., Petrosian, V., eds. 1986. *Gamma-Ray Bursts*. New York: Am. Inst. Phys. 206 pp.
Lindblom, L., Detweiler, S. L. 1983. *Ap. J. Suppl.* 53: 73–92
Lingenfelter, R. E., Hudson, H. S., Worrall, D. M., eds. 1982. *Gamma Ray Transients and Related Astrophysical Phenomena*. New York: Am. Inst. Phys. 500 pp.
Lingenfelter, R. E., Hueter, G. J. 1984. See Woosley 1984a, pp. 558–67
Livio, M., Taam, R. E. 1987. *Nature* 327: 398–400
Loredo, T. J., Epstein, R. I. 1989. *Ap. J.* 336: 896–919
Loznikov, V. M., Melioranskii, A. S.,

Kudryavtsev, M. I., Savenko, I. A., Shamolin, V. M. 1980. *Sov. Astron. Lett.* 6: 321–22
Manchester, R. N., Taylor, J. H. 1981. *Astron. J.* 86: 1953–73
Matz, S. M. 1986. PhD thesis. Univ. N. H., Durham. 132 pp.
Matz, S. M., Forrest, D. J., Vestrand, W. T., Chupp, E. L., Share, G. H., et al. 1985. *Ap. J. Lett.* 288: L37–40
Mazets, E. P. 1985. *Proc. Int. Cosmic Ray Conf., 19th, La Jolla* 9: 415–30
Mazets, E. P., Golenetskii, S. V. 1981. *Astrophys. Space Phys. Rev.* 1: 205–66
Mazets, E. P., Golenetskii, S. V. 1982. *Astrophys. Space Sci.* 88: 247–51
Mazets, E. P., Golenetskii, S. V. 1987. *Astronomia* 32: 16–42
Mazets, E. P., Golenetskii, S. V., Aptekar, R. L., Guryan, Yu. A., Ilyinskii, V. N. 1981b. *Nature* 290: 378–82
Mazets, E. P., Golenetskii, S. V., Guryan, Yu. A. 1979b. *Sov. Astron. Lett.* 5: 343–44
Mazets, E. P., Golenetskii, S. V., Guryan, Yu. A., Aptekar, R. L., Ilyinskii, V. N., et al. 1983. In *Positron-Electron Pairs in Astrophysics*, ed. M. L. Burns, A. K. Harding, R. Ramaty, pp. 36–53. New York: Am. Inst. Phys.
Mazets, E. P., Golenetskii, S. V., Guryan, Yu. A., Ilyinskii, V. N. 1982a. *Astrophys. Space Sci.* 84: 173–89
Mazets, E. P., Golenetskii, S. V., Ilyinskii, V. N., Aptekar', R. L., and Guryan, Yu. A. 1979a. *Nature* 282: 587–89
Mazets, E. P., Golenetskii, S. V., Ilyinskii, V. N., Guryan, Yu. A., Aptekar, R. L., et al. 1982b. *Astrophys. Space Sci.* 82: 261–82
Mazets, E. P., Golenetskii, S. V., Ilyinskii, V. N., Panov, V. N., Aptekar, R. L., et al. 1979c. *Cosm. Res.* 17: 674–80
Mazets, E. P., Golenetskii, S. V., Ilyinskii, V. N., Panov, V. N., Aptekar, R. L., et al. 1981a. *Astrophys. Space Sci.* 80: 3–143
Meegan, C. A., Fishman, G. J., Wilson, R. R. 1984. See Woosley 1984a, pp. 422–25
Meegan, C. A., Fishman, G. J., Wilson, R. R. 1985. *Ap. J.* 291: 479–85
Melia, F. 1988. *Ap. J. Lett.* 324: L21–25
Melia, F., Rappaport, S., Joss, P. C. 1986. *Ap. J. Lett.* 305: L51–55
Meszaros, P., Bagoly, Z., Riffert, H. 1989. *Ap. J. Lett.* 337: L23–27
Mitrofanov, I. G., Chernenko, A. M., Dolidze, V. S., Dyatchkov, A. V., Khavenson, N. G., et al. *Rep. No. PR-1497*, Inst. Space Res., Moscow. 66 pp.
Mitrofanov, I. G., Dolidze, V. S., Barat, C., Vedrenne, G., Niel, M., et al. 1984. *Sov. Astron. AJ* 28: 547–49
Motch, C., Pedersen, H., Ilovaisky, S. A., Chevalier, C., Hurley, K., et al. 1985. *Astron. Astrophys.* 145: 201–5
Murakami, T. 1988. In *Physics of Neutron Stars and Black Holes*, ed. Y. Tanaka, pp. 405–12. Tokyo: Univers. Acad. Press
Murakami, T., Fujii, M., Hayashida, K., Itoh, M., Nishimura, J., et al. 1988. *Nature* 335: 234–35
Murakami, T., Fujii, M., Hayashida, K., Itoh, M., Nishimura, J., et al. 1989. *Publ. Astron. Soc. Jpn.* 41: 405–26
Nolan, P. 1986. See Liang & Petrosian 1986, pp. 102–4
Nolan, P. L., Share, G. H., Chupp, E. L., Forrest, D. J., Matz, S. M. 1984. *Nature* 311: 360–62
Norris, J. P. 1983. PhD thesis. Univ. Md., College Park. 257 pp.
Norris, J. P., Cline, T. L., Desai, U. D., Teegarden, B. J. 1984. *Nature* 308: 434–35
Norris, J. P., Share, G. H., Messina, D. C., Dennis, B. R., Desai, U. D., et al. 1986. *Ap. J.* 301: 213–19
Paczynski, B. 1986. *Ap. J. Lett.* 308: L43–46
Paczynski, B. 1990. *Ap. J.* 348: 485–94
Paczynski, B., Long, K. 1988. *Ap. J.* 333: 694–99
Pavlov, G. G., Golenetskii, S. V. 1986. *Astrophys. Space Sci.* 128: 341–54
Pedersen, H., Danziger, J., Hurley, K., Pizzichini, G., Motch, C., et al. 1984. *Nature* 312: 46–48
Pedersen, H., Motch, C., Tarenghi, M., Danziger, J., Pizzichini, G., et al. 1983. *Ap. J. Lett.* 270: L43–47
Pedersen, H., Pizzichini, G., Schaefer, B., Hurley, K. 1986. See Liang & Petrosian 1986, pp. 39–46
Petrosian, V., Harding, A. K. 1986. See Liang & Petrosian 1986, pp. 108–21
Pizzichini, G., Gottardi, M., Atteia, J.-L., Barat, C., Hurley, K. 1986. *Ap. J.* 301: 641–49
Preece, R., Harding, A. K. 1989. *Ap. J.* 347: 1128–40
Puget, J. L. 1981. *Astrophys. Space Sci.* 75: 109–16
Ramaty, R., Bonazzola, S., Cline, T., Kazanas, D., Lingenfelter, R. E. 1980. *Nature* 287: 122–24
Ramaty, R., Lingenfelter, R. E., Bussard, R. W. 1981. *Astrophys. Space Sci.* 75: 193–203
Ramaty, R., Meszaros, P. 1981. *Ap. J.* 250: 384–88
Rappaport, S., Joss, P. C. 1985. *Nature* 314: 242–45
Ricker, G., Doty, J., Rappaport, S., Hurley, K., Fenimore, E., et al. 1988. In *Nuclear Spectroscopy of Astrophysical Sources*, ed. N. Gehrels, G. Share, pp. 407–16. New York: Am. Inst. Phys.

Rothschild, R. E., Lingenfelter, R. E. 1984. *Nature* 312: 737–39
Ruderman, M. 1975. *Ann. NY Acad. Sci.* 262: 164–80
Ruderman, M. 1987. *Proc. Tex. Symp. Relativ. Astrophys.*, *13th*, ed. M. Ulmer, pp. 448–59. Singapore: World Sci.
Ruderman, M., Cheng, K. S. 1988. *Ap. J.* 335: 306–18
Schaefer, B. E. 1981. *Nature* 294: 722–24
Schaefer, B. E. 1990. Submitted for publication
Schaefer, B. E., Bradt, H. V., Barat, C., Hurley, K., Neil, M., et al. 1984. *Ap. J. Lett.* 286: L1–4
Schaefer, B. E., Cline, T. L. 1985. *Ap. J.* 289: 490–93
Schaefer, B. E., Desai, U. D. 1988. *Astron. Astrophys.* 195: 123–28
Schaefer, B. E., Pedersen, H., Gouiffes, C., Poulsen, J. M., Pizzichini, G. 1987. *Astron. Astrophys.* 174: 338–43
Schaefer, B. E., Seitzer, P., Bradt, H. V. 1983. *Ap. J. Lett.* 270: L49–52
Schmidt, M. 1968. *Ap. J.* 151: 393–409
Schmidt, M., Higdon, J. C., Hueter, G. 1988. *Ap. J. Lett.* 329: L85–87
Schmidt, W. K. H. 1978. *Nature* 271: 525–27
Shklovskii, I. S., Mitrofanov, I. G. 1985. *MNRAS* 212: 545–51
Sturrock, P. 1986. *Nature* 321: 47–49
Taam, R. E. 1987. *Proc. Tex. Symp. Relativ. Astrophys.*, *13th*, ed. M. Ulmer, pp. 546–52. Singapore: World Sci.
Taam, R. E., Lin, D. N. C. 1984. *Ap. J.* 287: 761–68
Teegarden, B. J. 1982. See Lingenfelter et al 1982, pp. 123–42
Teegarden, B. J. 1984. See Woosley 1984a, pp. 352–56
Teegarden, B. J., Cline, T. L. 1980. *Ap. J. Lett.* 236: L67–70
Terrell, J., Evans, W. D., Klebesadel, R. W., Laros, J. G. 1980. *Nature* 285: 383–85
Tremaine, S., Zytkow, A. N. 1986. *Ap. J.* 301: 155–63
Tsygan, A. I. 1975. *Astron. Astrophys.* 44: 21–24
Van Horn, H. M. 1980. *Ap. J.* 236: 899–903
Wang, J. C. L., Lamb, D. Q., Loredo, T. J., Wasserman, I. M., Salpeter, E. E., et al. 1989. *Phys. Rev. Lett.* 63: 1550–53
Wheaton, W. A., Ulmer, M. P., Baity, W. A., Datlowe, D. W., Elcan, M. J., et al. 1973. *Ap. J. Lett.* 185: L57–61
Wood, K. S., Byram, E. T., Chubb, T. A., Friedman, H., Meekins, J. F., et al. 1981. *Ap. J.* 247: 632–38
Wood, K., Desai, U., Schaefer, B., Pizzichini, G., Norris, J., et al. 1986. See Liang & Petrosian 1986, pp. 4–23
Woosley, S. E., ed. 1984a. *High Energy Transients in Astrophysics*. New York: Am. Inst. Phys. 714 pp.
Woosley, S. E. 1984b. See Woosley 1984a, pp. 485–511
Woosley, S. E., Taam, R. E. 1976. *Nature* 263: 101–3
Woosley, S. E., Wallace, R. K. 1982. *Ap. J.* 258: 716–23
Yamagami, T., Nishimura, J. 1986. *Astrophys. Space Sci.* 121: 241–53
Yoshimori, M. 1978. *Aust. J. Phys.* 31: 189–94
Yoshimori, M., Okudaira, K., Hirasima, Y., Kondo, I. 1984. *Astrophys. Space Sci.* 105: 379–92
Zdziarski, A. A. 1980. *Acta Astron.* 30: 371–91
Zdziarski, A. A. 1987. *Proc. Tex. Symp. Relativ. Astrophys.*, *13th*, ed. M. Ulmer, pp. 553–62. Singapore: World Sci.
Zdziarski, A. A., Lamb, D. Q. 1986. *Ap. J. Lett.* 309: L79–82
Zdziarski, A. A., Coppi, P., Lamb, D. Q. 1990. *Ap. J.* In press
Zytkow, A. N. 1990. *Ap. J.* In press.

THE SPACE DISTRIBUTION OF QUASARS

F. D. A. Hartwick and David Schade

Department of Physics and Astronomy, University of Victoria,
P.O. Box 1700, Victoria, British Columbia V8W 2Y2, Canada

KEY WORDS: surface density of quasars, luminosity function of quasars, clustering of quasars

1. INTRODUCTION

It is now nearly 30 years since quasars were first identified by Matthews & Sandage (1963). In this review we summarize what has been learned about the space density (expressed in the form of luminosity functions) and clustering properties of these objects. Determination of the space distribution of quasars requires large homogeneous samples with well-understood selection biases. To put the problem in perspective, at $B = 22$ mag, one square degree of a high-latitude field contains $\sim 10,000$ objects, of which only ~ 100 are quasars! The rapid advances of the past decade in the field are due in part to the application of fast plate-scanning machines, automated measuring algorithms, multiaperture spectrographs, and a variety of novel selection techniques such as the CCD transit surveys.

1.1 Motivation

The most important reason for studying quasars is that they provide probes of conditions when the Universe was only 10–20% of its present age. Knowledge of the spatial distribution of quasars impacts on a large number of astrophysical problems, a sample of which is given below. Later sections of this review present in detail the data upon which the progress on these problems is based.

1.1.1 UNDERSTANDING THE CENTRAL "ENGINE" AND ITS EVOLUTION   The generally accepted model for the quasar phenomenon is accretion of matter onto a massive black hole as first suggested by Lynden-Bell (1969) [see

Rees (1984) for a review]. Blandford (1986) has shown how the quasar luminosity function and its evolution can constrain very simple black hole accretion models with admittedly somewhat arbitrary black hole birth rates and fueling rates. Ultimately, an observational handle on the fueling rates may come from the results of searches of regions around quasars to look for possible sources (companion galaxies?) and possible signatures (tidal distortions?) of the fueling process (see Section 1.1.2 below).

Other tests for self-consistency of the basic model come from two approaches to determining the black hole mass (M_{BH}) and, hence, the ratio of observed luminosity to the Eddington luminosity. One is a dynamical estimate based on the spectrum of the broad-line region that assumes the line widths to be gravitationally induced (Wandel & Yahil 1985, Padovani & Rafanelli 1988); the other involves fitting the spectrum of an accretion disk to the ultraviolet (UV) portion of the quasar spectrum (Malkan 1983, Wandel & Petrosian 1988). Both methods give results that are in accord: $M_{BH} \sim 10^{8-9.5}$ $M_\odot$ for quasars and $\sim 10^{7.5-8.5}$ $M_\odot$ for Seyfert galaxies, and ratios of UV luminosity to Eddington luminosity of ~ 1 to <0.01.

The problem of quasar evolution has remained unresolved since Schmidt (1968) first showed that the space density of quasars, ρ, was much higher in the past ($\rho_{z\sim 1} \sim 150 \times \rho_{z=0}$) under the assumption that the increase was independent of luminosity [now commonly referred to as pure density evolution]. However, as is discussed in Section 3, it can be argued (cf. Mathez 1978, Boyle et al 1988b) that the appearance of the luminosity functions up to $z \sim 2$ suggests an evolutionary picture referred to as pure luminosity evolution, whereby the space density of quasars remains constant while their luminosities decline with time. Both of the above limiting forms of evolution (pure density evolution and pure luminosity evolution) have served well as convenient mathematical formulations, but the actual evolution of quasars will only be determined after one is able to separate the evolution of an individual object from the birth/death rate of the population as a whole, i.e. to separate "physical" evolution from "statistical" evolution in the terminology of Lynds & Petrosian (1972) [see Cavaliere & Padovani (1989) for one recent approach to this problem].

Another problem, which is the focus for much of the current work on quasar luminosity functions, is the behavior of quasars at the highest redshifts. Does the density of quasars decline at $z > 3$? (cf. Schmidt 1970, Sandage 1972, Osmer 1982). A real decline such as that exhibited by the intrinsically brightest objects (cf. Schmidt et al 1988) might be signaling the birth not only of quasars but also of galaxies.

1.1.2 QUASAR-GALAXY ASSOCIATIONS AND EVIDENCE FOR INTERACTIONS/ MERGERS The idea that quasars might be "fueled" as a result of interactions with companion galaxies was first suggested by Toomre & Toomre (1972). Strong evidence exists that low-redshift quasars are found in regions where the density of galaxies is only moderately enhanced (Stockton 1978; reviewed by Stockton 1986). Furthermore, evidence for interactions between the quasar host galaxy (Kristian 1973) and nearby companions is also extensive (Wyckoff et al 1981, Stockton 1982, Hutchings et al 1984, Stockton & MacKenty 1987; but see Hutchings et al 1989, MacKenty 1989). The interaction/merger picture has been shown to be at least qualitatively consistent with the evolution of the luminosity function (De Robertis 1985, Roos 1985, Carlberg 1990).

1.1.3 QUASAR-GALAXY CORRELATIONS AND GRAVITATIONAL LENSING Stimulated by statistics that indicated an excess in the number of quasars associated with bright galaxies (Arp 1970, Arp & Sulentic 1979, Burbidge 1979), Canizares (1981) proposed, as an alternative to invoking the existence of noncosmological redshifts, that gravitational lensing caused by individual stars in foreground galaxies could give rise to such an effect. While the original statistics has been the subject of some debate (see Weedman 1986), more recent evidence for gravitational microlensing has been reported (Stocke et al 1987, Fugmann 1988, Webster et al 1988). In the case of the Webster et al study, the effect could manifest itself by bringing into a magnitude-limited sample of quasars fainter and (depending exactly on the shape of the $\log N$ vs. B relation) more numerous quasars. The magnitude of the effect thus depends sensitively on the slope of the count-magnitude relation [cf. Narayan (1989) for a clear discussion and earlier references]. A further complication with this interpretation is that Webster et al (1988) find a small positive quasar-galaxy correlation, whereas Boyle et al (1988a) find evidence for anticorrelations, although at larger separations.

1.1.4 THE CLUSTERING OF QUASARS AND THE GROWTH OF STRUCTURE IN THE UNIVERSE Observations of the cosmic microwave background indicate that the distribution of matter was extremely smooth at $z \sim 1000$. At the present epoch, the situation is clearly very different—the Universe appears to be lumpy on all scales up to ~ 60 Mpc [see Bahcall (1988) for a review]. Studies of the clustering of quasars offer the possibility of finding out what has happened at intermediate redshifts and look-back times. As is discussed later in this review, such studies are difficult, and conflicting results have been reported, owing in part to the low density of quasars. Clustering does appear to have been detected in some studies at $z \sim 1$ (Shanks et al 1988, Shaver 1988) at an amplitude comparable to that observed for the clustering of galaxies today. Furthermore, some workers find evidence for evolution in the quasar clustering amplitude (Iovino & Shaver 1988, Kruszewski 1988). So far, the theoretical predictions are scarce (but see Efstathiou & Rees 1988, Cole & Kaiser 1989).

1.1.5 COSMOLOGY Both the broad luminosity function and the current uncertainty of its evolution have so far prevented conclusive application of the classical cosmological tests to quasars (see Wampler 1987, and references therein). However, numerous count-magnitude determinations do allow limits to be placed on the isotropy of the Universe at large redshifts, and negative results for quasar clustering at very large scales (~ 1000 Mpc) allow limits to be placed on the homogeneity of the Universe (Shanks et al 1988).

1.1.6 UV AND X-RAY BACKGROUND Many years ago, Gunn & Peterson (1965) concluded that the Universe must have been highly ionized at early epochs (see also Bajtlik et al 1988). However, the source of the ionizing UV radiation has eluded detection. Quasars would appear to be natural candidates, and a number of authors (Bechtold et al 1987, Heisler & Ostriker 1988) have used the quasar luminosity functions to estimate their contribution to the UV background. The conclusion is that the presently observed numbers of quasars cannot provide the source of the ionizing radiation. Other possibilities are that high-redshift quasars are hidden from our view owing to absorption by dust (Heisler & Ostriker 1988; but see Fall & Pei 1989) or that the luminosity function for quasars may steepen at the faint end at high redshift (Steidel & Sargent 1989).

The situation with respect to the quasar contribution to the extragalactic X-ray background is complicated by the uncertain shape of the background spectrum at low (0.1–3 keV) energies and of the X-ray spectra of the dominant contributors. Using upper limits to the X-ray background at 0.1 keV, a spectral index of 1.4 for radio-quiet quasars (Canizares & White 1989), Fabian et al (1989) conclude that the quasars cannot contribute more than 13% of the 2–10 keV X-ray background.

1.2 The Operational Definition of a Quasar and the Scope of This Review

Quasars (we also use the term QSOs interchangeably) represent one part of the family of active galactic nuclei (AGN; see Osterbrock 1988). For the purposes of this review, quasars are defined to be those AGN with $M_B < -23$ when $H_0 = 50$ km s^{-1} Mpc^{-1}. Equivalently, we consider quasars to be those intrinsically luminous objects that have "quasarlike" spectra, i.e. broad permitted emission lines (widths ~ 5000 km s^{-1}), and narrower forbidden lines (widths ~ 500 km s^{-1}) superimposed on a featureless continuum. No distinction is necessarily made about whether the object was selected from optical, radio, X-ray, or infrared surveys because of the remarkable similarity of the optical spectra of optically selected quasars to those from radio surveys [but see Osterbrock (1988) for a discussion of subtle differences], X-ray surveys (Gioia et al 1984, Kruper & Canizares 1989), or infrared surveys (Low et al 1989). We do not discuss (*a*) the less luminous quasarlike AGN, which span a range from Seyfert galaxies and broad-line radio galaxies to possible "microquasars" that may reside in many otherwise normal galaxies [see Filippenko & Sargent (1989) for an extreme example]; (*b*) the BL Lacertae objects (cf. Stocke et al 1988); and (*c*) the possibly related ultraluminous infrared galaxies (Sanders et al 1988).

1.3 *Review Outline*

Significant progress in the subject of this review has come from an analysis of recent large, deep homogeneous quasar surveys. These surveys and a discussion of the relevant selection biases are the subject of Section 2. The data from these surveys are combined to form luminosity functions as a function of redshift in Section 3. Quasar associations are the subject of Section 4. There, we summarize what is known about quasar-galaxy correlations and quasar-quasar correlations (clustering). Some concluding remarks are made in Section 5.

The proceedings of a number of recent conferences and symposia contain reviews and contributions directly related to the subject of this review. These include IAU Symposium No. 119 (Swarup & Kapahi 1986), IAU Symposium No. 124 (Hewitt et al 1987), and a Workshop on Optical Surveys for Quasars (Osmer et al 1988b). Excellent discussions of the topics covered here can also be found in Weedman's (1986) book.

2. QUASAR SURVEYS

One of the goals of this section is to combine, for the first time, the data for over 1000 quasars from several different surveys. Assessing the completeness of each survey requires a rather detailed discussion of the various selection biases. The general reader, who may be more interested in the luminosity function, may wish to skip directly to Section 3.

2.1 *Techniques*

The ideal survey technique would produce a candidate list that included 100% of the objects of interest and contained no contaminating objects. In practice, one is faced with a trade-off between maximizing the level of completeness (missing as few true quasars as possible) while at the same time minimizing the number of contaminants. This latter aspect becomes more important as the objects get fainter and the follow-up work on the candidate list becomes more time consuming. Knowledge of the level of completeness and how it varies with luminosity and/or redshift is essential if reliable space densities are to be derived. The level of completeness can in theory be determined by comparing results from different survey techniques (e.g. multicolor selection vs. slitless spectroscopy).

The optical samples used here have been selected either on the basis of emission lines or peculiar colors. While quasars can also be discovered by their radio, X-ray, and infrared properties or by variability and lack of proper motion, it appears that samples selected by these other properties are subsets of the optically selected samples and not distinct classes of objects that have escaped optical detection (Weedman 1986). Optically selected samples are essentially complete samples of quasars, although all survey techniques have redshift- and, to a lesser extent, luminosity-dependent selection biases (Wampler & Ponz 1985). These biases are neither so severe nor so complex that they cannot be effectively dealt with (see Cavaliere et al 1989).

2.1.1 ULTRAVIOLET-EXCESS SURVEYS The first quasars were discovered by their radio emission. During the process of identifying the optical counterparts to radio sources, Sandage (1965) recounts how, when the telescope was set at the radio position, nearly every very blue (ultraviolet-excess) object near the field center turned out to be a correctly identified quasar. As the observations progressed it became clear that many other ultraviolet excess (UVX) objects existed that could not be identified with a common stellar population but that were not sources of radio emission. The earliest systematic quasar surveys (Sandage & Veron 1965, Sandage & Luyten 1967) identified numerous very blue objects. It was shown (Sandage 1965) that many of these objects were radio-quiet QSOs. Thus, an efficient discriminant of the QSO population had been found.

The UVX method and its multicolor variants were pioneered by the three-color Tonantzintla technique (Haro 1956) and the extensive three-color work by Luyten (1953–62) on the *Search for Faint Blue Stars*. Haro (1956) states that the three-color method—whereby a single plate is exposed through three filters with exposure times normalized to make it easy to pick out unusually blue stars—"will constitute a powerful and effective observational method that may be of great value in the statistical study of galaxies with bright lines and strong ultraviolet and blue colors." Haro & Luyten (1962) joined forces to produce the PHL catalog of blue objects using the Palomar 48" Schmidt telescope.

Quasar optical continua can be approximately described by a power law

$f_\nu \propto \nu^\alpha$, with α in the range -0.5 to -1.0. Such a spectrum results in a color much bluer in $U-B$ than most common stars (Matthews & Sandage 1963, Sandage & Luyten 1967) with the exception of white dwarfs. Good illustrations are given by Braccesi et al (1980), showing that the radio-selected quasars from the 3CR and 4C catalogues occupy a region of the two-color $(U-B, B-V)$ plane that is nearly devoid of stars. Selecting all objects with $U-B$ bluer than -0.4 gives a substantially complete list of quasar candidates. The success rate can be improved somewhat by using both colors and using a slightly more sophisticated selection criterion. Contamination of the candidate sample by hot stars was shown to be very small ($\sim 15\%$) at magnitudes fainter than $B = 18$.

A complication of the UVX method is that the colors are affected by emission lines as they are redshifted into and out of the filter passbands (Sandage 1966, Grewing 1967), an effect that is shown clearly in Figure 1 of Veron (1983) (or Figure 1 of Boyle et al 1987). If one knows the mean $[(U-B), z]$ relation and its dispersion, accurate completeness levels can be calculated by combining this information with the distribution of one's photometric errors. The UVX method is least complete when the $U-B$ colors are reddest because then the quasars scatter out of the color selection region. This, in fact, leads to a redshift dependence of completeness. In particular, at $0.5 < z < 1$, some surveys may only be 70% complete (Green 1986) or perhaps less. At redshifts greater than 2.2, Lα moves from the U passband to the B passband, and the ultraviolet excess decreases to the point where the method is ineffective.

The Baldwin effect (Baldwin 1977) is the inverse correlation found between QSO emission-line equivalent width and continuum luminosity. This correlation means that the effect of emission lines on quasar colors is most extreme for the least luminous objects. As a result, one would expect incompleteness in UVX surveys at a fixed redshift to be luminosity dependent to some extent—the least luminous objects are apparently fainter (and thus have larger photometric errors) *and* have redder colors than the more luminous objects.

The use of automated plate-measuring machines makes UVX selection a very efficient method of isolating large faint quasar samples at redshifts less than 2.2. UVX surveys (or extended UVX surveys, which include the $B-V$ color) used in our analysis include that of Boyle et al (1987; BFSP), the BF and AB samples (see Marshall et al 1984), and the Medium Bright Quasar Survey (MBQS; Mitchell et al 1984). The Palomar Bright Quasar Survey (BQS; Schmidt & Green 1983) is also a UVX selected sample. (For a listing of major quasar surveys and their acronyms, see Table 2.)

Based on comparisons with samples selected independently of color, the selection biases inherent in the UVX surveys appear to be well understood and manageable. It is particularly unfortunate that the UVX technique becomes ineffective near the redshift where a decline in space density of quasars may be taking place. Thus, it is necessary to rely on samples selected by other means to define the evolution at $z > 2.2$.

2.1.2 MULTICOLOR SURVEYS The extension of the UVX technique by the addition of a $B-V$ color (Braccesi et al 1980) is used by both the MBQS and Cristiani et al (1989), who claim that it is more effective than UVX alone. A further extension of the color selection method is the use of more colors (e.g. U, B, V, R, I) to find QSOs on the basis of segregation from normal stars in the multicolor space. This allows one to use color alone to find quasars at $z > 2.2$, where the (single-color) UVX method fails.

Koo & Kron (1982) used four-color plates (U, J, F, N, which give color indices similar to $U-B$, $B-V$, $R-I$) to a limiting magnitude of $B = 23$. Contaminating galaxies were eliminated by spatial image analysis. Power-law continua are well separated from thermal white dwarf spectra in the $U-J$, $F-N$ diagram. The most useful color plane for identifying likely QSOs turned out to be $U-J$, $J-F$. Candidate objects were selected because they are separated from the stellar locus of colors and not simply because of ultraviolet excess. In fact, many of the quasars with redshifts greater than 2 that were detected have $U-J > 0$ and thus are much too red to be selected by the standard UVX method. The method appears to be effective to $z \sim 3.2$. [Because moderate-redshift ($0.2 < z < 0.7$) compact galaxies turn out to be a major contaminant at faint magnitudes, it is useful to note (Koo & Kron 1988) that these objects are reasonably well separated in the two-color diagram from quasars. They have $-0.9 < U-J < -0.3$ and $J-F > 0.55$.]

The multicolor method has been very successful at finding high redshift objects. Shanks et al (1983) discovered a quasar of redshift 3.61 from U, B, V, and R plates. They examined quasars in their field that had been found by visual examination of objective-prism plates and used them to estimate the expected colors of high-redshift objects. Three of the 10 quasars currently known with $z > 4$ (Schneider et al 1989a) were selected on the basis of their colors. This includes the first object with $z > 4$ (Warren et al 1987a) and one of the highest redshift objects currently known (Warren et al 1987b). Warren et al (1988) describe their U, B, V, R, I photometry from UK Schmidt plates. The technique selects unusual objects regardless of color. A sample of 53 quasars with $z > 3$ has been compiled using this technique (Warren & Hewett 1990).

The major potential source of incompleteness in color-based surveys is the failure to detect QSOs that are embedded within the locus of normal stars in a multicolor diagram. This problem could be solved by comparing the survey results to those obtained by a method that is unbiased by color, e.g. a slitless survey or a radio survey. Marano et al (1988) combined U, J, F multicolor selection with the grism (grating-prism combination) technique and found (from a very small sample) that about 15% of the $z < 1$ objects selected by their colors were missed by the slitless spectroscopy, whereas a comparable number of $z > 2$ objects were missed by the color selection technique but picked up by the grism technique.

2.1.3 SLITLESS SPECTROSCOPIC SURVEYS It would seem natural to use a quasar's characteristically broad emission lines as a selection criterion. A strong-lined object appears strikingly different on a low-resolution spectroscopic plate than the common, usually red stars and galaxies from which one wishes to differentiate them. An excellent review of the development of this technique is provided by Smith (1978).

Slitless surveys use either wide-field Schmidt objective-prism plates or 4-m class grism or grens (grating-lens system) plates, which surpass the objective prism in finding faint, weak-lined objects because of their better spectral resolution. The first application of the slitless spectral technique to quasars was described by Hoag & Schroeder (1970), who used a transmission grating at the focus of the Kitt Peak 1-m telescope. Using an objective prism on the Curtis Schmidt telescope, Smith (1975) was able to take advantage of a much increased field of view, while Hoag (1976) used a transmission grating at the Cerro Tololo Inter-American Observatory 4-m telescope and was able to select fainter objects. It was quickly realized that such searches are capable of achieving high levels of completeness at redshifts where the strong Lα line is in the passband of the instrument. At other redshifts, completeness is more problematical. Clowes (1981) compared the distribution of equivalent widths and magnitudes of two samples of quasars—one selected on objective-prism plates, the other on 4-m grism plates. He demonstrated the relationship of detection probability, apparent magnitude, and equivalent width and assessed the effects of seeing. Gratton & Osmer (1987) have further quantified the problem of emission-line selection. Koo & Kron (1980) used the Kitt Peak 4-m telescope and a grism-CCD combination to search for high-redshift objects (with null results). Osmer (1982) tailored his instrument-emulsion combination to maximize sensitivity to Lα quasars with $3.7 < z < 4.7$. A number of objects were found but none in the redshift range of primary interest. This was a strong piece of evidence that the space density of optically selected QSOs declines at high redshift.

In most of the photographic surveys, plates are searched visually using a binocular microscope. Emission-line objects are selected along with objects with unusually blue continua. There are a number of uncertainties associated with this search technique. First, there is a bias in favor of strong-lined objects, which introduces the possibility of redshift-dependent selection effects due to strong lines being redshifted into and out of the passband. Second, the object selection criteria are subjective to some degree and may not be reproducible from worker to worker. A third complication, common to visual and automated search techniques, is that the resulting samples cannot be properly characterized in terms of a single limiting magnitude. The limiting continuum magnitude depends on line strength.

A number of slitless searches are currently underway or have recently been completed. The APM1 (automated plate measurement) survey [or Large Bright Quasar Survey (Foltz et al 1987)] and APM2 survey (Foltz et al 1989) use machine-scanned UK Schmidt telescope plates (direct and objective prism) to find QSO candidates with $16 < m_J < 18.5$. An important step forward is the use of objective automated selection procedures to identify candidates. The benefits of automation have been pointed out by others (Smith 1983), and such algorithms have been applied with success (Borra et al 1987). A large slitless survey has been undertaken at the Canada-France-Hawaii telescope (CFHT; Crampton & Rensing 1982, Crampton et al 1985, 1987, 1989). Good seeing on Mauna Kea and the high efficiency of the grens/IIIa-J emulsion combination have contributed toward making the CFHT probably the most efficient telescope for detecting faint quasars by slitless spectroscopy (Weedman 1985). The CFHT survey now consists of more than 300 objects, complete with follow-up spectroscopy (Crampton et al 1989), and is nominally complete to $m = 20.5$. This survey finds numbers of quasars comparable to—or higher than—any other method for redshifts up to $z = 3.4$. This success is attributable to the fact that relatively weak emission lines can be detected because of the excellent seeing. Furthermore, although quasars can be detected on the CFHT grens plates to $m \sim 21$, the adopted limit of $m = 20.5$ makes it very unlikely that substantial numbers of faint objects are missed. There appears to be no evidence of significant incompleteness in the CFHT survey in the range $0.3 < z < 3.4$, either from comparisons with other surveys or from examination of the equivalent width distribution of the detected emission lines (see Section 2.2.1.4), despite the fact that candidates are selected visually, which makes the sample susceptible to the possible biases discussed above.

Schmidt et al (1986a) have provided a complete characterization of the emission-line selection procedure. In order to obtain space densities from this type of data, one must know the intrinsic rest-frame equivalent width distribution of the quasar population that is being sampled. At present, there is no evidence that this distribution is dependent on redshift (see, e.g., Baldwin et al 1989, Schneider et al 1989a), and thus one may combine objects of all redshift to construct an empirical distribution function to use in the analysis. There is, however, ample evidence of a strong luminosity dependence of the equivalent width of the C IV line as well as a weaker dependence of Lα and other strong lines (Tytler et al 1987, Baldwin 1988, and references therein; Osmer et al 1988a).

Slitless spectroscopy has been very successful at finding high-redshift objects. Hazard et al (1986) describe results from a UK Schmidt objective-prism survey where they found many quasars with $z > 3$, including one at $z = 3.8$. Schneider et al (1989b) are continuing their CCD grism transit survey program with excellent results, including the discovery of the highest redshift quasar presently known ($z = 4.73$). Of 10 QSOs with $z > 4$ given by Schneider et al (1989a), 6 were found by slitless spectroscopy, including 5 by these authors.

In summary, the problems that have made slitless spectroscopic surveys difficult to use alongside the more easily quantifiable photometric surveys have now been largely eliminated through the use of machine searching and rigorous treatment of survey sensitivity. Unfortunately, some of the most valuable slitless surveys are subject to criticism because they are based on a visual search technique, although the original plates could be rescanned by machine.

2.1.4 VARIABILITY   Usher & Mitchell (1978) showed that nearly complete samples of quasars can be obtained by selecting for variability. Trevese et al (1989) carried out a variability survey to $B = 22.6$ for objects with dispersion in magnitude $\sigma > 0.1$ mag. This field (SA57) has been surveyed by other methods (Koo & Kron 1988), and thus a comparison of results is possible. About 70% of the confirmed QSOs are variable at the level $\sigma = 0.1$ mag (perhaps 30% at $\sigma = 0.2$ mag). Hawkins (1987) finds that $\sim 35\%$ of the UVX objects to $B = 21$ in his fields are variable at the $\sigma = 0.2$-mag level and $\sim 70\%$ at $\sigma < 0.3$ mag. In a small sample of known quasars (and three BL Lac objects) Barbieri et al (1988) find that 56% of the objects are variable at a level of $\sigma = 0.19$ mag. (Periodicity has not been detected in QSO light curves.) These three results suggest that, with good photometric precision, a substantial fraction of quasars could be detected by their variability. It is not clear, however, that any objects would escape detection by other methods. It is also not clear that this is an efficient method, although it should produce samples unbiased by color, line strength, or redshift (except to the extent that the time dilation effect gives a different effective time baseline to sample variability at high redshift than at low redshift).

2.1.5 PROPER MOTION   Proper motions were used by Sandage & Luyten (1967) to identify white dwarfs among their blue objects. Kron & Chiu

(1981) pointed out that the absence of proper motion can be used to detect compact extragalactic objects with no bias concerning spectral properties of the objects. They surveyed an area of 0.1 $\deg^2$ in SA57 to $V = 21.3$ with a baseline of 25 years. The same field had been surveyed by other methods, and a comparison of the methods showed that proper motion failed in 2 cases out of 7 in finding confirmed QSOs while identifying 2 subdwarfs as zero-proper-motion objects, a less serious and unavoidable problem. The UVX method would have failed in 1 case out of 10. Koo et al (1986) extended the survey in this field to $B \sim 23$ using a variety of techniques including proper motion, and showed that all the objects detected by Kron & Chiu (1981) on the basis of proper motion or variability were also detected by the multicolor technique. It is clear that proper motion is not an efficient search method, but the work discussed above is important in that it demonstrates that the conventional techniques do not miss any class of QSOs.

2.1.6 SELECTION AT OTHER WAVELENGTHS All of the selection techniques discussed thus far are carried out at optical wavelengths. Quasars selected in X-ray, infrared, or radio surveys have optical characteristics very similar to those of optically selected quasars (see Section 1.2). It is beyond the scope of this review to discuss the details of the surveys that have been carried out at these other wavelengths. Rather, the reader is referred to other reviews (cf. Smith 1983, 1986). The value of these other surveys for the purposes of this review is in providing samples of quasars that should be independent of the optical selection biases, such as effects of line contamination or color biases.

2.2 *The Surface Density of Optically Selected Quasars*

In this section we combine the available surveys and apply necessary corrections to the data to allow intercomparisons and to determine a combined surface density distribution.

2.2.1 CORRECTIONS TO THE RAW DATA Although many of the corrections that have been applied are small and are normally neglected, the goal here was to remove, as completely as possible, all sources of systematic error to provide a homogeneous data set.

2.2.1.1 *Extinction* The magnitude of each object has been corrected for galactic absorption using the Burstein & Heiles (1982) reddening maps. Reddenings for subareas of each survey were then derived. It is necessary to correct the limiting magnitude of each survey, as well as the magnitude of each object, for extinction. For the BQS survey a reddening was computed for each field (Green et al 1986) and for each object. For very small area surveys, a single reddening was computed for the entire area and all the objects. The vast majority of extinctions (using $A_B = 4.2E_{B-V}$) were negligible, with a small number in the range of 0.2–0.3 mag.

2.2.1.2 *Photometric error and the Eddington bias* In the case of a flat number-magnitude distribution (the differential counts $N(B)dB = $ constant), photometric error scatters as many objects toward brighter magnitudes as it does toward fainter magnitudes. But if the $N(B)$ relation is not flat, this symmetry is lost and the observed magnitude distribution is no longer an unbiased estimate of the true distribution. This problem is significant in the present case, where the number is a steep function of magnitude. A useful approximation exists and can be used to compute the size of the effect (Eddington 1940, Schmidt & Green 1983). The slope of the $\log N$ vs. B relation for quasars is about 0.9, leading to a correction (for overcompleteness in the observed counts) of 2% with photometric error $\sigma = 0.1$ mag and 30% with $\sigma = 0.3$ mag. Although the slope of the $\log N$ vs. B relation will be affected when errors are a function of magnitude (which is usually the case), this change in slope is a second-order effect in the present case and is ignored here. The total corrections to the counts, including the Eddington effect, are derived below.

2.2.1.3 *Variability* The quasar surface density is properly defined in terms of the time-average luminosity of the objects. Since periodic variability of QSOs has not been found, it is simplest to characterize the light curves by the dispersion in brightness. Variability behaves identically to photometric error and results in overcompleteness in a steeply rising number-magnitude relation.

A number of studies of quasar variability have been carried out (see Section 2.1.4). From an examination of the work of Hawkins (1987), Trevese et al (1989), and Barbieri et al (1988), the following estimates were made. It was assumed that 30% of QSOs are variable at the level of $\sigma = 0.15$ mag, 20% at $\sigma = 0.25$ and that 20% are variable with $\sigma = 0.35$ mag. The remaining quasars were assumed to be constant or variable at a level that is negligible for the purposes of producing a bias of the type under discussion. These assumptions lead to an estimate of overcompleteness of about 10% from variability alone. Extreme assumptions lead to estimates ranging from 7% to 13%.

Slitless surveys are selected differently but are nevertheless subject to the same bias, since objects are retained or rejected from a nominally magnitude-limited sample based on their measured magnitudes.

Variability can affect photometric surveys where plates have been taken at different epochs. Boyle et al (1987; BFSP) estimate that this is a small effect in their survey ($\sim 2\%$), and it is ignored here.

2.2.1.4 *Slitless surveys—the equivalent width distribution* In order to evaluate the level of completeness of spectroscopic surveys, it is necessary to know the true distribution of equivalent widths among quasars. One might expect that since slitless surveys select largely (or completely) on the basis of the presence of emission lines, QSO samples selected by other means might contain a population of weak-lined quasars that are missing from the slitless surveys. A correction could be made for this by using a weighting scheme. For example, it is clear intuitively that if only the strongest lined half of the QSO population were detectable by a spectroscopic survey while the weaker-lined half were beyond its sensitivity limit, one could compensate by counting each detected quasar twice.

Schmidt et al (1986a) have summarized quasar equivalent width properties derived from an examination of the literature. Data for radio-selected samples are available from Baldwin et al (1989) and Wilkes (1986). Savage et al (1988) present preliminary results from a radio-selected sample (which apparently overlaps that of Wilkes 1986). Equivalent widths are also available for the CFHT survey and the Osmer (1982) and Osmer (1980) samples. [Equivalent widths are included in the paper by Boyle et al (1990), which we received too late to include in this analysis.]

Peterson (1988) shows that quasar samples selected by variability and radio emission have much smaller mean Lα equivalent widths than the CFHT grens and APM objective-prism spectroscopic surveys. This implies that the spectroscopic surveys are incomplete for small equivalent width objects. Table 1 shows equivalent width measurements for the samples named above, along with the numbers taken from Peterson's (1988) Table 1 (for the APM, variable, and radio samples). The numbers in parentheses are equivalent widths measured for the Lα + N V blend. Guided by a comparison of the Wilkes (1986) and Savage et al (1988) results, we assume that the two Lα measurements without parentheses are deblended.

It appears to us that the apparent difference in mean Lα equivalent width between optical and radio samples is produced by comparing measurements that deblend Lα (1216Å) from N V (1240Å) with those where only the blend is measured. Given this assumption, there is no evidence in Table 1 of systematic differences in line strengths between these emission-line surveys and radio-selected samples. For example, the CFHT optical sample has a mean equivalent width larger than that of the Wilkes (1986) sample but smaller than the PKS radio sample (Baldwin et al 1989). There is a suggestion in this table that different workers systematically measure equivalent widths differently from one another. The PKS and Wilkes radio-selected samples differ significantly in mean equivalent width for all four major emission lines (Lα, C IV, C III], and Mg II), as judged by a U-test for equality of the means. It is less likely that they sample a

Table 1 Rest-frame equivalent widths (W in Å)

Sample	Lα(+N V)			C IV			C III]			Mg II		
	$\langle W \rangle$	σ	n	$\langle W \rangle$	σ	n	$\langle W \rangle$	σ	n	$\langle W \rangle$	σ	n
CFHT	(90)	45	108	46	33	187	33	29	203	39	25	125
Osmer1	(105)	64	42	53	38	56	34	14	15	47	15	2
Osmer2				47	25	5				37	21	2
APM1[a]	(95)	34	69									
Wilkes[b]	(83)	38	38	32	20	94	17	12	96	27	15	113
PKS[c]	(93)	46	15	49	41	29	22	20	31	43	38	32
Savage[a]	56	37	37									
Molonglo[d]	(78)	17	3	51	74	15	28	28	26	45	40	32
Variability[a]	61	41	19									
SSG1 (adopted)	(75)			35			20			40		
SSG[e] high z	(71)	16	10	25	10	10						

Angle brackets ($\langle \ \rangle$) indicate mean values.
References not included in Table 2:
[a] Peterson 1988
[b] Wilkes 1986
[c] Baldwin et al 1989
[d] Smith et al 1977
[e] Schneider et al 1989a.

different parent population than it is that the measuring procedure is biased. A further possibility is that the difference is a manifestation of the Baldwin effect, if there are differences in luminosity between the samples.

A comparison of the CFHT equivalent width distribution with that of the radio-selected samples provides no evidence that the CFHT survey is incomplete for weak equivalent width objects in any redshift range. This conclusion is supported by a comparison of the magnitude and redshift distributions of the CFHT and UVX samples.

2.2.1.5 *Emission-line correction* Two corrections were made in order to remove the effect of emission lines from the B magnitudes. After an examination of the line corrections derived for individual objects from slit spectra of CFHT survey objects, a line correction vs. redshift relation, similar to that given in graphical form by Koo & Kron (1988) (see also Veron 1983), was adopted. [The Koo & Kron (1988) relation is actually a *total K*-correction, but, since they assumed a model with power-law spectral index -1, it reduces to a line plus blue-bump correction.] This correction is reasonable for low-luminosity objects but ignores the Baldwin effect, thereby overestimating the line correction for higher luminosity quasars. Guided by an examination of the correlations of equivalent width with luminosity in the CFHT survey and in sources cited in Section 2.2.1.4, correlations for the four major lines Lα, C IV, C III], and Mg II were used to determine the change in the emission-line corrections as a function of luminosity. The line corrections are similar in magnitude to those given by Koo & Kron (1988) for low-luminosity objects, but they are smaller for higher luminosity objects.

2.2.1.6 *Application of the completeness corrections* For each survey we combined the authors' own estimates of completeness and photometric error with the adopted variability model outlined above and then computed a completeness value for each survey from these three quantities. In nearly all cases, after balancing the deficits and surpluses, the percentage completeness was found to be $100 \pm 8\%$. The BQS is 100% complete in the interval $0.6 < z < 0.8$ and 112% complete elsewhere. On the basis of these results, no correction factors were applied for photometric error or variability. Note that the surpluses introduced by photometric error and variability are significant only where the slope of the $\log N$ vs. B relation is steep. Beyond the turnover (at $B \sim 19$), the slope is much shallower. There the surplus disappears and the counts are likely $\sim 90\%$ complete on average. The same effect will occur for the luminosity functions derived later. If one wishes to make these small corrections, they should be applied only to the points at and beyond the turnovers in the individual luminosity functions.

2.2.2 COMBINING THE SAMPLES TO OBTAIN THE SURFACE DENSITY The UVX method is sensitive to quasars for $0 < z < 2.2$, and objects detected by this technique constitute the majority at these redshifts. On the other hand, spectroscopic surveys tend to be most sensitive at higher redshift and vary widely in their relative sensitivity at low redshift. This situation suggests that we construct surface densities in two redshift regimes: $0 < z < 2.2$ and $z > 2.2$. In each of these ranges, the surveys are compared and combined. The surface density results for that redshift interval are then compared with results of other workers. Table 2 gives a summary of major surveys with nominal redshift and magnitude limits. The actual limits used in this analysis differ from those in Table 2 in some details and are more complex for reasons discussed below.

2.2.2.1 *$0 < z < 2.2$* Detailed comparisons were made of various samples in an attempt to detect any evidence of incompleteness. A Kolmogorov-Smirnov (K-S) test was applied separately to the redshift and magnitude distributions to try to detect differences between the samples. Because the K-S test checks only the distribution shape, a method described by Gehrels (1986) was used to check that the normalizations agree. In a comparison of two surveys, one expects that the ratio of numbers of objects in the surveys (brighter than a given limit or in a

Table 2 Major quasar surveys and nominal sensitivity limits

Name	m_{lim}	z_{max}	Deg^2	n(QSOs)[a]	References
BQS	~16	2.2	10^4	114	Schmidt & Green 1983
MBQS	17.65	2.2	109	32	Mitchell et al 1984
AB	18.25	2.2	35.5	22	Marshall et al 1984
APM1	18.5	~3.3	102	192	Foltz et al 1987
APM2	18.75	~3.3	85	156	Foltz et al 1989
BF	19.8	2.2	1.72	35	Marshall et al 1984
Crisitani	19.8	2.2	10	99	Cristani et al 1989
Osmer1	~20	3.4	5.1	66	Osmer 1980
CFHT	20.5	3.4	9.40	268	Crampton et al 1989
SSG2	~20.5	4.5	7.84	8	Schmidt et al 1986b
BFSP	20.9	2.2	4.19	193	Boyle et al 1987, 1988b
Osmer2	~21	4.7	5	15	Osmer 1982
SSG1	~22	4.5	0.91	10	Schmidt et al 1986a
Marano	22	3.3	0.69	23	Marano et al 1988
KOKR	22.6	3.2	0.29	30	Koo & Kron 1988

[a] n is the number of QSOs in the survey.

magnitude interval) should equal the ratio of areas of sky coverage in the surveys. For example, when comparing survey 1 with survey 2 we would compute the ratio of objects detected (n_1/n_2) and use the method described by Gehrels (1986) to determine a two-sided confidence interval for the ratio. If the ratio of areas (A_1/A_2) was outside of this confidence interval, then we would take this to indicate that one of the samples was biased. If the area ratio was not included within the 99% confidence interval of the number ratio, we would say that the numbers detected differ from one another at the 1% level of significance. Implicit in the process of using these tests to identify incompleteness is the assumption of isotropy.

With certain exceptions, it was found that the surveys agreed with one another in numbers and distribution shapes. This finding must be tempered by the realization that the amount of overlap (in m-z space) between different surveys is less than ideal. Brighter than $B = 16$, the only survey that covers sufficient area to adequately sample the surface density is the BQS. A comparison of the bright end of the CFHT survey with a combination of the other moderately bright samples showed that, although the shapes of the redshift and magnitude distributions are similar, the CFHT sample contains more objects than the combined bright sample at $16 < B < 18.5$. The numbers of objects found are different at the 1% level of significance. Can this conflict be attributed to the different selection effects of the UVX and slitless techniques? The UVX technique tends to be less effective at redshifts $0.5 < z < 1.0$ for the reasons discussed above.

The redshift distribution of the CFHT sample has a sharp peak at about this redshift, whereas the UV excess samples show the expected deficit. The effect is large enough to entirely explain the significant difference in numbers. It appears unlikely that line identification errors could be responsible for the surplus of CFHT quasars in this range of redshift (D. Crampton, private communication, 1989). An alternative explanation for the surplus of CFHT quasars at bright magnitudes is the existence of a zero-point error in the CFHT magnitudes. (These are derived from PDS microdensitometer scans of Palomar Sky Survey plates calibrated by a number of CCD magnitude sequences.) The error would need to be 0.35 mag or greater to account for the anomaly, and such a large error is unlikely for two reasons. A reexamination of the photometric calibration appears to exclude such an error, and if one applies a ΔB to the CFHT magnitudes (to make the CFHT survey consistent with the other surveys at the bright end), one produces a conflict in the faint-end magnitude distributions that did not exist previously.

While it would be helpful to derive a completeness correction factor from the CFHT sample that could be applied to the UVX samples in the redshift range $0.5 < z < 1.0$, this cannot be done reliably. Other workers (e.g. Green 1986) have suggested that completeness might be as low as 70% in some part of this redshift range, while the CFHT data suggest completeness of less than 50% at bright magnitudes. At fainter magnitudes a comparison of the CFHT and BFSP samples does not reveal any significant deficit in the UVX-selected sample. Thus, a correction factor cannot be justified except, perhaps, at the bright end, where small numbers make the correction less relevant compared with statistical (counting) errors.

The fainter samples were combined after testing (as above) for differences in magnitude and redshift distributions. We limited the Cristiani et al (1989) and BF samples to $B \leq 19.5$. With this restriction, no significant differences were found among the BFSP (Boyle et al 1987), BF-AB (Marshall et al 1984), CFHT (Crampton et al 1989), and Cristiani et al (1989) samples. We adopted a conservative magnitude limit ($m = 20.0$) for the CFHT survey, despite the absence of evidence for incompleteness. The smaller, faint samples of Marano et al (1988) and Koo & Kron (1988; KOKR) (to $B = 21$) were also added. In order to estimate the surface density of the faintest quasars, the objects with $B > 21$ from Koo & Kron (1988) and Marano et al (1988) were used, even though follow-up work is not yet complete. In order to account for this incompleteness, the sky areas of these surveys were scaled by the fraction of objects in these surveys for which spectroscopic follow-up is complete.

Using the above data, the quasar population has been surveyed to

$z = 2.2$ and for the range $13 < B < 22.5$, with varying degrees of overlap between samples. To the extent that multiple surveys of the same parameter space strengthen the results, the weakest parts of the current sample are $B < 16$ (where only the BQS provides data) and $20 < B < 21$, where the CFHT survey is not yet complete and the BFSP survey is the only one that gives an appreciable number of objects. Very few quasars have been detected at $B > 21$.

Figure 1 shows the surface density of QSOs at $z < 2.2$. The turnover (i.e. the transition to a shallower slope) suggested by Koo & Kron (1982) is apparent at $B \sim 19$. Tables 3a and 3b give the differential and cumulative surface density, respectively, with approximate 1σ errors. These results agree in detail with the counts presented by Boyle et al (1988b). There is also good agreement with Koo & Kron (1982) and with Braccesi et al (1980) at the bright end. A fit to the differential counts for $14.75 < B < 18.75$ gives $\log N = 0.88(\pm 0.02)B - 15.8(\pm 0.5)$ (compare this to the slope of 0.75 first obtained by Sandage & Luyten 1969). At the faint end, the slope is 0.31, in agreement with Boyle et al (1988b).

2.2.2.2 *2.2 < z < 3.3* As noted above, the CFHT survey appears to be complete to $z = 3.4$ and $B = 20.5$ (although, as noted above, a limiting magnitude of $B = 20.0$ was adopted here). Because the CFHT survey is a large one, it is reasonable to use it as a benchmark against which to compare others. A magnitude limit for the Osmer (1980) sample (of objects

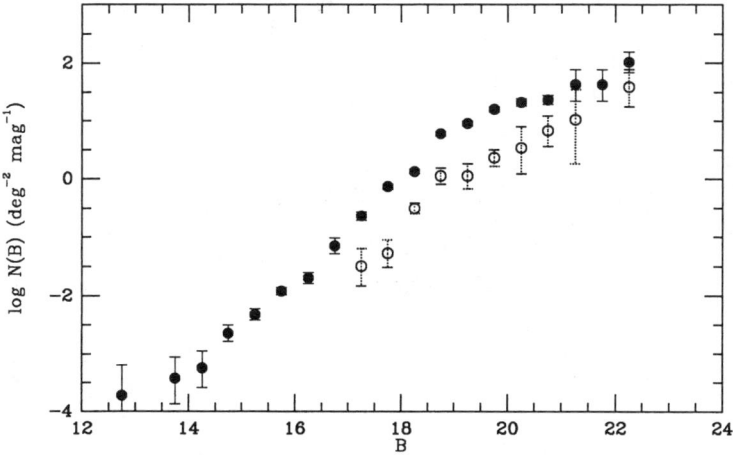

Figure 1 The surface density of quasars from the combined sample (*solid symbols*: $0 < z < 2.2$; *open symbols*: $2.2 < z < 3.3$).

Table 3a Differential quasar surface density

B	0 < z < 2.2			2.2 < z < 3.3		
	log N^a	σ	n	log N^a	σ	n
12.75	−3.7	±0.7	1			
13.75	−3.4	±0.4	2			
14.25	−3.25	±0.3	3			
14.75	−2.65	±0.15	12			
15.25	−2.33	±0.10	25			
15.75	−1.93	±0.06	54			
16.25	−1.70	±0.10	24			
16.75	−1.15	±0.14	13			
17.25	−0.64	±0.07	40			
17.75	−0.13	±0.05	99	−1.5	±0.33	3
18.25	0.13	±0.04	144	−1.3	±0.24	5
18.75	0.77	±0.04	119	−0.50	±0.09	26
19.25	0.95	±0.04	116	0.05	±0.14	13
19.75	1.20	±0.04	110	0.06	±0.21	6
20.25	1.32	±0.07	49	0.36	±0.15	12
20.75	1.36	±0.08	35	0.53	±0.4	2
21.25	1.6	±0.3	4	0.84	±0.3	4
21.75	1.6	±0.3	4	1.0	±0.7	1
22.25	2.0	±0.2	8	1.6	±0.3	3

[a] N has units QSOs deg^{-2} mag^{-1}.

Table 3b Cumulative quasar surface density

B	0 < z < 2.2		2.2 < z < 3.3	
	log N^a	σ	log N^a	σ
13.0	−4.0	±0.3		
13.5	−3.7	±0.5		
14.0	−3.55	±0.4		
14.5	−3.24	±0.2		
15.0	−2.77	±0.1		
15.5	−2.39	±0.07		
16.0	−1.99	±0.05		
16.5	−1.70	±0.07		
17.0	−1.23	±0.09		
17.5	−0.73	±0.06	−1.8	±0.4
18.0	−0.23	±0.04	−1.4	±0.2
18.5	0.12	±0.03	−0.69	±0.08
19.0	0.63	±0.03	−0.12	±0.09
19.5	0.93	±0.03	0.12	±0.10
20.0	1.21	±0.02	0.38	±0.09
20.5	1.40	±0.03	0.62	±0.16
21.0	1.52	±0.03	0.88	±0.14
21.5	1.73	±0.10	1.1	±0.25
22.0	1.87	±0.10	1.3	±0.25
22.5	2.11	±0.09	1.5	±0.25

[a] N has units QSOs deg^{-2} brighter than B (column 1).

found by Hoag & Smith 1977) was chosen by plotting the observed equivalent width (W) of the discovery emission lines against B magnitude. At bright apparent magnitudes one would expect all quasars to be detected, including those with weaker lines, whereas at fainter magnitudes, weak lines would be missing from the diagram (see Clowes 1981). A magnitude limit for the Osmer (1980) sample was selected where lines of all strengths were still detected using the distribution of W's from the CFHT survey for comparison. It appears that the Osmer sample includes objects to a fainter limit than the CFHT survey, although at the faintest magnitude all the Osmer (1980) quasars have $W(L\alpha) > 200$ Å. Since the CFHT survey includes objects with $W(L\alpha)$ as weak as 100 Å, we conclude that the Osmer sample becomes incomplete at the faint end, and thus we adopt a magnitude limit of $B = 20.25$.

Osmer (1982) presents results of a red survey capable of detecting $L\alpha$ at $3.7 < z < 4.7$ and shows that the survey goes 1.2 mag fainter than that of Hoag & Smith (1977). The highest redshift objects ($z < 3.4$) were found from the C IV lines. We adopt a limit of $B = 21$ for the redshift range where $L\alpha$ is within the passband, and a limit of $B = 20$ where C IV is the strongest line in the passband. These samples are not included in the surface density determination (because they do not have uniform magnitude limits) but are used to compute the luminosity function.

The samples that contribute to the high-redshift surface density are the CFHT survey, those of Marano et al (1988) and Koo & Kron (1988), APM1 (Foltz et al 1987), and APM2 (Foltz et al 1989). Figure 1 and Table 3 give the surface density of QSOs at $2.2 < z < 3.3$. If one gives all points in Figure 1 equal weight, the relation $\log N = 0.58(\pm 0.07)B - 8.23(\pm 1.4)$ is obtained for QSOs in this redshift interval alone. In general, the surface density of high-redshift quasars is about a factor of 4 lower than that of low-redshift objects at similar apparent magnitudes.

We conclude this section with a brief discussion concerning the fluctuations in surface density over large areas of the sky. Arp (1984) has called attention to apparently large variations in the number densities of radio-selected quasars over separations of tens of degrees. Subsequent work using carefully assembled samples of optically selected quasars with much larger numbers reports much smaller variations. The most authoritative study we know of is that by Shanks et al (1988), who report no deviations from isotropy that are greater than those expected for Poisson fluctuations (see also Hawkins 1985).

In the above section all the major surveys have been combined in a way that accounts for the main systematic differences. In general, our surface density results, summarized in Table 3 and Figure 1, confirm the results obtained by previous workers using subsets of these data.

3. THE LUMINOSITY FUNCTION

In this section we briefly describe how samples of quasars with well-defined selection criteria may be used to determine the luminosity function (LF) and then apply one of these methods to the combined sample defined in the previous section. For quasars, one must define both the luminosity and the comoving space density of objects as a function of redshift. In this review, the term "luminosity function" [designated $\Phi(M, z)$] refers to the space density of quasars per unit magnitude interval but in a restricted redshift range. Formulas appropriate to the quasar problem for the calculation of the K-correction, luminosity distance, and volume element for different cosmologies can be found in Schmidt & Green (1983). We adopt the usual method of determining the K-correction by assuming a power-law spectrum $f_\nu \propto \nu^\alpha$. Although there is evidence that spectral index (α) is a weak function of luminosity (Elvis 1989), such an approximation will not introduce appreciable errors. A Hubble constant of $H_0 = 50$ km s^{-1} Mpc^{-1} is adopted in this section. Luminosity functions were computed for $q_0 = 0.5$ and 0.1 and for spectral indices $\alpha = -0.5$ and -1.0. Tabulated results are presented for the two sets of parameters $[q_0, \alpha] = [0.5, -0.5]$ and $[0.10, -1.0]$, which demonstrate the extremes of the possible evolution of the LF. The *qualitative* description of the evolution of the quasar LF is relatively insensitive to the choice of cosmology but shows a stronger dependence (through the K-correction) on the assumed choice of spectral index.

3.1 Computing the Luminosity Function

A magnitude-limited sample of a population that is distributed uniformly throughout (Euclidean) space possesses a simple relation between the cumulative number of objects found and the limiting magnitude m: $N_{\rm observed} \propto V(m) \propto 10^{0.6m}$, where V is the volume. This is true regardless of the shape of the luminosity function of the objects. To determine the luminosity function (which is a function of absolute rather than apparent magnitude), the properties of a magnitude-limited (as opposed to volume-limited sample) must be considered. Comparing the observed numbers at two luminosities, $N_1(M_1)dM$ (this is the number counted in an infinitesimal interval $[M_1, M_1+dM]$) and $N_2(M_2)dM$, one finds that the more luminous objects are overrepresented in a magnitude-limited sample compared with the less luminous ones because a larger volume of space has been surveyed for them. This is the origin of the Malmquist bias. If one has a volume-limited sample (all objects within a volume V_a are detected), then the space density is simple to compute:

$$\rho(M_1) = \sum_{\substack{j \\ M_j \in (M_1, M_1 + dM)}} \frac{1}{V_a} = \frac{N_1(M_1)}{V_a}, \qquad 1.$$

and similarly for $\rho(M_2)$.

In a magnitude-limited sample, the volume sampled varies with absolute magnitude and we have

$$\rho(M_1) = \sum_{\substack{j \\ M_j \in (M_1, M_1 + dM)}} \frac{1}{V_a^j} \left(= \frac{N_1(M_1)}{V_a(M_1)} \right). \qquad 2.$$

The equality in parentheses is true for an infinitesimal bin size only: A finite bin width must be accounted for (see below). Note that $V_a(M_1)$ will, in general, be very different from $V_a(M_2)$, so that the fainter objects (with a smaller V_a) will have larger weights than the brighter objects. One can say that the ith object contributes to the luminosity function with a relative weight $1/V_a(M_i)$ (or $1/V_a^i$). In the terminology of Avni & Bahcall (1980), V_a^i is the *accessible volume*: It is the volume throughout which the ith object could be detected, taking account of the sensitivity limits of the survey.

Abandoning the assumption of uniform space density and instead assuming a radial density function, one can compute the luminosity functions in discrete shells of space with arbitrarily chosen boundaries. Then V_a^i is the volume, within the shell under consideration, accessible to an object with magnitude M_i, and N_{observed}^i is the corresponding number counted in that shell with magnitude M_i. An object with magnitude M_i may be detectable throughout the shell, detectable within a portion of the shell, or not detectable at all within the shell under consideration. The usefulness of computing the LF in shells of space is in allowing one to compare shells at different radii with one another to find evidence of a change in the LF between the shells. (This assumes that the shell is narrow enough so that the luminosity function is a function only of absolute magnitude M throughout its width.) If one abandons the assumption of Euclidean space, one changes only the computation of the volume element according to the cosmological model adopted.

There is a small complication in the process of converting the discrete observed counts (with weights that differ according to the accessible volume for each quasar) into a luminosity function, which is normally considered to be a continuous function of magnitude. To obtain a cumulative LF (number per unit volume brighter than M), one would compute the weighting factor $1/V_a^i$ for each object and sum the weighting factors of all objects brighter than M. There are at least two options for obtaining a differential luminosity function. One could bin the objects to obtain the (unweighted) counts between $M - dM/2$ and $M + dM/2$ and then compute a weighting factor at M (the middle of the bin). The LF estimate is the number within the bin multiplied by this mean weighting factor. This would yield a good estimate of the LF if small bins were used. However, it is better, especially when few objects are available, to compute V_a^j for each object (Schmidt 1968) and then sum:

$$\Phi(M,z)dM = \sum_{\substack{j \\ z_j \in (z_1, z_2) \\ M_j \in (M_1, M_2)}} \frac{1}{V_a^j} \begin{cases} M_1 = M - dM/2, \\ M_2 = M + dM/2, \\ z_1 = z - dz/2, \\ z_2 = z + dz/2. \end{cases} \qquad 3.$$

Equation 3 gives the LF (assumed to be a function of M only) for the shell within the redshifts z_1 and z_2.

When combining many samples, the accessible volume for an object is computed over all samples and not just the sample in which the object originated (the "coherent" method of Avni & Bahcall 1980).

The procedure described above is completely general, assuming that, within the shell of space under consideration, the LF is a function of magnitude alone (and not of redshift). Except for this, the computed LF is exact within statistical errors. Only recently have enough data become available to allow the use of this straightforward scheme.

Schmidt (1968) introduced the statistic V/V_{max} as a measure of the uniformity of the space distribution of a sample of radio sources from the 3CR catalogue. The test statistic is the average formed from the ratios of the volume (V) enclosed at the redshift of an object, divided by the volume that would be enclosed at the maximum redshift at which the object would be detectable (V_{max}, which is identical to V_a above). The maximum redshift is determined by the luminosity of the object and the limiting magnitude (and sometimes other properties) of the survey. One computes V/V_{max} for each object in a complete sample. If the objects are drawn from a population distributed uniformly in space, then the V/V_{max} values will be uniformly distributed between 0 and 1 with a mean of 0.5, whereas a skewed distribution of V/V_{max} indicates that the density of objects is a function of redshift.

The V/V_{max} statistic can be used as a measure of the acceptability of a hypothesized evolutionary model in the following way (Schmidt 1968). One defines a density function $\rho(z)$ that represents the density enhancement factor of the population over its local ($z = 0$) value. If one guesses the correct function and defines a new "weighted volume element"

$dV' = \rho(z)dV$, then it is the distribution of sample values of V'/V'_{max} that ought to be uniformly distributed between 0 and 1. One can try a particular analytical form for the density law and change parameters to obtain the desired distribution of V'/V'_{max}. Schmidt (1968) used this method to derive a density law that was consistent with the behavior of the 3CR radio sources. Because the density enhancement is a function of redshift only, this method assumes that objects of all luminosities increase in density at the same rate as one another. This is pure density evolution. Schmidt (1968) found that the mean values of V'/V'_{max} were different for optically weak and radio-weak subclasses. More generally, one could define a density function $\rho(z, L)$, where L is the radio or optical luminosity. The local ($z = 0$) luminosity function is obtained by binning the values of $1/V'_{max}$. The LF at redshift z is the local function multiplied by $\rho(z)$. A disadvantage of this method is that the solutions may lack uniqueness: A wide variety of models can be proposed that are acceptable. Each model will predict identical behavior (by construction) in the observed region of the M-z plane, but they may be quite different in their predictions of future observations. It should be noted that this approach is not equivalent to simple model fitting, a process where one would assume an analytical form $\Phi(M, z)$ for the luminosity function and its evolution. Then the parameters would be adjusted to obtain the best fit between model predictions and observations using, for example, a χ^2 statistic. The V'/V'_{max} method assumes $\Phi(M, z) = \phi(M)\rho(z)$, where $\phi(M)$ is obtained as a numerical function with no assumed form. However, $\rho(z)$ is described by a model, and the fit is obtained by adjusting parameters to give $\langle V'/V'_{max} \rangle = 0.5$. (Angle brackets denote the mean value.) Use of the V'/V'_{max} statistic is an excellent method of testing for evolution in space density but is a poor way to test for goodness of fit of an assumed model.

There are alternatives to guessing an analytical form for the density function $\rho(z)$ and adjusting parameters to obtain a best fit. Turner (1979b) devised a method of determining first the shape of the LF independent of density evolution and then the form of the density function, where both $\phi(M)$ and $\rho(z)$ are explicit numerical estimates. This is a distinct improvement over methods that require one to guess the correct analytic forms. If luminosity evolution is believed to be occurring, it is possible to use a method of estimating the luminosity evolution law (Turner 1979a) that depends only upon the weak assumption that the LF shape is preserved. (There may be density evolution.) A disadvantage of these procedures is that errors are difficult to accurately evaluate. Choloniewski (1987) showed how to derive the C-method of Lynden-Bell (1971) in a simple way. It must be assumed when applying this method that pure density evolution properly describes the behavior of the LF. As with Turner's method, explicit numerical values of $\phi(M)$ and $\rho(z)$ are obtained, and multiple samples of objects can be dealt with. In addition, it is shown that one obtains maximum likelihood estimates of the LF, and that the errors are well defined. The mathematical formalism presented is very useful, even when the method itself is not applicable, because it correctly takes account of the discrete nature of the observed counts.

The use of emission-line-selected quasar samples introduces further complications into the computation of the LF [see Schmidt et al (1986a, 1988), Schmidt (1987), and Section 3.2.2 below for discussions of this problem].

Of the several methods available to compute the luminosity function of quasars, the $1/V_a$ method (Equation 3) is the most general. When applied to discrete shells of space, it requires no assumptions about the form of the luminosity function or its evolution (nor does it provide any predictions of the behavior of the LF in regions of M-z space that have not been sampled—as other methods do). Methods that require the assumption of pure density evolution are not useful because, as shown in the following section, this fundamental requirement is not satisfied. The method used by Schmidt (1968) is more general but is most useful as a test for the presence of evolution.

3.2 *Luminosity Function Results*

There are two fundamental evolutionary models. In the pure luminosity evolution (PLE) scenario the shape of the luminosity function is preserved while some characteristic luminosity L^* changes with redshift: The entire LF shifts only along the luminosity axis. Pure density evolution (PDE) also maintains the shape of the LF but allows it to shift only along the density axis. Thus, with PDE, the relative space density of quasars at any two luminosities remains constant while the total number increases or decreases with redshift. If the entire LF is sampled, one can clearly distinguish between the two models, since PDE requires the total number of objects to change with time while PLE requires that the total number must remain constant. A clear description of the PLE and PDE models and variations of them is given by Koo (1986) (particularly his Figure 7).

The LF represents the distribution of luminosities (per unit volume) at a given redshift, whereas its evolution is the change in this distribution from one redshift to another. Various physical paths are possible to get from the LF at z_1 to that at z_2. It could be that all quasars were born simultaneously and are becoming less luminous at the same rate as time goes on. Alternatively, the birthrate of the most luminous objects may have been highest in the early Universe and is steadily decreasing, while the birthrate of the least luminous objects has been roughly constant. These two models require very different individual lifetimes but could produce the same apparent LF evolution. One can say, however, that if quasar lifetimes are short, then the change in the LF reflects changes in the QSO birthrate. If lifetimes are long, the change reflects the evolution of individuals. The fitting of PLE and PDE models thus has a physical motivation in addition to the fact that these models simplify the analysis considerably.

Schmidt (1968) used the V'/V'_{max} procedure to derive an evolution law that showed the space density of radio-selected QSOs to be about 2 orders of magnitude higher at $z = 1$ than at $z = 0$. Although based on very few data compared with that currently available, this estimate is approximately correct for optically selected quasars. The first optical quasar luminosity function calculations were done by Schmidt (1970). Mathez (1978) derived a luminosity function assuming pure luminosity evolution and showed it to fit the observations as well as PDE models.

Braccesi et al (1980) concluded that either luminosity or density evolution models could provide a satisfactory fit to the observations. Similarly, Marshall et al (1983) considered PLE and PDE models and showed that it was possible to fit either model successfully, although it was found that the PDE models had to be rejected because they implied an excess of 2-keV X-ray background. In addition, the PDE model predicted faint number counts, in conflict with the observations of Koo & Kron (1982), who showed that their counts were consistent with the PLE model of Braccesi et al (1980).

When the Palomar Bright Quasar Survey (BQS) became available, Schmidt & Green (1983) combined it with other data to determine the LF and its evolution. They concluded that—because the values of $\langle V/V_{max}\rangle$ were larger for more luminous than for less luminous objects—pure density evolution was unlikely. Instead, they proposed "luminosity-dependent density evolution," in which the evolution model is more general than either PDE or PLE. The shape of the LF is not preserved in such a model. This important paper is strongly recommended as a source of information on the issues and methods involved in the analysis of quasar survey data. Crampton et al (1987) proposed some modifications to the Schmidt & Green model when they found their new data for $z > 2$ in conflict with the earlier model.

Also of interest are models with a more specific physical basis. Koo & Kron (1988) devised a simple model where the lifetime of a quasar is dependent on its luminosity, with the brightest objects burning out fastest, and they showed that this model could successfully describe the observations. Blandford (1986) developed another simple model showing that suitable choices of fueling and birth-rate functions result in qualitatively realistic descriptions of luminosity function evolution.

The most successful and most fully constrained description of the luminosity function at $z < 2.2$ was provided by Boyle et al (1987, 1988b), who used a sample of ~ 400 UVX-selected objects. Pure luminosity evolution was found to fit the data well. In this model, the LF is fit by two power laws. The characteristic magnitude M^* (where the two power laws meet) evolves according to $M^*(z) = M_0^* - 2.5 \times k \times \log(1+z)$. The value of k is found to be ~ 3–4.

Recently, analyses of large, combined samples (similar to that in this paper) have been done by Schade (1988) and Green (1989). Both noted a decrease in space density (integrated to a fixed magnitude) beyond $z = 2$ for the brightest objects. Warren et al (1988) and Warren & Hewett (1990) provide new luminosity functions at high redshift based upon automated multicolor surveys.

3.2.1 $0 < z < 2.2$ To derive the luminosity function in this redshift range, we used all surveys in Table 2 that are sensitive at low redshifts. Tables 4 and 5 give numerical results for $q_0 = 0.5$, $\alpha = -0.5$ and $q_0 = 0.10$,

Table 4 Luminosity functions for $z \leq 2.2$—(QSOs Gpc^{-3} mag^{-1}), $H_0 = 50$ km s^{-1} Mpc^{-1}, $q_0 = 0.5$, $\alpha = -0.5$

	$0.16 < z < 0.4$			$0.4 < z < 0.7$			$0.7 < z < 1.0$			$1.0 < z < 1.2$		
M_B	Φ	σ	n	Φ	σ	n	Φ	σ	n	Φ	σ	n
-23.25	358	77	22	826	216	15	1103	373	9	1457	660	5
-23.75	38	18	5	364	90	18	1267	291	20	1459	557	7
-24.25	42	12	13	142	34	18	462	135	12	1192	324	14
-24.75	13	3.9	11	95	22	18	276	69	19	729	189	15
-25.25	3.3	1.4	6	37.8	13	9	88	22	16	318	104	11
-25.75	3.8	1.2	10	20.6	7.2	9	39.2	11.8	11	125	33	16
-26.25	0.33	0.33	1	0.98	0.75	2	14.7	6.6	5	48	15	10
-26.75				0.64	0.32	4	2.55	2.55	1	11.8	6.8	3
-27.25							0.35	0.35	1			
-27.75							0.18	0.13	2	0.32	0.32	1
-28.25										0.59	0.3	4
-28.75										0.11	0.11	1
-29.25												
-29.75												
$\langle V/V_{\max} \rangle$	0.65±0.032			0.60±0.029			0.57±0.031			0.53±0.030		
N_{total}	68			93			96			87		
$\int_{-\infty}^{-26} \Phi(M)dM$	0.16±0.16			0.81±0.41			8.9±3.5			30.3±8.3		
$\int_{-\infty}^{-24} \Phi(M)dM$	36.2±7.4			148±22			442±77			1213±196		

$\alpha = -1$. These results are similar to those in Boyle et al (1988b) (except that they used only $\alpha = -0.5$). Figure 2 shows the LF for four redshift intervals (with $q_0 = 0.5$). Also plotted is the Seyfert 1 luminosity function of J. Huchra & R. Burg (private communication, 1989) and the local ($z < 0.2$) luminosity function of quasars. The integrated density of quasars more luminous than $M_B = -23$ clearly increases strongly from its local value up to $z = 2$. It is also evident that pure luminosity evolution (a shift along the magnitude axis) is *sufficient to describe the change in density*. There is an apparent deviation from the PLE model at the faint end of the highest redshift interval ($1.9 < z < 2.2$), where the density is greater than would be expected under the assumption of PLE, although this rests on a single point (at $M = -24.25$). The LFs for $q_0 = 0.10$, $\alpha = -1$ (not plotted) show, qualitatively, the same behavior shown in Figure 2: Pure luminosity evolution appears to be a reasonable description of the change in the LF with redshift.

3.2.2 $z > 2.2$ The CFHT, Marano et al (1988), KOKR (Koo & Kron 1988), APM1, APM2, and Osmer (1980, 1982) samples are included in this redshift range. Also included are the surveys of Schmidt et al (1986a;

Table 4 (*continued*)

	$1.2 < z < 1.4$			$1.4 < z < 1.7$			$1.7 < z < 1.9$			$1.9 < z < 2.2$		
M_B	Φ	σ	n	Φ	σ	n	Φ	σ	n	Φ	σ	n
−23.25										4808	4808	1
−23.75	1081	484	5	652	493	2						
−24.25	812	340	6	1275	426	9	3303	1494	8	14970	7169	6
−24.75	1006	272	14	1229	339	14	2225	671	11	1147	406	8
−25.25	616	165	14	701	169	18	1132	306	14	1630	419	17
−25.75	244	71	14	414	104	16	618	174	13	576	157	14
−26.25	110	25	20	252	49	32	529	133	18	391	105	14
−26.75	31.0	11.8	7	45.0	12.1	14	107	30	15	209	53	20
−27.25	17.7	7.9	5	13.8	6.2	5	32.5	12.3	7	53.8	13.5	16
−27.75	3.35	3.35	1	9.1	4.5	4				8.5	4.9	3
−28.25	0.91	0.53	3									
−28.75	0.11	0.11	1	0.26	0.19	2				2.9	2.1	2
−29.25	0.11	0.11	1				0.31	0.22	2			
−29.75	0.11	0.11	1							0.15	0.10	2
$\langle V/V_{\max}\rangle$	0.56 ± 0.032			0.55 ± 0.026			0.55 ± 0.031			0.52 ± 0.028		
N_{total}	92			116			88			103		
$\int_{-\infty}^{-26}\Phi(M)dM$	81 ± 14			160 ± 26			334 ± 68			333 ± 59		
$\int_{-\infty}^{-24}\Phi(M)dM$	1421 ± 236			1969 ± 291			3973 ± 841			9496 ± 3598		

Table 5 Luminosity functions for $z \leq 2.2$—(QSOs Gpc^{-3} mag^{-1}), $H_0 = 50$ km s^{-1} Mpc^{-1}, $q_0 = 0.10$, $\alpha = -1$

	0.16 < z < 0.4			0.4 < z < 0.7			0.7 < z < 1.0			1.0 < z < 1.2		
M_B	Φ	σ	n	Φ	σ	n	Φ	σ	n	Φ	σ	n
−23.25	204	59	12	511	184	8	2196	819	8			
−23.75	173	45	15	470	126	15	587	249	6	5321	3850	2
−24.25	41	13	11	192	49	16	807	201	17	1024	396	7
−24.75	17	5	10	104	25	18	454	111	17	685	252	8
−25.25	7.0	2.4	9	66	15	18	162	46	14	515	139	14
−25.75	4.0	1.2	11	27	9	10	80	19	18	326	87	14
−26.25	1.1	0.6	4	10.0	3.5	9	34	9	15	139	38	17
−26.75				0.2	0.2	1	7.6	3.8	4	33	10	11
−27.25				0.4	0.2	3	5.8	2.9	4	21	7	9
−27.75				0.1	0.1	1	0.2	0.2	1	1.9	1.9	1
−28.25							0.1	0.07	2			
−28.75										0.4	0.2	4
−29.25										0.06	0.06	1
−29.75										0.06	0.06	1
$\langle V/V_{\max} \rangle$	0.67 ± 0.031			0.60 ± 0.028			0.59 ± 0.028			0.54 ± 0.030		
N_{total}	91			114			107			91		
$\int_{-\infty}^{-26} \Phi(M) dM$	0.56 ± 0.28			50.3 ± 1.7			24 ± 5			98 ± 20		
$\int_{-\infty}^{-25} \Phi(M) dM$	6.1 ± 1.4			52 ± 9			145 ± 25			519 ± 85		

SSG1) and Schmidt et al (1986b; SSG2), although they are treated separately. Using the information given in those two papers, and an assumed equivalent width distribution, it is possible to precisely determine an equivalent accessible volume V_a, taking into account the selection effects discussed by SSG1. The contribution of each object to the luminosity function in the interval (z_1, z_2) is given by

$$\psi_j = \frac{1}{V_a^j}, \qquad 4.$$

with

$$V_a^j = \sum_k \omega_k \int_{z_1}^{z_2} \int_{W_{\min}(L_k, z, M_j)}^{\infty} \chi(W) dW \frac{dV}{dz} dz. \qquad 5.$$

Adopting notation similar to SSG1, the sum is over sky areas ω_k with corresponding limiting magnitudes L_k. $W_{\min}$ is the minimum equivalent width (W) that would be included in the sample; its value can be calculated, for a given z and L, from equations given by SSG1. The equivalent width distribution $\chi(W)$ is assumed to be known and normalized to unity. If the

Table 5 (*continued*)

	$1.2 < z < 1.4$			$1.4 < z < 1.7$			$1.7 < z < 1.9$			$1.9 < z < 2.2$		
M_B	Φ	σ	n	Φ	σ	n	Φ	σ	n	Φ	σ	n
−23.25	3197	3197	1									
−23.75				3776	2672	2				9449	9449	1
−24.25				1637	1637	1				2142	2142	1
−24.75	516	231	5	214	214	1				1650	1650	1
−25.25	406	157	7	327	148	5	735	464	3	1366	1366	1
−25.75	461	121	15	740	185	17	712	237	9	2932	1534	6
−26.25	245	71	12	271	71	15	683	203	13	535	161	11
−26.75	115	31	17	204	48	18	392	97	17	425	107	17
−27.25	44	10	18	108	22	30	184	53	12	174	45	15
−27.75	14	5	7	24	6	16	91	24	17	113	31	14
−28.25	6.5	3.3	4	9.6	3.4	8	27	7	15	49	11	24
−28.75	1.6	1.6	1	3.8	1.9	4				9.8	3.3	9
−29.25	0.5	0.3	3							0.9	0.9	1
−29.75	0.1	0.07	2	0.12	0.09	2						
$\langle V/V_{max} \rangle$	0.58 ± 0.030			0.56 ± 0.026			0.55 ± 0.031			0.53 ± 0.028		
N_{total}	95			119			88			105		
$\int_{-\infty}^{-26} \Phi(M) dM$	213 ± 39			310 ± 45			689 ± 116			654 ± 101		
$\int_{-\infty}^{-25} \Phi(M) dM$	646 ± 107			844 ± 127			1412 ± 285			2803 ± 1032		

W distribution is correctly specified, this gives the exact weight for each object (within the usual assumptions of the $1/V_a$ method). We have used an equivalent width distribution similar to that given by SSG1 but have taken into account data from radio and other optical samples, assuming a redshift-independent distribution. The Baldwin effect, which has been ignored in this calculation, could be accounted for explicitly by the use of a W distribution appropriate for the luminosity of each object.

Work by Warren & Hewett (1990) and Warren et al (1988) provide additional independent information on the high-redshift luminosity function. Results from the above sources, including SSG1 and SSG2, are given in Table 6. Table 7 gives results for $q_0 = 0.10$, $\alpha = -1$. Figure 3 shows the LFs for three redshift slices.

3.2.2.1 $q_0 = 0.5$, $\alpha = -0.5$ It appears, from a careful inspection of Figure 3a, that PLE does not describe the behavior of the LF at high redshifts. The density of quasars at $z > 2.5$ is shown as computed from the combined sample compiled here and by Warren & Hewett (1990). Both data sets show that the space density of the bright ($M < -26$) QSOs declines compared with the density at $z \sim 2$ (for the assumed parameters

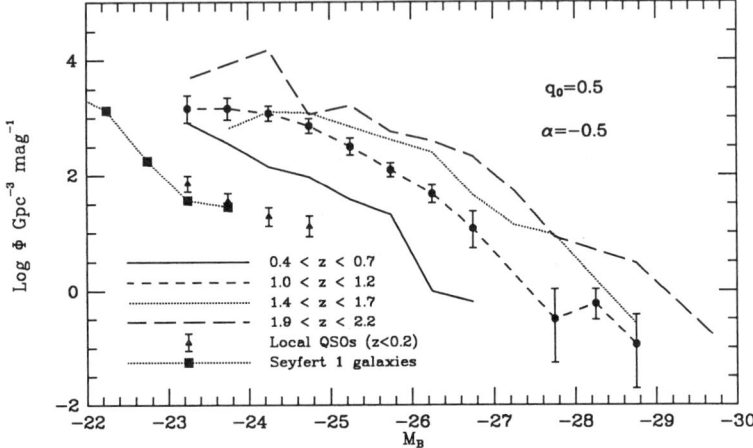

Figure 2 The luminosity functions for $z \leq 2.2$ for four redshift intervals. Error bars are included for $1.0 < z < 1.2$. (Errors for the other intervals appear in Table 4.) The *square symbols* connected by *dotted lines* show the Seyfert 1 data of J. Huchra & R. Burg (private communication, 1989). The *triangles* represent the local space density of quasars (predominantly from the BQS sample). There is reasonable agreement between the local quasar and Seyfert luminosity functions. The calculations have been made assuming $q_0 = 0.5$, $H_0 = 50$ km s^{-1} Mpc^{-1}, and power-law spectral index $\alpha = -0.5$.

q_0 and α). This result is supported by the SSG1 and SSG2 data. Thus, the decline in space density at high redshift suggested by Sandage (1972) and Osmer (1982) seems to be confirmed. Recent work by Green (1989) and Schade (1988) agrees that the space density of bright quasars declines at high redshift, and Schmidt et al (1988) estimate that the density of quasars with Lα luminosity greater than 10^{45} ergs s^{-1} (corresponding to $M_B \sim -25.8$) declines by a factor of 7 between redshifts 2.2 and 3.3, with the peak density occurring around $z = 2$ or 2.5. Warren et al (1988) find a decrease in density at $M_B = -26$ of a factor of 10 from $z = 2$ to $z = 4$. The present sample shows a decrease of a factor of 3 from the redshift slice centered at $z = 2.35$ to that centered at $z = 2.9$. Figure 4 shows the integrated space density of quasars with $M_B < -26$. This behavior is similar to that found for flat-spectrum compact radio sources by Peacock & Miller (1988). It is not clear where the peak density of quasars occurs. There is a conflict at $z \sim 3$ between the Warren & Hewett (1990) results and those obtained from the present combined sample (see Table 6). The present sample gives a lower density than the Warren & Hewett model.

Although all of the sources mentioned above present a consistent picture of a declining integrated space density of the brightest quasars, it is useful

Table 6 Luminosity functions for $z > 2.2$ —(QSOs Gpc^{-3} mag^{-1}), $H_0 = 50$ km s^{-1} Mpc^{-1}, $q_0 = 0.5$, $\alpha = -0.5$

M_B	2.2 < z < 2.5			2.2 < z < 2.5[a]			2.5 < z < 3.3			2.5 < z < 3.3[a]			z = 3[b]	z = 4[b]
	Φ	σ	n	Φ	σ	n	Φ	σ	n	Φ	σ	n	Φ	Φ
−23.25	4912	4912	1											
−23.75							8848	8848	1					
−24.25				5541	5541	1	1803	1803	1	15908	15908	1		
−24.75	2063	1193	3				1418	1004	2					
−25.25	2094	980	6				571	404	2					
−25.75	269	121	5	1059	1059	1	274	150	4	196	196	1	663	110
−26.25	241	108	5				127	48	7				440	73
−26.75	463	138	12	439	439	1	94	38	6	81	81	1	262	43
−27.25	69	24	10				20	12	3				120	20
−27.75	12.0	6.9	3	50	50	1	27	8	13	21	21	1	36	6
−28.25							7.8	3.5	5				8	1.3
−28.75							1.6	1.6	1				1.5	0.25
$\langle V/V_{\max} \rangle$	0.52 ± 0.04						0.41 ± 0.04							
N_{total}	45						45							
$\int_{-\infty}^{-26} \Phi(M)dM$	393 ± 88						139 ± 32						438	72
$\int_{-\infty}^{-25} \Phi(M)dM$	1574 ± 502						561 ± 218							
$\int_{-\infty}^{-24} \Phi(M)dM$	2606 ± 779						2172 ± 1055							

[a] Values from SSG1, SSG2 samples.
[b] Values computed from Warren & Hewett (1990) model.

Table 7 Luminosity functions for $z > 2.2$—(QSOs Gpc^{-3} mag^{-1}), $H_0 = 50$ km s^{-1} Mpc^{-1}, $q_0 = 0.10$, $\alpha = -1$

M_B	$2.2 < z < 2.5$			$2.2 < z < 2.5^a$			$2.5 < z < 3.3$			$2.5 < z < 3.3^a$		
	Φ	σ	n	Φ	σ	n	Φ	σ	n	Φ	σ	n
-24.25										5705	5705	1
-24.75	1567	1567	1				2934	2934	1			
-25.25				1696	1696	1						
-25.75							368	260	2			
-26.25	722	419	3	332	332	1	504	504	1	55	55	1
-26.75	557	261	6				466	273	3			
-27.25	85	38	5	140	140	1	133	90	3	23	23	1
-27.75	77	34	5				57	23	7			
-28.25	135	40	13				26	10	6	6	6	1
-28.75	17	6	9	16	16	1	13	7	4			
-29.25	4	2	3				8.3	2.5	11			
-29.75							3.6	1.3	8			
$\langle V/V_{max} \rangle$	0.54 ± 0.04						0.42 ± 0.04					
N_{total}	45						45					
$\int_{-\infty}^{-26} \Phi(M)dM$	799 ± 249						606 ± 291					

[a] Values from SSG1 and SSG2 samples.

to construct a formal statistical significance for this result. The only such result we can provide, at present, is through an examination of the distribution of the V/V_{max} statistics for the sample that has been compiled here. As discussed above, we expect the distribution to have a mean value of $0.5 [\pm (12n)^{-1/2}]$. Furthermore, a Kolmogorov-Smirnov (K-S) test comparing the observed distribution to the uniform distribution can be done and ought to be more sensitive than an examination of the mean value. It is found, using a single slice of space from $z = 2.2$ to $z = 3.3$ and only those objects brighter than $M_B = -26$, that the distribution of V/V_{max} differs from uniform (and thus indicates a change in integrated density) at roughly the 10% significance level; this is not a very strong result. The mean value found is $\langle V/V_{max} \rangle = 0.43 \pm 0.036$, again not an overwhelmingly positive indication of a decline in density. Although these numbers are unconvincing by themselves, the fact that a decline in space density is found from two other independent samples, each of which consists of more objects than the present sample and each of which has its own (very different) selection effects, argues strongly that a decline in space density of the most luminous objects is taking place.

The behavior of the LF at low luminosity does not necessarily mimic

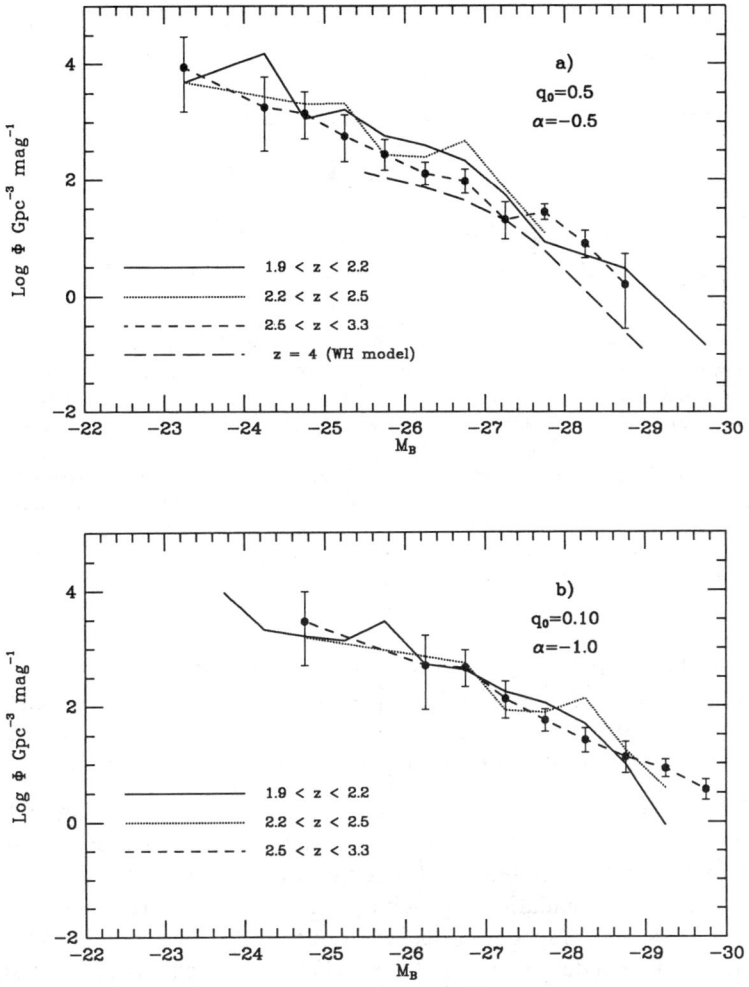

Figure 3 The high-redshift luminosity functions ($1.9 < z < 4$). (*a*) $q_0 = 0.5$. The $z = 4$ luminosity function is from the Warren & Hewett (1990) model. Error bars are shown for the interval $2.5 < z < 3.3$. A decline in density is evident at high redshift. (*b*) $q_0 = 0.10$ and $\alpha = -1$. There is no indication from the present sample that the luminosity function declines toward higher redshift for these assumed parameters.

the bright end. The density of the very faintest objects ($M_B \sim -23$) at $z \sim 3$ is consistent with its being as high or higher than at $z \sim 2$: There is no indication in the present data of a decline in density. Admittedly, there is considerable uncertainty in this result, as an examination of Tables 6

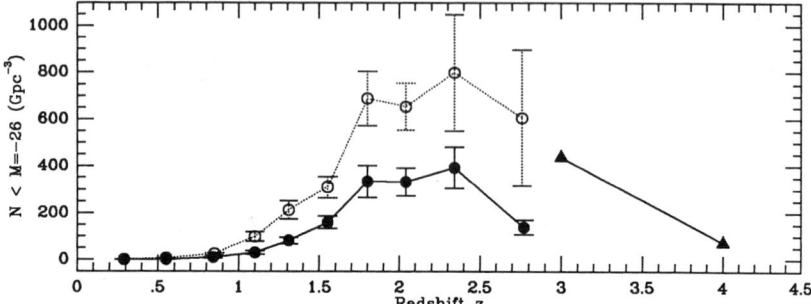

Figure 4 The integrated space density of luminous ($M_B < -26$) quasars versus redshift. We show results derived from the sample compiled in Section 2 for two parameter sets: $q_0 = 0.5$ and $\alpha = -0.5$ (*filled circles*), and $q_0 = 0.10$ and $\alpha = -1$ (*open circles*). The *triangles* are from the Warren & Hewett (1990) model, which assumes $q_0 = 0.5$ and $\alpha = -0.5$. There is a clear increase in density from the present ($z = 0$) up to $z \sim 2$, independent of cosmology or spectral index. There evidently exists, for $q_0 = 0.5$ and $\alpha = -0.5$, a drop in the space density somewhere between $z = 2$ and $z = 4$. For $q_0 = 0.10$ and $\alpha = -1$ there is no suggestion of such a decline up to $z \sim 3$ (the limit of the sample compiled here). Note that the existence of a decline depends fundamentally on the choice of assumed spectral index α and depends very little on cosmology. See Section 3.2 for a fuller discussion.

and 7 will confirm. The best estimate that can be made of the integrated space density of low-luminosity quasars (to $M_B = -24$) suggests a decrease similar to that for the most luminous quasars, but this result is less convincing than it is for the bright objects.

3.2.2.2 $q_0 = 0.10, \alpha = -1$ An examination of Figure 3b, Figure 4, or the data in Tables 5 and 7 shows no evidence that the space density of quasars declines at high redshift (assuming these values of the parameters q_0 and α). The calculation by Schmidt et al (1988) referred to above (showing a decline in space density) was done for $q_0 = 0.5$, but they suggest that the decline in density at $z > 2$ would be even more marked for smaller values of q_0. Similarly, Warren & Hewett (1990), noting the aparent drop in density (at $M_B \sim -26$), claim that the space density behaves much the same for $q_0 = 0$ as it does for $q_0 = 0.5$. It is unfortunate that hard numbers are not available to compare with the results obtained from the present sample. An examination of the distribution of the $V/V_{\max}$ statistics ($2.2 < z < 3.3$ and $M_B < -26$) for the present sample shows that the K-S test fails to support a decline at even the 20% significance level and $\langle V/V_{\max} \rangle = 0.47 \pm 0.032$. There is thus a conflict that we are unable to resolve with the presently available information. In particular, although both Schmidt et al (1988) and Warren & Hewett (1990) claim a decrease in density independent of the assumed cosmology, it is not explicitly stated that the effects of using a different spectral index have been accounted for. This is an important point, since the "correct" choice of spectral index α used to make the K-correction is subject to considerable uncertainty (see, e.g., Elvis 1989).

3.2.2.3 *Summary* The evolution of the luminosity function for $z < 2.2$ can be described in rather simple terms. The LF itself shifts systematically toward higher luminosity with little change in shape from $z = 0$ up to $z \sim 2$. This result, which is independent of the choice of q_0 and α, conforms to the pure luminosity evolution scenario, although it is not possible to decide whether this trend reflects the behavior of individual objects or simply the statistical properties of the ensemble. At higher redshift the evolution is less clear. We know from the analysis of recent data that were not available to us (i.e. Warren & Hewett 1990) that by $z = 4$ the comoving space density of luminous ($M \sim -26$) quasars has dropped by a factor of about 10 from its value at $z = 2$. A question that the data compiled here does allow us to address is what happens at $2 < z < 3.3$. The answer is moderately dependent on the value of spectral index adopted and relatively independent of cosmology. For $q_0 = 0.5$ and $\alpha = -0.5$ a decrease in integrated density of the bright objects ($M < -26$) by a factor of 3 from $z = 2$ to $z \sim 3$ is observed, whereas for $q_0 = 0.10$ and $\alpha = -1$ we see no decline at all (see Figure 4). Koo & Kron (1988) found no evidence for a redshift cutoff. Schmidt et al (1988), in a preliminary analysis, report a decrease of a factor of 7 from $z = 2.2$ to $z = 3.3$ in the integrated density of $M < -25.9$ objects. [A private communication (1989) from M. Schmidt indicates that more recent results suggest that the decline may be more gradual than this.]

We conclude that the comoving space density of luminous quasars has a relatively flat maximum at $2 < z < 3$, and that the decline becomes more pronounced at $z > 3$.

Presently no quasars are known beyond $z = 4.73$ (Schneider et al 1989b). As it now stands, we may be seeing the birth of quasars beginning when the Universe was $\sim 10\%$ of its present age, the peak in space density at a look-back time of $\sim 25\%$ of the age of the Universe, and a gradual decline in space density to the present.

4. QUASAR ASSOCIATIONS

Historically, interest in quasar-galaxy associations was directed toward establishing the cosmological nature of quasar redshifts. The first observational evidence that the redshifts are cosmological (as is assumed in this review) was based on the continuity of certain radio properties between quasars and radio galaxies (Heeschen 1966, Ryle 1968). The evidence that has accumulated since then makes the case in favor of cosmological redshifts seem strong to most astronomers (see Weedman 1986) but not to all (see Arp 1987). At the present time, therefore, the emphasis of work on quasar-galaxy associations has shifted toward studying the environments of quasars in order to learn more about the quasar phenomenon itself, as well as toward studying other unrelated physical phenomena, such as gravitational lensing. Quasar-quasar associations might be expected to provide an important observational link in the evolution of structure in the Universe between the apparently smooth state at $z \sim 1000$ (as revealed from observations of the cosmic microwave background) to the rich variety of structures that we observe around us today.

This subject would appear to be on a much more solid footing now that several relatively large homogeneous surveys are available for statistical analysis. This is in contrast to the situation as it stood in the late 1970s [see Wills (1978) for an excellent review].

4.1 Quantitative Measures of Association or Clustering

A convenient statistic to quantify the tendency of a sample of objects to cluster is the two-point correlation function (cf. Peebles 1980). The angular function $w(\theta)$ measures the ratio of the excess number of objects within $d\theta$ at angle θ away from a specified center to the expected number within the same solid angle, assuming that the objects are randomly distributed. The spatial function $\xi(r)$ is defined in an analogous manner, except that r represents three-dimensional separation. For galaxies it is found that both functions take the form of power laws—i.e. $w_{gg} \propto \theta^{1-\gamma}$ and $\xi_{gg} \propto r^{-\gamma}$, where $\gamma = 1.77$ (Groth & Peebles 1977). The auto-correlation function $w_{xx}(\theta)$ is computed using each object x in turn as center, while the cross-correlation function $w_{xy}(\theta)$ uses objects y as center but measures the relative excess of objects x at distance θ away. Other methods for detecting clustering that have been applied to the problems discussed below include the nearest neighbor test (cf. Osmer 1981) and power spectrum analysis (PSA; Webster 1976, 1982). A recent discussion of these and other statistical tests applied to the spatial clustering of quasars can be found in Gosset et al (1988).

4.2 Quasar-Galaxy Correlations

The first strong evidence for a galaxy being associated with a quasar was found by Gunn (1971). Later, the first comprehensive survey (i.e. obtaining redshifts for all galaxies brighter than the sky survey limit within 45" of a sample of 27 quasars) was carried out by Stockton (1978), who found roughly one half of the 29 galaxies satisfying the survey criteria had redshifts within 1000 km s^{-1} of the quasar in the field. French & Gunn (1983), on the basis of galaxy counts around 25 low-redshift quasars, found a strong tendency for the faintest galaxies to cluster around the quasars, a result in qualitative agreement with that of Stockton (1978) but with a lower amplitude and over larger areas. These surveys and results from other investigators (cf. Wyckoff et al 1981, Stockton 1982, Heckman et al 1984, Hintzen 1984, Hutchings et al 1984, Yee & Green 1984, Gehren et al 1984) firmly established the tendency for low-redshift quasars to be associated with loose aggregates of galaxies.

As mentioned above, quantification of quasar-galaxy associations is conveniently expressed using the cross-correlation functions $w_{gq}(\theta)$ or $\xi_{gq}(r)$ or their amplitudes A and B, where $w_{gq}(\theta) = A_{gq}\theta^{1-\gamma}$ and $\xi_{gq}(r) = B_{gq}r^{-\gamma}$. Yee & Green (1984, 1987) have evaluated B_{gq} using the projection procedure of Longair & Seldner (1979), which involves knowledge of the galaxy luminosity function and its evolution. They normalize B_{gq} to the present galaxy auto-correlation amplitude B_{gg}. Yee & Green (1987) find (a) evolution of the galaxy luminosity function (a brightening of 0.90 ± 0.5 mag at $z \sim 0.6$), (b) the power-law slope of the quasar-galaxy relation to be the same as the galaxy-galaxy function slope, and (c) between $z = 0.3$ and $z = 0.6$ an increase by a factor of 3 in the cross-correlation amplitude for radio-loud quasars that apparently does not occur for optically selected objects (Boyle et al 1988c). A summary of the determination of $\langle B_{gq}/B_{gg} \rangle$ for the Yee & Green surveys as well as for the earlier surveys of Stockton (1978) and French & Gunn (1983) is given in Table 8. The later work

Table 8 Spatial quasar-galaxy cross-correlation amplitudes

Source	Radio/optical (R/O)	$\bar{z}$ or Δz	$\left\langle \dfrac{B_{gq}}{B_{gg}} \right\rangle$	n
Stockton 1978	R+O	0.33	6 ± 3[a]	27
	—	—	4 ± 2[b]	—
French & Gunn 1983	R+O	0.24	1.2 ± 0.50	25
	O	0.23	1.7 ± 0.8	11
	R	0.26	0.9 ± 0.6	14
Yee & Green 1987	O	0–0.15	1.9 ± 0.6	17
	O	0.15–0.3	2.1 ± 0.7	16
	O	0.3–0.5	1.2 ± 0.7	7
	R	0.15–0.30	3.1 ± 1.1	9
	R	0.3–0.5	2.3 ± 1.1	10
	R	0.55–0.65	8.0 ± 2.1	9
Boyle et al 1988c	O	0.55–0.65	2.3 ± 1.3	8

[a] From analysis by Weymann et al 1978.
[b] From analysis by French & Gunn 1983.

reinforces the conclusions of the earlier studies that low-redshift ($z < 0.5$) quasars do associate with galaxies, but that the galaxy environment is not rich. By $z \sim 0.6$, radio-loud quasars are found in much richer environments (approaching Abell richness class 1 clusters), whereas the optically selected quasar environment does not show a similar change near the quasar. Recent work reported by Yee (1990) indicates that relatively high values of $\langle B_{gq}/B_{gg} \rangle$ can also occur for radio-selected quasars with $z < 0.5$ [see Ellingson et al (1989) for an example of a radio-loud quasar with $z = 0.198$ in an Abell richness class 1 cluster]. Furthermore, while $\langle B_{gq}/B_{gg} \rangle$ for high-redshift, optically selected quasars may be small when evaluated relatively near the quasar, values that are 3 times larger may occur farther away (~ 1–2 Mpc) from the quasar, prompting the suggestion that such quasars may be located near the edges of rich clusters.

Calculation of B_{gq} requires knowledge of the galaxy luminosity function and its evolution. To avoid this complication we have calculated the angular cross-correlation amplitude A_{gq} [assuming the same power-law slope $\gamma - 1 = 0.77$, which is justified from the work of Yee & Green (1987) and Tyson (1986)]. Calculating A_{gq} also allows useful comparisons with several additional investigations where B_{gq} has not or cannot be calculated (because, for example, the quasar redshift is sufficiently large that one would not expect to see the associated galaxies owing to their faintness). We have normalized A_{gq} to A_{gg}, where allowance has been made for the differing background galaxy counts by using the relation $\log A_{gg} = -0.7 \log n - 2.39$ based on the data from Stevenson et al (1985), where n is the background galaxy count in galaxies per square arcminute. Operationally, A_{gq} is obtained by integrating the expression that defines w_{gq}, which yields

$$A_{gq} = \frac{3-\gamma}{2} \theta_0^{\gamma-1} \frac{N_{obs}(<\theta_0) - N_{exp}(<\theta_0)}{N_{exp}(<\theta_0)}. \qquad 6.$$

Here, $N_{obs}(<\theta_0)$ is the observed number of galaxies within angle θ_0 of a particular quasar, and $N_{exp}(<\theta_0)$ is the expected number of galaxies based on the background surface density. The ratio $\langle A_{gq}/A_{gg} \rangle$ computed at either $\bar{z}$ or over range Δz for each of the surveys is given in Tables 9a and 9b and is shown plotted as a function of z in Figures 5a and 5b.

At low redshift ($z \lesssim 0.6$) the behavior of $\langle A_{gq}/A_{gg} \rangle$ is qualitatively similar to $\langle B_{gq}/B_{gg} \rangle$. At high redshift, where the quasars might be expected to serve as randomly placed centers, we would expect $\langle A_{gq}/A_{gg} \rangle$ to be zero. It is somewhat surprising then to find that among the radio-loud objects (Figure 5b) both the Tyson (1986) and Fugmann (1988) data are several sigmas greater than zero. These results seem especially puzzling, since Hintzen's (1984) imaging study to even fainter magnitudes ($R = 22.5$)

Table 9a Angular quasar-galaxy cross-correlation amplitudes for optically selected quasars

Survey	Plotting symbol	Δr (arcsec)	Galaxy background (galaxies min^{-2})	$\bar{z}$ or Δz	$\left\langle \frac{A_{qg}}{A_{gg}} \right\rangle$	n
Stockton 1978	S	45	0.3	0.29	1.2 ± 1.6	6
French & Gunn 1983	FG	~300	0.41	0.23	2.8 ± 1.8	11
Yee & Green 1984	YG84	~51	0.7	0.07	5.6 ± 2.1	16
Yee & Green 1984	YG84	~51	0.7	0.22	3.8 ± 1.9	12
Gehren et al 1984	GFWW	34–192	1.56	0.12	−1.7 ± 6.2	2
Tyson 1986	T	30	0.39	0.21	10.8 ± 4.5	4
Yee & Green 1987	YG87	~51	3.9	0.44	2.7 ± 1.8	7
Boyle et al 1988c	BSY	750 h^{-1} kpc		~0.6	5.2 ± 3.0[a]	8
Boyle et al 1988b	BSP	265	0.13	<2.2	−0.28 ± 0.08	447
Webster et al 1988	WHHW	6	0.28	>0.5	1.5 ± 0.6	296

[a] Scaled from $\langle B_{gq}/B_{gg} \rangle$.

Table 9b Angular quasar-galaxy cross-correlation amplitudes for radio-selected quasars

Survey	Plotting symbol	Δr (arcsec)	Galaxy background (galaxies min^{-2})	$\bar{z}$ or Δz	$\left\langle \frac{A_{qg}}{A_{gg}} \right\rangle$	n
Stockton 1978	S	45	0.3	0.34	2.5 ± 1.0	21
French & Gunn 1983	FG	≤300	0.41	0.26	1.4 ± 1.0	14
Yee & Green 1984	YG84	~51	0.7	0.22	6.5 ± 2.7	9
Gehren et al 1984	GFWW	<192	1.56	0.34	7.9 ± 2.4	13
Tyson 1986	T	30	0.39	0.28	10.0 ± 2.3	15
—	T	30	0.39	1.21	6.8 ± 1.5	23
Yee & Green 1987	YG87	~52	3.9	0.41	5.9 ± 2.8	10
—	YG87	~52	3.9	0.64	7.3 ± 2.8	11
Fugmann 1988	F	90	0.48	~0.75	12.1 ± 3.3	12
—	F	90	0.48	>1.7	10.3 ± 1.9	13

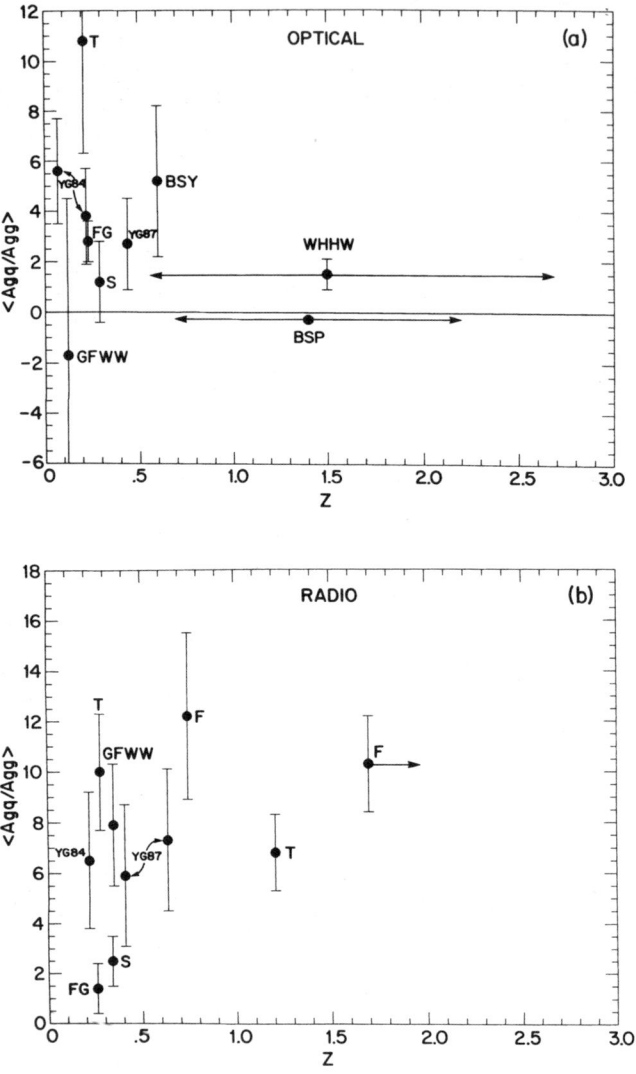

Figure 5 The angular quasar-galaxy cross-correlation amplitude, normalized to the galaxy autocorrelation amplitude for the same background galaxy density, versus redshift. (*a*) Optically selected quasars. (*b*) Radio-selected quasars. Plotting symbols are defined in Tables 9*a* and 9*b*.

showed a sharp drop in excess galaxies around five radio-loud quasars with $z > 0.65$. Furthermore, a result similar to that found by Hintzen was obtained by Yee & Green (1984) among their highest redshift objects but at a brighter limiting magnitude for the background galaxies. Among optically selected objects, Boyle et al (1988a) find a small anticorrelation (which they attribute to dust associated with the unseen cluster in which the quasar is embedded), and Webster et al (1988) find a small positive correlation [which they attribute to the effects of gravitational microlensing (Canizares 1981, Vietri & Ostriker 1983, Schneider 1986)]. According to a number of authors (cf. Kovner 1989, and references therein), the Webster et al (1988) result as it stands is too large to be explained by the lensing hypothesis. The startingly large correlations found by Tyson (1986) and Fugmann (1988) have yet to be subjected to the same scrutiny. Tyson, in fact, attributes his result to quasar-induced luminosity evolution of the surrounding galaxies. Fugmann (1988) emphasizes that his sample consists of flat-spectrum radio quasars in which, he argues, the effects of gravitational microlensing might be enhanced. Another case where gravitational microlensing may be operative comes from Stocke et al (1987). These authors showed that X-ray-selected quasars near galaxies have a different magnitude and redshift distribution than those that are not near galaxies. Rix & Hogan (1988) have discussed these observations within the gravitational-lensing context.

Is there a precedent for the Fugmann (1988) and Tyson (1986) results? Arp (1970), Burbidge et al (1971), Arp & Sulentic (1979), and Burbidge (1979) have argued that some quasars with large redshifts were associated with bright galaxies of much lower redshift. In addition, Seldner & Peebles (1979) found statistically significant evidence for a correlation between the angular positions of a list of 382 quasars and the Lick counts of galaxies. Nieto & Seldner (1982) later reexamined the Seldner & Peebles result and concluded that strong selection effects were present in the quasar list, and that if a correlation with galaxies did exist it was with respect to radio-loud quasars. Even then, the significance depended heavily on how far away from the quasar the galaxy background was measured. Chu et al (1984) performed a cross correlation between quasars in the Hewitt & Burbidge (1980) catalogue and the de Vaucouleurs et al (1976) catalogue of bright galaxies and found strong statistical evidence for association (strongest again in the case of radio-loud quasars). The early statistics referred to above has been criticized (see Wills 1978, Weedman 1986), and as the Nieto & Seldner (1982) result shows, analyses of inhomogenous catalogues can be misleading. Explaining the large correlations found by Fugmann (1988) for the highest redshift radio-selected group appears to be a problem that requires further work.

In summary, much has been learned from imaging surveys about the environments of low-redshift quasars ($z < 0.7$, or low enough that galaxies with the same redshift as the quasars are brighter than the limiting magnitude of the survey). Both the spatial and angular quasar-galaxy cross-correltion amplitudes indicate that *quasars are more likely to have companion galaxies than the average galaxy itself*. However, the galaxy environment in general is not as rich as Abell richness class 1 clusters, although some radio-selected quasars are found in such galaxy-rich environments. The situation with respect to quasar-galaxy correlations at high redshift ($z > 0.7$, or high enough that galaxies at the same redshift as the quasars would normally be fainter than the survey limit) remains unclear. In particular, among radio-selected objects, the strong correlations found by Tyson (1986) and Fugmann (1988) appear to be at odds with the results of Hintzen (1984) and Yee & Green (1984). Similarly, from optical samples, Webster et al (1988) find small positive correlations, while Boyle et al (1988a) report anticorrelations. It may be that some as yet unrecognized selection effects are responsible for these apparently contradictory results. Whatever the cause, the ambiguities will have to be resolved before the relevance of the gravitational microlensing phenomenon (cf. Schneider 1989) can be properly assessed.

4.3 *Quasar-Quasar Correlations*

Two of the earliest important papers concerned with the clustering of quasars took different approaches and arrived at different conclusions. One paper by Oort et al (1981) compiled lists of close quasar pairs gleaned from a variety of sources, including the Hewitt & Burbidge (1980) catalogue, and presented statistics evaluating the likelihood of the individual associations occurring by chance. These authors concluded that the data were compatible with clustering on scales of ~ 5–$30\ h^{-1}$ Mpc, where h represents the Hubble constant in units of 100 km s^{-1} Mpc^{-1}. In the other paper, Osmer (1981) applied two statistical tests to the two homogeneous surveys conducted at CTIO and found no evidence for clustering—a result confirmed later by Webster (1982) using a third test. These apparently contradictory conclusions persisted until the mid-1980s with the homogeneous surveys of Kunth & Sargent (1986), Clowes et al (1987), Crampton et al (1987), Drinkwater (1988), and Osmer & Hewett (1988), which all produced null results. However, analysis of catalogues (Shaver 1984, 1986b, Kruszewski 1987, Anderson et al 1988) gave a positive clustering signal for separations $< 10\ h^{-1}$ Mpc. The one homogeneous survey that did indicate the existence of clustering is the Durham UVX survey (Boyle 1986, Shanks et al 1987, 1988). One reason why the other surveys failed to show evidence for clustering involves the statistics of small numbers, as the following simple calculation shows. Following Anderson et al (1988), who (after ignoring the effects of the survey boundary, clustering, and density variations with redshift) approximated the expected number N_p of independent pairs of quasars with separation less than r as

$$N_p = \frac{N}{2}\left(\frac{4}{3}\pi r^3 \frac{N}{V}\right), \qquad 7.$$

where N is the number of quasars in the survey, and V is the volume under consideration. If one assumes a $q_0 = 1/2$ cosmology, the number of pairs with separation less than $10\ h^{-1}$ Mpc within $0 < z < 2.2$ is $N_p = 0.00111\sigma^2\Omega$, where σ is the surface density of quasars (per square degree) within this range of z, and Ω is the solid angle (in square degrees) covered by the survey. For $2.2 < z < 3.4$ the corresponding expression is $N_p = 0.00165\sigma^2\Omega$. From Section 2, $\sigma_{0-2.2}(B < 21) = 38$ and $\sigma_{2.2-3.4}(B < 21) = 8$, which gives $N_p(r < 10\ h^{-1}$ Mpc) of 1.6 deg^{-2} for $z < 2.2$ and 0.11 deg^{-2} for $z > 2.2$. Further, in order to achieve 3σ significance for $\xi = 1$, one would have to survey 11 deg^2 and 164 deg^2, respectively! For reference, the full Durham UVX survey covers 12 deg^2 and is only complete up to $z = 2.2$. It is clear from these numbers that reliable clustering results (especially at high redshifts) require an extensive effort! The situation with respect to radio-loud quasars is much more pessimistic. According to Figure 1 of Machalski (1987), the surface density of radio-loud quasars at $B \sim 21$ is down by a factor of ~ 60 compared with that for optically selected quasars with $1 < z < 2.2$.

Shaver (1984) was the first to systematically exploit for clustering analysis the large number of quasars available in a catalogue (in this case the catalogue of Veron-Cetty & Veron 1984) in a manner that is claimed to be independent of the inherent inhomogeneities in such a compilation. His method of "normalization to large scales" is based on the argument that because clustering is expected to show up among close pairs (both in projection and redshift) and not in pairs with large separations, the ratio of these two groups would signal true clustering independent of the multitude of selection effects (which should affect both groups in a similar manner). Using this method, Shaver concluded that clustering of quasars did exist with amplitude comparable to that of the present galaxy autocorrelation function. Kruszewski (1987) analyzed the same data using a different correction for inhomogeneities and confirmed Shaver's result but also found evidence for evolution with redshift—an effect hinted at in the work of Fang et al (1985). Anderson et al (1988) analyzed a later version of the Veron-Cetty & Veron (1987) catalogue using a modification of Osmer's (1981) scrambling technique for dealing with inhomogeneities and confirmed the previous results of Shaver (1984). Most recently, using the latest catalogue of Hewitt & Burbidge (1987) containing 3600 quasars and Shaver's (1984) method, Chu & Zhu (1988) find that radio-selected quasars may be more strongly clustered than optically selected objects, and that a sharp cutoff in clustering amplitude exists beyond $z = 2$ (Chu & Zhu 1989). In all of the above cases, the clustering signal is due to a relatively small number of pairs, and despite the precautions taken in allowing for effects of inhomogeneities in the catalogues, these results are still susceptible to systematic errors. For this reason, we return to a discussion of the homogeneous surveys referred to above.

In order to allow a fair comparison between the various surveys, we have computed the clustering amplitude B_{qq}, where $\xi(R) = B_{qq}R^{-\gamma}$ in an analogous manner to the computation of A_{gq} in the previous section. This amplitude was then normalized to the present galaxy-galaxy correlation amplitude under the assumption that $B_{gg,0} = (5\,h^{-1}\,\text{Mpc})^\gamma$ and $\gamma = 1.8$, i.e.

$$\frac{B_{qq,z}}{B_{gg,0}} = \left(\frac{R_0}{5}\right)^\gamma \frac{3-\gamma}{3} \frac{N_{\text{obs}}(<R_0) - N_{\text{exp}}(<R_0)}{N_{\text{exp}}(<R_0)}, \qquad 8.$$

where R_0 is the three-dimensional radius of the sphere within which $N_{\text{obs}}(<R_0)$ quasars are observed and $N_{\text{exp}}(<R_0)$ quasars are expected if randomly distributed. Justification for the use of $\gamma = 1.8$ comes from the work of Shanks et al (1988).

It is not always obvious how best to produce the random sample on which the clustering analysis is based. Nonuniformities creep in through selection, edge, and field effects related to the shape of the sampling area and the nature of sensitivity variations within that area [cf. Osmer (1981) and Anderson et al (1988) for a discussion of scrambling algorithms]. If the selection and field effects are small, as they are likely to be in relatively small survey areas, a convenient method for allowing for edge effects is to cross correlate the sample with a much larger randomly chosen data set (i.e. randomly chosen coordinates within the survey area or areas with redshifts chosen randomly from the observed redshift distribution). The number of pairs at the specified separations[1] are then normalized to the total quasar count in the catalogue (cf. Shanks et al 1987).

---

[1] Three-dimensional separations can be computed in a $q_0 = 0.5$ cosmology as follows (cf. Osmer 1981). Suppose that quasar A has redshift z_A and lies at an angle θ away from quasar B with redshift z_B. From the comoving coordinates of each quasar, $r_A = 1 - (1+z_A)^{-1/2}$ and $r_B = 1 - (1+z_B)^{-1/2}$, the comoving coordinate separation r_{AB} is given by the plane-triangle relation, i.e. $r_{AB} = (r_A^2 + r_B^2 - 2r_A r_B \cos\theta)^{1/2}$. The real distance between A and B is then $R_{AB} = 2r_{AB}(c/H_0)$ Mpc.

The average values of $B_{qq,z}/B_{gg,0}$ for each survey are given in Table 10 and plotted in Figure 6. (Note that distances in Table 10 are calculated assuming $h = 1$.) Where practical, certain samples have been divided into two redshift ranges, and $\langle B_{qq,z}/B_{gg,0} \rangle$ is given for each range. Note that two points have been plotted for the low-redshift portion of the Crampton et al (1989) survey. This was done in order to show the effects of excluding the 23 quasars within a narrow range of $z = 1.1$ in their 13-hr field, which these authors argue may represent a large group that produces a clustering signal out to $40\ h^{-1}$ Mpc. The reason for treating these data differently here is that in some sense they are biased in that the presence of a group was suggested in the previous work (Crampton et al 1987), and in the later work an effort was made to find more group members. The problems of assessing the significance of such a group after the fact are well known, and it would seem that the true significance of this group will only be known if and when similar structures are found in later surveys. Both the table and the figure show that in any individual survey except the Durham one (SBP), the amplitude of the clustering signal is not significantly different from zero for any z. However, when the results are averaged with weights inversely proportional to the square of the errors, the result at $z \sim 1$ is significant at greater than 3–4σ, whereas at $z \sim 2$ the average differs from zero at only 2σ. Similar conclusions have been reached by Kruszewski (1988) and Iovino & Shaver (1988). What is the situation near $z = 0$? The local density of quasars is much too low to give an answer to this queston. The low space density of the related Seyfert 1 galaxies also precludes a direct determination of their clustering properties, but a very indirect estimate can be made as follows. Hamilton (1988) has recently used the Harvard-Smithsonian Center for Astrophysics galaxy survey data base (Huchra et al 1983) to show that the most luminous spiral galaxies (i.e. those with $M_B < -22$ for $h = 0.5$) cluster more strongly than lower luminosity spirals by $3.5e^{\pm 0.4}$. According to Osterbrock (1988), one may be justified in identifying all spiral galaxies more luminous than $M \sim -22$ as Seyfert 1 galaxies. This leads to the extremely tentative result that $\langle B_{ss}/B_{gg} \rangle = 3.5e^{\pm 0.4}$. The symbol (S) denotes this point in Figure 6 near $z = 0$. Note that this point is ~ 0.5 times the amplitude found for radio galaxies (denoted RG in Figure 6) by Peacock & Miller (1988). How does B_{gg} evolve? For illustrative purposes, we show a curve in Figure 6 based on the stable clustering picture, i.e. $B_{qq} \propto (1+z)^{-1.2}$. [In this highly simplified picture, bound structures that include quasars are formed early and the clustering strength increases with time as the rest of the Universe expands away from these structures (cf. Phillipps et al 1978).] Cole & Kaiser (1989) have discussed the predictions for quasar clustering in the cold dark matter (CDM) model and conclude that the present observations are not incon-

Table 10 Spatial quasar autocorrelation amplitudes

Source	Plotting symbol	No. of QSOs	B_{lim}	Δz	Cosmology (q_0/H_0)	R (Mpc)	$\xi(<R)$	$\left\langle \dfrac{B_{qq,z}}{B_{gg,0}} \right\rangle$
Osmer 1981	O	71	~20.9	0.3–3.5	0/50	30 (9.4)[a]	1.2±1.2	1.5±1.5
Shanks et al 1988		392	~21	0.3–2.2	0.5/100	10	1.2±0.4	1.7±0.6
	SBP		~21	0.3–1.4	0.5/100	10	1.3±0.4	1.8±0.5
	SBP		~21	1.4–2.2	0.5/100	10	1.0±0.4	1.4±0.6
Osmer & Hewett 1988	OH	127	≥20	0.3–3.4	0.5/100	10	−0.1±0.6	−0.1±0.8
Drinkwater 1988	D	862	~20	1.8–2.6	0/100	25 (15.7)[a]	0.13±0.25	0.41±0.80
Crampton et al 1989		270	~20.3	0.3–3.2	0.5/100	10	0.68±0.53	0.95±0.74
(with group)	CCHg	145	~20.3	0.3–1.5	0.5/100	10	1.5±0.81	2.2±1.1
(without group)	CCHng	122	~20.3	0.3–1.5	0.5/100	10	0.43±0.71	0.60±0.99
	CCH	125	~20.3	1.5–3.2	0.5/100	20	0.07±0.26	0.36±1.26
Average (low z)	ALg				0.5/100			1.9±0.5
Average (without group)	ALng				0.5/100			1.6±0.4
Average (high z)	AH				0.5/100			0.8±0.4

[a] Adjusted to 0.5/100 cosmology.

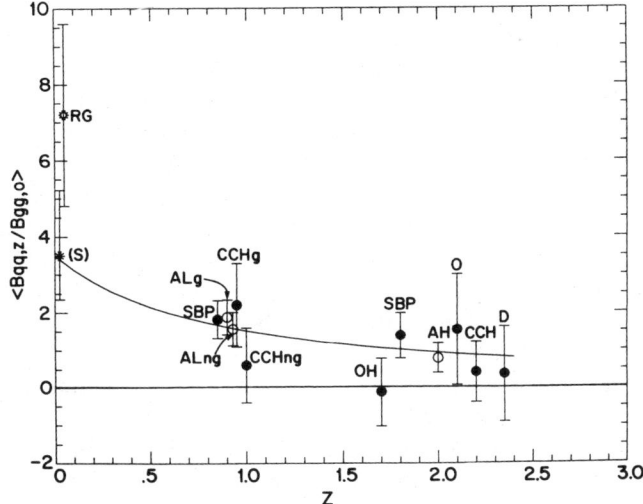

Figure 6 The normalized quasar-clustering amplitude as a function of redshift. RG is the result for radio galaxies from Peacock & Miller (1988) and (S) is an estimate for Seyfert 1 galaxies based on the work of Hamilton (1988). The other symbols are defined in Table 10. The *curved line* is an example of one possible evolutionary model, i.e. $B_{qq} \propto (1+z)^{-1.2}$.

sistent with the CDM picture. Clearly, the large observational uncertainties prevent unambiguous conclusions regarding the evolution of clustering to be drawn at this time.

Finally, are close quasar pairs with very different redshifts associated, as Burbidge et al (1985), following earlier work, argue? Shaver (1985, 1986a) counters that they are not by plotting the number of pairs of given separation versus separation from several (optically selected) surveys and finding that the distributions show no evidence for excess pairs over those expected assuming a random distribution.

In summary, the present evidence for quasar clustering comes mainly from the Durham survey, although the large uncertainties associated with the negative results from other surveys do not preclude a clustering strength of 1–2 times that of galaxies today for $z < 2$.

5. CONCLUSIONS AND OUTLOOK

Our present knowledge of the quasar luminosity function for $z < 2.2$ would appear to be quite secure. Furthermore, the statistics of the luminosity function for $z > 3$ are improving rapidly due to several large surveys currently underway [see Foltz & Osmer (1988) for a summary of some ongoing surveys]. Deep surveys are required in order to determine whether the decline in space density at $z > 3$ observed for the most luminous quasars also applies to the intrinsically faintest objects. With the luminosity functions becoming well defined, it might be expected that future work will concentrate on explaining the observed evolution in terms of physical models, such as the black hole accretion model.

The situation with respect to the clustering of quasars is, unfortunately, not so far advanced. The limiting factor would appear to be small numbers. Presently there is a very strong indication that the clustering strength at $z \sim 1$ is 1–2 times that of galaxies today, primarily on the basis of the Durham UVX survey. Large deep surveys are needed to strengthen this result and to clearly define the evolution of quasar clustering. The related problem of the existence of quasar groups is also in its infancy.

Finally, it can be expected that another focus of future work will be on determining the relationship between the classical (naked?) quasars discussed in this review and their less luminous and/or obscured relatives (see Section 1.2).

ACKNOWLEDGMENTS

We would like to thank David Crampton and Anne Cowley for sharing their quasar observations with us and for their help with this review. We are grateful to David Koo and Allan Sandage for many useful comments on an earlier version of this paper. We also wish to thank John Huchra for sending us his unpublished data on the Seyfert galaxy luminosity function. We acknowledge financial support from the Natural Sciences and Engineering Research Council of Canada in the form of Operating Grants (FDAH) and a Post-Graduate Scholarship (DS).

Literature Cited

Anderson, N., Kunth, D., Sargent, W. L. W. 1988. *Astron. J.* 95: 644
Arp, H. C. 1970. *Astron. J.* 75: 1
Arp, H. 1984. *Ap. J. Lett.* 227: L27
Arp, H. 1987. *Quasars, Redshifts and Controversies.* Berkeley, Calif: Interstellar Media
Arp, H., Sulentic, J. W. 1979. *Ap. J.* 229: 496
Avni, Y., Bahcall, J. N. 1980. *Ap. J.* 235: 694
Bahcall, N. A. 1988. *Annu. Rev. Astron. Astrophys.* 26: 631
Bajtlik, S., Duncan, R. C., Ostriker, J. P. 1988. *Ap. J.* 327: 570.
Baldwin, J. A. 1977. *Ap. J.* 214: 679
Baldwin, J. A. 1988. See Osmer et al 1988b, p. 346
Baldwin, J. A., Wampler, E. J., Gaskell, C. M. 1989. *Ap. J.* 338: 630

Barbieri, C., Cappellaro, E., Romano, G., Turatto, M., Szuszkiewicz, E. 1988. *Astron. Astrophys. Suppl.* 76: 477
Bechtold, J., Weymann, R. J., Lin, Z., Malkan, M. A. 1987. *Ap. J.* 315: 180
Blandford, R. D. 1986. See Swarup & Kapahi 1986, p. 359
Borra, E. F., Edwards, G., Petrucci, F., Beauchemin, M., Brousseau, D., et al. 1987. *Publ. Astron. Soc. Pac.* 99: 535
Boyle, B. J. 1986. PhD thesis. Univ. Durham, Engl.
Boyle, B. J., Fong, R., Shanks, T. 1988a. *MNRAS* 231: 897
Boyle, B. J., Fong, R., Shanks, T., Peterson, B. A. 1987. *MNRAS* 227: 717 (BFSP)
Boyle, B. J., Fong, R., Shanks, T., Peterson, B. A. 1990. *MNRAS* 243: 1

Boyle, B. J., Shanks, T., Peterson, B. A. 1988b. *MNRAS* 235: 935
Boyle, B. J., Shanks, T., Yee, H. K. C. 1988c. In *Large Scale Structures of the Universe, IAU Symp. No. 130*, ed. J. Audouze, M.-C. Pelletan, A. Szalay, p. 576. Dordrecht: Kluwer
Braccesi, A., Zitelli, V., Bonoli, F., Formiggini, L. 1980. *Astron. Astrophys.* 85: 80
Burbidge, E. M., Burbidge, G. R., Solomon, P. M., Strittmatter, P. A. 1971. *Ap. J.* 170: 233
Burbidge, G. 1979. *Nature* 282: 451
Burbidge, G. R., Narlikar, J. V., Hewitt, A. 1985. *Nature* 317: 413
Burstein, D., Heiles, C. 1982. *Astron. J.* 87: 1165
Canizares, C. R. 1981. *Nature* 291: 620
Canizares, C. R., White, J. L. 1989. *Ap. J.* 339: 27
Carlberg, R. G. 1990. *Ap. J.* 350: 505
Cavaliere, A., Giallongo, E., Vagnetti, F. 1989. *Astron. J.* 97: 336
Cavaliere, A., Padovani, P. 1989. *Ap. J. Lett.* 340: L5
Choloniewski, J. 1987. *MNRAS* 226: 273
Chu, Y., Zhu, X. 1988. *Astron. Astrophys.* 205: 1
Chu, Y., Zhu, X. 1989. *Astron. Astrophys.* 215: 14
Chu, Y., Zhu, X., Burbidge, G., Hewitt, A. 1984. *Astron. Astrophys.* 138: 408
Clowes, R. G. 1981. *MNRAS* 197: 731
Clowes, R. G., Iovino, A., Shaver, P. 1987. *MNRAS* 227: 921
Cole, S., Kaiser, N. 1989. *MNRAS* 237: 1127
Crampton, D., Cowley, A. P., Hartwick, F. D. A. 1987. *Ap. J.* 314: 129
Crampton, D., Cowley, A. P., Hartwick, F. D. A. 1989. *Ap. J.* 345: 59 (CFHT)
Crampton, D., Rensing, M. 1982. *Publ. Astron. Soc. Pac.* 94: 440
Crampton, D., Schade, D. J., Cowley, A. P. 1985. *Astron. J.* 90: 987
Cristiani, S., Barbieri, C., Iovino, A., LaFranca, F., Nota, A. 1989. *Astron. Astrophys. Suppl.* 77: 161
De Robertis, M. M. 1985. *Astron. J.* 90: 998
de Vaucouleurs, G. H., de Vaucouleurs, A. P., Corwin, H. G. 1976. *Second Reference Catalogue of Bright Galaxies*. Austin: Univ. Tex. Press
Drinkwater, M. 1988. *MNRAS* 235: 1111
Eddington, A. S. 1940. *MNRAS* 100: 354
Efstathiou, G., Rees, M. J. 1988. *MNRAS* 230: 5P
Ellingson, E., Yee, H. K. C., Green, R. F., Kinman, T. D. 1989. *Astron. J.* 97: 1539
Elvis, M. 1989. *Comments Astrophys.* 14: 137
Fabian, A. C., Canizares, C. R., Barcons, X. 1989. *MNRAS* 239: 15P
Fall, S. M., Pei, Y. C. 1989. *Ap. J.* 337: 7
Fang, L. Z., Chu, Y. Q., Zhu, X. F. 1985. *Astrophys. Space Sci.* 115: 99
Filippenko, A. V., Sargent, W. L. W. 1989. *Ap. J. Lett.* 342: L11
Foltz, C. B., Chaffee, F. H., Hewett, P. C., MacAlpine, G. M., Turnshek, D. A., et al. 1987. *Astron. J.* 94: 1423 (APM1)
Foltz, C. B., Chaffee, F. H., Hewett, P. C., Weymann, R. J., Anderson, S. F., MacAlpine, G. M. 1989. *Astron. J.* 98: 1959 (APM2)
Foltz, C. B., Osmer, P. S. 1988. See Osmer et al 1988b, p. 361
French, H. B., Gunn, J. E. 1983. *Ap. J.* 269: 29
Fugmann, W. 1988. *Astron. Astrophys.* 204: 73
Gehrels, N. 1986. *Ap. J.* 303: 336
Gehren, T., Fried, J., Wehinger, P. A., Wyckoff, S. 1984. *Ap. J.* 278: 11
Gioia, I. M., Maccacaro, T., Schild, R., Stocke, J. T., Liebert, J. W., et al. 1984. *Ap. J.* 283: 495
Gosset, E., Surdej, J., Swings, J. P. 1988. See Osmer et al 1988b, p. 281
Gratton, R. G., Osmer, P. S. 1987. *Publ. Astron. Soc. Pac.* 99: 899
Green, R. 1986. See Swarup & Kapahi 1986, p. 429
Green, R. F. 1989. In *The Epoch of Galaxy Formation*, ed. C. Frenk, R. Ellis, T. Shanks, A. Heavens, J. Peacock, p. 121. Dordrecht: Kluwer
Green, R., Schmidt, M., Liebert, J. 1986. *Ap. J. Suppl.* 61: 305
Grewing, M. 1967. *Ap. J.* 148: 705
Groth, E. J., Peebles, P. J. E. 1977. *Ap. J.* 217: 385
Gunn, J. E. 1971. *Ap. J. Lett.* 164: L113
Gunn, J. E., Peterson, B. A. 1965. *Ap. J.* 142: 1633
Hamilton, A. J. S. 1988. *Ap. J. Lett.* 331: L59
Haro, G. 1956. *Bol. Obs. Tonantzintla y Tacubaya* 2(14): 16
Haro, G., Luyten, W. J. 1962. *Bol. Obs. Tonantzintla y Tacubaya* 3(22): 37
Hawkins, M. R. S. 1985. *MNRAS* 216: 589
Hawkins, M. R. S. 1987. See Hewitt et al 1987, p. 691
Hazard, C., Morton, D. C., McMahon, R. G., Sargent, W. L. W., Terlovich, R. 1986. *MNRAS* 223: 87
Heckman, T. M., Bothun, G. D., Balick, B., Smith, E. P. 1984. *Astron. J.* 89: 958
Heeschen, D. S. 1966. *Ap. J.* 146: 517
Heisler, J., Ostriker, J. P. 1988. *Ap. J.* 332: 543
Hewitt, A., Burbidge, G. 1980. *Ap. J. Suppl.* 43: 57
Hewitt, A., Burbidge, G. 1987. *Ap. J. Suppl.* 63: 1
Hewitt, A., Burbidge, G., Fang, L. Z., eds.

1987. *Observational Cosmology*. *Proc. IAU Symp. No.* 124. Dordrecht: Reidel
Hintzen, P. 1984. *Ap. J. Suppl.* 55: 533
Hoag, A. A. 1976. *Publ. Astron. Soc. Pac.* 88: 860
Hoag, A. A., Schroeder, D. J. 1970. *Publ. Astron. Soc. Pac.* 82: 1141
Hoag, A. A., Smith, M. 1977. *Ap. J.* 217: 362
Huchra, J., Davis, M., Latham, D., Tonry, J. 1983. *Ap. J. Suppl.* 52: 89
Hutchings, J. B., Crampton, D., Campbell, B. 1984. *Ap. J.* 280: 41
Hutchings, J. B., Janson, T., Neff, S. G. 1989. *Ap. J.* 342: 660
Iovino, A., Shaver, P. A. 1988. *Ap. J. Lett.* 330: L13
Koo, D. C. 1986. In *Structure and Evolution of Active Galactic Nuclei*, ed. G. Giuricin, F. Mardirossian, M. Mezzetti, M. Ramella, p. 317. Dordrecht: Reidel
Koo, D. C., Kron, R. G. 1980. *Publ. Astron. Soc. Pac.* 92: 537
Koo, D. C., Kron, R. G. 1982. *Astron. Astrophys.* 105: 107
Koo, D. C., Kron, R. G. 1988. *Ap. J.* 325: 92 (KOKR)
Koo, D. C., Kron, R. G., Cudworth, K. M. 1986. *Publ. Astron. Soc. Pac.* 98: 285
Kovner, I. 1989. *Ap. J. Lett.* 341: L1
Kristian, J. 1973. *Ap. J. Lett.* 179: L61
Kron, R. G., Chiu, L.-T. G. 1981. *Publ. Astron. Soc. Pac.* 93: 397
Kruper, J. S., Canizares, C. R. 1989. *Ap. J.* 343: 66
Kruszewski, A. 1987. Preprint.
Kruszewski, A. 1988. *Proc. Workshop of the Max-Planck-Ges. and Acad. Sin.*, 2nd, ed. G. Borner, p. 255. Berlin: Springer-Verlag
Kunth, D., Sargent, W. L. W. 1986. *Astron. J.* 91: 761
Longair, M. S., Seldner, M. 1979. *MNRAS* 189: 433
Low, F. J., Cutri, R. M., Kleinmann, S. G., Huchra, J. P., 1989. *Ap. J. Lett.* 340: L1
Luyten, W. J. 1953–62. *A Search for Faint Blue Stars*. Minneapolis: The Obs., Univ. Minn.
Lynden-Bell, D. 1969. *Nature* 223: 690
Lynden-Bell, D. 1971. *MNRAS* 155: 95
Lynds, R., Petrosian, V. 1972. *Ap. J.* 175: 591
Machalski, J. 1987. See Hewitt et al 1987, p. 649
MacKenty, J. W. 1989. *Ap. J.* 343: 125
Malkan, M. A. 1983. *Ap. J.* 268: 582
Marano, B., Zamorani, G., Zitelli, V. 1988. *MNRAS* 232: 111
Marshall, H. L., Avni, Y., Braccesi, A., Huchra, J. P., Tananbaum, H., Zamorani, G., Zitelli, V. 1984. *Ap. J.* 283: 50 (BFAB)
Marshall, H. L., Avni, Y., Tananbaum, H., Zamorani, G. 1983. *Ap. J.* 269: 35
Mathez, G. 1978. *Astron. Astrophys.* 68: 17
Matthews, T. A., Sandage, A. R. 1963. *Ap. J.* 138: 30
Mitchell, K. J., Warnock, A. III, Usher, P. D. 1984. *Ap. J. Lett.* 287: L3 (MBQS)
Narayan, R. 1989. *Ap. J. Lett.* 339: L53
Nieto, J.-L., Seldner, M. 1982. *Astron. Astrophys.* 112: 321
Oort, J. H., Arp, H. C., de Ruiter, H. 1981. *Astron. Astrophys.* 95: 7
Osmer, P. S. 1980. *Ap. J. Suppl.* 42: 523
Osmer, P. S. 1981. *Ap. J.* 247: 762
Osmer, P. S. 1982. *Ap. J.* 253: 28
Osmer, P. S., Hewett, P. C. 1988. See Osmer et al 1988b, p. 273
Osmer, P. S., Porter, A. C., Green, R. F. 1988a. *Bull. Am. Astron. Soc.* 20: 968
Osmer, P. S., Porter, A. C., Green, R. F., Foltz, C. B., eds. 1988b. *Proc. Workshop Opt. Surv. for Quasars*. Provo, Utah: Brigham Young Univ. Print Serv.
Osterbrock, D. E. 1988. *Astrophysics of Gaseous Nebulae and Active Galactic Nuclei*. Mill Valley, Calif: Univ. Sci. Books
Padovani, P., Rafanelli, P. 1988. *Astron. Astrophys.* 205: 53
Peacock, J. A., Miller, L. 1988. See Osmer et al 1988b, p. 194
Peebles, P. J. E. 1980. *The Large-Scale Structure of the Universe*. Princeton, NJ: Princeton Univ. Press
Peterson, B. A. 1988. See Osmer et al 1988b, p. 23
Phillipps, S., Fong, R., Ellis, R. S., Fall, S. M., MacGillivray, H. T. 1978. *MNRAS* 182: 673
Rees, M. J. 1984. *Annu. Rev. Astron. Astrophys.* 22: 471
Rix, H.-W., Hogan, C. J. 1988. *Ap. J.* 332: 108
Roos, N. 1985. *Ap. J.* 294: 486
Ryle, M. 1968. In *Highlights of Astronomy*, ed. L. Perek, p. 33. Dordrecht: Reidel
Sandage, A. 1965. *Ap. J.* 141: 1560
Sandage, A. 1966. *Ap. J.* 146: 13
Sandage, A. 1972. *Ap. J.* 178: 25
Sandage, A., Luyten, W. J. 1967. *Ap. J.* 148: 767
Sandage, A., Luyten, W. J. 1969. *Ap. J.* 155: 913
Sandage, A., Veron, P. 1965. *Ap. J.* 142: 412
Sanders, D. B., Soifer, B. T., Elias, J. H., Madore, B. F., Matthews, K., et al. 1988. *Ap. J.* 325: 74
Savage, A., Jauncey, D. L., White, G. L., Peterson, B. A., Peters, W. L., et al. 1988. See Osmer et al 1988b, p. 204
Schade, D. 1988. MSc thesis. Univ. Victoria, Can.
Schmidt, M. 1968. *Ap. J.* 151: 393

Schmidt, M. 1970. *Ap. J.* 162: 371
Schmidt, M., 1987. See Hewitt et al 1987, p. 619
Schmidt, M., Green, R. 1983. *Ap. J.* 269: 352 (BQS)
Schmidt, M., Schneider, D. P., Gunn, J. E. 1986a. *Ap. J.* 306: 411 (SSG1)
Schmidt, M., Schneider, D. P., Gunn, J. E. 1986b. *Ap. J.* 310: 518 (SSG2)
Schmidt, M., Schneider, D. P., Gunn, J. E. 1988. See Osmer et al 1988b, p. 87
Schneider, D. P., Schmidt, M., Gunn, J. E. 1989a. *Astron. J.* 98: 1507
Schneider, D. P., Schmidt, M., Gunn, J. E. 1989b. *Astron. J.* 98: 1951
Schneider, P. 1986. *Ap. J. Lett.* 300: L31
Schneider, P. 1989. *Astron. Astrophys.* 221: 221
Seldner, M., Peebles, P. J. E. 1979. *Ap. J.* 227: 30
Shanks, T., Boyle, B. J., Peterson, B. A. 1988. See Osmer et al 1988b, p. 244
Shanks, T., Fong, R., Boyle, B. J. 1983. *Nature* 303: 156
Shanks, T., Fong, R., Boyle, B. J., Peterson, B. A. 1987. *MNRAS* 227: 739
Shaver, P. A. 1984. *Astron. Astrophys.* 136: L9
Shaver, P. A. 1985. *Astron. Astrophys.* 143: 451
Shaver, P. A. 1986a. *Nature* 323: 185
Shaver, P. A. 1986b. See Swarup & Kapahi 1986, p. 475
Shaver, P. A. 1988. In *Large Scale Structures of the Universe, IAU Symp. No. 130*, ed. J. Audouze, M.-C. Pelletan, A. Szalay, p. 359. Dordrecht: Kluwer
Smith, H. E., Burbidge, E. M., Baldwin, J. A., Tohline, J. E., Wampler, E. J., et al. 1977. *Ap. J.* 215: 427
Smith, M. 1975. *Ap. J.* 202: 591
Smith, M. 1978. In *Vistas in Astronomy*, ed. A. Beer, P. Beer, 22: 321. Oxford: Pergamon
Smith, M. 1983. In *Quasars and Gravitational Lenses. Proc. Liège Int. Astrophys. Colloq., 12th*, p. 4. Liège: Univ. Liège
Smith, M. 1986. See Swarup & Kapahi 1986, p. 17
Steidel, C. C., Sargent, W. L. W. 1989. *Ap. J. Lett.* 343: L33
Stevenson, P. R. F., Shanks, T., Fong, R., MacGillivray, H. T. 1985. *MNRAS* 213: 953
Stocke, J. T., Morris, S. L., Gioia, I. M., Maccacaro, T., Schild, R. E., Wolter, A. 1988. See Osmer et al 1988b, p. 311
Stocke, J., Schneider, P., Morris, S. L., Gioia, I. M., Maccacaro, J., Schild, R. 1987. *Ap. J. Lett.* 315: L11
Stockton, A. 1978. *Ap. J.* 223: 747
Stockton, A. 1982. *Ap. J.* 257: 33
Stockton, A. 1986. *Astrophys. Space Sci.* 118: 487
Stockton, A., MacKenty, J. W. 1987. *Ap. J.* 316: 584
Swarup, G., Kapahi, V. K., eds. 1986. *Quasars. Proc. IAU Symp. No. 119*. Dordrecht: Reidel
Toomre, A., Toomre, J. 1972. *Ap. J.* 178: 623
Trevese, D., Pittella, G., Kron, R. G., Koo, D. C., Bershady, M. 1989. *Astron. J.* 98: 108
Turner, E. 1979a. *Ap. J.* 230: 291
Turner, E. 1979b. *Ap. J.* 231: 645
Tyson, J. A. 1986. *Astron. J.* 92: 691
Tytler, D., Bokensberg, A., Sargent, W. L. W., Young, P., Kunth, D. 1987. *Ap. J. Suppl.* 64: 667
Usher, P. D., Mitchell, K. J. 1978. *Ap. J.* 223: 1
Veron-Cetty, M.-P., Veron, P. 1984. *ESO Sci. Rep. No. 4*
Veron-Cetty, M.-P., Veron, P. 1987. *ESO Sci. Rep. No. 5*
Veron, P. 1983. In *Quasars and Gravitational Lenses. Proc. Liège Int. Astrophys. Colloq., 12th*, p. 210. Liège: Univ. Liège
Vietri, M., Ostriker, J. P. 1983. *Ap. J.* 267: 488
Wampler, E. J. 1987. *Astron. Astrophys.* 178: 1
Wampler, E. J., Ponz, D. 1985. *Ap. J.* 298: 448
Wandel, A., Petrosian, V. 1988. *Ap. J. Lett.* 329: L11
Wandel, A., Yahil, A. 1985. *Ap. J. Lett.* 295: L1
Warren, S. J., Hewett, P. C. 1990. Preprint
Warren, S. J., Hewett, P. C., Irwin, M. J., McMahon, R. G., Bridgeland, M. T., et al. 1987a. *Nature* 325: 131
Warren, S. J., Hewett, P. C., Osmer, P. S. 1988. See Osmer et al 1988b, p. 96
Warren, S. J., Hewett, P. C., Osmer, P. S., Irwin, M. J. 1987b. *Nature* 330: 453
Webster, A. 1976. *MNRAS* 175: 61
Webster, A. 1982. *MNRAS* 199: 683
Webster, R., Hewett, P. C., Harding, M. E., Wegner, G. A. 1988. *Nature* 336: 358
Weedman, D. W. 1985. *Ap. J. Suppl.* 57: 523
Weedman, D. W. 1986. *Quasar Astronomy*. Cambridge: Univ. Press
Weymann, R. J., Boroson, T. A., Peterson, B. M., Butcher, H. R. 1978. *Ap. J.* 226: 603
Wilkes, B. 1986. *MNRAS* 218: 331
Wills, D. 1978. *Phys. Scr.* 17: 333
Wyckoff, S., Wehinger, P. A., Gehren, T. 1981. *Ap. J.* 247: 750
Yee, H. K. C. 1990. *Proc. Hubble Centen. Symp. Evol. of the Universe of Galaxies*, ed. R. Kron. Provo, Utah: Brigham Young Univ. Print Serv. In press
Yee, H. K. C., Green, R. F. 1984. *Ap. J.* 280: 79
Yee, H. K. C., Green, R. F. 1987. *Ap. J.* 319: 28

EQUILIBRIUM AND DYNAMICS OF CORONAL MAGNETIC FIELDS

B. C. Low

High Altitude Observatory, National Center for Atmospheric Research[1], Boulder, Colorado 80307

KEY WORDS: hydromagnetics, flares, coronal mass ejections, shocks

1. INTRODUCTION

The solar corona as a hot, highly ionized, magnetized gas displays a rich variety of directly observable hydromagnetic behaviors. At the 10^6-K temperature of the solar corona, the hydrostatic density scale height is of the order of a tenth of a solar radius. An interplay among pressure, gravity, and magnetic field organizes the large-scale corona into regions where the solar wind escapes along open magnetic fields and complementary regions where closed magnetic fields trap relatively high-density plasmas in near-static equilibrium. These structures are observed as fully resolved features down to several arcseconds, both above the solar limb in Thomson-scattered light (at eclipses or by the use of coronagraphs) and against the solar disk in EUV, XUV, and X-ray images. In the low corona, over smaller scales down to below a hundredth of a solar radius, the magnetic field dominates in a myriad of structures: plasma loops and condensations of various sizes and thermodynamic states, including the magnetically levitated, cool (10^4 K) long filaments called prominences. A body of phenomenological knowledge has accumulated over the last 30 years from observational studies of coronal structures and their dynamics. This review describes our theoretical understanding of these structures to the extent that it can be based upon hydromagnetic principles.

The vantage of being able to observe solar phenomena as spatially

---

[1] The National Center for Atmospheric Research is sponsored by the National Science Foundation.

resolved events makes it very desirable to develop numerical simulations sophisticated enough for direct comparison with observations. This is a way of checking theory and is one of the ultimate goals of theory development. However, a long way has to be traveled to identify the basic physical effects in a particular phenomenon, through posing the right kind of theoretical questions and constructing simple but appropriate models, before the undertaking of sophisticated numerical modeling can be meaningful and fruitful. We review in the next three sections some of these theoretical questions and their possible answers as provoked by the magnetic field in the coronal environment. Specifically, we treat the equilibrium and stability of magnetic fields (Section 2), the mechanics of onset of eruption (Section 3), and the large-scale organization and time-dependent, ordered dynamics of the corona (Section 4). The emphasis in this review is on physical understanding. Mathematical details are kept to a minimum, and our attention is focused instead upon the critical examination of the basic assumptions of models and the physical interpretation of results. Some small amount of mathematics is necessary in order not to lose sight of the precise content of specific hydromagnetic principles. For this reason, Section 2 contains a simple physical description of the governing equations. This description brings out various elementary hydromagnetic properties in terms of which the physical discussions in the rest of Section 2 and much of Sections 3 and 4 can be understood. The questions to be addressed arise from coronal observations, but no attempt is made to give a comprehensive review of current observations, for which readers are referred to other sources (e.g. Zirker 1977, Sturrock 1980, Orrall 1981) as well as to previous volumes of this series.

The physical processes discussed here have their origin in the corona, but they are interesting in their own right, with possible relevance to other astrophysical systems. What we learn in the case of the corona provides a basis for what we can only speculate about for other systems not accessible to direct observation.

2. EQUILIBRIUM AND STABILITY

The simple one-fluid ideal hydromagnetic equations are the basis of much of our discussion and are given by

$$\rho\left[\frac{\partial \mathbf{v}}{\partial t} + (\mathbf{v}\cdot\nabla)\mathbf{v}\right] = \frac{1}{4\pi}(\nabla \times \mathbf{B}) \times \mathbf{B} - \nabla p - \rho\nabla\Phi, \qquad 2.1$$

$$\frac{\partial \mathbf{B}}{\partial t} = \nabla \times (\mathbf{v} \times \mathbf{B}), \qquad 2.2$$

$$\frac{\partial \rho}{\partial t} + \nabla \cdot (\rho \mathbf{v}) = 0, \qquad 2.3$$

$$\frac{D}{Dt}\left(\frac{1}{\gamma-1}\frac{p}{\rho}\right) + p\frac{D}{Dt}\left(\frac{1}{\rho}\right) = \sigma, \qquad 2.4$$

in standard forms with $\mathbf{B}$, $\mathbf{v}$, Φ, p, ρ, and γ being the magnetic field, velocity field, solar gravitational potential, pressure, density, and the adiabatic index, respectively (Roberts 1964, Friedberg 1982). Equation 2.1 is the momentum equation. The plasma medium is approximated to be electrically neutral, so that only the (magnetic) Lorentz force appears in addition to the pressure and gravitational forces. Equation 2.2 is basically Faraday's law of induction. Equations 2.3 and 2.4 describe mass and energy conservation, respectively. In the latter equation, D/Dt is the Lagrangian derivative, and the term σ represents departures from adiabaticity.

2.1 Magnetostatic Equilibrium

In the low corona, within about one solar radius of the solar surface, long-lived magnetically closed structures may be approximated to be in static equilibrium. The last two decades saw considerable activity in generating static solutions to the hydromagnetic equations and in studying their mechanical stability to linear perturbations. The classical subject of equilibrium and stability in the context of solar physics differs from a similar pursuit in the study of laboratory plasma devices in several important ways. In the solar atmosphere, the plasmas we deal with are open systems without rigid walls. Coronal structures are constantly exchanging energy and mass with their environment, including the dense atmosphere below. Gravity is an important force, playing a dual role of stabilizing magnetic fields by anchoring them with dense plasma and destabilizing them through Rayleigh-Taylor-type instabilities. Finally, the physical length scales of interest are usually very large compared with the characteristic plasma length scales, so that dissipative processes like electrical resistivity and viscosity can often be neglected in the corona as a first approximation.

Let us review the rudiments of magnetostatic equilibrium, which are the basis for several important theoretical questions that are posed and discussed in this and subsequent sections. We first survey the known theoretical equilibrium solutions, pointing out their basic physical properties, and then treat the question of mechanical stability in the next subsection. Static equilibrium is described by setting the left side of Equation 2.1 to zero, i.e.

$$\frac{1}{4\pi}(\nabla \times \mathbf{B}) \times \mathbf{B} - \nabla p - \rho \nabla \Phi = 0, \qquad 2.5$$

requiring a balance of the Lorentz force against the plasma pressure gradient and weight. This relation is especially simple if we note that the Lorentz force is perpendicular to the magnetic field. Along each narrow tube of magnetic flux, a hydrostatic relation obtains between the pressure gradient and gravity. The magnetostatic atmosphere is thus made up of elemental magnetic flux tubes of hydrostatic plasmas stacked in global equilibrium, with the Lorentz force playing a role only in lateral force balance. The simple case of a system with an ignorable Cartesian coordinate x helps to fix ideas. The divergence-free magnetic field can be written as

$$\mathbf{B} = \left(Q, \frac{\partial A}{\partial z}, -\frac{\partial A}{\partial y} \right), \qquad 2.6$$

associated with the electric current density

$$\mathbf{J} = \frac{c}{4\pi}\left(-\nabla^2 A, \frac{\partial Q}{\partial z}, -\frac{\partial Q}{\partial y} \right), \qquad 2.7$$

in terms of two functions A and Q; c is the speed of light. In this symmetric system, the pressure and gravitational forces have no components in the x direction. For the Lorentz force to vanish in that direction, $\mathbf{B}$ and $\mathbf{J}$ must have parallel projections in the y-z plane, requiring Q to be a strict function of A. Force balance in the y-z plane is then described by

$$\frac{\partial}{\partial z} p(A,z) + \rho g = 0, \qquad 2.8$$

$$\nabla^2 A + 4\pi \frac{\partial}{\partial A} p(A,z) + Q(A) \frac{dQ}{dA} = 0, \qquad 2.9$$

where, for simplicity, gravity is taken to have a uniform acceleration g in the $-\hat{z}$ direction (Low 1975). The scalar function A describes magnetic flux surfaces and is constant along each magnetic line of force. Equation 2.8 is thus the hydrostatic relation along the lines of force, while Equation 2.9 describes the balance of forces across the magnetic field. In effect, Equation 2.9 determines the electric current density (see Equation 2.7) under the condition of static equilibrium.

The nearly static condition of long-lived structures in the low corona is maintained by two properties: the extremely high electrical conductivity of the corona, and the anchoring of the coronal magnetic fields by the dense photosphere. The former allows electric currents to persist in a static state in the corona for exceedingly long times. For structures of sizes 10 km and up, a Spitzer-type conductivity appropriate for the corona as an ionized gas gives resistive diffusion times of years, so that the electrical conductivity may be taken to be infinite for most purposes. The electric current density in the hydromagnetic approximation is related to the magnetic field by $\mathbf{J} = c/4\pi \nabla \times \mathbf{B}$ (Ampere's law). While it is sometimes useful to think physically in terms of the electric current, it is the magnetic flux rather than the electric current that is conserved or "frozen" in the perfect-conductor approximation (e.g. Parker 1979). For this reason, it is often convenient to think in terms of $\mathbf{B}$ alone and to take the curl operation on $\mathbf{B}$ to determine $\mathbf{J}$ whenever the electric current density is needed.

The field topology of any given state is accounted for deterministically by Faraday's law (Equation 2.2) in terms of the braiding and twisting of the magnetic lines of force by the fluid motion during the system's evolution from some initial state. The physical description of a magnetic field at any given time must therefore include a specification of its field topology as endowed by its evolutionary history. Take a magnetic field $\mathbf{B}$ in some volume V with all lines of force anchored at both ends on the boundary ∂V, a situation that we return to many times throughout this review. To describe the physical state of the magnetic field, we must specify independently the normal flux distribution at ∂V and the field topology in V, noting that magnetic fields of quite different field-line connectivities can have the same boundary flux distribution. This consideration leads to a physical but untraditional way of posing problems in magnetostatic equilibrium, which is to seek equilibrium states endowed with prescribed field topologies. Such kinds of problems give rise to integro–partial differential equations as the result of the nonlocal nature of the topology of magnetic field lines (Sakurai 1979, Low 1982a). It is important to realize that the static state as a mathematical solution to Equation 2.5 does not a priori have anything to do with Faraday's law expressed by Equation 2.2, since the momentum equation and Faraday's law are quite independent. There can arise situations where no smooth solutions to Equation 2.5 exist to satisfy the demand of a prescribed field topology, an interesting subject to be treated in the next section. For the present, we concentrate on the more familiar situations where the solutions are everywhere smooth.

Given almost any field topology, the twist and shear of a magnetic field imply the unavoidable presence of electric currents. The photosphere by its high inertia serves to lock these magnetic twists and shears in the corona, so that generally electric currents exist in the corona. In the quiescent state, the coronal magnetic field assumes a field geometry such that its electric currents account for both Ampere's law and the balance of forces. It is the so-called drift currents perpendicular to the magnetic field that generate the Lorentz force needed to balance the other body forces. The current parallel to the magnetic field produces no force but gives rise to magnetic twists and shears to account for the magnetic field's endowed topology.

Many particular magnetostatic solutions have been reported in the literature, both for the planar system described above and the axisymmetric system in cylindrical and spherical coordinates. The reader is referred to the original sources for applications of these solutions to the study of the large-scale corona, the support of prominences, and the confinement of sunspots and related structures (e.g. Tandberg-Hanssen 1974, Lerch & Low 1980, Hundhausen et al 1981, Pizzo 1986). Equation 2.5 needs to be closed in some suitable physical manner with a statement on static energy transport. In the corona, especially for the low-lying plasma loops, there is a static balance between thermal conduction, radiative loss, and heating. The magnetostatic problem by itself is formidably nonlinear, and thus few attempts have been made to treat the full problem. An exception is the thermal balance of the Kippenhahn-Schlüter prominence model, where tractability has been achieved by exploiting the simple magnetic geometry of this model (e.g. Milne et al 1979, Low & Wu 1981). Most previous works are based on closing Equation 2.5 by ad hoc mathematical assumptions to keep the problem simple. Issues of energy balance tend to be addressed as a separate concern in terms of one-dimensional atmospheres along prescribed, rigid magnetic lines of force (e.g Withbroe 1981). While such an approach seems reasonable tactically, the coupling between force balance and energy transport may be what is needed but is neglected in the consideration of certain important questions. The formation of prominences and the physics of low-lying plasma loops of various thermodynamic states probably cannot be understood adequately without accounting for this coupling explicitly.

The desire to model realistic solar structures is a natural motivation for the construction of equilibria without the type of strong symmetries associated with ignorable coordinates. Fully three-dimensional equilibria are distinct from symmetric ones in that they incur certain physical requirements on magnetic flux surfaces that are trivially satisfied and never in question under the assumption of strong symmetry. The point is best appreciated in terms of the magnetic lines of force expressed as the intersections between a pair of families of local magnetic flux surfaces given by the constant values of two distinct Euler potentials U and V (e.g. Stern 1966, Rosner et al 1989). By a proper choice of these flux surfaces, we can have the representation

$$\mathbf{B} = \nabla U \times \nabla V, \qquad 2.10$$

which is manifestly divergence free. In general, the gravitational potential Φ is independent of U and V, and therefore these three functions constitute a set of local curvilinear spatial coordinates. In terms of these coordinates, the magnetostatic equation (Equation 2.5) resolves into the three components

$$\frac{\partial}{\partial \Phi} p(U,V,\Phi) + \rho = 0, \qquad 2.11$$

$$\left[\nabla \times (\nabla U \times \nabla V)\right] \cdot \nabla V - \frac{\partial}{\partial U} p(U,V,\Phi) = 0, \qquad 2.12$$

$$\left[\nabla \times (\nabla U \times \nabla V)\right] \cdot \nabla U + \frac{\partial}{\partial V} p(U,V,\Phi) = 0. \qquad 2.13$$

Equation 2.11 is the hydrostatic relation along a magnetic line of force for which U and V are constant (cf. Equation 2.8), whereas the other two equations describe force balance in two independent directions perpendicular to the magnetic field. The assumption of a special symmetry can reduce these two equations to a single relation, such as given by Equation 2.9 for the system invariant in x. For the general case, a nontrivial (necessary) compatibility condition on the flux functions U and V arises from the demand that $\partial^2 p/\partial U \partial V = \partial^2 p/\partial V \partial U$ in order that a smoothly distributed equilibrium pressure is single valued in space. If the magnetic field is prescribed with a fixed topology, a global constraint is then imposed upon the U and V surfaces, and there may arise situations in which this global constraint does not permit the above compatibility condition (which is a local differential condition) to be satisfied everywhere in space. The equilibrium pressure cannot then be everywhere single valued. Pressure discontinuity is implied and electric current sheets must be admitted, an interesting property to be treated in Section 3.

On the other hand, a different approach can be pursued. Without first restricting the magnetic field topology, we can solve the compatibility condition for just those U and V flux surfaces that are compatible with a smooth distribution of equilibrium pressure. This is a difficult nonlinear problem. Particular families of this class of flux surfaces have been found, giving an intriguing first glimpse of fully three-dimensional static models of coronal structures, sunspots, prominences, and inverted-U plasma loops of finite loop diameters (Low 1982d, 1984a, 1985a, Hu 1983, Bogdan & Low 1986). Notable among these particular magnetostatic solutions are those reported by Low (1985a), which are versatile enough to accommodate the input of an arbitrary normal field at the boundary of the physical domain of interest. Thus, the magnetic field measurements provided by a magnetograph can be used as an input to model global three-dimensional solar magnetic fields. The plasma structures predicted by such a model can be compared directly with real observations, a prospect currently being pursued in the observational interpretation of the solar corona. Solving the compatibility condition for other families of U and V is a worthwhile undertaking. It is a step toward realistic modeling of solar structures, and from the theoretical point of view, there is also much to learn from the three-dimensional solutions that have yet to be discovered.

2.2 Linear Mechanical Stability

Observation of plasma structures in the corona suggests that coronal magnetic fields can persist with remarkable stability (e.g. Vaiana & Rosner 1978). The solar atmosphere is never truly quiescent or static, of course. Outside of violent eruptions, such as flares or coronal mass ejections, motions of various small magnitudes and over a broad range of time scales are ever present. To first order, these motions may be neglected, and the physics underlying the apparent stability of long-lived structure may be captured by the stability properties of idealized static states. What seemed surprising until recently, when the existence of stable equilibria could be demonstrated explicitly, was that stable structures were at all possible, since a great variety of hydromagnetic instabilities abound for the laboratory plasmas. In hindsight, there should have been no surprise, because coronal plasmas are open natural systems, as opposed to artificial engineering constructions, and are thus free to interact with their environments and discharge their free energies in favor of minimum energy states over the typically short Alfvén transit time. This statement is an oversimplification, but it seems that the coronal plasma naturally seeks out a stable quiescent existence until its environment, by slow changes, forces it to a dynamical transition.

Classical plasma physics associates a pressure-driven (exchange) instability with drift currents flowing perpendicular to the magnetic field (e.g. Friedberg 1982). A simple rule states that a dense plasma confined by a magnetic field concave to it is unstable because of the magnetic curvature force. There is a tendency for a destabilizing exchange between the dense plasma and the confining field. Conversely, a dense plasma confined by a magnetic field convex to it is stable. In the presence of gravity, a magnetic field behaves like a weightless gas with an isotropic (magnetic) pressure. The magnetic field therefore manifests a Rayleigh-Taylor-type instability that tends to upturn magnetic fields over the embedding plasma. Coupled with the frozen-in condition of perfect electrical conductivity, this effect usually proceeds with the drainage of plasma along the magnetic field to the valleys of an undulating flux rope. The instability runs away, with the evacuated part of the flux rope becoming more buoyant and the weighted valleys more heavy (Parker 1966).

It was by studying explicit examples of static equilibrium that it was found that stability may obtain through the competition of the above two effects associated with drift currents. Low (1982d) presented a three-dimensionsal, inverted-U plasma loop of finite thickness containing an untwisted magnetic field confined in equilibrium by a (drift) sheet current flowing in the loop surface. The interior and exterior atmospheres are stably stratified, embedding potential magnetic fields. No instability can arise from these regions. The stability of the whole system then hinges on whether the loop's boundary surface is also stable. It was found that everywhere on the loop boundary, an exchange displacement between the interior and the exterior is destabilizing to one but is exactly compensated for by being stable to the other of the magnetic curvature and buoyancy forces. Take the case of a U-loop of higher density than the surrounding plasma and confined by strong magnetic fields in the exterior. The boundary at the top side of the loop apex sees a dense (interior) plasma confined by a concave (exterior) magnetic field. The magnetic curvature force is thus destabilizing but is compensated for by having the dense loop plasma lying stably below the less dense exterior. Conversely, the curvature force is stabilizing and compensates for the otherwise Rayleigh-Taylor unstable stratification at the bottom side of the loop apex. Stability deriving from such a mutual canceling of these two competing effects associated with the drift currents has also been demonstrated for the Kippenhahn-Schlüter prominence (Zweibel 1982, Migliuolo 1982). The reader is referred to Low (1984c) for a stability study of an axisymmetric $\gamma = 6/5$ polytropic magnetostatic corona, which was shown to be absolutely stable to axisymmetric perturbations but unstable to nonaxisymmetric perturbations.

In these and most previous works, interest centers on the mechanical aspects of the problem, and an oversimplification is adopted in which the chromosphere is ignored and the lower coronal boundary is taken to be the photosphere obeying the rigid boundary conditions. These boundary conditions play a tacit but principal role in providing stability. It is the relative rigidity of the photosphere that, in the first instance, locks the magnetic twist and its associated electric currents in the equilibrium corona and at the same time prevents certain plasma displacements that would otherwise destabilize the system. While rigid boundary conditions are not unreasonable given that the photosphere is at least 7 orders of magnitude denser than the low corona, these are strong assumptions. Attempts to relax these assumptions in a meaningful way have not been successful; see the discussions in Low (1985b) and Rosner et al (1985). Here we continue to use the rigid boundary conditions, but in Section 3.3 we discuss these conditions critically.

Magnetic shear associated with parallel electric currents can suppress the exchange instabilities, although it introduces a separate class of instabilities if the shear is excessive (e.g Friedberg 1982). Considerable efforts have been made to establish the stability of anchored force-free magnetic fields. The magnetic intensity in the low corona over a solar active region is of the order of 10–10^2 G, so that the magnetic pressure far exceeds the typical local gas pressure. To a first approximation, these magnetic fields may be taken to be force free, characterized by the electric current density being everywhere parallel to the magnetic field (Gold 1964). If we drop the terms for pressure gradient and plasma weight, Equation 2.5 becomes

$$(\nabla \times \mathbf{B}) \times \mathbf{B} = 0. \qquad 2.14$$

For the planar system, the hydrostatic equation (Equation 2.8) can be ignored, and Equation 2.9 takes the form

$$\nabla^2 A + Q(A)\frac{dQ}{dA} = 0 \qquad 2.15$$

for a two-dimensional force-free field. It has been conjectured that the two-dimensional force-free fields in the infinite half-plane $z > 0$ generated by Equation 2.15 are all linearly stable if all lines of force of the magnetic field are rigidly anchored at the boundary $z = 0$ (Hood & Priest 1981, Hood 1983). Unfortunately, it has not been possible to prove this plausible conjecture rigorously because the stability analyses of these force-free fields are hindered by the question of completeness of the perturbations treated. Establishing the stability of force-free magnetic fields is important, since it is believed that the free energy of the order of 10^{31-32} ergs liberated by a flare can be stored in the form of electric currents parallel to the magnetic field. This energy may be built up by a slow evolution, during which a magnetic field is increasingly twisted or sheared by the photospheric displacements of the magnetic footpoints (Gold 1964). For this process to be viable, it is important that force-free magnetic fields of considerable shear or twist can be globally mechanically stable, at least in the linear regime, in order to give physical meaning to the idea of energy storage.

The preoccupation with two-dimensional force-free fields generated by Equation 2.15 is motivated largely by a tactical consideration to keep the stability problem simple. It has turned out that examples of stable force-free fields can be more readily established for more complex magnetic fields. There exists a class of three-dimensional force-free fields lying in parallel flux planes (Chang & Carovillano 1981). If we take these planes to be the constant x planes, the fields are generated by

$$B_y - iB_z = F(y+iz)\exp(i\zeta), \qquad 2.16$$

where $i = \sqrt{-1}$, F is a free analytic function of $y+iz$, and ζ is a free (real) function of x. The field pattern varies from one constant x plane to another according to $\zeta(x)$. This magnetic shear is associated with an electric current density everywhere parallel to the magnetic field. The special case of constant ζ corresponds to an unsheared potential field invariant in x. It can be shown by direct proof that any force-free magnetic field of this class is stable to all perturbations subject to the rigid boundary conditions (Low 1988a). A related class of force-free fields model, with a striking degree of geometric realism, the three-dimensional field configurations associated with two photospheric regions of opposite magnetic polarities intruding into each other (Low 1982b). These force-free fields have also been shown to be absolutely stable (Low 1988b). These few explicit examples provide a definite, though limited, basis (i.e. within the concern of linear mechanical stability) for believing that force-free magnetic fields of various degrees of twist and shear may exist in the solar corona.

3. ONSET OF MAGNETIC ERUPTION

Coronal magnetic fields evolve with time in response to heating, forcing by the lower dense atmosphere, and injections of mass and magnetic flux during the course of the solar cycle. Here we refer to quasi-steady changes due to forcing over time scales long compared with the typical dynamical time scales at which the coronal plasma adjusts from one equilibrium state to another. This quasi-steady evolution does not last indefinitely, of course. During the lifetime of a coronal structure, there are typically long periods of quasi-steady changes (days to weeks), interrupted by explosive dynamical transitions (minutes to hours) during which considerable amounts of energy (of the order of 10^{31-32} ergs) are liberated through intense heating, as in a flare, or in bulk motions, as in a coronal mass ejection (e.g. Sturrock 1980, Hundhausen 1987, Haisch & Rodono 1989). The energy liberated is believed to be built up (and stored) by the stressing of the magnetic field during the quasi-steady evolution. The manner in which quasi-steady evolution would reach a point where some form of eruption sets in is a fascinating question that we now critically assess. Specifically, we discuss the spontaneous formation of electric current sheets in coronal magnetic fields and the idea of the loss of equilibrium resulting from the unavailability of stable equilibrium states in the course of quasi-steady evolution.

3.1 *Spontaneous Formation of Electric Current Sheets*

Subject to stresses under the frozen-in condition, the plasma behaves as though the individual magnetic flux tubes are coherent entities with a tendency for discontinuous or tearing displacements along magnetic flux surfaces. These displacements can bring into contact parts of magnetic flux tubes initially separate in space and form magnetic rotational discontinuities or infinitely thin electric current sheets. In this manner, electric current sheets appear spontaneously in an otherwise smooth magnetic field, a process to be expected in fully three-dimensional systems with the full degrees of freedom for a magnetic flux tube to slip relative to its surrounding magnetic field. This property was first pointed out by Parker (1972) and also independently by Arnol'd (1974). General theoretical developments can be found in Parker (1972, 1986a,b, 1989a,b,c,d), who introduced the now well-known problem of constructing the magnetostatic states for a magnetic field in the space between two infinite parallel plates. The magnetic field is given a fixed topology created by interweaving the lines of force as a result of a prescribed but arbitrary smooth displacement of the footpoints, beginning with an initial uniform field perpendicular to the two plates. Parker concluded that with the fixed topology locked in by a rigid anchoring of the magnetic footpoints at the boundary plates, the deformed magnetic field cannot in general be in equilibrium without the formation of electric current sheets. An independent theoretical development due to Moffatt (1985, 1986) is also interesting, studying in particular the analogy between magnetostatic states and steady Euler flows. In the following, we take an intuitive approach to the process of current sheet formation based on simple explicit examples.

To be sure, electric current sheets can be forced upon any smooth magnetic field by imposing discontinuous displacements in a direct manner. A magnetic field in the infinite half-space $z > 0$ subject to a discontinuous footpoint displacement at the boundary $z = 0$ will develop an electric current sheet extending into the interior from the point of discontinuous displacement on the boundary. What is interesting about Parker's result is that smooth footpoint displacements at the boundary can also force an initially smooth interior field to seek an equilibrium state obtainable only through interior, discontinuous plasma displacements, resulting in the formation of current sheets. These current sheets have infinitesimal thickness under the frozen-in condition. The assumption of an infinite electrical conductivity is only approximate in realistic plasma, and thus when an electric current density peaks to form a thin sheet, it would dissipate away at some point in time when the width of the density peak becomes small enough for resistive dissipation to be no longer negligible. This process provides a means of converting volume electric currents, which are subject to negligible dissipation under normal coronal conditions, into a form that readily dissipates despite the high electrical conductivity of the corona. It is remarkable that high electrical conductivity itself can, through such a process, promote the rapid dissipation of electric currents that otherwise flow freely in the plasma.

Let us consider the formation of current sheets in force-free magnetic fields. The formation of an electric current sheet in association with a magnetic X-type neutral point has been known for many years (e.g. Syrovatskii 1971). When a two-dimensional magnetic field with an X-type neutral point is perturbed, the coherent behavior of the four magnetic lobes about the neutral point has the tendency for a pair of opposite magnetic lobes to squeeze the other pair out of the way. The neutral point is thus torn into a current sheet along which opposite fields meet. It was recognized only recently that an X-type neutral point is not essential to this process (Aly 1987, Low 1987, Moffatt 1987). This is illustrated in Figure 1. A quadrupolar potential field in the y-z plane is shown in Figure 1a. There is no neutral point in the domain of interest ($z > 0$), but a separatrix line separates three sets of magnetic flux: two (shaded) bipolar lobes lying under and kept apart by a third distinct lobe. Let the footpoints at $z = 0$ of this initial field be displaced in a continuous compressive motion in the y-z plane toward the origin and then anchored rigidly in their new locations. For finite (as opposed to infinitesimal) footpoint displacements, the equilibrium available to the deformed magnetic field under the frozen-in condition is in general one with a vertical current sheet (shown in Figure 1b). There are actually special finite footpoint displacements that can take the deformed magnetic field to an equilibrium state without an electric current sheet, but, in general, the current sheet is unavoidable. In seeking out its equilibrium, the two shaded magnetic lobes tend to squeeze toward each other with the complete expulsion of the intervening flux from the third (overlying) lobe, so that opposite fields touch along the thick vertical line (in Figure 1b) to form the current sheet. The general inevitability of this outcome follows from first recognizing that the electric current density of any magnetic field in the y-z plane is always in the x direction. The only way for the Lorentz force to vanish in this case to establish the force-free state is to require that the electric current density vanishes, except, possibly, at current sheets where the field reverses discontinuously. If a smooth force-free state is available to the magnetic field, this state must therefore be the unique potential field in the half-plane $z > 0$ associated with the normal flux distribution of the magnetic field at the boundary, quite irrespective of any concern over the field topology in the interior. For a fixed boundary normal flux distribution, the quadrupolar magnetic field

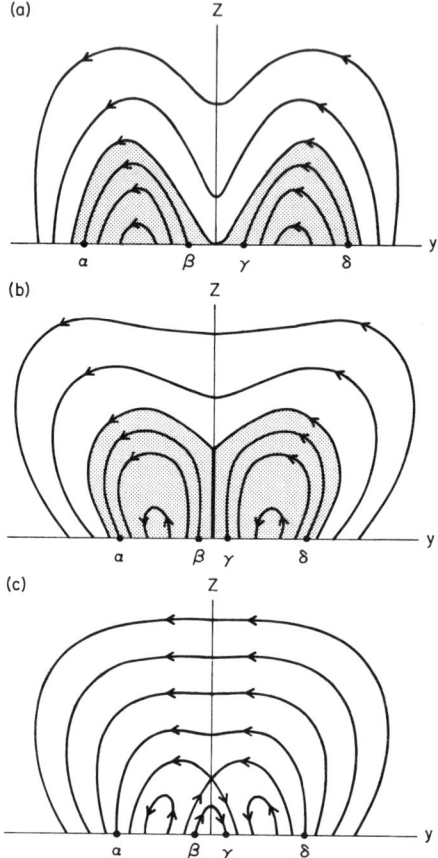

Figure 1 The field-line topologies of quadrupolar magnetic fields lying in the infinite half-plane $z > 0$.

can have an infinite number of topologies, depending on how pairs of footpoints at $z = 0$ are to be connected by the lines of force in the $z > 0$ half-plane. In general, the deformed magnetic field may have a topology different from the unique potential field in the $z > 0$ half-plane, although the two fields have the same boundary normal flux. In particular, the latter may in fact possess a neutral point in the $z > 0$ region such as is sketched in Figure 1c, whereas the deformed field does not have a neutral point. The difference in field topologies in these two states is indicated in Figure 1a and 1c by the field-line connectivities between the four footpoints labeled α, β, γ, and δ. The deformed magnetic field is then forbidden, under the frozen-in condition, to relax to this everywhere potential state. It must settle instead into the state sketched in Figure 1*b*, which is topologically compatible and is everywhere potential except on the vertical current sheet, a conclusion verifiable by direct construction (Low 1987).

The imposition of a specific field topology is additional to and independent of the requirement of force balance, as pointed out in Section 2.1. Magnetic singularities in the form of current sheets must be admitted when the two requirements are incompatible. The essence of the process forming these singularities is the interaction of three distinct magnetic flux bundles—namely, the squeezing of a third bundle out of the way of two others to allow the latter to come into contact. Once this essential point is recognized, the special role played by the separatrix line in the two-dimensional quadrupolar field of Figure 1 loses its significance. The condition for three-bundle interaction readily obtains in three-dimensional systems. Any magnetic field is locally made up of a family of parallel flux surfaces. Each flux surface contains a two-dimensional magnetic field. Let there be a third degree of variation—namely, a variation of the field patterns across the flux surfaces. This variation or shear in the magnetic field implies the presence of electric currents and therefore the availability of free energy. If this magnetic field is initially continuous but is subject to some stressing that pushes the flux surfaces together, there is a general tendency for the magnetic flux between a pair of flux surfaces to slip and to be expelled out of the way, creating a gap in space through which the two inward-pushing flux surfaces come into contact. The magnetic fields on these two surfaces are uncorrelated, so that in general a rotational discontinuity forms at the contact surface. To treat this process directly requires a three-dimensional model, and we refer the reader to Low (1989) for a demonstration using the force-free fields generated by Equation 2.16. The existence of gaps in equilibrium magnetic flux surfaces has also been demonstrated with elegance by the use of an optical analogy (Parker 1989b). The creation of such gaps is an intrinsically three-dimensional process, otherwise completely suppressed in symmetric systems. The mathematical problems associated with the formation of electric current sheets are exceedingly difficult, leaving room for some doubts (e.g. van Ballegooijen 1985, 1988, Zweibel & Li 1987, Field 1990). A notable attempt to prove that electric current sheets cannot form spontaneously is that of van Ballegooijen (1985), which we discuss in the next subsection. Actually, the reality of this process first pointed out by Parker is not questionable, for there are several explicit examples (Vainshtein & Parker 1986, Parker 1987, 1989a,d, Aly 1987, Low 1987, Moffatt 1987, Low & Wolfson 1988, Strauss & Otani 1988). On the other hand, many properties of this process are not yet fully understood, and working them out is a truly exciting development previously not expected in the classical study of magnetostatic equilibrium.

Coronal magnetic fields are subject to slow twisting on small scales due to random photospheric granular motions. Parker suggested that over these small scales, electric current sheets form readily in the corona and dissipate in response to random stresses. Order-of-magnitude estimates by Parker suggest that this process may heat and maintain the 10^6-K temperature in the corona over an active region (Parker 1986c). This part of the corona tends to be an order of magnitude denser than elsewhere, requiring an order of magnitude higher heating rate of some 10^7 ergs cm^{-2} s^{-1} (e.g. Withbroe & Noyes 1977). It is also the region of the corona where the magnetic fields are closed and anchored to the photosphere, conditions ideal for the working of Parker's mechanism. Elsewhere in open-field coronal regions, Parker's mechanism is not expected to work easily, since the twists in the field may propagate away as Alfvén waves along the open fields as fast as they are created by the relatively slow photospheric motions. The only known plausible way of heating the open-field regions requiring a lower heating rate of about 10^6 ergs cm^{-2} s^{-1} is the dissipation of hydromagnetic waves, an interesting subject for which the reader is referred to Leer et al (1982).

The sudden creation of a large-scale electric current sheet in the low corona may be the origin of such explosive processes as the flare (e.g. Canfield et al 1974, Low & Wolfson 1988). We have seen how energy can be stored in stable force-free magnetic fields in the form of volume electric currents. The creation of the current sheet would convert a fraction (probably a significant fraction in the case of a flare) into another form that can be rapidly and explosively dissipated away. There are some subtle and intriguing properties to examine in this connection. Let us return to the process of current sheet formation shown in Figures 1a and 1b. The initial field in Figure 1a is everywhere potential. A finite compressive displacement of the magnetic footpoints along the boundary $z = 0$ deforms the field so that it relaxes to the state shown in Figure 1b, which is everywhere potential except at the vertical current sheet shown. Now the initial potential field has no free energy. The free energy in the current sheet created can only come from the work done in displacing the magnetic footpoints. If the initial field is force free with considerable electric currents, the situation is conceivably quite different. To be specific, now regard the sketch in Figure 1a to represent the projection of a force-free field on the y-z plane generated from Equation 2.15 with $B_x = Q(A) \neq 0$. In particular, bear in mind that this field has a finite component out of the plane of projection. Subject this magnetic field to a continuous footpoint displacement in the x-direction on $z = 0$. In our consideration below, we do not restrict the footpoint displacement to be finite or infinitesimal. To make our point, take this displacement on $z = 0$ to be zero on the side

$y < 0$ and smoothly match it to a nonzero shearing displacement on the side $y > 0$. This displacement leaves the magnetic footpoints of the left shaded lobe unchanged. The footpoint separations of all lines of force outside this lobe would have changed, and for our purpose we take these changes to have increased the field component B_x everywhere outside the left shaded lobe. The enhanced field component tends to result in the left shaded lobe becoming compressed. A careful geometric consideration of the quadrupolar topology shows that although the footpoint displacement is continuous on $z = 0$, there is a discrete jump in the shear across the separatrix line that translates into a discrete jump in $B_x = Q(A)$ in the force-free state available to the displaced magnetic field. Following Low & Wolfson (1988), the final state is as sketched in Figure 2, possessing a complex current sheet formed with a three-way bifurcation out of the original separatrix line. In this case, the current sheet has a length as long as the original separatrix line, even if the imposed footpoint displacement is infinitesimal. Moreover, the energy (or the intensity) of the discrete current in the sheet is no longer to be bounded by the work done in displacing the footpoints. The free energy in the discrete current can include a major contribution from energy stored as volume currents in the initial state. This conversion of energy and the dissipation of current sheets created can help explain the common phenomenon of the almost spontaneous eruption of a flare in the corona without an obvious photospheric signal.

The foregoing analysis shows that current sheets may form with varying energy contents, depending on circumstances. The dissipation of a large-scale, highly energized current sheet obviously has catastrophic dynamical consequences. The consequences of the dissipation of a weak current sheet, on the other hand, are more subtle. If there is little free energy in the magnetic field that contains the current sheet, we expect no more than a

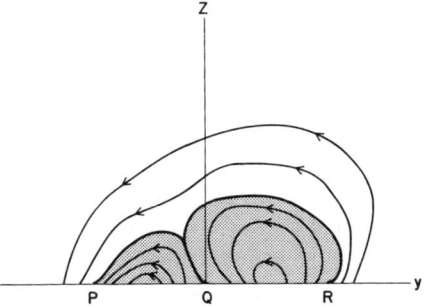

Figure 2 A sheared quadrupolar magnetic field with a complex electric current sheet intersecting the boundary $z = 0$ tangentially at points P, Q, and R.

weak heating from the dissipation of the current sheet. However, for a magnetic field containing a substantial amount of free energy, much depends on the change of magnetic topology brought about by the dissipation of the weak current sheet. If the changed topology allows the magnetic field to settle into a smooth equilibrium, again no more than a weak heating occurs. More interesting is the case where the changed magnetic topology under the nearly frozen-in condition requires the formation of new current sheets. This may initiate a runaway turbulent process in which more and more sheets form and dissipate, fueled by the free energy of the global magnetic field, in which case we have an explosive situation somewhat distinct from the sudden appearance of a large-scale current sheet (Parker 1988). These intriguing theoretical possibilities await further quantitative study.

3.2 Loss of Equilibrium

Quasi-steady evolution may also terminate at a point where instead of the spontaneous appearance of electric current sheets, the magnetic field is forced into a fully developed dynamical state of motion (Low 1982a, Birn & Schindler 1981). In this case, magnetic free energy is converted into ordered motion that may, perhaps, rapidly become turbulent and dissipative. I proposed an early model for such a loss of equilibrium that merits some discussion in the context of some recent criticisms of the model (Low 1977a,b, 1980, Jockers 1978, Aly 1984, Klimchuk & Sturrock 1989). The model is based on a set of solutions to the force-free equation (Equation 2.15) with

$$Q(A) = \lambda \exp(A), \qquad 3.1$$

which fixes the profile of Q over A while allowing the parameterized amplitude λ to be varied at will. These solutions describe bipolar force-free fields with lines of force anchored at both ends on the boundary $z = 0$. They share the same distribution of the normal field B_z on $z = 0$ but differ in the distribution of B_x in the half-plane $z > 0$. The continuous sequence of force-free states generated by increasing values of λ beginning with $\lambda = 0$ can be interpreted to represent a quasi-static evolution of a magnetic field from an initial potential state subject to a slow photospheric displacement in the x direction of the magnetic footpoints at $z = 0$. The initial potential field given by $\lambda = 0$ lies in the y-z plane, but an increasingly stronger x component develops as the shearing displacement takes the pair of magnetic footpoints of each line of force in opposite directions on the photosphere. It was found that the λ sequence terminates at a critical point $\lambda = 2$, beyond which no solutions that are topologically admissible under the frozen-in condition could be found. Based upon this result, I suggested that a dynamical transition out of the quasi-steady phase would set in at that critical point.

Further studies of the λ sequence provided two possible theoretical bases for the suggestion. An analysis of the linear stability of each of the $\lambda \leq 2$ force-free fields as a static state showed that absolute stability obtains for $\lambda < 2$, whereas the $\lambda = 2$ terminal state is marginally stable (Low 1977b). This analysis was performed for two-dimensional perturbations (i.e. $\partial/\partial x \equiv 0$), and the medium was taken to be cold (i.e. an absence of pressure). It was then argued that realistic force-free fields are never in a completely cold medium, and that the presence of pressure, however weak relative to the magnetic pressure, would introduce instability as the $\lambda = 2$ state is approached from below. In the language of the energy principle, we denote by δW the minimum change of the potential energy of the system due to infinitesimal perturbations (Bernstein et al 1958). For the λ sequence force-free fields in a cold medium, it was found that $\delta W > 0$ for $\lambda < 2$, implying absolute stability, and that $\delta W = 0$ at $\lambda = 2$, implying marginal stability. If the presence of weak pressure in the equilibrium state is allowed for, the resulting small correction of δW may dip the total minimum change in potential energy below zero, implying an unstable state, in the neighborhood $\lambda \to 2$. The analysis in Low (1977b) was based on a minimization over variations in the magnetic field rather than in the plasma displacement. It was realized later that the result $\delta W = 0$ at $\lambda = 2$ was based on magnetic field variations not compatible with plasma displacements that satisfy the rigid boundary conditions at $z = 0$ (Low 1985b). Correction of this technical flaw led to $\delta W > 0$ at $\lambda = 2$. It would be interesting to see if the result can be recovered by examining stability in response to three-dimensional perturbations. In any case, while the mechanism for the onset of instability proposed in Low (1977b) is not without interest for the quasi-static approach to any force-free state that is truly marginally stable in the limit of a cold medium, its relevance to the termination of the λ sequence by a loss of equilibrium is vitiated at this point of the development.

The second basis for suggesting a loss of equilibrium in the λ sequence is more promising (Low 1980). A role is again played by the weak equilibrium pressure expected in realistic environments. We remind ourselves that a static magnetic field in a low-β plasma is taken to be force-free only in the sense of the lowest order of approximation; β is the ratio between plasma and magnetic pressures. This approximation tacitly assumes that the higher order departures in the magnetic field from the exact force-free state can be readily computed if the presence of weak plasma pressure in static equilibrium is to be accounted for. These higher order corrections were shown to be calculable for the $\lambda < 2$ force-free fields but not for the

$\lambda = 2$ terminal state. The mathematical nonexistence of the magnetic field corrections for the $\lambda = 2$ force-free field shows that this field exists only in a completely cold medium. If an arbitrary pressure is introduced, however weak, the field cannot in general adjust by infinitesimal amounts under the frozen-in condition to accommodate the pressure gradient in equilibrium. Hence, as the $\lambda = 2$ state is approached from below, the presence of weak pressure that has negligible influence in the regime $\lambda < 2$ would become significant in initiating a loss of equilibrium.

There is a simple virial consideration that provides physical insight for understanding the absence of solutions to Equations 2.15 and 3.1 when λ exceeds a critical value (Aly 1984). If one integrates for the total Lorentz force in the half-plane $z > 0$ and sets it to zero for a force-free field, an application of Gauss' theorem yields, for the two-dimensional system, the necessary condition

$$\int_{z=0} (B_z^2 - B_x^2 - B_y^2) \, dy = 0, \qquad 3.2$$

where the field is assumed to vanish sufficiently rapidly at infinity in $z > 0$. The λ sequence of force-free fields has been constructed so that the normal field component B_z at $z = 0$ is independent of λ. By Equation 2.6, we have

$$A = -\int_0^y B_z \, dy$$

on $z = 0$, and thus A on $z = 0$ is known and also independent of λ. It then follows from setting $B_x = Q(A)$ with Equation 3.1 that condition 3.2 is violated if λ exceeds a certain upper bound beyond which no force-free state exists. A point of caution is worth noting, as first pointed out by Jockers (1978) and, more recently, by Klimchuk & Sturrock (1989; see also Low 1982a, Aly 1984). The failure of solutions to Equations 2.15 and 3.1 for $\lambda > 2$ does not preclude the possibility of continuing the λ sequence at $\lambda = 2$ smoothly into another sequence of force-free states generated by a different functional form of $Q(A)$. This is of course true but is not of physical significance, since (as pointed out in Low 1982a) loss of equilibrium is expected as $\lambda \to 2$ *from below*. In this sense, the rejection of the idea of loss of equilibrium by Klimchuk & Sturrock (1989) seems to be based on a narrow definition of the idea—namely, a restriction to the case where no pressure is admissible. Two important points should not be overlooked. The first is that the λ sequence involves footpoint displacements on $z = 0$ whose shear becomes infinite in the limit $\lambda \to 2$, when the two photospheric regions carrying opposite magnetic polarities behave as though they have slipped rigidly and discontinuously along their common boundary (Low 1977a). This is an extreme condition, and it is hardly surprising that some form of instability would set in prior to reaching the terminal state. The second point is that the quasi-static evolution modeled would be completely reversible if not for the question of departure from the quasi-static state via instability or loss of equilibrium in the neighborhood of the $\lambda = 2$ state. Smooth sequences of force-free fields with different forms of $Q(A)$ that lead smoothly to the $\lambda = 2$ force-free state merely imply the possibility of several different quasi-static paths leading to the same extreme state. It is the loss of equilibrium setting in near the extreme state that renders physically unlikely the continuation from one sequence to another through the extreme state (that connects them).

The above theory is incomplete in that no quantitative calculations have been made to show the precise nature of the dynamical state following the loss of equilibrium. It was suggested that this dynamical state may create current sheets and lead to reconnection of the magnetic field (Low 1982a, Aly 1987). It is also plausible that the dynamical state is actually uninteresting physically because it is rapidly quenched, whereupon the system attains another equilibrium state with a finite but insignificant change of energy, as Biskamp & Welter (1989) have suggested from their numerical simulations. An important consequence of the loss of equilibrium is the conversion of a magnetic field from a closed to an open configuration. This type of dynamical transition is believed to underlie the phenomenon of coronal mass ejections (discussed in Section 4.1). The effect is difficult to demonstrate with the planar model invariant in x because it takes an infinite amount of energy to open up a magnetic field in this system. In the y-z plane, a magnetic field opened to infinity declines with radial distance $R = (y^2+z^2)^{1/2}$ like $1/R$, and therefore its total energy, even if the integration is restricted over the plane, is unbounded. In three-dimensional space, it takes a finite energy to open up an initially closed magnetic field in all directions. Barnes & Sturrock (1972) have suggested that a highly twisted force-free magnetic field might have an energy exceeding that of a fully open untwisted field and in such a case would undergo a dynamical transition to open up its lines of force. This attractive mechanism is complicated by a plausible but unproven conjecture of Aly (1984). From particular examples it was suggested that an initially closed force-free magnetic field in an infinite space but anchored at some inner boundary (for example, an axisymmetric dipolar magnetic field anchored on the unit sphere) cannot have an energy exceeding that of an untwisted field that is everywhere open and having the same boundary flux distribution (Yang et al 1987). The open field may be visualized to be the result of taking the top of each bipolar line of force of the initial field to infinity while holding the magnetic footpoints anchored on the boundary, creating an electric current sheet surface separating magnetic fluxes of opposite signs. While we should bear in mind Aly's conjecture, it is important to realize that in an eruption, there is no compelling reason to demand the opening up of all magnetic lines of force. For example, take the dipolar magnetic field anchored on a unit sphere. It is conceivable that a highly twisted magnetic field may have enough free energy to transit to a lower energy state in which only a fraction of its total magnetic flux is opened to infinity. For the dipolar field, it would be the polar magnetic lines of force extending far from the unit-sphere boundary that open out to infinity, forming an infinite equatorial current sheet with some inner edge lying outside the unit sphere. The existence of this lower energy end state awaits proper direct demonstrations for specific initial closed force-free fields. A persuasive case for this possibility can be made by the analysis of the twisting of a dipolar magnetic field, which shows a loss of equilibrium for certain classes of magnetic footpoint displacements (Low 1986).

Loss of equilibrium has also been considered for sequences of equilibria in which pressure, gravity, and drift currents together play a role. The simplest example is provided by the equilibrium of a slender magnetic flux tube anchored at the base of a stratified, otherwise field-free atmosphere (Parker 1979). The thin flux tube is magnetically buoyant but can be held in equilibrium by the magnetic tension force, provided that the separation of the two magnetic footpoints is not greater than a critical distance related to the hydrostatic density scale height. The equilibrium shape of the flux tube is determined through equilibrating the magnetic pressure at each point of the thin flux tube with the external confining hydrostatic pressure. The magnetic intensity so fixed, together with the flux tube curvature, determines the magnetic tension force required to balance the buoyancy force. If the footpoint separation is increased to the critical distance from below, buoyancy rapidly dominates over the magnetic tension force, so that the heights of the loop apex of successive equilibrium states tend to infinity. As a quasi-static process, equilibrium is lost as the critical footpoint separation is approached because the adjustment to the next available equilibrium would involve finite upward displacements of the loop apex not bounded by the footpoint displacements. More sophisticated models involving continuous two-dimensional magnetic fields have been studied in connection with the initiation of coronal mass ejections (discussed in Section 4.1; Low 1982c, Low et al 1982, Wolfson 1982).

3.3 *Rigid Anchoring of Magnetic Fields*

It is instructive at this point to discuss subtle aspects of the role of the photosphere in anchoring the coronal magnetic field, which appear not to have been discussed in the literature. If we treat the photosphere as a rigid perfect conductor threaded by the magnetic field, all components of the plasma motion must vanish at the photosphere (e.g. Roberts 1964). The normal component vanishes because of rigidity, and the transverse components vanish because of the continuity of the tangential electric field. It has been suggested that these rigid boundary conditions may be relaxed by keeping the transverse plasma displacement zero but allowing a flow across the photosphere subject to a variety of artificial conditions in order to keep the corona energetically conserved (e.g. van Hoven et al 1981). These models have unclear physical meaning, as discussed by Low (1985b) and Rosner et al (1985). The principal criticism is that if the photosphere is not rigid, its interaction with the corona naturally leads to an exchange of energy, and the model becomes physically ambiguous if that interaction is not given a clear physical statement. Perhaps the problem of the photosphere-corona interaction is meaningless without an explicit consideration of the chromosphere that separates the two regions—that is, without treating the complexity of energy processes that maintain the transition between these two regions. We have no suggestion on how to treat this formidable problem.

Taking a different approach to this concern as a first step away from the rigid boundary condition, we may think of the photosphere by its own slow motion as a deformable wall across which no flow is allowed. This deformable wall by its high inertia may change in time. Therefore, given a static coronal magnetic structure anchored on the photosphere, we should be interested in more than just the linear mechanical stability of that static structure with the photosphere in a fixed configuration, the traditional stability problem. We should also investigate the whole spectrum of static equilibrium states available to that coronal structure associated with the photosphere being deformed (linearly) into different configurations in the neighborhood of some fixed configuration. The interesting question is, Which of these deformed photospheric configurations admit stable smooth equilibrium, and which others do not? The idea of structural stability arises naturally here. In this broader physical consideration, the criteria for structural stability are (a) that equilibrium exists and is individually linearly stable for *all* admissible photospheric configurations, and (b) that the equilibrium state is a continuous function of the photospheric configuration. The absence of smooth equilibrium or the presence of a linearly unstable equilibrium for any particular photospheric configuration implies structural instability. It is easy to show that the potential field in Figure 1a is mechanically (linearly) stable in the traditional sense but is structurally unstable to infinitesimal shearing displacements of magnetic footpoints in the x-direction (Low & Wolfson 1988).

Another aspect worth investigating arises from within the rigid boundary model. If we introduce an isotropic Ohmic resistivity with a constant coefficient η into Faraday's law (Equation 2.2), it can be shown that the scalar function A describing the two-dimensional magnetic field (Equation 2.6) evolves in time according to

$$\frac{\partial A}{\partial t} + \mathbf{v} \cdot \nabla A - \eta \nabla^2 A = 0. \qquad 3.3$$

In the absence of resistivity, i.e., $\eta = 0$, the rigid boundary condition sets $\mathbf{v} = 0$ at the photosphere $z = 0$ with the normal flux $B_z = -\partial A/\partial y$, and hence A is fixed in time at $z = 0$. Applying this condition rigorously is mathematically crucial when stability is established in the ideal approximation. With a small but finite resistivity, the slippage of plasma across magnetic field lines can be a small but significant perturbation of the boundary condition, resulting in the introduction of resistive instabilities (characterized by hybrid growth times) that are physically important for equilibrium states otherwise quite stable in the ideal approximation. These types of instabilities have not received attention previously. This idea is analogous to that of the resistive tearing instabilities of Furth et al (1963).

Finally, we present a case where the rigid boundary model itself must physically break down in the photosphere. We return to Figure 2, showing a force-free field that has developed a complex current sheet as the result of structural instability (Low & Wolfson 1988). Bear in mind that we are looking at the magnetic field of the form given by Equation 2.6, projected onto the y-z plane. There is a discrete jump in $B_x = Q(A)$ across the current sheet, and for equilibrium this jump must be compensated by appropriate jumps in B_y and B_z such that the total magnetic pressure $B_x^2 + B_y^2 + B_z^2$ is continuous across the sheet. We recall that this end state has been created by a smooth footpoint displacement at $z = 0$. Thus, B_z on $z = 0$ is everywhere continuous in the end state, and this condition has the following interesting implication. Use the symbol Δ to represent the jump of a quantity across the current sheet, and let $W^2 = B_y^2 + B_z^2$. Equilibrium therefore requires that $\Delta Q^2 + \Delta W^2 = 0$ along the current sheet. Take one of the three points P, Q, R in Figure 2 where current sheets intersect the boundaries. If the current sheet intersects the boundary $z = 0$ not tangentially as sketched in Figure 2 but at some nonzero angle θ from the boundary, then $B_y = W \cos \theta$ and $B_z = W \sin \theta$. At this point of intersection, B_z is continuous. Therefore, if W is finite, it is also continuous and it follows that B_y is continuous. Thus, $\Delta W^2 = 0$ at the point of intersection. But $B_x = Q(A)$ is constant along any line of force, and in particular the jump in B_x across the current sheet is constant along the current sheet, so that $\Delta B^2 = \Delta Q^2 + \Delta W^2 \neq 0$ at the intersection point on the boundary, where we would then not have equilibrium. To avoid this paradox within the rigid boundary model, we must let θ tend to zero and W tend to infinite values at different rates on the two sides of the current sheet as we approach the point of intersection, such that B_z remains finite and continuous whereas B_y is infinite and discontinuous but just such that $\Delta Q^2 + \Delta W^2 = 0$. The current sheet would then thread the boundary tangentially with $\theta = 0$, as sketched in Figure 2. The infinity in the magnetic field is integrable in a suitable manner, so that there is no infinite energy involved. However, there is an infinite Lorentz force acting on the boundary where the current sheet intersects it. This infinite force density does not present a problem within the logic of the mathematical model because it is taken up by the rigid boundary. Applied to the photosphere, however, a point will be reached during the formation of the electric current sheet, well before the Lorentz force becomes infinite, when the infinite rigidity of the boundary must be abandoned. The photosphere would then yield to the build up of stress.

Two points follow from this analysis. The first is that the strong field gradient developing at the footpoints of the electric current sheet suggests that these footpoints are among the first parts of the magnetic structure to manifest resistive heating if, for example, a flare is thus initiated. This process may contribute to the brightening of magnetic footpoints during the impulsive phase of a flare. This brightening effect has hitherto been interpreted to be caused by particle beams and thermal conduction fronts (Dennis & Schwartz 1989). The other point is that any assumption of regularity at the intersection between current sheets and rigid boundaries should be made with caution if paradoxical results are to be avoided. This second point has an implication for the interesting method due to van Ballegooijen (1985) for the construction of force-free magnetic fields lying between a pair of infinite parallel plates (the Parker problem). These magnetic fields are obtained by twisting an initial uniform magnetic field that threads these plates perpendicularly through some prescribed smooth boundary footpoint displacement. Van Ballegooijen's method is one of extrapolating for the interior field from one plate to the other, starting with initial data prescribed on one of the plates. Using a multiscale expansion, it was found that an infinite class of force-free fields can be so constructed. In particular, solutions exist for force-free fields varying with distance from one plate to the other, thus disproving an early result of Parker (1972) that only those solutions independent of this distance are admissible. Based on this construction, van Ballegooijen rejected the idea that electric current sheets would form in this system as the result of the random twisting of the magnetic field. It seems clear that van Ballegooijen's position cannot be supported. The crucial issue is whether the infinite set of solutions generated by van Ballegooijen's method admits all the possible field topologies that one can create by twisting the initial uniform field between the two plates. Short of imposing the necessary global constraints explicitly to fix field topologies, van Ballegooijen's method as a differential extrapolation simply generates those field topologies compatible with smooth solutions. This procedure is of the same spirit as that under which the three-dimensional magnetostatic solutions described in Section 2.1 are generated. We saw in the example shown in Figure 2 that where the current-sheet intersects the boundary, singularities may be expected. This suggests that if van Ballegooijen's method is to recover those force-free fields with electric current sheets claimed by Parker, the extrapolation must start with the proper singularities in the initial data to allow for current sheets that extend to one of the plates. It is also possible that the expansion procedure underlying this particular extrapolation scheme does not work at all in the presence of such singularities (Rosner & Knobloch 1982). Much further work needs to be done to address these open questions.

4. TIME-DEPENDENT OUTFLOWS

A good-quality eclipse photograph taken during solar activity minimum is the simplest way to see the large-scale ordering of the corona into closed and open magnetic field regions (Newkirk 1967). The corona at this point in the solar cycle has a predominantly dipolar field, open in extensive areas over the poles and closed over the magnetic equator. In Thomson-scattered light, the open regions are conspicuously dark ("the coronal holes") because of their low density, while the closed regions appear as bright helmet-shaped features that extend from opposite limbs of the Sun out to connect with the interplanetary current sheet (Parker 1963, Hundhausen 1977, Holzer 1979). The physics of these coronal features is still not fully understood (Pizzo et al 1987). Much work needs to be done on the structure and acceleration of the solar wind along the open magnetic fields, the equilibrium and stability of the closed field regions, and the interaction between these regions of contrasting dynamical states that controls the global atmosphere. The corona evolves quasi-steadily as a whole, driven by the solar dynamo that continually injects new magnetic flux to eventually reverse the solar magnetic polarity every half–solar cycle. This results in a change from the dipolar magnetic configuration at activity minimum to a much more complex configuration dominated by high magnetic moments during field reversal at activity maximum. Despite its complexity, the corona in the latter case is still ordered into the basic components of closed and open regions (Hundhausen 1977). In the early 1970s, it became possible to observe rapid changes in the dim outer corona by orbiting coronagraphs in space. Direct observations revealed that the quasi-steady change of the corona is interrupted as often as once a day by large-scale dynamical reconfigurations, with outward expulsions of mass of the order of 10^{15} g per event. The study of these dynamical processes became a primary interest in coronal physics. Once their existence became known, observations of these phenomena were also made from the ground. Data accumulated from three major orbiting coronagraphs on the Skylab (1973–74), the *Solwind* satellite (1979–85), and the *Solar Maximum Mission* (*SMM*) satellite (1980–89) and from the High Altitude Observatory coronagraphs at Hawaii have changed radically our perception of the corona, from the passive system deceptively suggested by static eclipse photographs to a highly active system capable of initiating major eruptions of its own (MacQueen 1980, Hundhausen et al 1984b, Hundhausen 1987, Kahler 1987).

4.1 *Coronal Mass Ejections*

While coronal mass ejections come in a broad variety of shapes, a great majority of them have the appearance in white light of a leading bright loop surrounding a dark cavity with a dense filamentary core (see, for example, Hundhausen 1987). An early suggestion was that these ejections were caused by the energy of a flare liberated impulsively at the coronal base (e.g. Maxwell & Dryer 1981). This idea was found to be untenable on two counts, leading to an eventual recognition that mass ejections are coronal eruptions in their own right. Firstly, many mass ejections originate from the dynamical breakup of helmet features, without being accompanied by flares (Munro et al 1979, Webb & Hundhausen 1987). The eruption of a flare is therefore not essential to the mass ejection phenomenon. Secondly, for those mass ejections accompanied by flares, a careful examination of the timing of the various components of these not uncommon events showed that the mass ejection would almost always precede the associated flare by some 20 min (Harrison 1986). That most mass ejections should originate from helmet features is interesting (Illing & Hundhausen 1985, 1986). This fact allows one to trace the origin of the three-part structure of the mass ejection to a similar structure in the preeruption helmet: the dense helmet dome overlying a low-lying cavity in which a quiescent prominence resides (Hundhausen et al 1984a, Low 1984b). At an average rate of occurrence of one event a day and an average mass of 10^{15} g per event, the contribution to the solar mass loss rate is well less than 10% of that estimated for the steady solar wind. The astrophysical significance of mass ejections lies not in this meager contribution but in the opening up of otherwise closed coronal magnetic fields to enable mass to escape in the solar wind.

The speeds at which mass ejections travel through the corona are surprising in that the leading edge of the loop can have a speed in a broad range, from 10 to 10^3 km s^{-1} (Hundhausen et al 1987). The corona is characterized by a sound speed of 120 km s^{-1}, an estimated Alfvén speed of about 600 km s^{-1} (at $3R_\odot$), and a gravitational escape speed of 550 km s^{-1} in the low corona. The median mass ejection speed is about 350 km s^{-1}, so that most mass ejections are supersonic but sub-Alfénic and, interestingly, tend to travel below the gravitational escape speed in the inner corona. In fact, there is a common tendency for the mass ejection loop to travel with little acceleration in the low corona. The total mechanical energy in an event is about 10^{31-32} ergs, the kinetic energy being comparable to the work done against gravity to lift 10^{15} g of matter away from the Sun. The magnetic field is believed to be the source of this energy. Unfortunately, the magnetic field cannot be measured directly, but it has been suggested from radio observations that some 10^{31-32} ergs of magnetic energy are carried away with the expelled mass (MacQueen 1980).

A simple way of understanding the mass ejection phenomenon is to realize that the corona maintained at a 10^6-K temperature would be everywhere expanding in the solar wind if not for the magnetic field. (Actually, it is the magnetic field that heats the corona, but let us for the purposes of this discussion not dwell on the origin of the heating.) The magnetic field through the Lorentz force introduces two things: an isotropic (magnetic) pressure and a tension force. The magnetic pressure cannot provide confinement; in fact, it enhances the tendency of the embedding hot gas to expand. The magnetic tension force can provide gas confinement in a closed field geometry, giving rise to the helmet features. When that confinement fails in the course of stressing during quasi-steady evolution, such as discussed in Section 2, the entire structure breaks lose and expands as a mass ejection. A credible scenario can be argued from theory and observation that the innocuous prominence (magnetic) cavity under the helmet dome can become buoyantly unstable and serve to trigger a mass ejection (Low 1982c, Low et al 1982, Wolfson 1982, Wolfson & Gould 1985).

When it is recognized that the initiation of the mass ejection need not be impulsive, the broad range of speeds it can have in the low corona becomes less of a surprise. For dynamical reasons not fully understood yet, the time-dependent outflow of a mass ejection tends to align the Lorentz force, gravity, and pressure gradient so that there is little acceleration of the mass ejection during its fully developed stage in the corona. Theoretically, we are assured that such outflows are indeed possible, as demonstrated by a class of two- and three-dimensional, self-similar, time-dependent solutions to the full hydromagnetic equations (Equations 2.1–2.4) with $\sigma = 0$ for a $\gamma = 4/3$ polytrope (Low 1984b). In these solutions the velocity has a certain dispersion in space, but individual plasma parcels tend to move inertially at constant Lagrangian speeds. Particularly interesting is that a completely free parameter allows the modeling of a mass ejection loop moving at any prescribed speed. A simple interpretation of these outflows is that they arise from a heated magnetized atmosphere that maintains just enough energy to moderately clear the gravitational potential barrier of the Sun at all points during its motion. The significance of the requirement that the polytropic index γ be 4/3 is that this polytrope has an internal energy that, together with the magnetic energy, degrades at the same rate with the gravitational potential energy in a homologous expansion. In this manner, the self-similarity of the outflow, once established, can be preserved. While these self-similar theoretical outflows do capture some essential aspects of the large-scale motions of mass ejections, significant departures from self-similar behavior are found in the detailed study of the evolution of the individual mass ejection structures (Illing 1984, Low & Hundhausen 1987).

4.2 Hydromagnetic Shocks

Mass ejections are finite-amplitude motions in the corona, and the possibility exists for these motions to generate hydromagnetic shocks. Let us recall that there are three hydromagnetic waves with anisotropic propagation speeds: the fast, the slow, and the Alfvén waves. The first two modes are compressive, coupling the plasma and magnetic pressures. When these two pressures oscillate in phase we have the fast mode, and when they oscillate out of phase we have the slow mode, with the former having the faster propagation speed, as their names imply. The Alfvén wave is incompressible, and it arises from the magnetic lines of force behaving like elastic strings. Shocks associated with these waves are generated whenever appropriate disturbances are created faster than the respective waves can respond to them.

In the corona, the Alfvén speed is generally higher than the sound speed. In this case a mass ejection can drive a fast-mode shock only if it travels faster than the local characteristic fast-mode speed, which by definition exceeds the Alfvén speed. The Alfvén speed in the corona is about 600 km s^{-1} at $3R_\odot$. There are not many mass ejections at any solar cycle epoch having such high speeds. An examination of the coronagraph data has found a few examples that are consistent with the interpretation that the loop's leading edge is a fast shock (Sime & Hundhausen 1987).

Since most mass ejections travel at sub-Alfvénic speeds, they cannot be expected to generate fast shocks in the corona. However, since the median speed is 350 km s^{-1}, many of these mass ejections are supersonic, meeting the physical requirement for them to generate slow shocks (Whang 1987, Hundhausen et al 1987). Slow shocks are peculiar, and thus a little digression is necessary (Sears & Resler 1961, Tamada 1964, Spreiter et al 1970, Kennel et al 1985, Wolfson 1987). Fast shocks are rather like the familiar gas-dynamic shock in that they would drape around the driver as sketched in Figure 3a (from Hundhausen et al 1987), which shows the

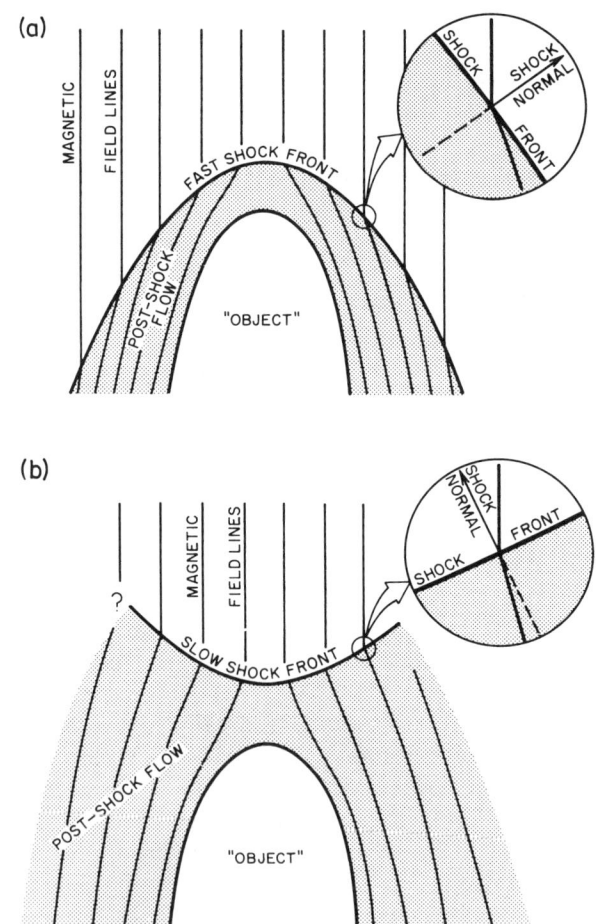

Figure 3 The shapes of steady fast and slow hydromagnetic shocks. The question mark in (*b*) indicates the finite but uncertain lateral extent of the slow shock front.

driver as an object moving steadily in the direction of the uniform magnetic field in the ambient medium. In a fast-mode wave, the plasma and magnetic pressure change in phase. Hence, the magnetic pressure increases with the plasma pressure across the fast shock. The normal component of the magnetic field is continuous across the shock. The increase of magnetic pressure across the shock therefore implies that the ambient magnetic field is deflected away from the shock normal. Such a magnetic configuration fits the drape-around geometry of the shock surface, in that the ambient magnetic field is deflected out of the way of the driver, as sketched in Figure 3a. With slow shocks, the drape-around geometry presents a problem. The changes of plasma and magnetic pressures are out of phase for a slow-mode wave. Across a slow shock, the magnetic pressure decreases with the increase of plasma pressure. The continuity of the normal magnetic field component in this case implies a deflection of the ambient magnetic field toward the shock normal. If the slow shock surface is draped around the driver as sketched in Figure 3a, the ambient magnetic field would be deflected into the way of the driver and cause a problem. The surface of a slow shock in such a configuration would evolve rapidly into a front geometry concaved the other way around, as shown in Figure 3b. This is a stable configuration, in that it allows the ambient magnetic field to be deflected out of the way of the driver. Moreover, theory shows that the lateral extent of the slow shock is limited, as sketched.

The coronagraph cannot determine the presence of a shock unambiguously; we need to treat the Rankine-Hugoniot relations to do that. But the unconventional shape of the slow shock front is something that one can hope to observe with a coronagraph. In the corona the ambient magnetic field in the open field region is radial, in spherical geometry. A slow shock front propagating into this magnetic field may not have a concave-outward geometry, as sketched in Figure 3b, but may instead take on a flattened shape as opposed to a convex or spherical front. Evidence of a conspicuous flattening of the leading edge of a mass ejection loop has been known since the Skylab observations and has also been found in the *SMM* coronagraph data set (Hundhausen et al 1987). What until recently has been regarded as a mere geometric curiosity of certain mass ejection loops now may possibly indicate a naturally occurring slow hydromagnetic shock. This development has led to renewed interest in hydromagnetic shocks. Some interesting questions have arisen on the interaction between the slow shock front with the medium ahead via the fast-mode waves and on the generation of the intermediate shocks, to name just two. These questions are actively being pursued in numerical studies (Steinolfson & Hundhausen 1989a,b, Hu et al 1990).

5. CONCLUSION

The previous three sections have surveyed various hydromagnetic processes whose study has been stimulated by the observed behavior of the magnetic field in the solar corona. The theoretical constructions used to study these processes teach us as much about the basic hydromagnetic principles as about the coronal magnetic field itself. This is as it should be, because there is no ready hydromagnetic knowledge waiting to be simply applied for an understanding of the corona. Thus, coronal phenomena extend our physical knowledge of hydromagnetic processes. The behavior of open plasma systems in the corona has motivated us to think of models that identify the right kind of theoretical questions to ask and to be critical about the physical interpretation of these models as we proceed. It seems a fair assessment that we have, through the study of equilibrium and its stability, of onset mechanisms for eruption, and of fully developed time-dependent flows, accomplished a qualitative physical understanding of the basic behavior of the coronal plasma. Several other important topics on the corona have not been discussed here, partly because of limitation of space and partly because there are others more competent than I to treat them. Notable among these omitted topics are plasma kinetic processes (especially those complex processes that make up the flare) and the many yet unresolved issues concerning the solar wind. Within the limited scope of this review, there is still much analytic exploration to be done on many outstanding questions, some of which have been raised or discussed in the preceding three sections, notably those concerning the onset of eruption. On the other hand, many basic physical issues have become more sharply focused, and it thus will be worthwhile to exploit numerical tools combined with careful interpretation to further our understanding of these issues.

ACKNOWLEDGMENTS

I thank Tom Holzer for comments on the manuscript, and Keith Moffatt for kind hospitality at Cambridge University during the writing of this review.

Literature Cited

Aly, J. J. 1984. *Ap. J.* 283: 349
Aly, J. J. 1987. *Proc. Workshop Interstellar Magn. Fields*, ed. R. Beck, R. Gräve, p. 240. New York: Springer-Verlag
Arnol'd, V. 1974. *Proc. Summer Sch. Differ. Equ., Erevan.* Armen. SSR Acad. Sci. (In Russian)
Barnes, C. W., Sturrock, P. A. 1972. *Ap. J.* 174: 659
Bernstein, I. B., Frieman, E. A., Kruskal, M. D., Kulsrud, R. M. 1958. *Proc. R. Soc. London Ser. A* 244: 17
Birn, J., Schindler, K. 1981. In *Solar Flare Magnetohydrodynamics*, ed. E. R. Priest, p. 337. New York: Gordon & Breach
Biskamp, D., Welter, H. 1989. *Sol. Phys.* 120: 49
Bogdan, T. J., Low, B. C. 1986. *Ap. J.* 306: 271
Canfield, R. C., Priest, E. R., Rust, D. M.

1974. In *Flare-Related Magnetic Field Dynamics*, ed. Y. Nakagawa, D. Rust, p. 361. Boulder, Colo: NCAR
Chang, H. M., Carovillano, R. L. 1981. *Bull. Am. Astron. Soc.* 13: 309
Dennis, B. R., Schwartz, R. A. 1989. *Sol. Phys.* 121: 75
Field, G. B. 1990. In *Topological Fluid Mechanics*, ed. H. K. Moffatt, p. 244. Cambridge: Univ. Press
Friedberg, J. P. 1982. *Rev. Mod. Phys.* 54: 801
Furth, H. P., Killen, J., Rosenbluth, M. N. 1963. *Phys. Fluids* 6: 45
Gold, T. 1964. In *Physics of Solar Flares. NASA SP-50*, ed. W. N. Hess, p. 389. Washington, DC: NASA
Haisch, B. M., Rodono, M., eds. 1989. *Solar and Stellar Flares. Proc. IAU Colloq. No. 104* (*Sol. Phys.*, Vol. 121)
Harrison, R. A. 1986. *Astron. Astrophys.* 162: 283
Holzer, T. E. 1979. In *Solar System Plasma Physics*, ed. C. F. Kennel, L. J. Lanzerotti, E. N. Parker, 1: 101. Amsterdam: North-Holland
Hood, A. W. 1983. *Sol. Phys.* 87: 279
Hood, A. W., Priest, E. R. 1981. *Geophys. Astrophys. Fluid Dyn.* 17: 297
Hu, Y. Q. 1983. *Chin. J. Space Sci.* 3: 261
Hu, Y. Q., Zhu, Z. W., Hundhausen, A. J., Holzer, T. E., Low, B. C. 1990. *Sci. Sin.* 333(3): 332
Hundhausen, A. J. 1977. See Zirker 1977, p. 225
Hundhausen, A. J. 1987. See Pizzo et al 1987, p. 181
Hundhausen, A. J., Holzer, T. E., Low, B. C. 1987. *J. Geophys. Res.* 92: 11,173
Hundhausen, A. J., MacQueen, R. M., Sime, D. G. 1984a. *Eos, Trans. Am. Geophys. Union* 65: 1069
Hundhausen, A. J., Burlaga, L. F., Feldman, W. C., Gosling, J. T., Hildner, E., et al. 1984b. In *Solar Terrestrial Physics: Present and Future*, ed. D. M. Butler, K. Papadopoulos (NASA Publication)
Hundhausen, J. R., Hundhausen, A. J., Zweibel, E. G. 1981. *J. Geophys. Res.* 86: 11,117
Illing, R. M. E. 1984. *Ap. J.* 280: 399
Illing, R. M. E., Hundhausen, A. J. 1985. *J. Geophys. Res.* 90: 275
Illing, R. M. E., Hundhausen, A. J. 1986. *J. Geophys. Res.* 91: 10,951
Jockers, K. 1978. *Sol. Phys.* 50: 405
Kahler, S. 1987. *Rev. Geophys.* 25: 663
Kennel, C. F., Edmiston, J. P., Hada, T. 1985. In *Collisionless Shocks in the Heliosphere. Geophys. Monogr.*, ed. R. G. Stone, B. T. Tsyrutani, 34: 1. Washington, DC: Am. Geophys. Union

CORONAL MAGNETIC FIELDS 523

Klimchuk, J., Sturrock, P. A. 1989. *Ap. J.* 345: 1034
Leer, E., Holzer, T. E., Fla, T. 1982. *Space Sci. Rev.* 33: 161
Lerche, I., Low, B. C. 1980. *Sol. Phys.* 67: 229
Low, B. C. 1975. *Ap. J.* 197: 251
Low, B. C. 1977a. *Ap. J.* 212: 234
Low, B. C. 1977b. *Ap. J.* 217: 988
Low, B. C. 1980. *Ap. J.* 239: 377
Low, B. C. 1982a. *Rev. Geophys. Space Sci.* 20: 145
Low, B. C. 1982b. *Sol. Phys.* 77: 43
Low, B. C. 1982c. *Ap. J.* 251: 352
Low, B. C. 1982d. *Ap. J.* 263: 952
Low, B. C. 1984a. *Ap. J.* 277: 415
Low, B. C. 1984b. *Ap. J.* 281: 392
Low, B. C. 1984c. *Ap. J.* 286: 772
Low, B. C. 1985a. *Ap. J.* 293: 31
Low, B. C. 1985b. *Sol. Phys.* 100: 309
Low, B. C. 1986. *Ap. J.* 307: 305
Low, B. C. 1987. *Ap. J.* 323: 358
Low, B. C. 1988a. *Ap. J.* 330: 992
Low, B. C. 1988b. *Sol. Phys.* 115: 269
Low, B. C. 1989. *Ap. J.* 340: 558
Low, B. C., Hundhausen, A. J. 1987. *J. Geophys. Res.* 92: 2221
Low, B. C., Wolfson, R. 1988. *Ap. J.* 324: 574
Low, B. C., Wu, S. T. 1981. *Ap. J.* 248: 335
Low, B. C., Munro, R. H., Fisher, R. R. 1982. *Ap. J.* 254: 335
MacQueen, R. M. 1980. *Philos. Trans. R. Soc. London Ser. A* 297: 605
Maxwell, A., Dryer, M. 1981. *Sol. Phys.* 73: 313
Migliuolo, S. 1982. *J. Geophys. Res.* 87: 8057
Milne, A. M., Priest, E. R., Roberts, B. 1979. *Ap. J.* 232: 304
Moffatt, H. K. 1985. *J. Fluid Mech.* 159: 359
Moffatt, H. K. 1986. *J. Fluid Mech.* 166: 359
Moffatt, H. K. 1987. In *Advances in Turbulence*, ed. G. Comte-Bellot, J. Meathieu, p. 228. New York: Springer-Verlag
Munro, R. H., Gosling, J. T., Hildner, E., MacQueen, R. M., Poland, A. I., Ross, C. L. 1979. *Sol. Phys.* 61: 201
Newkirk, G. Jr. 1967. *Annu. Rev. Astron. Astrophys.* 5: 213
Orrall, F. Q., ed. 1981. *Solar Active Regions*. Boulder: Colo. Assoc. Univ. Press
Parker, E. N. 1963. *Interplanetary Dynamical Processes*. New York: Interscience
Parker, E. N. 1966. *Ap. J.* 145: 811
Parker, E. N. 1972. *Ap. J.* 174: 499
Parker, E. N. 1979. *Cosmical Magnetic Fields: Their Origin and Their Activity*. Oxford: Oxford Univ. Press
Parker, E. N. 1986a. *Geophys. Astrophys. Fluid Dyn.* 34: 243
Parker, E. N. 1986b. *Geophys. Astrophys. Fluid Dyn.* 35: 277
Parker, E. N. 1986c. In *Coronal and Prom-*

inence Plasmas. *NASA CP-2442*, ed. A. I. Poland, p. 9. Washington, DC: NASA
Parker, E. N. 1987. *Ap. J.* 318: 376
Parker, E. N. 1988. *Ap. J.* 330: 474
Parker, E. N. 1989a. *Geophys. Astrophys. Fluid Dyn.* 45: 159
Parker, E. N. 1989b. *Geophys. Astrophys. Fluid Dyn.* 45: 169
Parker, E. N. 1989c. *Geophys. Astrophys. Fluid Dyn.* 46: 105
Parker, E. N. 1989d. *Geophys. Astrophys. Fluid Dyn.* In press
Pizzo, V. 1986. *Ap. J.* 302: 785
Pizzo, V. J., Sime, D. G., Holzer, T. E., eds. 1987. *Proc. Solar Wind Six Conference. NCAR TN-306*. Boulder, Colo: NCAR
Roberts, P. H. 1964. *An Introduction to Magnetohydrodynamics*. Am. Elsevier
Rosner, R., Knobloch, E. 1982. *Ap. J.* 262: 349
Rosner, R., Low, B. C., Holzer, T. E. 1985. In *Physics of the Sun*, ed. P. A. Sturrock, T. E. Holzer, D. Mihalas, R. Ulrich, 2: 135. Dordrecht: Reidel
Rosner, R., Low, B. C., Tsinganos, K., Berger, M. 1989. *Geophys. Astrophys. Fluid Dyn.* 48: 251
Sakurai, T. 1979. *Publ. Astron. Soc. Jpn.* 31: 209
Sears, E. R., Resler, E. L. Jr. 1961. *Adv. Aerospace Sci.* 3-4: 637
Sime, D. G., Hundhausen, A. J. 1987. *J. Geophys. Res.* 92: 1049
Spreiter, J. R., Marsh, M. C., Summers, A. L. 1970. *Cosmic Electrodyn.* 1: 5
Steinolfson, R. S., Hundhausen, A. J. 1989a. *J. Geophys. Res.* 94: 1222
Steinolfson, R. S., Hundhausen, A. J. 1989b. *J. Geophys. Res.* In press
Stern, D. P. 1966. *Space Sci. Rev.* 6: 147
Strauss, H. R., Otani, N. F. 1988. *Ap. J.* 326: 418
Sturrock, P. A., ed. 1980. *Solar Flares*. Boulder: Colo. Assoc. Univ. Press
Syrovatskii, S. I. 1971. *Sov. Phys. JETP* 33: 933
Tamada, K. 1964. In *Symposium Transsonicum*, ed. K. Oswatitsch. New York: Springer-Verlag
Tandberg-Hanssen, E. 1974. *Solar Prominences*. Dordrecht: Reidel
Vaiana, G. S., Rosner, R. 1978. *Annu. Rev. Astron. Astrophys.* 16: 393
Vainshtein, S. I., Parker, E. N. 1986. *Ap. J.* 304: 821
van Ballegooijen, A. A. 1985. *Ap. J.* 298: 421
van Ballegooijen, A. A. 1988. *Geophys. Astrophys. Fluid Dyn.* 41: 181
van Hoven, G., Ma, S. S., Einaudi, G. 1981. *Astron. Astrophys.* 97: 232
Webb, D. F., Hundhausen, A. J. 1987. *Sol. Phys.* 383: 108
Whang, Y. C. 1987. *J. Geophys. Res.* 92: 4349
Withbroe, G. L. 1981. See Orrall 1981, p. 199
Withbroe, G. L., Noyes, R. W. 1977. *Annu. Rev. Astron. Astrophys.* 15: 363
Wolfson, R. L. T. 1982. *Ap. J.* 255: 774
Wolfson, R. L. T. 1987. *J. Geophys. Res.* 92: 9875
Wolfson, R. L. T., Gould, S. A. 1985. *Ap. J.* 296: 287
Yang, W. H., Sturrock, P. A., Antiochos, S. K. 1987. *Ap. J.* 309: 383
Zirker, J. B., ed. 1977. *Coronal Holes and High Speed Wind Streams*. Boulder: Colo. Assoc. Univ. Press
Zweibel, E. G. 1982. *Ap. J. Lett.* 258: L53
Zweibel, E. G., Li, H. S. 1987. *Ap. J.* 312: 423

EXTRAGALACTIC H II REGIONS

G. A. Shields

Department of Astronomy, University of Texas, Austin, Texas 78712

KEY WORDS: nebulae, ionizing stars, abundances, galaxies

1. INTRODUCTION

Located 500 pc from Earth, in the plane of our Galaxy, the Orion Nebula has fascinated astronomers for many years. Roughly 50 $M_\odot$ of ionized gas at temperature $T \approx 8000$ K gas occupies a core of diameter ~ 0.5 pc and a more dilute halo of diameter ~ 5 pc (Aller 1984). A single hot star supplies most of the ultraviolet radiation that heats and ionizes the gas. This ionized nebula is embedded in the edge of a larger cloud of neutral molecular gas, comprising $\sim 10^3 \, M_\odot$, in which new stars are forming. The Orion complex is situated in an atomic hydrogen cloud of diameter ~ 70 pc and mass $\sim 10^5 \, M_\odot$. The ionized gas has a chemical composition roughly similar to that of the Sun, and the most abundant ions include H^+, He^+, O^+, and O^{++}. Excitation of bound energy levels of the various ions produces the characteristic emission-line spectrum, and bound-free and free-free processes give the continuous emission. Such diffuse, ionized nebulae are called "H II regions" after the spectroscopic notation for ionized hydrogen. A historical review of nebular studies has been given by Aller (1986). Properties of some galactic and extragalactic H II regions are summarized in Table 1.

Conspicuous because of its proximity to Earth and its high surface brightness, the Orion Nebula (M42) is a diminutive H II region by galactic standards. Throughout the disk of the Milky Way, H II regions of various sizes are found, with the largest occurring near the spiral arms and somewhat closer to the galactic center than the Sun's location. The more remote H II regions from Earth, which are heavily obscured by interstellar dust, are best observed at radio and infrared (IR) wavelengths. The giant H II region complex W49, roughly 150 pc across, lies 14 kpc from Earth. Some

Table 1 Properties of selected H II regions[a]

Object	Galaxy	Distance	Type	Diameter	N_e	EM	$Q(H^0)$	$M(H^+)$	N(O5 V)
Orion	MWG	0.5	C	5	3.5	6.2	48.85	1.5	0.2
W49	MWG	15	H	150	(2.0)	5.4	51.20	4.6	27
N70	LMC	50	S	110	—	3.0	—	—	1
30 Dor	LMC	50	H	370	2.5	5.5	52.05	5.8	230
N19	SMC	70	D	220	(1.2)	3.6	50.20	4.7	3
NGC 604	M33	800	H	400	≲1.8	5.1	51.50	5.8	65
NGC 5471	M101	6000	M	800	2.3	5.3	52.55	7.0	750

[a] Based on Kennicutt (1984). Distances are in kiloparsecs, diameters are in parsecs. The electron density N_e (in cm^{-3}), peak emission measure EM (in $pc\ cm^{-6}$), $Q(H^0)$ (in s^{-1}), and $M(H^+)$ (in $M_\odot$) are all given as $\log_{10}$. N(O5 V) is the equivalent number of O5 V stars needed to match the ionizing luminosity of the region. Electron densities: Orion core, Osterbrock (1989); W49 and M19, assumed filling factor; 30 Dor, Mathis et al (1985); NGC 604, Díaz et al (1987); NGC 5471, Torres-Peimbert et al (1989). Morphological types as follows: C, "classical" H II region; H, "high-surface-brightness" region; D, "diffuse" giant H II region; M, "multiple core complex"; S, "shell" or "ringlike" H II region.

200 times more luminous than M42, this object involves the equivalent of 30 luminous O5 V ionizing stars (Kennicutt 1984).

At a distance of ~60 kpc, two irregular galaxies orbit the Milky Way. Rich in interstellar gas and prolific at star formation, the Large and Small Magellanic Clouds (LMC, SMC) abound in H II regions. In the LMC lies the giant H II region 30 Doradus, the "Tarantula Nebula." This filamentary object is more than 300 pc across, and its ~$10^6\ M_\odot$ of ionized gas are excited by several hundred ionizing stars. Because this is the nearest example of a giant extragalactic H II region (GEHR), 30 Dor is well placed for observational studies of its detailed structure and kinematics.

The nearest spiral galaxy, M31 in Andromeda, has many H II regions but lacks giant complexes on the scale of 30 Dor. Its neighbor, the modest Scd galaxy M33, has prominent H II regions, dominated by NGC 604. Spirals of late Hubble type have many GEHRs clustered along their prominent spiral arms. The largest H II regions in these galaxies are supergiant complexes several times more luminous than 30 Dor or NGC 604. In the large Scd galaxy M101, the complexes NGC 5461 and NGC 5471, almost 10^4 times more luminous than Orion, have core and halo diameters of ~300 and ~1000 pc, respectively. Descriptions of these and other H II regions may be found in Aller (1984) and Kennicutt (1984). The latter reference gives an excellent photographic illustration of the range of sizes and morphologies of galactic and extragalactic H II regions. The distribution of H II regions over part of M101 is illustrated in Figures 1 and 2 (see also Melnick 1987).

GEHRs provide a variety of opportunities for study. Their diameters and luminosities give us a means of measuring distances to galaxies, and their luminosities indicate the rate of massive star formation (see Section

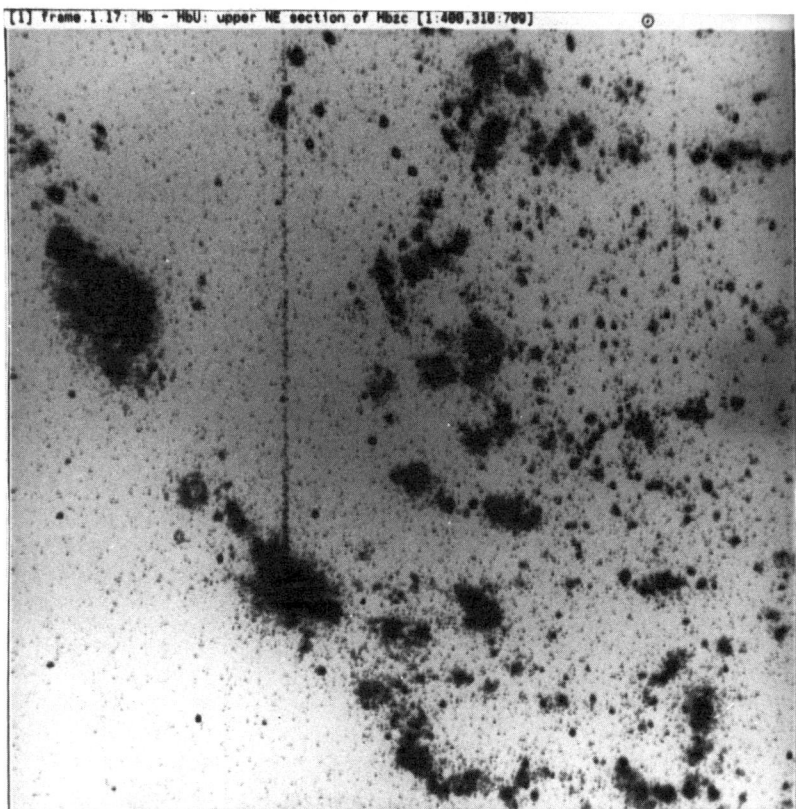

Figure 1 A logarithmic gray-scaled Hβ surface brightness image of M101 H II regions in an 8-arcmin diameter area ENE of the nucleus. The stellar continuum has been subtracted out, and the imagery is displayed in 256 gray logarithmic levels from a minimum (black) to maximum (white) surface brightness range of 1.0×10^{-18} to 2.0×10^{-15} erg cm^{-2} s^{-1} arcsec^{-2}. The imagery was taken with the Wide-Field PFUEUI camera on the Palomar 60-inch telescope by J. J. Hester and R. J. Dufour (2000-s exposure, 1.2" per pixel resolution). In Figures 1 and 2, north is to the top and east is to the left.

5). Their emission lines allow measurement of the internal motions of spiral and irregular galaxies (Rubin et al 1985) and provide an opportunity to study the chemical composition of the interstellar gas. The study of extragalactic H II regions has witnessed accelerating growth in recent years. Quantitative measurement of emission-line intensities has been facilitated by electronic spectrophotometers such as the image dissector scanners (IDS), sensitive in the optical and near ultraviolet. More recently, spectrographs with CCD detectors have allowed efficient, long-slit observations and have extended spectral coverage to the near infrared; and observations at longer infrared wavelengths are becoming increasingly

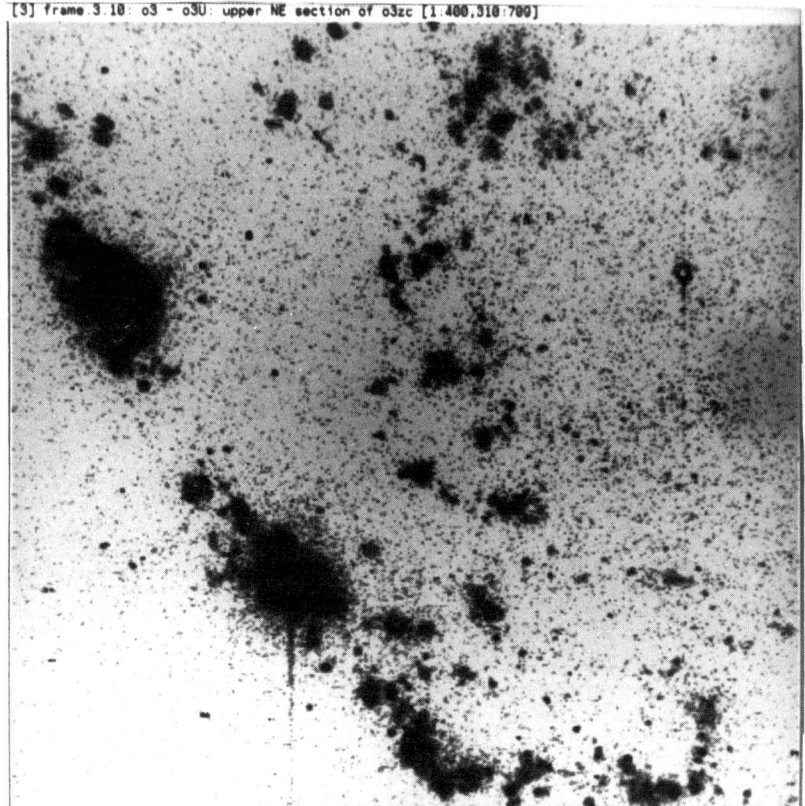

Figure 2 An [O III] $\lambda 5007$ logarithmic surface brightness image of M101 at identical field and gray-scale levels as in Figure 1 (3600-s exposure). Compared with the Hβ image of Figure 1, note that in [O III] the H II regions within 2 arcmin of the nucleus (*right side* of the photo) are largely invisible but become comparable with Hβ at $R \approx 4$ arcmin. The $\lambda 5007/\lambda 4861$ ratios exceed 5 in the cores of the more distant GEHRs, which are also much larger in size and complexity than the inner H II regions (photo courtesy of R. Dufour and P. Scowen, Rice University).

sensitive. The *International Ultraviolet Explorer* (*IUE*) has provided ultraviolet (UV) line intensities of H II regions and information on associated hot stars. Radio continuum studies with interferometers, notably the VLA, now achieve appropriate angular resolution to map the distribution of GEHRs in nearby galaxies. Coupled with theoretical advances in the modeling of stars and nebulae, this observational material is yielding valuable insights.

This review highlights some aspects of these observational and theoretical studies of extragalactic H II regions. In Section 2 I discuss the physical properties of H II regions, their distribution in galaxies, and extinction. Section 3 considers the physical processes in H II regions and the use of photoionization models. In Section 4 I describe the morphology and kinematics of GEHRs and their use as distance indicators. Section 5 considers the ionizing stars, and Section 6 deals with chemical abundance measurements. Other reviews of H II regions are given by Melnick (1987) and Dinerstein (1990). Nuclear starbursts (not considered here) are reviewed briefly by Shields (1989), and much information on them can be found in the volume edited by Thuan et al (1988). Emission lines of H II regions, starbursts, and active galactic nuclei are compared by Veilleux & Osterbrock (1987).

2. THE NATURE OF EXTRAGALACTIC H II REGIONS

2.1 Diagnostic Methods

Extragalactic H II regions are analyzed by the usual nebular techniques, as described, for example, in the books by Aller (1984) and Osterbrock (1989). These references and Mendoza (1983) summarize the relevant atomic data and literature. The commonly observed lines in GEHRs include the Balmer lines, He I $\lambda 5876$, and several other permitted lines and forbidden lines, including C III] $\lambda 1909$; [N II] $\lambda\lambda 6584, 6548$; [O II] $\lambda\lambda 3726$, 3729, 7325; [O III] $\lambda\lambda 5007, 4959, 4363$; [Ne III] $\lambda\lambda 3869, 3968$; [S II] $\lambda\lambda 6717$, 6731, 4072; [S III] $\lambda\lambda 9532, 9069, 6312$; and [Ar III] $\lambda\lambda 7751, 7136$.

A star or cluster emitting $Q(\mathrm{H}^0)$ ionizing photons per second produces a highly ionized nebula out to a fairly sharp cutoff at the "Strömgren radius" R_S, given by

$$Q(\mathrm{H}^0) = \int_{\nu_\mathrm{H}}^{\infty} L_\nu(h\nu)^{-1}\,d\nu = \frac{4\pi}{3} R_\mathrm{S}^3 f \alpha_\mathrm{B} N_\mathrm{e} N_\mathrm{p}, \qquad 1.$$

where L_ν is the stellar luminosity. Here ν_H is the ionization threshold of H^0, N_e is the electron density (cm^{-3}), N_p the H^+ density, $\alpha_\mathrm{B}(T)$ the excited-state radiative recombination coefficient (cm^3 s^{-1}) at the nebular temperature T (Osterbrock 1989), and $f \leq 1$ is the volume filling factor in "filamentary" models. The right-hand side is replaced by the corresponding integral if the gas distribution is nonuniform. If the gas extends beyond R_S, the nebula is called "radiation bounded," "ionization bounded," or "optically thick"; and a nebula truncated inside R_S is "matter bounded," "density bounded," or "optically thin." The local state of ionization and temperature is given by equilibrium equations involving photoionization and various recombination and cooling processes. Line emission from hydrogen and helium is mainly produced by radiative recombination, with a local emissivity j_l (ergs cm^{-3} s^{-1} sr^{-1}) given by

$$4\pi j_l = \varepsilon_l(A^{+i}, T, N_e)N_e N_{i+1}, \qquad 2.$$

where A^{+i} is the recombined ion ($i = 0, 1, 2, \ldots$), and ε_l is the emission coefficient (ergs cm^3 s^{-1}) for the transition. For radiative recombination, ε_l varies roughly as T^{-1}, as does α_B. Continuous emission (bound-free and free-free), mostly from H^0 and He^0, is given by

$$4\pi j_\nu = \gamma_\nu(A^{+i}, T)N_e N_{i+1}, \qquad 3.$$

where γ_ν (ergs cm^3 s^{-1} Hz^{-1}) is a continuous emission coefficient. The total luminosity in a recombination line of hydrogen, or in the radio free-free continuum, is proportional to $\frac{4}{3}\pi R_s^3 f N_e N_p$ and therefore to $Q(H^0)$ by Equation 1, and this provides a means of estimating the combined luminosity of the ionizing stars. The intrinsic Balmer decrement $j(H\alpha)/j(H\beta)$ and the line to radio-continuum ratio $j(H\alpha)/j_\nu^{ff}$ are relatively insensitive to electron density and temperature, providing useful indicators of extinction by interstellar dust.

The heavier elements, such as C, N, O, Ne, S, and Ar, emit forbidden lines following electron impact excitation of low-lying metastable energy levels. These lines are emitted according to Equation 2, with N_{i+1} replaced by N_i. Now ε_l involves the collision strength for exciting the upper level and an exponential temperature sensitivity $e^{-\chi_{12}/kT}$, where χ_{12} is the excitation potential of the upper level. For typical values $T \approx 10^4$ K and χ_{12} of a few eV, this means that an accurate temperature is required to analyze the line intensity in terms of the ionic abundances $N(A^{+i})/N(H^+)$. Favorable line ratios provide an opportunity to measure the electron temperature. For example, at $N_e \lesssim 10^4$ cm^{-3}, the [O III] line ratio is given by

$$R_{O\,III} \equiv \frac{I(\lambda\lambda 5007, 4959)}{I(\lambda 4363)} = 7.73 e^{3.29/t}, \qquad 4.$$

where $t \equiv T/10^4$ K; analogous equations hold for $R_{N\,II} \equiv I(\lambda\lambda 6548, 6584)/I(\lambda 5755)$ and some other ratios (Osterbrock 1989).

When collisional deactivation is important, the N_e dependence in ε_l of Equation 2 becomes significant. This again complicates the derivation of abundances but provides a diagnostic for N_e. In particular, the [O II] line ratio $r_{O\,II} \equiv I(\lambda 3729)/I(\lambda 3726)$ and the [S II] line ratio $r_{S\,II} \equiv I(\lambda 6717)/I(\lambda 6731)$ are sensitive to N_e in the range $\sim 10^2$–10^4 cm^{-3}. (Here I use R for a line ratio that measures temperature, and r for one that measures density.)

Extragalactic H II regions and other emission-line objects such as supernova remnants (SNRs) and active galactic nuclei (AGN) have characteristic differences in their emission-line spectra, mainly because of different physical conditions. This causes them to occupy different regions of certain diagnostic line-intensity ratio diagrams, such as a plot of [O I] $\lambda 6300$/[O III] $\lambda 5007$ versus [O II] $\lambda 3727$/[O III] $\lambda 5007$ (Heckman 1980, Baldwin et al 1981, Veilleux & Osterbrock 1987). Theoretical results for a variety of diagnostic diagrams are given by Evans & Dopita (1985).

2.2 Physical Properties

The properties of GEHRs in comparison with galactic H II regions have been summarized by Kennicutt (1984). Electron temperatures in galactic and extragalactic H II regions fall in the range ~ 5000–$\sim 20,000$ K, with the lower values corresponding to higher abundances of heavy elements (e.g. Smith 1975, Shaver et al 1983). The [S II] $\lambda\lambda 6717, 6730$ red doublet is at least partially resolved in typical spectra, and observed values of $r_{\text{S II}}$ usually imply densities $N_e \lesssim 300$ cm^{-3}. McCall et al (1985; hereinafter MRS) found a characteristic value $N_e \approx 140$ cm^{-3} for a sample of 100 GEHRs. A histogram of $r_{\text{S II}}$ and N_e in nuclear, hot spot, and disk H II regions has been given by Kennicutt et al (1989b). (However, because $r_{\text{S II}}$ approaches a low density limit of 1.4, the mean $r_{\text{S II}}$ will not give the mean N_e.) O'Dell & Castañeda (1984) measured $r_{\text{O II}}$ in several GEHRs and found densities typically in the range 50–150 cm^{-3}; their large value $N_e \approx 235$ cm^{-3} for NGC 5461 in M101 agrees with $r_{\text{S II}}$ for this object (MRS). GEHRs in blue compact galaxies have similar densities. For example, $r_{\text{S II}}$ implies that $N_e = 300$ and 100 cm^{-3} in Tol 1214−277 and UM 461, respectively (Pagel & Simonson 1989); and Dinerstein & Shields (1986) have found that $N_e = 140$ cm^{-3} for NGC 4861. These densities are lower than the $\sim 10^3$–10^4 cm^{-3} values that are characteristic of the cores of Orion and other small galactic H II regions (Aller 1984). From the diameters and radio or Hα luminosities (or surface brightnesses) of GEHRs, one finds rms electron densities of $N_{\text{rms}} \approx 1$–10 cm^{-3} (e.g. Searle 1971, Kennicutt 1984, Kaufman et al 1987, van der Hulst et al 1988). Because the line ratios give $N_e \gg N_{\text{rms}}$, the gas must be in clouds or filaments occupying a small fraction $f \approx 10^{-2}$ of the volume. (Note that $N_{\text{rms}} = N_e f^{1/2}$.) This was discovered in the case of Orion by Osterbrock & Flather (1959).

The internal structure of nearby GEHRs is observed to be complex (see Section 4). For example, Skillman (1985) made spatially resolved radio and optical observations of NGC 5471 in M101. He found substantial variations in T(O III) from $\sim 12,000$ to 14,000 K and in N_e from ~ 100 to 600 cm^{-3}. Skillman noted several "nuclei" of concentrated emission within the complex, as well as at least one SNR.

H II regions are ubiquitous in spiral and irregular galaxies. Atlases with Hα photographs of H II regions in many galaxies are given by Hodge

(1969, 1974) and Hodge & Kennicutt (1983b). The properties of first-ranked H II regions in 95 spiral and irregular galaxies were studied by Kennicutt (1988). For a typical first-ranked GEHR, the Hα luminosity is $\sim 10^{39}$ ergs s^{-1}, requiring $\sim 10^{51}$ ionizing photons per second from 100–200 O and B stars with a combined mass of several thousand solar masses. The ionized gas, amounting to 10^4–10^5 $M_\odot$, resides in an atomic and molecular gas cloud that is ~ 10 times more massive. Early-type spirals and low-luminosity irregulars may have a brightest H II region 10 to 100 times less luminous and less massive than these values, whereas in luminous Sc's and irregulars the brightest H II region may be 10 to 100 times larger.

Luminosity functions for extragalactic H II regions can be derived from radio or reddening-corrected Hα fluxes. They can be expressed as the cumulative number of H II regions brighter than luminosity L, in the form $N(L) \propto L^{-\beta}$. Kennicutt et al (1989a) measured Hα luminosity functions of H II regions in 30 spiral and irregular galaxies, finding $\beta \approx 2.0 \pm 0.5$. Earlier spirals have fewer H II regions at all luminosities, a steeper luminosity function, and a fainter cutoff at the bright end. Kennicutt et al suggested that the dependence on morphological type involves differences in the masses of interstellar clouds. Radio studies of various galaxies give similar values of β. The combined radio flux density of the giant H II regions in M51 is 16.7 mJy, compared with 49 mJy for the galaxy as a whole (Klein et al 1984, van der Hulst et al 1988). This difference presumably comes from smaller H II regions and diffuse H II, consistent with the extrapolation of the radio luminosity function to lower luminosities. Hodge et al (1989) studied H II regions in the tiny local group irregular GR8, finding $\beta = 2.0 \pm 0.2$ and a luminosity of $L(H\alpha) = 8 \times 10^{36}$ ergs s^{-1} for the brightest complex. The diameter function is $N(D) \propto \exp(-D/17$ pc), and the luminosity-diameter relation is $L(H\alpha) \propto D^{2.8}$, close to constant-volume emissivity.

Kaufman et al (1987) considered in some detail the spatial distribution of the H II regions in M81. Compared with optical H II region surveys including fainter objects (Hodge & Kennicutt 1983a), the giant H II regions dominating Kaufman et al's radio map are more concentrated along the spiral arms and on the inner H II ring at a radius of 300" (4.7 kpc at the adopted distance of 3.3 Mpc). The fainter regions are more uniformly distributed over the disk. The H II distribution, taken as an indicator of star formation rates, is not consistent with the predictions of Visser's (1980) density-wave theory unless molecular hydrogen is concentrated near the H I ring.

2.3 Reddening

Radio emission is not subject to interstellar extinction and thus provides a good basis for determining values of emission measure and $Q(H^0)$.

Measurements at two wavelengths, such as 6 and 20 cm with the VLA, allow a separation of thermal H II region continuum from nonthermal contamination. Comparisons of Hβ and radio luminosities with Hα or Hβ luminosities (or surface brightnesses) give a measure of the extinction at visual wavelengths, usually expressed as a visual ($\lambda 5500$) or Hβ ($\lambda 4861$) extinction (A_V or A_β) or as $C_{H\beta} \equiv \Delta \log_{10} I_{H\beta}$, where $\Delta \log_{10} I_{H\beta}$ refers to the observed intensity in comparison with the intensity that would be observed in the absence of extinction. This may be compared with the extinction implied by the reddening of the Balmer decrement. Assuming a normal galactic interstellar extinction curve as a function of wavelength, one can convert $E_{\beta-\alpha}$ to E_{B-V} and use $A_V \simeq 3E_{B-V}$. Israel & Kennicutt (1980) carried out this program for GEHRs in a number of galaxies and found A_V(radio) much greater than $3E_{B-V}$(Balmer) for most H II regions. Average values were A_V(radio) $\simeq$ 1.7 mag and $3E_{B-V}$(Balmer) $\approx$ 0.5 mag. This discrepancy is consistent with the idea that there is dust mixed with the line-emitting gas, or, alternatively, that there is patchy extinction in front of the nebula, so that the observed optical emission is weighted toward less obscured areas.

More recent results show a somewhat weaker effect. Caplan & Deharveng (1986) studied the extinction and reddening of H II regions in the LMC, finding that E_{B-V}(Balmer) correlates with A_V(radio), but that the ratio is typically ~ 5.1 rather than 3. The agreement between the reddening of the nebular Balmer lines and of the continuum of stars in the vicinity of the nebula suggests that some of the reddening occurs in the foreground, but additional dust ranging from within the H II region to several radii outside is also indicated. Opaque filaments are seen in images of H II regions such as N59, but the covering factor is small. Viallefond & Goss (1986) found that A_β(Balmer) $< A_\beta$(radio) for H II regions in M33, and they also find that A_β(Balmer) is larger for the core of an H II region than for its global emission. Van der Hulst et al (1988) observed that A_V(radio) exceeds A_V(Balmer) by ~ 0.5 mag in M51, and they suggested that a mismatch of radio and optical aperture sizes in earlier studies exaggerated the excess of A_V(radio) over A_V(Balmer). Skillman & Israel (1988) reached similar conclusions for several GEHRs in M101 on the basis of observations of the infrared Brackett-γ line, the Balmer lines, and the radio continuum. Kaufman et al (1987) found a mean A_V(radio) = 1.1 $\pm$ 0.4 in M81 and argued that most of this extinction occurs in M81 but outside the H II regions.

We discuss below the evidence for radial composition gradients across spiral galaxies, with O/H decreasing toward larger radii. Since dust grains are composed of heavy elements, one might expect the dust-to-gas ratio to vary with O/H. Sarazin (1976), Israel & Kennicutt (1980), and Viallefond & Goss (1986) found evidence for a radial decrease in A_V across M33 and

M101, consistent with this expectation. In M51, van der Hulst et al (1988) observed a radial gradient in $N_{H\,I}/A_V$ larger than the factor of 4 variation of O/H, and Walterbos & Schwering (1987) found a similar gradient of $N_{H\,I}/A_V$ in M31. Kaufman et al (1987) noted no radial trend in A_V in M81, in spite of a strong O/H gradient; however, the mean extinction for H II regions along the inner H I ring is greater than for the H II regions in the main arms.

3. MODELS OF H II REGIONS

3.1 *Physical Processes*

Numerical models involving the variation of ionization and temperature with location in an H II region are useful for determining physical parameters, ionizing star temperatures, and chemical abundances. The typical geometry involves a central "point" source of ionizing continuum with a spectrum based on stellar atmosphere models. Surrounding this point source is an effectively infinite expanse of gas with density N, filling factor f, and chemical abundances $N(A)/N(H)$. Starting at some radius R_{in}, one solves the equations of ionization and thermal equilibrium progressively farther from the star, taking account of the absorption of the ionizing continuum as well as its geometrical dilution. Once the edge of the Strömgren sphere is reached, as a result of exhaustion of the ionizing continuum, the emission-line luminosities are calculated by integrating Equation 2 over the ionized volume. Some codes include an iterative treatment of the diffuse radiation (e.g. Harrington 1968, Rubin 1985), and others use the "on-the-spot" (OTS) approximation or the "outward-only" approximation (e.g. Stasińska 1980, 1982, Evans & Dopita 1985, Dinerstein & Shields 1986). The differences are usually not important for applications to GEHRs.

The photoionization probability per second of hydrogen is

$$\Gamma_{pi} = \int_{v_H}^{\infty} a_v(L_v/4\pi R^2)e^{-\tau_v}(hv)^{-1}\,dv, \qquad 5.$$

where a_v is the photoionization cross section, and τ_v is the monochromatic optical depth between the source and the point in question. In a steady state, the rates (cm^{-3} s^{-1}) of photoionization and recombination are equal, so that, in OTS,

$$\Gamma_{pi}N_{H^0} = \alpha_B N_e N_p. \qquad 6.$$

Analogous equations for the ratios of neighboring ions of other elements also hold, but the inclusion of charge transfer and dielectronic recombination is important in some cases [see references in Osterbrock (1989) and Shields (1989)]. From the ionic ratios, the ionization fractions $X(A^{+i}) \equiv N(A^{+i})/N(A)$ are calculated. The ratio N_p/N_{H^0} from Equation 6 is proportional to ϕ_i/N_e, where $\phi_i = Q(H^0)/4\pi R^2$ is the ionizing photon flux. The degree of ionization in the interior of the nebula is thus governed by the dimensionless "ionization parameter"

$$U \equiv \frac{Q(H^0)}{4\pi R_S^2 Nc},\qquad 7.$$

where c is the speed of light. Typical values for GEHRs are $U \approx 10^{-2.5 \pm 0.5}$, corresponding to $X(H^0) \approx 10^{-3}$ at $R \approx 0.5 R_S$. The ionization of other elements in an H II region is governed by the ionization potentials of the various ions, the shape of the ionizing continuum, and the absorption of the continuum by H and He. The continua of the ionizing O and B stars are cut off at the He II Lyman edge at $v_{He^+} = 4v_H$. Ions with ionization potentials $v_{A^{+i}} > v_{He^+}$ cannot be further ionized. Thus He^+, C^{+3}, N^{+3}, O^{+2}, Ne^{+2}, and S^{+4} are the highest possible stages. The actual dominant ions may be lower, depending on U, on the stellar temperature T_*, and on the chemical composition; typically, C^{+2}, O^+ or O^{+2}, Ne^+ or Ne^{+2}, and S^{+2} or S^{+3} are most abundant.

3.2 Ionization Structure

Observed GEHRs are crudely categorized as "high excitation" or "low excitation." This terminology can be defined (Aller 1942, Searle 1971) in terms of large or small values of [O III]/H$\beta \equiv I(\lambda\lambda 5007, 4959)/I(\lambda 4861)$. Empirically, strong [O III] correlates with low O/H, high gas temperature, and large $\langle X(O^{+2}) \rangle$, where $\langle X \rangle$ is a nebular average (Searle 1971, Smith 1975).

For fixed composition and ionizing continuum shape (L_v/L_{v_H}), models with differing N, f, and $Q(H^0)$ but the same U are homologous. Although differences in relative abundances, such as C/O, N/O and He/H, have some effect on the ionization structure and temperature, the main defining parameters are T_*, O/H, and U. Photoionization models indicate a statistical decrease of T_* or U with increasing O/H among GEHRs (Shields & Searle 1978, Stasińska 1980, Dopita & Evans 1986, Campbell 1988).

Figure 3 illustrates the ionization structure of two models typical of low- and high-excitation GEHRs (models L and H, respectively). These were computed with the author's photoionization code (Dinerstein & Shields 1986, Garnett 1989b) and an outward-only treatment of the diffuse ionizing radiation of H and He. Both models have $Q(H^0) = 10^{51}$ s^{-1} and $N = 100$ cm^{-3}, with R_S and U adjusted by means of f to give typical O^+/O^{+2} ratios. Table 2 gives the model parameters and selected results, including line

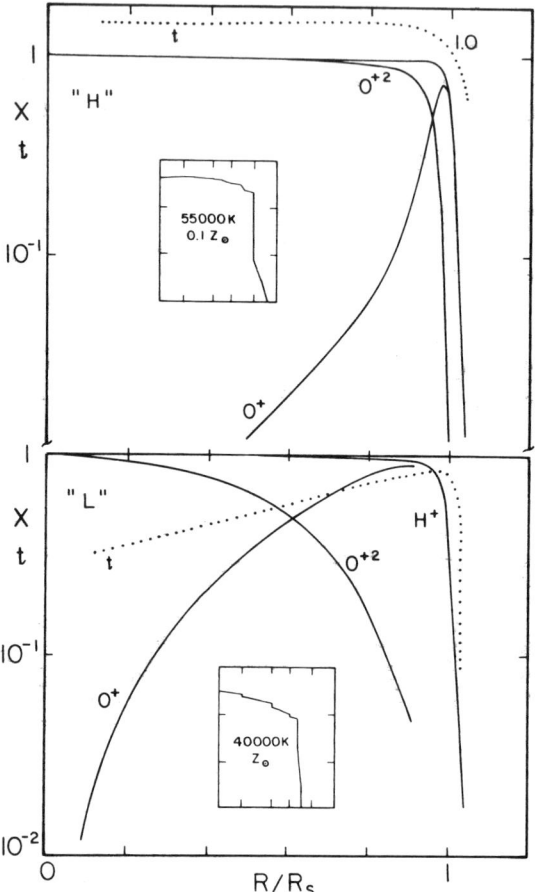

Figure 3 Ionization and temperature structure of photoionization models. *Upper* and *lower* panels are models H and L, respectively (see text and Table 2). *Solid curves* are fractional ionic abundances $X(A^{+i})$, and the *dotted curves* are the electron temperature $t \equiv T/10^4$ K. Insets show the ionizing continuum, with ticks on the ordinate at intervals of 2 dex in L_ν and on the abscissa at $\nu/\nu_H = 1.0$, 1.81, 2.58, and 4.0—the Lyman edges of H^0, He^0, O^+, and He^+, respectively.

intensities, fractional abundances, and ion weighted temperatures defined by Dinerstein & Shields (1986). The models use LTE stellar atmosphere fluxes, computed without line blanketing by D. R. Garnett using the ATLAS code (Kurucz 1979). Solar abundances (compiled by Anders & Grevesse 1989) were taken as $12 + \log A/H =$ (8.56, 8.04, 8.92, 8.08, 6.11, 7.26, 6.56, 6.04) for (C, N, O, Ne, Mg, S, Ar, Fe), including depletion of Mg and Fe into grains.

Table 2 Photoionization models[a]

Quantity	Model L		Model H		Quantity	Model L	Model H
T_*	4.0		5.5		H I 4861	0.00	0.00
$\log U$	−2.91		−2.76		H I 1215C	−2.23	1.07
[O/H]	0.0		−1.0		C II] 2326	−1.31	−1.16
[C/O]	0.0		−0.5		C III] 1909	−2.04	−0.33
[N/O]	0.0		−0.5		C IV 1550	−3.34	−0.43
He/H	0.100		0.075		[N II] 6584	−0.09	−1.11
f	0.03		0.10		[N II] 5755	−2.34	−2.64
	$\langle X \rangle$	$\langle t \rangle$	$\langle X \rangle$	$\langle t \rangle$	[N III] 57 μm	−0.30	−1.52
H$^+$	0.975	0.633	0.967	1.52	[O I] 6300	−1.63	−1.62
He$^+$	0.935	0.630	0.985	1.51	[O II] 3729	0.02	−0.14
C$^+$	0.435	0.679	0.169	1.32	[O II] 3726	−0.14	−0.28
C^{+2}	0.564	0.600	0.643	1.52	[O II] 7325	−1.85	−1.42
C^{+3}	0.000	0.850	0.188	1.65	[O III] 88 μm	0.40	−0.08
N$^+$	0.484	0.685	0.195	1.32	[O III] 52 μm	0.34	−0.18
N^{+2}	0.499	0.584	0.625	1.54	[O III] 5007	0.00	0.80
N^{+3}	0.000	0.744	0.167	1.65	[O III] 4363	−3.08	−0.88
O^0	0.025	0.699	0.030	1.08	[Ne II] 12.8 μm	−0.30	−2.26
O$^+$	0.577	0.672	0.161	1.34	[Ne III] 3869	−1.64	−0.23
O^{+2}	0.398	0.576	0.809	1.56	Mg II 2798	−3.07	−1.48
Ne$^+$	0.737	0.654	0.048	1.33	[S II] 6731	−0.87	−1.33
Ne^{+2}	0.247	0.574	0.949	1.52	[S II] 6717	−0.74	−1.20
Mg$^+$	0.029	0.699	0.204	1.38	[S II] 4072	−2.01	−2.14
S$^+$	0.129	0.700	0.072	1.22	[S III] 18.7 μm	−0.03	−0.98
S^{+2}	0.795	0.633	0.488	1.45	[S III] 9532	−0.04	−0.46
S^{+3}	0.076	0.539	0.361	1.59	[S III] 6312	−2.26	−1.85
Ar^{+2}	0.808	0.718	0.460	1.42	[S IV] 10.5 μm	−0.39	−0.54
Ar^{+3}	0.057	0.628	0.519	1.61	[Ar III] 7136	−0.10	−1.40
Fe^{+2}	0.291	0.670	0.093	1.38	t(N II)	0.70	1.33
Fe^{+3}	0.307	0.591	0.873	1.54	t(O III)	0.62	1.56

[a] $Q(H^0) = 10^{51}$ s^{-1}, $N = 100$ cm^{-3}. Abundances in brackets are logarithmic relative to solar values (see text). Ion-averaged temperatures $\langle t(A^{+i}) \rangle$ are given in units of 10^4 Kelvins. Wavelengths are in angstroms or microns; line intensities are given as $\log I(\lambda)/I(H\beta)$. H I 1215C is the collisional contribution.

Model H (top panel in Figure 3) represents high-excitation regions, such as NGC 5471. It has [O/H] ≡ $\log[(O/H)/(O/H)_\odot] = -1.0$, [C/O] = [N/O] = -0.5, $T_* = 55{,}000$ K, and $U = 10^{-2.76}$. The electron temperature is high and fairly uniform, averaging 15,200 K over the entire nebula (Table 2). Helium is He$^+$ throughout the nebula. As the H$^+ \to$ H^0 "transition zone" at R_S is approached, O^{+2} gives way to O$^+$ and then to O^0. Ne^{+2} fills most of the H$^+$ volume, and N$^+$ and S$^+$ occupy the transition zone in a fashion similar to O$^+$. The ions C^{+2} and S^{+2} occupy large fractions of the volume, although C^{+3} and S^{+3} are significant in the core.

The low-excitation case, such as regions S5 and H40 in M101 (Torres-

Peimbert et al 1989), is illustrated by model L (bottom panel in Figure 3). It has solar abundances, $T_* = 40{,}000$ K, and $U = 10^{-2.91}$. The temperature is lower than in model H and shows a radial increase. (This would be steeper in an OTS model.) The ionization structure is quite different from that of model H. He^+ gives way to He^0 slightly inside R_S. O^{+2} is confined to an inner core, and Ne^{+2} and S^{+3} are absent. The ions N^+, O^+, and S^{+2} dominate the ionized volume. In modeling S5 (an inner region of M101), Shields & Searle (1978) used the idea of Balick & Sneden (1976) that ionizing stars in metal-rich H II regions would have deep O II and Ne II Lyman edges in their spectra and hence low values of $Q(O^+)/Q(H^+)$, where

$$Q(O^+) \equiv \int_{\nu_{O^+}}^{\infty} L_\nu (h\nu)^{-1} d\nu.$$

This effect, coupled with a lower T_* in S5 than in NGC 5471, explains the combination of low $\langle X(O^{+2})\rangle$ and large observed $\langle X(He^+)\rangle$ in S5.

The electron temperature is much lower in model L, primarily because of the large heavy-element abundance and low value of T_*. Ironically, the oxygen-poor model has the stronger [O III] optical lines. Searle (1971) explained this situation in terms of the nebular cooling in the infrared fine-structure lines, including [O III] 52, 88 μm, as O/H increases. The gas temperature is forced to drop enough that the optical lines have diminished intensities, corresponding to the energy available from the ionizing continuum. However, the correlation of $\langle X(O^{+2})\rangle$ with O/H also contributes to the observed anticorrelation of [O III] and O/H.

The variation of the [O II] $\lambda 3727$ and [O III] $\lambda\lambda 5007$, 4959 lines (relative to Hβ) with model parameters is given in Figure 4. (Here $\lambda 3727$ refers to the sum of $\lambda 3726$ and $\lambda 3729$.) The models are characterized by T_*, U, and O/H. The inverse relationship of [O II]/[O III] as U varies is a simple consequence of photoionization equilibrium; the fractional volume of the transition zone varies roughly as U^{-1}. However, $R_{23} \equiv [I(\lambda 3727) + I(\lambda\lambda 5007, 4959)]/I(H\beta)$ is less sensitive to U. This supports the use of R_{23} as an empirical indicator of O/H (Edmunds & Pagel 1984, Stasińska et al 1981). However, it is clear that this quantity increases with T_* at fixed O/H, so that its utility in practice requires either constant T_* or a reliable trend in T_* with O/H. Below O/H $\approx 10^{-4}$, the [O II] and [O III] lines weaken with decreasing O/H as cooling by ultraviolet lines (including Lα) inhibits the further rise of the gas temperature (see Davidson & Kinman 1985, Dufour et al 1988). Skillman (1989) has shown that a useful empirical relation involving decreasing R_{23} with decreasing O/H holds for O/H $< 6 \times 10^{-5}$. This lower branch of the empirical relation depends only weakly on T_*, unlike the upper branch.

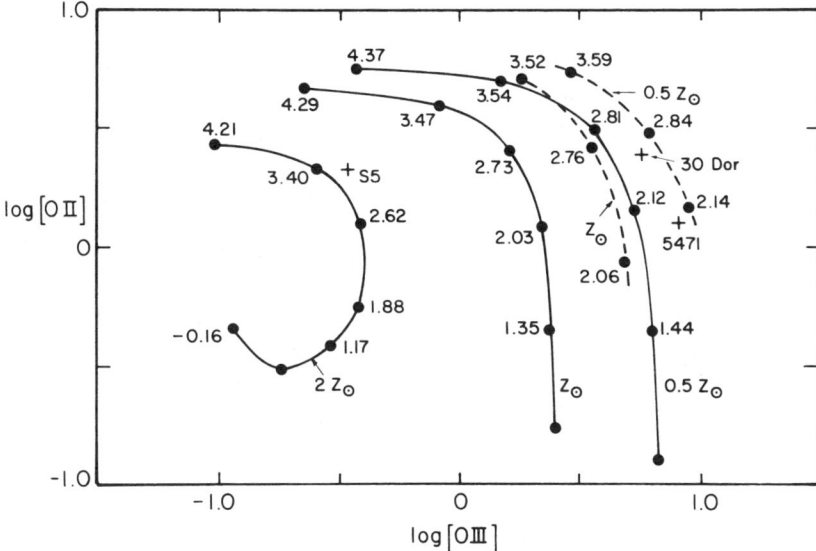

Figure 4 Model predictions of $\log_{10} I(\lambda\lambda 3726, 3729)/I(H\beta)$ and $\log_{10} I(\lambda\lambda 5007, 4959)/I(H\beta)$. *Solid* and *dashed* curves are for $T_* = 40{,}000$ K and $45{,}000$ K, respectively, with stellar continua calculated according to Shields & Searle (1978). Curves are labeled by heavy-element abundance relative to a solar value O/H = 7.4×10^{-4}. Points representing individual models are labeled by $-\log_{10} U$. Also shown are observed line intensities for 30 Dor (Mathis et al 1985) in the LMC and for S5 and NGC 5471 in M101 (Torres-Peimbert et al 1989).

Figure 4 is useful in constraining two of the parameters T_*, O/H, and U, but a third observational constraint is needed. Stasińska (1980) showed that an electron temperature, such as from [O III] $\lambda 4363$, is useful (see Section 5). Then O/H = $(O^+ + O^{+2})/H^+$ follows directly from the observations, and U and T_* follow from Figure 4 or an equivalent diagram. Alternatively, [O I], [Ne III], [S II], and [S III] are potentially useful in various regimes of excitation (Mathis 1985, Shields 1986).

From Equations 2 and 4, the ionic ratios A^{+i}/H^+ follow for the observed ions. For some elements, unobserved ionization stages constitute a considerable fraction of the element, integrated over the H II regions. This requires the use of an ionization correction factor (i_{CF}) to derive the total abundance of the element. Peimbert & Costero (1969) suggested some simple prescriptions based on ionization potentials. In some cases, such as N/O $\simeq$ N$^+$/O$^+$, models generally support these prescriptions. In other cases, such as He and S, a model tailored to the entire spectrum of an H II region is needed. Stasińska (1980) computed models with a composite density structure involving a low-density core surrounded by a higher density shell. Although real H II regions are indeed complex, abundances and stellar temperatures derived from composite models generally agree with simple models fit to the same data. Dopita & Evans (1986) argued that the weakly emitting, low-density regions of Stasińska's models mainly serve to increase the effective inner radius and thus alter the ionization parameter characterizing the dense shell. Model grids are given by Stasińska (1982), Rubin (1985), Evans & Dopita (1985), and Campbell (1988).

4. MORPHOLOGY AND KINEMATICS

The forms and sizes of GEHRs are diverse, and their internal structures complex. Their emitting gas occupies a small fraction of the nebular volume, and the gas moves at supersonic speeds. The underlying physics is controversial, with the debate centering on the relative importance of gravity, stellar winds, and hydrodynamical flows.

4.1 *Structural Properties*

Structural properties of GEHRs in nearby galaxies have been analyzed by Kennicutt (1984), who employed the morphological categories "classical" H II region, "high-surface-brightness" giant H II region, "diffuse" giant H II region, "multiple core complex," and "shell" or "ringlike" H II region. A few examples are listed in Table 1. The most luminous and well studied H II regions are the high-surface-brightness objects such as 30 Dor, NGC 604, and NGC 5471. These typically have a central surface brightness corresponding to an emission measure $EM = \int N_e^2 f ds \approx 10^{5.5}$ pc cm^{-6}, similar to the central EM of the largest galactic H II regions (e.g. W49). The higher luminosities of the brightest GEHRs result from their larger radii. There is a steep radial decrease in surface brightness passing from the "core" (perhaps ~ 50 pc in radius) to the "halo." Azimuthally averaged surface brightness profiles give little indication of a distinct physical separation of core and halo, but Kennicutt (1979) showed that the eye estimates of core and halo radii by Sandage & Tammann (1974) correspond roughly to photometric radii. Kennicutt uses core, halo sizes determined by the isophotes at 2×10^{-15}, 2×10^{-16} ergs cm^{-2} s^{-1} arcsec^{-2}, respectively. The mass of ionized gas depends on the electron density of the emitting gas, ranging up to $\sim 10^6 \, M_\odot$ for $N_e \approx 100$ cm^{-3} as indicated by forbidden-line ratios in the core. However, much of the luminosity comes from the halo at lower densities, so that the actual ionized mass may be several times larger. A filling factor of $\sim 10^{-2}$ is typical over a wide range of luminosities. The ionizing luminosities of these objects require the equivalent of up to 1000 O5 V stars. A Salpeter mass function extending down to 10 $M_\odot$ gives a corresponding stellar mass in the OB association of $\sim 10^{5.3}$ $M_\odot$ (Kennicutt 1984). The rms and actual electron densities are one or two orders of magnitude smaller than in compact galactic H II regions, and thus GEHRs cannot simply be superpositions of many individual H II regions; this also seems evident from photographs of nearby objects such as 30 Dor (Elliott et al 1977). However, some of the largest GEHRs, such as NGC 5471, have multiple cores with individual luminosities comparable to 30 Dor (Kennicutt 1984, Skillman 1985). The luminosities of the brightest H II regions in a galaxy increase strongly toward later Hubble type; those in the Milky Way are typical of an Sbc galaxy (Kennicutt 1984).

The diffuse giant H II regions have luminosities typical of the brightest galactic H II regions but a more diffuse and extended structure. On a smaller scale, the North America Nebula (NGC 7000) is a galactic analogue. The shell or ringlike H II regions, similar to galactic ring nebulae, have low surface brightnesses, luminosities, and densities (Gum & de Vaucouleurs 1953). Boulesteix et al (1974) illustrated H II rings in M33 and noted their increasing diameters at larger galactocentric distances. Chu (1983, and references therein) studied the kinematics of LMC ring nebulae associated with Wolf-Rayet (WR) stars. The four largest have diameters of ~ 100–200 pc, a size 10–20 times larger than a typical galactic counterpart. There may be an evolutionary sequence from diffuse to ringlike H II regions under the influence of winds from the massive stars (Castor et al 1975, McKee et al 1985).

The proximity of 30 Dor makes it a natural target for studies of the structure of GEHRs. The nebula surrounds R136, a cluster of early-type stars (Melnick 1985, Walborn 1986). Interference filter photographs of the emission-line structure of the core and halo have been made by Elliott et al (1977) and Meaburn (1979). On all scales, there is a complex structure of filaments and loops. Feast (1961) studied the densities and motions of the emitting filaments, finding values of $r_{\text{O II}}$ corresponding to densities of ~ 200–500 cm^{-3}, increasing to ~ 2000 cm^{-3} at R136. Elliott et al (1977) obtained long-slit spectra perpendicular to the bright east-west filament about 1.5 (25 pc) NE of R136, finding that the [N II] intensity peaks about 5" (1.4 pc) farther from the stars than does [O III]. This supports the idea that it is an optically thick ionization front eating into dense neutral material. (Time-dependent ionization is not the issue here; rather, the concentric O^{+2} and O^+ ionization zones of a spherical nebula are replaced by the corresponding layered structure.) The electron density in this filament is $N_e \approx 280$ cm^{-3} (Peimbert & Torres-Peimbert 1974, Mathis et al 1985). For a central $Q(H^0) = 10^{51.7}$ s^{-1}, this gives an ionizing flux $\phi_i = 7 \times 10^{10}$ cm^{-2} s^{-1} and an ionization parameter $U = 10^{-2.1}$. The predicted ionized thickness $l_s = \phi_i/(\alpha_B N_e^2) = 1.0$ pc is consistent with the observed filament. For $N(O)/N(H) = 3 \times 10^{-4}$, the [O II] and [O III] intensities of the filament (e.g. Peimbert & Torres-Peimbert 1974) are reproduced by a photoionization model with $U = 10^{-2.1}$ and $T_* = 43,000$ K (see Figure 4). For ionized gas flowing at the sound speed from an irradiated cylinder of dense neutral gas, the expected ionized density at the base of the flow is given by $\phi_i = \alpha_B N^2 a$, where a is the cylinder radius (Oort & Spitzer 1955). For $a = 1$ pc, we find $N_0 \approx 300$ cm^{-3}, in agreement with the observed N_e. The flux observed by Peimbert & Torres-Peimbert (1974) corresponds to $I(H\beta) = 10^{-12.8}$ erg s^{-1} cm^{-2} arcsec^{-2} if the filament fills half the length of their 75″ slit (position 30 Dor II). This agrees with the intensity for an irradiated cylinder. (An ionized flow from a neutral cloud of much larger radius than 1 pc would have a lower density and larger thickness than observed. A thin bright rim around such a large cloud would have to be confined by an external pressure such as a shocked stellar wind.) One problem with the idea that much of the emission-line luminosity of 30 Dor, or other GEHRs, comes from such ionized "champagne flows" (Tenorio-Tagle 1979) is the large accumulation of ionized gas (old champagne) in the intercloud volume (cf. Elmegreen 1976).

At lower surface brightnesses, images of 30 Dor show a system of giant loops extending over a diameter of 2000 pc (Meaburn 1979), including a "supergiant" shell of 1144 pc diameter to the north. Caulet et al (1982) made a kinematic study of a giant loop east of 30 Dor, fitting the observations with a portion of a shell of radius 475 pc and expanding at 30 km s^{-1}. The shell requires $\sim 10^{51.6}$ ionizing photons s^{-1}, possibly supplied by its own internal ionizing stars. The shell, which may result from the combined action of winds and supernovae, has an age of $\sim 10^7$ yr.

Smoothed models of the radial brightness profile of 30 Dor were derived from radio continuum measurements by Mills et al (1978) and fit to a core-plus-halo model. A radius of ~ 70 pc contains one half of the total flux. They suggested that the core is density bounded on the western side, facilitating ionization of more remote gas by the central cluster. Cersosimo & Loiseau (1984) added a third, more extended component of diameter 2100 pc. The three components together require $10^{52.5}$ ionizing photons s^{-1}, corresponding to ~ 500 O5 V stars. Melnick (1985) studied the spectral types of hot stars in the core of 30 Dor, which provide $\sim 10^{51.7}$ photons s^{-1}, enough to ionize at least the nebular core.

Spatially resolved spectra of NGC 604 by Díaz et al (1987) show a uniform chemical composition and reddening. Although [O II] and [O III] vary, their sum, R_{23}, is fairly constant, supporting the use of R_{23} as an empirical abundance indicator. Optical and radio line and continuum observations of the structure of NGC 5471 were made by Skillman (1985). He finds an H I mass of $6 \times 10^7 \, M_\odot$ and mean density $N_{HI} = 3.4$ cm^{-3}.

Several bright "cores" are seen, with heavier reddening than the surrounding regions. One location shows a factor of ~ 2 enhancement in N^+/O^+, and Skillman notes several indications of at least one supernova remnant (SNR), confirmed by Chu & Kennicutt (1986).

The subject of galactic H II regions lies beyond the scope of this review. Papers emphasizing the structure and kinematics of giant galactic H II regions in relation to GEHRs include the studies of NGC 3603 by Balick et al (1980) and Persi et al (1985). The Orion complex is reviewed by Genzel & Stutzki (1989), and star formation in the context of molecular clouds is reviewed by Evans (1989) and Shu et al (1987).

4.2 Kinematics

Kinematical studies of GEHRs focus on identifying the dynamical processes at work and on exploiting H II regions as standard candles for distance determinations. Underlying the kinematical studies is the fact that the emitting gas is concentrated in clouds, filaments, shells, or bright rims occupying a small fraction of the nebular volume. The observed line widths mostly result from the relative motion of these filaments, presumably as a result of gravity and stellar winds.

Feast (1961) obtained emission-line velocities for many positions across 30 Dor, finding an average velocity difference between pairs of points of ~ 11 km s^{-1} for separations of 9–60 pc, decreasing for separations of less than 4 pc. Smith & Weedman (1972) observed line profiles and velocities at 200 positions in 30 Dor and observed velocity differences of up to 80 km s^{-1}. The 30 Dor nebula as a whole is characterized by a line-of-sight velocity dispersion $\sigma \approx 25$ km s^{-1} (corrected for instrumental and thermal broadening), where $I(v) = I(0) \exp(-v^2/\sigma^2)$ and $v = c\Delta\lambda/\lambda$. This value exceeds the line widths of other LMC H II regions, but it resembles the line-width range of $\sigma \approx 22$–35 km s^{-1} found by Smith & Weedman (1970) for H II regions in M33 and M101. Long-slit spectrograms in the latter reference illustrate the differences in velocity between structures in large H II regions such as NGC 604. The individual filaments in 30 Dor move coherently and have narrow line widths ($\sigma \simeq 15$ km s^{-1}). This is consistent with an evaporation flow, as described above. Hydrodynamic models of a photoionized flow around a spherical cloud indicate that most of the line emission comes from gas moving at ~ 10–20 km s^{-1} (Bertoldi 1989).

Smith & Weedman (1970) argued that winds from the WR stars in 30 Dor could supply the $\sim 10^{51.5}$ ergs of kinetic energy of the emitting gas. A similar conclusion was reached by Rosa & D'Odorico (1982) for NGC 604, in which they observed numerous shell structures with a mean diameter of ~ 25 pc. Meaburn (1981, 1984) observed extensive sheets of ionized gas in 30 Dor and a variety of expanding shells presumably driven by stellar winds. Chu & Kennicutt (1986) found broad Hα line wings, indicating fast-moving gas, in H II regions of M101. The fast-moving gas involves only a small fraction of the line emission but possibly a large fraction of the nebular kinetic energy. However, Skillman (1985) noted that GEHRs with weaker WR features in their stellar continuum, such as NGC 5471, have equally large nebular line widths. Furthermore, Skillman & Balick (1984) found that GEHRs rarely show the line profile splittings expected for a wind-driven shell model (Dyson 1979), although narrow components and asymmetries are observed.

A systematic increase of "turbulent" velocity with radius for GEHRs ($R \propto \sigma^{2.3\pm0.2}$) was found by Melnick (1977), and a correlation of σ with $L(H\alpha)$ was obtained by Melnick (1978) and Terlevich & Melnick (1981). The latter argued that the same trends agree with the velocity dispersions of globular clusters and elliptical galaxies (Faber & Jackson 1976), supporting a gravitational origin for the motions.

Hippelein (1986) found $L(H\alpha) \propto \sigma^{6.6}$, but the metallicity dependence proposed by Terlevich & Melnick (1981) actually worsened the scatter. Melnick et al (1987) studied the problem with new data, employing core radii R_c (Sandage & Tammann 1974). They considered two theoretical models: "virial turbulence," giving $L \propto R\beta^2$; and a stellar wind model, predicting $L \propto R^3\sigma^2$. The least-squares fits to the data give $L(H\alpha) \propto (R\sigma^2)^{0.86}(O/H)^{-0.65}$ and $L(H\alpha) \propto (R^3\sigma^2)^{0.35}(O/H)^{0.75}$, favoring the gravitational model. Hippelein (1986) proposed three regimes to explain the σ-L correlation: (a) for $\sigma \lesssim 10$ km s^{-1}, "champagne" flows dominate with $\sigma \propto L^0$ (e.g. M42, M8); (b) for $10 \lesssim \sigma \lesssim 30$ km s^{-1} and $10^{38} \lesssim L(H\alpha) \lesssim 10^{40}$ ergs s^{-1}, stellar winds give $\beta \propto L^{1/5.5}$; and (c) for $\sigma > 30$ km s^{-1}, the motions obey a virial relation that extrapolates to the starburst galactic nuclei observed by Balzano (1983).

4.3 Distance Determinations

GEHRs have attracted interest as possible "standard candles" for distance measurements because they are large and bright enough to be observed in galaxies as remote as the Virgo cluster and beyond. The diameters and luminosities of the H II regions themselves are not useful, being "distance degenerate" because GEHRs have a characteristic surface brightness. Likewise, the diameters of the largest H II regions in a galaxy and its absolute magnitude satisfy a statistical relation $\Delta \log D/\Delta M_{pg} = -0.14$, uselessly close to the distance-blind value of -0.2 (Sandage & Tammann 1974). Sersic (1960) showed that the sizes of the largest H II regions depend on morphological type, measured by color index, peaking at Scd, and thus that they provide a potential standard candle. Sandage & Tammann (1974) showed a correlation of D with luminosity class, using subjective core and halo diameters; the halo diameters varied from ~ 550 pc for Sc I to 110 pc for luminosity class V. They made H II region diameters an important rung in the first formulation of their cosmic distance ladder. In a series of papers, Kennicutt (1981, and references therein) used quantitative Hα photometry of GEHRs in an attempt to refine the distance method. He found for 21 Virgo cluster galaxies that Hα luminosities as a function of Hubble type show less scatter than diameters, but the scatter is still discouraging. This appears to be a real dispersion rather than statistical sampling, since late-type galaxies often have a few anomalously large H II regions.

De Vaucouleurs (1983) argued that the diameters of H II rings provide an objective basis for distance determinations. The largest ring in a galaxy has a diameter D_1 that correlates only weakly with the galaxy's absolute magnitude, $\log \langle D_1 \rangle = 1.85 - 0.05(M_T^0 + 10)$, where D_1 is in parsecs. This is far from the "distance effect" coefficient $\Delta \log D_1 / \Delta M_T^0 = -0.2$, and the small dispersion offers the potential for distance moduli accurate to 0.25 mag.

Another approach involves the use of GEHR line widths to predict the diameters and luminosities of these regions, as described above. Roy & Arsenault (1986) find that $\langle \sigma \rangle$ for the three largest H II regions in a galaxy correlates better with the galaxy's absolute magnitude M_B than with the diameter of the H II regions (see also Melnick 1978, de Vaucouleurs 1979). However, the reality is that GEHRs are complex, inhomogeneous objects with ill-defined boundaries and much dispersion in their properties. In the absence of dramatic advances in their understanding, GEHRs may not be the most efficient distance indicators. Moreover, in the important case of the Virgo cluster, there is increasing evidence of important environmental effects on the interstellar medium of the spirals (Giovanelli & Haynes 1985, Warmels 1986). Other distance methods therefore seem preferable to the use of GEHRs. Prominent among these is the correlation of a galaxy's absolute magnitude and rotational velocity (Tully & Fisher 1977, Aaronson et al 1979), and the use of supergiant stars and Type I supernovae (Sandage & Tammann 1985).

5. IONIZING STARS

H II regions are heated and ionized by their luminous hot stars, which also play an important role in their dynamics. Star formation is reviewed by Scalo (1986, 1990), with the second paper emphasizing blue compact galaxies. The subject of ionizing stars of GEHRs is reviewed by Rosa & D'Odorico (1986). The underlying stellar continuum of GEHRs typically has an equivalent width of absorption in Hβ of 1.4 Å (MRS). In the case of nearby extragalactic H II regions, the individual stars can be studied. Melnick (1985) obtained spectral types of stars within 25 pc of the center of 30 Dor, finding 49 O-type stars, 12 WR stars, 8 B supergiants, and 1 M supergiant. At least 15 are earlier than O6, and 6 are O3. Excluding R136a, the stars provide $\sim 10^{51.5}$ ionizing photons s^{-1} (see also Walborn 1986). Taking R136a as the equivalent of 6 O3 If stars, the total increases to $\sim 10^{51.7}$ s^{-1}. Fitting non-LTE (NLTE) model atmospheres to spectra of O stars in the LMC, Gehren et al (1986) obtained values of $T_{\rm eff}$ up to $50,000 \pm 3000$ K. Massey et al (1989) found many hot O stars in NGC 346 in the SMC, with $T_{\rm eff}$ ranging up to 48,000 K (the limit of their spectral-type calibration). In NGC 604, D'Odorico & Rosa (1981) located ~ 50 WN7 and 50 early O-type stars. Wolf-Rayet stars in M33 H II regions were also studied by Conti & Massey (1981) and Massey & Conti (1983), who found that the ratio of WC to WN types decreases with increasing galactocentric radii as a result of changes in the initial mass function (IMF) or the chemical composition. Similar differences are seen between the Galaxy, the LMC, and SMC. For NGC 5471, however, Skillman (1985) sets upper limits on the $\lambda 4650$ WR feature, implying that less than 10% of the stellar continuum comes from WR stars, compared with 50% in NGC 604.

The large number of O3 stars in the LMC and SMC contrasts with their relative scarcity in the Milky Way (Melnick 1978). This may be related to other indications that hotter ionizing stars are present in H II regions with lower abundances. Shields & Tinsley (1976) suggested that such a trend would result from opacity effects on stellar structure for a given IMF upper mass limit $m_{\rm u}$, and in addition the physics of star formation might cause $m_{\rm u}$ to increase for lower Z. They noted, in support of this suggestion, that a radial increase in the equivalent width of the nebular Hβ emission, $W({\rm H}\beta)$, from ~ 65 Å in the inner H II arms of M101 to ~ 165 Å in the outer arms. The trend of T_* with O/H helps to explain the systematic dependence of nebular ionization on metallicity, as noted above. This trend remains somewhat controversial, however, especially as to the quantitative T_*-Z relation (Scalo 1986). MRS found little correlation of $W({\rm H}\beta)$ with R_{23}, the metallicity indicator.

Stasińska (1980) showed that T_* can be found from photoionization models and a plot of $R_{\rm O\,III}$ (or T) versus O/H, and he obtained a trend of increasing T_* with decreasing O/H. This method requires an observed electron temperature and so is restricted at present to metal-poor H II regions. In contrast, Evans & Dopita (1985) gave a grid of photoionization models and found from diagrams involving $\lambda 6300$, $\lambda 3727$, and $\lambda 5007$ a value of $T_* \approx 41,500$ K, with little systematic dependence on O/H. Nevertheless, Evans (1986) found lower T_* for higher O/H in models of individual H II regions in M101, in agreement with Shields & Searle (1978). Campbell et al (1986) obtained spectra of a sample of H II galaxies and found an increase of [O III]/[O II] with $W(H\beta)$ that they attributed at least in part to aging of the ionizing clusters. This suggests that the upper envelope of observed points in a T-O/H diagram represents the equivalent T_* of the IMF. The observed trend of this upper envelope implies a decrease in T_* from $\sim 55{,}000$ K for log O/H ≈ -4.5 to $\sim 40{,}000$ K for log O/H ≈ -3.7. A similar trend is found by Campbell (1988).

The T-O/H method for determining T_* is based on the idea that, for a given O/H, hotter stars, with their more energetic ionizing photons, will give a higher nebular temperature. This simple physics suggests that the method should give reliable results if the stellar atmosphere models give the correct dependence of the mean ionizing photon energy $\langle hv \rangle$ on T_* and O/H. Other methods, reviewed by Shields (1986), rely more heavily on the ionization of various elements in the nebula as influenced by heavy-element absorption edges in the stellar atmosphere. Garnett (1989b) has compared the effects of various stellar atmospheres in model H II regions. He found that line blocking is not very important for H II regions in LTE models computed with the ATLAS code (Kurucz 1979), but that heavy-element photoionization edges are critical. These edges tend to be weaker in NLTE than in LTE for given T_*, log g, and O/H, but models involving a fictitious "mean heavy element" fail to give the important edges at their true frequencies (Mihalas 1972, Borsenberger & Stasińska 1982). Needed for improved H II region modeling are grids of NLTE model atmospheres in the range $35{,}000 \lesssim T_* \lesssim 60{,}000$ K and $0.02 \lesssim (O/H)/(O/H)_\odot \lesssim 2$, including the actual absorption edges of the important heavy elements.

One method of determining T_* is to use the "radiation hardness" parameter $\eta \equiv (O^+/O^{+2})/(S^+/S^{+2})$, promoted by Vílchez & Pagel (1988) and Vílchez et al (1988). Garnett (1989b) showed that η varies somewhat with choice of stellar atmosphere and ionization parameter, but for a given type of stellar atmosphere, T_* and U can be determined from a diagram of O^+/O versus S^+/S^{+2} (as used by Mathis 1985). Unfortunately, some observed H II regions have values of S^+/S^{+2} that are so large, relative to O^+/O, as to require unreasonably high values of T_*, giving much larger intensities of He II $\lambda 4686$ than observed. Finding a similar problem for the H II galaxy NGC 4861, Dinerstein & Shields (1986) suggested incomplete atomic data, perhaps involving charge transfer.

The current status of T_* determinations is that there appears to be an increase of T_* with decreasing O/H, reflecting the need for higher $\langle hv \rangle$ and weaker absorption edges to fit the spectra of H II regions with lower O/H. Because of inadequate stellar atmospheres, absolute values of T_* are uncertain, as is the need for a variation in m_u. However, sequencing of

H II regions in T_* probably can be done reliably by a uniform application of existing model atmospheres.

Model H II regions also indicate a trend of increasing ionization parameter U with decreasing O/H. This can be seen in the models of M101 H II regions by Shields & Searle (1978) and Evans (1986), and it is generally found in analyses involving line-ratio diagrams (e.g. Dopita & Evans 1986). Vílchez et al (1988) found such a trend in M33, using the models of Stasińska (1980, 1982) and the parameters η and S^+/S^{+2} to fix T_* and U. A photoionization model sequence involving correlated changes in O/H, T_*, and U has been given by MRS. Nevertheless, U tends to fall in a fairly narrow range ($\sim 10^{-2.5 \pm 0.5}$) in the context of the much larger value that would occur if the gas were not confined in filaments with a small filling factor. Shields (1986) has considered the possibility that this "magic" ionization parameter range may involve the pressure of shocked stellar winds (Dyson 1979). Alternatively, the above discussion of filaments in 30 Dor raises the possibility that N and U are determined by the physics of ionized gas flows. Since N_e and U depend on the size of the filaments or clouds, there may be some regulating mechanism. This could involve the evaporation of clouds near the stars, the "rocket effect," and the accumulation of old "champagne" in the nebula (Elmegreen 1976, McKee et al 1985).

The Hα luminosities of H II regions are proportional to $Q(H^0)$ and thus to the rate of massive star formation, since most GEHRs are optically thick to the Lyman continuum (MRS). Tracing star formation this way, Kennicutt (1989) found good agreement with the idea of a critical surface density of disk gas above which gravitational instability occurs. For M83, Jensen et al (1981) compared star formation rates derived from Hα and from UBV surface photometry, finding that some clusters are born with a deficiency of ionizing stars.

6. CHEMICAL ABUNDANCES

One of the most important applications of GEHRs is the measurement of how the chemical composition of the interstellar gas varies with position in individual galaxies and from one galaxy to another. Such measurements, up to now based mostly on optical spectrophotometry, are valuable as clues to galactic evolution (Audouze & Tinsley 1976, Pagel & Edmunds 1981, Chiosi 1986, Matteucci 1986, Pagel 1987).

6.1 *Spiral Galaxies*

Aller (1942) noticed that the [O III]/Hβ ratios of GEHRs in M33 increase with galactocentric distance R, whereas [O II] remains relatively constant.

He suggested that the ionizing stars might be hotter at increasing radius. Searle (1971) obtained spectra of H II regions in M101 and M33 that showed this effect and interpreted it as a radial decrease in the O/H ratio. Searle & Sargent (1972) measured a high electron temperature for NGC 5471, an outlying H II region in M101. This implied an oxygen abundance only one third that in the Orion Nebula, which in turn has lower O/H than the Sun. Smith (1975) obtained spectrophotometry of numerous H II regions in several spiral galaxies that confirmed the presence of radial abundance gradients in O/H and a characteristic decrease in O^+/O^{+2} with decreasing O/H. The outermost H II regions of late-type galaxies had values as low as O/H $\approx$ 0.2 (O/H)$_\odot$; values in the inner disk were apparently greater than solar but were difficult to measure as a result of the faintness of [O III] $\lambda 4363$.

A theoretical understanding of GEHRs was needed to determine (a) O/H in the low-excitation, oxygen-rich regions and (b) any variations in other abundances, T_*, and U. Shields & Tinsley (1976) proposed an increase in T_* with decreasing O/H. The lower characteristic T_* in the inner H II regions helped to explain the large [O II]/[O III] ratio of these regions, as described above. Alternatively, Sarazin (1976) suggested that internal dust is more abundant in metal-rich H II regions, absorbing the ionizing radiation and giving rise to the observed gradients in O^+/O^{+2} and $W(H\beta)$.

A theoretical fit to the spectrum of the inner region S5 in M101 was achieved by Shields & Searle (1978), as described above. This led to an abundance by number of log O/H = -2.9 ± 0.2 at $R/R_{25} = 0.28$, where R_{25} is the photometric radius. The radial drop in log O/H—to -3.5 in NGC 5455 at $R/R_{25} = 0.55$, and to -3.9 in NGC 5471 at $R/R_0 = 0.99$—was steeper than could be accounted for by the "simple model" of chemical evolution described below. Modeling of H II regions in M101 by Evans (1986) gave a somewhat lower value of O/H in S5.

Spectrophotometric studies of abundance gradients have now been carried out for many galaxies. Many of these are referenced in the review by Pagel & Edmunds (1981), in the survey of 20 spiral and irregular galaxies by MRS, and in the study of M81 by Garnett & Shields (1987). Recent additional work includes the theoretical study by Evans (1986) of M101 and the observations of M33 by Vílchez et al (1988).

The issues addressed by these studies include (a) the dependence of the O/H gradient on galactic mass, luminosity, and type; (b) the variation of other elements relative to O; and (c) the implications for the evolution of stars and galaxies. The presence of radial gradients appears to be a universal property of spiral galaxies. For massive Scd galaxies like M101, O/H decreases by an order of magnitude from the inner to the outer disk.

The functional form of this O/H versus R relationship is unclear. Shields & Searle (1978) argued for O/H $\propto R^{-a}$ from three regions, but Evans (1986) found a form O/H $\propto e^{-r/b}$ for the same galaxy. Dufour et al (1980) argued that an exponential form log O/H $= A + BR$ fits the gradients of several spirals, and that the scale length $B^{-1} \approx -0.09$ dex kpc^{-1} is similar in galaxies of different luminosity and photometric radius R_0. However, in M33 the gradient becomes steeper in the inner disk in terms of $\Delta \log (O/H)/\Delta R$ (Vílchez et al 1988). McCall (1982) and Edmunds & Pagel (1984) found greater consistency when gradients are expressed in terms of R/R_e rather than R/R_{25}, where R_e is the exponential scale length of the disk.

One approach to this question involves the use of the local gas fraction $\mu = M_{gas}/M_{total}$ as an indicator of stellar processing. The "simple" closed-box model of chemical evolution with instantaneous recycling evolution predicts $z = y \ln \mu^{-1}$, where z is the mass fraction of one or more primary nucleosynthetic products, and y is the yield (Searle & Sargent 1972, Tinsley 1980). Matteucci & François (1989) and Pagel (1989) found that models with infall (Larson 1972, Tinsley & Larson 1978) can account for the gradients in μ and O/H in the Milky Way as well as for the distribution of stellar metallicities. Jensen et al (1976) analyzed abundance gradients across late-type spirals in terms of the density-wave picture of star formation; however, for instantaneous recycling, the detailed formation history does not affect the O/H-μ relation.

The dependence of the mean oxygen abundance averaged over the disk, $\langle O/H \rangle$, on galaxy type is controversial. Smith (1975) found evidence for a higher $\langle O/H \rangle$ and weaker gradients in earlier-type spirals, an impression supported by Hawley & Phillips (1980). Edmunds & Pagel (1984) found that O/H is about twice as large in early-type spirals than in Scd and Sd spirals at a given R/R_e. On the other hand, Garnett & Shields (1987) observed that the O/H gradient and $\langle O/H \rangle$ value of the Sab galaxy M81 agree with those of later-type spirals with the same mass, luminosity, and gas fraction. They argued that these physical parameters, rather than morphological type, determine chemical composition.

Barred spiral galaxies may have weaker gradients in O/H. Pagel et al (1979) found a weak gradient in the SBb galaxy NGC 1365 and suggested that mixing by radial gas flows in barred spirals may suppress composition gradients (cf. Tinsley 1980). However, Roy & Walsh (1988) noted a steep gradient in R_{23} at larger radii in NGC 1365.

The variation in the abundances of other heavy elements, relative to oxygen, is also controversial. This topic is discussed for oxygen-poor dwarf galaxies in Section 6.2. For spirals, the general impression is that the abundances of Ne, S, and Ar vary in fixed proportion to oxygen (e.g.

Smith 1975, Kwitter & Aller 1981, Fierro et al 1986, Rosa & Mathis 1987, Garnett 1989b). The variation of N/O appears to be complicated. Smith (1975) and Evans (1986) found that N/O does increase with O/H across M101, but that the variation is weaker than the relation N/O $\propto$ O/H predicted by the "simple model" of chemical evolution for secondary nitrogen nucleosynthesis (Tinsley 1980). Fierro et al (1986) observed a gentle N/O gradient across NGC 2403. However, other studies find little or no gradient in N/O across some spirals, including M33 and NGC 300 (Kwitter & Aller 1981, Pagel et al 1979, Vílchez et al 1988). M81 shows little N/O gradient in spite of a strong O/H gradient (Garnett & Shields 1987).

Studies of H II regions in the Milky Way show a radial gradient in O/H similar to those of other spiral galaxies (Peimbert et al 1978, Hawley 1978, Shaver et al 1983). Peimbert et al found a radial gradient in N/O, but Shaver et al did not. Infrared fine-structure lines of N III and O III suggest large N/O values toward the galactic center (Lester et al 1983, 1987), but Garnett (1989a) has argued that $N^{+2}/O^{+2} > N/O > N^+/O^+$ in low-excitation H II regions. Using photoionization models, Rubin et al (1988) argued for constant N/O across the galactic disk. Models show that $N/O = N^+/O^+$ is a fair approximation in high-excitation H II regions, even though O^+/O and N^+/N are both small.

The variation of S/O with O/H is controversial. Some studies have indicated that S/O increases as O/H decreases below solar (e.g. French 1981, Shaver et al 1983, Evans 1986, Vílchez et al 1988). Because of the low fractional abundance of S^+, results based on [S II] $\lambda\lambda 6717, 6731$ alone are unreliable. The S^{+2} ion is represented by the weak, temperature-sensitive $\lambda 6312$ line; however, modern CCD detectors can measure the stronger $\lambda\lambda 9069, 9532$ lines, subject to accurate correction for telluric water vapor absorption. From new measurements of $\lambda\lambda 9532, 9069$ analyzed with photoionization models, Garnett (1989b) found constant log S/O = -1.7 and argued that earlier work showing variable S/O was in error. Torres-Peimbert et al (1989) likewise found constant S/O with O/H.

Neon abundances can be measured only in high-excitation H II regions, in which Ne^{+2} predominates. Solar values of Ne/O are found. Observations of [Ne II] $\lambda 12.8$ μm would be especially valuable for low-excitation objects for which Ne^+ predominates. The few available measurements of Ar are consistent with solar Ar/O (e.g. Torres-Peimbert et al 1989).

Most measurements of composition gradients across spiral galaxies involve GEHRs, but some work on nearby galaxies also involves spectra of SNRs. For example, Blair et al (1982) compared abundances derived from SNR and H II regions in M31. They found that N/H and O/H decrease by factors of ~ 4 from $R = 4$ to 23 kpc; however, whereas the

N/H values agreed between results for H II regions and SNRs, there was a systematic discrepancy for O/H.

6.2 Dwarf Irregular Galaxies

Dwarf irregular galaxies have enjoyed increasing scrutiny since the discovery by Peimbert & Spinrad (1970) and Searle & Sargent (1972) that the abundances of N, O, and Ne in NGC 6822, II Zw 40, and I Zw 18 are only a small fraction of solar. These objects have the lowest interstellar abundances known, making them uniquely useful as probes of stellar nucleosynthesis and of the primordial helium abundance. The nearest examples are the Magellanic Clouds, reviewed by Dufour (1984) with recommended abundances based on several studies (e.g. Aller et al 1974, Dufour 1975). Blue compact galaxies (BCGs) are undergoing brief bursts of intensive star formation, as shown for II Zw 40 and I Zw 18 by Searle et al (1973). Much information on BCGs and related objects is contained in the volume edited by Kunth et al (1985). There is some controversy about the evolutionary connection between dwarf irregulars, BCGs, and dwarf ellipticals (e.g. Lin & Faber 1983, Silk 1987, Silk et al 1987).

Heavy-element abundances in dwarf irregulars are reviewed by Dufour (1984, 1986). One question is the relationship between size of the galaxy and oxygen abundance. An increase in O/H with mass was found by Lequeux et al (1979), Talent (1980), and Kinman & Davidson (1981). Recent work by Skillman et al (1988, 1989) clearly shows a trend, $12 + \log (O/H) = -0.153 M_B + 5.50$, where M_B is the blue absolute magnitude. This trend resembles the one obeyed by dwarf ellipticals. Dwarf irregular galaxies with large H I halos do not show lower O/H, as might be expected to result from accretion of primordial gas.

Dwarf irregulars have low N/O values as well as low O/H values. Pagel (1985) has illustrated the correlation of N/O with O/H in H II regions of spirals and irregulars. For $12 + \log O/H \lesssim 8.3$, there is much scatter but little systematic variation in N/O with O/H; for larger O/H, there is an increase in N/O with O/H. This resembles the results for metal-poor stars in the Galaxy (Tomkin & Lambert 1984). The large scatter in N/O, at a given O/H (French 1980, Pagel 1985), was confirmed by Garnett (1989a), who found a mean value $\log N/O = -1.47$. However, II Zw 40 and NGC 6822 have $\log N/O \approx -1.25$ and -1.75, respectively, at $12 + \log O/H = 8.2$. The difference is much greater than the estimated errors and the dispersion of independent measurements of each object. At a given O/H, spirals tend to have higher N/O values than do irregulars; the most extreme example is the outlying H II region Münch 1 in M81, with $\log N/O = -1.0$ at $12 + \log O/H = 8.2$ (Garnett & Shields 1987). Recent models of chemical evolution of dwarf irregulars have emphasized discrete bursts of star formation (Matteucci & Tosi 1985, Clayton & Pantelaki 1986, Matteucci 1986, Pantelaki 1988). In this spirit, Garnett (1989a) argues that the scatter in N/O for irregulars may result from bursts of star formation that produce oxygen promptly and nitrogen only later.

Models for spirals emphasize infall (Pagel 1989, Matteucci & François 1989) and radial flows (Lacey & Fall 1985). Infall may help explain high N/O at low O/H, as in Münch 1 (Serrano & Peimbert 1983, Garnett & Shields 1987); for example, sudden accretion of primordial gas would dilute O/H at fixed N/O.

Carbon abundances can be measured in oxygen-poor H II regions because the high temperature facilitates excitation of C III] $\lambda1909$. Measurements are summarized by Dufour (1986). The high C/O ratio for I Zw 18 by Dufour et al (1988) should be adjusted downward by a factor of 1.6 for aperture effects (Dufour & Hester 1990). The measured N/C ratios for dwarf irregulars range around the solar value or slightly below, showing little systematic variation with O/H. Tinsley (1980) noted that time delays in carbon enrichment could give C/O ratios mimicking the behavior of a secondary element like nitrogen. An interesting question is whether spirals with relatively high N/O have correspondingly high C/O.

Rosa & Mathis (1987) have studied the chemical homogeneity of 30 Dor, obtaining spectra at 10 positions in the outer regions. The elemental abundances are quite similar to those in the core of the nebula (Mathis et al 1985), with O/H $= 2.2 \times 10^{-4}$ and solar ratios Ne/O, S/O, Ar/O, and Cl/O. The measured value Fe/O $= 0.008$ is about 0.2 times solar, and Rosa & Mathis attribute this to incorporation of Fe in grains. However, in metal-poor stars Fe/C remains roughly solar (Wheeler et al 1989), so that the low value C/O ≈ 0.3 (C/O)$_\odot$ in 30 Dor may largely account for the low Fe/O. One region studied by Rosa & Mathis has high abundances (O/H $\approx 10.3 \times 10^{-4}$ and He/H $= 0.14$). This appears to be photoionized nebular gas whose low temperature results from the high abundances. The region is consistent with enrichment by about 10 $M_\odot$ of H-poor material lost by a star of ~ 80 $M_\odot$ during its late O star and WR evolutionary phases.

6.3 *The Primordial Abundance of Helium*

In the Big Bang model, primordial nucleosynthesis occurs during the first few minutes. A comprehensive review of primordial nucleosynthesis theory and measurements is given by Boesgaard & Steigman (1985). As the early Universe cools and expands, the neutron/proton ratio "freezes out" at its equilibrium value (n/p ≈ 0.2) at $t \approx 1$ s and $T \approx 10^9$ K. The neutron density drops slightly because of spontaneous decay, n $\to$ p+e$^-$+$\bar{\nu}$, until

T drops to $\sim 10^7$ K. At this point deuterons become stable against photodestruction, and all neutrons are incorporated into deuterons. Most further react to form ^{4}He, leaving a trace of ^{2}D and ^{7}Li. The abundance of helium by mass is $Y = 4y/(1+4y)$, where $y \equiv n(\text{He})/n(\text{H}) \approx 0.08$ in the standard Big Bang model. The primordial abundance value Y_p is a weak function of the nucleon-to-photon ratio $\eta \equiv N_n/N_\gamma$, the number of neutrino species N_ν, and the neutron half-life τ_n. A useful formula is

$$Y_p = 0.229 + 0.011 \ln \eta_{10} + 0.013(N_\nu - 3) + 0.014(\tau_n - 10.5), \qquad 8.$$

where $\eta_{10} = 10^{10}\eta$ (Boesgaard & Steigman 1985). Precise measurements of Y_p, to an accuracy of 5% or better, can provide important constraints on cosmology.

Measurements of Y in globular clusters, galactic stars, and nebulae are consistent with a universal minimum helium abundance augmented to varying degrees by nucleosynthesis in stars. The most promising method of measuring Y_p involves spectroscopy of H II regions in metal-poor dwarf irregular galaxies. The line-emission mechanisms are well understood, and radiative transfer is unimportant for most lines. These objects have the lowest interstellar abundances known, so that helium enrichment by stars presumably is minimized. Peimbert & Torres-Peimbert (1974) measured He/H and O/H in different objects to find $\Delta Y/\Delta Z$, the ratio of stellar enrichment in the helium and heavy-element mass fractions, and extrapolated to Y_p at $Z = 0$. Subsequent studies have used a similar approach, generally finding Y_p in the range 0.22–0.24. A review emphasizing nebular physics is given by Shields (1986), and a critical analysis of the systematic errors is made by Davidson & Kinman (1985).

High-excitation GEHRs have helium mostly in the form He$^+$. Recombination produces emission in several He I lines, including $\lambda\lambda 3889$, 4471, 5876, 6678, and 7065. The ionic ratio He$^+$/H$^+$ follows from Equation 2, with the effective recombination coefficients being given by Brocklehurst (1971, 1972). Several complications affect the line emission. Accumulation of atoms in the 2^3S metastable level leads to substantial optical depths in $\lambda 3889$, so that its intensity is diminished, whereas $\lambda 7065$ is enhanced, by fluorescence (Robbins 1968, Osterbrock 1989). Of the remaining lines, $\lambda 5876$ is the brightest, but it is affected by collisional excitation of 2^3S atoms by electrons. Ferland (1986) expressed this in terms of $\gamma_C \equiv I_{\text{coll}}/I_{\text{rec}}$. For various reasons, Ferland appears to have exaggerated the effect in GEHRs, but the effect is still significant. Clegg (1987) has given convenient formulas for γ_C. Because γ_C varies roughly as N_e for $N_e \lesssim 10^3$ cm^{-3}, the electron density must be known, or constrained to values $< 10^2$ cm^{-3}, for work of the required precision.

Also of concern is the ionization correction. Standard models of high-excitation nebulae with $T_* \gtrsim 40{,}000$ K give $i_{CF} = (\text{He}/\text{H})/(\text{He}^+/\text{H}^+) \approx 0.98$ or 0.99 (Stasińska 1982). However, Dinerstein & Shields (1986) warned that there could be pockets of gas with H^+, He^0, and O^+ around cooler ionizing stars. In the extreme, He^+/He could be almost as small as $1 - \text{O}^+/\text{O}$. Typical BCGs have $\text{O}^+/\text{O} = \text{O}^+/(\text{O}^+ + \text{O}^{+2}) \approx 0.1$–$0.2$, and thus this uncertainty is potentially devastating to cosmological applications. Peña (1986) finds that stars with $T_* \gtrsim 38{,}000$ K contribute relatively little ionizing radiation in the case of star formation described by a "reasonable" initial mass function, but it seems advisable to minimize this uncertainty by using H II regions with low O^+/O.

The value of $Y_p = 0.230 \pm 0.004$ (1σ) by Lequeux et al (1979) has proved to be quite durable. Pagel et al (1986) gave 0.237 ± 0.005. These authors preferred to plot Y versus N/H rather than O/H for extrapolation to Y_p, but the difference is not great (Torres-Peimbert et al 1989). The value of $Y = 0.243$ by Kunth & Sargent (1983) should be lowered to 0.234 because of problems with II Zw 40 (Pagel et al 1986). A recent study by Pagel & Simonson (1989) includes measurements of two BCGs (Tol 1214−277 and UM 461) with $\text{O}^+/\text{O} < 0.05$ and $\text{O}/\text{H} < 7 \times 10^{-5}$, respectively, ideal objects for Y_p determinations. Including selected observations from other workers, these authors find $Y_p = 0.229 \pm 0.004$ (1σ). The importance of electron density is illustrated by Tol 1214−277, for which Pagel & Simonson (1989) give $N_e = 330 \pm 270$ cm^{-3}. If N_e were only 50 cm^{-3}, collisional excitation of $\lambda 5876$ would be significantly reduced, and Y would be increased from 0.224 to 0.233 in this object.

The work currently available suggests a value $Y_p = 0.230 \pm 0.005$, where the error bar is a 1σ error based on the dispersion of measurements for different objects, in turn consistent with the uncertainties in measurements of individual line intensities (Pagel & Simonson 1989, Torres-Peimbert et al 1989). To this we may add a possible increase $\Delta Y_p \lesssim 0.01$ in the extreme case that 5% of the helium is neutral, corresponding to the O^+ fraction in the above named BCG.

A value of Y_p in the range 0.23–0.24 constrains the entropy of the Universe to $\eta \lesssim 2 \times 10^{-10}$ for $N_v = 3$ (see Equation 8 or Figure 2 of Boesgaard & Steigman 1985). Given the photon density of the 3 K background radiation, this corresponds to a present-day cosmic nucleon density $\rho_N < 2 \times 10^{-31}$ g cm^{-3}, much less than the critical density needed to bind the Universe. This has been the conclusion of Y_p studies since the early work of Peimbert & Torres-Peimbert (1974). Potential information on particle physics is possible by analyzing Y_p together with the cosmic deuterium abundance, which decreases with increasing η. Nucleosynthesis in stars destroys deuterium but returns some of it to the interstellar medium in the form of ^3He, so that the combined mass fraction of D and ^3He, Y_{23p}, is less vulnerable to uncertainties in galactic evolution. Using Boesgaard & Steigman's (1985) limit ($Y_{23p} < 10^{-4}$) and their Figure 3, we see that $Y_p \leq 0.24$ is marginally difficult to reconcile with three neutrino species. Any significant tightening of the upper limit on Y_p, as by ruling out significant He^0 corrections, would favor $N_v = 2$, possibly implying a rest mass for the τ neutrino.

7. CONCLUSIONS

Giant extragalactic H II regions present many challenges for future work. Their morphology and kinematics are substantially understood at a descriptive level, but the physics of their evolution is sketchy. What are the roles of stellar winds, hydrodynamic flows, and cloud acceleration and destruction? How does star formation propagate through the neutral gas? What determines the population of ionizing stars? Likewise, studies of chemical abundances have revealed the gross character of composition gradients across spiral galaxies and the range of abundances in irregulars. These results have yet to be fully exploited as constraints on the evolution of galaxies in various environments.

Morphological questions require further study of nearby GEHRs. The 30 Dor Nebula appears to be typical, so that a detailed understanding of its dynamics would seem to be the first priority. Observations of the velocity field across individual filaments and the sharpness of the filament edges may distinguish between evaporation flows and clouds confined by a medium. Measurements of the wind kinetic energy of the luminous stars are important for hydrodynamical models of the nebular evolution. A goal of such models should be to explain why the core surface brightnesses, densities, and ionization parameters of GEHRs fall in a fairly narrow range. Clarification of the U-T_*-O/H relationship will require photoionization models with improved stellar atmospheres and evolutionary tracks, coupled with good atomic data for a redundant set of nebular diagnostics. Giant telescopes will allow the measurement of weak lines, giving direct electron temperature determinations for low-excitation, oxygen-rich H II regions, aided by measurements of the far-IR fine-structure lines.

Outlying H II regions should be measured to probe the limits of abundance gradients and the nature of chemical evolution in locations dynamically dominated by dark matter. Correlations of line ratios measured azimuthally around a galactic disk may be able to distinguish between local abundance variations and variations in T_* and U, perhaps due to aging, at fixed abundances. Systematic observational studies of the environmental dependence of abundances are needed. Larger telescopes in space will permit better measurements of carbon abundances, and nebular C IV intensities will provide a measure of T_*.

Spectroscopy of irregular galaxies with improved sensitivity will allow precise measurements of N/O fluctuations and a search for correlated properties of the galaxies. Precise measurements of many He I lines will allow the effects of collisional excitation and fluorescence to be sorted out; the use of the helium lines as nebular diagnostics will be a fringe benefit of such observations. More objects with low O/H and low O^+/O must be found and studied, and spatially resolved observations of nearby GEHRs could set limits on the presence of neutral helium zones. Modest further improvements in Y_p measurements should give results of cosmological importance.

ACKNOWLEDGMENTS

I am grateful to F. Bash, H. Dinerstein, R. Dufour, D. Garnett, R. Kennicutt, C. McKee, D. Osterbrock, R. Pogge, and E. Skillman for valuable conversations and comments on the manuscript. This work was supported in part by the Robert A. Welch Foundation of Houston, Texas.

Literature Cited

Aaronson, M., Huchra, J., Mould, J. 1979. *Ap. J.* 229: 1–13
Aller, L. H. 1942. *Ap. J.* 95: 52–57
Aller, L. H. 1984. *Physics of Thermal Gaseous Nebulae*. Dordrecht: Reidel. 350 pp.
Aller, L. H. 1986. *Publ. Astron. Soc. Pac.* 98: 957–64
Aller, L. H., Czyzak, S. J., Keyes, C. D., Boeshaar, G. 1974. *Proc. Natl. Acad. Sci. USA* 71: 4496–4501
Anders, E., Grevesse, N. 1989. *Geochim. Cosmochim. Acta* 53: 197–214
Audouze, J., Tinsley, B. M. 1976. *Annu. Rev. Astron. Astrophys.* 14: 43–79
Baldwin, J. A., Phillips, M. M., Terlevich, R. 1981. *Publ. Astron. Soc. Pac.* 93: 5–19
Balick, B., Boeshaar, G. O., Gull, T. R. 1980. *Ap. J.* 242: 584–91
Balick, B., Sneden, C. 1976. *Ap. J.* 208: 336–45
Balzano, V. 1983. *Ap. J.* 268: 602–27
Bertoldi, F. 1989. PhD thesis. Univ. Calif., Berkeley
Blair, W. P., Kirshner, R. P., Chevalier, R. A. 1982. *Ap. J.* 254: 50–69
Boesgaard, A. M., Steigman, G. 1985. *Annu. Rev. Astron. Astrophys.* 23: 319–78
Borsenberger, J., Stasińska, G. 1982. *Astrophys.* 106: 158–62
Boulesteix, J., Courtès, G., Laval, A.,

Monnet, G., Petit, H. 1974. *Astron. Astrophys.* 37: 33–48
Brocklehurst, M. 1971. *MNRAS* 153: 471–90
Brocklehurst, M. 1972. *MNRAS* 157: 211–27
Campbell, A. 1988. *Ap. J.* 335: 644–57
Campbell, A., Terlevich, R., Melnick, J. 1986. *MNRAS* 223: 811–25
Caplan, J., Deharveng, L. 1986. *Astron. Astrophys.* 155: 297–313
Castor, J., McCray, R., Weaver, R. 1975. *Ap. J. Lett.* 200: L107–10
Caulet, A., Deharveng, L., Georgelin, Y. M., Georgelin, Y. P. 1982. *Astron. Astrophys.* 110: 185–97
Cersosimo, J. C., Loiseau, N. 1984. *Astron. Astrophys.* 133: 93–98
Chiosi, C. 1986. *Prog. Part. Nucl. Phys.* 17: 173–214
Chu, Y.-H. 1983. *Ap. J.* 269: 202–11
Chu, Y.-H., Kennicutt, R. C. 1986. *Ap. J.* 311: 85–97
Clayton, D. D., Pantelaki, I. 1986. *Ap. J.* 307: 441–48
Clegg, R. E. S. 1987. *MNRAS* 229: 31p–39p
Conti, P. S., Massey, P. 1981. *Ap. J.* 249: 471–80
Davidson, K., Kinman, T. D. 1985. *Ap. J. Suppl.* 58: 321–40

de Vaucouleurs, G. 1979. *Astron. Astrophys.* 79: 274–76
de Vaucouleurs, G. 1983. *Ap. J.* 224: 14–21
Díaz, A. I., Terlevich, E., Pagel, B. E. J., Vílchez, J. M., Edmunds, M. G. 1987. *MNRAS* 226: 19–37
Dinerstein, H. L. 1990. In *The Interstellar Medium of External Galaxies*, ed. H. A. Thronson, J. M. Shull. Dordrecht: Kluwer. In press
Dinerstein, H. L., Shields, G. A. 1986. *Ap. J.* 311: 45–57
D'Odorico, S., Rosa, M. 1981. *Ap. J.* 248: 1015–20
Dopita, M. A., Evans, I. N. 1986. *Ap. J.* 307: 431–40
Dufour, R. J. 1975. *Ap. J.* 195: 315–22
Dufour, R. J. 1984. In *Structure and Evolution of the Magellanic Clouds, IAU Symp. No. 108*, ed. S. van den Bergh, K. S. de Boer, pp. 353–62. Dordrecht: Reidel
Dufour, R. J. 1986. *Publ. Astron. Soc. Pac.* 98: 1025–31
Dufour, R. J., Garnett, D. R., Shields, G. A. 1988. *Ap. J.* 332: 752–61
Dufour, R. J., Hester, J. J. 1990. *Ap. J.* 350: 149–54
Dufour, R. J., Talbot, R. J. Jr., Jensen, E. B., Shields, G. A. 1980. *Ap. J.* 236: 119–34
Dyson, J. E. 1979. *Astron. Astrophys.* 73: 132–36
Edmunds, M. G., Pagel, B. E. J. 1984. *MNRAS* 211: 507–19
Elliott, K. H., Goudis, C., Meaburn, J., Tebbutt, N. J. 1977. *Astron. Astrophys.* 55: 187–201
Elmegreen, B. G. 1976. *Ap. J.* 205: 405–18
Evans, I. N. 1986. *Ap. J.* 309: 544–52
Evans, I. N., Dopita, M. A. 1985. *Ap. J. Suppl.* 58: 125–42
Evans, N. J. 1989. In *Frontiers of Stellar Evolution*, ed. D. L. Lambert. San Francisco: Astron. Soc. Pac. In press
Faber, S. M., Jackson, R. E. 1976. *Ap. J.* 204: 668–83
Feast, M. W. 1961. *MNRAS* 122: 1–16
Ferland, G. J. 1986. *Ap. J. Lett.* 310: L67–70
Fierro, J., Torres-Peimbert, S., Peimbert, M. 1986. *Publ. Astron. Soc. Pac.* 98: 1032–40
French, H. B. 1980. *Ap. J.* 240: 41–59
French, H. B. 1981. *Ap. J.* 246: 434–43
Garnett, D. R. 1989a. PhD thesis. Univ. Tex., Austin
Garnett, D. R. 1989b. *Ap. J.* 345: 282–97
Garnett, D. R., Shields, G. A. 1987. *Ap. J.* 317: 82–101
Gehren, T., Husfeld, D., Kudritzki, R. P., Conti, P. S., Hummer, D. G. 1986. In *Luminous Stars and Associations in Galaxies, IAU Symp. No. 116*, ed. C. W. H. de Loore, A. J. Willis, P. G. Laskarides, pp. 413–14. Dordrecht: Reidel

Genzel, R., Stutzki, J. 1989. *Annu. Rev. Astron. Astrophys.* 27: 41–85
Giovanelli, R., Haynes, M. P. 1985. *Ap. J.* 292: 404–25
Gum, C. S., de Vaucouleurs, G. 1953. *Observatory* 73: 152–55
Harrington, J. P. 1968. *Ap. J.* 152: 943–62
Hawley, S. A. 1978. *Ap. J.* 224: 417–36
Hawley, S. A., Phillips, M. M. 1980. *Ap. J.* 235: 783–92
Heckman, T. M. 1980. *Astron. Astrophys.* 87: 142–51
Hippelein, H. H. 1986. *Astron. Astrophys.* 160: 374–84
Hodge, P. W. 1969. *Ap. J. Suppl.* 18: 73–83
Hodge, P. W. 1974. *Ap. J. Suppl.* 27: 113–19
Hodge, P. W., Kennicutt, R. C. Jr. 1983a. *Ap. J.* 267: 563–70
Hodge, P. W., Kennicutt, R. C. Jr. 1983b. *Astron. J.* 88: 296–328
Hodge, P. W., Lee, M. G., Kennicutt, R. C. Jr. 1989. *Publ. Astron. Soc. Pac.* 101: 640–48
Israel, F. P., Kennicutt, R. C. 1980. *Astrophys. Lett.* 21: 1–9
Jensen, E. B., Strom, K. M., Strom, S. E. 1976. *Ap. J.* 209: 748–69
Jensen, E. B., Talbot, R. J. Jr., Dufour, R. J. 1981. *Ap. J.* 243: 716–35
Kaufman, M., Bash, F. N., Kennicutt, R. C. Jr. 1987. *Ap. J.* 319: 61–75
Kennicutt, R. C. Jr. 1979. *Ap. J.* 228: 696–703
Kennicutt, R. C. Jr. 1981. *Ap. J.* 247: 9–16
Kennicutt, R. C. Jr. 1984. *Ap. J.* 287: 116–30
Kennicutt, R. C. Jr. 1988. *Ap. J.* 334: 144–58
Kennicutt, R. C. Jr. 1989. *Ap. J.* 344: 685–703
Kennicutt, R. C. Jr., Edgar, B. K., Hodge, P. W. 1989a. *Ap. J.* 337: 761–81
Kennicutt, R. C. Jr., Keel, W. C., Blaha, C. A. 1989b. *Astron. J.* 97: 1022–35
Kinman, T. D., Davidson, K. 1981. *Ap. J.* 243: 127–39
Klein, U., Wielebinski, R., Beck, R. 1984. *Astron. Astrophys.* 135: 213–24
Kunth, D., Sargent, W. L. W. 1983. *Ap. J.* 273: 81–98
Kunth, D., Thuan, T. X., Tran Thanh Van, J., eds. 1985. *Star-Forming Dwarf Galaxies*. Gif sur Yvette, Fr: Ed. Front. 517 pp.
Kurucz, R. 1979. *Ap. J. Suppl.* 40: 1–340
Kwitter, K. B., Aller, L. H. 1981. *MNRAS* 195: 939–57
Lacey, C. G., Fall, S. M. 1985. *Ap. J.* 290: 154–70
Larson, R. B. 1972. *Nature Phys. Sci.* 236: 7–10
Lequeux, J., Peimbert, M., Rayo, J. F., Serrano, A., Torres-Peimbert, S. 1979. *Astron. Astrophys.* 80: 155–66

Lester, D. F., Dinerstein, H. L., Werner, M. W., Watson, D., Genzel, R. L. 1983. *Ap. J.* 271: 618–24
Lester, D. F., Dinerstein, H. L., Werner, M. W., Watson, D., Genzel, R., Storey, J. W. V. 1987. *Ap. J.* 320: 573–85
Lin, D. N. C., Faber, S. M. 1983. *Ap. J. Lett.* 266: L21–25
Massey, P., Conti, P. S. 1983. *Ap. J.* 273: 576–89
Massey, P., Parker, J. W., Garmany, C. D. 1989. *Astron. J.* 98: 1305–34
Mathis, J. S. 1985. *Ap. J.* 291: 247–59
Mathis, J. S., Chu, Y.-H., Peterson, D. E. 1985. *Ap. J.* 292: 155–63
Matteucci, F. 1986. *Publ. Astron. Soc. Pac.* 98: 973–78
Matteucci, F., François, P. 1989. *MNRAS* 239: 885–904
Matteucci, F., Tosi, M. 1985. *MNRAS* 217: 391–405
McCall, M. L. 1982. PhD thesis. Univ. Tex., Austin (*McDonald Obs. Publ. No. 20*)
McCall, M. L., Rybski, P. M., Shields, G. A. 1985. *Ap. J. Suppl.* 57: 1–62 (MRS)
McKee, C. F., Van Buren, D., Lazareff, B. 1985. *Ap. J. Lett.* 278: L115–18
Meaburn, J. 1979. *Astron. Astrophys.* 75: 127–32
Meaburn, J. 1981. *MNRAS* 196: 19P–27P
Meaburn, J. 1984. *MNRAS* 211: 521–33
Melnick, J. 1977. *Ap. J.* 213: 15–17
Melnick, J. 1978. *Astron. Astrophys.* 70: 157–62
Melnick, J. 1985. *Astron. Astrophys.* 153: 235–44
Melnick, J. 1987. In *Observational Evidence of Activity in Galaxies, IAU Symp. No. 121*, ed. E. Ye. Khachikian, K. J. Fricke, J. Melnick, pp. 545–65. Dordrecht: Reidel
Melnick, J., Moles, M., Terlevich, R., Garcia-Pelayo, J.-M. 1987. *MNRAS* 226: 849–66
Mendoza, C. 1983. In *Planetary Nebulae, IAU Symp. No. 103*, ed. D. R. Flower, pp. 143–72. Dordrecht: Reidel
Mihalas, D. 1972. *Non-LTE Model Atmospheres for B and O Stars. NCAR-TN/STR-76*
Mills, B. Y., Turtle, A. J., Watkinson, A. 1978. *MNRAS* 185: 263–76
O'Dell, C. R., Castañeda, H. O. 1984. *Ap. J.* 283: 158–64
Oort, J. H., Spitzer, L. 1955. *Ap. J.* 121: 6–23
Osterbrock, D. E. 1989. *Astrophysics of Gaseous Nebulae and Active Galactic Nuclei*. Mill Valley, Calif: Univ. Sci. Books. 408 pp.
Osterbrock, D. E., Flather, E. 1959. *Ap. J.* 129: 26–33
Pagel, B. E. J. 1985. In *Production and Distribution of the CNO Elements*, ed. J. Danziger, F. Matteucci, K. Kjär, pp. 155–70. Garching: Eur. South. Obs.
Pagel, B. E. J. 1987. In *The Galaxy*, ed. G. Gilmore, B. Carswell, pp. 341–64. Dordrecht: Reidel
Pagel, B. E. J. 1989. *Rev. Mex. Astron. Astrofis.* 18: 161–72
Pagel, B. E. J., Edmunds, M. G. 1981. *Annu. Rev. Astron. Astrophys.* 19: 77–113
Pagel, B. E. J., Edmunds, M. G., Blackwell, D. E., Chun, M. S., Smith, G. 1979. *MNRAS* 189: 95–113
Pagel, B. E. J., Simonson, E. A. 1989. *Rev. Mex. Astron. Astrofis.* 18: 153–60
Pagel, B. E. J., Terlevich, R. J., Melnick, J. 1986. *Publ. Astron. Soc. Pac.* 98: 1005–8
Pantelaki, I. 1988. PhD thesis. Rice Univ., Houston, Tex.
Peimbert, M., Costero, R. 1969. *Bol. Obs. Tonantzintla Tacubaya* 5: 3–22
Peimbert, M., Spinrad, H. 1970. *Astron. Astrophys.* 7: 311–17
Peimbert, M., Torres-Peimbert, S. 1974. *Ap. J.* 193: 327–33
Peimbert, M., Torres-Peimbert, S., Rayo, J. F. 1978. *Ap. J.* 220: 516–24
Peña, M. 1986. *Publ. Astron. Soc. Pac.* 98: 1061–65
Persi, P., Tapia, M., Roth, M., Ferrari-Toniolo, M. 1985. *Astron. Astrophys.* 144: 275–81
Robbins, R. 1968. *Ap. J.* 151: 511–29
Rosa, M., D'Odorico, S. 1982. *Astron. Astrophys.* 108: 339–43
Rosa, M., D'Odorico, S. 1986. In *Luminous Stars and Associations in Galaxies, IAU Symp. No. 116*, ed. C. W. H. de Loore, A. J. Willis, P. G. Laskarides, pp. 355–68. Dordrecht: Reidel
Rosa, M., Mathis, J. S. 1987. *Ap. J.* 317: 163–72
Roy, J.-R., Arsenault, R. 1986. *Ap. J.* 302: 579–84
Roy, J.-R., Walsh, J. R. 1988. *MNRAS* 234: 977–91
Rubin, R. H. 1985. *Ap. J. Suppl.* 57: 349–87
Rubin, R. H., Simpson, J. P., Erickson, E. F., Haas, M. R. 1988. *Ap. J.* 327: 377–88
Rubin, V. C., Burstein, D., Ford, W. K. Jr., Thonnard, N. 1985. *Ap. J.* 289: 81–104
Sandage, A., Tammann, G. A. 1974. *Ap. J.* 190: 525–38
Sandage, A., Tammann, G. A. 1985. In *Supernovae as Distance Indicators. Lect. Notes Phys.*, ed. N. Bartel, 224: 1–13. Berlin: Springer-Verlag
Sarazin, C. L. 1976. *Ap. J.* 208: 323–45
Scalo, J. M. 1986. *Fundam. Cosm. Phys.* 11: 1–278
Scalo, J. M. 1990. In *Windows on Galaxies*, ed. A. Renzini, G. Fabbiano, J. S. Gallagher, pp. 125–40. Dordrecht: Kluwer
Searle, L. 1971. *Ap. J.* 168: 327–41

Searle, L., Sargent, W. L. W. 1972. *Ap. J.* 173: 25–33
Searle, L., Sargent, W. L. W., Bagnuolo, W. G. 1973. *Ap. J.* 179: 427–38
Serrano, A., Peimbert, M. 1983. *Rev. Mex. Astron. Astrofis.* 8: 117–31
Sersic, J. L. 1960. *Z. Astrophys.* 50: 168–77
Shaver, P. A., McGee, R. X., Newton, L. M., Danks, A. C., Pottasch, S. R. 1983. *MNRAS* 204: 53–112
Shields, G. A. 1986. *Publ. Astron. Soc. Pac.* 98: 1072–75
Shields, G. A. 1989. In *Molecular Astrophysics*, ed. T. W. Hartquist, pp. 461–72. Cambridge: Univ. Press
Shields, G. A., Searle, L. 1978. *Ap. J.* 222: 821–32
Shields, G. A., Tinsley, B. M. 1976. *Ap. J.* 203: 66–71
Shu, F. H., Adams, F. C., Lizano, S. 1987. *Annu. Rev. Astron. Astrophys.* 25: 23–81
Silk, J. 1987. In *Stellar Evolution and Dynamics in the Outer Halo of the Galaxy. ESO Conf. Workshop Proc. No. 27*, ed. M. Azzopardi, F. Matteucci, pp. 653–63
Silk, J., Wyse, R. F. G., Shields, G. A. 1987. *Ap. J. Lett.* 322: L59–65
Skillman, E. D. 1985. *Ap. J.* 290: 449–61
Skillman, E. D. 1989. *Ap. J.* 347: 883–90
Skillman, E. D., Balick, B. 1984. *Ap. J.* 280: 580–91
Skillman, E. D., Israel, F. 1988. *Astron. Astrophys.* 203: 226–32
Skillman, E. D., Melnick, J., Terlevich, R. J., Moles, M. 1988. *Astron. Astrophys.* 196: 31–38
Skillman, E. D., Terlevich, R., Melnick, J. 1989. *MNRAS* 240: 563–72
Smith, H. E. 1975. *Ap. J.* 199: 591–610
Smith, M. G., Weedman, D. W. 1970. *Ap. J.* 161: 33–40
Smith, M. G., Weedman, D. W. 1972. *Ap. J.* 172: 307–17
Stasińska, G. 1980. *Astron. Astrophys.* 84: 320–28
Stasińska, G. 1982. *Astron. Astrophys. Suppl.* 48: 299–304
Stasińska, G., Alloin, D., Collin-Souffin, S., Joly, M. 1981. *Astron. Astrophys.* 93: 362–69
Talent, D. 1980. PhD thesis. Rice Univ., Houston, Tex.
Tenorio-Tagle, G. 1979. *Ap. J.* 233: 85–96
Terlevich, R., Melnick, J. 1981. *MNRAS* 195: 839–51
Thuan, T. X., Montmerle, T., Tran Thanh Van, J., eds. 1988. *Starbursts and Galaxy Evolution*. Gif sur Yvette, Fr: Ed. Front. 607 pp.
Tinsley, B. M. 1980. *Fundam Cosm. Phys.* 5: 287–388
Tinsley, B. M., Larson, R. B. 1978. *Ap. J.* 221: 554–61
Tomkin, J., Lambert, D. L. 1984. *Ap. J.* 279: 220–24
Torres-Peimbert, S., Peimbert, M., Fierro, J. 1989. *Ap. J.* 345: 186–95
Tully, R. B., Fisher, J. R. 1977. *Astron. Astrophys.* 54: 661–73
van der Hulst, J. M., Kennicutt, R. C., Craine, D. C., Rots, A. H. 1988. *Astron. Astrophys.* 195: 38–52
Veilleux, S., Osterbrock, D. E. 1987. *Ap. J. Suppl.* 63: 294–310
Viallefond, F., Goss, W. M. 1986. *Astron. Astrophys.* 154: 357–69
Vílchez, J. M., Pagel, B. E. J. 1988. *MNRAS* 231: 257–67
Vílchez, J. M., Pagel, B. E. J., Díaz, A. I., Terlevich, E., Edmunds, M. G. 1988. *MNRAS* 235: 633–53
Visser, H. C. D. 1980. *Astron. Astrophys.* 88: 159–74
Walborn, N. R. 1986. In *Luminous Stars and Associations in Galaxies, IAU Symp. No. 116*, ed. C. W. H. de Loore, A. J. Willis, P. G. Laskarides, pp. 185–98. Dordrecht: Reidel
Walterbos, R. A. M., Schwering, P. B. W. 1987. *Astron. Astrophys.* 180: 27–49
Warmels, R. H. 1986. PhD thesis. Univ. Groningen, Neth.
Wheeler, J. C., Sneden, C., Truran, J. W. Jr. 1989. *Annu. Rev. Astron. Astrophys.* 27: 279–349

RADIO PROPAGATION THROUGH THE TURBULENT INTERSTELLAR PLASMA

B. J. Rickett

Department of Electrical and Computer Engineering,
University of California, San Diego, La Jolla, California 92093

KEY WORDS: scattering, scintillation, radio-source variation, interstellar medium

1. INTRODUCTION

In recent years there has been a surge of interest among radio astronomers in the effects of wave propagation in the inhomogeneous interstellar plasma. At the time of my earlier review (Rickett 1977; hereinafter R77) the subject was of interest to observers of pulsars and to a lonely few theoreticians of plasma turbulence in the interstellar medium (ISM). The reasons for the recent interest were by no means foreseen in 1977. They derive partly from new observations, partly from theoretical advances (particularly, the recognition of refractive effects in strong scintillations), and partly from developments in VLBI. The various studies of scattering phenomena lead both to an increased awareness of the "radio seeing" conditions in the Galaxy and to an increased knowledge of the irregular (and apparently turbulent) interstellar plasma.

Wave propagation topics have figured in several recent conferences—in particular, the Beijing Union Radio Scientifique International/ International Astronomical Union (URSI/IAU) seminar in May 1989 on "Radio Astronomical Seeing," which included effects due to the neutral atmosphere, the ionosphere, the solar wind, and the interstellar medium. A 2-day workshop on "Radio Wave Scattering in the Interstellar Medium" was held in January 1988 at the University of California, San Diego (UCSD). I have drawn on the proceedings of the latter meeting

(Cordes et al 1988a) as well as on the general literature in preparing this review. The subject has now matured to the point where a review is indeed appropriate. The main part of this paper (Section 2) reviews the observations and advances in interstellar scattering and their impact on radio astronomy, with little discussion of wave propagation theory. However, advances in propagation theory have been very important and are described more fully in Appendix A. A brief discussion of interstellar plasma turbulence is given in Section 3.

2. INTERSTELLAR SCATTERING

There has been an important interplay between scattering theory and observation in recent years. However, here I focus on the astronomical implications, and readers interested in the theoretical development should follow Appendix A in parallel.

2.1 *Scattering Regimes and Source Diameters*

The various regimes of radio-wave scattering and scintillation are introduced here. There are many observable quantities that are perturbed by interstellar scattering (e.g. Rickett 1988). However, it is most fruitful to divide the regimes of scattering into weak and strong according to whether the rms fractional intensity fluctuation (= scintillation index) is much smaller or greater than unity, respectively, for a point source on a chosen line of sight. A typical radio observation (through more than, say, 200 pc of the galactic scattering medium and at frequencies of less than a few gigahertz) is in strong scattering. At frequencies above, say, 10 GHz the scattering is typically weak. Essentially all radio observations are very strong in the alternative sense that the rms phase is greater than 1 radian. The rms phase deviation is, however, hard to measure and to predict theoretically, since it depends on the largest scales present along a light of sight and on whether these largest structures are to be thought of as random or deterministic. Here I use the word scintillation when referring to intensity fluctuations, and the word scattering for more general perturbations of the wave field.

THE ELECTRON DENSITY SPECTRUM   The different phases of the ISM have received considerable attention in recent years (e.g. McKee & Ostriker 1977, McCray & Snow 1979). We are concerned here with the distribution of electron density, which is greatest in the "warm intercloud phase" and is less in the "hot coronal phase." We assume that the plasma density irregularities can be characterized by a spatial power spectrum $P_{3N}(\kappa)$, where κ is the (three-dimensional) wave number. This function is the three-dimensional Fourier transform of the spatial correlation function of electron density deviations from the mean. It is this "electron density spectrum" that we hope to determine from measurements of scintillation and scattering. We model it as a power law (Equations A1, A2), with a strength parameter (C_N^2) that is uniform in a galactic disk, but find evidence (Section 2.4) for localized enhancements distributed randomly in the inner Galaxy.

Lovelace (1970) was the first to propose a power-law electron density spectrum. Lee & Jokipii (1976) suggested a power law over many decades of wavenumber, as for neutral turbulence, and they extrapolated from the parsec scale of interstellar clouds all the way to the "microscales" responsible for radio scattering, suggesting that the cloud size was an outer scale of a "turbulent" process with a spectral exponent (β) equal to that of Kolmogorov turbulence in a neutral gas. These ideas were explored further by Armstrong et al (1981). Their implications for the interstellar plasma are the subject of Section 3.

WEAK INTERSTELLAR SCINTILLATIONS (WISS) In weak scattering, a point source will show intensity scintillations that can be characterized by a scintillation index ($\ll 1$) and a time scale. For power-law models of the density spectrum, the spatial scale for weak intensity variations is approximately the Fresnel scale r_f (time scale t_f) (see Appendix A for the formal description). The weak scintillation spatial and temporal scales can be approximated as

$$r_f \equiv (L/k)^{0.5} \approx 1.2 \times 10^9 \sqrt{L_{kpc}/f_{GHz}} \text{ m,} \qquad 2.1$$

$$t_f \equiv r_f/V \approx 6.7 \sqrt{L_{kpc}/f_{GHz}} \, (50/V_{km\,s^{-1}}) \text{ hr.} \qquad 2.2$$

Here L is the effective distance through the interstellar scattering medium, k is the radio wave number at frequency f_{GHz}, and V is the velocity of the diffraction pattern with respect to the Earth (50 km s^{-1} is typical of interstellar motion relative to the Earth, except for pulsars with larger proper motions). The point-source scintillation index for a given line of sight should increase a little faster than linearly with wavelength, until saturation near a value of unity. The reason that scintillations are not seen for all radio sources is that their angular extent is usually greater than r_f/L, so that scintillation patterns from different parts of the source overlap and smear each other out, eliminating a detectable variation. This, of course, is the same reason that planets do not twinkle, whereas stars do. Thus, sources with diameters greater than about θ_{weak} will have reduced scintillation:

$$\theta_{weak} \approx r_f/L \approx 8 \times 10^{-6} \sqrt{L_{kpc}/f_{GHz}} \text{ arcsec.} \qquad 2.3$$

Almost all sources exceed this diameter, and their index will be reduced and the time scale increased by the ratio of the intrinsic diameter to θ_{weak} (Rickett 1986; hereinafter R86). However, pulsars are small enough (i.e. smaller than θ_{weak}) and at centimeter wavelengths show WISS (e.g. Backer 1975). It may also be that centimeter-wavelength source variations occurring over times shorter than about 1 day (Quirrenbach et al 1989a,b) are also WISS. If so, studies of these variations have the power to detect radio source diameters at the microarcsecond level.

STRONG INTERSTELLAR SCINTILLATIONS The "strength of scintillation" increases with wavelength λ and with distance L (i.e. the galactic pathlength). Thus, as we go from a weak scintillation condition to greater λ or L, the point-source scintillation index (m_{point}) will approach unity. Further increases in λ or L will cause m_{point} to increase above unity, saturate, and then decrease asymptotically toward unity. However, there is an important change in the spatial characteristics as the scintillation becomes strong. The diffraction pattern takes on a two-scale character, by which is meant that the spatial power spectrum of the intensity pattern exhibits two regimes. Such spectra are sketched in Figure 1. The higher wave numbers are due to diffraction, while the bump at lower wave number is due to refraction [see Spangler (1988) for a review of diffractive interstellar scintillations (DISS)]. If these regimes are characterized by scales s_d and s_r, respectively, it follows that $s_d s_r \approx r_f^2$. Further into strong scattering s_d decreases, and so s_r increases. A physically intuitive understanding of these two components comes from the concepts of the scattering angle and the scattering disk.

We first define the field coherence scale s_0 as the lateral separation, at an observing plane, across which there is a 1-radian rms difference in the phase calculated along a straight-line path from the source to the observer. From this, the effective scattering angle θ_s can be defined:

$$\theta_s = 1/(ks_0). \qquad 2.4$$

At an observing location, the summing and interference of waves arriving from the angular spectrum (of width θ_s) cause the diffractive amplitude variations. A lateral displacement of s_0 substantially changes the relative phases of these interfering waves and, thus, the amplitude. Hence, the diffractive amplitude scale is nearly equal to the field coherence scale ($s_d \approx s_0$). It is useful to define a strength-of-scattering parameter u (strong scattering is $u > 1$):

$$u = r_f/s_0. \qquad 2.5$$

The concept of the scattering disk is also useful; it is the largest transverse

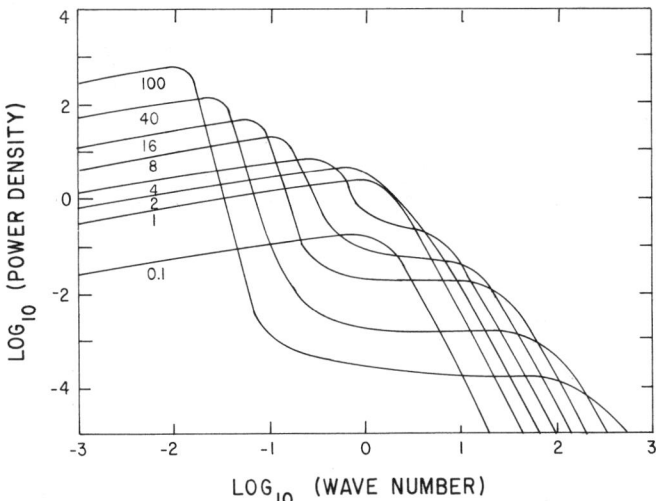

Figure 1 The power spectrum of intensity fluctuations for a screen (with Kolmogorov spectrum) plotted logarithmically versus the magnitude of the two-dimensional wave number, which is normalized to r_f^{-1}. The curves have the proper slopes and widths, and, though not precise, they show how the spectral width increases with the strength of scattering (u), which is marked against each curve. For $u > 1$, the peak at low wave numbers (u^{-1}) is due to refraction, and the plateau, which extends to high wave numbers (u), is due to diffraction. The variance, given by a two-dimensional integration over wave number, has contributions from both components. The diffractive variance is unity independent of u, so the plateau level decreases as u^{-2}. The refractive variance decreases only slowly with u. In weak scintillation ($u < 1$) there is a single peak at r_f^{-1}, and the variance and peak level vary as $u^{5/3}$.

scale that influences the signal arriving at a single observing point. At a typical distance to the scattering medium, its radius is $L\theta_s$. In an extended scattering medium, we should think of $L\theta_s$ as the typical cross-sectional radius of a scattering volume, which in fact will be cigar shaped. Since the medium causes increasing phase perturbations with scale, there are large phase differences across the scattering disk. These cause partial focusing or defocusing, giving "refractive" amplitude variations. A lateral displacement of the observer by $L\theta_s$ corresponds to a new scattering volume, and so to a change in the refractive modulation. Thus, the refractive scale s_r is approximately equal to the radius of the scattering disk:

$$s_r \approx L\theta_s = ur_f. \qquad 2.6$$

Using Equation 2.4, we quickly obtain the result $s_d s_r \approx r_f^2$. The refractive scale is a factor of u greater, and the diffractive scale a factor of u smaller, than r_f. In a medium with a "steep" spectrum that is enhanced at large scales, the above discussion avoids some of the subtleties. However, it remains accurate provided the scattering angle is defined in terms of an instantaneous angular spectrum, as opposed to an ensemble average angular spectrum, which includes deviations in angle of arrival over extremely long scales that do not modulate the amplitude. This point is discussed in Appendix A and in Section 2.5 under "Image Wander."

For an ideal point source the total variance in intensity can be greater than 1, particularly near $u \approx 1$. In the limit of strong scattering the refractive variations can be modeled as multiplying the diffractive variations. Then a modulation index m_d and m_r can be defined for a total variance:

$$m_{\text{point}}^2 \approx m_d^2 + m_r^2 + m_d^2 m_r^2. \qquad 2.7$$

The diffractive interstellar scintillations (DISS) have $m_d^2 = 1$. Typically, observed refractive interstellar scintillations (RISS) have m_r^2 of about 0.1, decreasing slowly to zero as u increases.

The influence of source diameter can be considered in the same fashion as for weak scintillation. A source must be smaller than the angular sizes s_d/L and s_r/L, respectively, to show diffractive and refractive scintillation. Since s_d is smaller than r_f, sources showing DISS have to be even smaller than is necessary for WISS. Dennison & Condon (1981) observed a sample of 22 sources that were suspected of being extremely compact on the basis of low-frequency variability. They found none to exhibit DISS and so set lower limits on their angular diameters in the range of microarcseconds. Their conclusions were in agreement with earlier searches for DISS in nonpulsar sources (Condon & Backer 1975, Armstrong et al 1977, Condon & Dennison 1978).

However, the limit for RISS is less stringent; using Equation 2.6, we see that if the intrinsic angular size $\theta_{\text{source}} < \theta_s$, then RISS will be observable at a level characterized by m_r. This limit is of the order of milliarcseconds at 1 GHz and $L \approx 0.5$ kpc. Such refractive variations are discussed in the following sections. The refractive regime was overlooked in the radio literature until Rickett et al (1984; hereinafter RBC) proposed it as an explanation for slow variations in pulsars and in other classes of radio sources. Several authors have now studied a wide variety of refractive effects [e.g. Cordes et al 1986 (hereinafter CPL), Romani et al 1986 (hereinafter RNB)]. It is interesting to note that even though the theory of the two regimes was already developed in the optical literature (e.g. Prokhorov et al 1975), we overlooked the application to interstellar scattering until Sieber (1982) published his study of slow-pulsar amplitude variations. Our wave propagation group at UCSD had even discussed the theory (e.g. Rumsey 1976) and observed the two regimes in optical measurements (Coles & Frehlich 1982).

2.2 RISS of Pulsars

Sieber's (1982) recognition that the slow amplitude variations of pulsars are due to interstellar propagation was the essential first step leading to the identification of RISS. Refractive effects from discrete structures in the interstellar medium had been proposed earlier as an explanation of the problematical low-frequency variability of some radio sources (Shapirovskaya 1978), but Sieber's work led RCB to propose refractive scintillation from the same "turbulence" spectrum that causes diffractive scintillation. Most of the data analyzed by Sieber (1982) were from early pulsar investigations (Cole et al 1970, Rankin et al 1974, Helfand et al 1977). Sieber noticed a marked increase in time scale with dispersion measure and hence distance. Clearly this flagged the variations as extrinsic rather than intrinsic. As he noted, the sense of the relation is exactly opposite to that known for DISS, in which s_0 decreases with distance. However, it is in good agreement with Equation 2.6 for RISS. An important corollary was also recognized by RCB: namely, that RISS might account for the low-frequency radio-source variations that had lacked a reasonable intrinsic interpretation. Dennison & Condon (1981) had also noted that a propagation process could explain the low-frequency variables.

The depth of the slow pulsar amplitude variation was, however, shown to be greater than that expected for RISS (m_r) in the standard Kolmogorov model. This led several investigators to question the Kolmogorov model. Most notably, Blandford & Narayan (1985) proposed that the plasma density power spectrum for the ISM was a power-law function with an exponent $\beta \geq 4$. This proposal was driven partly by concern for the theoretical difficulties in supporting a turbulent cascade in the interstellar plasma. The Kolmogorov and related power-law models have attracted interest because of the analogy that can be made with neutral gas turbulence; by contrast, the steep spectra do not need a turbulent explanation. Blandford & Narayan and coworkers have since explored their proposal in detail. These "steep-spectrum" models are strongly refracting, and the scattering analysis thus requires extra care. The shape for the density spectrum is still debated, and its determination continues to be a central part of ISS investigations [see Narayan (1988) for a summary of how the various observations constrain our knowledge of it]. He emphasizes the need to solve the inverse problem of determining the density spectrum from the observations. Unfortunately, the problem is neither linear nor invertible, and thus one must resort to model fitting to the data, a process that by no means has a unique solution.

Following Sieber's paper, there have been some new RISS observations (Stinebring & Condon 1990, Rickett & Lyne 1990). Figure 2, reproduced

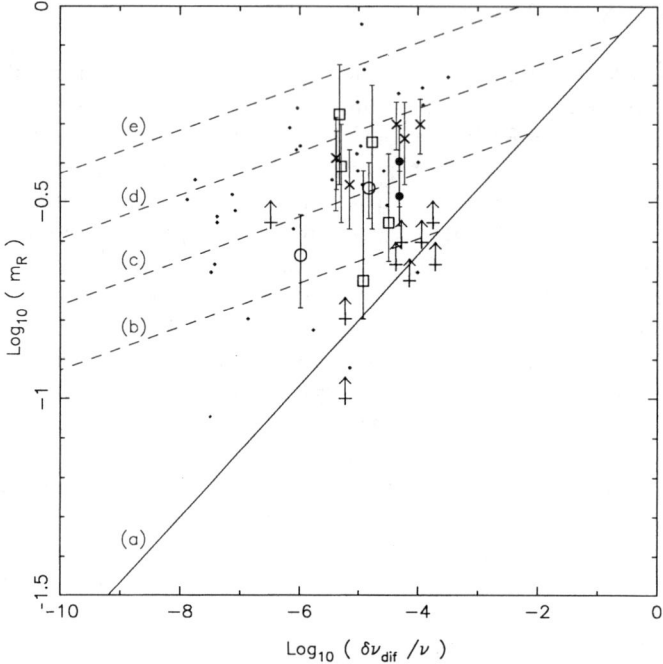

Figure 2 Refractive scintillation index versus $\delta v_d/v$, which varies inversely with u. For a Kolmogorov spectrum, we have $\delta v_d/v \sim 2u^{-2}$. The observational estimates are plotted with their error bars [see Rickett & Lyne (1990) for detailed references]. The lines give the first-order theory for a uniform medium/Kolmogorov spectrum and for inner scales of (*a*) 0, (*b*) 10^7 m, (*c*) 10^8 m, (*d*) 10^9 m, and (*e*) 10^{10} m. The observations are all above line (*a*) and substantially below unity, which is predicted for steep-spectrum models; they are thus consistent with an inner scale of 10^8–10^9 m.

from the latter paper, summarizes our state of knowledge on m_r. In brief, the values, though poorly estimated, are definitely higher than the predictions of the Kolmogorov model but are significantly below the 100% predicted for the steep power-law models. The dashed lines in Figure 2 show approximate predictions for a Kolmogorov density spectrum with a high-wave-number cutoff due to an inner scale (Coles et al 1987). There is no consensus, although the model of Shapirovskaya & Sieber (1984) now seems unlikely, and more observations of pulsar RISS would be helpful.

2.3 RISS of Other Radio Sources

Interstellar scintillation is well established for pulsars; this means that similar effects must be occurring for all radio sources. Though there appear to be no cases in which sources other than pulsars have small enough angular diameters to exhibit DISS, refractive variability of other galactic and extragalactic sources can be important. In some cases RISS may explain all of the observed variations and remove the need for the sometimes exotic source models invoked to explain the compactness of objects based on light travel-time arguments. The various categories of variable sources were considered by R86 in terms of a galactic disk distribution of scattering and are only briefly reviewed here.

LOW-FREQUENCY VARIABLE SOURCES   Sources with 3–30% variations over periods of months observed below 1 GHz have been labeled as low-frequency variables. The high brightness temperatures implied by such intrinsic variations were a serious enough astrophysical dilemma to be the topic of a National Radio Astronomy Observatory (NRAO) workshop (Cotton & Spangler 1982). It is now clear that RISS is at least partly responsible for the variations (R86, Shapirovskaya 1985).

Spangler et al (1990) have statistically analyzed the long data set observed at 408 MHz at Bologna with a view toward examining a RISS explanation. They found difficulties with a completely RISS explanation and concluded that there were often two characteristic time scales (for 20 out of 39 sources studied), suggesting that the longer of them was intrinsic and the shorter was due to RISS. They did not find the close relations between time scale, modulation index, and galactic latitude, expected in the simplified RISS model of R86. However, their data definitely show that RISS contributes to the variations, in accord with Cawthorne & Rickett (1985) and Gregorini et al (1986), who found an increasing fraction of low-frequency variables at low latitudes.

Refinements in the analysis and in the RISS model are needed. Mantovani et al (1990) obtained better agreement with the RISS model for a subset of 24 Bologna sources with VLBI data at nearby frequencies. They corrected the index values of Spangler et al (1990) for extended (nonscintillating) components and found an average latitude dependence in accord with the disk distribution of scattering, but with an angular broadening at the galactic pole that was a factor of ~ 3 less than the nominal value. The uniform disk distribution is probably an oversimplification, needing random enhancements in scattering along the line of sight (Section 2.4). The conclusion from these analyses is that there are both RISS and intrinsic variations at 408 MHz. The latter are slower and typically below the 10^{12}-K limit, and so the chief puzzle associated with low-frequency variables is removed. However, Slee & Siegman (1988) have reported an analysis of 412 sources observed with the Culgoora array at 80 and 160 MHz in which they found that nearly half of their sources were

10–30% variable over times of 1 month or 1 yr. However, the expected latitude dependences were barely significant, and thus they concluded that normal RISS is not a satisfactory explanation.

GALACTIC PLANE VARIABLES Sources at low galactic latitudes have a greater chance of enhanced scattering, and so a greater fraction may show RISS with time scales of tens of days at centimeter wavelengths. However, the random distribution of enhanced scattering precludes detailed predictions. Dennison et al (1987) posed RISS as the cause of variations in the low-latitude sources 2013+370 and 0355+208, observed at 2.7 and 8.3 GHz during 1979–86. Hjellming & Narayan (1986) studied the inverted spectrum variable source 1741−038 over a year with time resolution down to a few days at frequencies of 1.49, 4.9, 15, and 22 GHz. Their results showed a 300-day intrinsic variation at 22 GHz, a largely RISS 20-day variation at 1.49 GHz, and variation at 4.9 GHz in which both effects were important. These findings are in agreement with the measured angular broadening and imply a level of scattering 100 times stronger than for the simple galactic disk model for a line of sight at $13°$ galactic latitude. The 20-day time scale at 1.49 GHz implies a distance of about 500 pc for the scattering—considerably closer than for the disk model but possibly compatible with a clumpily distributed enhanced scattering (see Section 2.4).

Gregory & Taylor (1986) have made careful surveys of the galactic plane for variable radio sources at frequencies of 1.4 and 5 GHz. They classify their variable sources as short-term (1–20 days) or long-term variables (~ 1 yr). Many of these sources may be extragalactic, and so the distance through the galactic plane may be quite large and strong scattering conditions are likely to apply even at 5 GHz. However, RISS was not considered as part of the interpretations by Gregory (1987), Duric & Gregory (1988), and Duric et al (1987); rather, they concluded that they were observing intrinsic variations from a variety of galactic and extragalactic sources. Similarly, Duric et al (1989) observed strong variations in the cores of two extended double radio sources at both 1.45 GHz and 4.87 GHz, which they interpreted as intrinsic. Though RISS complicates the interpretation, sources as compact as is implied by an intrinsic variability must also exhibit RISS.

RAPID VARIATIONS (FLICKERING) In a phenomenon that he called flickering, Heeschen (1984) found variations of a few percent at 3.3 GHz over times of about 10 days for a large fraction of flat-spectrum radio sources. As for low-frequency variables, intrinsic flickering would imply brightness temperatures well above 10^{12} K. Although a RISS explanation was rejected in the two-frequency study by Simonetti et al (1985), it was supported by

Heeschen & Rickett (1987), who found an increasing level of flicker with decreasing galactic latitude. Once again we have the problem of determining whether the variations are intrinsic, extrinsic (RISS), or a mixture.

Even faster variability has now been reported in other high-latitude sources. Heeschen et al (1987) observed 15 compact sources at 2.7 GHz at 2–4 hr intervals for 3–4 days at three epochs and detected variations of a few percent, which they classified in two types. The first they identified as RISS, and the second as either intrinsic variations or variations due to refraction from discrete regions of the ISM. Dreher et al (1986) observed a 1.8% decline in flux over 7 hr and a 0.5% variation in 15 min in the 5-GHz flux of OJ 287; these variations could also be RISS, although Dreher et al posed an intrinsic explanation. Quirrenbach et al (1989a,b) observed rapid variations from 0917+624, with peak-to-peak variations of 20–30% at both 2.7 GHz and 5 GHz, with partial correlation. The time scales were in the range 0.5–2 days, with the variations noticeably faster at 5 GHz than at 2.7 GHz. If intrinsic, the variations imply brightness temperatures of 10^{17}–10^{19} K. Quirrenbach et al also concluded that the simple slab model of RISS required an unrealistically close distance for the scattering. However, in applying the same RISS model, I disagree, finding a time scale of 18 hr at 2.7 GHz, which is compatible with a source diameter of 0.08 mas (milliarcseconds), a velocity of 50 km s^{-1}, and a 500-pc distance for the scattering. The associated scintillation index would be about 30%, which could be reduced to the observed 7% (23% ÷ 3) by an extended nonscintillating component. At 5 GHz, the time scale would require a still smaller diameter. At 2.7 GHz, it requires 0.09 Jy in, say, a 80-μas (microarcsecond) component, implying a brightness temperature of about 3×10^{12} K, which is only slightly above the inverse Compton limit. The later observations of Quirrenbach et al (1989b) showed remarkable variations of the linearly polarized flux by a factor of 3, which were anticorrelated with the total flux. Refractive interstellar scintillation does not cause significant variability of polarization, and thus a RISS explanation appears to be ruled out. However, as the authors point out, the superposition of orthogonally polarized core and jet emission could give the 1–2% average polarization, and RISS modulation of the compact core could give anticorrelated variations of total and polarized flux. VLBI polarimetry could test this possibility.

In generalizing, R86 considered sources with a brightness of $T_{12} \times 10^{12}$ K (i.e. a diameter proportional to wavelength). Under strong scattering conditions, RISS at latitudes $b \geq 10°$ is constrained by a rate of change of flux:

$$dS/dt \lesssim S_{\rm rms}/\tau \sim 30 T_{12}(\sin b)^{0.5} \, {\rm mJy\,day}^{-1}.$$

This relation shows whether an RISS interpretation requires a diameter that infringes the Compton limit.

In spite of questions, the combination of RISS and the idea that component sizes decrease with wavelength explains an important property common to many of the observations. As the frequency is increased, the high-latitude sources vary with smaller amplitudes and shorter time scales. Figure 1 of R86 illustrates this for a latitude of 45°. There is, however, another scenario that should be explored—weak interstellar scintillation. As the frequency is increased, the strength of scattering decreases, and above ~ 10 GHz it will become weak. Given the uncertainties in the model, the frequency of this transition is poorly known. In WISS the time scale is given by Equation 2.2 for a plane wave source and is lengthened by the ratio of the intrinsic source diameter to θ_{weak} (given in Equation 2.3); this same factor governs the decrease of scintillation index over a point source, which by postulate is already less than 1.0. The WISS behavior consistent with the slab model would apply above a frequency of ~ 10 GHz (Equation 2 in R86). However, for a density spectrum with an inner scale, WISS becomes a more important possibility for the high-latitude variations discussed here. It may also be important for flare stars and might even contribute to the frequency structure observed in recent flare star studies (Bastian & Bookbinder 1987, Jackson et al 1987).

OH AND H_2O MASER SOURCES The importance of angular broadening in maser sources has been recognized for some time (see Reid & Moran 1981). This in turn indicates intrinsic diameters smaller than the scattered diameter, and so to the expectation of RISS variability as a point source. Since the sources are typically at low galactic latitudes, they are likely to be influenced by the clumpy enhanced scattering, making it difficult to predict quantitatively. Typical angular diameters at 1.6 GHz (OH) are a few mas and are smaller by roughly the square of the frequency ratio at 22 GHz; corresponding time scales depend on L and V, but they will be many years for 1.6 GHz and tens of days for 22 GHz. The latter time scale could thus be studied readily, and there are indeed a number of reports of short-term variations of H_2O masers (e.g. White & Macdonald 1979, Rowland & Cohen 1986). The masers often have many features at differing velocities and positions. Whether the RISS modulation should be independent from feature to feature depends on whether the features are separated by transverse distances bigger than the scattering disk. Once again, we need to consider both intrinsic and RISS variations.

DISCRETE PROPAGATION EVENTS A remarkable new type of interstellar variability has been discovered by Fiedler et al (1987). Figure 3 reproduces their observations of source 0945+658 from 1980.5 to 1981.7. This source

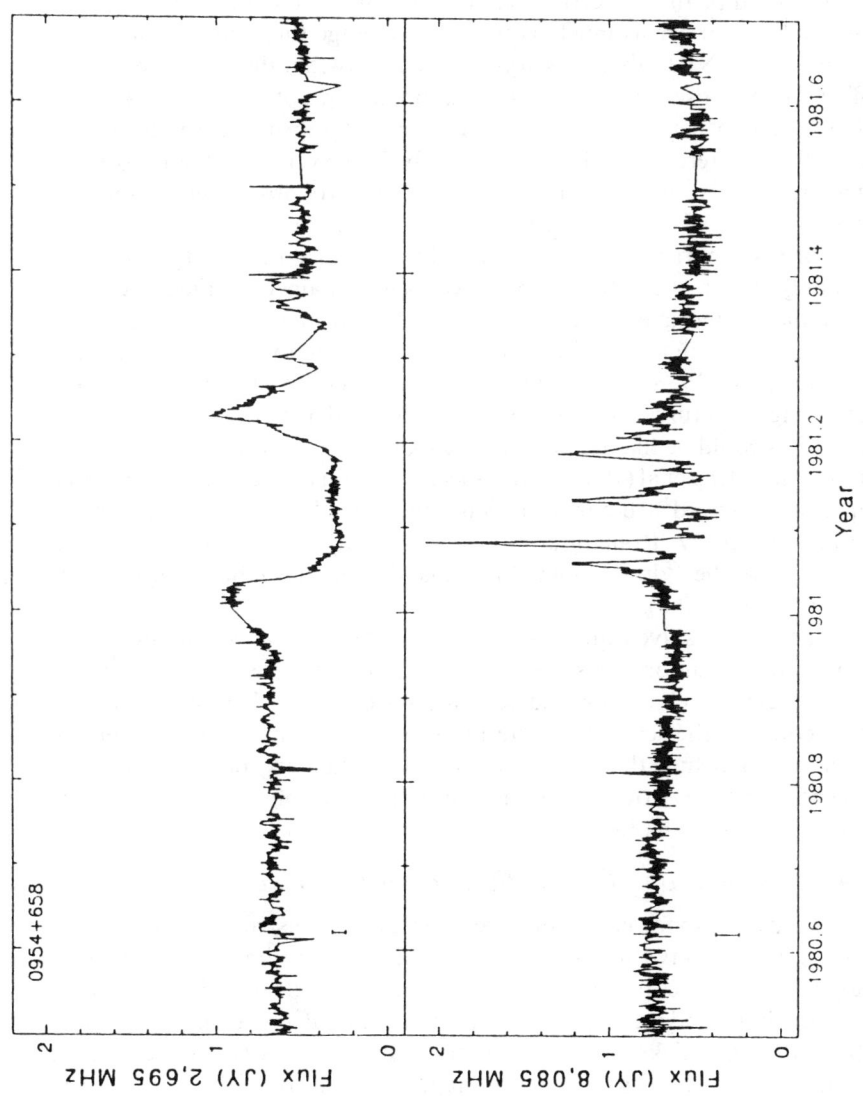

Figure 3 Intensity series at 2.7 and 8.1 GHz for the discrete propagation event on 0945+658 observed by Fiedler et al (1987). Note the pronounced dip in the intensity, preceded and followed by local maxima that are several times the average source flux.

was observed daily at Green Bank as part of the Naval Research Laboratory monitoring program. The 2.7-GHz flux shows a notable $\sim 50\%$ dip lasting for about 60 days, preceded and followed by maxima of $\sim 30\%$; during the dip, the 8.1-GHz flux is also reduced but has superimposed four spikes that are as much as twice the average flux and that last for a few days. The NRL observers argue persuasively that this is a propagation effect and propose a particular explanation in terms of a discrete scattering object passing between us and the source. On approach, we receive the extra flux scattered out of the direct path. This event was the most spectacular, but they reported three events in 395 days of observing 160 sources. In a reanalysis of all the Green Bank data, Fiedler et al (1988) reported a total of 8 comparable events, although for the other 7 events the variation was only strong at 2.7 GHz. These events were named by Fiedler et al as "extreme scattering events," in accord with their model outlined above.

Others have referred to them as "enhanced refraction events," following Romani et al (1988), who proposed an explanation in terms of discrete refracting structures; Romani et al emphasized the possibility that the structures could be filamentary or sheetlike rather than spherical. Romani (1988) and Clegg et al (1988) have concentrated on refraction from possible large-scale ionized structures, particularly interstellar shocks, which seems a likely geometry. Romani et al also point out that the same refraction could cause the "double-imaging" episodes in the dynamic spectra of pulsars (Section 2.5).

A major unresolved question is whether the structures responsible are related to the random density irregularities that cause ISS or whether they are discrete, unrelated ionized regions. As discussed below, we already must consider the scattering medium to be very clumpy, and it would be attractive to include these new structures as particularly dense clumps. We also need to know how these structures fit into our physical picture of the different parts of the ISM.

2.4 Distribution of Galactic Scattering Material

The discussion so far has emphasized questions of distribution of electron density versus wave number, i.e. the spatial spectrum. Our zero-order model has been a power-law dependence of spectral density on wave number, independent of position in a uniform galactic disk, taken to be about 1000 pc thick. This was the model studied in R77; the conclusion there was that, whereas the spectral shape could well be of the Kolmogorov form, the spatial distribution was not uniform, tending to stronger scattering density with increasing galactic pathlength.

A major study of the galactic distribution of scattering material was completed by Cordes et al (1985; hereinafter CWB). They observed the diffractive scintillation of 31 pulsars, and for each they estimated the characteristic frequency width δv_d of the scintillations. They also assembled observations from as many frequencies as possible on 5 pulsars and demonstrated that the frequency scaling laws for δv_d were close to those expected for the Kolmogorov spectrum. They then combined these observations with other measurements from the literature and investigated their dependence on galactic coordinates. In strong scattering, the diffractive scintillations become very narrowband (see Section 2.6), varying randomly on frequency separations δv_d and on a time scale t_d ($\approx s_0/V$). The measurements of δv_d can be used to estimate the average scattering parameter (C_N^2, defined in Equation A1) under an assumption of a uniform distribution of scattering electrons along the line of sight to each pulsar (see CWB for detailed formulas). Their results showed a very wide range in the derived values of C_N^2, in disagreement with the assumption of uniformity. They showed that the variation of C_N^2 with galactic coordinates is compatible with a two-part distribution of scattering material: A uniform disk with a scale height of ± 500 pc and $C_N^2 \approx 10^{-3.5}$ m$^{-6.67}$ and a superimposed clumpy distribution with a scale height of 100 pc and C_N^2 in the range 10^{-3}–1 m$^{-6.67}$. The clumpy material has a filling factor in the range 10^{-3}–10^{-4} and a mean free path of 1–10 kpc for a line of sight to intersect a region of enhanced scattering. These parameters relied on assuming a scale of about 1 pc for the clumps. Note that if a 5-kpc line of sight has an increase of 10^3 in distance-averaged C_N^2, the local C_N^2 in the clump would have to be increased by the ratio of pathlength to clump size (i.e. a further 5×10^3), making it 10^6 times the background. It becomes of great interest to identify what galactic objects or phases of the medium cause such enormously enhanced scattering.

The CWB model of the scattering distribution has been the starting point of several further studies. Rao & Ananthakrishnan (1984) used the method of interplanetary scintillations to demonstrate the presence of enhanced scattering for lines of sight within about 5° of the galactic center. Dennison et al (1984) surveyed 29 low-latitude ($<5°$) compact extragalactic sources using VLBI at 408 MHz. They found that many of the sources with longitudes close to the galactic center were resolved. This they interpreted as enhanced angular broadening due to the same distribution of irregularities. Their combined results were summarized by Cordes et al (1984). Alurkar et al (1986) reported temporal broadening of 33 pulsars at frequencies of 410, 160, and 80 MHz. They combined these measurements with other published scattering observations and proposed a single distribution of density inhomogeneities that is highly clumped and has a scale height at least as great as that of the pulsars. Their proposed distribution is strongly peaked near the galactic center, with a fall-off by a factor of 1000 as the galactocentric distance increases from 1 to 10 kpc. Thus, they suggest that the diffuse uniform component proposed by CWB is not separate but is part of a single clumpy distribution with a large scale height (>0.5 kpc). They confirmed the wide range of scattering on sources within a few degrees.

Figures 4a and 4b are from Cordes et al (1988b), who recently summarized our knowledge of this subject and reaffirmed the two distributions. For each object there is an estimate of distance L, mostly derived from pulsar dispersion measures using the electron distribution model of Lyne et al (1985). In Figure 4a, C_N^2 from high latitudes ranges from 10^{-4} to 10^{-2} m$^{-6.67}$ with a mean near $10^{-3.5}$ m$^{-6.67}$. Cordes et al concluded that the high-latitude objects are consistent with an exponential z-distribution with scale height 0.5–1 kpc and a dependence on galactocentric distance, which they model as a Gaussian with a $1/e$ radius of 7 ± 2 kpc. Low latitudes show an even larger range ($C_N^2 \sim 10^{-4}$–$10^{0.5}$ m$^{-20/3}$) and a higher mean. However, since the low-latitude objects do not sample as great a range of heights as the higher latitude (and nearer) objects, the extremely variable strength of C_N^2 only weakly constrains the scale height of the enhanced scattering. Attempts to find the galactic objects or regions responsible for enhanced scattering remain inconclusive (see Section 3). Meanwhile, evidence for specific, very strongly scattered lines of sight continues to accumulate.

2.5 Angular Broadening

THEORY In principle, angular broadening provides one of the most direct ways to measure scattering in the ISM; of course, the conditions for such measurements to be successful are just those where scattering causes the worst interstellar "seeing." The synthesis of an image in radio astronomy usually relies on correlations from multiple interferometer baselines. A single interferometer measurement (on baseline σ) provides an estimate of the average visibility $E(\mathbf{s})E^*(\mathbf{s}+\boldsymbol{\sigma})$ (see Appendix A). Equation A3 shows that (given sufficient integration) the visibility leads to a direct estimate of the structure function for geometric phase ϕ. However, if there is no averaging in time, space, or frequency, the visibility will be highly variable—the analog of speckle in the image of a filled-aperture telescope. Thus, in a synthesis observation, the applicability of Equation A3 depends on receiver bandwidth, coherent integration time, intrinsic source structure, and method used for calibration (self-calibration or otherwise).

Assuming the conditions for strong scattering, there are independent fluctuations of visibility over a distance s_0 and frequency interval equal to that for intensity scintillations (δv_d in Equation A14). A typical observing bandwidth B may include many such independent "scintles," reducing the

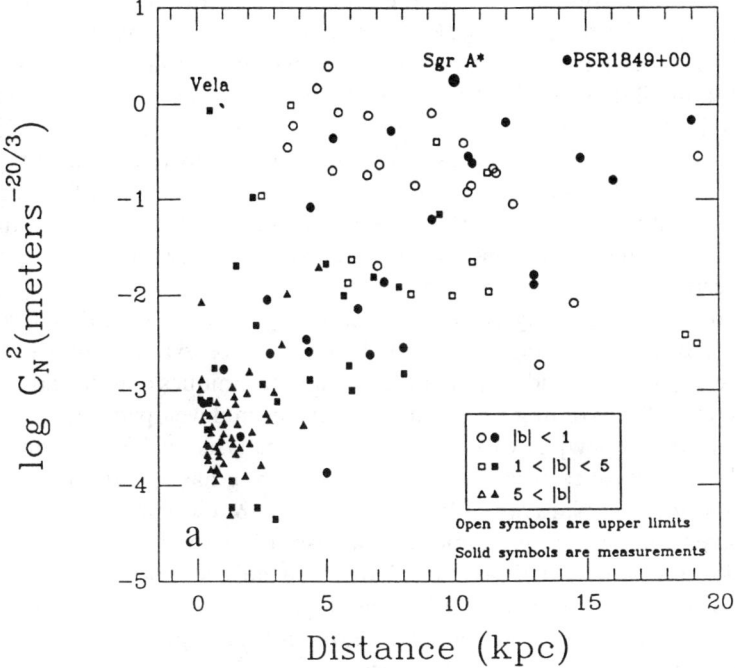

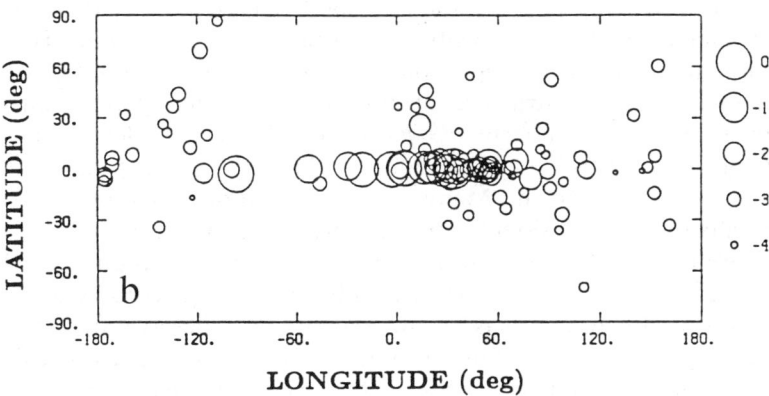

Figure 4 Plots (from Cordes et al 1988b) of the line-of-sight-averaged C_N^2 values versus (*a*) distance from the Earth and (*b*) galactic coordinates. Plot (*a*) shows the erratic and increasing scattering at low latitudes and great distances, and (*b*) demonstrates that these increases are strongest toward the inner part of the galactic disk. The origin of the localized, strongly enhanced "turbulence" is actively being sought.

visibility variations. Similarly, the coherent integration time T may be longer than the time t_d ($\sim s_0/V$) for a scintle to cross. And, finally, if the source is extended sufficiently to smear the pattern spatially ($\theta_{source} > s_0/L$) but too small to be resolved on the baseline, there will be further averaging. These ideas were discussed qualitatively in an important paper by Cohen & Cronyn (1974). In practice, source smearing is likely to dominate typical VLBI observations (Rickett & Coles 1988). Narayan et al (1989) discuss the possibility of estimating an image with very high angular resolution (s_0/L); the same resolution is achievable from intensity scintillation, but it only estimates the squared visibility.

A quantitative analysis in weak scattering was presented by Cronyn (1972), but it has been little used. In optical observations atmospheric twinkling is a well-studied phenomenon, and techniques (e.g. speckle interferometry) for avoiding its limitations have been developed [see Coulman (1985) for a review]. A careful quantitative analysis of the variations in a single visibility measurement in strong scattering has recently been made by Goodman & Narayan (1989) and Narayan & Goodman (1989). They classified the averaging regimes as "snapshot," with 100% variation due to diffractive speckles; "average," with partially smoothed speckle and full refractive variability; and "ensemble average," which also smoothes refractive variability and for which Equation A3 applies. They did not address the effects of self-calibration.

There have been several recent VLBI observations of angular broadening in the ISM. Results have been presented in plots of angular diameter versus wavelength or as a scattered image (restored with the usual synthesis techniques) or as the magnitude of visibility versus baseline. The latter form allows a direct comparison with Equation A3, from which one can estimate the phase structure function D_ϕ. A particular point of interest is to estimate the exponent $\alpha(=\beta-2)$. The inner-scale density spectrum (Coles et al 1987) implies $\alpha = 2$ for baselines shorter than their proposed inner scale near 10^9 m. Many of the observations discussed below give results that are consistent with the Kolmogorov value of $\alpha = 5/3$, in disagreement with this inner scale. However, the interpretation of the data is still a matter of debate. The first problem, as discussed by Goodman & Narayan (1989), is a bias at low visibility. The observations are in the "average" regime, and if the magnitude of the visibility is estimated, there is an upward bias that decreases as the inverse square root of the number of independent scintles in time T and bandwidth B. At low visibility this is the same bias that occurs in estimating the magnitude of the correlation coefficient for any pair of uncorrelated random quantities. Such a bias reduces the apparent exponent α but typically has been ignored. A second problem is the effect of intrinsic source structure; the intrinsic source visibility multiplies the scattered visibility, steepening the decorrelation versus baseline—a bias in the opposite direction. The third problem is the influence of self-calibration, which essentially centers the image, removing a linear phase gradient from the field. Since a D_ϕ exponent of 2 corresponds to a linear phase gradient, self-calibration may cause a bias, but no proper study has been made. The VLBI results for the exponent α should thus be viewed with caution. Note, however, that Spangler & Gwinn (1990) conclude that the published values demonstrate an inner scale of 10^6 m.

OBSERVATIONS In recent years several groups have applied the powerful angular resolution of VLBI to interstellar broadening, particularly at meter wavelengths. The criterion for sufficient resolution is that the baseline is greater than s_0, which is more readily satisfied at larger wavelengths, since s_0 decreases as $\lambda^{-1.2}$. Conversely, the conditions under which VLBI will be uncorrupted by ISS effects favor shorter wavelengths and baselines. Many heavily scattered objects have been found, which confirms the ideas of Cordes et al (1984) that there is low-latitude enhanced, clumped scattering.

VLBI observers have found a number of radio sources whose angular extent vary approximately as the wavelength squared, a pattern characteristic of plasma scattering. The low-latitude survey at 408 MHz by Dennison et al (1984) revealed strong but variable levels of scattering on nearby lines of sight. Fey et al (1988) studied the angular broadening in the Cygnus region of the Galaxy, finding similarly large variations over only a few degrees. Both groups noted the large scattered diameter of 2048+313 (CL 4) which lies at $8°$ latitude behind the Cygnus Loop. Spangler et al (1986) measured the angular scattering in the vicinity of supernova remnants. Whereas they found several heavily scattered sources, they were unable to establish the supernova association. Spangler & Cordes (1988) further investigated the question for the remnant G33.6+0.1. Using the VLA at 333 MHz, they observed an anisotropic (3:1) and very heavily scattered image for 1849+005. They also observed 45 other nearby sources at 1.46 GHz and concluded that the scattering increased between $42°$ and $22°$ longitude, and rejected the supernova (G33.6+0.1) as the cause. Presumably, the heavy scattering of pulsar PSR 1849+00 (Clifton 1986) has the same cause. The elliptical image of 1849+005 is the strongest evidence for anisotropic scattering; such a large value implies a truly anisotropic density structure, rather than the smaller randomly changing anisotropy predicted by RNB due to differing degrees of phase curvature in orthogonal directions. If verified, the large anisotropy will provide important evidence for interstellar turbulence elongated by local magnetic fields (e.g. Higdon 1986; see Section 3). Narayan & Hubbard (1988) should be consulted for details of propagation theory with substantial anisotropy.

Wilkinson et al (1988) measured the strongly scattered image of Cyg X3 during its 1987 outburst. They found a 2.8-arcsec diameter at 408 MHz, in close agreement with a 1972 measurement. In a useful clue to the cause of enhanced scattering, Moran et al (1990) observed very heavy scattering for NGC 6334B, which they suggest is due to one of the lobes of NGC 6334A—a bipolar H II region.

Mutel & Lestrade (1988, 1989) used sensitive Mark III VLBI recordings at 5 GHz to study the heavily scattered source 2005+403. They found a visibility decreasing with baseline, as expected, but they also found that it saturated at about 1% beyond 5000 km. Although Mutel & Lestrade initially attempted an interpretation using Equation A3, their 1989 analysis suggested that the saturation is due to refractive bias in visibilities (Goodman & Narayan 1989). Alternatively, the cause may be the bias due to insufficient averaging of the diffractive speckle, as discussed above for estimates of visibility magnitude; I estimate this to be a significant bias ($\sim 0.6\%$), but it needs further details of the processing.

Backer (1988) reviewed the VLBI observations of the compact source Sag A* at the galactic center. The λ^2 scaling of its angular diameter shows clearly that it is dominated by a very heavy scattering. He noted an apparent exponent steeper than 2 for the structure function, estimated via Equation A3 for the data of Lo et al (1985). However, Lo et al did not properly account for the slight anisotropy in the visibility function. Armstrong et al (1990) dealt with this problem for the highly anisotropic broadening (10:1) in the inner solar wind by plotting visibility as a function of the quadratic combination of baseline components that characterizes the fitted elliptical correlations. The location of Sag A* makes it an intriguing object, but scattering hinders its study. Backer concluded that it is intrinsically very compact and variable.

Interstellar scattering is evident in distant maser sources. This causes a number of OH/H_2O masers to show diameters in a λ^2 ratio between 18 and 1.35 cm (Reid & Moran 1981). Diamond et al (1988) observed the diameters of 12 OH maser sources (at 1665 MHz) to increase with distance (to about 50 mas at 10 kpc). This increase is similar to the extra broadening of low-latitude extragalactic sources.

Gwinn et al (1988) reported VLBI observations of PSR 1933+16 at 326 and 608 MHz. They found a moderate level of scattering, measuring a 50% visibility on a baseline of about 10^7 wavelengths at 326 MHz. Their result can be combined with the temporal pulse-broadening to estimate the distance for an equivalent scattering screen—about 75% of the pulsar distance, which is a reasonable result.

IMAGE WANDER   Scattering can cause a (slowly variable) shift in apparent source position, in addition to broadening it. The magnitude of such a position wander θ_r compared with the scattering angle θ_s is a potential discriminator for the interstellar density spectrum. In particular, the "steep" power-law models predict strong refractive shifts. The refractive shift is determined by the gradient in phase, averaged over the dimensions of the scattering disk, $s_r = L\theta_s$. Thus, it is given by

$$k\theta_r \approx D_\phi(s_r)^{0.5}/s_r, \qquad 2.8$$

which (if $0 < \alpha < 2$) can be expressed as

$$\theta_r \approx \theta_s[s_0/s_r]^{1-\alpha/2} = \theta_s u^{\alpha-2}. \qquad 2.9$$

For $\alpha = 5/3$, this gives $\theta_r \approx \theta_s u^{-1/3}$. In strong scattering ($u > 1$), this shows that $\theta_r < \theta_s$. In contrast, for steep spectra ($2 < \alpha < 4$) $\theta_r \gg \theta_s$ and θ_r depends on the outer scale (see RNB, CPL).

Unfortunately, absolute position measurements are difficult to make. The apparent separation of two sources (of true separation $\Delta\theta$) is also subject to an error $\delta\theta$ due to a relative refractive shift. Since relative positions can be measured more accurately, repeated observations should reveal how $\delta\theta$ varies with time and frequency. For a baseline v, $\delta\theta$ depends on the second difference of the phase front, which is characterized by $F(v, L\Delta\theta)$ in Equation A12. This function also controls the development of intensity scintillations. Thus, $\delta\theta$ becomes a less valuable discriminant between spectral models, since a square-law term in the phase structure function contributes nothing to F, and it is the square-law term that is dominant for the steep spectra (see Goodman & Narayan 1985, Rickett & Coles 1988).

Gwinn et al (1988) made an interesting VLBI study of refractive shifts in a cluster of H_2O masers in Sgr B2. They determined that over the course of 6 months, the rms wander of the individual maser "spots" was smaller than 18 μas. This is substantially smaller than the measured angular broadening of 300 μas for Sgr B2. They used the theories of RNB and CPL and corrected for differential position errors, concluding that $\beta < 3.67$ on spatial scales up to 10^{11} m. They also reported similar preliminary results from W49. With a view to determining whether Sgr A* is truly at the galactic center, Backer (1988) made proper-motion measurements and estimated that any refractive position error was 10 times smaller than the broadened image. Mutel & Hodges (1986) and Mutel & Lestrade (1989) observed the compact extragalactic double source 2050+364, whose components are broadened as λ^2. They found a refractive error $\delta\theta < 1$ mas at 610 MHz, which again implies that $\theta_r < \theta_s$ and thus that $\alpha < 2$ ($\beta < 4$).

2.6 Other ISS Phenomena

DYNAMIC SPECTRA When observed between, say, 100 and 1000 MHz, pulsar spectra show deep modulations in time and frequency (Rickett 1969). This phenomenon was early identified as diffractive scintillation (e.g. Scheuer 1968) and has since proved a rich source of information on the interstellar propagation process [for example, see CWB's important study of the dynamic spectra of many pulsars (Section 2.4)]. It is, however, only pulsars that have sufficiently small diameters to exhibit these effects.

The theory of DISS and the related phenomenon of pulse broadening are introduced in Appendix A. In the limit of fully saturated scintillations, there are independent variations over time scale s_0/V and frequency scale δv_d with a 100% modulation index. As the strength of scattering decreases, the refractive scintillations become more important and modulate the DISS in various ways. Refraction may be the cause of various apparently organized patterns that have been observed in pulsar dynamic spectra since the early observations of Ewing et al (1970). The observed phenomena are

1. drifting bands, in which a dynamic spectrum shows features drifting in frequency-time;
2. changes in the estimated frequency decorrelation bandwidth δv_d over days to months;
3. criss-cross patterns, in which there are overlapping slopes of opposite sign; and
4. periodic patterns, in which there are repeating bands at an angle in the frequency-time plane, with from a few to tens of "fringes" modulating a single diffractive intensity burst.

Figure 5 shows examples of these phenomena. These features can be explained qualitatively by refractive steering of the diffractive pattern. However, the quantitative tests have not been conclusive, and scattering by specific deterministic structures is also possible. In particular, it is difficult to explain (d) on the basis of random RISS. Refraction is caused by phase structure on the spatial scale of the scattering disk ($s_r \approx L\theta_s$). Whereas intensity modulation is caused by a curvature in the wave front, a local refraction angle is caused by a linear phase gradient over the scale s_r.

Drifting bands and variable frequency width The angles of refraction are given by $k\theta_{rx} = \partial\phi/\partial x$ and $k\theta_{ry} = \partial\phi/\partial y$, where the phase gradients are averaged over the scale s_r; typical values are given by Equations 2.8 and 2.9 for $0 < \alpha < 2$. Shishov (1974) gave a quantitative analysis in terms of refraction by a separate large-scale component of the medium. The angles are independent in the x and y directions and refractively shift (or steer) the diffractive pattern by distances $X \sim L\theta_{rx}$ and $Y \sim L\theta_{ry}$. The shifts are

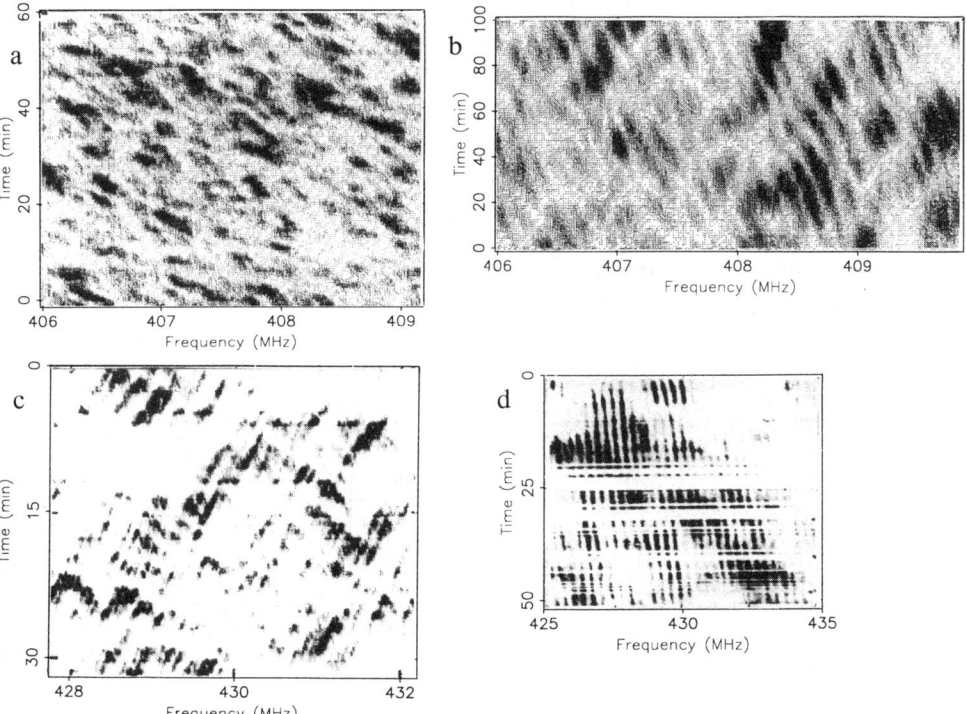

Figure 5 Dynamic spectra from DISS of pulsars, showing a range of phenomena. Spectra, averaged over many pulses, are plotted versus time in a gray scale (plot density increasing linearly with amplitude). The spectra are for (*a*) PSR 1642−03, (*b*) PSR 2016+28, (*c*) PSR 0832+26, and (*d*) PSR 1237+25. Plots (*a*) and (*b*) are from Jodrell Bank (see Gupta et al 1988), and (*c*) and (*d*) are from Arecibo (J. M. Cordes, private communication; Wolszczan & Cordes 1987).

frequency dependent, approximately as v^{-2}, so that an intensity peak will appear to drift in frequency-time. An observer at a fixed site sees temporal changes due to motion of the pattern (at velocity V, which is chosen as the x direction) and so sees a sloping feature (Figure 5*a*) with

$$dt/dv \sim (dX/dv)/V \sim (-2X/v)/V \sim -2L\theta_r/(vV). \qquad 2.10$$

Measurements of such drifting bands have been made by Smith & Wright (1985), who estimated the "drift rate" dv/dt from observations of 32 pulsars and found θ_r/θ_s to be consistent with a Kolmogorov spectrum. However, it is not a very precise estimate, since for each pulsar there was only a single estimate of θ_r, which is expected to vary randomly about a zero mean over the refractive time scale. Systematic series of observations of dynamic spectra have been made at Arecibo (J. M. Cordes & A. Wolszczan, private communication) and Jodrell Bank (Gupta et al 1988), but they are still being analyzed. A practical point is that in estimating the slope of a drifting band, it is more satisfactory to report dt/dv rather than dv/dt, since according to Equation 2.10 it has an average of zero (and so an average dv/dt of infinity). As mentioned above, steep-spectrum models predict larger θ_r/θ_s ratios than does the Kolmogorov spectrum; thus, these observations can potentially discriminate between the models.

The same observations show fluctuations of the apparent decorrelation bandwidth δv (e.g. Roberts & Ables 1982). Simple refraction has the effect of narrowing the frequency structure when X or Y are nonzero. A corollary is that observations of δv will vary and will typically underestimate δv_d (e.g. CPL). If one assumes normal statistics for X and Y, the fractional bias and rms variation in δv will be about 0.5 (θ_r/θ_s). Thus, the variations should be very pronounced for a steep power-law spectrum. For a Kolmogorov spectrum, they should be most noticeable near $\theta_r/\theta_s \sim 1$ (which is near the transition from weak to strong scintillation) and decrease slowly as u increases.

One of the parameters that can be estimated from the dynamic spectra is δv; others include the slope, the time scale, and the apparent average intensity. There are correlated variations expected between these parameters (see RNB for specific predictions). A preliminary analysis reported by Gupta et al (1988) showed significant deviations from this first-order theory, and more work is needed. For a discussion including a curvature as well as a gradient in the wave front, see CPL.

As can be seen in Figure 5c there can be opposite slopes overlapping in a single spectrum. The refractive shift interpretation implies two angles of arrival. It is not known whether this is compatible with a random distribution or whether instead it requires discrete density structures. The latter is an unattractive option, since the phenomenon is quite common, requiring widespread discrete structures. However, for propagation through an extended region of random density, it initially seems surprising to find only two angles of arrival contributing at a given point. Such effects seem more likely near $u \sim 1$, where focusing and interference are first occurring. Goodman et al (1987) proposed an interesting explanation for spectra with an inner scale; such spectra are strongly influenced by caustics, of which the commonest case is a "fold" with two dominant angles of arrival.

Periodicities Dramatic periodicities are evident in some dynamic spectra; examples can be seen in Figures 5b and 5d. This phenomenon has been a subject of increasing interest. Hewish (1980), Roberts & Ables (1982), and Cordes & Wolszczan (1986) emphasized periodicities, which Hewish et al (1985) explained in terms of interference between waves arriving from two (or three) directions. They interpreted their observations as requiring ratios $\theta_r/\theta_s > 1$, which in turn implies a spectral exponent $\beta > 4$ ($\alpha > 2$), a conclusion also reached by Roberts & Ables (1982). This seems to be in disagreement with much of the other evidence on spectral shape. An inner scale and the caustics of Goodman et al (1987) seem to be the best resolution of this inconsistency.

Wolszczan & Cordes (1987) observed a remarkable episode of periodic structure from PSR 1237+25 at 430 MHz. Figure 5d shows one of their spectra from 9 December 1986; observations 13 days later also showed fringes but with a larger period, and 19 days later the fringes had essentially disappeared. More than 10 fringes can be seen across a single diffractive scintle, which makes $\theta_r/\theta_s \sim 10$. Strong periodicities, also referred to as "multiple imaging events," have been observed from other pulsars, but as yet there is no statistical analysis of their occurrence.

"Interferometry" In observing PSR 1237+25, Wolszczan & Cordes (1987) kept spectra resolved according to phase within the pulsar period. They found systematic shifts in the periodic spectra from different parts of the average pulse profile. This they interpreted as due to a lateral separation of the emitting regions across the pulse profile, which gives an exciting resolution of the pulsar emitting region. In reanalyzing their results, Cordes & Wolszczan (1988) concluded that emission is from regions near the "light cylinder" of the pulsar, in surprising disagreement with the standard hollow-cone pulsar models. They note that the apparent angular resolution is equivalent to an interferometer of 1 AU baseline. Wolszczan et al (1988) have reported a similar resolution of PSR 1133+16, except that the periodicity was predominantly in the time domain. Further applications of this technique should be very fruitful, although much observing time is needed to catch an event.

VELOCITY DETERMINATIONS The diffractive scintillation of pulsars has been used to extract a typical time scale t_d as well as a typical frequency scale δv_d. The time scale is related to the spatial scale via the velocity V of the diffraction pattern with respect to the Earth (i.e. $t_d \approx s_0/V$), and s_0 can be related to δv_d and the pulsar distance L; thus a velocity estimate V_{iss} is possible (e.g. Rickett 1970). Lyne & Smith (1982) used this technique to measure V_{iss} for 20 pulsars and found statistically reasonable agreement with independent measurements of pulsar proper-motion velocities. Cordes (1986) made ISS observations of 65 pulsars and combined them with earlier data, giving a list of 71 pulsars with estimates of δv_d, t_d, and derived V_{iss}. He found good agreement in V_{iss} with the values of Lyne &

Smith. The velocity estimate V_{iss} depends both on the velocity of the pulsar relative to the Earth and on the distribution and motion of scattering electrons along the line of sight, for which there is no precise model. Since pulsar velocities tend to dominate, Cordes was able to use his results to investigate the spatial distribution of pulsars and their velocities.

In a creative use of this technique Lyne (1984) measured the apparent V_{iss} for the binary pulsar PSR 0655+64 as a function of phase in its binary orbit. He found that V_{iss} changed systematically around the orbit, providing a measure of the transverse speed; he thus estimated the systematic velocity, the orbital inclination, and the mass of the companion ($<0.8\ M_\odot$). Dewey et al (1988) made similar observations of PSR 1855+09 and constrained the inclination of the orbit, which together with the known mass function gave the pulsar companion's mass ($\sim 0.25\ M_\odot$). For PSR 1913+16, Dewey et al observed a variation in V_{iss} around the orbit but could not constrain the parameters significantly.

TIMING VARIABILITY The intrinsic timing accuracy of most pulsars is about a milliperiod (see Cordes & Downs 1985), which is much greater than delay variations expected from the irregular densities of the ISM (R77). However, some millisecond pulsars now have timing accuracy better than a microsecond, and thus the ISM has become an important limit.

Rawley (1986) and Rawley et al (1988) report on the extraordinary timing stability of PSR 1937+21. From Arecibo observations at 1.4 and 2.5 GHz they estimated the contribution ($\sim 1\ \mu s$) due to varying dispersion delay. Cordes et al (1990) made time-of-arrival measurements at frequencies down to 0.32 GHz in order to investigate propagation effects. They too found very stable intrinsic behavior and slowly varying time differences between 0.43 and 1.4 GHz. The inferred dispersion variation showed only partial agreement with Rawley's. The difference could be partly explained as an intrinsic frequency dependence of pulse shape or emission time, but the dispersion variations have become a little uncertain. However, Cordes et al note that other influences may be at work, such as time-of-arrival variations due to wander in angle of arrival, a phenomenon that scales more steeply with frequency (see also Blandford et al 1984). Clearly, this subject needs further work. Both Rickett (1988) and Cordes et al used the variable dispersion delay to infer the geometric phase (1 μs of delay corresponds to $> 10^3$ radians), and they estimated D_ϕ at scales up to 10^{12} m.

The ultimate timing accuracy from millisecond pulsars requires techniques for correcting dispersion variations (see Foster & Cordes 1988). At a lower level there are other propagation effects that cannot be corrected. Some effects depend on the method of dispersion removal used in the receiver (e.g. postdetection, predetection, swept local oscillator). There is a diffractive effect that has not been considered, which could be important in timing PSR 1937+21. In the observations of Rawley et al (1988) and the swept local oscillator measurements of Biraud & Bourgois (1988), a total bandwidth of about 6–8 MHz was used. This includes only a few independent diffractive scintles in the band. Associated with each scintle there is an unavoidable timing uncertainty of $(2\pi\delta v_{\rm d})^{-1}$ equal to the diffractive pulse broadening. This error should change on the time scale of the diffractive intensity scintillations. For a typical measurement reported by these two groups, this contributes ~ 0.1 μs, which may become a significant timing limit, although it could be reduced by a larger bandwidth and integration time.

FARADAY ROTATION FLUCTUATIONS   Multifrequency observations of a linearly polarized radio source allow the Faraday rotation measure (RM) to be determined. This quantity is proportional to the integral of the line-of-sight component of the magnetic field times the local electron density. Simonetti & Cordes (1988) summarized their measurements as the structure function of RM for separations from less than 1' to several degrees in the sky. It is not possible to separate the effects of variation in density and field, but Simonetti & Cordes found that irregularities in either one or the other extend to scales as large as 1 pc. In an interesting extension of this work Lazio et al (1990) measured RM variations across eight extended extragalactic radio sources using the VLA. The sources all lie in the heavily scattered Cygnus region of the Galaxy and showed significantly higher RM and RM variance than comparable sources at high galactic latitudes. Changes of RM by 40 rad m^{-2} over an arcminute were typical. Lazio et al concluded that they were seeing the effect of the density fluctuations responsible for the enhanced scattering. Using C_N^2 estimates from VLBI of nearby sources and assuming a uniform field, they showed that their results are consistent with a density spectrum continuous from scales of 10^7 m to 10^{15}–10^{17} m. Evidently, this is a powerful method for probing scales not previously accessible to radio techniques. However, interpretations in which magnetic rather than density fluctuations are responsible may also be possible, especially given the magnetic irregularities proposed for cosmic-ray confinement (Jokipii 1988).

3. PLASMA TURBULENCE IN THE ISM

The possibility of a turbulent plasma in the interstellar medium has emerged in the last two decades as a serious astrophysical question. The subject is beyond my ability to evaluate critically, and thus I summarize the literature here with little critical evaluation. It is exciting to note that serious attempts are under way to understand the plasma physics of the density structures that cause ISS.

3.1 The Turbulence Hypothesis

Turbulence has long been discussed for the neutral interstellar gas, particularly concerning the need to dissipate turbulent energy before a gas cloud can collapse gravitationally towards "stardom." The theoretical questions of turbulence in the partially ionized interstellar gas clouds (the cool phase) have been addressed by various workers over the years (e.g. Kaplan 1966, pp. 103–22; Larson 1979, Fleck 1981). However, there is no satisfactory theory for turbulence in a magnetized plasma. Furthermore, there will be differing physics in the cool (~ 100 K), warm ($\sim 10^4$ K), and hot ($\sim 10^6$ K) phases of the ISM (McKee & Ostriker 1977, McCray & Snow 1979). Lee & Jokipii (1976) suggested that the microscale density structures responsible for radio-wave scattering (scales of $\sim 10^9$ m or smaller) are part of a spectrum of density irregularities that is continuous up to the scale of interstellar clouds or even to the thickness of the disk (10^{17}–10^{19} m). For power-law spectra (Equation A1) with $\beta > 3$, the variance of density (the integrated density spectrum) is determined both by C_N^2 and by the low-wave-number cutoff. The associated rms density should then be less than or comparable to the local mean density. The results are consistent with a Kolmogorov spectrum and suggest a turbulent process with energy input from the parsec scales of the clouds and a turbulent cascade over 8–9 decades in wave number. However, there remain very serious problems with such an idea, and it is these that motivate the proposals for "steep" power-law spectra.

The first major question is whether we can even infer the presence of turbulence from density observations alone. The idea of turbulence in a neutral fluid is centered on the *dynamics* of the motion as a function of scale. Density perturbations are not directly involved in the dynamics, and so they provide no evidence for turbulence (or even randomness) in the velocity field or magnetic field. Higdon (1984) proposed that though the density structures are not central to the physics of the motion, they are nevertheless passive tracers. Similarly, Montgomery et al (1987) invoked an equation of state for the plasma to calculate weak density perturbations from pressure perturbations, obtaining a Kolmogorov form for the density spectrum. These interpretations support the idea of plasma turbulence down to the small radio-scattering scales. With this assumption, I now discuss the three parts of a turbulent spectrum: the power input range, the inertial (power-law) range, and the dissipation range.

3.2 Power Input

The power input at the largest scales could be random motion of cool clouds or H II regions, or expanding shock fronts from supernovae or large-scale stellar wind flows. The source could be one or a combination of these very different processes, and no consensus exists. In a recent paper Bykov (1988) calls for a mixture of these processes from scales up to 100 pc and a power input of 10^{-27} watts m^{-3}. Altunin (1981) argues against a cascade from such large scales and proposes instead that instabilities in shock propagation directly generate turbulence on smaller magnetoacoustic wavelengths of, say, 10^9 m. Max et al (1988) also discuss a source mechanism at shocks. They discuss how a Fermi-type cosmic-ray acceleration mechanism may also generate large-amplitude Alfvén waves and associated density fluctuations. Spangler et al (1988) and Spangler & Cordes (1988) discuss the magnetohydrodynamic (MHD) waves at the Earth's solar wind bow shock, suggesting that similar wave generation may occur in interstellar shocks. Pimenov (1985) emphasizes the role of a shock passage in amplifying an existing spectrum of small-scale irregularities. It seems that shock generation sites cannot easily account for the diffuse scattering medium, but they might be responsible for the clumpy sites of strongly enhanced scattering.

3.3 Nonlinear Energy Exchange—The Inertial Range

Nonlinear interactions are necessary for a turbulent cascade from one scale to another. Possible wave modes include the Alfvén waves and fast and slow magnetoacoustic waves, but there is little theory for their nonlinear interaction. Matthaeus & Montgomery (1981) and Shebalin et al (1983) have made analytical and numerical assaults on plasma turbulence under the incompressible MHD approximation. They obtain predominantly two-dimensional turbulent flow normal to the local B field; the velocity and magnetic field spectra take on the Kolmogorov form in two dimensions. On smaller scales, the local B field includes the random components on all larger scales and so may not remain aligned with a large uniform field. Recently, Shebalin & Montgomery (1988) numerically modeled the effects of slight compressibility, finding results in agreement with those of Montgomery et al (1987). Note that Kraichnan (1965) has argued that the power-law exponent for a turbulent cascade in a hydromagnetic plasma should be 3.5 rather than 3.67 (see also McIvor 1977, Bykov 1988, Tu 1988).

Several theoretical studies emphasize dissipation mechanisms, which might interrupt an energy cascade down to the small scales. For example,

McIvor (1977) and Cesarsky (1980) both discussed this question for the three phases of the medium. They concluded that important dissipation should occur at scales much larger than the radio-scattering scales, although they disagreed on some details; particularly for the hot phase, McIvor found thermal conduction effects to suppress any cascade, whereas Cesarsky argued that a cascade could exist from scales of 10^{17} to 10^{10} m. Zweibel et al (1988) considered waves generated in the hot ISM by the interaction of clouds and supernovae remnants, concluding that viscous damping precluded a turbulent cascade to small scales. They also noted, as had McIvor and Cesarsky, that collisions with neutrals would be an important damping mechanism in the partially ionized warm phase.

Higdon (1984, 1986) proposed that stationary constant-pressure density structures (entropy structures and tangential pressure balances) are convected and sheared by the turbulent velocity field and acquire the same spectrum as the velocity field. His arguments rely on analogies with turbulence in both a neutral gas and in an incompressible plasma but are a thoughtful consideration of this difficult problem. He suggests that the two-dimensional nature of the plasma turbulence (see above) may avoid the damping mechanisms discussed by McIvor and Cesarsky.

Jokipii (1988) has recently suggested that the magnetic irregularities that are involved in the confinement of cosmic rays in the Galaxy may be part of the same turbulent cascade. The diffusion of cosmic rays is determined by magnetic irregularities on a scale of the associated proton gyroradius, which for a magnetic field of 3 μgauss covers, say, 10^{10}–10^{17} m. Jokipii's analysis for Fermi-acceleration of protons in strong interstellar shocks and a Kolmogorov spectrum of magnetic irregularities over this range of scales gives a proton spectrum in excellent agreement with the observations. This is an interesting new argument in support of turbulence at intermediate scales. Alfvénic fluctuations have been measured in the solar wind (Belcher & Davis 1971, Roberts et al 1987, Bavassano & Bruno 1989), and there has been a long-standing debate about whether these are noninteracting incompressible Alfvén waves generated in or near the Sun or whether they represent an input power at large scales and a nonlinear turbulent cascade to small scales (see reviews by Hollweg 1978, Barnes 1979). The picture that now emerges (e.g. Tu et al 1989) is a solar origin for Alfvénic turbulence with sufficient amplitude to cause nonlinear interaction between scales and associated density fluctuations. These studies certainly clarify the plasma physics of the fluctuations, and although the ISM lacks a spherical outflow, the suggestion of Alfvénic turbulence of sufficient amplitude to drive density fluctuations is in accord with many of the ideas posed for the ISM.

3.4 Dissipation

In a turbulent neutral gas the inertial range extends to an inner scale where viscous forces dissipate the turbulent energy. As discussed above, there has been no lack of dissipation mechanisms; indeed, the problem has been to avoid dissipation and find what wave modes might interact and transfer energy down to the small radio-scattering scales. Magnetohydrodynamic and magnetoacoustic modes can extend to length scales shorter than the mean free path for collisions. However, there should be no waves smaller than the gyroradius for the thermal protons. For the warm intercloud medium, which is most commonly proposed as the site for the turbulence, this scale is $\sim 10^4$–10^5 m, in agreement with the results of Spangler & Gwinn (1990) but much smaller than the 10^9 m proposed by Coles et al (1987). Bieber et al (1988) discuss the dissipation scale for magnetic turbulence and its influence on cosmic-ray propagation. Harmon (1989) discusses density spectra and dissipation for Alfvén waves, associated with the inner scale, measured in the solar wind by Coles & Harmon (1989). These authors find enhanced density spectra just above a cutoff; if a similar enhancement exists in the ISM, there may be an associated enhancement in the focusing properties that might explain the double-imaging events in pulsar spectra.

A further theoretical difficulty surrounding the turbulence hypothesis is that active turbulence implies a power flux of, perhaps, 10^{-26} watts m^{-3} from very large to very small scales, which is comparable to other heating and cooling rates for the ISM (see Spitzer 1978). In other words, the turbulence itself could be important in the energy balance of the plasma, supplying heat to the plasma at the dissipation scale and absorbing energy at the outer scale. Gibson (1988) has introduced the notion of fossil turbulence into this discussion. He argues, by analogy with ocean turbulence, that the density spectrum may be a "fossil" remnant of a turbulent process that is no longer active. The Kolmogorov density spectrum could persist well after the plasma has ceased being actively stirred by turbulent motion. This avoids the large steady power flux for active turbulence. A formal examination of this idea must address the lifetime of density structures in a collisionless plasma.

3.5 Sites for Enhanced Scattering

As discussed above, shock generation or amplification of plasma fluctuations has been proposed by several authors. Thus, supernova remnants have been an obvious candidate for the location of the regions of enhanced "turbulence." Several groups of VLBI observers have looked for evidence of an excess of source broadening around supernova remnants, where interstellar shocks are expected. The results have not been conclusive. An alternative site of enhanced scattering is proposed by Anantharamaiah & Narayan (1988) as the low-density outer envelopes of H II regions, particularly those responsible for galactic ridge recombination lines (see Anantharamaiah 1986, Kassim 1989). In addition, Spangler & Gwinn's (1990) proposed inner scale is near the proton gyroradius for this medium. The enhanced scattering of NGC 6334B by a bipolar H II region (Moran et al 1990) is also a valuable clue.

3.6 Nonturbulent Models

Models that do not involve turbulence have also been proposed. A spectrum with $\beta \sim 4$ is consistent with a random superposition of discontinuities, as might be due to shock fronts; the constraint would be that the shock thickness was at least as small as the radio-scattering scales of $\sim 10^9$ m. In contrast, Hall (1980, 1981) has argued that a narrow spectral range in the density spectrum near 10^9 m would result from the "mirror instability" in the hot phase of the ISM. Though ISS observations suggest a wider spectrum, it is interesting that this proposed peak is near the inner scale proposed at 10^9 m.

4. SUMMARY

Interstellar scattering has now become important in several branches of astronomy and astrophysics, and the present situation can be summarized as follows:

1. *The bad news.* Scattering causes a seeing limitation in radio observations. This limitation was already evident for pulsars, but it is now clear that variation due to RISS is likely to be important for several classes of variable sources, particularly low-frequency variables and centimeter-wave flickering; RISS must also be included in the interpretation of sources small enough to vary intrinsically. However, even here the news is partly good, in that we now need fewer astrophysical explanations of extraordinary brightness temperatures. Interstellar scattering also broadens the image of many sources at low galactic latitude and provides a limit to VLBI resolution at low latitudes and long wavelengths (e.g. Dennison & Booth 1987).

2. *The good news.* By studying the flux variability of pulsars and extragalactic sources and the VLBI visibility curves, observers have new techniques for probing the ISM. The existence of a density spectrum covering many orders of magnitude in wave number is now established. There is a

hard question as to whether this spectrum is steeper or shallower than the exponent 4. There are conflicting interpretations of pulsar dynamic spectra, VLBI data, and image wander. Much of the evidence points to a shallower spectrum, with an inner-scale cutoff, which creates caustics. In spite of the great strides in wave propagation theory, particularly with regard to refractive effects, there are some unsolved questions. We lack quantitative interpretations of the periodic behaviors in pulsar dynamic spectra and in VLBI visibility curves obtained with self-calibration. It is likely that numerical simulation of these observations will resolve these questions.

Notable recent discoveries include the dramatic enhancements of scattering in the inner Galaxy and models for its distribution; the application of pulsar dynamic spectra to pulsar astronomy (binary orbit determination and resolution of a pulsar emitting region from episodes of "double imaging"); and the discrete propagation events, which may be due to deterministic structures.

3. The associated questions for theories of the ISM are now being addressed. The possibility of plasma turbulence in the various interstellar regions is under serious consideration. There are conflicting conclusions that support and deny a turbulent cascade over the proposed 8–10 orders of magnitude in scale. Location of the sites where "turbulence" is enhanced by a factor of 10^6 will be an important clue. The fact that the solar wind has a similar density spectrum with the appearance of a turbulent cascade adds weight to the turbulence hypothesis. In situ solar wind measurements may help resolve the nature of the interstellar as well as interplanetary density irregularities.

APPENDIX A. WAVE PROPAGATION THEORY

The Density Spectrum

As in R77, I assume here that the plasma density irregularities may be characterized by a spatial power spectrum $P_{3N}(\kappa)$, where κ is the three-dimensional wave number. It is this function that we hope to determine from measurements of scintillation and scattering. The following form is useful:

$$P_{3N}(\kappa) = C_N^2(r)(\kappa^2 + \kappa_{\text{outer}}^2)^{-\beta/2} \exp(-\kappa^2/\kappa_{\text{inner}}^2). \qquad \text{A1.}$$

For wave numbers between κ_{outer} and κ_{inner}, the power-law region is given by

$$P_{3N}(\kappa) = C_N^2(r)\kappa^{-\beta}, \qquad \kappa_{\text{outer}} \ll \kappa \ll \kappa_{\text{inner}}. \qquad \text{A2.}$$

For $\kappa < \kappa_{\text{outer}}$ the power spectrum saturates at a constant value; for $\kappa > \kappa_{\text{inner}}$ it falls rapidly to zero. The wave number κ_{outer} is the reciprocal of an outer scale, and κ_{inner} is the reciprocal of an inner scale; between them they define an "inertial range" for the plasma turbulence. Unless specified otherwise, we assume the following simplified model: the wave numbers relevant to the scintillation are far removed from κ_{inner} and κ_{outer}; the exponent β is equal to 11/3, corresponding to the Kolmogorov spectrum for turbulence in a neutral gas; the factor $C_N^2(r)$ is taken to vary slowly with position, and we define $\alpha = \beta - 2$. In the slab model C_N^2 it falls to zero outside the galactic disk of thickness L_0. The theory is stated for a general form of $P_{3N}(\kappa)$, where convenient. Good reviews of propagation through such a weakly scattering medium have been given by Prokhorov et al (1975) and Fante (1975, 1980).

Second Moment of the Scattered Field

Spatial deviations in the refractive index cause phase modulations as the waves travel through the ISM. The resulting wave front can be analyzed into a spectrum of plane waves, the width of which we call the scattering angle θ_s. A point or plane-wave source thus suffers angular broadening. Consider a pair of antennas at vector positions $\mathbf{s}$ and $\mathbf{s}+\boldsymbol{\sigma}$; the resulting interferometer fringes have electric field (E) amplitude and phase given by $E(\mathbf{s})E^*(\mathbf{s}+\boldsymbol{\sigma})$. If this complex quantity is averaged sufficiently, we obtain the ensemble average visibility function, for which there is a simple and generally valid result:

$$\langle E(\mathbf{s})E^*(\mathbf{s}+\boldsymbol{\sigma})\rangle = \exp[-0.5 D_\phi(\boldsymbol{\sigma})]. \qquad \text{A3.}$$

This equation is independent of the position coordinate $\mathbf{s}$ if the medium is statistically stationary. The quantity $D_\phi(\boldsymbol{\sigma})$ is the structure function of geometric phase—called the "wave structure function" in the optical literature; it is defined as

$$D_\phi(\boldsymbol{\sigma}) = \langle [\phi(\mathbf{s}) - \phi(\mathbf{s}+\boldsymbol{\sigma})]^2 \rangle, \quad \text{where } \mathbf{s} = (x, y, L), \ \boldsymbol{\sigma} = (\xi, \eta, 0), \quad \text{A4.}$$

$$\phi(x, y, L) = r_e \lambda \int_0^L N_e(x, y, z)\, dz. \qquad \text{A5.}$$

Here ϕ is the phase deviation calculated on a straight-line path (z axis) from the source to the observer, assuming that the observing frequency is everywhere much greater than the local plasma frequency; $N_e(x, y, z)$ is the deviation in electron density from its mean; λ is the wavelength (in the mean plasma density); and r_e is the classical electron radius (2.82×10^{-15} m). Under astronomical conditions the pathlength $L \gg 1/\kappa_{outer}$, and the wave structure function can be related to the density spectrum:

$$D_{\phi p}(\sigma) = \int_0^L D'_\phi(z,\sigma)\,dz \quad \text{and} \quad D_{\phi s}(\sigma) = \int_0^L D'_\phi(z,\sigma z/L)\,dz. \qquad \text{A6.}$$

The quantity D_ϕ is basic to the wave propagation process, and different forms are needed for plane-wave ($D_{\phi p}$) and spherical-wave sources ($D_{\phi s}$). Its derivative $D'_\phi(z,\sigma)$ in Equation A6 is related to the spectrum by

$$D'_\phi(z,\sigma) = 4\pi\lambda^2 r_e^2 \int_{-\infty}^{\infty} [1-\cos(\boldsymbol{\kappa}\cdot\boldsymbol{\sigma})] P_N(\kappa_x,\kappa_y,\kappa_z=0)\,d^2\boldsymbol{\kappa}. \qquad \text{A7.}$$

Consider the spectral form of Equation A1 with $2 < \beta < 4$ (or with $\alpha = \beta - 2$, $0 < \alpha < 2$) and $\sigma \ll 1/\kappa_{\text{outer}}$:

$$D'_\phi(z,\sigma) = [\Gamma(1-\alpha/2)/\Gamma(1+\alpha/2)](8\pi^2/\alpha 2^\alpha)\lambda^2 r_e^2 C_N^2(z)\sigma^\alpha$$

$$\text{if } \sigma > 1/\kappa_{\text{inner}}, \qquad \text{A8.}$$

$$D'_\phi(z,\sigma) = \Gamma(1-\alpha/2)\pi^2\lambda^2 r_e^2 C_N^2(z)(\kappa_{\text{inner}})^{2-\alpha}\sigma^2 \quad \text{if } \sigma < 1/\kappa_{\text{inner}}. \qquad \text{A9.}$$

Here D_ϕ increases with σ from zero at the origin, and for $\sigma > 1/\kappa_{\text{outer}}$ it saturates at a value equal to twice the phase variance; we define the outer scale by the value of σ at which it crosses half that value. For strongly scattering media, we can define a scale where $D_\phi(\sigma)$ equals unity. This defines the field coherence scale s_0, which is the spatial separation across which an rms phase difference of 1 radian exists and also at which the visibility function (Equation A3) falls to $e^{-1/2}$. The field coherence scale s_0 is often smaller than the outer scale by many orders of magnitude. By using Equations A6 and A8 or A9, it is straightforward to express $D_\phi(\sigma) \approx (\sigma/s_0)^\alpha$ and so obtain s_0 in terms of the wavelength and the average of C_N^2 over distance L for spherical- or plane-wave sources (see, for example, Coles et al 1987).

Since the visibility function in Equation A3 is a two-dimensional Fourier transform of the scattered source brightness distribution, the uncertainty relation between widths in the two domains suggests that we define the width of the scattered image (scattering angle) by $\theta_s = 1/(ks_0)$ (i.e. Equation 2.4). This gives the width of an image averaged over very many realizations or over a spatial region much larger than the outer scale. The coherent integration times typical of a VLBI observation are not necessarily this long; in particular, if the density spectrum is "steep" (with $\beta > 4$), the scattered image will wander by an angular extent much larger than its instantaneous angular size, which is accordingly much smaller than θ_s defined by Equation 2.4. This situation can complicate the interpretation of VLBI observations (Section 2.5).

Weak Scattering

The scintillation index (m) is the rms variation in intensity normalized by the mean intensity. In weak scintillation ($m \ll 1$) there is a linear solution (Born solution) for the spectrum of the intensity variations as a Fresnel filter function times the phase spectrum for each scattering layer of thickness dz. For a plane-wave source, the wave-number spectrum for the intensity variations is

$$P_I(\kappa_x, \kappa_y) = 8\pi r_e^2 \lambda^2 \int_0^L \sin^2(\kappa^2 \lambda z/4\pi) P_N(\kappa_x, \kappa_y, \kappa_z = 0) \, dz. \qquad \text{A10.}$$

Here the medium extends from the observer to a distance L. For a point source at distance L, the z in the argument of the $\sin^2$ function is replaced by $\zeta(z, L) = z(L-z)/L$. For the density spectrum (Equation A1) with $2 < \beta < 4$ and a uniform slab of scattering material, the resulting intensity spectrum increases with wave number up to an effective maximum wave number about equal to the reciprocal of the Fresnel radius r_f (Equation 2.1). The scintillation index is found as the square root of the integral of the appropriate $P_I(\kappa_x, \kappa_y)$. When the index is substantially smaller than unity, the scintillations are weak and we find $s_0 > r_f$. Weak scintillations are correlated over a wide frequency range, of the order of an octave [see R77 and Scott et al (1980) for the cross correlation of intensity variations in weak scintillations].

Strong Scattering

In strong scattering (i.e. a large scintillation index calculated from the Born solution), the linear relationship breaks down and no simple result is available. Then it also follows that $s_0 < r_f$, which suggests the choice of $r_f/s_0 = u$ as a parameter (Equation 2.5). For scattering concentrated in a thin "screen," the intensity spectrum (arbitrary u) from a plane-wave source can be written (e.g. Equation 4.9 of Prokhorov et al 1975) as

$$P_I(\kappa) = \int_{-\infty}^{\infty} \exp\{-F(\mathbf{r}, \kappa L/k) + i\boldsymbol{\kappa} \cdot \mathbf{r}\} \, d^2\mathbf{r}/(4\pi^2), \qquad \text{A11.}$$

where

$$F(\mathbf{r}, \mathbf{t}) = D_\phi(\mathbf{r}) + D_\phi(\mathbf{t}) - 0.5 D_\phi(\mathbf{r}+\mathbf{t}) - 0.5 D_\phi(\mathbf{r}-\mathbf{t}). \qquad \text{A12.}$$

For a spherical-wave source (at distance L from observer) seen through a screen (at distance z from the observer), Equation A11 applies with the effective distance $\zeta(z, L)$ in place of L.

For an extended medium, the solution is much more awkward and a great deal of effort has been devoted to various regimes of approximation. Prokhorov et al (1975) gave approximations valid for high and low wave numbers (e.g. their Equations 4.40 and 4.41; see also Frehlich 1987). There is another interesting approximate solution (Uscinski 1982), that can be recognized as expressing the intensity spectrum as for an equivalent screen. Uscinski's Equation 102 for a plane-wave source, generalized to three dimensions, is identical to Equation A11, but with an equivalent function F_{eq} in place of F:

$$F_{eq}(\mathbf{r}, \mathbf{t}) = \int_0^L \{D'_\phi(z, \mathbf{r}) + D'_\phi(z, \mathbf{t}z/L)$$
$$- 0.5 D'_\phi(z, \mathbf{r} + \mathbf{t}z/L) - 0.5 D'_\phi(z, \mathbf{r} - \mathbf{t}z/L)\} \, dz. \quad \text{A13.}$$

The equivalent-screen method for a point source in an extended medium has been presented by Uscinski et al (1982) and Macaskill (1983). Whitman & Beran (1985) explored the range of u for which the equivalent-screen solutions are valid, concluding that they apply for both weak and strong scattering, provided only that as a wave travels through the medium, large-phase fluctuations develop more rapidly than intensity fluctuations; this process can be described as extreme forward scatter. The solution is Equations A11 and A13 with a substitution in Equation A13 of $L-z$ in place of z wherever z occurs in the second argument of D'_ϕ, and with the substitution of $\mathbf{r}z/L$ in place of $\mathbf{r}$.

Unfortunately, none of these expressions (Equations A11/A12 or A11/A13 or their spherical-wave equivalents) are particularly simple, and the form of the spatial spectrum of intensity is hard to visualize. We thus resort to approximation techniques.

In weak scattering we obtain Equation A10. In asymptotically strong scattering, the solutions take on a two-scale form. This can be seen from low- and high-wave-number expansions of Equation A11. The first-order high-wave-number expansion leads to the diffractive (small-scale) component. The first term of the low-wave-number expansion leads to the refractive (large-scale) component. It is described by a modified Equation A10, in which there is an additional cutoff at wave numbers above the reciprocal of the scattering disk size. The refractive term is actually of the second order compared with the first diffractive term, and the second diffractive term needs also to be considered, since it makes a contribution to the total variance equal to that of the first refractive term. This leads to the approximate result expressed in Equation 2.7. The expansions have been discussed by various authors since the review of Prokhorov et al (1975). For example, Codona et al (1986a,b) discuss the various techniques starting from a full wave treatment for spectra with exponents $\beta < 4$.

(They also show the equivalence of path integral and moment equation methods of analysis.)

Other groups have made useful alternative approximations to the diffractive and refractive regimes, two of which have concentrated on interstellar applications. Blandford & Narayan (1985) and RNB used a mixture of wave (for diffraction) and ray (for refraction) theories to examine the variability of many observable quantities. In a parallel investigation, CPL separated the phase perturbation of a screen into slowly varying (refractive) and rapidly varying (diffractive) components; by approximating the former by a Taylor expansion, they estimated the refractive and diffractive perturbations for many observable quantities. (Their separation of the phase into refractive and diffractive components remains an ad hoc procedure that has to be rechecked after the analysis is complete.) Similar results were obtained from both approaches and included many potentially observable parameters that had not been addressed in the full wave treatments. Both papers have tables of expressions relating observables to C_N^2, distance, and frequency. Although their numerical factors differ, the basic agreement of the two methods gives confidence in their results, and these studies have been used in several comparisons of observation and theory (e.g. for image wander by Gwinn et al 1988).

The Caltech group had earlier demonstrated an interesting solution to the scattering problem for media with exponents $\beta > 4$ (Blandford & Narayan 1985, Goodman & Narayan 1985). These spectra have more energy in the low wave numbers as compared with the Kolmogorov spectrum and cause stronger refractive effects. The importance of this solution is that if such an electron density spectrum can explain the observations, there is no need to invoke turbulence in the interstellar plasma on the microscales responsible for interstellar scattering (see Section 3).

The following sections discuss the refractive and diffractive regimes of strong scattering for density spectra with $\beta < 4$. For steeper spectra the structure function has a square-law term that cancels in the second difference F (Equation A11), and thus a modified expansion method is needed (see below).

Diffractive Scintillation

The asymptotic high-wave-number expansion for the intensity spectrum $P_I(\kappa)$ results in the Fourier transform of the square of the spatial covariance of the electric field, as in Equation A3. The resulting spatial scale for DISS is s_0, with a decorrelation time $\sim s_0/V$. There is also an important decorrelation in frequency, δv_d, which is typically in the range of 10 kHz to 10 MHz. There are thus independent diffractive scintles in frequency-time cells of δv_d by s_0/V. The depth of modulation is 100%.

In the strong scattering limit, the intensity covariance, at position s_1 and frequency v_1 and position s_2 and frequency v_2, is approximated by the square of the second moment for the electric field. Consider the particular case of coincident receivers at different frequencies. The square of the second moment includes a term depending on the dispersion delay; as discussed by Lee & Jokipii (1975), a more useful approximation is obtained by removing the dispersive term. (I avoid calling it the refractive term in order to distinguish it from refractive scintillation.) However, there was no satisfactory derivation of how an intensity observation removes this dispersive term until the work of Codona et al (1986c). Their result is a modified dispersive term that arises from dispersion differences across the scattering disk. This term is important for frequency differences above about 0.1%. In typical interstellar observing conditions the term is unimportant, since diffractive decorrelation occurs over even narrower differences. Thus, as has been commonly assumed, diffractive decorrelation in frequency is governed by the square of the *diffractive* second moment versus frequency, which in turn is the temporal Fourier transform of the diffractive impulse response. Geometrical ray-path considerations give the typical broadening time for a pulse as

$$\tau_d = aL\theta_s^2/c = 1/2\pi\delta v_d. \qquad \text{A14.}$$

Here θ_s is the apparent angular broadening at the observer, and $a = 0.5$ for a plane-wave incident on a thin screen at distance L (and for a point source at distance L with a screen halfway in between). For extended scattering, similar expressions result, with L being the total pathlength and the factor a depending on the distribution of irregularities. There is, regrettably, no simple exact result for the scattered pulse shape or the second moment of a point source in a uniform medium. Williamson (1974) made ray calculations for the scattered pulse shape in a uniform extended medium and showed that the rising edge of the pulse was particularly sensitive to the distribution along the line of sight of the scattering irregularities, which he assumed had a Gaussian spectrum. This is an area of research where fairly straightforward wave simulations could be done that would advance our ability to interpret the scattered profiles of pulsars. I use $a = 0.5$ in Equation A14 unless otherwise noted.

As explained above, the decorrelation function for diffractive scintillation is the Fourier transform of the scattered pulse shape, and so its width δv_d is related to the pulse width τ_d as in Equation A14. This result has been tested by a number of authors comparing observations of pulse broadening with decorrelation of diffractive scintillations for the same pulsars. Evidently, its validity depends in detail on the definition of frequency and time scales; a half-power half-width in frequency and a $1/e$

width in time have often been used (see CWB, Slee et al 1980, Roberts & Ables 1982).

Refractive Scintillation

In asymptotically strong scattering, a low-wave-number expansion for the intensity spectrum yields the refractive scintillation spectrum. It is given by a modified Born solution, in which there is an additional cutoff at wave numbers above the reciprocal of the scattering-disk size—Equation A10 for $P_1(\kappa)$ multiplied by $\exp[-\int_0^z D'_\phi(z', \kappa z'/k)\, dz']$ [see Frehlich (1987) for the spherical-wave form]. The refractive variance is given by integration over κ_x and κ_y, and the Fourier transform gives the intensity correlation and hence the structure function needed to predict the depth of variation and time scales of RISS. It is straightforward to show that the bandwidth of RISS is about an octave, as for weak scintillation. The above results are only valid when the refractive variance $m_r^2 \ll 1$. This excludes the interesting region near $u = 1$, where we need numerical techniques.

Inner-Scale Effects

Most of the foregoing expressions apply for density spectra with $\alpha < 2$ with or without an inner-scale cutoff. However, the integrals then depend on the inner scale as well as on the strength of scattering. Coles et al (1987) showed how the presence of a cutoff in the spectrum can enhance the refractive variance over that of the Kolmogorov spectrum, adjusted to give the same strength of scattering. This provides a possible explanation of the RISS of pulsars, in which m_r is greater than the Kolmogorov value (Figure 2). The relative importance of the inner scale is determined by the ratio $s_{\text{inner}}/L\theta_s$. The greatest enhancements in m_r occur where this ratio is unity.

Goodman et al (1987) went further in examining the "optics" near the focusing condition $s_{\text{inner}}/L\theta_s = 1$. Their analysis of caustics in this regime provides important insight. At distances somewhat beyond the focus there will be common occurrences of a "fold," which corresponds to two distinct directions of arrival. This is a promising explanation of the double-imaging events (Figure 5d). In the limit of saturated scintillations there will be many directions of arrival, and the scintillations will take on a more stationary random appearance. The focusing condition for the inner scale of 10^9 m proposed by Coles et al (1987) probably corresponds to observations above 1 GHz at distances of 100–1000 pc and is represented by the heavy dashed line in Figure A1. This figure is for a uniform slab of scattering material, and large local deviations would be expected in the inner Galaxy. The focusing of a 10^6-m inner scale would be less dramatic, since it is smaller than r_f.

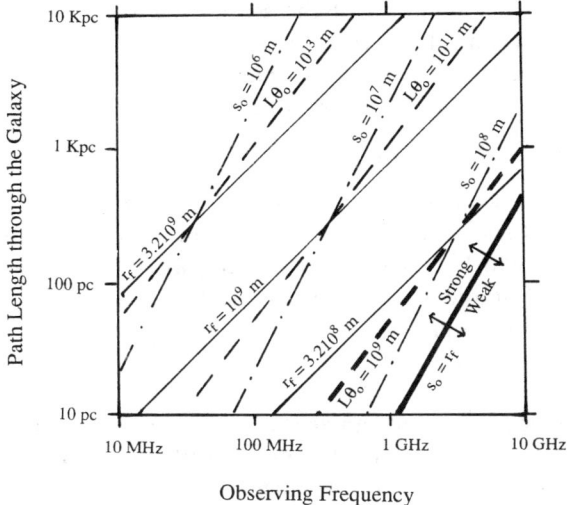

Figure A1 Lines of constant scale in "observing coordinates" for a uniform distribution of scattering material. Dash-dot lines are for constant diffractive scale s_0. Dashed lines are for constant refractive scale s_r. Solid lines are for constant Fresnel scale r_f. The heavy line divides weak from strong scattering. The heavy dashed line is where an inner scale of 10^9 m would fill the scattering disk and so focus. Conditions near and to the left of this line might lead to double imaging. This plot only represents the diffuse scattering medium; large localized increases in scattering strength are to be expected, particularly toward the inner part of the Galaxy.

Steep Spectra

Goodman & Narayan (1985) have analyzed scintillation from a screen with $\beta > 4$ in Equation A1; the spectrum is then said to be steep. There are important modifications to Equations A8 and A9. Ignoring the effect of an inner scale, the dominant term for $D'_\phi(z, \sigma)$ at small σ is a square law. This gives $D_\phi(\sigma) \approx (\sigma/s_0)^2$, with the s_0 determined by the outer scale. The geometrical phase is dominated by linear gradients, and in terms of the scattered image the angle $1/ks_0$ corresponds to image wander ($\theta_r \sim 1/ks_0$). The width of the instantaneous scattered image comes from deviations of $D_\phi(\sigma)$ from a square law. Equations A11 and A12 still describe the intensity spectrum. However, in Equation A12 F is completely independent of square-law terms in D_ϕ. The next term in a D_ϕ expansion can be written as $-(\sigma/s_\alpha)^\alpha$; for $4 > \alpha > 2$ the scale s_α is independent of the outer scale, depending only on C_N^2 (as for $2 > \alpha > 0$).

The scale s_α determines both the development of intensity scintillations and the instantaneous width of the scattered image. Evaluation of the integral in Equation A11 gives a "two-scale" intensity spectrum, but the negative sign and the fact that $\alpha > 2$ changes the behavior. The results can be understood by considering a modified D_ϕ, namely

$$D_{\phi\alpha} = (\sigma/s_0)^2 - D_\phi(\sigma) \approx (\sigma/s_\alpha)^\alpha, \qquad \text{A15.}$$

which represents deviations from linear phase gradients. The instantaneous scattering disk (of scale s_r) is determined by the interference condition that $s_r \approx L\theta(s_r)$, in which $\theta(s)$ is the scattering angle determined from the modified phase gradient averaged over scale s {i.e. $k\theta(s) = [D_{\phi\alpha}(s)]^{0.5}/s$}. Inserting the form Equation A15 gives

$$s_r \approx r_f(r_f/s_\alpha)^{\alpha/(4-\alpha)}.$$

The instantaneous scattering angle is $\theta_s = s_r/L$, and the diffractive scale s_d is $1/k\theta_s$. This result was originally found by Blandford & Narayan (1985) and refined in several subsequent papers.

Literature Cited

Altunin, V. I. 1981. *Sov. Astron AJ* 25: 304–8

Alurkar, S. K., Slee, O. B., Bobra, A. D. 1986. *Aust. J. Phys.* 39: 433–38

Anantharamaiah, K. R. 1986. *J. Astrophys. Astron.* 7: 131–39

Anantharamaiah, K. R., Narayan, R. 1988. See Cordes et al 1988a, pp. 185–89

Armstrong, J. W., Coles, W. A., Kojima, M., Rickett, B. J. 1990. *Ap. J.* In press

Armstrong, J. W., Cordes, J. M., Rickett, B. J. 1981. *Nature* 291: 561–64

Armstrong, J. W., Spangler, S. R., Hardee, P. E. 1977. *Astron. J.* 82: 785–90

Backer, D. C. 1975. *Astron. Astrophys.* 43: 395–404

Backer, D. C. 1988. See Cordes et al 1988a, pp. 111–16

Barnes, A. 1979. In *Solar System Plasma Physics*, ed. C. F. Kennel, L. J. Lanzerotti, E. N. Parker, pp. 249–319. Amsterdam: North-Holland

Bastian, T. S., Bookbinder, J. A. 1987. *Nature* 326: 678–80

Bavassano, B., Bruno, R. 1989. *J. Geophys. Res.* 94: 11,977–82

Belcher, J. W., Davis, L. 1971. *J. Geophys. Res.* 76: 3534–63

Bieber, J. W., Smith, C. W., Matthaeus, W. H. 1988. *Ap. J.* 334: 470–75

Biraud, F., Bourgois, G. 1988. Presented at Symp. Pulsar Timing, Nancay, Fr.

Blandford, R. D., Narayan, R. 1985. *MNRAS* 213: 591–611

Blandford, R. D., Narayan, R., Romani, R. W. 1984. *J. Astrophys. Astron.* 5: 369–88

Bykov, A. M. 1988. *Sov. Astron. Lett.* 14: 60–62

Cawthorne, T. J., Rickett, B. J. 1985. *Nature* 315: 40–42

Cesarsky, C. J. 1980. *Annu. Rev. Astron. Astrophys.* 18: 289–319

Clegg, A. W., Chernoff, D. F., Cordes, J. M. 1988. See Cordes et al 1988a, pp. 174–78

Clifton, T. R. 1986. PhD thesis. Univ. Manchester, Engl.

Codona, J. L., Creamer, D. B., Flatte, S. M., Frehlich, R. G., Henyey, F. S. 1986a. *J. Math. Phys.* 27: 171–77

Codona, J. L., Creamer, D. B., Flatte, S. M., Frehlich, R. G., Henyey, F. S. 1986b. *Radio Sci.* 21: 929–48

Codona, J. L., Creamer, D. B., Flatte, S. M., Frehlich, R. G., Henyey, F. S. 1986c. *Radio Sci.* 21: 805–14

Cohen, M. H., Cronyn, W. M. 1974. *Ap. J.* 192: 193–97

Cole, T. W., Hesse, H. K., Page, C. G. 1970. *Nature* 225: 712–13

Coles, W. A., Frehlich, R. G. 1982. *J. Opt. Soc. Am.* 72: 1042–48

Coles, W. A., Frehlich, R. G., Rickett, B. J., Codona, J. L. 1987. *Ap. J.* 315: 666–74

Coles, W. A., Harmon, J. K. 1989. *Ap. J.* 227: 1023–34

Condon, J. J., Backer, D. C. 1975. *Ap. J.* 197: 31–38

Condon, J. J., Dennison, B. 1978. *Ap. J.* 224: 835–40
Cordes, J. M. 1986. *Ap. J.* 311: 183–96
Cordes, J. M., Ananthakrishnan, S., Dennison, B. 1984. *Nature* 309: 689–91
Cordes, J. M., Downs, G. S. 1985. *Ap. J. Suppl.* 59: 343–82
Cordes, J. M., Pidwerbetsky, A., Lovelace, R. V. E. 1986. *Ap. J.* 310: 737–67 (CPL)
Cordes, J. M., Rickett, B. J., Backer, D. C., eds. 1988a. *Radio Wave Scattering in the Interstellar Medium. AIP Conf. Proc. No. 174.* New York: Am. Inst. Phys.
Cordes, J. M., Spangler, S. R., Weisberg, J. M., Clifton, T. R. 1988b. See Cordes et al 1988a, pp. 180–84
Cordes, J. M., Weisberg, J. M., Boriakoff, V. 1985. *Ap. J.* 288: 221–47 (CWB)
Cordes, J. M., Wolszczan, A. 1986. *Ap. J. Lett.* 307: L27–31
Cordes, J. M., Wolszczan, A. 1988. See Cordes et al 1988a, pp. 212–16
Cordes, J. M., Wolszczan, A., Dewey, R. J., Blaskiewisicz, M., Stinebring, D. R. 1990. *Ap. J.* 349: 245–61
Cotton, W. D., Spangler, S. R., eds. 1982. *Proc. NRAO Workshop Low-Freq. Var. of Extragalact. Radio Sources.* Green Bank, W. Va: NRAO
Coulman, C. E. 1985. *Annu. Rev. Astron. Astrophys.* 23: 19–57
Cronyn, W. M. 1972. *Ap. J.* 174: 181–200
Dennison, B., Booth, R. S. 1987. *MNRAS* 224: 927–34
Dennison, B., Condon, J. J. 1981. *Ap. J.* 246: 91–99
Dennison, B., Fiedler, R., Johnston, K. J., Spencer, J. H., Waltman, E. B., et al. 1987. *Ap. J.* 313: 141–45
Dennison, B., Thomas, M., Booth, R. S., Brown, R. L., Broderick, J. J., Condon, J. J. 1984. *Astron. Astrophys.* 135: 199–212
Dewey, R. J., Cordes, J. M., Wolszczan, A., Weisberg, J. M. 1988. See Cordes et al 1988a, pp. 217–21
Diamond, P. J., Martinson, A., Booth, R. S., Winnberg, A. 1988. See Cordes et al 1988a, pp. 195–99
Dreher, J. W., Roberts, D. H., Lehar, J. 1986. *Nature* 320: 239–42
Duric, N., Gregory, P. C. 1988. *Astron. J.* 95: 1149–58
Duric, N., Gregory, P. C., Taylor, A. R. 1987. *Astron. J.* 93: 890–97
Duric, N., Gregory, P. C., Tsutsumi, T. 1989. *Nature* 337: 143–45
Ewing, M. S., Batchelor, R. A., Friefeld, R. D., Price, R. M., Staelin, D. H. 1970. *Ap. J. Lett.* 162: L169–72
Fante, R. L. 1975. *Proc. IEEE* 63: 1669–92
Fante, R. L. 1980. *Proc. IEEE* 68: 1424–43
Fey, A. L., Spangler, S. R., Mutel, R. L. 1988. See Cordes et al 1988a, pp. 190–94

Fiedler, R. L., Dennison, B., Johnston, K. J., Hewish, A. 1987. *Nature* 326: 675–78
Fiedler, R. L., Simon, R., Johnston, K. J., Dennison, B., Hewish, A. 1988. See Cordes et al 1988a, pp. 150–55
Fleck, R. C. 1981. *Ap. J. Lett.* 246: L151–54
Foster, R. S., Cordes, J. M. 1988. See Cordes et al 1988a, pp. 205–11
Frehlich, R. G. 1987. *J. Opt. Soc. Am. A* 4: 360–66
Gibson, C. H. 1988. See Cordes et al 1988a, pp. 74–79
Goodman, J., Narayan, R. 1985. *MNRAS* 214: 519–37
Goodman, J., Narayan, R. 1989. *MNRAS* 238: 995–1028
Goodman, J. J., Romani, R. W., Blandford, R. D., Narayan, R. 1987. *MNRAS* 229: 73–102
Gregorini, L., Ficarra, A., Padrielli, L. 1986. *Astron. Astrophys.* 168: 25–31
Gregory, P. C. 1987. *Proc. NRAO Workshop Large-Scale Surv. of the Sky*, ed. J. J. Condon, F. J. Lockman. Green Bank, W.Va: NRAO
Gregory, P. C., Taylor, A. R. 1986. *Astron. J.* 92: 371–411
Gupta, Y., Rickett, B. J., Lyne, A. G. 1988. See Cordes et al 1988a, pp. 140–44
Gwinn, C. R., Cordes, J. M., Bartel, N. H., Wolszczan, A. 1988. See Cordes et al 1988a, pp. 106–10
Gwinn, C. R., Moran, J. M., Reid, M. J., Schneps, M. H. 1988. *Ap. J.* 330: 817–27
Hall, A. N. 1980. *MNRAS* 190: 371–83
Hall, A. N. 1981. *MNRAS* 195: 685–96
Harmon, J. K. 1989. *J. Geophys. Res.* 94: 15,399–405
Heeschen, D. S. 1984. *Astron. J.* 89: 1111–23
Heeschen, D. S., Krichbaum, Th., Schalinski, C. J., Witzel, A. 1987. *Astron. J.* 94: 1493–1507
Heeschen, D. S., Rickett, B. J. 1987. *Astron. J.* 93: 589–91
Helfand, D. J., Fowler, L. A., Kuhlman, J. V. 1977. *Astron. J.* 82: 701–5
Hewish, A. 1980. *MNRAS* 192: 799–804
Hewish, A., Wolszczan, A., Graham, D. A. 1985. *MNRAS* 213: 167–79
Higdon, J. C. 1984. *Ap. J.* 285: 109–23
Higdon, J. C. 1986. *Ap. J.* 309: 1342–61
Hjellming, R. M., Narayan, R. 1986. *Ap. J.* 310: 768–72
Hollweg, J. V. 1978. *Rev. Geophys.* 16: 689–720
Jackson, P. D., Kundu, M. R., White, S. M. 1987. *Ap. J. Lett.* 316: L85–90
Jokipii, J. R. 1988. See Cordes et al 1988a, pp. 48–59
Kaplan, S. A. 1966. *Interstellar Gas Dynamics.* Oxford: Pergamon
Kassim, N. E. 1989. *Ap. J.* 347: 915–24

Kraichnan, R. H. 1965. *Phys. Fluids* 8: 1385–87
Larson, R. B. 1979. *MNRAS* 186: 479–90
Lazio, T. J., Spangler, S. R., Cordes, J. M. 1990. Preprint (Cornell Univ., Ithaca, N.Y.)
Lee, L. C., Jokipii, J. R. 1975. *Ap. J.* 201: 532–43
Lee, L. C., Jokipii, J. R. 1976. *Ap. J.* 206: 735–43
Lo, K. Y., Cohen, M. H., Readhead, A. C. S., Backer, D. C. 1985. *Ap. J.* 249: 504–12
Lovelace, R. V. E. 1970. PhD thesis. Cornell Univ., Ithaca, N.Y.
Lyne, A. G. 1984. *Nature* 310: 300–2
Lyne, A. G., Manchester, R. N., Taylor, J. H. 1985. *MNRAS* 213: 613–39
Lyne, A. G., Smith, F. G. 1982. *Nature* 298: 825–27
Macaskill, C. 1983. *Proc. R. Soc. London Ser. A* 386: 461–74
Mantovani, F., Fanti, R., Gregorini, L., Padrielli, L., Spangler, S. 1990. *Astron. Astrophys.* In press
Matthaeus, W. H., Montgomery, D. 1981. *J. Plasma Phys.* 25: 11–41
Max, C. E., Zachary, A., Arons, J. 1988. See Cordes et al 1988a, pp. 61–65
McCray, R., Snow, T. P. Jr. 1979. *Annu. Rev. Astron. Astrophys.* 17: 213–40
McIvor, I. 1977. *MNRAS* 178: 85–99
McKee, C. F., Ostriker, J. P. 1977. *Ap. J.* 218: 148–69
Montgomery, D., Brown, M. R., Matthaeus, W. H. 1987. *J. Geophys. Res.* 92: 282–84
Moran, J. M., Rodriguez, L. F., Greene, B., Backer, D. C. 1990. *Ap. J.* 348: 147–52
Mutel, R. L., Hodges, M. W. 1986. *Ap. J.* 307: 472–77
Mutel, R. L., Lestrade, J.-F. 1988. See Cordes et al 1988a, pp. 122–28
Mutel, R. L., Lestrade, J.-F. 1989. Preprint (Univ. Iowa, Iowa City)
Narayan, R. 1988. See Cordes et al 1988a, pp. 17–31
Narayan, R., Cornwell, T. J., Goodman, J., Anantharamaiah, K. R. 1989. *Proc. URSI/IAU Symp. Radio Astron. Seeing, Beijing.* In press
Narayan, R., Goodman, J. 1989. *MNRAS* 238: 963–94
Narayan, R., Hubbard, W. B. 1988. *Ap. J.* 325: 503–18
Pimenov, S. F. 1985. *Sov. Astron. Lett.* 10: 218–20
Prokhorov, A. M., Bunkin, F. V., Gochelashvily, K. S., Shishov, V. I. 1975. *Proc. IEEE* 63: 790–811
Quirrenbach, A., Witzel, A., Krichbaum, T., Hummel, C. A., Alberdi, A., Schalinski, C. 1989a. *Nature* 337: 442–44
Quirrenbach, A., Witzel, A., Qian, S. J., Krichbaum, T., Hummel, C. A., Alberdi, A. 1989b. *Astron. Astrophys.* 226: L1–4
Rankin, J. M., Payne, R. R., Campbell, D. B. 1974. *Ap. J. Lett.* 193: L71–74
Rao, A. P., Ananthakrishnan, S. 1984. *Nature* 312: 707–11
Rawley, L. A. 1986. PhD thesis. Princeton Univ., N.J.
Rawley, L. A., Taylor, J. H., Davis, M. M. 1988. *Ap. J.* 326: 947–53
Reid, M. J., Moran, J. M. 1981. *Annu. Rev. Astron. Astrophys.* 19: 231–76
Rickett, B. J. 1969. *Nature* 221: 158–59
Rickett, B. J. 1970. *MNRAS* 150: 67–91
Rickett, B. J. 1977. *Annu. Rev. Astron. Astrophys.* 15: 479–504 (R77)
Rickett, B. J. 1986. *Ap. J.* 307: 564–74 (R86)
Rickett, B. J. 1988. See Cordes et al 1988a, pp. 2–16
Rickett, B. J., Coles, W. A. 1988. In *The Impact of VLBI on Astrophysics and Geophysics, IAU Symp. No. 129*, ed. M. J. Reid, J. M. Moran, pp. 287–94. Dordrecht: Kluwer
Rickett, B. J., Coles, W. A., Bourgois, G. 1984. *Astron. Astrophys.* 134: 390–95 (RCB)
Rickett, B. J., Lyne, A. G. 1990. *MNRAS.* In press
Roberts, D. A., Goldstein, M. L., Klein, L. W., Matthaeus, W. H. 1987. *J. Geophys. Res.* 92: 12,023–35
Roberts, J. A., Ables, J. G. 1982. *MNRAS* 201: 1119–38
Romani, R. W. 1988. See Cordes et al 1988a, pp. 156–62
Romani, R. W., Blandford, R. D., Cordes, J. M. 1988. *Nature* 328: 324–26
Romani, R. W., Narayan, R., Blandford, R. D. 1986. *MNRAS* 220: 19–49 (RNB)
Rowland, P. R., Cohen, R. J. 1986. *MNRAS* 220: 233–51
Rumsey, V. H. 1976. *Radio Sci.* 11: 545–49
Scheuer, P. A. G. 1968. *Nature* 218: 920–22
Scott, S. L., Rickett, B. J., Armstrong, J. W. 1980. *Astron. Astrophys.* 123: 191–206
Shapirovskaya, N. Y. 1978. *Sov. Astron. AJ* 22: 544–47
Shapirovskaya, N. Y. 1985. *Sov. Astron. Lett.* 11: 289–91
Shapirovskaya, N. Y., Sieber, W. 1984. *Astron. Astrophys.* 136: 171–74
Shebalin, J. V., Matthaeus, W. H., Montgomery, D. 1983. *J. Plasma Phys.* 29: 525–47
Shebalin, J. V., Montgomery, D. 1988. *J. Plasma Phys.* 39: 339–67
Shishov, V. I. 1974. *Sov. Astron. AJ* 17: 598–602
Sieber, W. 1982. *Astron. Astrophys.* 113: 311–13
Simonetti, J. H., Cordes, J. M. 1988. See Cordes et al 1988a, pp. 134–38

Simonetti, J. H., Cordes, J. M., Heeschen, D. S. 1985. *Ap. J.* 296: 46–59

Slee, O. B., Dulk, G. A., Otrupcek, R. E. 1980. *Proc. Astron. Soc. Aust.* 4: 100–6

Slee, O. B., Siegman, B. C. 1988. *MNRAS* 235: 1313–41

Smith, F. G., Wright, W. C. 1985. *MNRAS* 214: 97–107

Spangler, S. R. 1988. See Cordes et al 1988a, pp. 32–46

Spangler, S. R., Cordes, J. M. 1988. See Cordes et al 1988a, pp. 66–69

Spangler, S. R., Fanti, R., Gregorini, L., Padrielli, L. 1990. *Astron. Astrophys.* In press

Spangler, S. R., Fuselier, S., Fey, A., Anderson, G. 1988. *J. Geophys. Res.* 93: 845–57

Spangler, S. R., Gwinn, C. 1990. Preprint (Univ. Iowa, Iowa City)

Spangler, S. R., Mutel, R. L., Benson, J. M., Cordes, J. M. 1986. *Ap. J.* 301: 312–19

Spitzer, L. 1978. *Physical Processes in the Interstellar Medium*. New York: Wiley

Stinebring, D., Condon, J. J. 1990. *Ap. J.* 352: 207–21

Tu, C. 1988. *J. Geophys. Res.* 93: 7–20

Tu, C., Marsch, E., Thieme, K. M. 1989. *J. Geophys. Res.* 94: 11,739–59

Uscinski, B. J. 1982. *Proc. R. Soc. London Ser. A* 380: 137–68

Uscinski, B. J., Macaskill, C., Ewart, T. E. 1982. *J. Acoust. Soc. Am.* 74: 1474–99

White, G. J., Macdonald, G. H. 1979. *MNRAS* 188: 745–64

Whitman, A. M., Beran, M. J. 1985. *J. Opt. Soc. Am. A* 2: 2133–43

Wilkinson, P. N., Spencer, R. E., Nelson, R. F. 1988. In *The Impact of VLBI on Astrophysics and Geophysics, IAU Symp. No. 129*, ed. M. J. Reid, J. M. Moran, pp. 305–6. Dordrecht: Kluwer

Williamson, I. P. 1974. *MNRAS* 166: 499–512

Wolszczan, A., Bartlett, J. E., Cordes, J. M. 1988. See Cordes et al 1988a, pp. 145–49

Wolszczan, A., Cordes, J. M. 1987. *Ap. J. Lett.* 320: L35–39

Zweibel, E. G., Ferriere, K. M., Shull, J. M. 1988. See Cordes et al 1988a, pp. 70–73

RAPIDLY OSCILLATING Ap STARS

D. W. Kurtz

Department of Astronomy, University of Cape Town, Rondebosch 7700, South Africa

KEY WORDS: asteroseismology, magnetic stars, pulsating stars

1. INTRODUCTION AND BACKGROUND

The rapidly oscillating Ap (roAp) stars are hydrogen core-burning stars of mass $M \approx 2M_\odot$ which have global dipole magnetic fields with effective strengths of a few hundred to a few thousand gausses and which pulsate in high-overtone, low-degree, nonradial p-modes with periods in the range 4–15 min. Studies of these pulsations can potentially determine the stars' rotation periods, rotational inclinations, magnetic geometries, internal magnetic field strengths, radii, masses, luminosities, and ages. In addition, they can help constrain the possible mechanisms that give rise to the peculiarities of roAp stars. Much of this information comes from asteroseismology, a field that derives from helioseismology [see (35, 45, 73) for overviews of helio- and asteroseismology]. Although the luminosity amplitudes of the oscillations in the roAp stars are about a factor of 10^3 greater than those of the 5-min oscillations in the total solar irradiance (184), the similarity of the periods is intriguing.

Before discussing the pulsation of the roAp stars, I draw a qualitative picture of the general properties of Ap stars. References to the extensive literature on Ap stars and an excellent discussion of the properties of all of the A stars are given in the monograph by Wolff (181). An earlier review of the Ap stars by Preston (137) can be found in this series. Up-to-date reviews of most aspects of the Ap stars, both observational and theoretical, are presented in the proceedings of IAU Colloquium No. 90 (41). Other reviews of the roAp stars themselves are given by Kurtz (85, 91, 92), Shibahashi (150), Weiss (175), and Matthews (117).

1.1 General Properties of the Ap Stars and the Oblique Rotator Model

The Ap stars have been recognized as peculiar stars since the late nineteenth century. Their spectra show enhanced lines of the Fe-peak elements and greatly enhanced lines of the rare earth elements compared with spectra of normal stars. Attempts have been made to explain these line strength anomalies by invoking abnormal atmospheric structure, but this cannot account completely for the observed spectra. Thus, the line strength anomalies are generally accepted to be caused predominantly by atmospheric abundance anomalies. These anomalies must be confined to a thin layer in the atmosphere for two reasons: (a) Ap stars are observed to have only main sequence and subgiant luminosities; by the time an Ap star becomes a giant, its spectral abnormalities are no longer apparent. (b) In some Ap stars the overabundance of certain rare earth elements (e.g. Eu, which is overabundant by a factor of 10^4 in some cases) is so great that a significant fraction of the supply of such elements in the Universe would be contained in Ap stars if the observed abundances extended throughout the star.

It has been known since the first decade of this century that some Ap stars show periodic spectrum variability; in the second decade, light variability was discovered which is in phase (or antiphase) with the spectrum variations. The cause of this variability remained unexplained until the late 1940s, when H. W. Babcock discovered in some Ap stars global magnetic fields which vary in phase with the spectrum and light variations. The surface magnetic fields in the Ap stars are (to first order) dipolar, and their effective strengths range from the lower limit of detectability of a few hundred gausses up to several tens of kilogausses (see ref. 23 for a general review of magnetic stars).

Stibbs (158) developed the oblique rotator model of the Ap stars, which accounts well for the form of the magnetic, spectrum, and light variations. This model assumes that the magnetic field is frozen-in to the stellar atmosphere and has an axis which is inclined by an angle β to the rotation axis, which is itself inclined to the line of sight by an angle i. Figure 1 shows this geometry.

With the exception of a few cases where the magnetic field is strong enough to cause the Zeeman components of the spectral lines to be resolved, only the effective magnetic field strength B_{eff} (the longitudinal component of the field integrated over the visible surface) can be measured. This is related to the polar field strength H_p by

$$B_{\text{eff}} \propto H_p P_l(\cos \alpha). \qquad 1.$$

Here $P_l(\cos \alpha)$ is the Legendre polynomial appropriate to the magnetic field configuration, and $\cos \alpha$ can easily be derived from Figure 1 to be

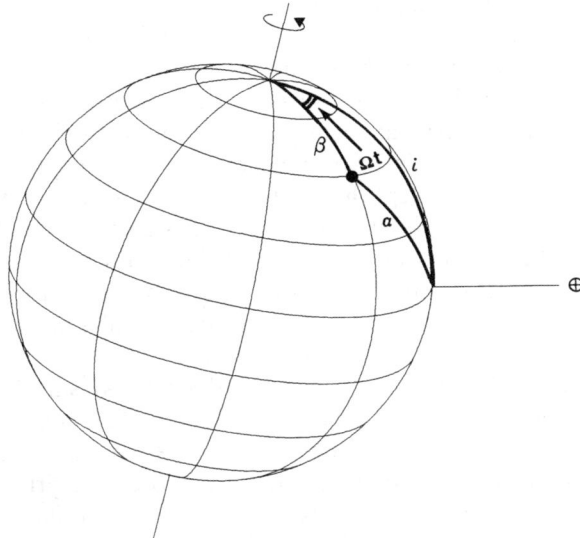

Figure 1 The geometry of an oblique rotator. The rotational inclination is i, the magnetic obliquity is β, and the rotation frequency is $\Omega = 2\pi\nu_{\rm rot}$, where phase zero is taken to be at magnetic maximum.

$$\cos \alpha = \cos i \cos \beta + \sin i \sin \beta \cos \Omega t, \qquad 2.$$

where $\Omega = 2\pi\nu_{\rm rot}$ is the angular rotation frequency, and phase zero is at the time of magnetic maximum. For dipolar magnetic fields $P_l(\cos \alpha) = \cos \alpha$, so that $B_{\rm eff}$ varies with the rotation period of the star about a zero point proportional to $\cos i \cos \beta$ and with an amplitude proportional to $\sin i \sin \beta$.

The oblique rotator model supposes that the abundance anomalies are not uniform over the surface of the star; many elements and ions are overabundant or deficient in spots or rings usually centered on the magnetic poles. This patchy abundance distribution gives rise directly to the observed spectrum variations and, supposedly by flux redistribution due to line blocking (although see Section 7.2), to the rotational light variations. Thus, studies of the magnetic field variations, spectrum variations, and/or rotational light variations yield the rotation period and magnetic phase of an oblique rotator and place constraints on i and β. Rotation periods are observed to range from just under a day to many decades. This is consistent with the observation that almost all Ap stars are slow rotators; in general, the cooler Ap stars have $v \sin i \leq 100$ km s^{-1}, and many of them have $v \sin i < 10$ km s^{-1}.

The origin of the line strength anomalies in the Ap stars is not completely settled. Attempts have been made to explain the line strength anomalies with abnormal model atmospheres, accretion of planetesimals, interior nuclear processes with mixing, surface nuclear processes, magnetic accretion from the interstellar medium, and radiative diffusion. Cowley (39) briefly describes a multiple working hypothesis approach to this problem. Most workers now accept the diffusion hypothesis as the most viable mechanism to explain the abundance anomalies and the patchy surface distribution of the elements, although knowledge of $T_{\rm eff}$, L, $v \sin i$, magnetic field strength, pulsation amplitude (if any), and binary nature are insufficient to predict quantitatively the abundance anomalies in any particular star.

The diffusion model was initially developed for the Ap stars by Michaud (127); more recent reviews are given by Michaud (128) and Alecian (2). The basic hypothesis is that the atmospheres of the Ap stars are sufficiently stable against turbulent mixing that radiation pressure drives outward those elements that have many absorption lines near flux maximum, whereas elements that have few absorption lines near flux maximum or that are cosmically abundant, so that their lines are saturated, sink gravitationally under their own weight in a sea of hydrogen.

This hypothesis (a) can account qualitatively for the observed abundance anomalies, (b) is consistent with the requirement that the abundance anomalies be confined to the surface layers, (c) can account for the patchy abundance distribution by invoking various magnetic field geometries, and (d) explains why Ap stars rotate slowly. Rotation distorts a star's spherical shape, which drives meridional circulation currents which increase in speed with increasing rotational velocity and finally become turbulent at some critical rotational velocity [supposedly $v \sin i \approx 100$ km s^{-1}, although a few Ap Si stars rotate substantially faster (see 181)]. Turbulence inhibits diffusion by mixing. Diffusion is also consistent with the observation that A stars with abnormal abundances generally do not pulsate, a point which is discussed in detail in Section 1.3.

Problems with diffusion are that it has too many free parameters to explain the observations quantitatively, and that the diffusion velocities predicted are in the range 10^{-4}–1 cm s^{-1}. These small diffusion velocities require that in the diffusion zones, the star must be exceedingly stable against turbulence.

1.2 The Range of the Ap Phenomenon

It is now clear that the magnetic Ap phenomenon (meaning magnetic stars which are oblique rotators) extends from effective temperatures of 7400 K (the SrCrEu stars) up to 23,000 K (the He-strong stars) [see (18, 22, 24) for discussions of the various subclasses of Ap stars and their magnetic field strengths]. Hence, the appellation "Ap" is too narrow, but I use it here for traditional reasons. The coolest magnetic "Ap" stars are the SrCrEu stars (which can be as cool as early F). The subjects of this review, the roAp stars, are cool, magnetic oblique rotators of the SrCrEu subclass of Ap stars.

1.3 Metallicism and Pulsation: the δ Scuti Stars

About 30% of the stars in the lower instability strip (where it crosses the main sequence among the A stars) pulsate with amplitudes greater than 0.01 mag (see 27, 181) and with periods between about 30 min and 6 hr. These are the δ Scuti stars, which pulsate in low-overtone radial and nonradial p-modes. None pulsates with an amplitude large enough to saturate the driving mechanism, and it is not known what limits the amplitude (27a, 53).

Observationally, the two most important factors which correlate with pulsation amplitude are rotation and metallicism (27). All rapidly rotating δ Scuti stars have small amplitudes, and all large-amplitude δ Scuti stars are slow rotators. However, most slowly rotating A stars are not δ Scuti stars—they are Am or Ap stars which either do not pulsate or pulsate with very small amplitudes. Breger (25) originally suggested that the Am stars do not pulsate. Later, some evolved Am stars (80), marginal Am stars (81, 88), and finally a classical Am star (94) were found that do pulsate. There are also a few Ap stars that are known to be low-amplitude δ Scuti stars (77, 78, 173, 174). Finally, there is HD 188136.

Wegner (170) found that the rare earth element abundances in HD 188136 are characteristic of the cool magnetic Ap stars. He compared them to those in the most extreme Ap star, HD 101065 (Section 1.4). HD 188136 is also a multiperiodic δ Scuti star which pulsates with peak-to-peak light variations of about 0.05 mag (82). Thus this star shows that extreme abundance peculiarities can coexist with typical, low-overtone, multiperiodic δ Scuti pulsation [see Kurtz (94) for a more extensive discussion of this topic and the terminology].

The importance of all of this lies in the use of the diffusion hypothesis to explain the line strength anomalies of the Am and Ap stars. Pulsation in the δ Scuti stars is primarily driven by the κ-mechanism operating in the He II ionization zone. The general exclusion between the δ Scuti stars and the Am and Ap stars has been taken as strong support for the diffusion hypothesis: It is supposed that in a slowly rotating A star diffusion occurs, helium drains from the He II ionization zone, and a nonpulsating Am or Ap star results. In a rapidly rotating A star, diffusion is inhibited by turbulent meridional circulation, resulting in a chemically normal A star which often (and possibly always) becomes a δ Scuti star. It is this near, but not complete, exclusion between the abnormal abundance A stars and the δ Scuti stars that led to the discovery of the roAp stars.

1.4 The Discovery of the roAp Stars

Prior to 1978 it appeared that Ap stars do not pulsate. They have strong magnetic fields which should inhibit pulsation, and they also have strong abundance peculiarities which, according to the diffusion hypothesis, should be accompanied by He depletion in the He II ionization zone. Even if an Ap star should pulsate, there seemed to be no need to search for periods shorter than 30 min; periods very much shorter than this are only associated with very high overtone pulsation, and in most harmonic oscillators high overtones are only weakly excited and/or heavily damped.

In 1978 Gary Wegner and I were arguing about the temperature of Przybylski's star, which he claimed to be about 7400 K while Przybylski preferred about 6300 K. Przybylski's star, HD 101065, is one of the most peculiar nondegenerate stars in the sky. Its exceedingly complex spectrum is dominated by the lines of the rare earth elements (40, 138–140) and has fueled a long-standing controversy regarding its nature (40, 141–143, 169, 171, 172). Particularly contentious issues have been T_{eff} (estimates range from 6000 to 7500 K) and the question of whether the Fe abundance is normal or not (estimates range from normal to deficient by orders of magnitude). Because of the discovery of qualitatively similar spectra in the known magnetic Ap stars HD 51418 (75) and HR 465 (59) and the discovery of a -2200-G magnetic field by Wolff & Hagen (182), Wegner and I believed HD 101065 to be an extreme Ap star. As such, it seemed most unlikely to be a δ Scuti star.

I tested HD 101065 for light variability in April 1978 using differential photometry with two comparison stars giving a time resolution of about 8 min. The appearance of its light curve seemed to suggest that variations were occurring on a time scale too short to see with the differential photometry, so the next month I observed the star through a Johnson B filter with continuous 20-s integrations. The 12.14-min variations were immediately (and unbelievably) obvious. Figure 2 shows part of a recent light curve of HD 101065 to illustrate how obvious these variations are. In the belief that HD 101065 is an Ap star, I began a search for rapid pulsation in the magnetic Ap stars.

"And thick and fast, they came at last/And more, and more, and more" (46).

1.5 Fundamental Data for the roAp Stars

Fourteen roAp stars are known at the time of this writing. Table 1 lists them and gives their Strömgren photometric indices, and Table 2 describes

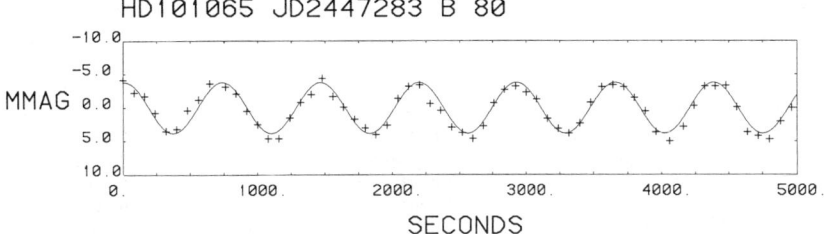

Figure 2 A section of a light curve of HD 101065 obtained with the *SAAO* 1.0-m telescope, showing the principal 12.14-min variation. The data were captured with 10-s integrations, which were coadded to 80-s integrations. Some low-frequency noise has been filtered from the data, but that has not affected the apparent noise level here, which is predominantly caused by scintillation. Variations of this amplitude are obvious at the telescope.

what is known about their magnetic field strengths and gives their spectral types and positions. The magnetic fields in the roAp stars are, in general, not well studied; there is a strong need for magnetic studies of these stars. They are all field stars, and twelve of them have southern declinations. Table 3 gives the rotation frequencies and periods derived from studies of the magnetic field variations, spectrum variations, and/or mean light variations. ("Mean light variation" is used to describe the photometric variability with rotation to distinguish such variability from the pulsational light variability.) Table 3 also lists the principal pulsation frequencies known for each star and gives terse comments about the frequencies and the conclusions derived from them.

Table 1 The roAp stars and their Strömgren indices

HD	HR	Name	V	$b-y$	m_1	c_1	β	δm_1	δc_1	Ref.
6532			8.45	0.084	0.237	0.846		−0.032	−0.050	17, 60, 61
24712	1217		5.99	0.183	0.212	0.634	2.744	−0.029	−0.034	17, 60, 61
60435			8.889	0.132	0.234	0.843		+0.002	+0.035	165
80316			7.808	0.122	0.295	0.657		−0.089	−0.169	165
83368	3831		6.174	0.146	0.203	0.796	2.827	+0.004	−0.035	60, 61, 165
101065		Przbylyski's star	8.004	0.448	0.368	−0.014	2.641	−0.185	−0.386	109
128898	5643	α Cir	3.198	0.152	0.195	0.760	2.831	+0.012	−0.077	60, 61, 165
134214			7.479	0.211	0.288	0.597	2.766	−0.098	−0.115	131, 132
137949		33 Lib	6.674	0.188	0.321	0.584	2.833	−0.114	−0.256	60, 61, 165
166473			7.953	0.213	0.311	0.538		−0.131	−0.101	165
176232	7167	10 Aql	5.89	0.150	0.208	0.829	2.809	−0.004	+0.031	60, 61
201601	8097	γ Equ	4.68	0.147	0.238	0.760	2.819	−0.032	−0.058	17, 60, 61
203932			8.820	0.169	0.196	0.736	2.814	+0.009	−0.072	165,[a]
217522			7.520	0.289	0.215	0.487	2.701	−0.043	−0.046	165,[a]

[a] W. W. Weiss (private communication).

Table 2 Magnetic field strengths, spectral types, right ascensions, and declinations for the roAp stars

HD	B_{eff} (G)	Ref.	Spectral type	Ref.	α (2000)	δ (2000)
6532			Ap SrCrEu	70	01 05 56	−26 44
24712	+400 to +1300	104, 136	A5p, Ap SrCr(Eu)	68, 71	03 55 16	−12 06
60435	$\|B_{eff}\| < 1000$	108, 123	Ap SrEu, Fp SrEu	15, 71	07 31 00	−58 00
80316			Ap SrEu[a]	71	09 18 25	−20 22
83368	−700 to +700	102, 161	Ap SrCrEu[b]	15, 69	09 36 25	−48 45
101065	−2200	182	Controversial	Section 1.4	11 37 37	−46 43
128898	−300 (variable)	21, 22, 183	Ap SrCrEu, Ap SrEu	15, 72	14 42 30	−64 59
134214			Ap SrEu(Cr), F0 SrEu	15, 71	15 09 09	−14 00
137949	+1400 to +1800	180	Ap SrCrEu, Fp SrCrEu	15, 71	15 29 35	−17 26
166473			Ap SrEu	70	18 12 26	−37 45
176232			F0p SrEu	68	18 58 47	+13 54
201601	+500 to −800	7, 20, 147	F0p	68	21 10 21	+10 08
203932			Ap SrEu	15, 70	21 26 04	−29 56
217522			Ap SiCr,[c] Fp SrEu	15, 69	23 01 47	−44 50

[a] Very strong Sr.
[b] Strong Sr.
[c] May be Eu rather than Si.

2. HR 3831 (HD 83368) AND THE OBLIQUE PULSATOR MODEL

HR 3831 is the "best behaved" of the roAp stars: Its frequency spectrum is stable over 5.3 yr and is relatively simple. The frequencies in this star originally led to the oblique pulsator model, which suggests that the roAp stars are pulsating in nonradial modes with the pulsation axis and magnetic axis aligned; thus, HR 3831 is a good object with which to illustrate that model. There are also some intriguing features in the frequency spectrum of HR 3831 that are currently not understood. The frequency analysis of HR 3831 and much of the discussion of the frequencies in this section are a synopsis of work by D. W. Kurtz, H. Shibahashi, and P. R. Goode (in preparation).

2.1 Basic Data for HR 3831

HR 3831 is a southern naked-eye SrCrEu Ap star. From Breger's (26) calibration of the Strömgren indices, it is found to have an effective temperature $T_{eff} = 8200$ K. The luminosity of HR 3831 cannot be estimated reliably using the Strömgren δc_1 index (Section 7.1) but can be determined by reference to its companion. HR 3831 is a visual binary with a separation of 3.1 arcsec and a magnitude difference of $\Delta V = 2.84$. Hurly & Warner (74) obtained UBV photometry for both components using an area scanner. Assuming the secondary is on the main sequence gives $M_V = 2.1$

for HR 3831. For $B-V = 0.25$, one finds that $M_V(\text{ZAMS}) = 2.6$. HR 3831 is thus $\Delta M_V = 0.5$ mag above the ZAMS. Given $R_{\text{ZAMS}} = 1.6\,R_\odot$ at $T_{\text{eff}} = 8200$ K (3), the radius of HR 3831 is then about $R = 2.0\,R_\odot$.

The rotation period of HR 3831 is accurately known from observations of the mean light variations over a time span of 12 yr. Kurtz & Marang (102) give an ephemeris for the time of mean B-light minimum of

$$\text{HJD(mean } B \text{ minimum)} = 2440000.35(\pm 0.02)$$

$$+ 2.851962(\pm 0.000014)\,E. \qquad 3.$$

HR 3831 has two maxima per cycle in the mean light curve, indicating that both magnetic poles are seen.

The magnetic field of HR 3831 was discovered by Thompson (161), who measured its strength nine times over a time span of 12 days. He found a sinusoidal polarity-reversing field. If a function of the form $B_{\text{eff}} = B_0 + B_1 \times \cos[(2\pi/P_{\text{rot}})(t-t_0) + \phi]$ is fitted to Thompson's data, then $B_0 = -22 \pm 28$ G, $B_1 = 737 \pm 84$ G, and $t(B_{\text{eff}} \text{ maximum}) = \text{HJD } 2440000.21 \pm 0.06$. Thus the magnetic ephemeris for HR 3831 is

$$\text{HJD(magnetic maximum)} = 2440000.21(\pm 0.06)$$

$$+ 2.851962(\pm 0.000014)\,E. \qquad 4.$$

Magnetic extrema coincide with mean B-light minimum, which is typical of Ap stars; the abundance spots or rings are centered on the magnetic poles.

Magnetic fields in Ap stars are often characterized by the parameter $r = B_{\text{eff}}(\min)/B_{\text{eff}}(\max)$ $(-1 \leq r \leq +1)$, where (from Equations 1 and 2) $\tan i \tan \beta = (1-r)/(1+r)$. From B_0 and B_1 given above, we have $r = -0.94 \pm 0.16$ for HR 3831. In terms of the oblique rotator model, this indicates that either $i > 80°$ or $\beta > 80°$.

2.2 Rapid Oscillations in HR 3831

I discovered rapid oscillations in HR 3831 in December 1980 (85). The principal period is near 11.67 min; its amplitude varies with rotation from zero to about 10 mmag peak-to-peak through a Johnson B filter. Figure 3 is a portion of the light curve obtained on JD 2446150, when HR 3831 was near amplitude maximum. Figure 4 is the amplitude spectrum of the entire light curve on that night and shows that HR 3831 pulsates nonsinusoidally with a single period; the first and second harmonics of the fundamental are visible. Figure 5 is a portion of the light curve obtained two days before that shown in Figure 3; it demonstrates the rotational amplitude modulation.

Table 3 Frequencies for the roAp stars[a]

HD	ν_{rot} (μHz) P_{rot} (day)	Ref.	ν_{pul} (mHz)	Ref.	Comments
6532	5.9513 1.9448	[b], 100	2.396215 2.402165 2.408120 4.804299	98, 99	$\nu_{1,2,3}$ are equally split by ν_{rot}; dipole mode; $\nu_4 = 2\nu_2$; 1H; $\Delta B = 4.8$ mmag
24712	0.92911 12.4572	101	2.61965 2.65294 2.68760 2.72085 2.75569 2.80568	104–106, 125	$\nu_2, \nu_3, \nu_4, \nu_5, \nu_6$ have rotational sidelobes split by ν_{rot}; $A_{rv} = 200$ m s^{-1}; $\Delta_1 = \nu_2 - \nu_1 = \nu_4 - \nu_3 = 33.27$ μHz; $\Delta_2 = \nu_3 - \nu_2 = \nu_5 - \nu_4 = 34.75$ μHz; ν_2, ν_4 are dipole modes; ν_3, ν_5 are not spherical harmonics; ν_6 is an enigma; $\Delta\nu_0 = 34$ or 68 μHz; $\Delta M_V(\Delta\nu_0) = 0.7$ or 1.9 mag; $M_V(\Delta\nu_0) = 2.3$ or 1.1; tan i tan $\beta = 0.52$; $\Delta B = 10$ mmag
60435	1.5072 7.6793	108	0.9–4.2	120–123	Many frequencies; mode lifetimes less than P_{rot}; $\Delta\nu_0 = 52$ μHz; $\Delta M_V(\Delta\nu_0) = 1.1$ mag; $M_V(\Delta\nu_0) = 1.7$; 1H; 2H; $\Delta B = 16$ mmag
80316	5.6? 2.1?	96, 113	2.2516 2.2460	96	Dipole mode?; tan i tan $\beta = 1.23$? $\Delta B = 2$ mmag
83368	4.05829 2.851962	102	1.42395445 1.42801257 1.43207097 2.84790845 2.85602534 2.86414166 4.28809652	Section 2.2, 85, 107	$\nu_{1,2,3}$ is a dipole mode; tan i tan $\beta = 9.4$; $\nu_{4,5,6}, \nu_7$ unexplained—see Section 2.2; ΔM_V(binary) = 0.5 mag; $M_V = 2.2$; 1H; 2H; $\Delta B = 10$ mmag

HD	ν_{rot}	P_{rot} ref	ν_{pul}	Comments ref	Comments
101065	2.94? 3.94?	115	1.3728660 1.3699260 1.3150340 2.7457355	115, 84, 83, 109	Evidence for small amplitude instability; $\dot{P}$ measured—possible low-mass companion; $\delta\nu = 2.94$ μHz?; 1H; $\Delta\nu_0 = 57.8$ μHz; $\Delta M_V(\Delta\nu_0) = 0.9$ mag; $M_V(\Delta\nu_0) = 2.1$; $\Delta B = 13$ mmag
128898	?	33, 34	2.442041 2.4395	97, 145, 176	$\delta\nu = 2.5$ μHz?; $\pi = 0.056''$; $\Delta M_V(\pi) = 0.7$ mag; $M_V(\pi) = 1.9$; $\Delta B = 4.8$ mmag;
134214			2.94960	79	$\Delta B = 6.4$ mmag
137949	?	33, 34	2.01482 4.02964	c, 6, 85	1H; $\Delta B = 2.8$ mmag
166473			1.892 1.824 1.928	103	$\Delta\nu_0 = 68$ μHz?; $\Delta M_V(\Delta\nu_0) = 0.7$ mag?; $M_V(\Delta\nu_0) = 2.2$; $\Delta B = 2.0$ mmag
176232	?	33, 34	1.435996 1.385373 1.239261	64, 65	$\Delta\nu_0 = 50.6$ μHz?; $\Delta M_V(\Delta\nu_0) = 1.1$ mag?; $M_V(\Delta\nu_0) = 1.6$
201601	?	33, 34	1.339 1.366 1.427	86, 112, 177	$A_{rv} \approx 20$ m s^{-1}; $\Delta\nu_0 = 58$ μHz?; $\Delta M_V(\Delta\nu_0) = 0.9$ mag; $M_V(\Delta\nu_0) = 1.8$; $\pi = 0.028''$; $\Delta M_V(\pi) = 0.8$ mag; $M_V(\pi) = 1.9$; $\Delta B = 2.8$ mmag mode lifetimes short?;
203932			2.8051	d, 93, 89	Other frequencies present but uncertain; $\Delta B = 2.0$ mmag
217522			1.2151	87	Other frequencies inferred; $\Delta B = 4.2$ mmag

[a] Notes: The column marked ν_{rot}/P_{rot} is the rotational frequency in microhertzes (above) and the rotational period in days (below). A question mark in the ν_{rot} column means that some estimates of ν_{rot} are available, but that they are uncertain and/or contradictory. See (33, 34) for a guide to the literature on P_{rot}. A blank entry in the ν_{rot} column means that there is no information. The column marked ν_{pul} lists the pulsation frequencies; by necessity this list is cursory—see the references for error estimates, discussions, and additional frequencies. In the Comments column 1H means that the first harmonic of one of the frequencies is present; 2H means that the second harmonic is also present; $\Delta M_V = M_V - M_V(\mathrm{ZAMS})$; and ΔB is the maximum peak-to-peak amplitude observed in the light variations in the Johnson B filter; A_{rv} is the radial velocity semiamplitude.
[b] D. W. Kurtz, F. Marang & F. van Wyk, in preparation.
[c] D. W. Kurtz, unpublished.
[d] P. Martinez, D. W. Kurtz & C. H. Heller, in preparation.

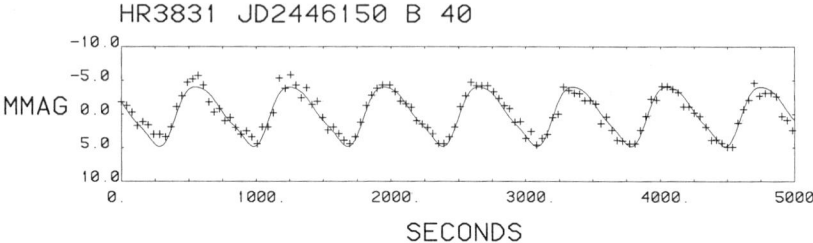

Figure 3 A portion of the light curve of HR 3831 near pulsation maximum (40-s integrations). Note the obvious nonlinear shape of the light curve.

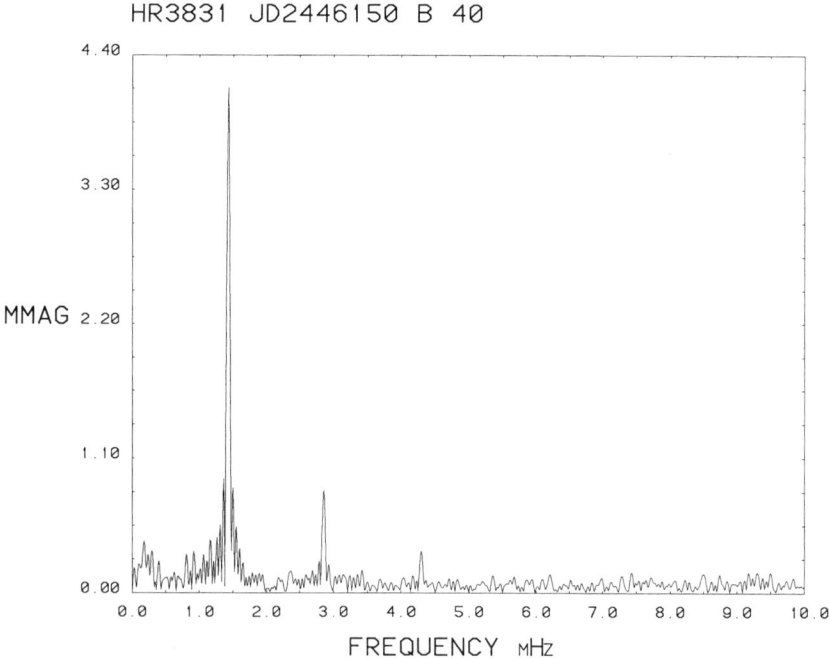

Figure 4 The amplitude spectrum of the entire 5.8-hr light curve obtained on JD 2446150 (a portion of which is shown in Figure 3). The principal frequency and its first two harmonics are clear; there are no other significant frequencies. Note that at higher frequencies, where the noise is scintillation limited, the highest noise peaks are less than 0.2 mmag for this single light curve obtained with the *SAAO* 0.75-m telescope.

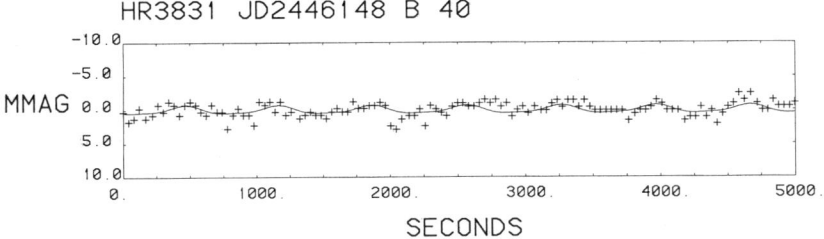

Figure 5 A portion of the light curve obtained two days prior to that shown in Figure 3. This demonstrates the amplitude modulation that occurs in the roAp stars with rotation.

2.2.1 FREQUENCIES I now have 238 hr of high-speed photometric observations of HR 3831 obtained over a time span of 5.3 yr from 1980 to 1986. A frequency analysis of these data gives the seven frequencies listed in Table 4 and shown schematically in Figures 6 and 7. There is a low-frequency triplet, v_{123}, where the frequencies are equally split by $v_{\rm rot}$ (Figure

Table 4 A nonlinear least-squares fit of the seven frequencies v_1–v_7 determined for the entire 1980–86 data set for HR 3831[a]

	Frequency (mHz)	Amplitude (mmag)	Phase (radians)
v_1	$1.42395445 \pm 0.00000002$	$A^{(1)}_{-1} = 2.095 \pm 0.017$	1.166 ± 0.011
v_2	$1.42801257 \pm 0.00000009$	$A^{(1)}_{0} = 0.403 \pm 0.017$	-0.830 ± 0.060
v_3	$1.43207097 \pm 0.00000002$	$A^{(1)}_{+1} = 1.707 \pm 0.017$	1.168 ± 0.014
v_4	$2.84790845 \pm 0.00000019$	$A^{(2)}_{-2} = 0.189 \pm 0.017$	1.147 ± 0.126
v_5	$2.85602534 \pm 0.00000009$	$A^{(2)}_{0} = 0.419 \pm 0.017$	1.046 ± 0.057
v_6	$2.86414166 \pm 0.00000019$	$A^{(2)}_{+2} = 0.188 \pm 0.017$	1.353 ± 0.126
v_7	$4.28809652 \pm 0.00000035$	$A^{(3)}_{+1} = 0.104 \pm 0.017$	1.764 ± 0.230

$\sigma = 1.6150$ mmag
$v_2 - v_1 = 4.05812 \pm 0.00009$ μHz
$v_3 - v_2 = 4.05840 \pm 0.00009$ μHz
$v_5 - v_4 = 8.11689 \pm 0.00021$ μHz $= 2(4.05845 \pm 0.00011$ μHz$)$
$v_6 - v_5 = 8.11632 \pm 0.00021$ μHz $= 2(4.05816 \pm 0.00011$ μHz$)$
$(v_2 - v_1) - (v_3 - v_2) = -0.28 \pm 0.13$ nHz
$[(v_5 - v_4) - (v_6 - v_5)]/2 = 0.29 \pm 0.15$ nHz
$v_5 - 2v_2 = 0.20 \pm 0.20$ nHz
$v_7 - 3v_2 = 4.05881 \pm 0.00044$ μHz
$\langle \Delta v \rangle = [(v_2 - v_1) + (v_3 - v_2) + (v_5 - v_4)/2 + (v_6 - v_5)/2]/4 = 4.05828 \pm 0.00005$ μHz
$v_{\rm rot} = 4.05829 \pm 0.00002$ μHz (102)

[a] These parameters fit the relation $\Delta B = \Sigma A_i \cos[2\pi v_i(t - t_0) + \phi_i]$, where $t_0 =$ JD 2444598.96 ± 0.03.

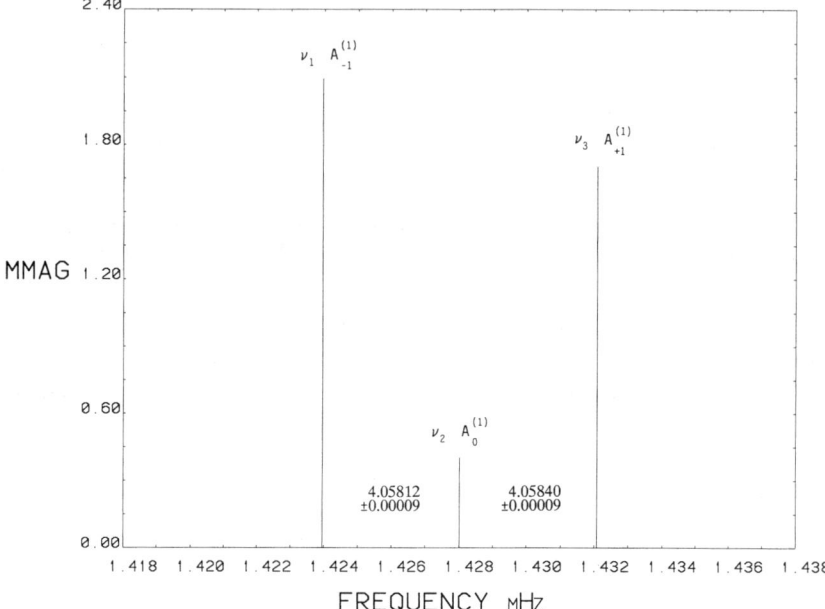

Figure 6 A schematic amplitude spectrum of the low-frequency triplet in HR 3831. This figure and the next graphically show the relative amplitudes and spacings of the frequencies given in Table 4.

6). There is also a high-frequency triplet, v_{456}, where the frequencies are equally split by $2v_{rot}$ (Figure 7). The seventh frequency, v_7, equals $3v_2 + v_{rot}$.

2.2.2 THE TIME OF PULSATION MAXIMUM   A time of pulsational amplitude maximum for HR 3831 is t(amp max) = HJD 2444598.96 ± 0.03. From Equation 4, we have $t(-B_{eff}$ extremum) = HJD 2444599.00. Hence, the pulsational amplitude maximum (which occurs twice per rotation, because v_1 and v_3 are separated by $2v_{rot}$) coincides with the magnetic extremum. This coincidence of pulsational and magnetic maxima is also observed in HR 1217 (Section 3.2).

2.3 *Rotationally Perturbed* m-*modes*

Assuming that the pulsation modes in the roAp stars can be described by spherical harmonics of radial overtone n, spherical degree l, and azimuthal order m, the frequency patterns in Figures 6 and 7 strongly suggest that

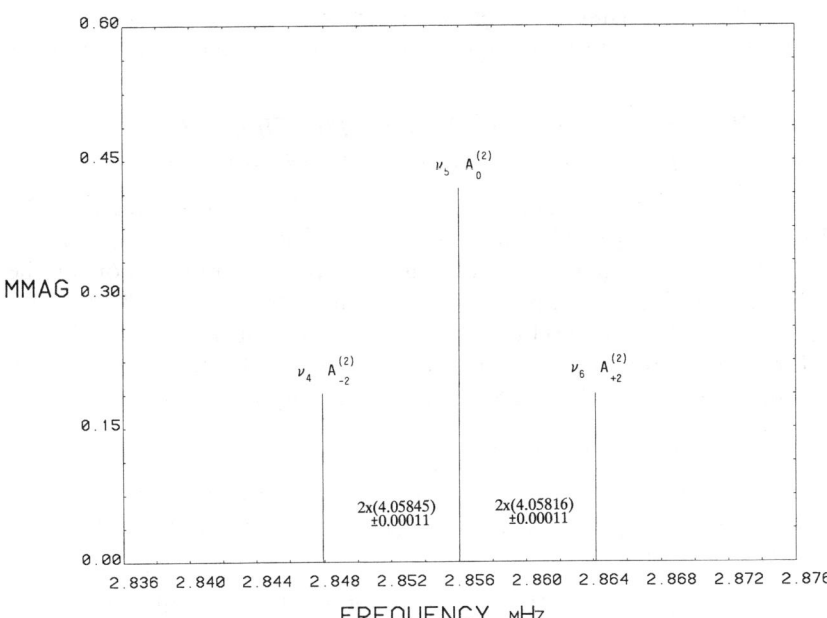

Figure 7 A schematic amplitude spectrum of the high-frequency triplet in HR 3831.

rotationally perturbed m-modes are excited. The frequencies of such modes are (111)

$$\nu_m = \nu_0 + m(1 - C_{nl})\nu_{\rm rot}, \qquad 5.$$

where C_{nl} is a constant that depends on stellar structure (see ref. 43 for a physical interpretation of m-mode splitting). Shibahashi & Saio (151) calculate for A star models that C_{nl} is in the range 10^{-2}–10^{-3} for $10 \leq n \leq 50$ appropriate to the short periods of the roAp stars. If the frequency triplets shown in Figures 6 and 7 result from excited, rotationally perturbed m-modes, then from their separations and the knowledge of $\nu_{\rm rot}$ from the magnetic and mean light measurements, C_{nl} can be calculated from Equation 5. Such a calculation gives $C_{nl} = (0.0 \pm 1.0) \times 10^{-5}$ for HR 3831.

Thus, C_{nl} measured in HR 3831 is at least a factor of 100 smaller than expected from model calculations. Furthermore, if C_{nl} is not exactly zero, then the time of magnetic maximum and the time of pulsation maximum are coincident only by happenstance: They must drift apart on a time scale proportional to $1/C_{nl}$. That we both find these two times equal and find

C_{nl} at least two orders of magnitude smaller than expected argues that $C_{nl} = 0.0$ exactly. That is, it argues that a model of rotationally perturbed m-modes is not correct for v_{123}. A better model is the *oblique pulsator model*.

2.4 The Oblique Pulsator Model: The Simple Case Neglecting Effects of the Magnetic Field and Rotation

The oblique pulsator model assumes that each pulsation mode can be described by a single spherical harmonic with its pulsation axis aligned with the magnetic axis in an oblique rotator. The modulation of the pulsation amplitude is thus naturally explained as an aspect effect analogous to the rotational modulation of the magnetic field strength (85).

If one assumes that the v_{123} triplet in HR 3831 is due to an $l = 1, m = 0$ oblique pulsation mode, this leads to a luminosity variation of

$$\Delta L/L \propto P_1(\cos \alpha) \cos(\omega t + \phi), \qquad 6.$$

where ω is the pulsation frequency, and α is the angle between the pulsation pole (the magnetic pole) and the line of sight. From Figure 1 and Equation 2 we can then see that

$$\Delta L/L \propto A_{-1} \cos[(\omega - \Omega)t + \phi] + A_0 \cos[\omega t + \phi] + A_{+1} \cos[(\omega + \Omega)t + \phi], \quad 7.$$

where

$$A_{\pm 1} = \tfrac{1}{2} \sin i \sin \beta, \qquad 8.$$

$$A_0 = \cos i \cos \beta, \qquad 9.$$

$$2A_{\pm 1}/A_0 = \tan i \tan \beta. \qquad 10.$$

The notation $A_{\pm 1}$ means either A_{+1} or A_{-1}. Since we are considering a dipole pulsation that is amplitude modulated exactly like the dipole magnetic field, it follows that

$$r_{\text{pul}} = (1 - 2A_{\pm 1}/A_0)/(1 + 2A_{\pm 1}/A_0). \qquad 11.$$

From Section 2.1, we have $r_{\text{mag}} = -0.94 \pm 0.16$ for HR 3831, where $r_{\text{mag}} = B_{\text{eff}}(\min)/B_{\text{eff}}(\max)$. Taking $A_{\pm 1} = (A_{+1} + A_{-1})/2$ and using the data in Table 4, we calculate from Equation 11 that $r_{\text{pul}} = -0.81 \pm 0.05$, which agrees with r_{mag} within 1σ. Thus, to first order, v_{123} is due to an axisymmetric dipole mode with the pulsation and magnetic axes aligned. This conclusion is supported by the frequency separations, which are equal to v_{rot}, and it is also supported by the relative amplitudes of $A_{\pm 1}$ to A_0, as shown by $r_{\text{pul}} = r_{\text{mag}}$.

This argument has so far discussed only the amplitudes of v_{123} and not their phases. Note that if the oblique pulsator model is correct and the

v_{123} triplet is due to a single oblique dipole mode, then, since we see polarity reversal of the magnetic field at quadrature, we should also see a phase reversal of π radians in the pulsation phase of v_{123} at quadrature—i.e. the phase ϕ of the oscillation described by Equation 6 will be constant except for a reversal of π radians when $P_1 (\cos \alpha)$ reverses sign.

To test this conclusion, v_2 was fitted to sections of the data that are two cycles (23.34 min) long by the least-squares method. Figure 8 shows the phases determined from these fits, plotted as a function of magnetic phase for the entire 1937-day (5.3-yr) time span of the data. Magnetic phase zero was taken from the ephemeris in Equation 4 to be at magnetic maximum. When one pulsation pole is facing us, we see a constant pulsation phase; when the other pole faces us, the phase is reversed by π radians. The reversals occur at magnetic quadrature when $\cos \alpha = 0$ (Equation 2); therefore, we have $\cos \Phi_{quad} = -1/(\tan i \tan \beta) = -A_0/2A_{\pm 1}$, and hence $\Phi_{quad} = 0.23$ and 0.77. These Φ_{quad} phases are marked by vertical lines in Figure 8; the horizontal lines are separated by π radians. This is clearly

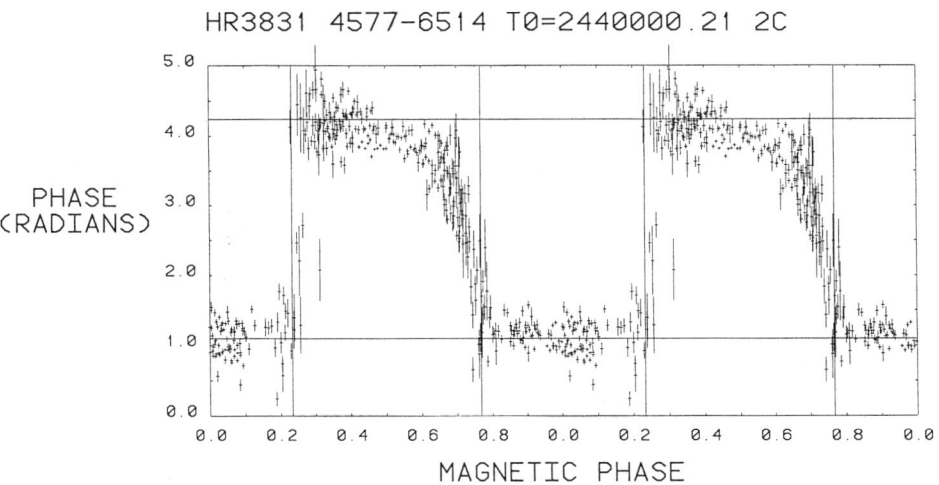

Figure 8 A phase diagram for HR 3831, showing the pulsation phase ϕ vs. the magnetic (rotation) phase Φ. The pulsation phase was obtained by fitting v_2, the central frequency of v_{123}, to two-cycle (23.34-min) lengths of data by the least-squares method. The magnetic phase was calculated from Equation 4 independently. The horizontal lines are separated by π radians. The vertical lines are the calculated times of magnetic quadrature; these are very near phases 0.25 and 0.75 because $i > 80°$. This diagram clearly shows the phase reversal characteristic of an oblique dipole pulsator. The ϕ error bars are $\pm 1\sigma$; they are inversely dependent on amplitude, and hence amplitude can approximately be seen in this diagram also. Note the low errors (high amplitude) near the magnetic extrema and the large errors near quadrature, where the amplitude goes through zero.

the signature of a dipole oblique pulsator. The data span 679 rotation cycles. Magnetic polarity is positive for $0.77 < \Phi < 0.23$ and negative for $0.23 < \Phi < 0.77$.

2.5 A Dipole Mode in HD 6532

HD 6532 is another roAp star that shows the clear signature of an oblique dipole pulsation mode (98). It has a frequency triplet with spacings equal to the rotation frequency. Figure 9 shows the phase diagram for HD 6532, where the magnetic phase has been assumed to be equal to the rotation phase determined from the mean light variations. Kurtz & Cropper (98) found that $\tan i \tan \beta = 2.14$; this implies that $\Phi_{quad} = 0.33$ and 0.67, and these phases are marked by vertical lines in Figure 9. This diagram predicts that HD 6532 has a polarity-reversing magnetic field with $r_{mag} = -0.36$, which is consistent with the double-wave nature of the mean light curve (100). No magnetic field measurements of HD 6532 have been published.

2.6 Oblique Pulsation and the RR Lyrae Stars

Cousens (38) proposed the *obliquely oscillating magnetic rotator model* (oblique pulsator model) as a possible explanation of the Blazhko effect

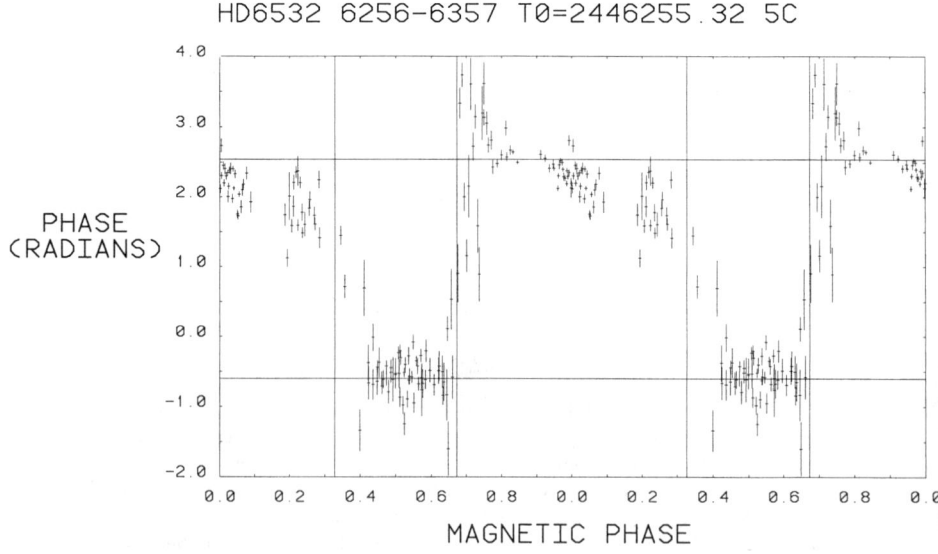

Figure 9 The phase diagram for HD 6532, showing the phase reversal characteristic of an oblique dipole pulsator. The geometry of this star is such that one pole is seen for a larger fraction of the rotation cycle than the other. That pole comes closer to the line of sight than the other pole, and hence the phase errors are smaller (amplitudes higher) for $0.67 \leq \Phi \leq 0.33$ than for $0.33 \leq \Phi \leq 0.67$.

(the long-term periodic modulation of amplitude) in some RR Lyrae stars. Although the Blazhko effect has been known since the first decade of this century, it is still unexplained (see 159). Moskalik (129), for example, has recently suggested that it may be caused by resonances in double-mode pulsators. This problem is mentioned to point out that the oblique pulsator model may have applications beyond the magnetic Ap stars.

2.7 Two Difficulties With the Simple Oblique Pulsator Model

Dolez & Gough (47) pointed out that an axisymmetric dipole mode aligned with the magnetic axis is equivalent to a linear sum of $m = -1, 0,$ and $+1$ modes aligned with the rotation axis. This is a simple consequence of the well-known coordinate rotation relation for spherical harmonics:

$$Y_l^m(\theta_B, \phi_B) = \sum_{m'=-l}^{l} d_{mm'}^{(l)}(\beta) Y_l^m(\theta_R, \phi_R), \qquad 12.$$

where the matrix $d_{mm'}^{(l)}$ can be found in Edmonds (54) [see Kurtz & Shibahashi (107) for an explicit form of $d_{mm}^{(l)}$ for $l = 1$ and Kurtz et al (104) for an explicit form for $l = 2$]. The coordinates (θ_B, ϕ_B) refer to the magnetic axis; (θ_R, ϕ_R) refer to the rotation axis.

The simple consequence of Equation 12 is that the oblique dipole mode aligned with the magnetic axis, which described the data so well in Sections 2.4 and 2.5, looks just like a linear sum of m-modes aligned with the rotation axis. But the latter should obey Equation 5, and hence the pulsation pattern should drift with respect to the magnetic field—i.e. the frequency splitting should equal $(1 - C_{nl})v_{\rm rot}$ with $C_{nl} \neq 0$. That the pulsation maximum and the magnetic maximum coincide in both HR 3831 (Section 2.4) and HR 1217 (Section 3.2) argues that C_{nl} does equal zero.

Dolez & Gough (47) suggested that the growth times for the observed high-overtone modes are short with respect to the drift time between the pulsation and magnetic poles, and thus only modes with pulsation axes currently aligned with the magnetic axis are excited to observable amplitudes. If this were true, however, then as some modes drifted into phase with the magnetic field and were excited while others drifted out of alignment and were damped, phase coherence would not be maintained. Figure 8 demonstrates that phase coherence *is* maintained over 679 rotation cycles in HR 3831. Figure 9 shows that it is maintained over 57 rotation cycles in HD 6532. Hence, the pulsation and magnetic axes are locked in these stars.

The second problem is that $A_{+1} \neq A_{-1}$, as expected in Equation 8. This same problem is reflected in Figure 8, where although the basic pattern is that of π-radian phase reversal, there is a clear deviation from that pattern, especially between magnetic phases 0.23 and 0.77, where the pulsation phase ϕ falls off instead of remaining constant. It is also related to the inequality of the phase of v_2 with those of v_1 and v_3 (Table 4) at the time of pulsation maximum; all three should have the same phase at that time according to Equation 7.

2.8 A Generalized Oblique Pulsator Model

Following Dziembowski & Goode (50) and Gough & Taylor (58), Dziembowski & Goode (51, 52) developed a generalization of the oblique pulsator model in which the effects of both the magnetic field and rotation were taken into account. They assumed that the perturbation to the eigenfrequencies by the magnetic field dominates perturbation due to rotation and showed that this leads naturally to the expectation that the pulsation axis should be rigidly locked to the magnetic axis. Including the effects of the Coriolis force and Lorentz force, which were neglected in the simple oblique pulsator model, explains the inequality $A_{+1} \neq A_{-1}$. The difference in these amplitudes leads to a measure of the magnetic field strength integrated over the volume of the star. An application of the generalized oblique pulsator model to HR 3831 is now given. Derivations of relations presented and discussion of the formalism are given by Shibahashi (149) and Kurtz & Shibahashi (107).

It is assumed that the eigenfunctions in the presence of an axisymmetric magnetic field can be approximately described by the spherical harmonics $Y_l^m(\theta_B, \phi_B)$, where $\theta_B = 0, \pi$ are the magnetic poles. Since the magnetic field is symmetric with respect to prograde and retrograde m-modes, the perturbation to the eigenfrequencies depends on $|m|$ and is given by $\omega_{|m|}^{(1)\text{mag}}$. The perturbed frequencies are, therefore

$$\omega = \omega^{(0)} + \omega_{|m|}^{(1)\text{mag}}. \qquad 13.$$

It is assumed that the magnetic field dominates the Coriolis force, i.e. $\omega_{|m|}^{(1)\text{mag}} \gg C_{nl}\Omega$. (This assumption locks the pulsation axis to the magnetic axis; it is justified below.) The observed luminosity variation for a single mode is then (107, 149)

$$\frac{\Delta L}{L} \propto \sum_{m'=-l}^{l} (-1)^{m'} \left\{ d_{mm'}^{(l)}(\beta) + C_{nl}\Omega \sum_{k=-l}^{l}{}'' \frac{d_{km'}^{(l)}(\beta)}{\omega_{|m|}^{(1)\text{mag}} - \omega_{|k|}^{(1)\text{mag}}} \right.$$
$$\left. \times \left[\sum_{p=-l}^{l} p d_{kp}^{(l)}(\beta) d_{mp}^{(l)}(\beta) \right] \right\} d_{m'0}^{(l)}(i) \cos \{[\omega^{(0)} + \omega_{|m|}^{(1)\text{mag}}$$
$$+ m C_{nl}\Omega \cos(\beta) - m'\Omega] t + \phi\}, \qquad 14.$$

where Σ'' means summation over k except for $k = \pm m'$.

2.8.1 APPLICATION TO THE LOW-FREQUENCY TRIPLET IN HR 3831   For an $l = 1$, $m = 0$ mode, Equation 14 becomes

$$\frac{\Delta L}{L} \propto \frac{1}{2}\sin i \sin \beta \left(1 - \frac{C_{nl}\Omega}{\omega_1^{(1)\mathrm{mag}} - \omega_0^{(1)\mathrm{mag}}}\right) \cos[(\omega^{(0)} + \omega_0^{(1)\mathrm{mag}} - \Omega)t + \phi]$$

$$+ \cos i \cos \beta \cos[(\omega^{(0)} + \omega_0^{(1)\mathrm{mag}})t + \phi]$$

$$+ \frac{1}{2}\sin i \sin \beta \left(1 + \frac{C_{nl}\Omega}{\omega_1^{(1)\mathrm{mag}} - \omega_0^{(1)\mathrm{mag}}}\right) \cos[(\omega^{(0)} + \omega_0^{(1)\mathrm{mag}} + \Omega)t + \phi]. \quad 15.$$

Note that when the Coriolis force is neglected ($C_{nl} = 0$) and the magnetic field is neglected ($\omega_{|m|}^{(1)\mathrm{mag}} = 0$), then Equation 15 reduces to the simple case in Equation 7.

Now, however, $A_{+1}^{(1)} \ne A_{-1}^{(1)}$, so

$$(A_{+1}^{(1)} + A_{-1}^{(1)})/A_0^{(1)} = \tan i \tan \beta \quad 16.$$

and

$$(A_{+1}^{(1)} - A_{-1}^{(1)})/(A_{+1}^{(1)} + A_{-1}^{(1)}) = C_{nl}\Omega/(\omega_1^{(1)\mathrm{mag}} - \omega_0^{(1)\mathrm{mag}}). \quad 17.$$

Note that Equation 16 is equivalent to averaging $A_{+1}^{(1)}$ and $A_{-1}^{(1)}$ and applying Equation 10, as was done in Section 2.4, so the constraint on i and β remains the same as previously derived. From Equation 17 and the data in Table 4 we find that $C_{nl}\Omega/(\omega_1^{(1)\mathrm{mag}} - \omega_0^{(1)\mathrm{mag}}) = -0.102 \pm 0.006$, a value that justifies the assumption that $\omega_{|m|}^{(1)\mathrm{mag}} \gg C_{nl}\Omega$. Futher justification comes from the dipole mode in HD 6532, where $C_{nl}\Omega/(\omega_1^{(1)\mathrm{mag}} - \omega_0^{(1)\mathrm{mag}}) = -0.25 \pm 0.02$ (98).

2.8.2 COMPLICATIONS   The inequality of the phases of v_{123} at the time of magnetic maximum is still not explained; also, in independent data sets the phase diagram shown in Figure 8 shows slight differences that appear to be real. This probably means either that v_{123} in HR 3831 is not simply an oblique dipole mode but rather has small contributions from other modes which are not yet resolved, or that our assumption that each pulsation mode can be described by a single spherical harmonic is not quite correct.

In the first case, if $m = \pm 1$ modes are excited at low amplitude, then the frequency spacing of v_{123} should not be exactly equal. This could lead to small changes in the phase diagram over the 5.3-yr time span of the data. It is not likely that a small-amplitude ($n-1, l = 3$) mode is present because the time span of the data is long enough that modes of ($n, l = 1$) and ($n-1, l = 3$) should be resolved (see Section 3.1).

2.9 Measuring the Internal Magnetic Field Strength

For a dipole magnetic field, Dziembowski & Goode (51) give

$$\omega_{|m|}^{(1)\mathrm{mag}} \propto \frac{l(l+1)-3m^2}{4l(l+1)-3} K^{\mathrm{mag}},\qquad 18.$$

where K^{mag} follows from a $|Y_l^m|^2$-weighted integration over the P_2 distortion of the matter caused by the field. The ratio of the magnetic pressure to the gas pressure is $K^{\mathrm{mag}}/\omega^{(0)}$, where $\omega^{(0)}$ is the unperturbed pulsation frequency. Combining Equations 17 and 18 gives (for $l=1$, $m=0$)

$$K^{\mathrm{mag}} = \frac{-5C_{nl}\Omega}{3}\frac{A_{+1}^{(1)}+A_{-1}^{(1)}}{A_{+1}^{(1)}-A_{-1}^{(1)}}.\qquad 19.$$

Assuming that $C_{nl} = 0.01$, then from the data in Table 4 for HR 3831, we have $K^{\mathrm{mag}} = 4.2 \pm 0.3$ μHz, and for HD 6532 (98) we have $K^{\mathrm{mag}} = 2.5 \pm 0.2$ μHz. This implies that the internal magnetic field strength of HR 3831 is greater than that of HD 6532. It is not known whether the internal magnetic field strengths have some simple relation to the measured surface fields; study of the roAp stars may be able to answer this question. If there is a simple relation, then when the magnetic field in HD 6532 is measured, it should be found to be weaker than that in HR 3831. If measurements of K^{mag} for many modes of different l, m can be made in a single star, then in principle the internal field structure can be modeled.

2.10 The High-Frequency Triplet in HR 3831

Interpretation of v_{456} in HR 3831 is an interesting problem. To very high accuracy, one finds that $v_4 = 2v_1$, $v_5 = 2v_2$, and $v_6 = 2v_3$. In spite of this, Kurtz (85) suggested that v_{456} is not the first harmonic of v_{123} but rather is due to pulsation in an oblique quadrupole ($l=2$, $m=0$) mode which is driven by a small, undetectable first harmonic of v_{123}, hence forcing the exact 2:1 ratio between the frequencies.

The constraint provided by the dipole interpretation of v_{123}—i.e. $(A_{+1}+A_{-1})/A_0 \equiv x = 9.4 \pm 0.4$—along with the additional constraint provided by the amplitude modulation of the supposed quadrupole mode require that $i = 86°$, $\beta = 36°$, or vice versa (85). In Section 2.1 the radius of HR 3831 was found to be $R = 2.0\ R_\odot$. For $P_{\mathrm{rot}} = 2.851962$ days this gives an equatorial rotational velocity of $v = 36$ km s^{-1}. Carney & Peterson (32) measured $v \sin i = 33 \pm 3$ km s^{-1} for HR 3831, a value requiring that $i \approx 90°$; this was taken to support the quadrupole interpretation of v_{456} which gave $i = 86°$.

H. Shibahashi (private communication) pointed out that the quadrupole interpretation of v_{456} is not correct. D. W. Kurtz, H. Shibahashi, and

P. R. Goode (in preparation) have considered this problem and find the following. The three frequencies v_{456} are in phase at magnetic maximum, so their amplitudes sum to a maximum at that time. At quadrature, $A^{(2)}_{+2}$ and $A^{(2)}_{-2}$ are again in phase with each other but are π radians out of phase with $A^{(2)}_0$. Since $A^{(2)}_{+2} + A^{(2)}_{-2} = A^{(2)}_0$, this means that the amplitude of v_{456} is zero at quadrature. An $m = 0$ quadrupole has a maximum at quadrature, so v_{456} cannot be due to an $m = 0$ quadrupole mode. This is the behavior of a mode that is modulated by $\cos^2 \alpha$: It is zero at quadrature and maximum at both magnetic extrema, but without the phase reversal of the dipole ($\cos \alpha$-modulated) mode. Therefore, the only constraint on i and β for HR 3831 comes from v_{123} and Equation 16. As we have previously seen, this requires that i or $\beta > 80°$; Carney & Peterson's (32) measurement of $v \sin i$ removes the ambiguity: It is $i > 80°$.

The 2:1 ratio between the frequencies v_{456} and v_{123} strongly suggests that v_{456} is the first harmonic of v_{123}. Since v_{123} is a dipole mode with a surface distortion proportional to $\cos \theta$, where θ is the colatitude, our theoretical expectation is that the first harmonic should show a surface distortion of the form $\cos^2 \theta$, since the harmonic arises from the second-order terms in the wave equation. A $\cos^2 \theta$ distortion is not observed to have zero apparent amplitude at quadrature; hence, our theoretical expectation and the observations are in clear disagreement. The reason for this is not known.

2.11 The Second Harmonic Frequency v_7

The second harmonic frequency lies at $3v_2 + v_{\text{rot}}$ (Table 4). It should arise from third-order terms and hence, for a dipole fundamental oscillation, should have a surface distortion proportional to $\cos^3 \theta$. This can be written as a linear sum of $P_1(\cos \theta)$ and $P_3(\cos \theta)$, but the P_3 term disappears as seen from the observer's viewpoint when the limb-darkening $I(\theta) = 1$; otherwise, the P_3 term contributes very little, so the second harmonic should look like a dipole. This means that the amplitude of $A^{(3)}_{-1}$ should be greater than that of $A^{(3)}_{+1}$, as is the case for v_{123}. It is therefore a surprise that it is the $A^{(3)}_{+1}$ component that is observed and not the $A^{(3)}_{-1}$ component. The reason for this is also not known.

2.12 Conclusions About HR 3831 and the Oblique Pulsator Model

HR 3831 is pulsating in an ($l = 1$, $m = 0$) dipole mode in which the pulsation axis and magnetic axis are aligned. The pulsation amplitude modulation and the magnetic field strength modulation are in phase and have the same form; they require the same i, β geometry (tan i tan β = 9.4 ± 0.4), so either i or $\beta > 80°$. The observed value of $v \sin i$ removes this ambiguity, requiring that $i > 80°$. From photometry of its companion, HR 3831 lies 0.5 mag above the ZAMS and has a radius of $R = 2.0\ R_\odot$. Pulsation maximum, magnetic extremum, and mean light extremum coincide.

The oblique pulsator model accounts for the observed splitting of the pulsation frequencies, which is equal to v_{rot} (or $2v_{rot}$) to high accuracy. It accounts for the relative amplitudes of v_{123} and the phases of v_1 and v_3; the phase of v_2 is still unexplained. The first harmonic frequencies, v_{456}, and second harmonic frequency, v_7, are not yet understood.

3. HR 1217 (HD 24712) AND ASTEROSEISMOLOGY

HR 1217 is one of the coolest of the Ap SrCrEu stars. It has sinusoidal magnetic field variations from $+400$ to $+1300$ G, mean light variations, and spectrum variations all with the rotation period of 12.4572 days. It also pulsates in at least six modes with periods near 6 min and has an amplitude spectrum reminiscent of the solar 5-min p-mode spectrum. Kurtz et al (104) have recently discussed this star in great detail, so only a synopsis of their results and some background material are presented here.

3.1 Asteroseismology

The asymptotic relation for high-overtone p-modes which applies to the Sun and roAp stars is (36, 160)

$$v_{nl} \approx \Delta v_0(n+l/2+\varepsilon)+\delta v, \qquad 20.$$

where ε is a constant which depends on stellar structure. The parameter $\Delta v_0 = [2\int_0^R dr/c]^{-1}$, where c is the sound speed, is the return sound travel time from the surface to the core. Because higher degree modes do not penetrate as deeply into the star as lower degree modes, the second term, δv, lifts the degeneracy in the first term for modes of (n,l), $(n-1,l+2)$, etc. Thus, measurement of both Δv_0 and δv can in principle allow the determination of stellar mass and age (36, 163, 164) by constraining the interior density distribution.

3.2 HR 1217

Kurtz et al (104) obtained 365 hr of observations of HR 1217 at eight observatories during three months in 1986. They found that the 6-min light variations are amplitude modulated in phase with the magnetic variations, with amplitude maximum and magnetic maximum coinciding (as in HR 3831), but with mean light minimum occurring at a slightly (but probably significantly) different time. A frequency analysis of 324 hr of these observations over a 46-day time-span yielded six principal frequencies, all of which are amplitude modulated with the rotation of the star. Figure 10 shows a schematic amplitude spectrum, which bears a striking resemblance to the solar p-mode spectrum. Compare it, for example, with the power spectrum of the solar whole-disk radial velocity variations in Figure 1 of Gelly et al (56) or in Figure 5 reproduced by Deubner & Gough (45), or with the power spectrum of the solar total irradiance variations shown in Figure 1 of Woodard & Hudson (184).

From Figure 10 and other data, Kurtz et al (104) made the following conclusions:

1. HR 1217 is an oblique rotator with a centered dipole magnetic field. The geometry of the field requires that $\tan i \tan \beta = 0.52 \pm 0.03$.

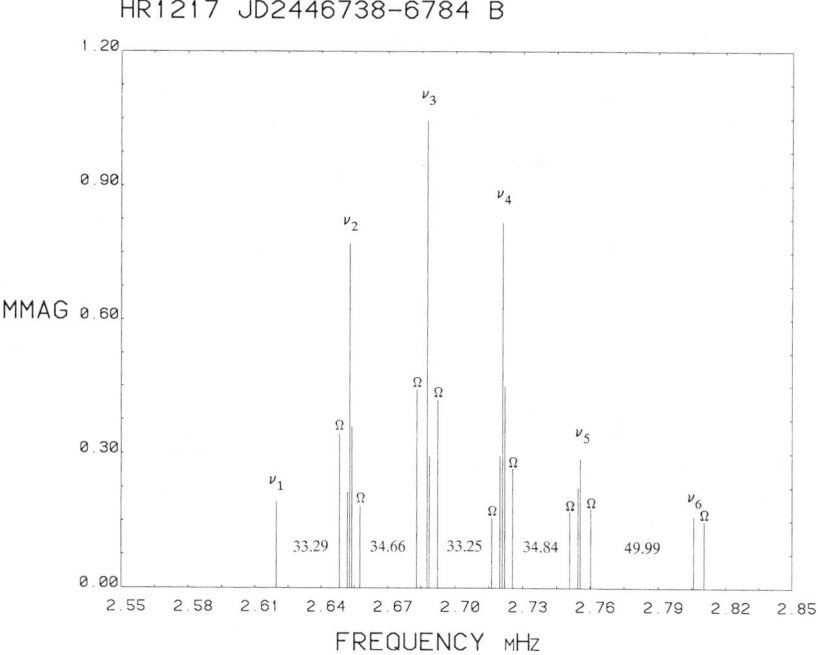

Figure 10 A schematic amplitude spectrum for HR 1217, showing its striking similarity to the solar low-l p-mode amplitude spectrum. The spacing of the principal frequencies is to scale; the separations are in microhertz. The rotational sidelobes are marked Ω; their separations and the separations of the secondary frequencies from the principal frequencies have been exaggerated for clarity (from 104).

2. HR 1217 is an oblique pulsator with the pulsation axis and the magnetic axis aligned. This is required by the coincidence of the times of amplitude maximum and magnetic maximum and by the phases of the frequency triplets.
3. There are six principal frequencies of pulsation in HR 1217; v_2 and v_4 are dipole modes ($l = 1$, $m = 0$); v_3 and v_5 cannot be described by single spherical harmonics—they look similar to dipole modes, but they have amplitude minima which are lower than expected.
4. The frequencies v_1–v_5 have an alternating frequency spacing: $\Delta_1 = v_2 - v_1 = v_4 - v_3 = 33.27 \pm 0.08$ μHz; $\Delta_2 = v_2 - v_3 = v_5 - v_4 = 34.75 \pm 0.08$ μHz.
5. It is not possible to discriminate between the hypothesis that v_1, v_2, v_3, v_4, and v_5 are basically due to alternating even and odd l-modes with $\Delta v_0 = 68$ μHz, and the hypothesis that they are all basically due to dipole modes with $\Delta v_0 = 34$ μHz (see Equation 20). The hypothesis of alternating even and odd l-modes seems incorrect because the amplitude ratios of the rotational sidelobes to the central frequencies are incorrect for l = even, although theoretically the appearance of modes with $l \neq 1$ is expected to be altered by the magnetic field (16). The hypothesis of $l = 1$ for all modes seems incorrect because the alternating Δ_1, Δ_2 frequency spacing is then difficult to explain and seems more consistent with alternating even and odd l-modes.
6. If $\Delta v_0 = 68$ μHz, then HR 1217 lies 0.7 mag above the ZAMS and has a radius of $R = 1.9\ R_\odot$; if $\Delta v_0 = 34$ μHz, then HR 1217 lies 1.9 mag above the ZAMS and has a radius of $R = 3.4\ R_\odot$ (assuming that $R_{ZAMS} = 1.4\ R_\odot$ at F0).
7. The sixth principal frequency, v_6, could be an $l = 2$ mode if $\Delta v_0 = 34$ μHz, but then the alternating Δ_1, Δ_2 frequency spacing is difficult to explain. If $\Delta v_0 = 68$ μHz, then the Δ_1, Δ_2 spacing makes sense, but the $v_6 - v_5 = 1.503\Delta_1$ spacing is inexplicable; v_6 is an enigma.
8. There are secondary frequencies which indicate that the principal frequencies are amplitude modulated on a time scale of months. This is not likely to be due to pulsation in higher degree modes (which would produce amplitude modulation on a much shorter time scale), and hence it indicates that the amplitudes of the principal frequencies are not completely stable.

The study of HR 1217 by Kurtz et al (104) demonstrates how much information can be gleaned from an intensive study of the roAp stars. It also demonstrates how little we understand of their pulsation, and it shows the large observing effort needed to overcome alias problems in their frequency analysis.

3.3 Radial Velocity Variations in HR 1217

Matthews et al (125) discovered radial velocity variations in HR 1217. They found a peak-to-peak radial velocity amplitude of $2K = 400 \pm 50$ m s^{-1} on one night when the peak-to-peak light variation in B was 6.8 mmag. On the previous night the light variation was only 1.8 mmag, and they found an upper limit to the radial velocity variations of 130 m s^{-1}. Thus, they demonstrated conclusively that the light variations in HR 1217 are due to pulsation. The amplitude modulation of the radial velocity also supports the oblique pulsator model in preference to another, the spotted pulsator model, which we now discuss.

4. THE SPOTTED PULSATOR MODEL

The Ap stars have mean light variations because of inhomogeneous abundance distributions. If the roAp stars pulsated in nonradial modes with the pulsation axes and rotation axes aligned, we would still see some amplitude modulation due to the changing background mean light caused by the spots. However, the mean light variations are typically less than 20% in Ap stars; in HR 1217 they are 6% peak-to-peak in B, and in HR 3831 they are 1% peak-to-peak in B. Hence, if we consider only the mean light variations, the maximum modulation of the light amplitude of pulsation which could occur would be equivalent to those percentages and thus would not be comparable to the amplitude modulations actually seen. Kurtz (85) rejected this model by this argument.

Mathys (116) proposed and developed the formalism for the *spotted pulsator model*. He pointed out that the surface inhomogeneities may cause more than the mean light variations; they may also cause an inhomogeneous distribution in the ratio of the flux to radius variation, f, and in the phase lag between the flux and radius variations, ψ [see (13, 14, 168) for a discussion of these parameters]. By allowing f and ψ to vary over the surface within reasonable limits, Mathys showed that the spotted pulsator model can describe the frequency spectrum of HR 3831 with a single dipole mode aligned with the rotation axis. However, it is now possible to discriminate between the spotted pulsator model and the oblique pulsator model—in favor of the latter.

The spotted pulsator model predicts that the radial velocity amplitude of a single pulsation mode will not be modulated, but that the light amplitude will be modulated. The oblique pulsator model predicts that both will be modulated in the same way with changing aspect. The observation of the modulation of the radial velocity amplitude in HR 1217 by Matthews et al (125) thus favors the oblique pulsator model. This is not unequivocal evidence, however. The radial velocity amplitude in the spotted pulsator model *will* be modulated by beating between multiple frequencies, and HR 1217 has at least six pulsation frequencies. Nevertheless, Matthews et al found the greater radial velocity amplitude at magnetic maximum, which is consistent with the oblique pulsator model, and they observed for 5 hr on each of two nights, which should have averaged over many of the beat cycles of the six frequencies.

A second argument favoring the oblique pulsator model is that the time of pulsation maximum is consistent with the time of magnetic maximum in HR 1217 but differs from the time of mean light minimum by 3σ. In the spotted pulsator model it is the abundance distribution which is supposed to give rise to the variations in f and ψ, so there the pulsation maximum should coincide with mean light extremum.

The spotted pulsator model still applies, at least partially, to the roAp stars because of the effect of the mean light variations on the pulsation amplitude, an effect that has not yet been quantitatively examined for any star. In HR 1217 the modes of largest amplitude have $A_B \approx 1$ mmag; with a mean B-light variation of 6%, this means that the rotational sidelobes will have about a 0.06-mmag contribution due to this effect, which is only slightly less than the noise level in the amplitude spectrum of the current observations. Models of future, more accurate observations will have to account for the effect of the spots.

5. HD 60435 AND THE LIFETIMES OF THE PULSATION MODES IN THE roAp STARS

HD 60435 has the most complex frequency spectrum of any of the roAp stars. It pulsates with many frequencies which have a basic separation of 25.8 μHz and appear to be alternating even and odd l-modes, but the lifetimes of at least some of the modes seem to be as short as a few days (120–123).

A rotation period of 7.6793 ± 0.0006 days has been determined for HD 60435 from its mean light variations (108). The mean light curves show a double wave of unequal depth, so both magnetic poles are seen (probably from different aspects), indicating that neither i nor β are near $90°$. The time of mean light extremum coincides with pulsation maximum, indicating that HD 60435 is an oblique pulsator. Although HD 60435 is only a mild spectrum variable, a study of the line strengths made simultaneously with high-speed photometric observations suggests that the Sr II maximum coincides with the pulsation maximum (124). Since the Sr II maximum is usually coincident with the magnetic maximum in Ap stars, this also suggests that HD 60435 is an oblique pulsator. Two upper-limit measurements of the magnetic field strength [J. D. Landstreet & D.A. Bohlender—reported by Matthews et al (123)] give $|B_{\text{eff}}| < 1000$ G using the rotation ephemeris of Kurtz et al (108).

Amplitude modulation of the pulsation frequencies over one rotation cycle can be seen in many of the data sets, but during the next rotation cycle, the same frequencies may not be present. If they do return, they do so at their original frequencies. This behavior is qualitatively similar to the Sun's; it indicates that HD 60435 is pulsating in many high-overtone p-modes with lifetimes that can be shorter than one rotation cycle. This conclusion is based on over 400 hr of high-speed photometric observations, many of which were obtained contemporaneously from Chile and South Africa; frequency resolution and aliases do not confuse the analysis.

Other roAp stars also show evidence of mode lifetimes on various time scales. Libbrecht (112) suggested from his radial velocity measurements and from the photometry of Kurtz (86) that the lifetimes of the pulsation modes in γ Equ may also be as short as those in HD 60435, but γ Equ is not a well-studied roAp star. HR 1217 (Section 3) has six pulsation frequencies that have constant amplitudes over one rotation period of 12.4572 days, but that show variable amplitudes on a time scale of four rotation periods (104). HD 101065 has one large-amplitude (about 10 mmag peak-to-peak in B) pulsation mode and at least one much smaller amplitude pulsation mode (115); the large-amplitude mode shows evidence of some variation in amplitude over the 10 yr it has been observed, but on a shorter time scale it appears stable. In HR 3831 (Section 2) the principal dipole oscillation is stable with a lifetime on the order of the 5.3 yr for which it has been observed. This is shown by Figure 8, although slight changes in that phase diagram from year to year may indicate very small changes in the amplitude of the principal oscillation. HR 3831 is singly periodic (with harmonics).

There is a qualitative pattern here: It appears that the lifetimes of the pulsation modes are short in those roAp stars that pulsate in many modes, whereas the lifetimes are longer in the stars with few or only one mode. This is not a problem of resolution; the separation between modes, both theoretically and observationally, is in the range 20–80 μHz, and thus the modes can, in principle, be resolved over a single night of observation. On the other hand, the number of stars from which this conclusion is made is few.

Many fundamental data are lacking for the roAp stars, so it is not yet possible to say whether pulsation amplitudes, mode lifetimes, number of modes excited, or anything else is correlated with fundamental physical parameters. Only a few roAp stars have luminosity estimates, and most of these are from secondary methods; accurate space-based parallaxes for these stars are needed. Temperatures are somewhat uncertain, in most cases due to the spectral peculiarities; in the worst case, HD 101065, the temperature dispute has ranged over 1500 K (Section 1.4). Of the roAp stars for which magnetic field measurements are listed in Table 2, only HR 1217 (HD 24712) and γ Equ (HD 201601) could be described as well studied, and γ Equ has not yet gone through one magnetic cycle in the 40 yr it has been observed—there is some dispute as to whether the magnetic field variations arise from oblique rotation with a period of about 70 yr or whether they are due to a stellar magnetic cycle (e.g. see 76). Rotation periods are also not known for most roAp stars (Table 3).

These fundamental data are needed for asteroseismology to be used to its fullest extent to probe the interior structure and interior magnetic field configuration of these stars, to determine their masses and ages, and to discover what governs the mode lifetimes.

6. PERIOD CHANGES IN THE roAp STARS

Observations of period changes, $\dot{P} \equiv dP/dt$, in pulsating stars provide a powerful tool for the following reasons: (a) Secular $\dot{P}$ values are caused by changes in stellar structure and hence measure stellar evolution in real time. (b) Measurement of a periodic variation in $\dot{P}$ due to Doppler shifts is an accurate way to measure binary orbits when the pulsation frequencies are accurately known. The pulsation frequencies in the roAp stars can be determined to very high accuracy for the stars that pulsate in modes with long lifetimes. Heller & Kawaler (63) have calculated $\dot{P}$ for evolutionary A star models. They show that for HR 1217 (Section 3), $\dot{P}$ should be detectable with the 6 yr of observations available if $\Delta v_0 = 34$ μHz and $\Delta M_V = 1.9$ mag, whereas it should not be detectable if $\Delta v_0 = 68$ μHz and $\Delta M_V = 0.7$ mag. Unfortunately, alias problems in the early single-site data for HR 1217 preclude making this test yet, but in a few years time this test may help to distinguish between $\Delta v_0 = 34$ μHz and 68 μHz and hence to determine whether the pulsation modes in HR 1217 are basically all dipole modes or whether they are alternating even and odd l-modes.

Martinez & Kurtz (115) have found that $\dot{P}$ in the principal frequency of HD 101065 is much larger than expected from evolutionary changes in a H-core-burning A star. This could be due to a very late evolutionary state for HD 101065 if it is a unique object (rather than an extreme example of an Ap star), or it could be due to a Doppler shift. In the latter case the secondary has a mass less than 0.1 $M_\odot$ unless the orbit is within 2° of pole-on. Future observations will test this possibility.

7. EXCITATION MECHANISMS

7.1 The κ-Mechanism in the He II Ionization Zone and Position in the HR Diagram

The roAp stars have effective temperatures in the range of about 7000 K to 8300 K based on their photometric properties in the Strömgren system (Table 1) and in the Geneva system (62). These temperatures are more uncertain than those of normal stars; the continuum flux distributions in the cool Ap stars can deviate substantially from those of normal stars (see 1, 8, 9, 181). Particularly uncertain is the temperature of HD 101065 (Section 1.4); it is possible that the continuum is not seen in the visible part of the spectrum of this star owing to line blanketing.

The luminosities of the roAp stars are not well known either. The Balmer jump is smaller than normal and hence is not a good measure of luminosity (57, 126, 137). This can be seen in Table 1, where a naive interpretation of the Strömgren δc_1 index indicates that most of the roAp stars lie below the ZAMS. Other luminosity criteria, including parallaxes for two stars (Table 3), show that this is not the case. The roAp stars lie within about 2 mag of the ZAMS and are Population I H-core-burning stars (see 10, 181).

Because the c_1 index in a roAp star is smaller than in a normal star of the same luminosity, δc_1 is a principal criterion for finding these stars: If $\Delta M_V \approx 8\delta c_1$ is much smaller than normal for an Ap SrCrEu star, then that star is probably a roAp star. This effect is caused by a depression of the apparent continuum by line blanketing on the long-wavelength side of the Balmer jump, thus decreasing the size of the jump and mimicking a star of higher surface gravity. The effect is most pronounced in cooler stars and in stars with extreme abundance anomalies, which is why it highlights roAp stars. In practice, δc_1 is negative for most of the roAp stars because negative δc_1 indices are easily recognized. Higher luminosity roAp stars may have positive δc_1 indices that are too low for their luminosities, but in the absence of other luminosity criteria, these values will not appear abnormal and hence may escape notice. Matthews (118a) has recently shown that the location of the cool Ap stars in the $[c_1]$-$[u-b]$ and $[c_1]$-$[b-y]$ planes provides a better indicator of the roAp phenomenon than does the δc_1 index alone.

The observed δ Scuti instability strip lies within 3 mag of the ZAMS between about 7000 and 9000 K (27); hence, the roAp stars lie within the instability strip. This suggests that the excitation mechanism for their pulsation is the κ-mechanism in the He II ionization zone, which drives δ Scuti pulsation. An objection to this conclusion is that most Ap stars seem to be stable against δ Scuti pulsation, as theoretically expected within the diffusion hypothesis (Section 1.3). But it is clear that a variety of peculiar stars, both magnetic and nonmagnetic, are δ Scuti stars (Section 1.3); if those stars can be driven by the κ-mechanism in the He II ionization zone, then the same may be true of the roAp stars. Cox et al (42) suggest that residual He in the He II ionization zone is sufficient to drive pulsation in cool, slightly evolved Am (δ Del) stars; this may also be true for the roAp stars.

The pulsation velocities are probably not a problem for diffusion or for any other hypothesis for the abundance anomalies. Surface pulsation velocities in the roAp stars may all be similar to the 20–200 m s^{-1} range seen in HR 1217 (125; Section 3.3 herein) and γ Equ (112). Velocity amplitude drops rapidly with depth in the atmosphere of a pulsating star, and since we know that the abundance anomalies are confined to the surface (Section 1.1), there can be no turbulent mixing of the surface layers with lower layers.

Hence, there is no major objection to the κ-mechanism in the He II ionization zone being the driving mechanism for the roAp stars. A roAp star lying well outside of the instability strip would conflict with this hypothesis, but no there is no such star (unless HD 101065 has $T_{\rm eff} \approx 6300$ K—see Section 1.4). Detection of nonpulsating Ap stars in the instability strip would be a problem, although that same problem is unexplained for the δ Scuti stars. There is also a possibility that the known roAp stars lie in the instability strip because of a selection effect: I have discovered most of them, and that is where I have been looking. This is coupled with the problem of detection and, especially, nondetection of these stars, which is discussed in Section 10.

7.2 Magnetic Overstability

For the case where He is depleted from the He II ionization zone, and to explain the coincidence of the pulsation and magnetic axes, Shibahashi (148; see also 44, 66, 67) suggested that the oscillations in the roAp stars may be excited by magnetic overstability. For a dipolar magnetic geometry, convection is unstable at the magnetic equator but is overstable at the magnetic poles. In superadiabatic layers at the magnetic poles the magnetic field provides an additional restoring force to convective motion, so that convective cells return to their original positions faster than they move out. This suppresses convective energy transport while causing the convective motions to become oscillatory. Shibahashi calculates that the time scale of the convective oscillation is on the order of the observed pulsation periods in the roAp stars; he suggests that resonance between global eigenmodes and overstable convective oscillation excites the few modes seen.

This suggestion explains why the pulsation axes and magnetic axes should be aligned: driving is at the poles [although Dolez et al (48) argue that this is an insufficient criterion for alignment—see Section 7.3]. It explains the short periods (high overtones) of the observed pulsation modes (Section 8.1), because the star is resonating with the short time scale of the convection. It suggests that axisymmetric modes ($m = 0$) should be excited, as is observed, due to driving at the poles. And it suggests that growth times for the modes may be as small as days and hence may account for the short mode lifetimes seen in some stars.

Shibahashi (148) also pointed out that magnetic overstability will cause the energy transport to be less efficient at the magnetic poles than at the equator. (This is the well-known Biermann mechanism for the darkness of sunspots.) This effect should cause an increased temperature gradient at the pole and hence may explain the mean light variations in Ap stars, which are not completely understood (see 181). The magnetic poles of the well-studied magnetic Ap star α^2 CVn are about 400–1000 K cooler than the magnetic equator (19, 143a), and perhaps magnetic overstability should be considered as a contributing factor to this temperature difference and to the mean light variations.

If magnetic overstability is the excitation mechanism in the roAp stars, then it might be expected that they should not be confined to the instability strip. However, certain constraints may conspire to keep them there. Their cool border would not change significantly compared with the κ-mechanism excitation, since the magnetic Ap phenomenon cuts off at about the same temperature as the δ Scuti red edge. The magnetic phenomenon cuts off because of the onset of large-scale convective envelopes in the early F stars; δ Scuti pulsation cuts off because of the increased depth of the He II ionization zone.

For hotter Ap stars, the temperature gradient at the magnetic poles is less superadiabatic than in the cooler stars; thus, magnetic overstability may have a blue edge similar to the observed δ Scuti blue edge. There is an additional complication, however. Hotter δ Scuti stars tend to pulsate in higher overtones with smaller amplitudes (27). It is possible that the observed blue border of the δ Scuti instability strip is an observational cutoff where the light amplitudes drop below about 0.01 mag. If this is true, then high-overtone δ Scuti pulsation might be expected to have a blue border much hotter than the presently observed one. None of the known roAp stars is hotter than the present blue border, but again, selection effects weaken any conclusion that may be drawn from this (a point discussed further in Section 10).

7.3 *The κ-Mechanism Operating Only at the Magnetic Poles*

Dolez & Gough (47) calculated that diffusion should cause He to be preferentially depleted at the magnetic poles of an Ap star compared with the equator because of the suppression of convection at the poles. This implies that modes aligned with the magnetic axis should be preferentially damped—just the opposite of what is observed. Dolez et al (48) modified their diffusion calculations to include stellar winds, which led them to the opposite conclusion that He accumulates at the magnetic poles. Thus, they suggest that the κ-mechanism operating in the He II ionization zone only at the poles is responsible for the driving in roAp stars, although they argue that driving at the poles is, of itself, insufficient to force alignment of the magnetic and pulsation axes.

They further argue that the model of Dziembowski & Goode (see Section 2.8) requires internal magnetic field strengths of megagausses which are unlikely given that only kilogauss surface fields are seen. They suggest instead that lateral inhomogeneity of the surface abundance distributions causes a perturbation to the pulsation frequencies, Δ_l, which is greater than the Coriolis perturbation, $C_{nl}\Omega$, and that this causes the alignment of the magnetic axis and pulsation axis. In their model, they have $\Delta_l \gg C_{nl}\Omega$ instead of $\omega_{|m|}^{(1)\mathrm{mag}} \gg C_{nl}\Omega$ (see Section 2.8); otherwise, the formalism remains the same. In this model, the inequality $A_{+1} \neq A_{-1}$ is a measure of the surface abundance distribution, which in principle can be deduced from observations. As an example, see Landstreet's (110) model of the magnetic field and abundance distribution geometry in 53 Cam. In this paper it is also demonstrated how the magnetic field geometry in an Ap star can be far from dipolar; the implications of this complication remain unexplored for the roAp stars.

7.4 The κ-Mechanisn for Si IV

Matthews (118) suggested that pulsation might be driven in the roAp stars by the κ-mechanism operating on enhanced abundances of Si IV, even if He should be depleted in the surface layers of these stars. He argues that Si enhancement will be greater at the magnetic poles, the driving will be preferentially at the poles, and hence the pulsation and magnetic axes will be aligned.

To test this model Matthews suggests that very accurate spectroscopic surveys of roAp stars and other cool, nonpulsating Ap stars should find differences in the surface Si abundances. For this to be done, however, it must first be shown that cool, nonpulsating Ap stars exist (Section 10).

7.5 Alfvén Modes

Campbell & Papaloizou (31) studied pulsation in the presence of a magnetic field and found that substantial pulsational energy is lost to Alfvén modes. Excitation calculations for the roAp stars must, therefore, include this extra damping.

Shore et al (152) suggested that the roAp stars might provide a unique tool to study heating of the outer stellar atmosphere by Alfvén waves in main sequence stars, but they note that ultraviolet spectra of HR 1217, HR 3831, and HD 137949 show no evidence of chromospheric emission lines. This probably rules out Alfvénic heating in these stars.

8. MODE SELECTION

8.1 *The Radial Overtone*, n

The radial overtones of the pulsation modes in the roAp stars are very high. This can be estimated from Equation 20 for stars in which consecutive overtones are observed. For HR 1217 (Section 3), $n \approx 80$ if consecutive overtones of $l = 1$ are excited, and $n \approx 40$ if alternating even and odd l-modes are excited. In HD 60435, a range of values is found: $13 \le n \le 28$ (Section 5). In general, the very short periods of the roAp stars imply that $n > 10$. Shibahashi (148) suggests that these high overtones are selected by resonance of the frequencies of the high-overtone eigenmodes with the time scale of the oscillatory motion of the overstable magnetic convection (Section 7.2). Dolez & Gough (47) find maximum driving in simple A star models for $n \approx 15$.

It is not known whether all cool A stars are unstable to many high-overtone modes. If a large range of modes is excited (as in the Sun) at low amplitude (say, less than a few millimagnitudes), then such oscillations are unlikely to be detected due to the interference of the many modes. If high-overtone modes are excited in δ Scuti stars, for example, they would not have been detected by any survey yet done. Because of the relatively long periods of δ Scuti stars (30 min to 6 hr), differential photometry is used to observe them. This means that the time resolution of such photometry is 5–10 min, which precludes the detection of periods less than 10–20 min. On the other hand, high-speed photometry of δ Scuti stars by itself may not detect high-overtone modes because any low-amplitude, high-frequency signal would get lost in the noise of the window pattern of the lower frequency variations. Most δ Scuti stars have frequency patterns so complex that it is not possible to predict the light curve from past observations; hence, the low-frequency window pattern cannot be accurately removed from the amplitude spectrum. Two-channel high-speed photometry of selected δ Scuti stars and comparison stars might determine whether high-overtone modes are excited in these stars.

Given the above, it is possible that the roAp stars are the only A stars with a mode *suppression* mechanism—i.e. they are the only A stars for which so few high-overtone modes are excited that they can be detected. If this is the case, then it is likely that the magnetic field plays a direct role in mode selection. If the abundance distribution were the mode selection mechanism, then Am stars might also be expected to be observed to oscillate rapidly. I have searched for rapid oscillations in some classical Am stars (unpublished) and have not detected any, although I have found one that is a low-amplitude δ Scuti star (94).

On the subject of the abundance distributions, it is interesting to note that pulsating DA white dwarf stars, the ZZ Ceti stars, have pulsation frequencies which are very long compared with the frequency of the fundamental mode; the modes are thus inferred to be high-overtone g-modes. This has been explained by mode trapping caused by the layered structure of white dwarfs (see 178, 179). Although the Ap stars have abnormal abundance distributions, they do not have the density discontinuities of the white dwarfs, so this mechanism is unlikely to be at work in the roAp stars.

8.2 *The Spherical Degree* l *and the Azimuthal Order* m

It was shown in Sections 2 and 3 that dipole ($l = 1$) modes are excited in HD 6532, HR 1217, and HR 3831. It seems probable that dipole modes are predominantly excited in the roAp stars because the match of the dipole pulsation geometry with that of the magnetic field minimizes the resistance of the magnetic field to the pulsation. Also, damping by energy loss to Alfvén modes increases with increasing l and thus preferentially allows the excitation of modes with lower l (31).

In HD 60435 and possibly HR 1217, both even and odd l-modes are excited. The amplitude modulation of these modes argues in favor of $l = 2$ quadrupole modes rather than $l = 0$ radial modes. This is complicated, however, by the deduction that some of the modes in HR 1217 cannot be explained by single spherical harmonics. Theoretically, this is not surprising; probably when $l \neq 1$, the surface appearance of the mode is distorted by the $l = 1$ (or more complicated) geometry of the magnetic field (16).

In general, higher l-modes will produce less easily detectable light or radial velocity variations because the integral of P_l over the visible surfaces goes to, or tends toward, zero regardless of excitation amplitude. Kurtz & Balona (97) suggested that two modes of (n, l) and $(n-1, l+2)$ are excited in α Cir. The separation of these two frequencies gives $\delta v = 2.5$ μHz. Martinez & Kurtz (115) also suggested the possible presence of modes of (n, l) and $(n-1, l+2)$ in HD 101065 with $\delta v = 2.92$ μHz. Models of A stars suggest that δv should be perhaps an order of magnitude larger than this (36, 55, 151); models including the effects of the magnetic field are needed to resolve this problem.

Finally, there is no indication that $m \neq 0$ modes are excited in the roAp stars. Such modes would be detectable if they were excited. The amplitudes and phases of their frequencies and rotational sidelobes are very different from those of the $m = 0$ modes (Section 2; Equation 14).

8.3 The Harmonics

Five of the roAp stars show harmonics for at least one of their principal pulsation frequencies (Table 3). The problem of understanding the well-observed first and second harmonics in HR 3831 is discussed in detail in Section 2.9. The five roAp stars that show harmonics are basically larger amplitude pulsators; it is therefore possible that all roAp stars will show harmonics when better signal-to-noise observations are obtained. The photometric accuracy obtainable at the higher harmonic frequencies can be improved by observing with larger telescopes. For frequencies over 2.5 mHz, the photometric noise decreases roughly in proportion to telescope aperture (90) for bright stars.

The theoretical implications of the harmonics have not been studied. It is clear that the millimagnitude pulsations in the roAp stars are nonlinear, in stark contrast to δ Scuti stars, where nonlinearities in the light curves are only seen for stars with amplitudes of tenths of a magnitude.

8.4 The Critical Frequency

In a real star the nonzero surface temperature limits the number of eigenmodes. The energy of acoustic waves is reflected over a finite length at the stellar surface; when the frequency of a pulsation mode becomes too high, its wavelength becomes short with respect to the length of the surface reflection layer, and thus phase coherence in a standing wave cannot be maintained. Therefore, there is a critical frequency v_{crit} above which standing acoustic waves should not be observed (unless very strongly driven).

Shibahashi & Saio (151) have calculated v_{crit} for standard equilibrium A star models of various masses and chemical compositions. They find that $v_{\text{crit}} < 2$ mHz for $\Delta v_0 < 80$ μHz appropriate to the roAp stars; many of the roAp stars pulsate with frequencies higher than 2 mHz. Refinements to Shibahashi & Saio's models might increase the calculated v_{crit}. Their models were homogeneous, whereas Ap stars tend to be cooler at the magnetic poles; v_{crit} increases with lower temperature. Also, line blanketing by the overabundant elements at the magnetic poles increases the temperature gradient, which sharpens the surface reflective layer and increases v_{crit}.

Shibahashi & Saio (151) find that one way to increase v_{crit} in their models to the frequencies observed in HR 1217 is to lower $T_{\text{surf}}/T_{\text{eff}}$ compared with standard models. This implies that the temperature in the line-forming regions is lower than is normally assumed. On the other hand, Dolez et al

(48) find modes excited above the adiabatic v_{crit} for their models with enhanced He driving at the magnetic poles; they caution that the lower T_{surf} suggested by Shibahashi & Saio for HR 1217 may not be necessary to explain the observed frequencies.

Shibahashi & Saio (151) also note that the direct effect of the magnetic field on the oscillations will be to lower v_{crit}, although indirectly the effect of magnetic pressure is to decrease the gas pressure, which will decrease the surface temperature and increase v_{crit}. Studies of inhomogeneous A star models with magnetic fields are needed. The comparison of v_{crit} in such models with the observed frequencies in the roAp stars will provide an additional constraint on the models. It was pointed out in Section 1.1 that nonstandard model atmospheres cannot entirely explain the line strength anomalies in the Ap stars. This led to the conclusion that the line strength anomalies imply abundance anomalies. Nevertheless, the above arguments about v_{crit} suggest that the roAp stars (and thus, by implication, other Ap stars) may have nonstandard atmospheres. If the temperatures in the abundance patches at the magnetic poles are lower than previously thought, then the abundances are still anomalous but perhaps by factors smaller than previously calculated. Studies of the pulsation frequencies in the roAp stars may lead to better model atmospheres and hence to improved abundance determinations.

9. MODE IDENTIFICATION

It has been shown in Sections 2, 3 and 8 that n and l can be determined, or at least estimated, for some oblique pulsators from a study of the amplitude spectra of the photometric light variations alone. There are other methods for mode identification which have been used successfully for other types of stars but not yet for the roAp stars.

9.1 *Line Profile Variations*

Line profile variations can identify modes in nonradially pulsating stars. In a classic paper, Ledoux (111) used this fact to deduce that some β Cep stars pulsate in nonradial modes. Osaki (133) showed how profiles are affected by rotation, inclination, and l, and Balona (12) developed a quantitative method for making mode identifications from the line profile observations.

Intensive observations by Myron Smith of the late-B 53 Per variables show spectacular line profile variations (see, e.g., 155). These are interpreted as nonradial pulsations, and the time-dependent character of the variations can constrain or identify l and m. Line profile variations due to nonradial pulsation have also been detected in δ Scuti stars (153) and Be stars (see 5, 134). Smith (154) reviews nonradial pulsation in OB stars. As an example, in an interesting paper on ζ Oph, a rapidly rotating ($v \sin i = 370$ km s^{-1}) O9.5 Ve star, Vogt & Penrod (166) interpreted the line profile variations as nonradial pulsation from an ($l = 8, m = -8$) mode. In this case, the rapid rotation Doppler shifts the various slices of the high-m mode so that bumps march across the line profile with rotation. This is a case where high l, m can be detected in the line profile variations, even though the luminosity variations from such a mode may be undetectable due to averaging over the many surface elements.

The successes, both observational and theoretical, of line profile studies of the O and B stars have led to several investigations of line profile variations in roAp stars. Theoretical calculations of expected line profile variations in an oblique pulsator have been made by Odell & Kreidl (130), Baade & Weiss (6), and Pesnell (135). Schneider et al (146) and Weiss & Schneider (177) searched unsuccessfully for line profile variations in γ Equ and α Cir. Schneider et al (146) put an upper limit of 120 m s^{-1} on the velocity variations in α Cir, whereas Weiss & Schneider (177) find an upper limit for γ Equ of 30–100 m s^{-1}, which is consistent with Libbrecht's (112) observation of 20 m s^{-1}.

Detection of line profile variations in the roAp stars is difficult; large telescopes and bright stars are needed to achieve the required signal-to-noise ratio. The brightest roAp stars, α Cir and γ Equ, have been observed first, but they do not have the largest photometric amplitudes. HR 3831 is an excellent candidate because it has a relatively large photometric amplitude and an accurately known rotation period, and because its principal pulsation mode is known to be a dipole (Section 2). Thus, it could provide a good test and calibration of the theoretical line profile calculations. It is, however, faint for this type of work ($V = 6.2$), so the largest telescopes in the Southern Hemisphere are needed to make the observations.

9.2 The Phase Shift Between the Light and Color Curves

The spherical harmonic l in a nonradially oscillating star can be identified in some cases from the phase difference between the light and color curves (11, 13, 14, 29, 49, 157, 168). Application of this technique has had some success for β Cep, 53 Per, δ Scuti, Cepheid, and ZZ Ceti variables, but it has been spectacularly unsuccessful for the roAp stars. No consistent interpretation of the spherical harmonic index l and the phase shift $\Delta\phi(B-V, V) = \phi(B-V) - \phi(V)$ is yet possible for the roAp stars which have been observed simultaneously in more than one color (see 168).

Those stars are HD 6532 [*BV* (99)], HD 24712 [*VBLUW* (175); *uvby* and *BV* (104)], HD 83368 [*UBV* (85)], HD 101065 [*UBV* (83, 109)], HD

128898 [BV (97, 176)], and HD 134214 [BV (79)]. Although the phase parameter normally used in the linear theory for identifying l does not produce consistent results for these stars, the multicolor observations do provide some important information: In general, the amplitude of the light variations is greatest for observations made with blue filters and drops substantially for observations made with longer wavelength filters. The phase of the oscillations for some stars is similar in all colors, but in others the phase of the light and color curves can differ by 100°. These factors have important implications in the search for, and study of, roAp stars.

10. THE SEARCH FOR roAp STARS

To find the answers to the many questions posed by the roAp stars, many more of these stars must be found. It is also important to determine which Ap stars do not pulsate. Finding more pulsators will provide more objects on which to test asteroseismological theory; finding the constant stars will determine what physical parameters control the pulsation in the variables.

In practice, proving which stars are constant is the more difficult task. Because the roAp stars are oblique pulsators, their observed amplitudes may go to zero at certain rotation phases. As some Ap stars have rotation periods of decades, e.g. HR 465 [$P \approx 21$ yr (144)] and, arguably, γ Equ [$P \approx 70$ yr (see 76)], the observation that a particular Ap star does not show rapid variations only indicates that it was not variable at the time of observation. Unless the rotation period and magnetic geometry are known, it is impossible to conclude that it is not a roAp star, no matter how many times it is observed to be constant.

It is therefore important to observe Ap stars with well-studied magnetic fields, so that they can be observed at the times of magnetic maximum. One such star is β CBr, which is an F0p SrCrEu star; its Strömgren photometric indices are very similar to those of the known roAp stars in Table 1, and its magnetic field is well studied and has an accurate ephemeris (95). Several groups have searched for rapid light variations in this star without finding any (64, 177; T. J. Kreidl, G. Wegner, J. M. Matthews, private communications); these observations are not yet definitive, however, because they have not been made under excellent conditions through blue filters exactly at magnetic maximum.

It is clear from the data in Tables 1 and 2 that an Ap SrCrEu spectral classification and a negative δc_1 index are good indicators of roAp stars. As was pointed out in Section 7.1, however, Matthews (118a) has shown that the location of the cool Ap stars in the $[c_1]$-$[u-b]$ and $[c_1]$-$[b-y]$ planes provides a better indicator of the roAp phenomenon than does the δc_1 index alone.

I have searched for rapid light variability in many hotter magnetic stars, up to the He-strong stars (unpublished). While these surveys have not been systematic, they do suggest that if hotter rapid pulsators exist, their apparent amplitudes may be less than those of the cooler stars already discovered. It is interesting to note, however, that if the absolute amount of energy driving the rapid pulsations were approximately constant for all roAp stars, then the observed amplitude would drop with increasing temperature and increasing luminosity, making such pulsation more difficult to detect in hotter stars. Observations of the highest accuracy are obviously required.

10.1 *Photometric Search Techniques*

Most of the work done on the roAp stars has used single-channel high-speed photometry. At a good photometric observing site there is little to be gained by using a second channel on a comparison star for these types of observations. One reason for this is that most of the roAp stars are bright; finding a comparison star of similar brightness within the limited field of view of the second channel is usually not possible, and normalizing to a much fainter star will increase the noise. Another reason can be seen in the amplitude spectrum of the noise caused by sky transparency variations, scintillation, and photon statistics. On good photometric nights at good sites the dominant noise at the frequencies of the roAp stars is scintillation; the stars are bright, so photon statistics is not a limiting factor, and low-frequency sky transparency variations generally merge with the scintillation noise for frequencies higher than 1 mHz or so. Thus, on good photometric nights the use of a comparison star will reduce the noise level only if the scintillation of the two stars is correlated. This appears to be the case at the frequencies of the roAp stars for stars within a few arcminutes of each other (167), but then area detectors are needed to make the observations.

HR 3831 and γ Equ both have fairly bright secondaries which could be used as comparison stars; observations of these stars with a CCD might produce more accuracy than heretofore obtained, but such observations are currently limited by the slow readout time of most CCDs and the difficulty of handling the large amount of data gathered this way. These problems are not insuperable; high-speed photometry with CCDs will probably become standard.

That the observations of roAp stars are often scintillation limited means that higher accuracy is obtained with larger telescope aperture. To first order, the number of cells averaged over goes as the collecting area. Since noise goes down as the square root of the number of cells, the scintillation noise goes down roughly inversely proportional to aperture. Thus, there is a clear need for large aperture to observe roAp stars—not in order to collect additional photons, but to reduce scintillation noise.

Another interesting technique that has not been tried is to observe a roAp star with a two-channel photometer with a beam splitter. It was pointed out in Section 9.2 that the roAp stars have much larger amplitudes in the blue than in the visual; their amplitudes in the far-red and near-infrared are unstudied. It is possible that some or all of them are essentially constant in the red, so that observations in one channel in the blue would show the variations, whereas observations in the other channel in the red could serve for purposes of comparison. Sky transparency and scintillation variations would be highly correlated between the two channels, and hence the residual noise after normalizing the blue channel by the red channel might approach the photon statistical limit.

Most observations of roAp stars have been made through readily available standard filters to make it easy for many observers to cooperate. Because there are often large phase shifts from filter to filter, it is clear that the signal in any one broadband filter is phase smeared to some extent, and that a carefully selected narrow-band blue filter could produce a better signal-to-noise ratio than has yet been obtained. Such a filter could be very narrow for the brighter roAp stars (consider that α Cir is $V = 3.2$), but thought must be given to the bandpass. This is because the Ap stars have abnormal spectra, and narrow bandpasses containing pathological spectral lines might produce results different from bandpasses containing mostly continuum.

10.2 Some Photometric Searches

I have conservatively listed 14 known roAp stars in Tables 1–3. There have been claims of other members of the class: Some of these may be correct but in my opinion are not yet proven, some are based on poor data, and some are overinterpretations of noise. Some published searches are mentioned here, but publishing null results is not generally exciting, and, because null results for potential roAp stars are usually inconclusive (Section 10.1), much of the work which has been done remains unpublished.

Matthews et al (119) searched for rapid light variability in three Ap stars in the cluster NGC 2516 and one nearby field star which, from Strömgren photometry, looked like promising candidates. Discovery of roAp stars in this cluster of known distance and age would be important to our understanding of these stars. Of the four stars, the three cluster members were nonvariable, while the field star, HD 166473, appeared to have periods of 7.1 and 19.8 min. Martinez (114) made further observations of HD 166473 but could find no variability. He suggested that the peaks in

Matthews et al's amplitude spectra may have been "false alarms," and he discussed this problem in general for these stars. Because the roAp stars are oblique pulsators, however, the nonvariability in the three stars is inconclusive, and Martinez's nonconfirmation of variations in HD 166473 does not necessarily disprove variability in this star. Another search was conducted by Heller & Kramer (64), who found 11.4-min variability in 10 Aql but failed to find variations in two other stars previously reported as variable. These two searches are illustrative of the kind of work being done, and they point out the difficulty of the observations and the controversial nature of the interpretation of the amplitude spectra.

10.3 *Radial Velocity Studies*

Very high accuracy radial velocity measurements of sharp-lined stars are now being made with several techniques by many groups. Some examples are (*a*) the search for substellar companions to stars by Campbell et al (30), who quote an external error for their measurements of ± 13 m s^{-1} per observation; (*b*) the discovery of a 2-day periodicity in the radial velocity of Arcturus with an amplitude of 200 m s^{-1} by Cochran (37), who quotes an error of 15–20 m s^{-1} per observation; and (*c*) the discovery of 2.5-hr oscillations in Aldeberan and Pollux with amplitudes of about 5 m s^{-1} by Smith & McMillan (156), who quote an error of ± 3 m s^{-1} per observation.

With accuracies as high as these, radial velocity observations are an excellent method for studying the roAp stars. The accuracies will not be quite as good as those quoted above, since even the brightest of the roAp stars are not as bright as the stars in these studies, and the roAp stars have slightly higher rotational velocities. The very short periods of the roAp stars also necessitate short integration times, which will mean larger standard deviation per observation, but this noise can be reduced by analyzing many cycles, just as it is for the photometric variations.

In Section 9.1 line profile observations for some roAp stars were discussed which gave radial velocities accurate to several tens to hundreds of meters per second. In Section 3.3, radial velocity variations in HR 1217 were discussed; Matthews et al (125) found $2K = 400 \pm 50$ m s^{-1} at pulsation maximum. Libbrecht (112) reported rapid radial velocity variations in γ Equ with amplitudes of about 15 m s^{-1} at periods consistent with those determined photometrically by Kurtz (86). Ando et al (4) searched for rapid radial velocity variations in HR 1217 and the cool magnetic Ap star β CBr. They report the possible detection of rapid variations in these stars with errors of 50–200 m s^{-1}.

These studies show that radial velocity observations are a viable method to study the roAp stars. Matthews et al (125) estimate the velocity-amplitude to light-amplitude ratio to be $2K/\Delta m_B = 59 \pm 12$ km s^{-1} mag^{-1} for HR 1217, and Libbrecht (112) estimates $2K/\Delta m_B = 26$ km s^{-1} mag^{-1} for γ Equ. In Section 10.1 it was pointed out that the B amplitudes of the roAp stars are generally about twice those in V, so we can estimate for HR 1217 and γ Equ that $2K/\Delta m_V \approx 50$–100 km s^{-1} mag^{-1}; this is similar to the 50–125 km s^{-1} mag^{-1} range found for the δ Scuti stars (28).

The implication of these results is that pulsation frequencies for the roAp stars may in some cases be more accurately determined from radial velocity measurements than from photometry. Consider α Cir: It has a peak-to-peak amplitude of about $\Delta B = 0.0048$ mag, which implies that $2K = 120$–240 m s^{-1}. This star has $V = 3.2$ and is sharp lined, so integration over many cycles could give a very good signal-to-noise ratio. In addition, because of the patchy abundance distribution in Ap stars, radial velocities measured using lines of elements that are concentrated near the magnetic poles may give even better signal-to-noise values, since pulsation amplitude is largest at the poles for axisymmetric nonradial modes.

11. SUMMARY

11.1 What is Known About the roAp Stars?

1. (Section 1) They are Ap SrCrEu hydrogen-core-burning stars which are cool magnetic oblique rotators. They pulsate in high-overtone, low-degree nonradial p-modes with periods in the range 4–15 min.
2. (Section 2) They are oblique pulsators whose magnetic and pulsation axes are aligned. HR 3831 and HD 6532 show the clear signature of oblique dipole pulsation. Application of the oblique pulsator model allows the rotational inclination i and the magnetic obliquity β to be constrained (and potentially to be determined). It also gives a measure of the internal magnetic field strength.
3. (Section 3) Asteroseismology potentially can determine the luminosity, mass, and age of the roAp stars from analyses of their frequency spectra.
4. (Section 4) The spotted pulsator model does not describe the pulsations in the roAp stars as well as the oblique pulsator model. However, it will be needed in the future when more accurate observations will have to account for the small effect caused by the mean light variations.
5. (Section 5) The lifetimes of some pulsation modes in some stars are years; in others, they are as short as days.
6. (Section 6) Measurement of $\dot{P}$ values in roAp stars potentially can be used to determine their evolutionary time scale. Some frequencies are very accurately known; Doppler shifts in these frequencies can detect the presence of low-mass companions.

7. (Section 7) They lie within the observed δ Scuti instability strip.
8. (Section 8) They pulsate in high-overtone modes with $n \geq 10$; in some cases, n may be as large as 80. They pulsate in axisymmetric ($m = 0$) low-l modes; $l = 1$ has been identified in some stars (Section 2). Some stars pulsate in alternating even and odd l-modes simultaneously (Section 3).
9. (Section 9) Mode identification has been made only from analyses of the frequency spectra. Line profile studies potentially can provide additional mode identifications.
10. (Section 10) Cool Ap SrCrEu stars with Strömgren c_1 indices too small for their luminosities are excellent candidates for roAp stars. Radial velocity studies, particularly for the longer period stars, may provide better signal-to-noise ratios than photometric studies.

11.2 What is not Known?

1. (Section 2) The nature of the nonlinearities in the light curves is not understood. The amplitude of the first harmonic in HR 3831 is modulated as $\cos^2 \alpha$, for reasons not yet known. There is some dispute about whether the magnetic field strengths in the roAp stars are strong enough to lock the pulsation and magnetic axes (Section 7.3).
2. (Section 3) The factor that alters the surface distortions of some of the pulsation modes from spherical harmonics is probably the magnetic field (16, 31). There seems to be no consistent interpretation of all six principal frequencies in HR 1217 in terms of the high-overtone, asymptotic p-mode relation.
3. (Section 5) It is not known what governs mode lifetimes in the roAp stars. This question is coupled to that of the excitation mechanism and growth rates (Section 7).
4. (Section 7) The excitation mechanism for the roAp stars is not known. The κ-mechanism in the He II ionization zone, or at the magnetic poles only, or in Si IV at the magnetic poles only have all been suggested, as has magnetic overstability.
5. (Section 8) The mode selection mechanism is unknown. Suggestions have been made based on magnetic overstability and on growth rates in simple A star models. The magnetic field probably selects modes of low degree (16, 31). Some frequencies seem to be above the critical frequency; this may indicate abnormal atmospheric structure, which would affect abundance analyses.
6. (Section 9) The phase shift between the light and color curves in many pulsating stars provides information on the pulsation mode. In the roAp stars these phase shifts scatter inconsistently over a wide range; the reason for this is unknown.

7. (Section 10) No systematic surveys for roAp stars have been made; their temperature and luminosity ranges are uncertain. It is not known for certain whether cool magnetic Ap stars exist which are stable against high-overtone pulsation.

Literature Cited

1. Adelman, S. J., Cowley, C. R. 1985. See Ref. 41, p. 305
2. Alecian, G. 1986. See Ref. 41, p. 381
3. Allen, C. W. 1973. *Astrophysical Quantities*. London: Athlone. 3rd ed.
4. Ando, H., Watanabe, E., Yutani, M., Shimizu, Y., Nishimura, S. 1988. *Publ. Astron. Soc. Jpn.* 40: 249
5. Baade, D. 1987. In *Physics of Be Stars, IAU Colloq. No. 92*, ed. A. Slettebak, T. P. Snow, p. 361. Cambridge: Univ. Press
6. Baade, D., Weiss, W. W. 1987. *Astron. Astrophys. Suppl.* 67: 147
7. Babcock, H. W. 1958. *Ap. J. Suppl.* 3: 141
8. Babu, G. S. D., Shylaja, B. S. 1981. *Astrophys. Space Sci.* 79: 243
9. Babu, G. S. D., Shylaja, B. S. 1982. *Astrophys. Space Sci.* 81: 269
10. Babu, G. S. D., Shylaja, B. S. 1983. *Astrophys. Space Sci.* 89: 341
11. Balona, L. A. 1981. *MNRAS* 196: 159
12. Balona, L. A. 1986. *MNRAS* 219: 111
13. Balona, L. A., Stobie, R. S. 1979. *MNRAS* 187: 217
14. Balona, L. A., Stobie, R. S. 1979. *MNRAS* 189: 649
15. Bidelman, W. P., MacConnell, D. J. 1973. *Astron. J.* 78: 687
16. Biront, D., Goosens, M., Cousens, A., Mestel, L. 1982. *MNRAS* 201: 619
17. Blanco, V. M., Demers, S., Douglass, G. G., Fitzgerald, M. P. 1970. *Publ. US Nav. Obs.* 11: 1
18. Bohlender, D. A., Brown, D. N., Landstreet, J. D., Thompson, I. B. 1987. *Ap. J.* 323: 325
19. Böhm-Vitense, E., Van Dyk, S. D. 1987. *Astron. J.* 93: 1527
20. Bonsack, W. K., Pilachowski, C. A. 1974. *Ap. J.* 190: 327
21. Borra, E. F., Landstreet, J. D. 1975. *Publ. Astron. Soc. Pac.* 87: 961
22. Borra, E. F., Landstreet, J. D. 1980. *Ap. J. Suppl.* 42: 421
23. Borra, E. F., Landstreet, J. D., Mestel, L. 1982. *Annu. Rev. Astron. Astrophys.* 20: 191
24. Borra, E. F., Landstreet, J. D., Thompson, I. 1983. *Ap. J.* 53: 151
25. Breger, M. 1970. *Ap. J.* 162: 597
26. Breger, M. 1975. *Dudley Obs. Rep.* 9: 31
27. Breger, M. 1979. *Publ. Astron. Soc. Pac.* 91: 5
27a. Breger, M. 1982. *Publ. Astron. Soc. Pac.* 94: 845
28. Breger, M., Hutchins, J., Kuhi, L. V. 1976. *Ap. J.* 210: 163
29. Buta, R. J., Smith, M. A. 1979. *Ap. J.* 232: 213
30. Campbell, B., Walker, G. A. H., Yang, S. 1988. *Ap. J.* 331: 902
31. Campbell, C. G., Papaloizou, J. C. B. 1986. *MNRAS* 220: 577
32. Carney, B. W., Peterson, R. C. 1985. *MNRAS* 212: 33P
33. Catalano, F. A., Renson, P. 1984. *Astron. Astrophys. Suppl.* 55: 371
34. Catalano, F. A., Renson, P. 1987. *Astron. Astrophys. Suppl.* 72: 1
35. Christensen-Dalsgaard, J. 1988. See Ref. 36a, p. 3
36. Christensen-Dalsgaard, J. 1988. See Ref. 36a, p. 295
36a. Christensen-Dalsgaard, J., Frandsen, S., eds. 1988. *Advances in Helio- and Asteroseismology. Proc. IAU Symp. No. 123*. Dordrecht: Reidel
37. Cochran, W. D. 1988. *Ap. J.* 334: 349
38. Cousens, A. 1983. *MNRAS* 203: 1171
39. Cowley, C. R. 1981. In *Upper Main Sequence Chemically Peculiar Stars. Proc. Liège Colloq. No. 23*, p. 5. Liège: Inst. Astrophys., Univ. Liège
40. Cowley, C. R., Cowley, A. P., Aikman, G. C., Crosswhite, H. M. 1977. *Ap. J.* 216: 37
41. Cowley, C. R., Dworetsky, M. M., Mégessier, C., eds. 1986. *Upper Main Sequence Stars With Anomalous Abundances. Proc. IAU Colloq. No. 90*. Dordrecht: Reidel
42. Cox, A. N., King, D. S., Hodson, S. W. 1979. *Ap. J.* 231: 798
43. Cox, J. P. 1984. *Publ. Astron. Soc. Pac.* 96: 577
44. Cox, J. P. 1984. *Ap. J.* 280: 220
45. Deubner, F.-L., Gough, D. 1984. *Annu. Rev. Astron. Astrophys.* 22: 593
46. Dodgson, C. L. 1871. In *Through the Looking Glass and What Alice Saw There*

47. Dolez, N., Gough, D. O. 1982. In *Pulsations in Classical and Cataclysmic Variables*, ed. J. P. Cox, C. J. Hansen, p. 248. Boulder, Colo: JILA
48. Dolez, N., Gough, D. O., Vauclair, S. 1988. See Ref. 36a, p. 291
49. Dziembowski, W. 1977. *Acta Astron.* 27: 201
50. Dziembowski, W., Goode, P. R. 1984. *Mem. Soc. Astron. Ital.* 55: 185
51. Dziembowski, W., Goode, P. R. 1985. *Ap. J. Lett.* 296: L27
52. Dziembowski, W., Goode, P. R. 1986. In *Seismology of the Sun and Distant Stars*, ed. D. O. Gough, p. 441. Dordrecht: Reidel
53. Dziembowski, W., Krolikowska, M., Kosovitchev, A. 1988. *Acta Astron.* 38: 61
54. Edmonds, A. R. 1957. *Angular Momentum in Quantum Mechanics*. Princeton, NJ: Princeton Univ. Press
55. Gabriel, M., Noels, A., Scuflaire, R., Mathys, G. 1985. *Astron. Astrophys.* 143: 206
56. Gelly, B., Fossat, E., Grec, G., Pomerantz, M. 1988. See Ref. 36a, p. 21
57. Gerbaldi, M., Hauck, B., Morguleff, N. 1974. *Astron. Astrophys.* 30: 105
58. Gough, D. O., Taylor, P. P. 1984. *Mem. Soc. Astron. Ital.* 55: 215
59. Hartoog, M. R., Cowley, C. R., Cowley, A. P. 1973. *Ap. J.* 182: 847
60. Hauck, B., Mermilliod, M. 1980. *Astron. Astrophys. Suppl.* 40: 1
61. Hauck, B., Mermilliod, M. 1985. *Astron. Astrophys. Suppl.* 60: 61
62. Hauck, B., North, P. 1982. *Astron. Astrophys.* 114: 23
63. Heller, C. H., Kawaler, S. D. 1988. *Ap. J. Lett.* 329: L43
64. Heller, C. H., Kramer, K. S. 1988. *Publ. Astron. Soc. Pac.* 100: 583
65. Heller, C. H., Kramer, K. S. 1990. *MNRAS* 244: 372
66. Hermans, D., Goosens, M. 1987. *Astron. Astrophys.* 172: 85
67. Hermans, D., Goosens, M., Kerner, W., Lerbinger, K. 1988. See Ref. 36a, p. 395
68. Hoffleit, D., Jaschek, C. 1982. *Bright Star Catalogue*. New Haven, Conn: Yale Univ. Obs. 4th ed.
69. Houk, N. 1978. *Michigan Spectral Catalogue*, Vol. 2. Ann Arbor: Dep. Astron., Univ. Mich.
70. Houk, N. 1982. *Michigan Spectral Catalogue*, Vol. 3. Ann Arbor: Dep. Astron., Univ. Mich.
71. Houk, N. 1988. *Michigan Spectral Catalogue*, Vol. 4. Ann Arbor: Dep. Astron., Univ. Mich.
72. Houk, N., Cowley, A. P. 1975. *Michigan Spectral Catalogue*, Vol. 1. Ann Arbor: Dep. Astron., Univ. Mich.
73. Hudson, H. S. 1988. *Annu. Rev. Astron. Astrophys.* 26: 473
74. Hurly, P. R., Warner, B. 1983. *MNRAS* 202: 761
75. Jones, T. J., Wolff, S. C., Bonsack, W. K. 1974. *Ap. J.* 190: 579
76. Krause, F., Scholz, G. 1981. In *Upper Main Sequence Chemically Peculiar Stars. Proc. Liège Colloq. No. 23*, p. 323. Liège: Inst. Astrophys., Univ. Liège
77. Kreidl, T. J. 1985. *MNRAS* 216: 1017
78. Kreidl, T. J. 1986. *Lect. Notes Phys.* 274: 134
79. Kreidl, T. J., Kurtz, D. W. 1986. *MNRAS* 220: 313
80. Kurtz, D. W. 1976. *Ap. J. Suppl.* 32: 651
81. Kurtz, D. W. 1978. *Ap. J.* 221: 869
82. Kurtz, D. W. 1980. *MNRAS* 193: 29
83. Kurtz, D. W. 1980. *MNRAS* 191: 115
84. Kurtz, D. W. 1981. *MNRAS* 196: 61
85. Kurtz, D. W. 1982. *MNRAS* 200: 807
86. Kurtz, D. W. 1983. *MNRAS* 202: 1
87. Kurtz, D. W. 1983. *MNRAS* 205: 3
88. Kurtz, D. W. 1984. *MNRAS* 206: 253
89. Kurtz, D. W. 1984. *MNRAS* 209: 841
90. Kurtz, D. W. 1984. *Proc. Workshop Improv. to Photom., NASA CP-2350*, ed. W. J. Borucki, A. Young, p. 56. Washington, DC: NASA
91. Kurtz, D. W. 1986. In *Seismology of the Sun and Distant Stars*, ed. D. O. Gough, p. 417. Dordrecht: Reidel
92. Kurtz, D. W. 1988. In *Multimode Stellar Pulsations*, ed. G. Kovács, L. Szabados, B. Szeidl, p. 107. Budapest: Konkoly Obs.-Kultura
93. Kurtz, D. W. 1988. *MNRAS* 233: 565
94. Kurtz, D. W. 1989. *MNRAS* 238: 1077
95. Kurtz, D. W. 1989. *MNRAS* 238: 261
96. Kurtz, D. W. 1990. *MNRAS* 242: 489
97. Kurtz, D. W., Balona, L. A. 1984. *MNRAS* 210: 779
98. Kurtz, D. W., Cropper, M. S. 1987. *MNRAS* 228: 125
99. Kurtz, D. W., Kreidl, T. J. 1985. *MNRAS* 216: 987
100. Kurtz, D. W., Marang, F. 1987. *MNRAS* 228: 141
101. Kurtz, D. W., Marang, F. 1987. *MNRAS* 229: 285
102. Kurtz, D. W., Marang, F. 1988. *MNRAS* 231: 1039
103. Kurtz, D. W., Martinez, P. 1987. *MNRAS* 226: 187
104. Kurtz, D. W., Matthews, J. M., Martinez, P., Seeman, J., Cropper, M. S., et al. 1989. *MNRAS* 240: 881
105. Kurtz, D. W., Schneider, H., Weiss, W. W. 1985. *MNRAS* 215: 77

106. Kurtz, D. W., Seeman, J. 1983. *MNRAS* 205: 11
107. Kurtz, D. W., Shibahashi, H. 1986. *MNRAS* 223: 557
108. Kurtz, D. W., van Wyk, F., Marang, F. 1990. *MNRAS* 243: 289
109. Kurtz, D. W., Wegner, G. 1979. *Ap. J.* 232: 510
110. Landstreet, J. D. 1988. *Ap. J.* 326: 967
111. Ledoux, P. 1951. *Ap. J.* 114: 373
112. Libbrecht, K. G. 1988. *Ap. J. Lett.* 330: L51
113. Manfroid, J., Mathys, G. 1986. *Astron. Astrophys. Suppl.* 64: 9
114. Martinez, P. 1989. *MNRAS* 238: 439
115. Martinez, P., Kurtz, D. W. 1990. *MNRAS* 242: 636
116. Mathys, G. 1985. *Astron. Astrophys.* 151: 315
117. Matthews, J. M. 1986. In *The Study of Variable Stars Using Small Telescopes*, ed. J. R. Percy, p. 117. Cambridge: Univ. Press
118. Matthews, J. M. 1988. *MNRAS* 235: 7p
118a. Matthews, J. M. 1990. *Lect. Notes Phys.* In press
119. Matthews, J. M., Kreidl, T. J., Wehlau, W. W. 1988. *Publ. Astron. Soc. Pac.* 100: 255
120. Matthews, J. M., Kurtz, D. W., Wehlau, W. H. 1986. *Ap. J.* 300: 348
121. Matthews, J. M., Kurtz, D. W., Wehlau, W. H. 1986. See Ref. 41, p. 239
122. Matthews, J. M., Kurtz, D. W., Wehlau, W. H. 1986. See Ref. 36a, p. 261
123. Matthews, J. M., Kurtz, D. W., Wehlau, W. H. 1987. *Ap. J.* 313: 782
124. Matthews, J. M., Slawson, R. W., Wehlau, W. H. 1986. In *Hydrogen Deficient Stars and Related Objects*, ed. K. Hunger, D. Schönberner, N. K. Rao, p. 313. Dordrecht: Reidel
125. Matthews, J. M., Wehlau, W. H., Walker, G. A. H., Yang, S. 1988. *Ap. J.* 324: 1099
126. Mégessier, C. 1988. *Astron. Astrophys. Suppl.* 72: 551
127. Michaud, G. 1970. *Ap. J.* 160: 641
128. Michaud, G. 1980. *Astron. J.* 85: 589
129. Moskalik, P. 1986. *Acta Astron.* 36: 333
130. Odell, A. P., Kreidl, T. J. 1984. In *Theoretical Problems in Stellar Stability and Oscillations. Proc. Liège Colloq. No. 25*, p. 148. Liège: Inst. Astrophys., Univ. Liège
131. Olsen, E. H. 1983. *Astron. Astrophys. Suppl.* 54: 55
132. Olsen, E. H., Perry, C. L. 1984. *Astron. Astrophys. Suppl.* 54: 229
133. Osaki, Y. 1971. *Publ. Astron. Soc. Jpn.* 23: 485
134. Percy, J. R. 1987. In *Physics of Be Stars, IAU Colloq. No. 92*, ed. A. Slettebak, T. P. Snow, p. 49. Cambridge: Univ. Press
135. Pesnell, W. D. 1989. *Ap. J.* 339: 1038
136. Preston, G. W. 1972. *Ap. J.* 175: 465
137. Preston, G. W. 1974. *Annu. Rev. Astron. Astrophys.* 12: 257
138. Przybylski, A. 1961. *Nature* 189: 739
139. Przybylski, A. 1963. *Acta Astron.* 13: 217
140. Przybylski, A. 1966. *Nature* 210: 20
141. Przybylski, A. 1977. *MNRAS* 177: 71
142. Przybylski, A. 1977. *MNRAS* 178: 735
143. Przybylski, A. 1982. *Ap. J. Lett.* 257: L83
143a. Rakosch, K. D., Sexl, R., Weiss, W. W. 1974. *Astron. Astrophys.* 31: 441
144. Rice, J. B. 1988. *Astron. Astrophys.* 199: 299
145. Schneider, H., Weiss, W. W. 1989. *Astron. Astrophys.* 210: 147
146. Schneider, H., Weiss, W. W., Kreidl, T. J., Odell, A. P. 1987. *Lect. Notes Phys.* 274: 138
147. Scholz, G. 1979. *Astron. Nachr.* 300: 213
148. Shibahashi, H. 1983. *Ap. J. Lett.* 275: L5
149. Shibahashi, H. 1986. In *Hydrodynamic and Magnetohydrodynamic Problems in the Sun and Stars*, ed. Y. Osaki, p. 195. Tokyo: Univ. Tokyo
150. Shibahashi, H. 1987. *Lect. Notes Phys.* 274: 112
151. Shibahashi, H., Saio, H. 1985. *Publ. Astron. Soc. Jpn.* 37: 245
152. Shore, S. N., Brown, D. N., Sonneborn, G., Gibson, D. M. 1987. *Astron. Astrophys.* 182: 285
153. Smith, M. A. 1982. *Ap. J.* 254: 242
154. Smith, M. A. 1986. In *Hydrodynamic and Magnetohydrodynamic Problems in the Sun and Stars*, ed. Y. Osaki, p. 145. Tokyo: Univ. Tokyo
155. Smith, M. A., Fullerton, A. W., Percy, J. R. 1987. *Ap. J.* 320: 768
156. Smith, P. H., McMillan, R. S. 1988. In *The Impact of Very High S/N Spectroscopy on Stellar Physics, IAU Symp. No. 132*, ed. G. Cayrel de Strobel, M. Spite, p. 291. Dordrecht: Kluwer
157. Stamford, P. A., Watson, R. D. 1981. *Astrophys. Space Sci.* 77: 131
158. Stibbs, D. W. N. 1950. *MNRAS* 110: 395
159. Szeidl, B. 1988. In *Multimode Stellar Pulsations*, ed. G. Kovács, L. Szabados, B. Szeidl, p. 45. Budapest: Konkoly Obs.-Kultura
160. Tassoul, M. 1980. *Ap. J. Suppl.* 43: 469

161. Thompson, I. B. 1983. *MNRAS* 205: 43P
162. Deleted in proof
163. Ulrich, R. K. 1986. *Ap. J. Lett.* 306: L37
164. Ulrich, R. K. 1988. See Ref. 36a, p. 299
165. Vogt, N., Faúndez, M. 1979. *Astron. Astrophys. Suppl.* 36: 477
166. Vogt, S. S., Penrod, G. D. 1983. *Ap. J.* 275: 661
167. Walker, A. R. 1984. *Proc. Workshop Improv. to Photom.*, *NASA CP-2350*, ed. W. J. Borucki, A. Young, p. 177. Washington, DC: NASA
168. Watson, R. D. 1988. *Astrophys. Space Sci.* 140: 255
169. Wegner, G. 1976. *MNRAS* 177: 99
170. Wegner, G. 1981. *Ap. J.* 247: 969
171. Wegner, G., Cummins, D. J., Byrne, P. B., Stickland, D. J. 1983. *Ap. J.* 272: 646
172. Wegner, G., Petford, A. D. 1974. *MNRAS* 168: 557
173. Weiss, W. W. 1983. *Astron. Astrophys.* 128: 152
174. Weiss, W. W. 1983. *Hvar Obs. Bull.* 7: 263
175. Weiss, W. W. 1986. See Ref. 41, p. 219
176. Weiss, W. W., Schneider, H. 1984. *Astron. Astrophys.* 135: 148
177. Weiss, W. W., Schneider, H. 1989. *Astron. Astrophys.* 224: 101
178. Winget, D. E. 1988. See Ref. 36a, p. 305
179. Winget, D. E., Fontaine, G. 1982. In *Pulsations in Classical and Cataclysmic Variables*, ed. J. P. Cox, C. J. Hansen, p. 46. Boulder, Colo: JILA
180. Wolff, S. C. 1975. *Ap. J.* 202: 127
181. Wolff, S. C. 1983. *The A-type Stars: Problems and Perspectives. NASA SP-463.* Washington, DC: NASA
182. Wolff, S. C., Hagen, W. 1976. *Publ. Astron. Soc. Pac.* 88: 119
183. Wood, H. J., Campusano, L. B. 1975. *Astron. Astrophys.* 45: 303
184. Woodard, M., Hudson, H. S. 1983. *Nature* 305: 589

THE SOFT X-RAY BACKGROUND AND ITS ORIGINS

Dan McCammon and Wilton T. Sanders

Physics Department, University of Wisconsin, Madison, Wisconsin 53706

KEY WORDS: X rays, diffuse background, interstellar medium, galaxies

1. Introduction

The sky in the soft X-ray region of the electromagnetic spectrum (photon energies between roughly 0.1 and 10 keV) is far from dark: Over this entire energy range, the diffuse flux is much greater than the contribution from identifiable discrete sources. It now appears that there are several diffuse X-ray backgrounds, and despite the title of this review, we are not all that sure of the origin of any of them.

Soft X rays are an energetic phenomenon. Thermal production over this energy range requires temperatures in the range 10^6–10^8 K. Nonthermal processes require 50–500 GeV electrons for synchrotron radiation in a typical galactic magnetic field, 5–50 MeV electrons for inverse Compton scattering of starlight photons, and 300–3000 MeV electrons for inverse Compton on microwave background photons.

The dominant interaction with matter for X rays below 10 keV is photoelectric absorption. This cross section increases rapidly at lower energies, scaling approximately as E^{-3}. As shown in Figure 1, this results in a range of mean free paths in an assumed average interstellar density of 1 H atom cm^{-3} that encompasses the entire range of galactic distance scales. At 0.1 keV, X rays would barely reach us from neighboring stars (although the use of an average density here is certainly not correct), whereas at 10 keV the entire diameter of the disk is transparent. Interstellar scattering can be ignored over the entire energy range. The sharp jumps

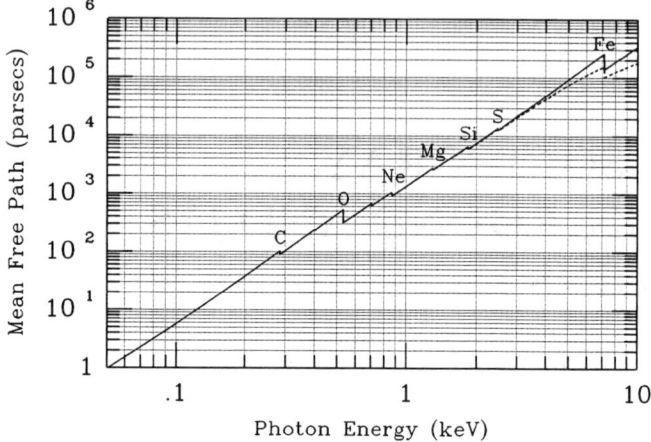

Figure 1 Interstellar mean free path at an assumed average density of 1 H atom cm^{-3} and assumed solar abundances. The solid line is photoelectric absorption only, whereas the dashed line includes Thomson scattering. Elements responsible for the major absorption edges are indicated. Data are from (88).

in the mean free path are caused by atomic absorption edges of the more abundant heavy elements. These provide an opportunity for direct measurement of interstellar column densities of particular species.

Diffuse X rays were one of the first discoveries of X-ray astronomy. In 1962, a sounding rocket carrying Geiger counters sensitive in the 2–6 keV range observed both the first "X-ray star," Scorpius X-1, and a copious isotropic flux (36). The observed isotropy implies that the origin is either very local or very distant, since all known material in the Galaxy is distributed quite anisotropically as viewed from the Sun. The lack of interstellar absorption at these energies means that there is nothing to restrict our view to a local source, so only an extragalactic origin seems possible. This was an exciting discovery, as it was the first cosmological diffuse flux ever observed. It was followed a few years later by the discovery of the cosmic microwave background by Penzias & Wilson (96), and these remain the only regions of the electromagnetic spectrum where extragalactic diffuse radiation is not overwhelmed by local sources. Further observations have shown that the photon flux extends to at least 100 MeV (31, 71), where it is dominated by the harder spectrum of galactic diffuse emission produced through interactions of cosmic rays with interstellar material (5).

There are a number of theories for the origin of the extragalactic component of the radiation. The integrated emission from active galactic nuclei probably contributes significantly to it and could conceivably account for all of the flux (37, 87a). On the other hand, thermal bremsstrahlung from a thin intergalactic plasma with $kT = \sim 40$ keV provides an excellent fit to the observed spectrum from 3 to 50 keV (79). There are difficulties with both of these models, however, and some unknown population of sources at large redshifts may be responsible for the bulk of the flux (116, and references therein).

Soon after the discovery of the soft X-ray background flux, it was speculated that when instruments capable of observing the extragalactic flux below 1 keV were developed, one could study the distribution of interstellar material in the Galaxy by looking at the patterns of absorption. In effect, the isotropic background could be used as a source to "x-ray" the interstellar medium. The first successful measurement in this energy range was reported by Bowyer et al in 1968 (10). The observed flux at low energies decreased as expected at low galactic latitudes, where the interstellar absorption of an extragalactic source should be greater. Shortly thereafter, it was noted that the absolute intensity of the flux far exceeded an extrapolation of the spectrum observed above 2 keV (49). At that time it seemed natural to interpret this excess as another component of the extragalactic flux. A particularly attractive proposition attributed it to thermal radiation from an intergalactic plasma at a density somewhat larger than the critical value required to close the Universe. The inferred temperature, about 10^6 K, was within the narrow allowed range for such a dense medium, bounded below by the lack of $L\alpha$ absorption continua in quasar spectra, and above by the requirement that its thermal radiation not exceed the X-ray flux observed above 2 keV (131). As discussed below, observational details never really favored this interpretation, but the theoretical attractiveness of the model and the lack of reasonable alternatives kept it alive for a number of years.

It is now generally thought that the excess below 0.25 keV is due to thermal radiation from a 10^6-K component of the interstellar gas in our Galaxy, and that most of it originates within a few hundred parsecs of the Sun. The degree of pervasiveness of such hot gas in the galactic disk is unknown, but the possibility that it is common has dramatically altered our view of the structure and evolution of the interstellar medium (ISM). Since 10^6-K gas can most easily be detected through its X-ray emission, diffuse soft X-ray observations have indeed become an important tool for studying the ISM, although not in the mode originally envisioned.

This review is concerned primarily with diffuse galactic emission. However, in the intermediate energy range (0.5–1 keV) it is not yet clear just how much of the observed radiation is galactic or extragalactic, so we cannot entirely isolate this subject. Even where it can be shown that almost all of the detected X rays are galactic, this is not the same thing as saying that there are no extragalactic X rays. The great intrinsic interest of extragalactic "background light" levels at any wavelength makes it worth using what understanding we have of the foreground emission to extract as much information as possible about the value of the extragalactic flux. Current limits and the prospects for improving them are discussed in Section 6.

A number of good reviews exist for observations of the extragalactic diffuse X-ray background and its interpretation (7, 37, 39, 75, 113, 116, 117). Of the previous reviews of the galactic diffuse emission (23, 105, 124), we particularly recommend the encyclopedic 1977 review by Tanaka & Bleeker (124). For many areas, it provides a more detailed discussion than we can include here, and it is a good reference to many of the important early experiments. In most cases Tanaka & Bleeker reached the same conclusions that we do here, although a good deal of intuition and taste were required on their part to extract these from the more limited and conflicting data available at that time.

2. Emission Mechanisms for the Galactic Diffuse Background

Several authors have considered whether the soft X-ray background could arise from the superposition of point sources or if it must be truly diffuse (12, 38, 76, 128). The low level of intensity fluctuations of the 0.25-keV background implies a space density of point source emitters of at least 0.2 pc^{-3}, greater than that of all stars. Stellar X-ray luminosity functions measured using the *Einstein* satellite (19, 47, 104; J. Schmitt & S. Snowden, private communication) predict that normal stars of all types produce less than a few percent of the observed 0.25-keV diffuse background and no more than 20% of the 0.5–1.0 keV flux.

Several diffuse nonthermal mechanisms have been considered to explain the soft X-ray background and have been rejected because they predict other effects that are not observed (40, 70, 132). The electron flux needed for soft X-ray production by bremsstrahlung on interstellar gas or inverse Compton scattering from starlight would produce much more ionization of the interstellar gas than is allowed. Electrons capable of producing the soft X-ray background by inverse Compton scattering from the 3-K microwave background or as transition radiation from collisions with dust would also produce a flux of 100-MeV gamma rays much larger than is observed. Electrons energetic enough to produce soft X-rays by the synchrotron process in the galactic field would lose half their energy on a time scale of 5000 yr.

Diffuse thermal emission at temperatures of $\sim 10^6$ K remains as the probable source of the 0.25-keV X-ray background, and at 2–4×10^6 K it may well be responsible for most of the galactic emission in the 0.5–1.0 keV range. Direct evidence for hot interstellar gas at slightly lower temperatures has come from the *Copernicus* observations of O VI absorption (59–61). Radiation from plasmas in the $0.5-4 \times 10^6$ K range is entirely dominated by collisionally excited emission lines of the partially ionized heavy elements ($Z \geq 5$). Figure 2 shows calculated X-ray emission spectra for gas with solar abundances in collisional equilibrium at temperatures in the range 10^5-10^7 K (98; see also 66, 87, 99, 99a, 123). Times required for such plasmas to reach collisional equilibrium can be long compared with heating and cooling time scales, so considerable work has been done

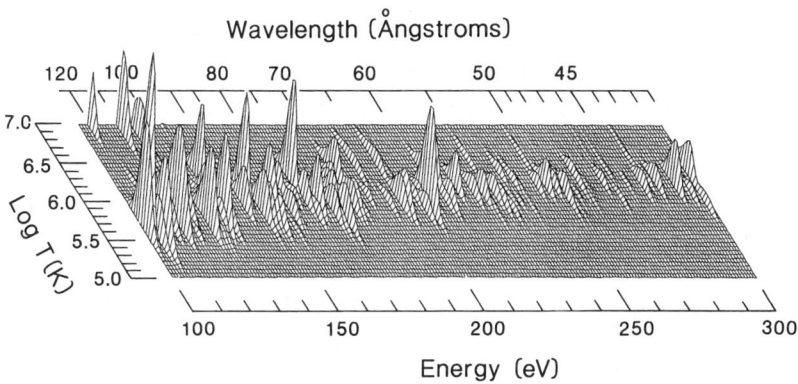

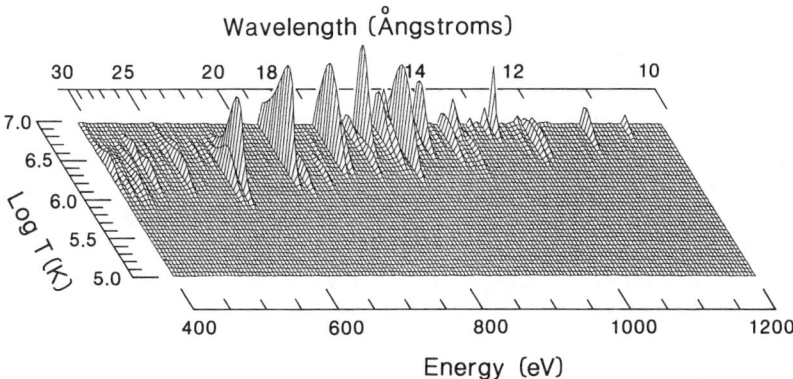

Figure 2 Calculated X-ray emission for an optically thin plasma with solar abundances in collisional equilibrium (98). Spectra are shown as a function of the gas temperature from 10^5 to 10^7 K. The vertical scales are linear in photon emissivity.

also on calculations for various nonequilibrium situations (1, 22, 29, 50, 119, 120a).

While high-resolution spectroscopy of the diffuse X-ray background is currently in its infancy, there is clearly a wealth of information potentially available on the detailed physical conditions in these hot regions of the ISM. Not only can temperature distributions and elemental abundances be measured with precision, but it should also be possible to determine much about the past history of the gas, including the time elapsed since it was heated.

3. Observations

INSTRUMENTATION The Geiger counters of the first rocket flights were soon replaced with proportional counters, for which the amplitude of the signal pulses is proportional to the energy of the absorbed photon. The spectral resolving power ($E/\Delta E$) of these detectors is limited to about 2.5 at 1 keV by fundamental statistical processes in the gas, and it scales as $E^{1/2}$. Variants exist that can double this resolving power (97, 115, 120), but so far only limited measurements of diffuse emission have been made with them (51, 67, 68).

In addition to information about the spectrum of the incident X rays, the pulse-height distribution provides invaluable discrimination against non-X-ray events. This is particularly important for diffuse background studies, where it is not possible to use the discrete source technique of estimating backgrounds by observing a nearby area of empty sky. Other valuable characteristics of proportional counters are a very low background rate and the ease of scaling to large areas with relatively low weight and cost. A more extensive discussion of their application to soft X-ray observations can be found in (55).

The revolution in sensitivity for discrete source measurements provided by imaging X-ray optics has largely bypassed diffuse background studies. The large focal ratios of X-ray telescopes give them less throughput than mechanically collimated detectors for angular resolutions coarser than a degree or so, and thus far no large telescope/detector combination has been flown that was sufficiently well optimized for extended sources to allow routine diffuse background measurements to be made with higher angular resolution. A few experiments, most notably SAS 3 (45), have successfully used nonimaging reflectors to improve the angular resolution, but most of the data discussed below were obtained with mechanically collimated proportional counters.

The limited energy resolution of proportional counters becomes particularly serious below 1 keV, and at 0.25 keV the resolving power is down to about unity (although the shape of the pulse-height distribution continues to provide valuable discrimination against non-X-ray counts). Additional spectral information is generally obtained through the use of atomic absorption-edge filters, either placed in front of the counter or incorporated into its entrance window. The bandpass resulting from these filters has a characteristic shape with a sharp cutoff on the high-energy side and a gradual drop toward low energies. This can be seen in the Wisconsin Be-, B-, and C-band response curves of Figure 3; the figure shows the energy discrimination that can be obtained below 2 keV with conventional proportional counters by using pulse-height discrimination at the higher energies and K-edge filters of beryllium, boron, and carbon at lower energies.

Future instruments will include imaging telescopes with fast optics optimized for energies of ~ 1 keV and below, such as the upcoming ROSAT observatory (125). With appropriate detectors, these will enable observations with greatly improved angular resolution. Bragg spectrometers are awkward for diffuse sources, but the Wisconsin Diffuse X-ray Spectrometer (111) will obtain spectra of the 0.25-keV background with ~ 10-eV resolution when it flies as an attached Space Shuttle payload. Microcalorimeters can have similar or better energy resolution with much higher throughput for a given detector size (89, 134).

MAJOR DATA SETS Several experiments have mapped diffuse soft X-ray emission from a large fraction of the sky: (*a*) A University of Wisconsin

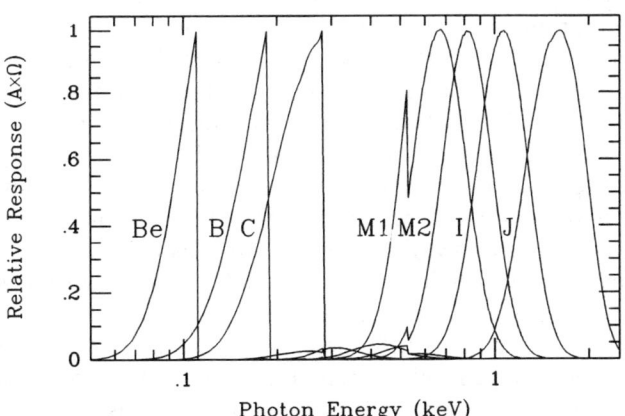

Figure 3 Band response functions obtainable from proportional counters below 2 keV. The M1, M2, I, and J bands are defined primarily by using the pulse-height resolution of the counter. The Be, B, and C bands are defined primarily by the transmission of a filter made of the corresponding element (from 4, 84).

sounding-rocket survey resulted in a set of seven maps covering the energy range 0.12–6 keV with 7° angular resolution and a typical exposure of 500 cm^2 s (84). These data are available on magnetic tape from the National Space Science Data Center or from the authors. (b) Marshall & Clark (80) have published a 0.25-keV (C-band) map using data from the *SAS 3* satellite. It has 4.5° resolution and an average exposure of $\sim$1000 cm^2 s. (c) The Penn State group (G. Garmire, J. A. Nousek, D. N. Burrows & R. L. Fink) has recently completed a set of maps from the A2 LED detectors on *HEAO 1*. Five maps cover the energy range 0.12–3 keV with 3° resolution and typical exposures of $\sim$8000 cm^2 s (32). The maps are available from those authors as FITS images. (d) The Goddard Space Flight Center X-ray group (E. Boldt et al) is preparing 2–20 keV maps from the *HEAO 1* A2 MED and HED detectors (106) and expects to publish them in 1990. These data are not yet distributed, but they are available at Goddard. Figures 4–7 compare sample maps from (a), (b), and (c) above.

CAVEATS: HOW RELIABLE ARE THE DATA? Extraneous signals are a serious problem for any diffuse background measurement. In principle, contamination levels can be estimated only by evaluating each physical process that is a potential contributor and then finding some way to estimate or limit its magnitude. There clearly can never be a guarantee that you have thought of everything.

The most important sources of non-X-ray contamination are cosmic rays, ultraviolet photons, and electrons in the 2–50 keV energy range that can penetrate thin detector windows. Noncosmic X rays may come from solar X rays scattered in the upper atmosphere, from thermalization of the solar wind, or from an intense flux of low-energy electrons that can produce X rays by bremsstrahlung from the atmosphere or from surfaces in the instrument itself. X-ray observations below 1 keV require thin detector windows and are the most susceptible to all of these effects. Electron contamination is particularly troublesome for satellite experiments, as it varies strongly with orbital position as well as with time. These mechanisms are discussed in detail in (81) and (82). Methods for estimating limits to their contributions are dealt with in (84).

Some general arguments help increase our confidence in the data. The most powerful of these result from multiple observations of the same part of the sky at different times and from different locations. We would expect any source associated with the Earth or the solar system to be variable in time on the many-year scale over which some of these observations have been repeated, and those associated with the upper atmosphere should look different from different altitudes or latitudes. Figures 4 and 5 show the $\sim$160–284 eV (C-band) maps from the Wisconsin and *SAS 3* sky surveys, respectively. A comparison of these is comforting, since, except for two areas in the Wisconsin survey that had been identified as likely to be contaminated by electrons (84), there is excellent feature-by-feature agreement between them. These maps were made with very different instruments (one had mechanical collimators, the other a reflective concentrator), from different vantage points (one was done from sounding rockets, the other from a satellite), and many years elapsed between the observations for some parts of the sky. A similar comparison can be made for the ~ 500–~ 1100 eV (M-band) maps from the Wisconsin and *HEAO 1* A2 LED surveys shown in Figures 6 and 7, respectively.

The band responses on the various surveys are slightly different, so a source spectrum must be assumed to make absolute flux comparisons. Using a $10^{6.0}$-K thermal emission spectrum that fits the average observed B/C-band rate ratio on the Wisconsin survey, the *SAS 3, HEAO 1*, and Wisconsin C-band average observed rates all agree within 10%. The M-band spectrum is more variable across the sky, but we again take the all-sky average M1/M2-band rate ratio from the Wisconsin survey to fix the temperature of a thermal emission model at $10^{6.5}$ K. With this source spectrum, the *HEAO 1* and Wisconsin M-Band (combined M1+M2 from the Wisconsin survey) average absolute fluxes fortuitously agree to better than 1%.

These comparisons are encouraging. However, large amounts of data were discarded in each of these surveys owing to internal disagreement of repeated observations or other indications of contamination, so time-variable contamination, usually from electrons, is a common occurrence. It is therefore prudent to regard with considerable caution any feature that has been observed only once. We also point out that while the possibility of an isotropic, time-invariant extraneous contribution seems remote, it cannot be ruled out by repeated observations, and many of the interpretations of the data discussed below would be profoundly affected by the removal of an isotropic component equal in magnitude to the minimum observed intensity.

MAJOR OBSERVATIONAL CHARACTERISTICS AT DIFFERENT ENERGIES For convenience in the following discussion, we divide the soft X-ray region into three energy ranges. These are not entirely arbitrary but are chosen because of evidence that different source mechanisms dominate in each.

2–10 keV Diffuse flux in this energy range comes largely from the extragalactic background. Between 3 and 10 keV, the spectrum can be fit by a $\sim 8E^{-0.4}$ keV (cm^2 s sr keV)$^{-1}$, power law [79, 113, 114; also personal communication of *Ginga* (126) results by Y. Tanaka]. There is some disagreement on the absolute intensity of this spectrum. Many sounding-

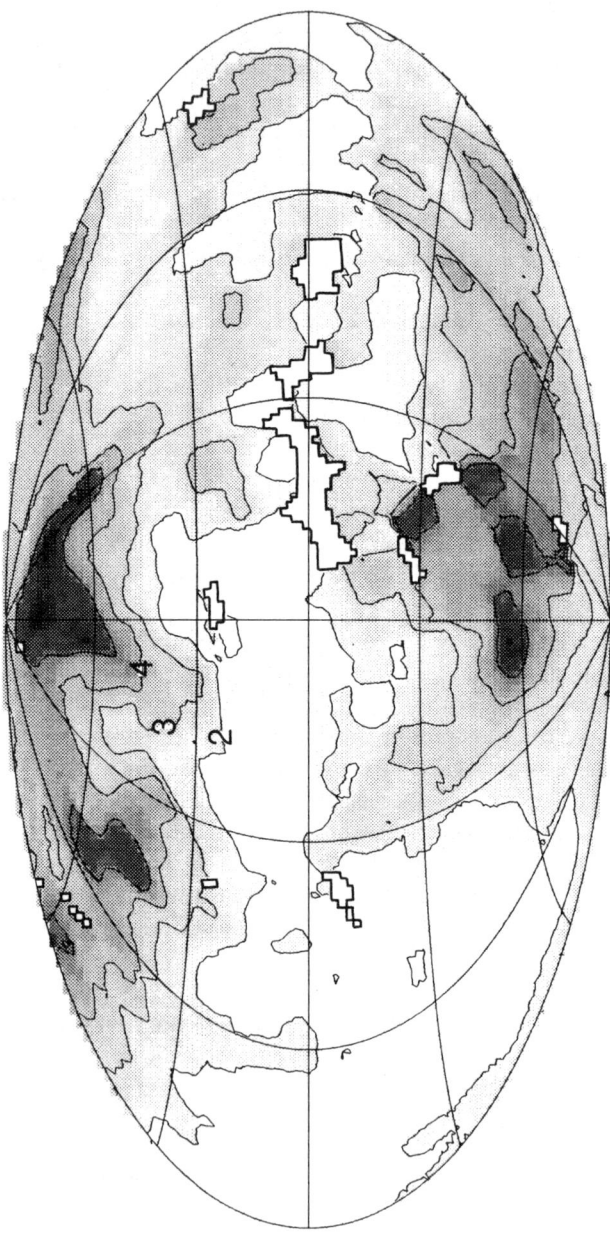

Figure 4 C-band (~160–284 eV) map from the Wisconsin sky survey (84). The map is in galactic coordinates, with $l = 0°$ at the center and increasing to the left. The contour unit is 0.001 cm^{-6} pc emission measure for a $10^{6.0}$-K equilibrium plasma with solar abundances (99a) and no interstellar absorption. Bright discrete sources have been removed from the Wisconsin maps.

THE SOFT X-RAY BACKGROUND   667

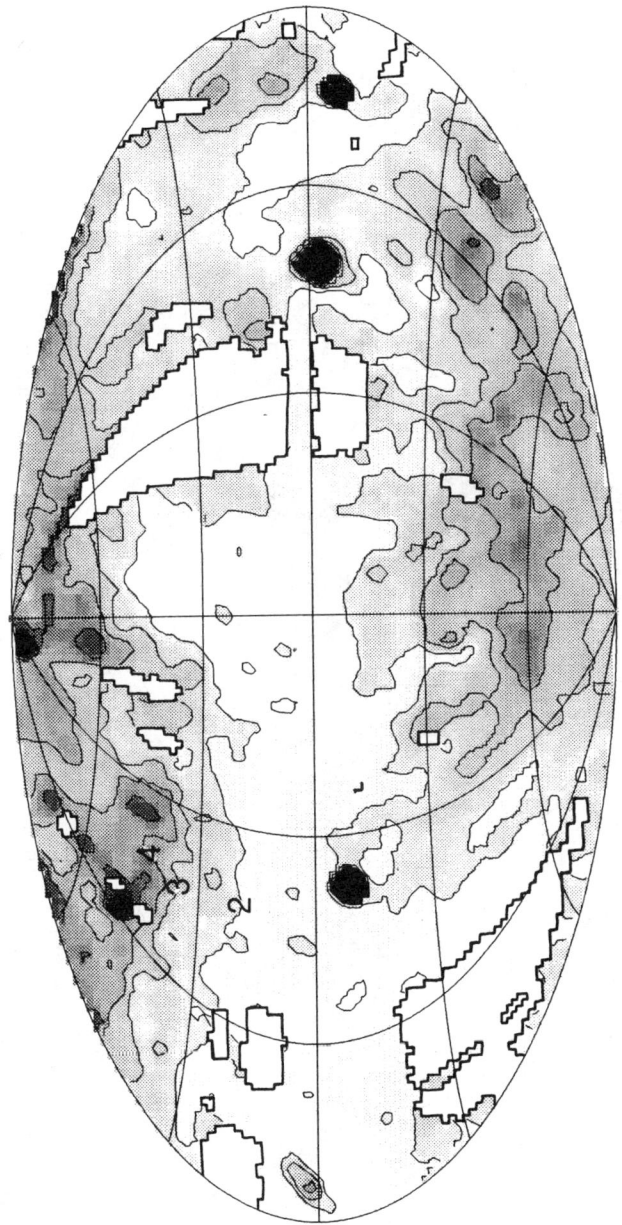

Figure 5 C-band (~130–284 eV) map from the *SAS 3* survey (80). Same projection and contours as in Figure 4.

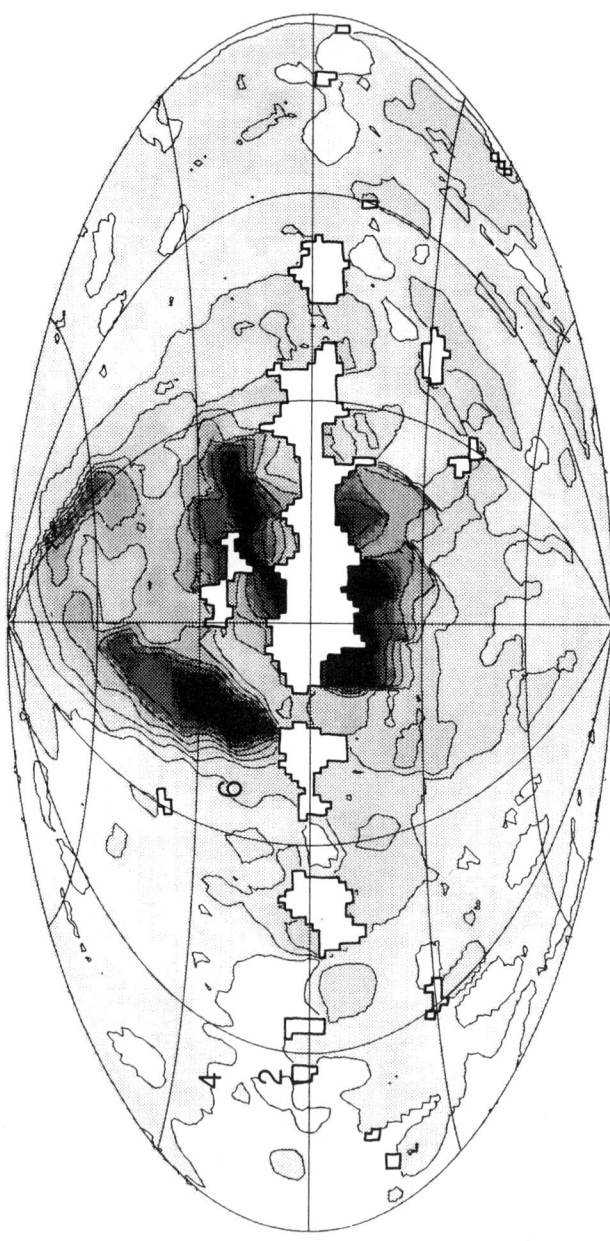

Figure 6 M-band (~490–~1090 eV) map from the Wisconsin sky survey (84). The map is in galactic coordinates, with $l = 0°$ at the center and increasing to the left. The contour unit is 0.001 cm^{-6} pc emission measure for a $10^{6.5}$-K equilibrium plasma with solar abundances (99a) and no interstellar absorption. Bright discrete sources have been removed from the Wisconsin maps.

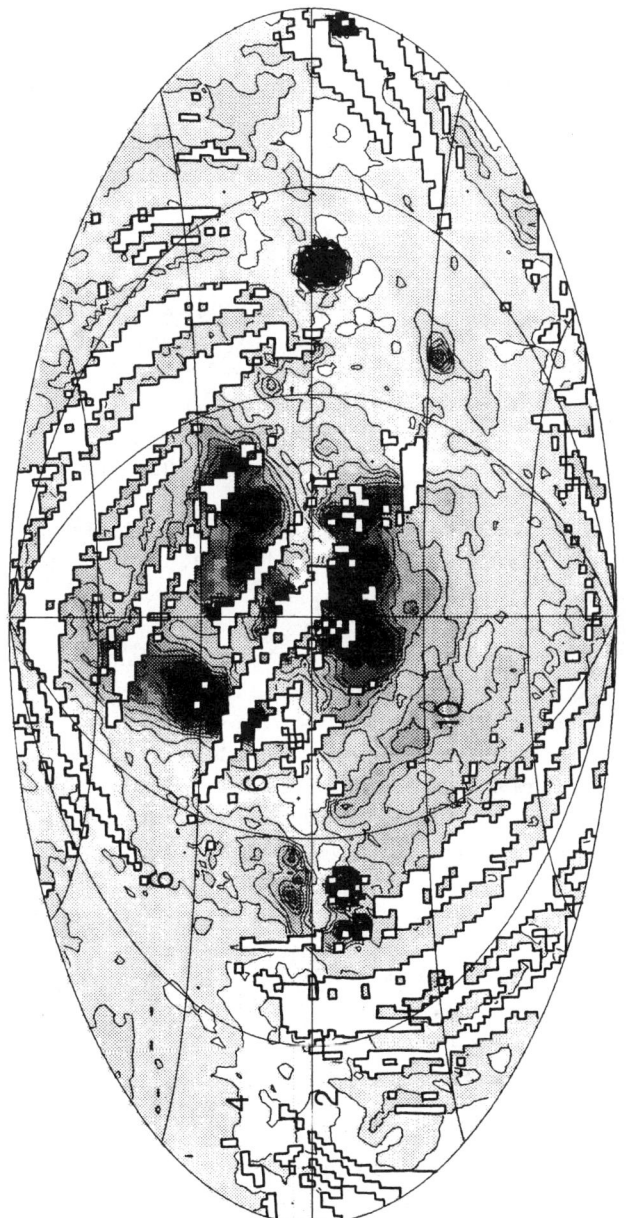

Figure 7 M-band (~480–~960 eV) map from the *HEAO 1* A2 LED survey (32). Same projection and contours as in Figure 6.

rocket experiments report fitting a power law closer to $11E^{-0.4}$ keV (cm² s sr keV)$^{-1}$ (13, 26, 33, 48, 94). The experiments giving the higher values in most cases had less sensitivity at the higher energies, and more sensitivity below 3 keV, so the obvious conclusion would be that the spectrum steepens at the low end of the 2–10 keV range. The results are not so easily reconciled, however, as this should also have resulted in a steeper power-law index in the thin-window experiments. The single most precise measurement in the 3–10 keV range seems to the *HEAO 1* A2 MED/HED survey (79), so we provisionally adopt the $8E^{-0.4}$ fit to their data, keeping in mind that the intensity (at least at 2 keV) may be as much as 30% above this spectrum.

This is an important point, since spectra for most of the discrete sources that are supposed to contribute significantly to the extragalactic background have been measured primarily in the 1–3 keV range. On the other hand, a galactic origin for the possible excess flux at 2 keV cannot be ruled out. Below 1 keV, the extragalactic flux is unknown: Here the poorly determined galactic emission could clearly be an important or dominant contributor.

In addition to the isotropic extragalactic flux, two relatively faint galactic emission components have been identified (54, 129, 130, 133). One is a narrow ridge confined to the galactic plane, with a scale height of ~ 250 pc and a radial extent of ~ 10 kpc. It has a spectrum that is considerably softer than the extragalactic background, being consistent with thermal plasma emission with kT varying from ~ 3 to 14 keV in different directions (67, 68). This spectral variation leads to the conclusion that if discrete sources are responsible, their number cannot be so large that variations are averaged out, implying a minimum luminosity of about 10^{33} ergs s^{-1} for the individual sources. The spectrum has been found to show strong emission lines of highly ionized iron, apparently confirming a thermal origin (68). The second emission component has a nominal scale height of several kiloparsecs (54). This component is much fainter but has a similar spectrum that recently has also been found to contain iron lines (67), although these were not seen by *HEAO 1* (54). The similar spectra indicate that both these components may have the same origin. It seems unlikely that either is truly diffuse, but no known class of discrete sources meets all of the requirements (127).

0.5–1 keV The spatial structure in this energy range is dominated by an irregular feature about 110° in diameter located in the general direction of the galactic center (see Figures 6 and 7) whose boundaries coincide roughly with the outline of radio Loop I [3; see also the H I map in the review by Dickey & Lockman (26a) in this volume]. The two brightest parts of the boundary of the northern half line up very well with the brightest sections of the radio ridge defining Loop I, and there can be little doubt that these are related. The bright knots in the interior appear to be associated with Loop I, but they might also represent emission from hot gas around the galactic center (35). Loop I and its relation to the X-ray emission have been discussed in a number of papers (8, 25, 42, 46, 53, 92). Two other, much smaller features that show up clearly on these maps are the Eridanus-Orion enhancement near 200°, −35° (90, 92, 101) and the Cygnus "superbubble" at 80°, +5° (20; this feature has been partially removed from Figure 6 along with the Cygnus Loop).

Aside from these features, the spatial distribution is quite smooth, with an intensity that exceeds an extrapolation of the $8E^{-0.4}$ keV (cm² s sr keV)$^{-1}$ extragalactic spectrum by more than a factor of 2. The intensity shows almost no variation in going across the galactic plane at most longitudes. This is quite curious, since an extragalactic or halo source should be completely absorbed in the galactic plane, while any source distributed throughout the disk should be much brighter near the plane than at the poles. Figure 8 shows the average latitude dependence of the M1+M2-band (~ 490–~ 1090 eV) data, excluding the quadrant $270° < l < 60°$ containing Loop I and a small region containing the Eridanus-Orion feature.

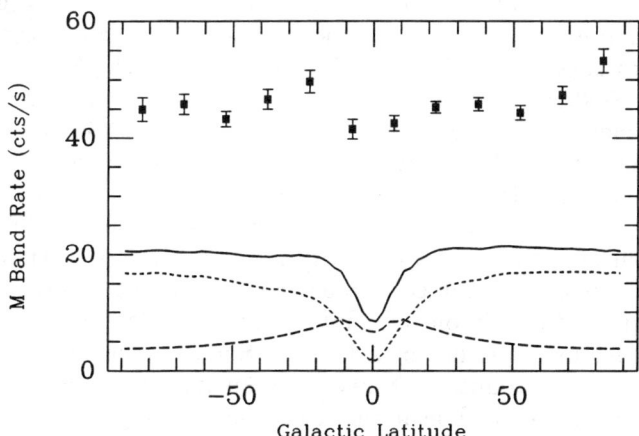

Figure 8 Average M-band (~ 490–~ 1090 eV) counting rate as a function of galactic latitude. Longitudes including Loop I ($270° < l < 60°$) have been excluded, as has a small region around the bright feature in Eridanus. The short-dashed line shows the calculated contribution from an assumed $8E^{-0.4}$ keV (cm² s sr keV)$^{-1}$ extragalactic source. The long-dashed line is the calculated contribution from dM stars (J. Schmitt & S. Snowden, personal communication). The solid line is the sum of these. Data are from (84).

A few observations of the 0.5–1 keV background have been made with instruments providing better than the usual proportional counter energy resolution. Inoue et al (51, 52) observed regions near the North Galactic Pole and North Polar Spur with a gas scintillation proportional counter with about 200-eV FWHM resolution at 600 eV. They report strong evidence for emission lines of O VII and O VIII. Schnopper et al (112) and Rocchia et al (103) observed essentially the same regions with a silicon solid-state detector ($\sim$150-eV FWHM resolution), and their data clearly show line emission due to O VII and good evidence for several other lines. This is a direct demonstration of the existence of an interstellar plasma in the $1–4 \times 10^6$ K temperature range in these directions, and these experiments represent the first step toward obtaining the wealth of information available in higher spectral resolution observations.

70–284 eV Proportional counters offer little useful energy resolution in this range, but K-edge filters allow it to be divided into three reasonably distinct bands, which we refer to as the Be, B, and C bands after the elements employed in the filters. Figure 3 shows the response of the Wisconsin version of these bands. Other thicknesses for the filters will change the low-energy cutoffs somewhat.

The spatial distribution at these energies is quite different from that in the M band, but all three of the low-energy bands appear quite similar to each other. As can be seen in Figures 4 and 5, the intensity is relatively low and uniform in the galactic plane and increases by a factor of 2 to 3 at high latitudes. There is a good deal of structure. The X-ray intensity below 284 eV shows a striking anticorrelation with the total column density of interstellar atomic hydrogen ($N_{H\,I}$). The anticorrelation between $N_{H\,I}$ and the Wisconsin C-band intensity is shown in Figure 9 and results from structure in longitude as well as from the general latitude variations. This is qualitatively the signature expected for an absorbed extragalactic flux, but it is difficult to fit quantitatively with an acceptable model. The effective absorption cross section must be much reduced from the normal atomic cross sections. In addition, attempts to find correlations with specific absorbing features have had mixed results. Some H I features that should be optically thick produce no X-ray variation at all, whereas others show clearly correlated reductions in the X-ray flux, although seldom with the expected quantitative behavior or energy dependence (16). This is in sharp contrast to the case for interstellar absorption of soft X rays from supernova remnants at known distances, which is found to be in good agreement with atomic cross sections (102, 107). As discussed in Section 5 below, it now seems probable that most, if not all, of the observed X rays below 284 eV come from closer than at least 90% of the interstellar gas.

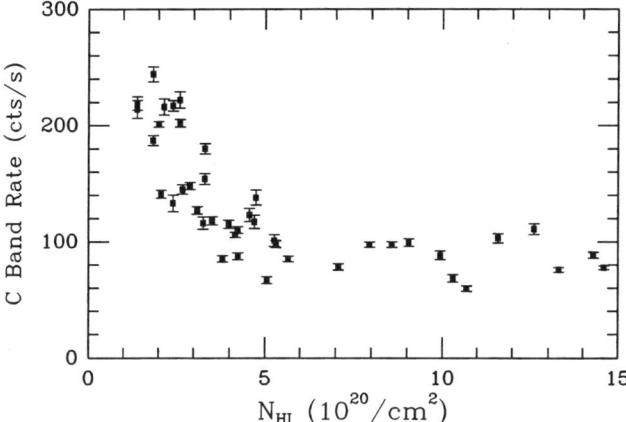

Figure 9 C band (~160–284 eV) rate from the Wisconsin survey (84) plotted against total H I column density (20a). Excluded regions of the sky are the same as in Figure 8. Plotted values are averages over 22.5° in galactic latitude by 30° in longitude. This X ray vs. H I anticorrelation has the qualitative appearance of an absorbed distant source, but this probably is not the case. See text.

The B band map (84) appears quite similar to the C-band map in Figure 4, although the ratio of counting rates varies significantly from one part of the sky to another. The B/C count-rate ratio is plotted against total H I column density in Figure 10, which shows that these variations have only a very slight correlation with $N_{H\,I}$. If one assumes that the X rays are produced by thermal emission from a collisional equilibrium plasma with solar abundances, the models of Raymond & Smith (99a; see also 98) require a temperature range of only $0.9–1.2 \times 10^6$ K to produce the observed range of ratios in the absence of significant interstellar absorption.

Sounding-rocket observations below the beryllium K-edge at 111 eV have sampled seventeen 15° × 15° fields (~10% of the sky) over a wide range of galactic latitudes and longitudes on two flights (4, 63). For almost all of these fields, the ratio of Be-band to B-band counting rates is the same to within the ~15% statistical precision of the individual measurements. (Figure 11 shows the as-yet unpublished data from the second flight.) This is considerably more constant than the B/C ratio: a surprising result, because the interstellar absorption cross sections for the C and B bands differ by only a factor of 2, whereas the Be-band effective cross section is about 6 times larger than that for the B band. Since an H I column density of only 1×10^{19} cm^{-2} (about 5% of the average high-latitude total) represents one optical depth for the Be band, very small amounts of

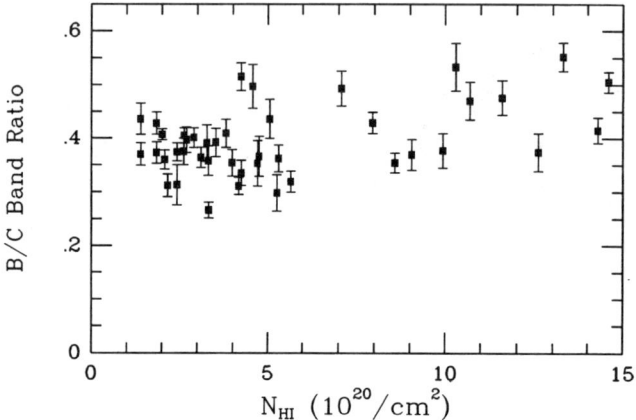

Figure 10 Ratio of counting rates in the B and C bands plotted against total H I column density (20a). The variations in this ratio are spatially very coherent: All points with B/C > 0.4 are contiguous in a region roughly centered on $l = 160°$, $b = +15°$. Excluded regions are the same as in Figure 8.

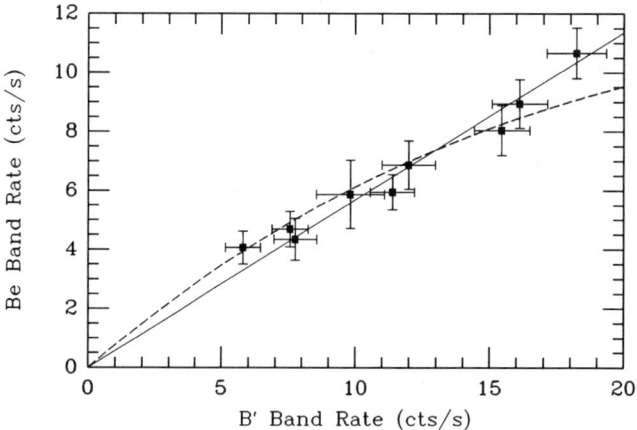

Figure 11 Be-band (~70–111 eV) vs. B'-band (~100–188 eV) rates for nine 15° × 15° fields widely distributed across the sky (62). The straight line shows the relation expected if there is no intermixed absorbing gas. The dashed curve shows the relation expected for uniformly intermixed absorbing material with a total column density of $\sim 1.4 \times 10^{19}$ cm^{-2} to the edge of the X-ray-emitting gas for the brightest point. This is a 2σ upper limit to what is allowed by the data.

foreground absorption should produce large changes in the Be/B ratio. The observed ratio is reproduced by the 1984 Raymond & Smith model for a 10^6-K plasma with no absorption, and the predicted ratio is almost independent of temperature over the small interval required to produce the observed range of B/C ratios (4). One should be cautious about regarding this as more than a happy coincidence, however, since the simple assumptions of this plasma emission model may be a poor approximation to reality.

A recent, partially analyzed third rocket flight observed an additional eight fields in the Be band, including some targets where the Be/B ratio might be expected to be different owing to the presence of a nearby absorbing cloud or a more distant source that should contribute mostly to the B band. The five random fields showed the same relation to the B band as the previous flights, while the other three were 25–50% lower (B. Edwards, personal communication). Analysis is proceeding to see if these results are consistent with interstellar absorption or with emission from much higher temperatures.

A grating spectrometer experiment has been flown on a rocket by the Berkeley group in a pioneering effort to obtain a high-resolution spectrum of the diffuse background in the 20–150 eV energy range (73). This first attempt produced marginal detections for two soft X-ray lines at 70 and 125 eV and demonstrates the promise of this type of observation.

Overview Figure 12 shows the total soft X-ray diffuse background spectrum from (84). The plotted points are the all-sky average fluxes for each of the energy bands, excluding the quadrant containing Loop I. The upper end of the vertical "error bar" is the average for $|b| > 60°$, and the lower end is the average for $|b| < 20°$. This figure should be interpreted with some caution. First, a cursory examination of the maps shows that this spectrum is a composite of independent components that dominate at different energies. Second, the differential fluxes have been plotted assuming continuous spectra, whereas thermal emission lines probably dominate for $E < 1$ keV. [Sufficient information for determining the corresponding thermal emission measures can be found in (84).] The horizontal error bars show the shift in the appropriate effective energy for the point in going from an assumed $E^{-0.5}$ to an $E^{-1.5}$ spectrum.

4. *Models of the 0.5–1 keV Emission*

The M-band sky consists of a few bright extended features superimposed on an otherwise nearly isotropic background. The most prominent features are radio Loop I, with the associated North Polar Spur, the Eridanus enhancement, and the Cygnus superbubble. Each is associated with struc-

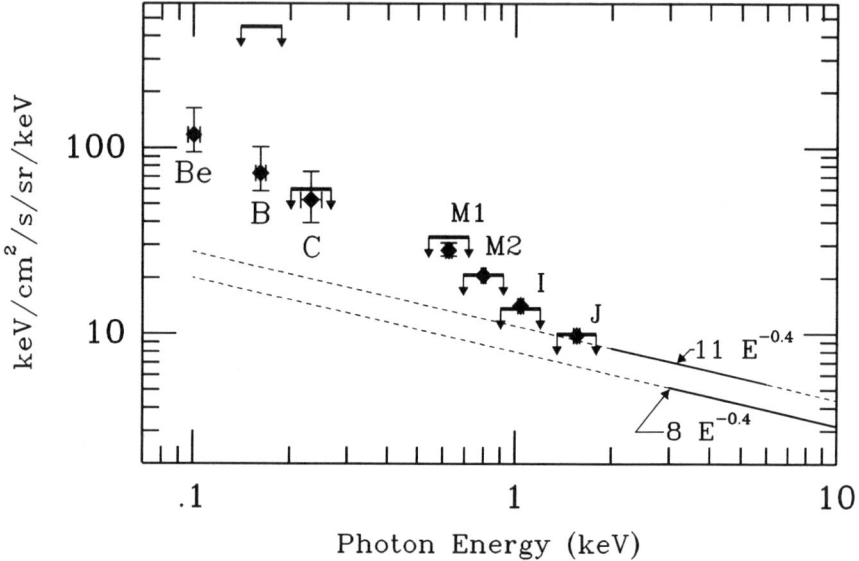

Figure 12 Average observed diffuse background intensity vs. energy. The plotted points are the all-sky averages (with the same exclusions as in Figure 8). The upper "error bar" is the average for $|b| > 60°$, while the lower "error bar" is for $|b| < 20°$. Be-band points assume that the unobserved part of the sky also tracks the B band. The intensities are plotted for an assumed E^{-1} power-law spectrum. The correct effective energy for each point is sensitive to the assumed spectral index: The horizontal error bars show the effect of a ± 0.5 change. The heavy bars represent the best upper limits that can be placed at each energy for the flux incident on the Galaxy from an extragalactic or halo source.

tures observed in the same directions at 21 cm or other wavelengths and has been modeled as a large cavity in the ISM, formed either by stellar winds from young associations or by a series of supernovae, and then heated at a large diameter by a subsequent supernova (8, 17, 20, 25, 42, 43, 46, 53, 91, 92, 101). These features are seen also in the C band where interstellar absorption permits.

The remaining flux mirrors none of the strong latitude and longitude dependence of the 0.25-keV background. Equilibrium thermal emission at 1.0×10^6 K that would provide all of the 0.25-keV flux in fact would contribute only a few percent to the M bands. It is possible to concoct emission mechanisms that produce the entire 0.25–1 keV spectrum in a particular direction (1, 28, 50), but given the disparate spatial structure above and below 0.5 keV, there is little motivation for doing so.

The minimal latitude dependence shown in Figure 8 is a strong observational constraint. The intermediate interstellar mean free path (800 pc at 0.78 keV) implies a strong latitude dependence for anything but a very local source. Two contributors that should exist at some level are an extrapolation of the observed extragalactic spectrum and the integrated contribution of dM stars (19, 47, 104; J. Schmitt & S. Snowden, private communication). Estimates for both of these are shown in Figure 8. Their combined latitude dependence is consistent with the observations, but an additional, nearly isotropic component is required for $\sim 55\%$ of the flux. So far, we know of no proposal for a local isotropic M-band source compatible with an anisotropic local 0.25-keV source. Several published attempts to model the M-band flux either fail to predict the observed latitude dependence near the plane (64, 92) or fail to reproduce the observed longitudinal isotropy (14, 109).

5. *Models of the 70–284 eV Emission*

Essential features of the data that must be addressed are (*a*) that the soft X-ray intensity is lowest in the galactic plane and increases by a factor of 2–3 at high latitudes; (*b*) that the intensity varies by up to a factor of 3 with longitude at some latitudes; (*c*) that the intensity is strongly but imperfectly anticorrelated with the column densities of atomic hydrogen measured at 21 cm; (*d*) that this anticorrelation exhibits no systematic energy dependence; and (*e*) that the sky maps in the B and C bands are quite similar to one another, with available Be-band observations suggesting a spatial distribution nearly identical to that of the B band.

In this section we describe three simple pictures that have been suggested for interpreting the Be-, B-, and C-band diffuse background. Each picture is clearly an oversimplification and only approximates the observations, but each has the virtue of a manageable number of free parameters. Following our discussion of the simple pictures, we briefly touch on more complete models of the ISM that attempt to include mechanisms for the formation and maintenance of their density and temperature structure in a self-consistent way.

ABSORBED EXTRAGALACTIC OR CORONAL EMISSION The observation that the soft X-ray intensity is highest at high galactic latitudes and lowest in the galactic plane leads easily to the suggestion of an extragalactic or halo origin for the soft X rays beyond the neutral gas of the Galaxy (10, 12, 49, 80, 118). The existence of a dominant extragalactic component is well established for the higher energy diffuse background, while support for an X-ray-emitting halo comes from the suggestion that a 3×10^6-K halo would provide enough pressure to confine high-latitude neutral clouds (122). In these pictures, intensity variations of the soft X rays are caused by absorption of this extragalactic flux by the intervening column of interstellar neutral gas, which is observed to be small at high galactic latitudes and large in the galactic plane. The opacity of this column is so high at low latitudes for X-ray energies below 1 keV that models using primarily an extragalactic or coronal component must include an additional local component to provide the flux observed in the galactic plane.

The strength of such a picture is that it produces a negative correlation of soft X-ray intensity with $N_{H\,I}$, but its major weaknesses are that the predicted falloff of soft X-ray intensity with increasing $N_{H\,I}$ is much steeper than is observed, and that an energy dependence of the falloff is predicted where none is seen. In addition, such a picture explains neither the large scatter in the correlation nor the lack of correlation toward some galactic $N_{H\,I}$ features. Clumping of the interstellar gas, typically into clouds of thickness several times 10^{20} H I cm^{-2}, could reduce the predicted falloff of X-ray intensity with increasing $N_{H\,I}$ to match that observed and eliminate the energy dependence of the falloff as well (9, 12), but 21-cm and other interstellar studies over all conceivably relevant angular scales have shown that the necessary clumping is not present in the galactic H I (27, 56, 57).

EMISSION FROM A LOCAL CAVITY ("DISPLACEMENT" MODEL) In this picture, the soft X-ray background originates entirely from a local emission region occupying an anisotropic H I cavity of suitable size and shape in which the Sun is also located (21, 44, 69, 108, 110, 124). Variations in the observed X-ray intensity are due to variations in the extent of the emission volume along the line of sight. In directions with a larger X-ray intensity, the emitting region extends farther along the line of sight, "displacing" more of the neutral material. The negative correlation between soft X rays and $N_{H\,I}$ that is generated by the competition for space between the hot X-ray-emitting component and the cooler absorbing material fits the observations at least as well as the absorbed extragalactic or coronal emission models. It also results in a reasonable value for the interstellar pressure, which is effectively the only free parameter in such a model when it is fit to the observed H I distribution (121). The required hot-gas cavity is also more or less consistent with the local H I–deficient region crudely outlined by interstellar ultraviolet absorption studies (95). Since the magnitude of the X-ray vs. $N_{H\,I}$ correlation is unrelated to atomic absorption cross sections in this displacement picture, it naturally should be independent of X-ray energy. Scatter in the correlation is expected because any structure in the H I outside the cavity would be unrelated to X-ray emission inside the cavity.

The emission is assumed to arise from a hot plasma that occupies the local cavity but does not necessarily fill it. This plasma has been modeled as an expanding adiabatic blast wave (1, 22, 28, 29) or as the interior of an older superbubble (50). Cox & Reynolds (23) have recently reviewed the local ISM, the local cavity, and its X-ray emission. The displacement picture seems to account for all of the observed behavior in the three soft X-ray bands and is the simplest of the simple pictures in that it involves the fewest free parameters.

Although this picture requires no absorption at all for a good fit, it is known from optical and UV observations that small amounts of cooler material exist within the local cavity (see ref. 23 for a summary), and solar Lα backscattering measurements show that the Sun is itself embedded in a low-density cloud of a few parsecs extent in which $n_H \sim 0.1$–0.2 cm^{-3}. Juda (62; Juda et al, in preparation) has examined the effects of optically thin absorbing clouds uniformly intermixed with the hot gas in the cavity. If there is more than a very small amount of included H I, the predicted intensity no longer increases proportionally to path length within the hot cavity but instead tends to saturate as the optical depth approaches unity at each energy. This introduces a curvature to the expected Be-band vs. B-band relation, as shown in Figure 11. Observational limits to this curvature place a 2σ upper limit of $\sim 7 \times 10^{18}$ H I cm^{-2} on the total column density along an average path to the edge of the emitting region (62). For the best-fit cavity with $\langle R \rangle = 100$ pc, this gives $\langle n_{H I} \rangle < 0.02$ cm^{-3} and allows a filling factor of up to 25% for cloudlets similar to the one surrounding the Sun.

ABSORBERS INTERMIXED WITH EMISSION   In this picture, the X-ray-emitting material is intermixed with the interstellar absorbing material (15, 18, 23, 26, 34, 41, 58, 65, 76). A strength of this model is that intermixture reduces the predicted magnitude of absorption, but the models so far have been unable to reproduce all of the major features of the data.

Jakobsen & Kahn (58) have developed a framework for exploring the effects of an embedded cloud geometry on the very soft X-ray background. They include explicitly in their analysis the effects of absorbing clouds randomly interspersed within the emitting medium. In this statistical idealization, they find that the properties of the background depend on two parameters: R, the ratio of the scale height of the absorbing clouds to that of the X-ray emission; and η, a parameter that describes the reduction of the effective cross section of the absorbing component caused by its being clumped into clouds. ($\eta = 1$ corresponds to no clumping; $\eta = 0$ corresponds to infinite clumping.) Jakobson & Kahn find that these parameters are not severely constrained by the observed smoothness of the background nor by the shape of the anticorrelation of soft X-ray intensity versus $N_{H I}$, but the galactic latitude dependence of the data is a significant constraint. The increase in soft X-ray brightness at high galactic latitudes would require values of $\eta \sim 0.5$ and $R \sim 0.1$, where most of the emission is behind most of the (highly clumped) absorbing gas. This model is essentially the same as the absorbed coronal model discussed above, including its requirement for more extreme clumping of the absorbing material than seems allowed by observations.

Since one is restricted to $\eta \sim 1$ by the observed lack of clumping, then R is also required to be about unity to reproduce the observed constancy of the B/C and Be/B count-rate ratios (65). This works because with equal scale heights for emission and absorption, the source function $S_v = \Lambda_v n_e^2/4\pi\sigma_v n_H$ is constant along a line of sight, and the observed intensity I_v saturates at this value. (Here $\Lambda_v n_e^2$ is the volume emissivity for electron density n_e, and σ_v is the frequency-dependent cross section per hydrogen atom.) This is not really a "model" in the usual sense of the word, since it predicts equal X-ray intensity in all directions. An obvious implication here is that one could introduce intensity variations and get an energy-independent anticorrelation of I_x with N_H by making n_H different in different directions. A little reflection reveals that there is no geometrically reasonable way of doing this, since n_H must remain approximately constant along the length of each line of sight. Instead, the requirements of homogeneity and uniform statistical properties must be relaxed, and large-scale structures introduced. The above authors are working on models of this sort; the challenge is to create one in which the observable part of it is significantly different from the local cavity model described above.

In one example of such a modification, Burrows (15) fit the Wisconsin C-, B-, and Be-band data to an embedded cloud model with the addition of a local unabsorbed component. He found that the model with appreciably intermixed emission and absorption could be rejected based on the χ^2 residual, but the formal application of such statistics to simple models that are known to be incomplete descriptions of reality is questionable.

MORE COMPREHENSIVE MODELS   Ultimately, we would like a successful model of the nature and origin of the soft X-ray background to be part of a more comprehensive model of the ISM that includes mechanisms for producing and maintaining its structure. Cox & Reynolds (23) have reviewed models of the local ISM, paying particular attention to the constraints afforded by the soft X-ray observations, and they conclude that the local ISM is not typical of the general ISM. Two competing pictures emerge from their discussion: that of the local ISM as an older ($\sim 10^7$ yr), quiescent, pressure-confined hot X-ray-emitting bubble (24, 50); and that of the local ISM as a younger ($\sim 10^5$ yr), active, hot X-ray-emitting supernova remnant (1, 22, 29). Although they acknowledge that the former picture may prove untenable if cloud evaporation provides significant cooling (11, 86), in the absence of evidence for the necessary clouds, they tentatively adopt the quiescent bubble scenario for the local ISM.

In their general ISM model, McKee & Ostriker (86) find a pervasive hot component in a self-consistent steady state where supernova heating is balanced by cloud evaporation, but its mean temperature is too low ($\sim 4 \times 10^5$ K) to produce the observed X rays. The local cavity seems deficient in evaporating clouds, however, which would give it a longer than average lifetime at its elevated temperature.

An elaborate picture of the local ISM has been assembled by Bochkarev (6), who incorporated extensive radio, IR, optical, UV, and X-ray observations of the solar neighborhood into a model with the Sun located near the edge of a giant cavern in the ISM centered on the Sco-Cen OB association. In this model, the X-ray background is produced by shock fronts from the OB association's stellar winds interacting with the cavern walls and the boundary of the small cloud in which the Sun is embedded.

6. Limits on the Extragalactic Flux

Below 0.25 keV it now seems likely that very little of the observed flux can come from outside the Galaxy. However, in placing the complementary upper limit on an extragalactic component, one must correct for the opacity of interstellar gas along a line of sight out of the Galaxy. At these energies, the opacity is high even near the galactic poles, and the correction factor is large.

The best limit for the C-band range comes from limits on spatial variation in the observed flux on and near the Small Magellanic Cloud (SMC; 83, 85). The analysis in (85) used an overly conservative criterion for placing confidence limits on parameter values, extending the allowed parameter range to the point where the absolute $P(\chi^2)$ for the model becomes small. Applying the now-standard criteria of Lampton et al (74) to the same χ^2 residuals results in somewhat lower limits for an extragalactic flux at 0.25 keV than were originally reported in (85): The 95% confidence upper limits are 30 keV (cm^2 s sr keV)$^{-1}$ for a flux originating beyond the SMC, or 38 keV (cm^2 s sr keV)$^{-1}$ for X rays originating closer than the SMC but beyond all of the galactic H I, as for emission from an extended galactic halo. These limits are respectively 2.1 and 2.7 times an extrapolation of the $8E^{-0.4}$ keV (cm^2 s sr keV)$^{-1}$ extragalactic spectrum observed above 3 keV.

It has recently been recognized that there is an ionized component of the ISM with a scale height of ~ 1 kpc and a total column density averaging $\sim 1 \times 10^{20}$csc(b) cm^{-2} (100). A high-z pulsar fortuitously located directly in front of the SMC has a dispersion measure corresponding to $N_{\mathrm{H\,II}} = 7.2 \times 10^{19}$ cm^{-2} (77a). The presumably neutral helium associated with this H II provides about 0.4 optical depth in the C band, since hydrogen normally is responsible for only 25% of the total absorption (88). This ionized component was not included in the previous analyses, so the above limits should be increased by about a factor of 1.5. The resulting extragalactic upper limits are comparable to the total observed intensity, but after absorption they can account for at most 13% of it in the direction of the SMC. Limits obtained using the Large Magellanic Cloud (LMC; 77) and M31 (78) also require most of the observed flux to be galactic but do not as effectively constrain the extragalactic flux owing to the high interstellar opacity in these directions.

The Wisconsin group also reports a similar limit based on lack of absorption by galactic H I in two other regions of the sky (16). This limit from the simultaneous B- and C-band analysis took advantage of an implicit (and unintended) assumption that the observed contribution of any extragalactic component would be the same for both bands, which is unreasonable. The weaker limits they obtained using the C band alone are valid (but should be corrected for absorption by the high-z H II).

It should be possible to improve these existing 0.25-keV limits by about a factor of 7 (to below a power-law extrapolation of the extragalactic spectrum) with high-angular-resolution observations of compact H I features in regions where the total absorption is particularly low. The ROSAT X-ray telescope is well-suited for such observations.

In the 0.5–1 keV range, it is not possible to use large-scale interstellar absorption to place limits on the extragalactic flux in a way that is independent of the rather unsatisfactory models for galactic emission at these energies. The best that can presently be done is to take the lowest observed intensity, far enough away from the galactic plane such that absorption is negligible, to be an upper limit to the extragalactic contribution. This limit goes from about 3 times the $8E^{-0.4}$ keV (cm^2 s sr keV)$^{-1}$ extrapolation at 0.5 keV to 2 times this extrapolation at 0.8 keV. Again it should be possible to improve these limits considerably with high-resolution observations of compact absorbing features. In this case, however, features with adequate optical depth will probably contain molecular material, which makes determination of the expected absorption more difficult and uncertain. If the local emission is entirely thermal (and therefore 99% in lines), it may also be possible to use high spectral resolution to look between the lines for an excess continuum due to an extragalactic component.

As discussed in Section 3 above, the total observed intensity in the 1–3 keV range is uncertain but may be about 30% above a simple bremsstrahlung or power-law fit to the uncontaminated extragalactic flux above

3 keV. If it is, this additional flux could be due either to a steepening of the extragalactic spectrum or to a galactic contribution. This creates an awkward situation for attempts to determine the contribution of active galactic nuclei (AGNs) to the extragalactic background, since their spectra are currently best determined in this energy range where the diffuse background spectrum is uncertain.

Galactic emission in the 3–10 keV range is quite faint away from the plane and almost certainly makes a negligible contribution to the extragalactic flux at high latitudes. However, structure in the large-scale-height component of the emission serves to confuse searches for the Compton-Getting effect in the extragalactic background, which should otherwise be marginally detectable with current data (K. Jahoda, private communication; 7, 117).

The best current upper limits to the extragalactic flux below 1 keV are shown in Figure 12. These include corrections for the opacity of galactic H I and the high-z H II layer discussed above. Even these crude upper limits provide useful constraints on contributions to the extragalactic background by AGNs with soft spectra (30).

7. Summary

Galactic diffuse X-ray emission goes from being a negligible contributor to the extragalactic background at 10 keV to completely dominating it at 0.25 keV and below. Several quite different sources of galactic emission are important at different energies.

In the 3–10 keV range, galactic emission is confined to a thin disk in the galactic plane and a much fainter thick disk. These have similar thermal spectra showing strong emission lines of iron. It is doubtful that either of these represents truly diffuse emission; both are more likely due to a population of discrete sources, although no known class of sources seems to have the right characteristics.

From about 1 to 3 keV, the observed diffuse flux is probably mostly extragalactic, but may be 25–30% higher than an extrapolation of the 3–10 keV spectrum. This excess could be either galactic or extragalactic, with some consequences for models of AGN contributions to the background.

The 0.5–1 keV range is probably the most poorly understood part of the diffuse X-ray background. A power-law extrapolation of the extragalactic spectrum above 3 keV would account for about 35% of the flux observed at latitudes above 20°, where interstellar absorption is negligible. Discrete sources can contribute at most another $\sim 15\%$ of the high-latitude flux. There are a small number of identifiable bright features, such as Loop I and the Cygnus and Eridanus "superbubbles," that appear to be pockets of hot gas near 3×10^6 K with radii of 100 pc or so. At this temperature and higher, we should be able to see all such features out to at least 1 kpc, and their filling factor appears to be less than 10%. Aside from these isolated features, the background in this energy range is quite smooth, showing only a slight tendency to decrease in the galactic plane (Figure 8).

The most straightforward way to account for this distribution is to postulate a very local isotropic source that provides about two thirds of the observed M-band photons. The expected contributions from an $8E^{-0.4}$ keV (cm^2 s sr keV)$^{-1}$ extragalactic spectrum and dM stars then fill in the remainder and give a reasonable fit to the average latitude dependence. An alternative is to assume that the extragalactic spectrum steepens below 2 keV to provide most of the observed M-band flux at high latitudes, or, equivalently, that there is a large-scale-height halo. A second source distribution is then needed in the galactic disk, whose scale height and intensity must be carefully adjusted to fill in for the interstellar absorption of the extragalactic component. In any of these cases, there must be some as-yet unidentified galactic component. High-resolution spectra could provide important clues for the physical origin of this flux, while high-spatial-resolution observations of potential absorbing features would place limits on its location.

The observed 70–284 eV background is almost entirely galactic. More than half of it must originate within a few hundred parsecs of the Sun; quite possibly all of it does. It is most easily (but not uniquely) explained as thermal emission from 10^6-K gas filling an irregular cavity around the Sun. The required cavity is consistent with other observations of a low-density cavity in the local ISM, although there are directions in which the X-ray emission would have to end well before a dense wall is encountered. No absorption within the emitting region is required by the X-ray data, but the small amounts of neutral material found by other observations can be accommodated.

Models in which a substantial fraction of the observed 70–284 eV flux comes from beyond all of the neutral hydrogen seem to be ruled out by stringent radio and optical limits on the clumping of the intervening material. Work is currently being done on the more general intermediate case, where hot gas is intermixed with the absorbing material, but it remains to be seen whether this can result in viable models that distribute the observed emission substantially differently from the simple cavity model. Any guidance that such models could give toward obtaining observational evidence for or against more widespread hot gas would be valuable.

In any case, the Sun *is* in an H I hole—i.e. the average density for ~ 100 pc in most directions, and much farther in some, is much less than the galactic average of 1 cm^{-3}. A similar path length of gas at 10^6 K will provide the observed 0.25-keV X-ray flux at a reasonable pressure ($\sim 10^4$

cm^{-3} K; 23). It seems logical enough to put the two together. The result can be interpreted in three ways: (a) The Sun is in an unusual location; (b) most of the H I in the Galaxy is organized into sheets on scales of a few hundred parsecs, with hot gas filling a large fraction of volume between; or (c) a situation in between (a) and (b) is the case, where we have a "normal" intermixture of clouds and intercloud medium, with some modest filling factor of hot bubbles. It is indicative of our current state of understanding of the global organization of the ISM that we cannot discriminate among these disparate pictures (72).

There is considerable reason to hope for an improvement in our understanding of galactic soft X-ray emission in the not-too-distant future. The German X-ray telescope on ROSAT is nearly an optimum design for studying diffuse emission at high angular resolution. This should allow detailed absorption studies to determine the location of the X-ray emission. Its all-sky survey program should provide maps of the global distribution in the 0.1–2 keV region with more than an order-of-magnitude improvement over existing data in both angular resolution and statistical precision. Observations of other galaxies will allow at least upper limits to be placed on the total amounts of 10^6-K gas in the disks of galaxies similar to our own. Measurements with Bragg spectrometers and microcalorimeter detectors will resolve the emission-line spectra of thermal sources, providing detailed information on the physical conditions in the source plasmas as well as indications of the time since heating. Observations of the individual emission lines should also make it possible to unambiguously distinguish galactic thermal emission from a truly extragalactic source.

ACKNOWLEDGMENTS

We thank D. P. Cox, W. L. Kraushaar, F. J. Lockman, and R. J. Reynolds for helpful discussions and critical readings of the manuscript; the Penn State group for permission to use their all-sky maps; J. J. Bloch for preparing Figure 2; and B. Edwards and M. Juda for their unpublished results. This work was supported in part by NASA grant NAG 5-629.

Literature Cited

1. Arnaud, M., Rothenflug, R. 1986. *Adv. Space Res.* 6: 119–28
2. Baity, W. A., Peterson, L. E., eds. 1979. *X-Ray Astronomy.* Oxford: Pergamon. 558 pp.
3. Berkhuijsen, E. M., Haslam, C. G. T., Salter, C. J. 1971. *Astron. Astrophys.* 14: 252–62
4. Bloch, J. J., Jahoda, K., Juda, M., McCammon, D., Sanders, W. T., Snowden, S. L. 1986. *Ap. J. Lett.* 308: L59–62
5. Bloemen, H. 1989. *Annu. Rev. Astron. Astrophys.* 27: 469–516
6. Bochkarev, N. G. 1987. *Astrophys. Space Sci.* 138: 229–302
7. Boldt, E. A. 1977. *Phys. Rep.* 146(4): 215–57
8. Borken, R. J., Iwan, D. C. 1977. *Ap. J.* 218: 511–20

9. Bowyer, C. S., Field, G. B. 1969. *Nature* 223: 573–75
10. Bowyer, C. S., Field, G. B., Mack, J. E. 1968. *Nature* 217: 32–34
11. Boehringer, H., Hartquist, T. W. 1987. *MNRAS* 228: 915–31
12. Bunner, A. N., Coleman, P. L., Kraushaar, W. L., McCammon, D., Palmieri, T. M., et al. 1969. *Nature* 223: 1222–26
13. Bunner, A. N., Coleman, P. L., Kraushaar, W. L., McCammon, D., Williamson, F. O. 1973. *Ap. J.* 179: 781–88
14. Burrows, D. N. 1982. PhD thesis. Univ. Wisc., Madison. 460 pp.
15. Burrows, D. N. 1989. *Ap. J.* 340: 775–85
16. Burrows, D. N., McCammon, D., Sanders, W. T., Kraushaar, W. L. 1984. *Ap. J.* 287: 208–18
17. Burrows, D. N., Nousek, J. A., Truax, R. J., Garmire, G. P., Singh, K. P. 1985. *Bull. Am. Astron. Soc.* 17: 883 (Abstr.)
18. Burstein, P., Borken, R. J., Kraushaar, W. L., Sanders, W. T. 1977. *Ap. J.* 213: 405–20
19. Caillault, J.-P., Helfand, D. J., Nousek, J. A., Takalo, L. O. 1986. *Ap. J.* 304: 318–25
20. Cash, W., Charles, P., Bowyer, S., Walter, F., Garmire, G., Riegler, G. 1980. *Ap. J. Lett.* 238: L71–76
20a. Cleary, M. N., Heiles, C., Haslam, C. G. T. 1979. *Astron. Astrophys. Suppl.* 36: 95–127
21. Cox, D. P. 1977. In *Topics in Interstellar Matter*, ed. H. van Woerden, pp. 17–25. Dordrecht: Reidel. 295 pp.
22. Cox, D. P., Anderson, P. A. 1982. *Ap. J.* 253: 268–89
23. Cox, D. P., Reynolds, R. J. 1987. *Annu. Rev. Astron. Astrophys.* 25: 303–44
24. Cox, D. P., Snowden, S. L. 1986. *Adv. Space Res.* 6(2): 97–107
25. Davelaar, J., Bleeker, J. A. M., Deerenberg, A. J. M. 1980. *Astron. Astrophys.* 92: 231–37
26. Davidsen, A., Shulman, S., Fritz, G., Meekins, J. F., Henry, R. C., Friedman, H. 1972. *Ap. J.* 177: 629–42
26a. Dickey, J. M., Lockman, F. J. 1990. *Annu. Rev. Astron. Astrophys.* 28: 215–61
27. Dickey, J. M., Salpeter, E. E., Terzian, Y. 1978. *Ap. J. Suppl.* 36: 77–114
28. Dyson, J. E., Hartquist, T. W. 1987. *MNRAS* 228: 453–61
29. Edgar, R. J. 1986. *Ap. J.* 308: 389–400
30. Fabian, A. C., Canizares, C. R., Barcons, X. 1989. *MNRAS* 239: 15P–18P
31. Fichtel, C. E., Hartman, R. C., Kniffen, D. A., Thompson, D. J., Oegelman, H. B., et al. 1977. *Ap. J. Lett.* 217: L9–13
32. Fink, R. 1989. PhD thesis. Pa. State Univ., Univ. Park. 286 pp.
33. Fried, P. M. 1978. PhD thesis. Univ. Wisc., Madison. 183 pp.
34. Fried, P. M., Nousek, J. A., Sanders, W. T., Kraushaar, W. L. 1980. *Ap. J.* 242: 987–1004
35. Garmire, G. P., Nugent, J. J. 1981. *Bull. Am. Astron. Soc.* 13: 786–87 (Abstr.)
36. Giacconi, R., Gursky, H., Paolini, F. R., Rossi, B. B. 1962. *Phys. Rev. Lett.* 9: 439–43
37. Giacconi, R., Zamorani, G. 1987. *Ap. J.* 313: 20–27
38. Gorenstein, P., Tucker, W. H. 1972. *Ap. J.* 176: 333–44
39. Hamilton, T. T., Helfand, D. J. 1987. *Ap. J.* 318: 93–102
40. Hayakawa, S. 1973. In *X- and Gamma-Ray Astronomy, IAU Symp. No. 55*, ed. H. Bradt, R. Giacconi, pp. 235–49. Dordrecht: Reidel. 323 pp.
41. Hayakawa, S. 1979. See Ref. 2, pp. 323–35
42. Hayakawa, S., Kato, T., Nagase, F., Yamashita, K. 1979. *Publ. Astron. Soc. Jpn.* 31: 71–86
43. Hayakawa, S., Kato, T., Nagase, F., Yamashita, K., Murakami, T., Tanaka, Y. 1977. *Ap. J. Lett.* 213: L109–13
44. Hayakawa, S., Kato, T., Nagase, F., Yamashita, K., Tanaka, Y. 1978. *Astron. Astrophys.* 62: 21–28
45. Hearn, D. R., Richardson, J. A., Bradt, H. V. D., Clark, G. W., Lewin, W. H. G., et al. 1976. *Ap. J. Lett.* 203: L21–24
46. Heiles, C., Chu, Y., Reynolds, R. J., Yegingil, I., Troland, T. H. 1980. *Ap. J.* 242: 533–40
47. Helfand, D. J., Caillault, J.-P. 1982. *Ap. J.* 253: 760–67
48. Henry, R. C., Fritz, G., Meekins, J. F., Chubb, T., Friedman, H. 1971. *Ap. J. Lett.* 163: L73–77
49. Henry, R. C., Fritz, G., Meekins, J. F., Friedman, H., Byram, E. T. 1968. *Ap. J. Lett.* 153: L11–18
50. Innes, D. E., Hartquist, T. W. 1984. *MNRAS* 209: 7–13
51. Inoue, H., Koyama, K., Matsuoka, M., Ohashi, T., Tanaka, Y., Tsunemi, H. 1979. *Ap. J. Lett.* 227: L85–88
52. Inoue, H., Koyama, K., Matsuoka, M., Ohashi, T., Tanaka, Y., Tsunemi, H. 1980. *Ap. J.* 238: 886–91
53. Iwan, D. 1980. *Ap. J.* 239: 316–27
54. Iwan, D., Marshall, F. E., Boldt, E. A., Mushotzky, R. F., Shafer, R. A.,

Stottlemyer, A. 1982. *Ap. J.* 260: 111–23
55. Jahoda, K., McCammon, D. 1988. *Nucl. Instrum. Methods A* 272: 800–13
56. Jahoda, K., McCammon, D., Dickey, J. M., Lockman, F. J. 1985. *Ap. J.* 290: 229–37
57. Jahoda, K., McCammon, D., Lockman, F. J. 1986. *Ap. J. Lett.* 311: L57–61
58. Jakobsen, P., Kahn, S. M. 1986. *Ap. J.* 309: 682–93
59. Jenkins, E. B. 1978. *Ap. J.* 219: 845–60
60. Jenkins, E. B. 1978. *Ap. J.* 220: 107–23
61. Jenkins, E. B., Meloy, D. A. 1974. *Ap. J. Lett.* 193: L121–25
62. Juda, M. 1988. PhD thesis. Univ. Wisc., Madison. 130 pp.
63. Juda, M., Bloch, J. J., McCammon, D., Sanders, W. T., Snowden, S. L. 1987. *Bull. Am. Astron. Soc.* 19: 722 (Abstr.)
64. Kahn, S. M., Caillault, J.-P. 1986. *Ap. J.* 305: 526–33
65. Kahn, S. M., Jakobsen, P. 1988. *Ap. J.* 329: 406–9
66. Kato, T. 1976. *Ap. J. Suppl.* 30: 397–449
67. Koyama, K. 1989. *Publ. Astron. Soc. Jpn.* 41: 665–78
68. Koyama, K., Makishima, K., Tanaka, Y., Tsunemi, H. 1986. *Publ. Astron. Soc. Jpn.* 38: 121–31
69. Kraushaar, W. L. 1976. *Bull. Am. Astron. Soc.* 8(4): 548 (Abstr.)
70. Kraushaar, W. L. 1979. See Ref. 2, pp. 293–308
71. Kraushaar, W. L., Clark, G. W., Garmire, G. P., Borken, R., Higbie, P., et al. 1972. *Ap. J.* 177: 341–63
72. Kulkarni, S. R., Heiles, C. 1987. In *Interstellar Processes*, ed. D. J. Hollenbach, H. A. Thronson, pp. 87–122. Dordrecht: Reidel. 807 pp.
73. Labov, S., Bowyer, S. 1988. *J. Phys.* 49: C1-63–66
74. Lampton, M., Margon, B., Bowyer, S. 1976. *Ap. J.* 208: 177–90
75. Leiter, D., Boldt, E. 1982. *Ap. J.* 260: 1–18
76. Levine, A., Rappaport, S., Halpern, J., Walter, F. 1977. *Ap. J.* 211: 215–22
77. Long, K. S., Agrawal, P. C., Garmire, G. P. 1976. *Ap. J.* 206: 411–17
77a. Manchester, R. N., Lyne, A. G., Johnston, S., D'Amico, N., Lim, J., Kniffen, D. A. 1989. *IAU Circ. No. 4892*
78. Margon, B., Bowyer, S., Cruddace, R., Heiles, C., Lampton, M., Troland, T. 1974. *Ap. J. Lett.* 191: L117–19
79. Marshall, F. E., Boldt, E. A., Holt, S. S., Miller, R. B., Mushotzky, R. F., et al. 1980. *Ap. J.* 235: 4–10
80. Marshall, F. J., Clark, G. W. 1984. *Ap. J.* 287: 633–52
81. Mason, I. M. 1981. PhD thesis. Univ. College London, Engl. 372 pp.
82. Mason, I. M., Culhane, J. L. 1983. *IEEE Trans. Nucl. Sci.* NS-30: 485–90
83. McCammon, D., Bunner, A. N., Coleman, P. L., Kraushaar, W. L. 1971. *Ap. J. Lett.* 168: L33–37
84. McCammon, D., Burrows, D. N., Sanders, W. T., Kraushaar, W. L. 1983. *Ap. J.* 269: 107–35
85. McCammon, D., Meyer, S. S., Sanders, W. T., Williamson, F. O. 1976. *Ap. J.* 209: 46–52
86. McKee, C. F., Ostriker, J. P. 1977. *Ap. J.* 218: 148–69
87. Mewe, R., Gronenschild, E. H. B. M., van den Oord, G. H. J. 1985. *Astron. Astrophys. Suppl.* 62: 197–254
87a. Morisawa, K., Matsuoka, M., Takahara, F., Piro, L. 1989. *Proc. ESLAB Symp. X-Ray Astron., 23rd.* In press
88. Morrison, R., McCammon, D. 1983. *Ap. J.* 270: 119–22
89. Moseley, S. H., Mather, J. C., McCammon, D. 1984. *J. Appl. Phys.* 56: 1257–62
90. Naranan, S., Shulman, S., Friedman, H., Fritz, G. 1976. *Ap. J.* 208: 718–26
91. Nousek, J. A., Burrows, D. N., Good, J., Singh, K. P. 1986. *Bull. Am. Astron. Soc.* 18: 1029 (Abstr.)
92. Nousek, J. A., Fried, P. M., Sanders, W. T., Kraushaar, W. L. 1982. *Ap. J.* 258: 83–95
93. Pallavicini, R., ed. 1988. *Hot Thin Plasmas in Astrophysics*. Dordrecht: Kluwer. 434 pp.
94. Palmieri, T. M., Burginyon, G. A., Grader, R. J., Hill, R. W., Seward, F. D., Stoering, J. P. 1971. *Ap. J.* 169: 33–39
95. Paresce, F. 1984. *Astron. J.* 89(7): 1022–37
96. Penzias, A. A., Wilson, R. W. 1965. *Ap. J.* 142: 419–21
97. Policarpo, A. J. P. L. 1977. *Space Sci. Instrum.* 3: 77–107
98. Raymond, J. C. 1988. See Ref. 93, pp. 3–20
99. Raymond, J. C., Smith, B. W. 1977. *Ap. J. Suppl.* 35: 419–39
99a. Raymond, J. C., Smith, B. W. 1984. Informally distributed update to Ref. 99
100. Reynolds, R. J. 1989. *Ap. J. Lett.* 339: L29–32
101. Reynolds, R. J., Ogden, P. M. 1979. *Ap. J.* 229: 942–53
102. Ride, S. K., Walker, A. B. C. Jr. 1977. *Astron. Astrophys.* 61: 347–52
103. Rocchia, R., Arnaud, M., Blondel, C., Cheron, C., Christy, J. C., et al. 1984. *Astron. Astrophys.* 130: 53–61

104. Rosner, R., Avni, Y., Bookbinder, J., Giacconi, R., Golub, L., et al. 1981. *Ap. J. Lett.* 249: L5–9
105. Rothenflug, R. 1988. See Ref. 93, pp. 197–211
106. Rothschild, R., Boldt, E., Holt, S., Serlemitsos, P., Garmire, G., et al. 1979. *Space Sci. Instrum.* 4: 269–301
107. Ryter, C., Cesarsky, C. J., Audouze, J. 1975. *Ap. J.* 198: 103–9
108. Sanders, W. T. 1976. PhD thesis. Univ. Wisc., Madison. 134 pp.
109. Sanders, W. T., Burrows, D. N., McCammon, D., Kraushaar, W. L. 1983. In *Supernova Remnants and Their X-Ray Emission, IAU Symp. No. 101*, ed. J. Danziger, P. Gorenstein, pp. 361–65. Dordrecht: Reidel. 614 pp.
110. Sanders, W. T., Kraushaar, W. L., Nousek, J. A., Fried, P. M. 1977. *Ap. J. Lett.* 217: L87–91
111. Sanders, W. T., Snowden, S. L., Edgar, R. J. 1990. In *High Resolution X-Ray Spectroscopy of Cosmic Plasmas*, ed. P. Gorenstein, M. V. Zombeck. Cambridge: Univ. Press. In press
112. Schnopper, H. W., Delvaille, J. P., Rocchia, R., Blondel, C., Cheron, C., et al. 1982. *Ap. J.* 253: 131–35
113. Schwartz, D. A. 1979. See Ref. 2, pp. 453–65
114. Schwartz, D. A., Gursky, H. 1974. In *X-Ray Astronomy*, ed. R. Giacconi, H. Gursky, pp. 359–88. Dordrecht: Reidel. 450 pp.
115. Schwartz, H. E., Mason, I. M. 1984. *Nature* 309: 532–34
116. Setti, G. 1989. *Proc. Yamada Conf. XX: Big Bang, Active Galactic Nuclei, and Supernovae.* In press
117. Shafer, R. A. 1983. PhD thesis. Univ. Md., College Park. *NASA Tech. Memo. 85029.* 471 pp.
118. Shapiro, P. R., Field, G. B. 1976. *Ap. J.* 205: 762–65
119. Shapiro, P. R., Moore, R. T. 1976. *Ap. J.* 207: 460–83
120. Siegmund, O. H. W., Clothier, S., Culhane, J. L., Mason, I. M. 1982. *IEEE Trans. Nucl. Sci.* NS-30: 350
120a. Slavin, J. D. 1989. *Ap. J.* 346: 718–27
121. Snowden, S. L., Cox, D. P., McCammon, D., Sanders, W. T. 1990. *Ap. J.* 354: 211–19
122. Spitzer, L. 1956. *Ap. J.* 124: 20–34
123. Stern, R., Wang, E., Bowyer, S. 1978. *Ap. J. Suppl.* 37: 195–222
124. Tanaka, Y., Bleeker, J. A. M. 1977. *Space Sci. Rev.* 20: 815–88
125. Truemper, J. 1986. In *Cosmic Radiation in Contemporary Astrophysics*, ed. M. M. Shapiro, pp. 241–47. Dordrecht: Reidel. 274 pp.
126. Turner, M. J. L., Thomas, H. D., Patchett, B. E., Reading, D. H., Makishima, K., et al. 1989. *Publ. Astron. Soc. Jpn.* 41: 345–72
127. van den Heuvel, E. P. J., Rappaport, S. 1987. In *Physics of Be Stars*, ed. A. Slettebak, T. P. Snow, pp. 291–308. Cambridge: Univ. Press. 557 pp.
128. Vanderhill, M. J., Borken, R. J., Bunner, A. N., Burstein, P. H., Kraushaar, W. L. 1975. *Ap. J. Lett.* 197: L19–22
129. Warwick, R. S., Pye, J. P., Fabian, A. C. 1980. *MNRAS* 190: 243–60
130. Warwick, R. S., Turner, M. J. L., Watson, M. G., Willingale, R. 1985. *Nature* 317: 218–21
131. Weymann, R. 1967. *Ap. J.* 147: 887–900
132. Williamson, F. O., Sanders, W. T., Kraushaar, W. L., McCammon, D., Borken, R., Bunner, A. N. 1974. *Ap. J. Lett.* 193: L133–37
133. Worrall, D. M., Marshall, F. E., Boldt, E. A., Swank, J. H. 1982. *Ap. J.* 255: 111–21
134. Zhang, J., Edwards, B., Juda, M., Kelley, R., Madejski, G., et al. 1990. In *High Resolution X-Ray Spectroscopy of Cosmic Plasmas*, ed. P. Gorenstein, M. V. Zombeck. Cambridge: Univ. Press. In press

SUBJECT INDEX

A

A stars
  in illumination of reflection nebulae, 53
  main sequence in, 611
  mode selection in, 642
Abell 41, 115
Abundances
  aluminum, 308
  argon, 536-37, 550-51, 553
  carbon, see Carbon, abundance of
  chlorine, 553
  in extragalactic H II regions, 536-40, 549-57
  of gases in planetary atmospheres, 369-70, 373-74, 376-80, 389-92
  helium, see Helium, abundance of
  in hot stars, 306, 308-11, 318-19, 327-28
  hydrogen, see Hydrogen, abundance of
  in interstellar matter, 241
  iron, see Iron, abundance of
  magnesium, see Magnesium, abundance of
  neon, see Neon, abundance of
  nitrogen, see Nitrogen, abundance of
  oxygen, see Oxygen, abundance of
  phosphorus, 66
  rare earth element, 608, 612
  in roAp stars, 608, 612, 637-40, 647
  silicon, see Silicon, abundance of
  solar, in collisional equilibrium, 661
  sulfur, 535-39, 550-51, 553
  in supernova remnants, 551, 552
  in white dwarf progenitors, 141-43, 145-46
Accretion
  black hole, 437-38, 486
  onto neutron star, 31
Accretion disks
  in low-mass X-ray binary stars, 192
  around neutron stars, 414
  in quasars, 438

Accretion-induced collapse, of white dwarfs, 185, 190-209
Active galactic nuclei, 683
AE Aur, 315
AM CVn, 110
Andromeda, H II regions in, 526
Aperture synthesis telescopes, H I studies at, 218
10 Aql, 613
Arago-Poisson light spot, 23
Arecibo Telescope, 220, 222, 584
ASTRO missions, 57
Asymptotic giant branch (AGB) luminosities, 122-23
Atomic clouds, small-scale structure in, 244-45
Aur OB1, 314
AzV 238, 316
AzV 243, 316
AzV 388, 316

B

B stars
  in illumination of reflection nebulae, 53
  line profiles in, 645
  superwinds of, 320
  ultraviolet extinction curves of, 311
BD +60°513, 314
Be stars
  line profile variations in, 644
  in X-ray binary systems, 185
Becklin-Neugebauer object, polarization in, 58
BG 2107+49, 243
Binary ion mixtures (BIMs), 194-95
Binary stars, high-mass X-ray, 185-86, 202, 208
Binary stars, low-mass X-ray, 185-86, 191-92, 202-8
Binary stars, massive X-ray, 31, 187-88, 208
Binary systems, close, white dwarfs in, 115, 126, 191-93
Binary systems, ultrashort-interaction, white dwarfs in, 110, 115
Binary systems, wide, white dwarfs in, 110, 126

Black holes, 33, 187, 437-38, 486
BL Lacertae objects, 441, 447
Born-again stars, 116, 122
Brown dwarfs, 142
Bubbles, 242-44
2175-Å bump, in optical/ultraviolet extinction, 42-44

C

C stars, ultraviolet bump in, 60
α Cam, 315
Cambridge Telescope, planetary radio imaging and, 357
Canada-France-Hawaii Telescope, 446
Canal rays, 6, 8
Carbon
  2175-Å bump due to, 43-44
  abundance of, 43
    in extragalactic H II regions, 535-37, 553, 557
    in hot stars, 308-9, 319
    in Large Magellanic Cloud, 45
    in Small Magellanic Cloud, 46
  in hot interstellar gas, 91-92, 98
  ignition of, lower limiting mass for, 103
  in interstellar dust, 45, 62-67
  in interstellar medium, 62
  red continuum emission and, 51
  in white dwarfs, 143, 145, 193-202
  in Wolf-Rayet stars, 340
  unidentified infrared band production and, 49-50
Carbon stars
  HR diagram of, 146
  interstellar dust formation in, 60
Carina cluster, 329
γ Cas, 51
Cas A, 74, 227
Cas OB6, 313, 315
β CBr, 646, 649
CD −38°0980, 107
α Cen A, 290
α Cen B, 290
Cen X-3, 185

689

SUBJECT INDEX

β Cep, 645
λ Cep, 314
μ Cep, 47
Cepheid variables, 645
Cep OB III, 40
Cerro Tololo Inter-American Observatory Telescope, 445
Cherenkov radiation, see Vavilov-Cherenkov radiation
Chromosphere, solar radio emission and, 21
α Cir, 613-14, 617, 645-46, 648, 650
Circumstellar dust, around oxygen-rich stars, 46
Circumstellar shells, ice band in, 60
Cirrus emission, galactic, 49, 239
CLEAN deconvolution technique, in radio imaging, 352-54
Cloud mass spectrum, 247-49
Cold dark matter, 483, 485
Colliding galaxies, hypothesis of, 24
Column density measurement, of H I, 224-25
Cometary dust, 60, 67
Copernicus satellite observations, 235, 237, 661
Cork evolution, in solar convection, 276-79
Coronal gas, galactic, 73, 90-92
Corrugations, 253-54
COS-B satellite observations, 28, 241-42
Cosmic radio emission, nonsolar, 25-29
Cosmic ray pressure, 95
Cosmic rays
  escape of from Galaxy, 97-98
  generation of, 28
  in interstellar medium, 95
Cosmic strings, 33, 401
Crab Nebula
  high magnetic field in, 77
  hot interstellar gas in, 74
  pulsar in, 184
Curtis Schmidt Telescope, 445
α² CVn, 639
Cyclotron radiation, in radio imaging of planets, 355
Cyclotron reprocessing, in neutron star magnetosphere, 414
VI Cyg 12, 39, 65
Cyg OB1, 314, 315
Cygnus, 72, 579, 587, 671, 675, 683
Cyg X-2, 206
Cyg X3, 580

D

DA stars, 104-8, 111, 129, 143-44, 158, 642
DA+dK2 pairs, 110
Dark matter, white dwarfs in, 129, 132, 171
DB stars
  evolution of, 128, 158
  masses of, 108-9, 111, 117, 127
DC stars, 108
dM stars, 671, 677
DQ stars, 108, 145, 158
Deconvolution techniques, in radio imaging, 352-354
Degenerate matter, 139-40, 152, 155-63, 177
δ Del stars, 638
DETAIL program, 308
Digital registration and de-stretching, 269
Dominion Radio Astrophysical Observatory, aperture synthesis telescope at, 218
30 Doradus, see Tarantula Nebula
Double-shell-burning phase, 103
Draco, 240
Dracula cloud, 240
DS1, 115

E

EGB 5, 115
Einstein satellite observations, 412, 660
Electric current sheets, in solar corona, 502-8
γ Equ, 613, 636, 645-47
40 Eri B, 107-8, 110, 115
Eridanus-Orion enhancement, 671, 675, 683
Extragalactic extinction, 45-46
Extragalactic objects, unidentified infrared bands in, 49
Extreme helium stars, 119
Extreme ultraviolet (EUV) emission
  from coronal magnetic fields, 491
  H I absorption in, 240-41
  hot star spectra in, 305, 309
  stellar winds and, 322
  unified model atmospheres and, 333

F

Far-infrared (FIR) emission, 49, 239-40
  of planets, 369
  unified model atmospheres and, 333
Far-ultraviolet extinction, 45
Filled-center remnants, hot interstellar gas from, 74
Fractals, 278, 280

G

Galactic bulge
  planetary nebulae in, 115
  sources of, 186, 204-206
Galactic center
  silicate absorption features in, 46-47
  X-ray globular cluster distance from, 205
Galactic disk
  age of, 129, 140, 169, 172-76
  finite size of, 233
  gamma-ray bursts in, 418
  hot interstellar gas in, 78-81, 83, 96-97
  low-mass X-ray binary stars in, 205
  supernovae in, 86
  warped, 90
Galactic field, subluminous OB stars in, 318
Galactic fountains, 88-90, 95
Galactic light, diffuse, 53-55
Galactic nuclei, active, emission-line spectra of, 530
Galactic plane
  H I emission spectra at, 219
  hot gas effects far from, 86-92
Galactic wind, hot interstellar gas in, 88, 96
Galaxies
  clustering of, 439
  cosmic rays in, 28
  evolution of, 106
  quasar associations with, 438-39, 441, 473-80, 485
  soft X-ray background in, 657-60, 663-75, 677-79, 681-85
Galaxies, blue compact, H II regions in, 531, 545, 552
Galaxies, colliding, 24
Galaxies, dwarf elliptical
  evolution of, 552
  H II regions in, 525, 542-43
Galaxies, dwarf irregular, H II regions in, 552-54
Galaxies, infrared, 441
Galaxies, irregular
  H II regions in, 527, 531-32, 549

SUBJECT INDEX 691

superbubbles in Local Group, 86
Galaxies, radio, 24, 28, 485
Galaxies, spiral
  H II regions in, 525-27, 541, 531-34, 548-56
  identification of, 483
Galaxy
  coronal gas in, 73, 90-92
  cosmic ray escape from, 97-98
  dark matter in, 171
  evolution of, 126, 129-132
  extinction laws in, 45
  formation of, models of, 28
  H I in, 215-55
  hot interstellar gas in, 71-73, 98
  interstellar dust in, 37
  interstellar radiation field in, 52
  ionizing stars in, 546
  luminosity of, 52
  soft X-ray background in, 658-59, 676-77, 681, 685
  superbubbles in, 86-88
  supernova rate in, 79
  unidentified infrared band emission in, 49
  X-ray sources in, 185-86
Gamma-ray bursts, 28, 207-8, 401-32
  classification of, 403
  continuum spectra of, 419-24
  detection of, 401-2
  distribution of on sky, 414-15
  durations of, 403-4
  emission processes of, 428-29
  nomenclature of, 402-3
  optical transients of, 413-14
  origins of, 430-32
  periodicities of, 408
  positions of, 411-12
  quiescent counterparts of, 412-13
  recurrence of, 409-11
  size-frequency distributions of, 415-19
  spectral features of, 424-28
  temporal structure of, 404-7
Gamma-ray emission
  from discrete sources, 28
  of H I, 241-42
  from neutron stars, 184
  from π^0 meson decay, 28
Gamma-ray lines, 28
Gamma Ray Observatory observations, 242, 402
Gamma-ray sources, see Gamma-ray bursts
GB780325, 420, 422-25, 428
GB781119, 413, 428
GB790107, 410
GB790113, 413
GB790305b, 402, 405, 407-10, 412
GB790324, 411
GB790325a, 411
GB790325b, 413-14, 420, 427-32
GB790327a, 411
GB790402b, 427-28
GB791105, 413
GB811231a, 426
GB820331, 405-6
GB820405, 403
GB830801b, 405-7, 420-21
GB840304, 403
GB841215, 405-6
GB870303, 422, 429-30
GB880205, 422, 425
GBS 0526 − 66, 409-10, 412, 414, 420, 427
GBS 1806 − 20, 409-10, 420-21
GBS 1900+14, 409-11, 420
Ginga observations, 402, 421, 425, 429, 665
Globular clusters
  disruption of, 205, 208-9
  hot subdwarfs in, 318
  low-mass X-ray binary stars in, 203-4
  white dwarfs in, 109, 118-21
  X-ray sources in, 186
Globules, optical properties of, 55
GONG observations, 296
Gould's Belt, 243
GP Com, 110
Granulation, solar, 263-66, 271-80, 283-89
Graphite
  2175-Å bump due to, 43-44
  in interstellar dust, 45, 63-64
Gravitational redshift, DA star masses from, 106-9
Gravitational waves, 5
Green Bank Telescope, 218, 222, 237, 574
Gum Nebula, 43
GX 1+4, 186, 206
G33.6+0.1, 579
G61 − 29, 110
G107 − 70, 110

H

H I regions
  hot interstellar gas in, 79-80
  indirect tracers of, 238-42
  in interstellar medium, 62
  observations of, 216-34, 237
  organization of, 242-49
  in soft X-ray background emission, 670, 672, 678-79, 682, 684-85
  vertical structure of, 249-55
H I sky, integral properties of, 215, 231-35
H II regions
  bipolar, 580, 592
  corrugations in, 254
  evolution of, 52
  extragalactic, 525-557
    chemical abundances in, 548-56
    distance determinations of, 544-45
    ionizing stars in, 545-48
    kinematics of, 543-44
    models of, 534-40
    nature of, 529-34
    structural properties of, 540-543
  galactic, 525, 541-43
  ground-state continuum of, 305
  hot interstellar gas in, 84
  ionization of, 45, 333
  low-excitation, 240
  optical/ultraviolet extinction in, 39-40
  random motion of, 589
  small-scale density fluctuations in, 244
  unidentified infrared bands in, 49-50
Hakucho observations, 402
Halo gas, structure of, 92-98
Hat Creek Observatory Telescope, 226, 397
HD 6532, 613-14, 616, 624-25, 627-28, 642-45, 650
HD 14633, 315
HD 15558, 314, 330
HD 15570, 314
HD 15629, 314, 330
HD 16429, 315
HD 18409, 315
HD 24712, 613-14, 616
HD 29647, 40, 43-44
HD 30614, 315
HD 34078, 315
HD 34656, 314, 330
HD 36486, 315
HD 37128, 315
HD 37742, 315
HD 38771, 315
HD 44179, 51
HD 46150, 314
HD 46223, 314
HD 46966, 315
HD 48099, 41
HD 48279, 315

HD 50896, 339-40
HD 51418, 612
HD 60435, 613-14, 616, 634-35, 641-42
HD 62542, 43
HD 66811, 314
HD 80316, 613-14, 616
HD 83368, 613-16, 618-30, 633, 635, 641-43, 645, 647, 650-51
HD 89137, 315
HD 90657, 339
HD 93028, 44
HD 93030, 315
HD 93128, 314
HD 93129A, 314, 330
HD 93222, 314
HD 93250, 314, 330
HD 101065, 612-14, 616, 635-38, 642, 645
HD 128898, 613-14, 617, 645-46, 648, 650
HD 134214, 613-14, 617, 646
HD 137949, 613-14, 617, 641
HD 149438, 315
HD 164794, 314
HD 166473, 613-14, 617, 648-49
HD 176232, 613-14, 617
HD 192639, 315, 330
HD 193514, 315, 330
HD 193682, 314
HD 201601, 613-14, 617
HD 203932, 613-14, 617
HD 210839, 314
HD 214680, 315
HD 217522, 613-14, 617
HD 303308, 330
HDE 268685, 316
HDE 269504, 316
HDE 303308, 314
HEAO (High Energy Astronomical Observatory) observations, 402, 425, 664-65, 669-70
Helios 2 observations, 402, 408
Helium, abundance of
 in H II regions, 535, 537-38, 553-56
 in hot stars, 309-10, 311, 318
 in Orion Nebula, 525
 in roAp stars, 612, 637-40, 647
 in Wolf-Rayet stars, 338-40
Her X-1, 186
Herschel 36, 40, 41
High Altitude Observatory, Hawaii, 517
Hinotori observations, 402
HR 465, 612, 646
HR 1217, 613, 620, 625, 630-33, 635-36, 638, 641-45, 649-51

HR 3831, 613-16, 618-30, 633, 635, 641-43, 645, 647, 650-51
HR 5643, 613
HR 7167, 613
HR 8097, 613, 636, 645-47
HR 24712, 613, 620, 625, 630-33, 635-36, 638, 641-45, 649-51
Hubble Space Telescope observations, 109, 235-36, 330
Hyades, 107, 120
Hydrogen
 abundance of
  in H II regions, 534-35, 537, 539, 550-53
  in hot stars, 306, 309-10
  in Orion Nebula, 525
 far-UV extinction and, 45
 in interstellar medium, 63
 unidentified infrared band production and, 49
Hydrogenated amorphous carbon (HAC), 65
β Hyi, 290
HZ 9, 110
HZ 29, 110
H40, 537
H1504+65, 118

I

IC 435, 51
ICE (*International Cometary Explorer*) observations, 402, 404, 410, 428
Induced emission, in radio wave propagation, 7
Infrared (IR) emission
 from H I regions, 239-40
 from H II regions, 525, 556
 of interstellar grains, 38, 49-53, 55-58, 65-66
 from planets, 362-63, 380
 unified model atmospheres and, 333-34
 from white dwarfs, 122-125
Intergalactic gas, heating of, 24-25
Interplanetary dust particles (IDPs), 60, 61, 67
Interstellar dust, see also Interstellar grains
 continuum emission from, 50-53
 diffuse, 41-42, 44-47, 63, 65
 evolution of, 62-63, 67
 H I in, 238-39
 inner-cloud, 38
 mass of, 52
 outer-cloud, 41-42, 44-46, 63, 65

 scattering from, 38, 53-55
 site of formation of, 60
 unidentified infrared band emission from, 49-50
Interstellar extinction
 continuous, 38-46
 galactic, 533
 mean extinction laws of, 40-41, 47-49
 silicate features of, 46, 47
Interstellar gas
 depletions in, 59-60
 diffuse, 83
 heating of, 62, 71-98
Interstellar grains, see also Interstellar dust
 composition of, 45-48, 64-66
 infrared emission of, 38, 49-53, 55-58
 modification of, 41-42
 origin of, 66
Interstellar medium (ISM)
 diffuse, 57
 dust in, 62-67
 H I in, 215-16, 233-34, 238-42, 250-54
 lines of sight sampling physical conditions in, 37-40, 44
 mass of, 52
 metal accretion from, 160, 162-163
 PAH ionization in, 50
 plasma turbulence in, 561, 587-93
 polarization in, 55
 properties of, 95
 soft X-ray background in, 658-59, 662, 676, 680-81, 684-85
 temperature of, 37
 unidentified infrared bands in, 49-50
Interstellar scattering
 angular broadening in, 576-87
 from dust, 38, 53-55
 galactic distribution of material for, 574-76
 in image wander, 580-81
 in pulsar amplitude variations, 567-68
 in radio source amplitude variations, 568-74
 regimes and source diameters for, 562-66
 sites for enhanced, 591-92
 strong, 596-98
Interstellar scintillation
 diffractive, 566-67, 569, 582-583, 598-600
 dynamic spectra in, 582-85
 refractive, 566-74, 582, 592, 600

SUBJECT INDEX 693

strong, 564-66
timing variability in, 586-87
weak, 563-64, 572
velocity determinations in, 585-86
Ionosphere, radio wave propagation in, 21, 22
IRAS (Infrared Astronomical Satellite) observations, 49-51, 115, 239-40
Iron, abundance of
  in H II regions, 536-37, 553
  in roAp stars, 608, 612
IRS 3, 65
IRS 7, 65
ISEE observations, 426
IUE (International Ultraviolet Explorer) observations, 39, 108, 235, 311, 528

J

Jet Propulsion Laboratory, Planetary Radio Astronomy group at, 372
Jodrell Bank Observatory, 584
Jupiter, radio images of, 347-49, 351-52, 354-63, 377-80

K

Kitt Peak Telescope, 445
K648, 118

L

10 Lacertae, 315
Large Magellanic Cloud
  abundances in, 45
  extragalactic extinction in, 45
  H II regions in, 526, 533, 539, 541
  N49 supernova remnant in, 401, 409, 412, 414, 430
  SN 1987A in, 184
  soft X-ray background emission from, 682
  optical/ultraviolet extinction in, 40
  O stars in, 316
Laziness, 116
LH −43°81, 316
33 Lib, 613
Local Bubble, 72
Local correlation tracking, 269, 275, 277
Local group of galaxies, superbubbles in, 86
Lynga 8, 65
Lynds-type clouds, 246
L151-8A/B, 111
L870-2, 110

M

Magellanic Clouds, see also Large Magellanic Cloud; Small Magellanic Cloud
  abundances in, 45-46
  central stars of planetary nebulae in, 112, 114
  extragalactic extinction in, 45
  H I emission from, 231
  white dwarfs in, 123-24
Magellanic Stream, 89
Magnesium
  abundance of
    in H II regions, 536-37
    in white dwarfs, 200-1, 209
  in interstellar extinction, 43
Magnetic stars, collapse of, 33
Magnetospheres
  of neutron stars, 414, 431
  of planets, 347
  of pulsars, 25
Mars, radio images of, 347, 350, 381-84
Maser sources, 572, 580-81
MEM (Maximum Entropy Method), in radio imaging, 353-54
Mercury, radio images of, 347, 350, 392, 394-97
Merope reflection nebula, red continuum in, 51
Mesoscale flows, in solar convection, 276, 278, 283, 285
Metallicity
  correlation of mass loss with, 123
  of ionizing stars, 546
  of roAp stars, 611
  stellar, in H II regions, 550
Meteorites, interstellar grains in, 60-61, 67
Meteoroids, interstellar grains in, 61
Microquasars, 441
Microwave emission
  from cosmic background, 439, 658
  from planets, 350
Milky Way
  H II regions in, 525, 541, 550-51
  infrared emission from, 240
Mini-black holes, 33
Molecular clouds
  biological grains in, 66
  cold, 81, 83, 89, 93, 97
  conducting envelopes of, 81-82
  extragalactic extinction in, 45-46

giant, in disruption of globular clusters, 205
gravitational energy removed from, 37
interstellar grain structures in, 61, 67
optical/ultraviolet extinction in, 39, 41-42
radiation at center of, 42
red continuum in, 51
scattering by, 54
silicate absorption features in, 46-49
star formation by, 52, 543
supernova remnant interaction with, 80-86
Moon, radio wave diffraction and, 23
Murray meteorite, interstellar grains in, 61
M3, 318
M5, 186
M7, 413
M15, 318
M31, 205-6, 526, 534, 551, 682
M33, 526, 533, 541, 548-51
M42, see Orion Nebula
M51, 533-34
M81, 532-34, 549, 551
M83, 548
M92, 318
M101, H II regions in, 526-28, 533-34, 537-39, 543-44, 546-49

N

Naŋay Telescope, 222
Near-infrared (NIR) continuous extinction, 39, 44-45, 51
Nebulae
  red continuum in, 50
  unidentified infrared bands in, 49-50
Neon, abundance of
  in H II regions, 535-37, 550-53
  in hot stars, 308
  in white dwarfs, 200-1, 209
Neptune, radio images of, 347-50, 354, 376-81
Neutrinos
  cooling of, 166, 170-71, 200
  in core collapse of massive stars, 184
  energy loss from, 151
  generation of, 29
  high-energy astrophysics of, 25, 28-29
Neutron stars
  accretion onto, 31

in binary systems, 142, 183-209
capture mechanisms for, 188-90
compact binary systems of, 142
core collapse of massive, 187-88
formation of through accretion-induced collapse, 190-202
gamma-ray bursts from, 401-2, 408, 422-24, 430-32
origin of, 184
NGC 246, 114
NGC 288, 318
NGC 300, 551
NGC 346, 316, 546
NGC 604, 526, 540, 542-43, 546
NGC 6334A, 580
NGC 6334B, 580, 592
NGC 1365, 550
NGC 1688, 122
NGC 1866, 123
NGC 2023, 50
NGC 2071, 243
NGC 2168, 121
NGC 2244, 40
NGC 2516, 108, 120-21, 648
NGC 3532, 121
NGC 3603, 543
NGC 4861, 531, 547
NGC 5455, 549
NGC 5461, 526, 531
NGC 5471, 526, 531, 537-42, 544, 546
NGC 6397, 318
NGC 6611, 314, 315
NGC 6624, 203
NGC 6752, 119, 318
NGC 6822, 552
NGC 7000, see North America Nebula
NGC 7027, 112
Nitrogen
abundance of
in H II regions, 535-38, 551-53
in hot interstellar gas, 91-92
in hot stars, 308-10, 319
in interstellar extinction, 43
North America Nebula, 541
North Celestial Pole, shell around, 234
North Polar Spur, H I sky and, 234
Novae, interstellar dust formation in, 60
N19, 526

N49 supernova remnant, 401, 409, 412, 414, 430
N59, 533
N70, 526

O

O stars
atmospheric parameters for, 312-16
in galactic field, 318
H II ground-state continuum in, 305
in ionization of hot coronal gas, 91
IUE spectra of, 311
line profiles in, 645
statistical equilibrium analyses of, 318
stellar winds of, 320, 328-29
supernovae produced from, 87
X-ray emission of, 332
OAO-2 space observatory observations, 235
OB stars
in galactic field, 318
nonradial pulsation in, 645
Observatoire de Pic du Midi, see Pic du Midi Observatory
OJ 287, 571
Old disk population stars, 186
ζ Oph, 645
Ophiuchus molecular cloud, optical/ultraviolet extinction in, 39
Optical extinction, 39-44
Optical transients, 413-14
Orbiting Astronomical Observatory 2, diffuse galactic light from, 44
Orion A, VLA map of, 243
Orion Molecular Cloud
Becklin-Neugebauer Object in, see Becklin-Neugebauer Object
optical/ultraviolet extinction in, 39
polarization in, 57
Orion Nebula
H I in, 227, 236
H II regions in, 525-26, 531, 549
Trapezium in, 47
OT 1901, 413
OT 1928, 413
OT 1944, 413
Owens Valley Radio Observatory, interferometer at, 356, 363, 397
Oxygen
abundance of

in H II regions, 531, 535-39, 549-53
in hot stars, 308-10, 319
in interstellar dust, 43
in Orion Nebula, 525
in white dwarf progenitors, 143, 146
atomic, in molecular cloud conducting envelopes, 82-85
in white dwarfs, 193-202, 209
in Wolf-Rayet stars, 340

P

P Cygni, 327, 329, 331
Palomar Telescope, 245, 442-43, 527
Parkes Radio Observatory, H I studies at, 222
Peak finding, for granulation recognition, 274
53 Per, 645
Phobos observations, 402
Pic du Midi Observatory, 264-66, 269, 275
PG 1159, 111, 117-118
PG 1346+082, 110
PHL 932, 115
Pioneer 10 observations, 53, 356
Pioneer 11 observations, 356, 358
Pioneer Venus Orbiter observations, 387, 402, 417, 426
Planetary atmospheres, radio emissions from, 348-50, 362-70, 373-74, 376-80, 383-85, 389-93
Planetary nebulae
central stars of, 305, 316-18, 327, 336
as white dwarf progenitors, 103-4, 111-17, 122-23, 127-29
ionization of, 333
nuclei of, in white dwarf development, 147, 151, 176
ultraviolet bump in, 60
unidentified infrared bands in, 49-50
Planets, radio emissions from, 347-97
Pleiades cluster, white dwarf in, 120, 141
Plerion remnants, hot interstellar gas from, 74
Pluto, radio images of, 347
Polarization, interstellar
continuum, 55-57

SUBJECT INDEX 695

in spectral features, 57-58
Polycyclic aromatic hydrocarbons (PAHs), 49-51, 63-65
Population I stars
 binary X-ray sources, 209
 roAp, 637
 RR Lyrae, 125
 white dwarfs, 118-19, 122-23, 145
Population II stars
 in globular clusters, 186
 white dwarfs, 109, 123
Procyon, 290
Procyon B, 110
Prognoz observations, 402
Proto-planetary nebulae, 111, 115-116
Przybylski's star, 612-14, 616, 635-38, 642, 645
PSR 0355+208, 570
PSR 0655+64, 186, 206-7, 586
PSR 0820+02, 207
PSR 0917+624, 571
PSR 0945+658, 572, 573
PSR 1133+16, 585
PSR 1237+25, 585
PSR 1516+02A, 186
PSR 1516+02B, 186
PSR 1620+26, 186
PSR 1831+00, 186
PSR 1849+00, 579
PSR 1849+05, 579
PSR 1855+09, 186, 586
PSR 1913+16, 186, 188, 206
PSR 1933+16, 580
PSR 1937+21, 586-87
PSR 1953+29, 186, 206-7
PSR 1957+20, 186, 207
PSR 2005+403, 580
PSR 2013+370, 570
PSR 2303+46, 186, 188, 206
Pulsars
 binary
  millisecond, 184, 186, 192, 206-9
  radio, 184
 diffractive interstellar scintillations in, 575, 583
 dynamic spectra of, 582-84
 hot interstellar gas and, 94-95
 magnetospheres of, 25
 observation of 21-cm absorption against, 220
 refractive interstellar scintillations in, 566-68
 weak interstellar scintillations in, 563-64
 X-ray, 183, 185
Pulsators, 117, 124, 167
ζ Puppis, 314, 322-23, 330, 334

Q

Quantitative spectroscopy, of hot stars, 303-41
Quasars
 associations of, 473-85
 clustering of, 437, 439-41, 474, 480-86
 gravitational lensing and, 439, 479-80
 quasar-galaxy, 438-39, 441, 473-80
 evolution of, 437, 438, 440
 luminosity function of, 437-38, 440-41, 458-473, 485-86
 operational definition of, 440-41
 surface density of, 448-57
 surveys of, 441-57
  at nonoptical wavelengths, 448
  multicolor, 444-45
  proper motion and, 447-48
  slitless spectroscopic, 445-47, 450, 453
  ultraviolet excess emission, 442-44, 447-48, 451, 453
  variability and, 447

R

R CrB stars, 119
21-cm radiation, for observation of interstellar atomic H I, 215-34
Radio emission
 cosmic synchrotron, 24-27, 33, 95
 from H II regions, 525, 532-33
 induced, 7
 from planets, 347-97
 from quasars, 476-80
 solar, 21-22, 24
Radio Loop I, 670-71, 675, 683
Radio source variation, 562-74, 586-87, 592-593
Radio star hypothesis, 26-27
Rapidly oscillating (roAp) stars
 asteroseismology of, 630-33, 646, 650-51
 classification of, 610-11
 diffusion model for, 609
 excitation mechanisms in, 637-41, 651
 frequencies for, 616-17
 general properties of, 608-9
 lifetimes of pulsation modes in, 634-36, 650-51
 mode identification in, 644-46, 651
 mode selection in, 641-44, 651
 oblique pulsator model for, 608-9, 622-30, 650-51
 period changes in, 636, 650
 search for, 646-52
 spotted pulsator model of, 633-34, 650
 Strömgren indices of, 613
Ratan 600 Telescope, 222
Red Rectangle, see HD 44179
Reflection nebulae
 illumination of, 53-54
 red continuum in, 50-51
 unidentified infrared bands in, 49
Relativity, general theory of, 33
ROB 162, 118, 318
ROSAT observations, 241, 663, 682, 685
RR Lyrae, 125, 624-25
Rye Canyon Observatory, 266
RY Sgr, 119
R136, 541
R136a, 546

S

2S 0921 − 63, 206
SA 57, 447-48
Sacramento Peak Observatory, 266
Sag A, 580
Sanduleak-Pesch pair, 110-11
SAS-2 observations, 241
SAS-3 observations, 662, 665, 667
Saturn, radio images of, 347-51, 354, 363-72, 377-80
Scattering, see Interstellar scattering
Schwarzschild black holes, 33
α Sco, 60
τ Sco, 315
Sco-Cen OB association, 681
Scorpius X-1, 658
Sco X-1, 206
δ Scuti stars, 611-12, 637-39, 641-45, 650-51
Self-absorption, H I emission and, 223-24, 245
Seyfert galaxies, 438, 441, 483, 485
9 Sgr, 314
Sgr A, 57
Sgr B2, 581
Shells, 242-44

SUBJECT INDEX

Silicates, in intrastellar grains, 45-48, 63-64
Silicon
  abundance of, 43
  in hot interstellar gas, 91-92, 98
  in hot stars, 308, 327-28
  in roAp stars, 640
  in white dwarfs, 107
  in interstellar extinction, 43
Single-dish absorption studies, for galactic H I spectra, 220-21
Sirius A, 107
Sirius B, 107-8, 110, 127
Sk 119, 316
Sk 159, 316
Sk $-70°69$, 316
Sk $-65°21$, 316
Sk $-66°172$, 316
Sk $-68°41$, 316
Sk $-69°213$, 316
Skylab observations, 521
Small Magellanic Cloud
  abundances in, 46
  H II regions in, 526, 546
  infrared emission from, 240
  O stars in, 316
  SM 79, 40
  soft X-ray background emission from, 681-82
SMM (Solar Maximum Mission) observations, 225-26, 402, 423, 517, 521
SN 1987A, 29, 184
Soft X-ray bands, H I absorption in, 240-41
SOHO observations, 296
Solar convection
  granulation in, 263-66, 271-80, 283-89, 292
  observational data for, 268-80
  theories of, 266-68, 280-96
Solar corona
  magnetic fields in, 491-522
    dynamics of, 516-21
    equilibrium and stability of, 492-501
    eruption from, 501-16
    radio emission from, 21-22
    time-dependent outflows from, 516-21
Solar photosphere, in anchoring coronal magnetic field, 512-16
Solar system, dust in, 60
Solar wind, from solar corona, 516
Solwind satellite observations, 517

SOUP (Solar Optical Universal Polarimeter) observations, 266, 269-70, 272, 275
Soviet Stratospheric Solar Observatory observations, 272
Space-time filtering, 269-70
Spacelab 2 observations, 266
Spiral galaxies, superbubbles in Local Group, 86
SrCrEu stars, 610-11, 614, 630, 637, 646, 650-51
SrEu stars, 614
Star quakes, 431
Stars
  A-type, see A stars
  abundances in, see Abundances
  B-type, see B stars
  binary, see Binary stars, high-mass X-ray; Binary stars, low-mass X-ray; Binary stars, massive X-ray
  blue
    luminous, 331
    subluminous, 111, 126, 318
  C-type, see C stars
  formation rate of, 62, 129-32
  evolution of
    late stage, 103, 106, 109, 111-32
    massive, HR diagram of, 326
    with mass loss, 121-22
  high-mass, 120, 185-86, 202, 205
  hot, quantitative spectroscopy of, 303-41
  intermediate-mass, 103, 120, 141
  interstellar dust formation in, 37, 60
  ionizing, of extragalactic H II regions, 545-48, 556
  Lα absorption measurements of, 236
  low-mass, 103, 114-15, 185-86, 191-92, 202-9, 414
  magnetic, see Magnetic stars
  massive, 206, 208, 184, 187-88
  in molecular clouds, 47, 52
  neutron, see Neutron stars
  O-type, see O stars
  oxygen-rich, 46
  red giant, 146, 188, 191
  structure of, 142-43
  subdwarf, hot, 111, 116-17, 318, 320
  supergiant, 116, 124, 319, 321, 545-46
  ultraviolet-bright, 118

Stein 2051B, 110
Stellar atmospheres
  of hot stars, 304-5, 312-18
  of roAp stars, 608, 641
  unified model of, 332-37
Stellar flares, 401
Stellar photospheres, of hot stars, 304-19
Stellar winds
  blanketing of, 309-10, 313
  in creation of planetary nebulae, 122
  from extragalactic H II regions, 543, 544
  from hot stars, 304-5, 309-10, 319-32, 336-41
  interstellar gas in, 62
  low-mass stars and, 103
  mass transfer and, 185
  opaque, 305
  optically thick, 336, 338-41
  optically thin, 305, 319-32
  from roAp stars, 640
  superbubbles and, 87
Sun
  gravitational potential barrier of, 519
  in H I sky, 233-34
  in Hertzsprung-Russell diagram, 289-90, 296
  interior of, 268-69, 291-96
  interstellar dust near, 62
  interstellar gas near, 62
  radio wave reflection and, 21-22, 24
  surface layers of, 269-72, 280-91, 296
  theory of sporadic emission by, 24
  white dwarfs in vicinity of, 127
Superbubbles, 86-88, 92, 95-96, 98, 671, 675, 679 683
Supergranulation, 276, 278, 280
Supernovae
  hot interstellar gas from, 71-73
  interstellar dust formation in, 60-61, 67
  Type I, 126, 191, 209, 545
  Type Ia, 191-193, 197, 209
  Type Ib, 86
  Type II, 79, 86, 203
Supernova flares
  neutron star formation and, 28
  in radio galaxies, 24
Supernova remnants
  abundances in, 551, 552
  cosmic rays in, 28
  emission-line spectra of, 530

SUBJECT INDEX 697

expansion of, 71-79
  magnetic fields in, 76-79
  spherical models of, 74-76
  thermal conduction in, 76-79
gas temperature in, 71-72
molecular cloud interaction with, 80-86
Supernova shock
  compression in, 77-78, 80, 82
  temperature of, 71
Supershells, 88, 242-44
Superwinds, 122, 145, 320
SURFACE program, 308
Swedish Vacuum Solar Telescope, 266, 269, 271-72, 275, 277
Synchrotron radio emission
  origin of, 95
  in radio imaging of planets, 349-50, 355-61, 376-77
  theory of, 24-27, 33
S5 region, 537-539, 549

T

T tauri stars, 43
Tarantula Nebula, 526, 539-43, 556
Taurus Molecular Cloud
  optical/ultraviolet extinction in, 39
  silicate absorption features in, 46
Thermal radiation, from planets, 355, 361-63, 381-82, 392-95
Thorne-Zytkov object, 188
Tidal capture, in binary system formation, 188-90, 204, 209
Tol 1214 − 277, 555
Trace particle plots, 283-84
Translucent molecular clouds, 245-46
Trapesium cluster, 47
Tycho, 74

U

4U 1626 − 67, 186
4U 1820 − 30, 203
Ultraviolet-bright stars, 118
Ultraviolet emission
  from anisotropic H I cavity, 679
  from H II regions, 528
  from planets, 371
  from quasars, 438, 440
  stellar winds and, 322, 327

Ultraviolet excess (UVX) emission, from quasars, 442-44, 447-48, 451, 453
Ultraviolet extinction, 37, 39-45, 311
Ultraviolet observations, see IUE observations
Ultraviolet spectra, of interstellar atomic H I, 216-31, 234-41, 246, 249
Ultraviolet stellar radiation, superbubbles and, 87
UM 461, 555
Unidentified infrared bands (UIBs), 49-53, 63-65
Universe
  closing of, 659
  growth of structure in, 439
  isotropy of, 440
Uranus, radio images of, 347-50, 352, 354, 372-76, 378-81
uvby observations, 105

V

V444 Cygni, 340
V471 Tau, 110
Valley finding, for granulation recognition, 274
Van Allen belt, 355
Vavilov-Cherenkov radiation, 8-9
Vela observations, 402
Venera observations, 402-4, 408, 411, 414-15, 417-19, 426-27
Venus, radio images of, 347, 350, 384-93, 396
Viking observations, 383
Virgo cluster, H II regions in, 544-45
VLA (Very Large Array) observations
  of extragalactic H II regions, 528, 533
  of galactic H I, 217, 220-22, 243
  of planets, 351-52, 357-58, 371-72, 382-83, 385-88
  of wave propagation, 579, 587
VLBI observations
  of galactic H I, 222
  phase closure technique in, 354
  of wave propagation, 561, 578-81, 591-93
Voyager observations, 108, 369, 373

W

W3, 227, 243
Warm neutral medium (WNM), in supernova remnants, 78-79
Warp, H I, 255
Wave propagation theory, 593-602
Westerbork Telescope, 227, 357-58
White dwarfs
  accretion-induced collapse of, 185, 190-209
  convection boundaries for, 159
  cooling of, 141, 164, 166-77
    by-products of, 171
    five stage, 166-69
    general features of, 164-65
    luminosity function and, 141, 171-76
    mass dependence of, 169-70
    pulsations and, 170-71
  energy release from, 139-41, 148-55
  evolution of, 103-4, 111-12, 116-32
  formation of, 141-48
  heat transfer in, 141, 155-64
  masses of, 103-32
  progenitors of, 141-48
    chemical composition of, 142-48
    masses of, 111-19
  pulsation frequencies of, 642
Wisconsin Diffuse X-ray Spectrometer, 663
WN5 stars, 338-39
WN5+O6, 340
WN7 stars, 546
Wolf-Rayet stars
  as central stars of planetary nebulae, 114
  in H II regions, 541, 543-44, 546
  interstellar dust and, 46-47
  quantitative spectroscopy of, 305, 338-40
WUPPE experiment, 57
W49 complex, 525-26, 540

X

X-ray binary stars, see Binary stars, high-mass

X-ray; Binary stars, low-mass X-ray; Binary stars, massive X-ray
X-ray bursts, 183, 410
X-ray emission, 25
  from coronal magnetic fields, 491
  from Cygnus Loop, 72
  from Galaxy, 185-186
  from hot interstellar gas, 91
  from neutron stars, 184
  from O stars, 332
  from quasars, 440, 479
  from supernovae, 72
  from supernova remnants, 74, 80
X-ray halo, around point sources, 45
X-ray photons, 183
X-ray soft background flux
  emission mechanisms for, 660-62
  limits on, 681-83
  models of, 675-81, 684
  observation of, 662-75
  origin of, 657-60
X-ray sources, 401

XUV emission, from coronal magnetic fields, 491

Z

Zeeman effect, at 21-cm radiation, 225-26
I Zw 18, 552
II Zw 40, 552, 555
ZZ Ceti, 642, 645

MISCELLANEOUS

3-30 μm continuum, 51

CUMULATIVE INDEXES

CONTRIBUTING AUTHORS, VOLUMES 18–28

A

Abbott, D. C., 25:113–50
Abt, H. A., 21:343–72
Adams, F. C., 25:23–81
Akasofu, S.-I., 20:117–38
Ambartsumian, V. A., 18:1–13
Angel, J. R. P., 18:321–61
Arnett, W. D., 27:629–700
Athanassoula, E., 23:147–68

B

Backer, D. C., 24:537–75
Bahcall, J. N., 24:577–611; 27:629–700
Bahcall, N. A., 26:631–86
Bai, T., 27:421–67
Balick, B., 20:431–68
Baliunas, S. L., 23:379–412
Beckwith, S., 20:163–90
Beichman, C. A., 25:521–63
Bertout, C., 27:351–95
Bignami, G. F., 21:67–108
Binggeli, B., 26:509–60
Binney, J., 20:399–429
Bloemen, H., 27:469–516
Bodenheimer, P., 26:145–97
Boesgaard, A. M., 23:319–78
Boggess, A., 27:397–420
Böhm-Vitense, E., 19:295–318
Borra, E. F., 20:191–220
Bosma, A., 23:147–68
Bradt, H. V. D., 21:13–66
Brault, J. W., 22:291–317
Bridle, A. H., 22:319–58
Brown, R. L., 22:223–65

C

Cameron, A. G. W., 26:441–72
Canal, R., 28:183–214
Carswell, R. F., 19:41–76
Caughlan, G. R., 21:165–76
Cesarsky, C. J., 18:289–319
Chapman, G. A., 25:633–67
Chincarini, G. L., 22:445–70
Chiosi, C., 24:329–75
Chupp, E. L., 22:359–87
Conti, P. S., 25:113–50
Coulman, C. E., 23:19–57
Cowie, L. L., 24:499–535

Cowling, T. G., 19:115–35; 23:1–18
Cox, A. N., 18:15–41
Cox, D. P., 25:303–44

D

D'Antona, F., 28:139–81
Davidson, K., 23:119–46
Davis, M., 21:109–30
de Pater, I., 28:347–99
Deubner, F.-L., 22:593–619
Dickey, J. M., 28:215–61
Djorgovski, S., 27:235–77
Dravins, D., 20:61–89
Dressler, A., 22:185–222
Dulk, G. A., 23:169–224
Dupree, A. K., 24:377–420

E

Edmunds, M. G., 19:77–113
Ellis, G. F. R., 22:157–84
Elson, R., 25:565–601

F

Fabbiano, G., 27:87–138
Feast, M. W., 25:345–75
Fesen, R. A., 23:119–46
Forman, W., 20:547–85
Fowler, W. A., 21:165–76
Freeman, K. C., 19:319–56; 25:603–32
Frogel, J. A., 26:51–92
Fujimoto, M., 24:459–97
Fusi Pecci, F., 26:199–244

G

Gallagher, J. S., 22:37–74
Garstang, R. H., 27:19–40
Gehrz, R. D., 26:377–412
Genzel, R., 25:377–423; 27:41–85
Gillett, F. C., 19:411–56
Gilmore, G., 27:555–627
Ginzburg, V. L., 28:1–36
Giovanelli, R., 22:445–70
Goldreich, P., 20:249–83
Golub, L., 23:413–52
Gough, D., 22:593–619

Greenstein, J. L., 22:1–35
Gustafsson, B., 27:701–56

H

Harris, M. J., 21:165–76
Hartmann, L. W., 25:271–301
Hartwick, F. D. A., 28:437–89
Haynes, M. P., 22:445–70
Heckman, T. M., 20:431–68
Hellings, R. W., 24:537–75
Hermsen, W., 21:67–108
Higdon, J. C., 28:401–36
Hillas, A. M., 22:425–44
Ho, P. T. P., 21:239–70
Hodge, P. W., 19:357–72; 27:139–59
Hollenbach, D. J., 18:219–62
Holt, S. S., 20:323–65
Holzer, T. E., 27:199–234
Houck, J. R., 25:187–230
Howard, R., 22:131–55
Hoyle, F., 20:1–35
Hudson, H. S., 26:473–507
Hummer, D. G., 28:303–45
Hunter, D. A., 22:37–74
Hurford, G. J., 20:497–516
Hut, P., 25:565–601

I

Iben, I. Jr., 21:271–342
Inagaki, S., 25:565–601
Ionson, J. A., 19:7–40
Isern, J., 28:183–214

J

Jones, C., 20:547–85
Joss, P. C., 22:537–92
Joyce, R. R., 19:411–56

K

Kaler, J. B., 23:89–117
Kellermann, K. I., 19:373–410
Kirshner, R. P., 27:629–700
Kleinmann, S. G., 19:411–56
Kondo, Y., 27:397–420
Kormendy, J., 27:235–77
Kudritzki, R. P., 28:303–45
Kuijken, K., 27:555–627

699

CONTRIBUTING AUTHORS

K
Kuperus, M., 19:7–40
Kurtz, D. W., 28:607–55

L
Labay, J., 28:183–214
Lada, C. J., 23:267–317
Landstreet, J. D., 20:191–220
Larson, H. P., 18:43–75
Léger, A., 27:161–98
Liebert, J., 18:363–98; 25:473–519
Lingenfelter, R. E., 28:401–36
Linsky, J. L., 18:439–88
Liszt, H. S., 22:223–65
Lizano, S., 25:23–81
Lockman, F. J., 28:215–61
Low, B. C., 28:491–524
Lubow, S. H., 19:227–93

M
Mackay, C. D., 24:255–83
Maeder, A., 24:329–75
Maran, S. P., 27:397–420
Margon, B., 22:507–36
Mariska, J. T., 24:23–48
Marsh, K. A., 20:497–516
Mathews, W. G., 24:171–203
Mathis, J. S., 28:37–70
Mazzitelli, I., 28:139–81
McAlister, H. A., 23:59–87
McCammon, D., 28:657–88
McClintock, J. E., 21:13–66
McCray, R., 20:323–65
McCrea, W. H., 25:1–22
McKee, C. F., 18:219–62
Mendis, D. A., 26:11–49
Mestel, L., 20:191–220
Miley, G., 18:165–218
Monet, D. G., 26:413–40
Moore, R., 23:239–66
Moran, J. M., 19:231–76
Morgan, W. W., 26:1–9
Morris, M., 20:517–45
Morrison, D., 20:469–95
Mould, J. R., 20:91–115

N
Narayan, R., 24:127–70
Ness, N. F., 20:139–61
Neugebauer, G., 25:187–230
Newkirk, G. Jr., 21:429–67
Nityananda, R., 24:127–70
Nordlund, Å., 28:263–301
Norris, J., 19:319–56
Noyes, R. W., 25:271–301

O
Oort, J. H., 19:1–5; 21:373–428
Osterbrock, D. E., 24:171–203

P
Pagel, B. E. J., 19:77–113
Pauliny-Toth, I. I. K., 19:373–410
Pearson, T. J., 22:97–130
Peebles, P. J. E., 21:109–30
Perley, R. A., 22:319–58
Phillips, T. G., 20:285–321
Pollack, J. B., 22:389–424
Popper, D. M., 18:115–64
Pringle, J. E., 19:137–62
Probst, R. G., 25:473–519
Puget, J. L., 27:161–98

R
Rabin, D., 23:239–66
Rappaport, S. A., 22:537–92
Raymond, J. C., 22:75–95
Readhead, A. C. S., 22:97–130
Rees, M. J., 22:471–506
Reid, M. J., 19:231–76
Renzini, A., 21:271–342; 26:199–244
Reynolds, R. J., 25:303–44
Rickard, L. J, 20:517–45
Rickett, B. J., 28:561–605
Ridgway, S. T., 22:291–317
Rood, H. J., 26:245–94
Rosner, R., 23:413–52

S
Saikia, D. J., 26:93–144
Salter, C. J., 26:93–144
Sandage, A., 24:421–58; 26:509–60, 561–630
Sanders, W. T., 28:657–88
Schade, D., 28:437–89
Schwartz, R. D., 21:209–37
Sellwood, J. A., 25:151–86
Shields, G. A., 28:525–60
Shu, F., 19:277–93; 25:23–81
Shull, J. M., 20:163–90
Smith, M. G., 19:41–76
Sneden, C., 27:279–349
Sofue, Y., 24:459–97
Soifer, B. T., 21:177–207; 25:187–230
Songaila, A., 24:499–535
Spicer, D. S., 19:7–40
Spinrad, H., 25:231–69
Spite, F., 23:225–38
Spite, M., 23:225–38
Spitzer, L. Jr., 27:1–17; 28:71–101
Spruit, H. C., 28:263–301
Sramek, R. A., 26:295–341
Steigman, G., 23:319–78
Stein, W. A., 21:177–207
Stern, D. P., 20:139–61
Stinebring, D. R., 24:285–327

Stockman, H. S., 18:321–61
Strömgren, B., 21:1–11
Sturrock, P. A., 27:421–67
Stutzki, J., 27:41–85
Sunyaev, R. A., 18:537–60
Syrovatskii, S. I., 19:163–229

T
Tammann, G. A., 26:509–60
Taylor, J. H., 24:285–327
Telesco, C. M., 26:343–76
Tenorio-Tagle, G., 26:145–97
Title, A. M., 28:263–301
Townes, C. H., 21:239–70; 25:377–423
Tremaine, S., 20:249–83
Trimble, V., 25:425–72
Truran, J. W. Jr., 27:279–349
Tsuji, T., 24:89–125

V
Vaiana, G. S., 23:413–52
van Altena, W. F., 21:131–64
van der Klis, M., 27:517–53
Vauclair, G., 20:37–60
Vauclair, S., 20:37–60
Vaughan, A. H., 23:379–412

W
Wagner, W. J., 22:267–89
Walker, A. R., 25:345–75
Wannier, P. G., 18:399–437
Weaver, T. A., 24:205–53
Weidemann, V., 28:103–37
Weiler, K. W., 26:295–341
Weiss, R., 18:489–537
Wetherill, G. W., 18:77–113
Weymann, R. J., 19:41–76
Wheeler, J. C., 27:279–349
Whitford, A. E., 24:1–22
Wielebinski, R., 24:459–97
Woody, D. P., 20:285–321
Woolf, N. J., 20:367–98
Woosley, S. E., 24:205–53; 27:629–700
Wynn-Williams, C. G., 20:587–618
Wyse, R. F. G., 27:555–627

Y
York, D. G., 20:221–48
Yorke, H. W., 24:49–87

Z
Zel'dovich, Ya. B., 18:537–60
Zimmerman, B. A., 21:165–76
Zuckerman, B., 18:263–88
Zwaan, C., 25:83–111

CHAPTER TITLES, VOLUMES 18–28

PREFATORY CHAPTER

On Some Trends in the Development of Astrophysics	V. A. Ambartsumian	18:1–13
Some Notes on My Life as an Astronomer	J. H. Oort	19:1–5
The Universe: Past and Present Reflections	F. Hoyle	20:1–35
Scientists I Have Known and Some Astronomical Problems I Have Met	B. Strömgren	21:1–11
An Astronomical Life	J. L. Greenstein	22:1–35
Astronomer by Accident	T. G. Cowling	23:1–18
A Half-Century of Astronomy	A. E. Whitford	24:1–22
Clustering of Astronomers	W. H. McCrea	25:1–22
A Morphological Life	W. W. Morgan	26:1–9
Dreams, Stars, and Electrons	L. Spitzer, Jr.	27:1–17
Notes of an Amateur Astrophysicist	V. L. Ginzburg	28:1–36

SOLAR SYSTEM ASTROPHYSICS

Infrared Spectroscopic Observations of the Outer Planets, Their Satellites, and the Asteroids	H. P. Larson	18:43–75
Formation of Terrestrial Planets	G. W. Wetherill	18:77–113
Planetary Magnetospheres	D. P. Stern, N. F. Ness	20:139–61
The Dynamics of Planetary Rings	P. Goldreich, S. Tremaine	20:249–83
The Satellites of Jupiter and Saturn	D. Morrison	20:469–95
Origin and History of the Outer Planets: Theoretical Models and Observational Constraints	J. B. Pollack	22:389–424
Comets and Their Composition	H. Spinrad	25:231–69
A Postencounter View of Comets	D. A. Mendis	26:11–49
Origin of the Solar System	A. G. W. Cameron	26:441–72
Radio Images of the Planets	I. de Pater	28:347–99

SOLAR PHYSICS

On the Theory of Coronal Heating	M. Kuperus, J. A. Ionson, D. S. Spicer	19:7–40
High Spatial Resolution Solar Microwave Observations	K. A. Marsh, G. J. Hurford	20:497–516
Variations in Solar Luminosity	G. Newkirk, Jr.	21:429–67
Solar Rotation	R. Howard	22:131–55
Coronal Mass Ejections	W. J. Wagner	22:267–89
High-Energy Neutral Radiations From the Sun	E. L. Chupp	22:359–87
Helioseismology: Oscillations as a Diagnostic of the Solar Interior	F.-L. Deubner, D. Gough	22:593–619
Radio Emission From the Sun and Stars	G. A. Dulk	23:169–224
Sunspots	R. Moore, D. Rabin	23:239–66
The Quiet Solar Transition Region	J. T. Mariska	24:23–48
Elements and Patterns in the Solar Magnetic Field	C. Zwaan	25:83–111
Variations of Solar Irradiance due to Magnetic Activity	G. A. Chapman	25:633–67
Observed Variability of the Solar Luminosity	H. S. Hudson	26:473–507
Interaction Between the Solar Wind and the Interstellar Medium	T. E. Holzer	27:199–234
Classification of Solar Flares	T. Bai, P. A. Sturrock	27:421–67
Solar Convection	H. C. Spruit, Å. Nordlund, A. M. Title	28:263–301
Equilibrium and Dynamics of Coronal Magnetic Fields	B. C. Low	28:491–524

701

STELLAR PHYSICS

Title	Authors	Reference
The Masses of Cepheids	A. N. Cox	18:15–41
Stellar Masses	D. M. Popper	18:115–64
Envelopes Around Late-Type Giant Stars	B. Zuckerman	18:263–88
White Dwarf Stars	J. Liebert	18:363–98
Stellar Chromospheres	J. L. Linsky	18:439–88
Mass, Angular Momentum, and Energy Transfer in Close Binary Stars	F. H. Shu, S. H. Lubow	19:277–93
The Effective Temperature Scale	E. Böhm-Vitense	19:295–318
Element Segregation in Stellar Outer Layers	S. Vauclair, G. Vauclair	20:37–60
Photospheric Spectrum Line Asymmetries and Wavelength Shifts	D. Dravins	20:61–89
Magnetic Stars	E. F. Borra, J. D. Landstreet, L. Mestel	20:191–220
The Search for Infrared Protostars	C. G. Wynn-Williams	20:587–618
The Optical Counterparts of Compact Galactic X-Ray Sources	H. V. D. Bradt, J. E. McClintock	21:13–66
Galactic Gamma-Ray Sources	G. F. Bignami, W. Hermsen	21:67–108
Herbig-Haro Objects	R. D. Schwartz	21:209–37
Asymptotic Giant Branch Evolution and Beyond	I. Iben, Jr., A. Renzini	21:271–342
Normal and Abnormal Binary Frequencies	H. A. Abt	21:343–72
Observations of Supernova Remnants	J. C. Raymond	22:75–95
High Angular Resolution Measurements of Stellar Properties	H. A. McAlister	23:59–87
Planetary Nebulae and Their Central Stars	J. B. Kaler	23:89–117
Radio Emission From the Sun and Stars	G. A. Dulk	23:169–224
The Composition of Field Halo Stars and the Chemical Evolution of the Halo	M. Spite, F. Spite	23:225–38
Stellar Activity Cycles	S. L. Baliunas, A. H. Vaughan	23:379–412
On Stellar X-Ray Emission	R. Rosner, L. Golub, G. S. Vaiana	23:413–52
Molecules in Stars	T. Tsuji	24:89–125
The Physics of Supernova Explosions	S. E. Woosley, T. A. Weaver	24:205–53
Recent Progress in the Understanding of Pulsars	J. H. Taylor, D. R. Stinebring	24:285–327
The Evolution of Massive Stars With Mass Loss	C. Chiosi, A. Maeder	24:329–75
Mass Loss From Cool Stars	A. K. Dupree	24:377–420
The Population Concept, Globular Clusters, Subdwarfs, Ages, and the Collapse of the Galaxy	A. Sandage	24:421–58
Pulsar Timing and General Relativity	D. C. Backer, R. W. Hellings	24:537–75
Star Formation in Molecular Clouds: Observation and Theory	F. H. Shu, F. C. Adams, S. Lizano	25:23–81
Wolf-Rayet Stars	D. C. Abbott, P. S. Conti	25:113–50
Rotation and Magnetic Activity in Main-Sequence Stars	L. W. Hartmann, R. W. Noyes	25:271–301
Very Low Mass Stars	J. Liebert, R. G. Probst	25:473–519
Tests of Evolutionary Sequences Using Color-Magnitude Diagrams of Globular Clusters	A. Renzini, F. Fusi Pecci	26:199–244
Supernovae and Supernova Remnants	K. W. Weiler, R. A. Sramek	26:295–341
The Infrared Temporal Development of Classsical Novae	R. D. Gehrz	26:377–412
Abundance Ratios as a Function of Metallicity	J. C. Wheeler, C. Sneden, J. W. Truran, Jr.	27:279–349
T Tauri Stars: Wild as Dust	C. Bertout	27:351–95
Quasi-Periodic Oscillations and Noise in Low-Mass X-Ray Binaries	M. van der Klis	27:517–53
Supernova 1987A	W. D. Arnett, J. N. Bahcall, R. P. Kirshner, S. E. Woosley	27:629–700
Chemical Analyses of Cool Stars	B. Gustafsson	27:701–56

CHAPTER TITLES

Masses and Evolutionary Status of White Dwarfs and Their Progenitors	V. Weidemann	28:103–37
Cooling of White Dwarfs	F. D'Antona, I. Mazzitelli	28:139–81
The Origin of Neutron Stars in Binary Systems	R. Canal, J. Isern, J. Labay	28:183–214
Quantitative Spectroscopy of Hot Stars	R. P. Kudritzki, D. G. Hummer	28:303–45
Gamma-Ray Bursts	J. C. Higdon, R. E. Lingenfelter	28:401–36
Rapidly Oscillating Ap Stars	D. W. Kurtz	28:607–55

DYNAMICAL ASTRONOMY

Astrometry	W. F. van Altena	21:131–64
Dynamical Evolution of Globular Clusters	R. Elson, P. Hut, S. Inagaki	25:565–601
The Galactic Spheroid and Old Disk	K. C. Freeman	25:603–32
Recent Advances in Optical Astrometry	D. G. Monet	26:413–40

INTERSTELLAR MEDIUM

Interstellar Shock Waves	C. F. McKee, D. J. Hollenbach	18:219–62
Cosmic-Ray Confinement in the Galaxy	C. J. Cesarsky	18:289–319
Nuclear Abundances and Evolution of the Interstellar Medium	P. G. Wannier	18:399–437
Interstellar Molecular Hydrogen	J. M. Shull, S. Beckwith	20:163–90
Herbig-Haro Objects	R. D. Schwartz	21:209–37
Interstellar Ammonia	P. T. P. Ho, C. H. Townes	21:239–70
Observations of Supernova Remnants	J. C. Raymond	22:75–95
The Influence of Environment on the H I Content of Galaxies	M. P. Haynes, R. Giovanelli, G. L. Chincarini	22:445–70
Planetary Nebulae and Their Central Stars	J. B. Kaler	23:89–117
Cold Outflows, Energetic Winds, and Enigmatic Jets Around Young Stellar Objects	C. J. Lada	23:267–317
The Dynamical Evolution of H II Regions—Recent Theoretical Developments	H. W. Yorke	24:49–87
High-Resolution Optical and Ultraviolet Absorption-Line Studies of Interstellar Gas	L. L. Cowie, A. Songaila	24:499–535
Star Formation in Molecular Clouds: Observation and Theory	F. H. Shu, F. C. Adams, S. Lizano	25:23–81
The Local Interstellar Medium	D. P. Cox, R. J. Reynolds	25:303–44
Large-Scale Expanding Superstructures in Galaxies	G. Tenorio-Tagle, P. Bodenheimer	26:145–97
Supernovae and Supernova Remnants	K. W. Weiler, R. A. Sramek	26:295–341
The Orion Molecular Cloud and Star-Forming Region	R. Genzel, J. Stutzki	27:41–85
A New Component of the Interstellar Matter: Small Grains and Large Aromatic Molecules	J. L. Puget, A. Léger	27:161–98
Interaction Between the Solar Wind and the Interstellar Medium	T. E. Holzer	27:199–234
Diffuse Galactic Gamma-Ray Emission	H. Bloemen	27:469–516
Interstellar Dust and Extinction	J. S. Mathis	28:37–70
Theories of the Hot Interstellar Gas	L. Spitzer, Jr.	28:71–101
Extragalactic H II Regions	G. A. Shields	28:525–60
Radio Propagation Through the Turbulent Interstellar Medium	B. J. Rickett	28:561–605

SMALL STELLAR SYSTEMS

Mass, Angular Momentum, and Energy Transfer in Close Binary Stars	F. H. Shu, S. H. Lubow	19:277–93
The Chemical Composition, Structure, and Dynamics of Globular Clusters	K. C. Freeman, J. Norris	19:319–56
Normal and Abnormal Binary Frequencies	H. A. Abt	21:343–72
Dynamical Evolution of Globular Clusters	R. Elson, P. Hut, S. Inagaki	25:565–601

The Galactic Nuclear Bulge and the Stellar Content of Spheroidal Systems	J. A. Frogel	26:51–92
Tests of Evolutionary Sequences Using Color-Magnitude Diagrams of Globular Clusters	A. Renzini, F. Fusi Pecci	26:199–244
Quasi-Periodic Oscillations and Noise in Low-Mass X-Ray Binaries	M. van der Klis	27:517–53

THE GALAXY

Cosmic-Ray Confinement in the Galaxy	C. J. Cesarsky	18:289–319
The Chemical Composition, Structure, and Dynamics of Globular Clusters	K. C. Freeman, J. Norris	19:319–56
Stellar Populations in the Galaxy	J. R. Mould	20:91–115
Gas in the Galactic Halo	D. G. York	20:221–48
The Optical Counterparts of Compact Galactic X-Ray Sources	H. V. D. Bradt, J. E. McClintock	21:13–66
Galactic Gamma-Ray Sources	G. F. Bignami, W. Hermsen	21:67–108
Sagittarius A and Its Environment	R. L. Brown, H. S. Liszt	22:223–65
Neutron Stars in Interacting Binary Systems	P. C. Joss, S. A. Rappaport	22:537–92
Star Counts and Galactic Structure	J. N. Bahcall	24:577–611
Physical Conditions, Dynamics, and Mass Distribution in the Center of the Galaxy	R. Genzel, C. H. Townes	25:377–423
The IRAS View of the Galaxy and the Solar System	C. A. Beichman	25:521–63
The Galactic Spheroid and Old Disk	K. C. Freeman	25:603–32
The Galactic Nuclear Bulge and the Stellar Content of Spheroidal Systems	J. A. Frogel	26:51–92
Large-Scale Expanding Superstructures in Galaxies	G. Tenorio-Tagle, P. Bodenheimer	26:145–97
Diffuse Galactic Gamma-Ray Emission	H. Bloemen	27:469–516
Kinematics, Chemistry, and Structure of the Galaxy	G. Gilmore, R. F. G. Wyse, K. Kuijken	27:555–627
H I in the Galaxy	J. M. Dickey, F. J. Lockman	28:215–61

EXTRAGALACTIC ASTRONOMY

The Structure of Extended Extragalactic Radio Sources	G. Miley	18:165–218
Optical and Infrared Polarization of Active Extragalactic Objects	J. R. P. Angel, H. S. Stockman	18:321–61
Absorption Lines in the Spectra of Quasistellar Objects	R. J. Weymann, R. F. Carswell, M. G. Smith	19:41–76
Abundances in Stellar Populations and the Interstellar Medium in Galaxies	B. E. J. Pagel, M. G. Edmunds	19:77–113
Compact Radio Sources	K. I. Kellermann, I. I. K. Pauliny-Toth	19:373–410
Dynamics of Elliptical Galaxies and Other Spheroidal Components	J. Binney	20:399–429
Extranuclear Clues to the Origin and Evolution of Activity in Galaxies	B. Balick, T. M. Heckman	20:431–68
Molecular Clouds in Galaxies	M. Morris, L. J Rickard	20:517–45
X-Ray-Imaging Observations of Clusters of Galaxies	W. Forman, C. Jones	20:547–85
Dust in Galaxies	W. A. Stein, B. T. Soifer	21:177–207
Superclusters	J. H. Oort	21:373–428
Structure and Evolution of Irregular Galaxies	J. S. Gallagher, III, D. A. Hunter	22:37–74
The Evolution of Galaxies in Clusters	A. Dressler	22:185–222
Extragalactic Radio Jets	A. H. Bridle, R. A. Perley	22:319–58
Black Hole Models for Active Galactic Nuclei	M. J. Rees	22:471–506
Shells and Rings Around Galaxies	E. Athanassoula, A. Bosma	23:147–68
Emission-Line Regions of Active Galaxies and QSOs	D. E. Osterbrock, W. G. Mathews	24:171–203

Global Structure of Magnetic Fields in Spiral Galaxies	Y. Sofue, M. Fujimoto, R. Wielebinski	24:459–97
The IRAS View of the Extragalactic Sky	B. T. Soifer, J. R. Houck, G. Neugebauer	25:187–230
Cepheids as Distance Indicators	M. W. Feast, A. R. Walker	25:345–75
Existence and Nature of Dark Matter in the Universe	V. Trimble	25:425–72
Polarization Properties of Extragalactic Radio Sources	D. J. Saikia, C. J. Salter	26:93–144
Voids	H. J. Rood	26:245–94
Enhanced Star Formation and Infrared Emission in the Centers of Galaxies	C. M. Telesco	26:343–76
The Luminosity Function of Galaxies	B. Binggeli, A. Sandage, G. A. Tammann	26:509–60
Observational Tests of World Models	A. Sandage	26:561–630
Large-Scale Structure in the Universe Indicated by Galaxy Clusters	N. A. Bahcall	26:631–86
X Rays From Normal Galaxies	G. Fabbiano	27:87–138
Populations in Local Group Galaxies	P. Hodge	27:139–59
Surface Photometry and the Structure of Elliptical Galaxies	J. Kormendy, S. Djorgovski	27:235–77
The Space Distribution of Quasars	F. D. A. Hartwick, D. Schade	28:437–89
Extragalactic H II Regions	G. A. Shields	28:525–60

OBSERVATIONAL PHENOMENA

Infrared Spectroscopic Observations of the Outer Planets, Their Satellites, and the Asteroids	H. P. Larson	18:43–75
Optical and Infrared Polarization of Active Extragalactic Objects	J. R. P. Angel, H. S. Stockman	18:321–61
Measurements of the Cosmic Background Radiation	R. Weiss	18:489–537
Preliminary Results of the Air Force Infrared Sky Survey	S. G. Kleinmann, F. C. Gillett, R. R. Joyce	19:411–56
Spectra of Cosmic X-Ray Sources	S. S. Holt, R. McCray	20:323–65
Galactic Gamma-Ray Sources	G. F. Bignami, W. Hermsen	21:67–108
The Evolution of Galaxies in Clusters	A. Dressler	22:185–222
Observations of SS 433	B. Margon	22:507–36
Recent Developments Concerning the Crab Nebula	K. Davidson, R. A. Fesen	23:119–46
The IRAS View of the Extragalactic Sky	B. T. Soifer, J. R. Houck, G. Neugebauer	25:187–230
Existence and Nature of Dark Matter in the Universe	V. Trimble	25:425–72
Polarization Properties of Extragalactic Radio Sources	D. J. Saikia, C. J. Salter	26:93–144
X Rays From Normal Galaxies	G. Fabbiano	27:87–138
Populations in Local Group Galaxies	P. Hodge	27:139–59
Astrophysical Contributions of the International Ultraviolet Explorer	Y. Kondo, A. Boggess, S. P. Maran	27:397–420

GENERAL RELATIVITY AND COSMOLOGY

Measurements of the Cosmic Background Radiation	R. Weiss	18:489–537
Microwave Background Radiation as a Probe of the Contemporary Structure and History of the Universe	R. A. Sunyaev, Ya. B. Zel'dovich	18:537–60
The Extragalactic Distance Scale	P. W. Hodge	19:357–72
Evidence for Local Anisotropy of the Hubble Flow	M. Davis, P. J. E. Peebles	21:109–30

Alternatives to the Big Bang	G. F. R. Ellis	22:157–84
Big Bang Nucleosynthesis: Theories and Observations	A. M. Boesgaard, G. Steigman	23:319–78
Pulsar Timing and General Relativity	D. C. Backer, R. W. Hellings	24:537–75
Existence and Nature of Dark Matter in the Universe	V. Trimble	25:425–72
Observational Tests of World Models	A. Sandage	26:561–630
Large-Scale Structure in the Universe Indicated by Galaxy Clusters	N. A. Bahcall	26:631–86
The Space Distribution of Quasars	F. D. A. Hartwick, D. Schade	28:437–89

INSTRUMENTATION AND TECHNIQUES

Millimeter- and Submillimeter-Wave Receivers	T. G. Phillips, D. P. Woody	20:285–321
High Resolution Imaging from the Ground	N. J. Woolf	20:367–98
Astrometry	W. F. van Altena	21:131–64
Image Formation by Self-Calibration in Radio Astronomy	T. J. Pearson, A. C. S. Readhead	22:97–130
Astronomical Fourier Transform Spectroscopy Revisited	S. T. Ridgway, J. W. Brault	22:291–317
Fundamental and Applied Aspects of Astronomical "Seeing"	C. E. Coulman	23:19–57
High Angular Resolution Measurements of Stellar Properties	H. A. McAlister	23:59–87
Maximum Entropy Image Restoration in Astronomy	R. Narayan, R. Nityananda	24:127–70
Charge-Coupled Devices in Astronomy	C. D. Mackay	24:255–83
The Art of N-Body Building	J. A. Sellwood	25:151–86
Recent Advances in Optical Astrometry	D. G. Monet	26:413–40
The Status and Prospects for Ground-Based Observatory Sites	R. H. Garstang	27:19–40
Astrophysical Contributions of the International Ultraviolet Explorer	Y. Kondo, A. Boggess, S. P. Maran	27:397–420

PHYSICAL PROCESSES

Interstellar Shock Waves	C. F. McKee, D. J. Hollenbach	18:219–62
Nuclear Abundances and Evolution of the Interstellar Medium	P. G. Wannier	18:399–437
The Present Status of Dynamo Theory	T. G. Cowling	19:115–35
Accretion Discs in Astrophysics	J. E. Pringle	19:137–62
Pinch Sheets and Reconnection in Astrophysics	S. I. Syrovatskii	19:163–229
Masers	M. J. Reid, J. M. Moran	19:231–76
Interaction Between a Magnetized Plasma Flow and a Strongly Magnetized Celestial Body With an Ionized Atmosphere: Energetics of the Magnetosphere	S.-I. Akasofu	20:117–38
Interstellar Molecular Hydrogen	J. M. Shull, S. Beckwith	20:163–90
The Optical Counterparts of Compact Galactic X-Ray Sources	H. V. D. Bradt, J. E. McClintock	21:13–66
Thermonuclear Reaction Rates, III	M. J. Harris, W. A. Fowler, G. R. Caughlan, B. A. Zimmerman	21:165–76
The Origin of Ultra-High-Energy Cosmic Rays	A. M. Hillas	22:425–44
Observations of SS 433	B. Margon	22:507–36
The Physics of Supernova Explosions	S. E. Woosley, T. A. Weaver	24:205–53
Quasi-Periodic Oscillations and Noise in Low-Mass X-Ray Binaries	M. van der Klis	27:517–53
The Origin of Neutron Stars in Binary Systems	R. Canal, J. Isern, J. Labay	28:183–214
Gamma-Ray Bursts	J. C. Higdon, R. E. Lingenfelter	28:401–36